T0202804

Lecture Notes in Computer Science   11480

Commenced Publication in 1973
Founding and Former Series Editors:
Gerhard Goos, Juris Hartmanis, and Jan van Leeuwen

More information about this series at http://www.springer.com/series/7407

Andrea Lodi · Viswanath Nagarajan (Eds.)

Integer Programming and Combinatorial Optimization

20th International Conference, IPCO 2019
Ann Arbor, MI, USA, May 22–24, 2019
Proceedings

 Springer

Editors
Andrea Lodi ⓘ
École Polytechnique de Montréal
Montreal, QC, Canada

Viswanath Nagarajan ⓘ
University of Michigan
Ann Arbor, MI, USA

ISSN 0302-9743   ISSN 1611-3349   (electronic)
Lecture Notes in Computer Science
ISBN 978-3-030-17952-6   ISBN 978-3-030-17953-3   (eBook)
https://doi.org/10.1007/978-3-030-17953-3

LNCS Sublibrary: SL1 – Theoretical Computer Science and General Issues

This Springer imprint is published by the registered company Springer Nature Switzerland AG
The registered company address is: Gewerbestrasse 11, 6330 Cham, Switzerland

Preface

This volume collects the 33 extended abstracts presented at IPCO 2019, the 20th Conference on Integer Programming and Combinatorial Optimization, held May 22–24, 2019, at Ann Arbor, Michigan, USA. IPCO is under the auspices of the Mathematical Optimization Society, and it is an important forum for presenting the latest results of theory and practice of the various aspects of discrete optimization. The first IPCO conference took place at the University of Waterloo in May 1990, and the University of Michigan is hosting the 20th such event.

The conference had a Program Committee consisting of 15 members. In response to the call for papers, we received 114 submissions, of which two were withdrawn prior to the decision process. The Program Committee met at Aussois, France, in January 2019. Each submission was reviewed by at least three Program Committee members. There were many high-quality submissions, of which the committee selected 33 to appear in the conference proceedings. We expect the full versions of the extended abstracts appearing in this volume of *Lecture Notes in Computer Science* to be submitted for publication in refereed journals, and a special issue of *Mathematical Programming Series B* is on the way.

This year, IPCO was preceded by a Summer School during May 20–21, 2019, with lectures by Nikhil Bansal, Samuel Burer and João Gouveia. We thank them warmly for their contributions. We would like to thank:

- The authors who submitted their research to IPCO
- The members of the Program Committee, who spent much time and energy reviewing the submissions
- The expert additional reviewers whose opinions were crucial in the paper selection
- The members of the local Organizing Committee: Alexander Barvinok, Jon Lee, Viswanath Nagarajan, Seth Pettie, Siqian Shen and Dan Steffy, who made this conference possible
- The Mathematical Optimization Society and in particular the members of its IPCO Steering Committee: David Williamson, Jens Vygen, Oktay Günlük and Jochen Koenemann, for their help and advice
- EasyChair for making paper management simple and effective
- EasyChair and Springer for their efficient cooperation in producing this volume

We would also like to thank the following sponsors for their financial support: FICO, Gurobi Optimization, IBM, LLamasoft, Microsoft, MOSEK, Springer, The Optimization Firm, The Office of Naval Research, The National Science Foundation, the Michigan Center for Applied and Interdisciplinary Mathematics, and the Department of Industrial and Operations Engineering (University of Michigan).

March 2019   Andrea Lodi
                                                            Viswanath Nagarajan

Organization

Program Committee

Amitabh Basu	Johns Hopkins University, USA
Jose Correa	Universidad de Chile, Chile
Sanjeeb Dash	IBM Research, USA
Volker Kaibel	Otto-von-Guericke Universität Magdeburg, Germany
Simge Küçükyavuz	Northwestern University, USA
Retsef Levi	Massachusetts Institute of Technology, USA
Andrea Lodi	École Polytechnique de Montréal, Canada
James Luedtke	University of Wisconsin-Madison, USA
Tom Mccormick	The University of British Columbia, Canada
Ruth Misener	Imperial College London, UK
Viswanath Nagarajan	University of Michigan, USA
Marc Pfetsch	TU Darmstadt, Germany
Martin Skutella	TU Berlin, Germany
Gautier Stauffer	Kedge Business School, France
Angelika Wiegele	Alpen-Adria-Universität Klagenfurt, Austria

Contents

Identically Self-blocking Clutters

Ahmad Abdi[(⊠)], Gérard Cornuéjols, and Dabeen Lee

Tepper School of Business, Carnegie Mellon University, Pittsburgh, USA
{aabdi,gc0v,dabeenl}@andrew.cmu.edu

Abstract. A clutter is *identically self-blocking* if it is equal to its blocker. We prove that every identically self-blocking clutter different from $\{\{a\}\}$ is nonideal. Our proofs borrow tools from Gauge Duality and Quadratic Programming. Along the way we provide a new lower bound for the packing number of an arbitrary clutter.

1   The Main Result

All sets considered in this paper are finite. Let V be a set of *elements*, and let $\mathcal{C}$ be a family of subsets of V called *members*. If no member contains another, then $\mathcal{C}$ is said to be a *clutter* over *ground set* V [12]. All clutters considered in this paper are different from $\{\}, \{\emptyset\}$. Let $\mathcal{C}$ be a clutter over ground set V. A *cover* is a subset of V that intersects every member. The *covering number*, denoted $\tau(\mathcal{C})$, is the minimum cardinality of a cover. A *packing* is a collection of pairwise disjoint members. The *packing number*, denoted $\nu(\mathcal{C})$, is the maximum cardinality of a packing. Observe that $\tau(\mathcal{C}) \geq \nu(\mathcal{C})$. A cover is *minimal* if it does not contain another cover. The family of minimal covers forms another clutter over ground set V; this clutter is called the *blocker of* $\mathcal{C}$ and is denoted $b(\mathcal{C})$ [12]. It is well-known that $b(b(\mathcal{C})) = \mathcal{C}$ [12,17]. We say that $\mathcal{C}$ is an *identically self-blocking* clutter if $\mathcal{C} = b(\mathcal{C})$. (This terminology was coined in [4].) Observe that $\{a\}$ is the only identically self-blocking clutter with a member of cardinality one.

Theorem 1 ([6]). *A clutter* $\mathcal{C}$ *is identically self-blocking if, and only if,* $\nu(\mathcal{C}) = \nu(b(\mathcal{C})) = 1$.

Consider for $w \in \mathbb{Z}_+^V$ the dual pair of linear programs

$$
\begin{array}{ll}
\min\ w^\top x & \max\ \mathbf{1}^\top y \\
\text{s.t.}\ \sum (x_u : u \in C) \geq 1\ \forall C \in \mathcal{C} & \text{s.t.}\ \sum (y_C : u \in C \in \mathcal{C}) \leq w_u\ \forall u \in V \\
\quad\ x \geq \mathbf{0} & \quad\ y \geq \mathbf{0},
\end{array}
$$

labeled $(P), (D)$, respectively. Denote by $\tau^\star(\mathcal{C}, w), \nu^\star(\mathcal{C}, w)$ the optimal values of $(P), (D)$, respectively, and by $\tau(\mathcal{C}, w), \nu(\mathcal{C}, w)$ the optimal values of $(P), (D)$ subject to the additional integrality constraints $x \in \mathbb{Z}^V, y \in \mathbb{Z}^{\mathcal{C}}$, respectively. Observe that by Strong Linear Programming Duality, $\tau(\mathcal{C}, w) \geq \tau^\star(\mathcal{C}, w) = \nu^\star(\mathcal{C}, w) \geq \nu(\mathcal{C}, w)$.

© Springer Nature Switzerland AG 2019
A. Lodi and V. Nagarajan (Eds.): IPCO 2019, LNCS 11480, pp. 1–12, 2019.
https://doi.org/10.1007/978-3-030-17953-3_1

Notice the correspondence between the 0–1 feasible solutions of (P) and the covers of $\mathcal{C}$, as well as the correspondence between the integer feasible solutions of (D) for $w = \mathbf{1}$ and the packings of $\mathcal{C}$. In particular, $\tau(\mathcal{C}, \mathbf{1}) = \tau(\mathcal{C})$ and $\nu(\mathcal{C}, \mathbf{1}) = \nu(\mathcal{C})$. We will refer to the feasible solutions of (P) as *fractional covers*, and to the feasible solutions of (D) for $w = \mathbf{1}$ as *fractional packings*.

$\mathcal{C}$ has the *max-flow min-cut property* if $\tau(\mathcal{C}, w) = \nu(\mathcal{C}, w)$ for all $w \in \mathbb{Z}_+^V$ [10]. $\mathcal{C}$ is *ideal* if $\tau(\mathcal{C}, w) = \nu^*(\mathcal{C}, w)$ for all $w \in \mathbb{Z}_+^V$ [11]. Clearly clutters with the max-flow min-cut property are ideal. The max-flow min-cut property is not closed under taking blockers, but

Theorem 2 ([18]). *A clutter is ideal if, and only if, its blocker is ideal.*

If $\mathcal{C}$ is an identically self-blocking clutter different from $\{\{a\}\}$, then $\tau(\mathcal{C}) \geq 2 > 1 = \nu(\mathcal{C})$ by Theorem 1, so $\mathcal{C}$ does not have the max-flow min-cut property. In this paper, we prove the following stronger statement:

Theorem 3. *An identically self-blocking clutter different from $\{\{a\}\}$ is non-ideal.*

For an integer $n \geq 3$, denote by Δ_n the clutter over ground set $\{1, \ldots, n\}$ whose members are $\{1, 2\}, \{1, 3\}, \ldots, \{1, n\}, \{2, 3, \ldots, n\}$. Notice that the elements and members of Δ_n correspond to the points and lines of a degenerate projective plane. Denote by $\mathbb{L}_7$ the clutter over ground set $\{1, \ldots, 7\}$ whose members are $\{1, 2, 3\}, \{1, 4, 5\}, \{1, 6, 7\}, \{2, 4, 7\}, \{2, 5, 6\}, \{3, 4, 6\}, \{3, 5, 7\}$. Notice that the elements and members of $\mathbb{L}_7$ correspond to the points and lines of the Fano plane. It can be readily checked that $\{\Delta_n : n \geq 3\} \cup \{\mathbb{L}_7\}$ are identically self-blocking clutters. There are many other examples of identically self-blocking clutters, and in fact there is one for every pair of blocking clutters ([4], Remark 3.4 and Corollary 3.6). Another example, for instance, is the clutter over ground set $\{1, \ldots, 6\}$ whose members are $\{6, 1, 2\}, \{6, 2, 3\}, \{6, 3, 4\}, \{6, 4, 5\}, \{6, 5, 1\}, \{1, 2, 4\}, \{2, 3, 5\}, \{3, 4, 1\}, \{4, 5, 2\}, \{5, 1, 3\}$.

Conjecture 4. *An identically self-blocking clutter different from $\{\{a\}\}$ has one of $\{\Delta_n : n \geq 3\}, \mathbb{L}_7, \{\{1, 2\}, \{2, 3\}, \{3, 4\}, \{4, 5\}, \{5, 1\}\}$ as minor.*

(Notice that the last clutter above is *not* identically self-blocking.) For disjoint $X, Y \subseteq V$, the *minor of $\mathcal{C}$* obtained after *deleting* X and *contracting* Y is the clutter over ground set $V - (X \cup Y)$ whose members are

$$\mathcal{C} \setminus X / Y := \text{the inclusionwise minimal sets of } \{C - Y : C \in \mathcal{C}, C \cap X = \emptyset\}.$$

It is well-known that $b(\mathcal{C} \setminus X / Y) = b(\mathcal{C})/X \setminus Y$ [21], and that if a clutter is ideal, then so is every minor of it [22]. It can be readily checked that the clutters in Conjecture 4 are nonideal. Thus Conjecture 4 – if true – would be a strengthening of Theorem 3.

The rest of the paper is organized as follows: We will present two proofs of Theorem 3, one will be short and indirect (Sect. 2) while the other will be a longer and direct proof that essentially unravels the first proof (Sect. 5). In Sect. 3, by using our techniques, we will provide a new lower bound for the packing number of an arbitrary clutter, and in Sect. 4, we will see a surprising emergence of *cuboids*, a special class of clutters. In Sect. 6 we will address the relevance of studying identically self-blocking clutters, a relatively narrow problem, and why it may be of interest to the community.

2   Gauge Duality

Here we present a short and indirect proof of Theorem 3. Take an integer $n \geq 1$ and let M be a matrix with n columns and nonnegative entries and without a row of all zeros. Consider the polyhedron $P := \{x \in \mathbb{R}^n_+ : Mx \geq \mathbf{1}\}$. The *blocker of P* is the polyhedron $Q := \{z \in \mathbb{R}^n_+ : z^\top x \geq 1 \ \forall x \in P\}$. Fulkerson showed that there exists a matrix N with n columns and nonnegative entries and without a row of all zeros such that $Q = \{z \in \mathbb{R}^n_+ : Nz \geq \mathbf{1}\}$, and that the blocker of Q is P [14,15]. In 1987 Chaiken proved the following fascinating result:

Theorem 5 ([8]). *Take an integer $n \geq 1$, let P, Q be a blocking pair of polyhedra in $\mathbb{R}^n$, and let R be a positive definite n by n matrix. Then $\min\{x^\top Rx : x \in P\}$ and $\min\{z^\top R^{-1}z : z \in Q\}$ have reciprocal optimal values.*

Theorem 5 exhibits an instance of *gauge duality*, a general framework introduced by Freund later the same year [13]. Theorem 5 in the special case of diagonal R's was also proved by Lovász in 2001 [19]. Both Freund and Lovász seem to have been unaware of Chaiken's result.

Let $\mathcal{C}$ be a clutter over ground set V. Define the *incidence matrix* of $\mathcal{C}$ as the matrix M whose columns are indexed by the elements and whose rows are the incidence vectors of the members, and define $Q(\mathcal{C}) := \{x \in \mathbb{R}^V_+ : Mx \geq \mathbf{1}\}$. Fulkerson showed that if $\mathcal{C}, \mathcal{B}$ are blocking *ideal* clutters then $Q(\mathcal{C}), Q(\mathcal{B})$ give an instance of blocking polyhedra [14,15]. Therefore Theorem 5 has the following consequence:

Theorem 6. *Let $\mathcal{C}, \mathcal{B}$ be blocking ideal clutters. Then $\min\{x^\top x : x \in Q(\mathcal{C})\}$ and $\min\{z^\top z : z \in Q(\mathcal{B})\}$ have reciprocal optimal values.*

We will need the following lemma whose proof makes use of concepts such as the *Lagrangian* and the *Karush-Kuhn-Tucker* conditions (see [7], Chapter 5):

Lemma 7 ([8]). *Let $\mathcal{C}$ be a clutter over ground set V, and let M be its incidence matrix. Then $\min\{x^\top x : Mx \geq \mathbf{1}, x \geq \mathbf{0}\}$ has a unique optimal solution $x^\star \in \mathbb{R}^V_+$. Moreover, there exists $y \in \mathbb{R}^{\mathcal{C}}_+$ such that $M^\top y = x^\star$, $\mathbf{1}^\top y = x^{\star\top}x^\star$ and $y^\top(Mx^\star - \mathbf{1}) = \mathbf{0}$.*

Proof. Notice that $\min\{x^\top x : Mx \geq \mathbf{1}, x \geq \mathbf{0}\}$ satisfies Slater's condition, that there is a feasible solution satisfying all the inequalities strictly. As $x^\top x$ is a strictly convex function, our quadratic program has a unique optimal solution $x^\star \in \mathbb{R}^V_+$. Denote by $L(x; \mu, \sigma) := x^\top x - \mu^\top(Mx - \mathbf{1}) - \sigma^\top x$ the Lagrangian of the program. Since Slater's condition is satisfied, there exist $\mu^\star \in \mathbb{R}^\mathcal{C}_+$ and $\sigma^\star \in \mathbb{R}^V_+$ satisfying the Karush-Kuhn-Tucker conditions:

$$\mathbf{0} = \nabla_x L(x^\star; \mu^\star, \sigma^\star) = 2x^\star - M^\top \mu^\star - \sigma^\star$$
$$0 = \mu^{\star\top}(Mx^\star - \mathbf{1})$$
$$0 = \sigma^{\star\top} x^\star.$$

Let $y := \frac{1}{2}\mu^\star$. Since $\sigma^\star$ and M have nonnegative entries, and the third equation holds, the first equation implies that $M^\top y = x^\star$. Multiplying the first equation by $x^{\star\top}$ from the left, and taking the next two equations into account, we get that $\mathbf{1}^\top y = x^{\star\top} x^\star$. As $y^\top(Mx^\star - \mathbf{1}) = \mathbf{0}$ clearly holds, y is the desired vector. □

We are now ready for the first, short and indirect proof of Theorem 3, stating that an identically self-blocking clutter different from $\{\{a\}\}$ is nonideal:

Proof (of Theorem 3). Let $\mathcal{C}$ be an identically self-blocking clutter over ground set V that is different from $\{\{a\}\}$, and let M be its incidence matrix. Suppose for a contradiction that $\mathcal{C}$ is ideal. Then Theorem 6 applies and tells us that $\min\{x^\top x : Mx \geq \mathbf{1}, x \geq \mathbf{0}\} = 1$. Let $x^\star, y$ be as in Lemma 7; so $x^\star = M^\top y$ and $1 = x^{\star\top} x^\star = \mathbf{1}^\top y$. As $\mathcal{C}$ is an identically self-blocking clutter different from $\{\{a\}\}$, every member has cardinality at least two, and by Theorem 1 every two members intersect, implying in turn that $MM^\top \geq J + I$.[1] As a result,

$$1 = x^{\star\top} x^\star = y^\top MM^\top y \geq y^\top(J+I)y = y^\top \mathbf{1}\mathbf{1}^\top y + y^\top y = 1 + y^\top y,$$

implying in turn that $y = \mathbf{0}$, a contradiction.   □

3   Lower Bounding the Packing Number

Here we present a lower bound on the packing number of an arbitrary clutter. We need the following lemma from 1965 proved by Motzkin and Straus:

Lemma 8 ([20]). *Let $G = (V, E)$ be a simple graph, and let L be its V by V adjacency matrix: for all $u, v \in V$, $L_{uv} = 1$ if $\{u,v\} \in E$ and $L_{uv} = 0$ otherwise. Then*

$$\max\{y^\top Ly : \mathbf{1}^\top y = 1, y \geq \mathbf{0}\} = 1 - \frac{1}{\omega(G)}$$

where $\omega(G)$ is the maximum cardinality of a clique of G.

Let $\mathcal{C}$ be a clutter over ground set V. Finding $\nu(\mathcal{C})$ can be cast as finding the maximum cardinality of a clique of a graph. This observation, combined with Lemma 8, has the following consequence:

---

[1] Throughout the paper, J is a square all ones matrix of appropriate dimension, and I is the identity matrix of appropriate dimension.

Lemma 9. *Let $\mathcal{C}$ be a clutter over ground set V, and let M be its incidence matrix. Then*

$$\min\left\{ y^\top MM^\top y - \sum_{C \in \mathcal{C}}(|C|-1)y_C^2 : \mathbf{1}^\top y = 1, y \geq \mathbf{0} \right\} = \frac{1}{\nu(\mathcal{C})}.$$

Proof. ($\leq$) Let $y \in \mathbb{R}_+^{\mathcal{C}}$ be the incidence vector of a maximum packing of $\mathcal{C}$. Then $\frac{1}{\nu(\mathcal{C})} \cdot y$ is a feasible solution whose objective value is $\frac{1}{\nu(\mathcal{C})}$, implying in turn that $\leq$ holds. ($\geq$) Let G be the graph whose vertices correspond to the members of $\mathcal{C}$, where two vertices are adjacent if the corresponding members are disjoint. Let L be the adjacency matrix of G. Then $MM^\top \geq \text{Diag}(|C| - 1 : C \in \mathcal{C}) + J - L$. Notice that there is a bijection between the packings in $\mathcal{C}$ and the cliques in G, and in particular that $\nu(\mathcal{C}) = \omega(G)$. Thus by Lemma 8, for any $y \in \mathbb{R}_+^{\mathcal{C}}$ such that $\mathbf{1}^\top y = 1$,

$$1 - \frac{1}{\nu(\mathcal{C})} \geq y^\top Ly \geq \sum_{C \in \mathcal{C}}(|C|-1)y_C^2 + y^\top Jy - y^\top MM^\top y$$

$$= \sum_{C \in \mathcal{C}}(|C|-1)y_C^2 + 1 - y^\top MM^\top y,$$

implying in turn that $\geq$ holds. □

As a consequence, after employing Carathéodory's theorem (see [9], §3.14) and the Cauchy-Schwarz inequality (see [23]), we get the following lower bound on the packing number of a clutter:

Theorem 10 ([5]). *Let $\mathcal{C}$ be a clutter over ground set V, and let M be its incidence matrix. Then*

$$\nu(\mathcal{C}) \geq \left(\frac{y^\top MM^\top y}{y^\top Jy} - \frac{\min\{|C|-1 : C \in \mathcal{C}\}}{\min\{|V|, |\mathcal{C}|\}} \right)^{-1} \quad \forall y \in \mathbb{R}_+^{\mathcal{C}}, y \neq \mathbf{0}.$$

Proof. Pick a nonzero $y \in \mathbb{R}_+^{\mathcal{C}}$. By Carathéodory's theorem there is a $y' \in \mathbb{R}_+^{\mathcal{C}}$ such that $M^\top y' \leq M^\top y$, $\mathbf{1}^\top y' = \mathbf{1}^\top y$ and $|\text{supp}(y')| \leq |V|$. Lemma 9 applied to $\frac{1}{\mathbf{1}^\top y'} \cdot y'$ implies that

$$\nu(\mathcal{C}) \geq \left(\frac{y'^\top MM^\top y' - \sum_{C \in \mathcal{C}}(|C|-1)y_C'^2}{y'^\top Jy'} \right)^{-1}$$

$$\geq \left(\frac{y^\top MM^\top y}{y^\top Jy} - \frac{\sum_{C \in \mathcal{C}}(|C|-1)y_C'^2}{y'^\top Jy'} \right)^{-1}.$$

By the Cauchy-Schwarz inequality applied to the nonzero entries of y',

$$\frac{\sum_{C \in \mathcal{C}}(|C|-1)y_C'^2}{y'^\top Jy'} \geq \frac{\left(\sum_{C \in \mathcal{C}} \sqrt{|C|-1} \cdot y_C' \right)^2}{|\text{supp}(y')| \cdot y'^\top Jy'} \geq \frac{\min\{|C|-1 : C \in \mathcal{C}\}}{\min\{|V|, |\mathcal{C}|\}}.$$

Combining the last two inequalities proves the theorem. □

This theorem was proved implicitly by Aharoni, Erdős and Linial in 1988. Given that M is the incidence matrix of $\mathcal{C}$, the authors explicitly proved Theorem 10 for y a maximum fractional packing of $\mathcal{C}$:

$$\nu(\mathcal{C}) \geq \left(\frac{\mathbf{1}^\top \mathbf{1}}{\nu^{*2}(\mathcal{C})} - \frac{\min\{|C| - 1 : C \in \mathcal{C}\}}{\min\{|V|, |\mathcal{C}|\}} \right)^{-1} \geq \frac{\nu^{*2}(\mathcal{C})}{|V|}.$$

But one can do better:

Theorem 11. *Let $\mathcal{C}$ be a clutter over ground set V, and let $\alpha := \min\{x^\top x : x \in Q(\mathcal{C})\}$. Then*

$$\nu(\mathcal{C}) \geq \left(\frac{1}{\alpha} - \frac{\min\{|C| - 1 : C \in \mathcal{C}\}}{\min\{|V|, |\mathcal{C}|\}} \right)^{-1}.$$

Proof. Let M be the incidence matrix of $\mathcal{C}$, and let $x \in \mathbb{R}_+^V$ be the point in $Q(\mathcal{C})$ such that $x^\top x = \alpha$. By Lemma 7, there exists $y \in \mathbb{R}_+^{\mathcal{C}}$ such that $x = M^\top y$ and $\mathbf{1}^\top y = \alpha$. By Theorem 10,

$$\nu(\mathcal{C}) \geq \left(\frac{x^\top x}{y^\top \mathbf{1}\mathbf{1}^\top y} - \frac{\min\{|C| - 1 : C \in \mathcal{C}\}}{\min\{|V|, |\mathcal{C}|\}} \right)^{-1}$$
$$= \left(\frac{\alpha}{\alpha^2} - \frac{\min\{|C| - 1 : C \in \mathcal{C}\}}{\min\{|V|, |\mathcal{C}|\}} \right)^{-1},$$

as required.   □

Let

$$\beta := \min\left\{ \frac{y^\top M M^\top y}{y^\top J y} : y \geq \mathbf{0}, y \neq \mathbf{0} \right\} \quad \text{and} \quad \alpha := \min\{x^\top x : Mx \geq \mathbf{1}, x \geq \mathbf{0}\}.$$

By Strong Conic Programming Duality (see [7], Chapter 5),

$$\frac{1}{\sqrt{\beta}} = \max\left\{ \mathbf{1}^\top y : \|M^\top y\| \leq 1, y \geq \mathbf{0} \right\} = \min\{\|x\| : Mx \geq \mathbf{1}, x \geq \mathbf{0}\} = \sqrt{\alpha},$$

so $\beta = \frac{1}{\alpha}$. As a result, the inequality given by Theorem 11 is the best lower bound derived from Theorem 10.

As an immediate consequence of Theorem 11, we get another new lower bound on the packing number of a clutter:

Theorem 12. *Let $\mathcal{C}$ be a clutter. Then $\nu(\mathcal{C}) \geq \min\{x^\top x : x \in Q(\mathcal{C})\}$.*

See ([1], Chapter 3, Theorem 3.2) for an alternative proof of this theorem.

4   Cuboids

Take an even integer $n \geq 2$. A *cuboid* is a clutter over ground set $\{1, \ldots, n\}$ where every member C satisfies $|C \cap \{1,2\}| = |C \cap \{3,4\}| = \cdots = |C \cap \{n-1,n\}| = 1$. Introduced in [2], cuboids form a very special class of clutters, to the point that some of the main conjectures in the field can be phrased equivalently in terms of cuboids [3]. Cuboids play a special role here also:

Theorem 13. *Let $\mathcal{C}$ be an ideal clutter over n elements whose members do not have a common element, and let $\alpha := \min\{x^\top x : x \in Q(\mathcal{C})\}$. Then $\alpha \geq \frac{4}{n}$. Moreover, the following statements are equivalent:*

(i) $\alpha = \frac{4}{n}$,
(ii) n is even, after a possible relabeling of the ground set the sets $\{1,2\}, \{3,4\}$, $\ldots, \{n-1,n\}$ are minimal covers, and the members of minimum cardinality form an ideal cuboid over ground set $\{1, \ldots, n\}$ whose members do not have a common element.

To prove this theorem we need a few preliminary results. Given a simple graph $G = (V, E)$, a *fractional perfect matching* is a $y \in \mathbb{R}_+^E$ such that $\mathbf{1}^\top y = \frac{|V|}{2}$ and for each vertex $u \in V$, $\sum (y_e : u \in e) = 1$. We need the following classic result:

Lemma 14 (folklore). *If a bipartite graph has a fractional perfect matching, then it has a perfect matching.*

Lemma 15. *Take an integer $n \geq 2$, and let $\mathcal{C}$ be a clutter over ground set $\{1, \ldots, n\}$. Then the following statements are equivalent:*

(i) $\mathcal{C}$ has a fractional packing of value 2 and $b(\mathcal{C})$ has a fractional packing of value $\frac{n}{2}$,
(ii) n is even, and in $\mathcal{C}$, after a possible relabeling of the ground set the sets $\{1,2\}, \{3,4\}, \ldots, \{n-1,n\}$ are minimal covers, and the members of minimum cardinality form a cuboid over ground set $\{1, \ldots, n\}$ with a fractional packing of value 2.

Proof. (ii) $\Rightarrow$ (i) is immediate. (i) $\Rightarrow$ (ii): Let M, N be the incidence matrices of $\mathcal{C}, b(\mathcal{C})$, respectively. Let $y \in \mathbb{R}_+^{\mathcal{C}}$ be a fractional packing of $\mathcal{C}$ of value 2; so $M^\top y \leq 1$ and $\mathbf{1}^\top y = 2$. Let $t \in \mathbb{R}_+^{b(\mathcal{C})}$ be a fractional packing of $b(\mathcal{C})$ of value $\frac{n}{2}$; so $N^\top t \leq 1$ and $\mathbf{1}^\top t = \frac{n}{2}$. Then $n = \mathbf{1}^\top \mathbf{1} \geq t^\top N M^\top y \geq t^\top J y = t^\top \mathbf{1}\mathbf{1}^\top y = \frac{n}{2} \cdot 2 = n$. Thus equality holds throughout, implying in turn that **(1)** $M^\top y = 1$ and $\mathbf{1}^\top y = 2$, **(2)** $N^\top t = 1$ and $\mathbf{1}^\top t = \frac{n}{2}$, **(3)** if $y_C > 0$ and $t_B > 0$ for some $C \in \mathcal{C}$ and $B \in b(\mathcal{C})$, then $|C \cap B| = 1$. Notice that $\tau(\mathcal{C}) \geq 2$ and $\tau(b(\mathcal{C})) \geq \frac{n}{2}$, so every member of $\mathcal{C}$ has cardinality at least $\frac{n}{2}$ while every member of $b(\mathcal{C})$ has cardinality at least 2. Together with (1) and (2), these observations imply that n is even, and **(4)** if $y_C > 0$ for some $C \in \mathcal{C}$, then $|C| = \frac{n}{2}$, **(5)** if $t_B > 0$ for some $B \in b(\mathcal{C})$, then $|B| = 2$. Let G be the graph over vertices $\{1, \ldots, n\}$ whose edges correspond to $\{B \in b(\mathcal{C}) : t_B > 0\}$. Pick $C \in \mathcal{C}$ such that $y_C > 0$. Then by (3) the vertex subset C intersects every edge of G exactly once, implying in turn that G is a bipartite graph. By (2) G has a fractional perfect matching, and as the graph is bipartite, there must be a perfect matching by Lemma 14, labeled as $\{1,2\}, \{3,4\}, \ldots, \{n-1,n\}$ after a possible relabeling of the ground set. As a consequence, the members of C of minimum cardinality form a cuboid over ground set $\{1,\ldots,n\}$ which by (1) and (4) has a fractional packing of value 2. Thus (ii) holds.   □

Remark 16. *Let C be a clutter over n elements, and let $\alpha, x^\star$ be the optimal value and solution of $\min\{x^\top x : x \in Q(C)\}$, respectively. Then the following statements hold:*

(i) $\alpha \geq \frac{\nu^{\star 2}(C)}{n}$. Moreover, equality holds if and only if $x^\star = \frac{\nu^\star(C)}{n} \cdot \mathbf{1}$.
(ii) Assume that every member has cardinality at least two. Then $\alpha \leq \frac{n}{4}$. Moreover, equality holds if and only if $x^\star = \frac{1}{2} \cdot \mathbf{1}$.

Proof. **(i)** The Cauchy-Schwarz inequality implies that $\alpha = x^{\star\top} x^\star \geq \frac{(\mathbf{1}^\top x^\star)^2}{n} \geq \frac{\tau^{\star 2}(C)}{n} = \frac{\nu^{\star 2}(C)}{n}$. Moreover, equality holds throughout if and only if the entries of $x^\star$ are equal and $\mathbf{1}^\top x^\star = \nu^\star(C)$, i.e. $x^\star = \frac{\nu^\star(C)}{n} \cdot \mathbf{1}$. **(ii)** is immediate.   □

We also need the following result proved implicitly in [2] (its proof can be found in the proof of Theorem 1.6, Claim 3 on p. 543):

Lemma 17 ([2]). *Take an even integer $n \geq 2$, and let C be an ideal clutter over ground set $\{1,\ldots,n\}$ where $\{1,2\}, \{3,4\}, \ldots, \{n-1,n\}$ are minimal covers. Then $\{C \in C : |C| = \frac{n}{2}\}$ is an ideal cuboid.*

We are now ready to prove Theorem 13:

Proof (of Theorem 13). By Remark 16 (i), $\alpha \geq \frac{\nu^{\star 2}(C)}{n} = \frac{\tau^2(C)}{n} \geq \frac{4}{n}$, where the equality follows from the fact that C is ideal, and the last inequality holds because the members have no common element. **(i)** $\Rightarrow$ **(ii)**: Assume that $\alpha = \frac{4}{n}$. Let $x^\star$ be the optimal solution of $\min\{x^\top x : x \in Q(C)\}$. Then by Remark 16 (i), $\tau(C) = 2$ and $x^\star = \left(\frac{2}{n}, \frac{2}{n}, \ldots, \frac{2}{n}\right)$. Let M be the incidence matrix of C. By Lemma 7, there is a $y \in \mathbb{R}_+^C$ such that $M^\top y = x^\star$ and $\mathbf{1}^\top y = \frac{4}{n}$, that is,

$\frac{n}{2} \cdot y$ is a fractional packing of C of value 2.

Let $\beta, z^\star$ be the optimal value and solution of $\min\{z^\top z : z \in Q(b(C))\}$. As C is an ideal clutter, it follows from Theorem 6 that $\beta = \frac{1}{\alpha} = \frac{n}{4}$. Thus by Remark 16 (ii), $z^\star = \left(\frac{1}{2}, \frac{1}{2}, \ldots, \frac{1}{2}\right)$. Let N be the incidence matrix of $b(C)$. By Lemma 7, there is a $t \in \mathbb{R}_+^{b(C)}$ such that $N^\top t = z^\star$ and $\mathbf{1}^\top z^\star = \frac{n}{4}$, that is,

$2t$ is a fractional packing of $b(C)$ of value $\frac{n}{2}$.

It therefore follows from Lemma 15 that n is even, after a possible relabeling of the ground set the sets $\{1,2\}, \{3,4\}, \ldots, \{n-1,n\}$ are minimal covers of C, and the members of C of minimum cardinality form a cuboid C_0 over ground set

$\{1, \dots, n\}$ with a fractional packing of value 2. In particular, the members of $\mathcal{C}_0$ do not have a common element. Moreover, since $\mathcal{C}$ is an ideal clutter, it follows from Lemma 17 that $\mathcal{C}_0$ is an ideal cuboid, thereby proving (ii). **(ii)** $\Rightarrow$ **(i)**: Observe that every member has cardinality at least $\frac{n}{2}$, so $\left(\frac{2}{n}, \frac{2}{n}, \dots, \frac{2}{n}\right) \in Q(\mathcal{C})$, implying in turn that $\alpha \leq \frac{4}{n}$. Since $\alpha \geq \frac{4}{n}$ also, (i) must hold. $\qquad\square$

We showed that among ideal clutters $\mathcal{C}$ whose members do not have a common element, it is essentially cuboids that achieve the smallest possible value for $\min\{x^\top x : x \in Q(\mathcal{C})\}$. Our proof relied on Lemma 15, which in itself has another consequence. Viewing clutters as *simple games*, Hof et al. [16] showed that given a clutter $\mathcal{C}$ over n elements, its *critical threshold value* is always at most $\frac{n}{4}$, and this maximum is achieved if, and only if, $\mathcal{C}$ has a fractional packing of value $\frac{n}{2}$ and $b(\mathcal{C})$ has a fractional packing of value 2. Thus by Lemma 15, it is essentially blockers of cuboids that achieve the largest possible critical threshold value.

5   Bypassing Gauge Duality

Here we present a longer and direct proof of the main result of the paper, Theorem 3. This proof will bypass the use of Theorem 6. We will need the following two lemmas:

Lemma 18. *Let $\mathcal{C}, \mathcal{B}$ be clutters over ground set V such that $|C \cap B| = 1$ for all $C \in \mathcal{C}, B \in \mathcal{B}$, for which there exist nonzero $y \in \mathbb{R}_+^{\mathcal{C}}$ and $t \in \mathbb{R}_+^{\mathcal{B}}$ such that $\sum_{C \in \mathcal{C}} y_C \chi_C = \sum_{B \in \mathcal{B}} t_B \chi_B$. Then*

$$\nu(\mathcal{C}) \geq \left(\frac{\mathbf{1}^\top t}{\mathbf{1}^\top y} - \frac{\min\{|C| - 1 : C \in \mathcal{C}\}}{\min\{|V|, |\mathcal{C}|\}} \right)^{-1}$$

$$\nu(\mathcal{B}) \geq \left(\frac{\mathbf{1}^\top y}{\mathbf{1}^\top t} - \frac{\min\{|B| - 1 : B \in \mathcal{B}\}}{\min\{|V|, |\mathcal{B}|\}} \right)^{-1}.$$

Proof. Due to the symmetry between $\mathcal{C}$ and $\mathcal{B}$, it suffices to prove the first inequality. After possibly scaling t, we may assume that $\mathbf{1}^\top t = 1$. Our hypotheses imply that for each $C' \in \mathcal{C}$,

$$\sum_{C \in \mathcal{C}} y_C |C' \cap C| = \sum_{C \in \mathcal{C}} y_C \chi_{C'}^\top \chi_C = \sum_{B \in \mathcal{B}} t_B \chi_{C'}^\top \chi_B = \sum_{B \in \mathcal{B}} t_B |C' \cap B| = 1.$$

Thus, given that M is the incidence matrix of $\mathcal{C}$, the equalities above state that $MM^\top y = \mathbf{1}$. And Theorem 10 applied to y implies that

$$\nu(\mathcal{C}) \geq \left(\frac{y^\top \mathbf{1}}{y^\top J y} - \frac{\min\{|C| - 1 : C \in \mathcal{C}\}}{\min\{|V|, |\mathcal{C}|\}} \right)^{-1},$$

therefore implying the first inequality. $\qquad\square$

Given a clutter, a fractional cover is *minimal* if it is not greater than or equal to another fractional cover. Given an ideal clutter, it is well-known that every minimal fractional cover can be written as a convex combination of the incidence vectors of minimal covers (see for instance [11]). We will use this fact below:

Lemma 19. *Let $\mathcal{C}, \mathcal{B}$ be blocking ideal clutters. Then there exist nonempty $\mathcal{C}' \subseteq \mathcal{C}$ and $\mathcal{B}' \subseteq \mathcal{B}$ such that $|C \cap B| = 1$ for all $C \in \mathcal{C}', B \in \mathcal{B}'$, for which there exist nonzero $y \in \mathbb{R}_+^{\mathcal{C}'}$ and $t \in \mathbb{R}_+^{\mathcal{B}'}$ such that $\sum_{C \in \mathcal{C}'} y_C \chi_C = \sum_{B \in \mathcal{B}'} t_B \chi_B$.*

Proof. Let M, N be the incidence matrices of $\mathcal{C}, \mathcal{B}$, respectively. Let $x^\star$ be the optimal solution of $\min\{x^\top x : x \in Q(\mathcal{C})\}$. Then $x^\star$ is a minimal fractional cover of $\mathcal{C}$. As $\mathcal{C}$ is an ideal clutter, $x^\star = N^\top t$ for some $t \in \mathbb{R}_+^{\mathcal{B}}$ such that $\mathbf{1}^\top t = 1$. Moreover, by Lemma 7, there exists $y \in \mathbb{R}_+^{\mathcal{C}}$ such that $M^\top y = x^\star$ and $y^\top (Mx^\star - \mathbf{1}) = \mathbf{0}$. Thus, $\mathbf{1}^\top y = x^{\star\top} M^\top y = t^\top N M^\top y \geq t^\top J y = t^\top \mathbf{1}\mathbf{1}^\top y = \mathbf{1}^\top y$, implying that $t^\top N M^\top y = t^\top J y$. Therefore, if $\mathcal{C}' := \{C \in \mathcal{C} : y_C > 0\}$ and $\mathcal{B}' := \{B \in \mathcal{B} : t_B > 0\}$, we have that $|C \cap B| = 1$ for all $C \in \mathcal{C}', B \in \mathcal{B}'$. Moreover, the equation $M^\top y = x^\star = N^\top t$ implies that $\sum_{C \in \mathcal{C}'} y_C \chi_C = \sum_{B \in \mathcal{B}'} t_B \chi_B$. As $\mathcal{C}'$ and $\mathcal{B}'$ are clearly nonempty, they are the desired clutters. □

Theorem 20. *Let $\mathcal{C}, \mathcal{B}$ be blocking ideal clutters. If $\tau(\mathcal{C}) \geq 2$ and $\tau(\mathcal{B}) \geq 2$, then $\nu(\mathcal{C}) \geq 2$ or $\nu(\mathcal{B}) \geq 2$.*

Proof. Assume that $\tau(\mathcal{C}) \geq 2$ and $\tau(\mathcal{B}) \geq 2$. By Lemma 19, there exist nonempty $\mathcal{C}' \subseteq \mathcal{C}$ and $\mathcal{B}' \subseteq \mathcal{B}$ such that $|C \cap B| = 1$ for all $C \in \mathcal{C}', B \in \mathcal{B}'$, for which there exist nonzero $y \in \mathbb{R}_+^{\mathcal{C}'}$ and $t \in \mathbb{R}_+^{\mathcal{B}'}$ such that $\sum_{C \in \mathcal{C}'} y_C \chi_C = \sum_{B \in \mathcal{B}'} t_B \chi_B$. As the members of $\mathcal{C}'$ and $\mathcal{B}'$ have cardinality at least two, we get from Lemma 18 that $\nu(\mathcal{C}') > \frac{\mathbf{1}^\top y}{\mathbf{1}^\top t}$ and $\nu(\mathcal{B}') > \frac{\mathbf{1}^\top t}{\mathbf{1}^\top y}$. As $\nu(\mathcal{C}) \geq \nu(\mathcal{C}')$ and $\nu(\mathcal{B}) \geq \nu(\mathcal{B}')$, it follows that $\nu(\mathcal{C}) \geq 2$ or $\nu(\mathcal{B}) \geq 2$, as required. □

We are now ready to prove Theorem 3 again, stating that an identically self-blocking clutter different from $\{\{a\}\}$ is nonideal:

Proof (of Theorem 3). Let $\mathcal{C}$ be an identically self-blocking clutter different from $\{\{a\}\}$. Then $\tau(\mathcal{C}) \geq 2$, and $\nu(\mathcal{C}) = 1$ by Theorem 1. Theorem 20 now applies and tells us that $\mathcal{C}$ cannot be ideal, as required. □

6   Concluding Remarks

Given a general blocking pair $\mathcal{C}, \mathcal{B}$ of ideal clutters, what can be said about them? This is an important research topic in Integer Programming and Combinatorial Optimization. We showed that if $\tau(\mathcal{C}) \geq 2$ and $\tau(\mathcal{B}) \geq 2$, then $\nu(\mathcal{C}) \geq 2$ or $\nu(\mathcal{B}) \geq 2$ (Theorems 3 and 20). Equivalently, if the members of $\mathcal{C}, \mathcal{B}$ have cardinality at least two, then one of the two clutters has a *bicoloring*, i.e. the ground set can be bicolored so that every member receives an element of each color. Next to Lehman's *width-length* characterization [18], this is the only other fact known about the structure of $\mathcal{C}$ and $\mathcal{B}$. As such, we expect the results as well as the tools introduced here to help us address the question in mind.

Our main result led us to two computable lower bounds – one weaker than the other – on the packing number of an arbitrary clutter (Theorems 11 and 12). We believe these lower bounds will have applications beyond the scope of this paper. We also characterized the clutters on which one of the lower bounds is at its weakest; we showed that these clutters are essentially cuboids (Theorem 13). Combined with evidence from [2,3], this only stresses further the central role of cuboids when studying ideal clutters.

Finally, we used techniques from Convex Optimization to prove the main result of the paper. A natural question is whether there is an elementary and discrete approach for proving the result? Conjecture 4 provides a potential approach and leads to an interesting research direction.

Acknowledgements. We would like to thank Fatma Kılınç-Karzan, Kanstantsin Pashkovich and Levent Tunçel for fruitful discussions about different parts of this work. We would also like to thank the referees; their feedback improved the presentation of the paper. This work was supported in part by ONR grant 000141812129 and NSF grant CMMI 1560828.

References

1. Abdi, A.: Ideal clutters. Ph.D. dissertation, University of Waterloo (2018)
2. Abdi, A., Cornuéjols, G., Pashkovich, K.: Ideal clutters that do not pack. Math. Oper. Res. **43**(2), 533–553 (2018)
3. Abdi, A., Cornuéjols, G., Guričanová, N., Lee, D.: Cuboids, a class of clutters. Submitted
4. Abdi, A., Pashkovich, K.: Delta minors, delta free clutters, and entanglement. SIAM J. Discrete Math. **32**(3), 1750–1774 (2018)
5. Aharoni, R., Erdős, P., Linial, N.: Optima of dual integer linear programs. Combinatorica **8**(1), 13–20 (1988)
6. Berge, C.: Hypergraphs: Combinatorics of Finite Sets. North Holland, Amsterdam (1989)
7. Boyd, S., Vandenberghe, L.: Convex Optimization. Cambridge University Press, Cambridge (2004)
8. Chaiken, S.: Extremal length and width of blocking polyhedra, Kirchhoff spaces and multiport networks. SIAM J. Algebraic Discrete Methods **8**(4), 635–645 (1987)
9. Conforti, M., Cornuéjols, G., Zambelli, G.: Integer Programming. Springer, Berlin (2014)
10. Cornuéjols, G.: Combinatorial Optimization. Packing and Covering. SIAM, Philadelphia (2001)
11. Cornuéjols, G., Novick, B.: Ideal 0, 1 matrices. J. Comb. Theor. Ser. B **60**(1), 145–157 (1994)
12. Edmonds, J., Fulkerson, D.R.: Bottleneck extrema. J. Comb. Theor. Ser. B **8**(3), 299–306 (1970)
13. Freund, R.M.: Dual gauge programs, with applications to quadratic programming and the minimum-norm problem. Math. Program. **38**, 47–67 (1987)
14. Fulkerson, D.R.: Blocking and anti-blocking pairs of polyhedra. Math. Program. **1**, 168–194 (1971)
15. Fulkerson, D.R.: Blocking polyhedra. In: Harris, B. (ed.) Graph Theory and Its Applications, pp. 93–112. Academic Press, New York (1970)
16. Hof, F., Kern, W., Kurz, S., Pashkovich, K., Paulusma, D.: Simple games versus weighted voting games: bounding the critical threshold value. Preprint, arXiv:1810.08841 (2018)

17. Isbell, J.R.: A class of simple games. Duke Math. J. **25**(3), 423–439 (1958)
18. Lehman, A.: On the width-length inequality. Math. Program. **17**(1), 403–417 (1979)
19. Lovász, L.: Energy of convex sets, shortest paths, and resistance. J. Comb. Theor. Ser. A **94**(2), 363–382 (2001)
20. Motzkin, T.S., Straus, E.G.: Maxima for graphs and a new proof of a theorem of Turán. Canad. J. Math. **17**, 533–540 (1965)
21. Seymour, P.D.: The forbidden minors of binary matrices. J. Lond. Math. Soc. **2**(12), 356–360 (1976)
22. Seymour, P.D.: The matroids with the max-flow min-cut property. J. Comb. Theor. Ser. B **23**(2–3), 189–222 (1977)
23. Steele, J.M.: The Cauchy-Schwarz Master Class: An Introduction to the Art of Mathematical Inequalities. Cambridge University Press, Cambridge (2004)

Min-Max Correlation Clustering
via MultiCut

Saba Ahmadi[1]([⊠]), Samir Khuller[2], and Barna Saha[3]

[1] University of Maryland, College Park, College Park, USA
saba@cs.umd.edu
[2] Northwestern University, Evanston, USA
samir.khuller@northwestern.edu
[3] College of Information and Computer Science,
University of Massachussetts Amherst, Amherst, USA
barna@cs.umass.edu

Abstract. Correlation clustering is a fundamental combinatorial optimization problem arising in many contexts and applications that has been the subject of dozens of papers in the literature. In this problem we are given a general weighted graph where each edge is labeled positive or negative. The goal is to obtain a partitioning (clustering) of the vertices that minimizes disagreements – weight of negative edges trapped inside a cluster plus positive edges between different clusters. Most of the papers on this topic mainly focus on minimizing total disagreement, a global objective for this problem.

In this paper we study a cluster-wise objective function that asks to minimize the maximum number of disagreements of each cluster, which we call min-max correlation clustering. The min-max objective is a natural objective that respects the quality of every cluster. In this paper, we provide the first nontrivial approximation algorithm for this problem achieving an $\mathcal{O}(\log n)$ approximation for general weighted graphs. To do so, we also obtain a corresponding result for multicut where we wish to find a multicut solution while trying to minimize the total weight of cut edges on every component. The results are then further improved to obtain an $\mathcal{O}(r^2)$-approximation for min-max correlation clustering and min-max multicut for graphs that exclude $K_{r,r}$ minors.

Keywords: Correlation clustering · Multicut ·
Approximation algorithms

A full version of this paper appears at http://cs.umd.edu/~samir/LCC.pdf
The first two authors are supported by NSF grant CNS 156019. Part of the research was done when the third author was visiting the Simons Institute of Theory of Computing and the author is supported by NSF CAREER 1652303, NSF CCF 1464310 and a Google faculty award.

A. Lodi and V. Nagarajan (Eds.): IPCO 2019, LNCS 11480, pp. 13–26, 2019.
https://doi.org/10.1007/978-3-030-17953-3_2

1   Introduction

Correlation clustering is a fundamental optimization problem introduced by Bansal, Blum and Chawla [3]. In this problem, we are given a general weighted graph where each edge is labeled positive or negative. The goal is to obtain a partitioning of the vertices into an arbitrary number of clusters that agrees with the edge labels as much as possible. That is a clustering that minimizes disagreements, which is the weight of positive edges between the clusters plus the weight of negative edges inside the clusters. In addition, correlation clustering captures some fundamental graph cut problems including min s-t cut, multiway cut and multicut. Correlation clustering has been studied extensively for more than a decade [1,2,6,7,10]. Most of the papers have focused on a global min-sum objective function, i.e. minimizing total number of disagreements or maximizing the total number of agreements.

In recent work, Puleo and Milenkovic [14] introduced a local vertex-wise min-max objective for correlation clustering which bounds the maximum number of disagreements on each node. This problem arises in many community detection applications in machine learning, social sciences, recommender systems and bioinformatics [8,13,16]. This objective function makes sure each individual has a minimum quality within the clusters. They showed this problem is NP-hard even on un-weighted complete graphs, and developed an $O(1)$ approximation algorithm for unweighted complete graphs. Charikar et al. [5] improved upon the work by Puleo et al. [14] for complete graphs by giving a 7 approximation. For general weighted graphs, their approximation bound is $O(\sqrt{n})$ where n is the number of vertices. Both these algorithms rely on LP rounding, based on a standard linear program relaxation for the problem. In contrast, for the global minimization objective an $O(\log n)$-approximation can be obtained [10]. Therefore, the local objective for correlation clustering seems significantly harder to approximate than the global objective.

In this work, we propose a new local cluster-wise min-max objective for correlation clustering – minimizing the maximum number of disagreements of each cluster. This captures the case when we wish to create communities that are harmonious, as global min sum objectives could create an imbalanced community structure. This new local objective guarantees fairness to communities instead of individuals. To name a few applications for this new objective, consider a task of instance segmentation in an image which can be modeled using correlation clustering [11,12]. A cluster-wise min-max objective makes sure each detected instance has a minimum quality. Another example is in detecting communities in social networks, this objective makes sure there are no communities with lower quality compared to the other communities. No hardness results are known for the cluster-wise min-max objective.

A similar objective was proposed for the multiway cut problem by Svitkina and Tardos [15]. In the min-max multiway cut problem, given a graph G and k terminals the goal is to get a partitioning of G of size k that separates all terminals and the maximum weight of cut edges on each part is minimized. Svitkina and Tardos [15] showed an $\mathcal{O}(\log^3 n)$ approximation algorithm for min-max multiway cut on general graphs (this bound immediately improves to $\mathcal{O}(\log^2 n)$ by using better bisection algorithms). Bansal et al. [4] studied a graph partitioning problem called min-max k-partitioning from a similar perspective. In this problem, given a graph $G = (V, E)$ and $k \geq 2$ the goal is to partition the vertices into k roughly equal parts $S_1, \cdots, S_k$ while minimizing $\max_i \delta(S_i)$. They showed an $\mathcal{O}(\sqrt{\log n \log k})$ approximation algorithm for this problem. They also improved the approximation ratio given by Svitkina et al. [15] for min-max multiway cut to $\mathcal{O}(\sqrt{\log n \log k})$. Bansal et al's seminal work [4] uses the concept of orthogonal separators introduced by Chlamtac et al. [9] to achieve their result.

2   Results and High Level Ideas

In this paper, we give an approximation algorithm for the problem of min-max correlation clustering.

Definition 1 *(Min-max Correlation Clustering). Let $G = (V, E)$ be an edge-weighted graph such that each edge is labeled positive or negative. The min-max correlation clustering problem asks for a partioning of the nodes (a clustering) where the maximum disagreement of a cluster is minimized. Disagreement of a cluster C is the weight of negative edges with both endpoints inside C plus the weight of positive edges with exactly one endpoint in C.*

We prove the following theorem for min-max correlation clustering.

Theorem 1. *Given an edge weighted graph $G = (V, E)$ on n vertices such that each edge is labeled positive or negative, there exists a polynomial time algorithm which outputs a clustering $\mathcal{C} = \{C_1, \cdots, C_C\}$ of G such that the disagreement on each $C_i \in \mathcal{C}$ is at most $\mathcal{O}(\log(n)) \cdot OPT$; where OPT is the maximum disagreement on each cluster in an optimal solution of min-max correlation clustering.*

In order to prove Theorem 1, we give a reduction from the problem of min-max correlation clustering to a problem which we call min-max multicut.

Definition 2 *(Min-max Multicut). Given an edge weighted graph $G - (V, E)$ and a set of source-sink pairs $\{(s_1, t_1), \cdots, (s_T, t_T\}$, the goal is to give a partitioning $\mathcal{P} = \{P_1, P_2, \cdots, P_{|\mathcal{P}|}\}$ of G such that all the source sink pairs are separated, and $\max_{1 \leq i \leq |\mathcal{P}|} \delta(P_i)$ is minimized.*

In min-max multicut, we do not force each part of the partitioning to have a terminal and there could be some parts without any terminals in the final solution. However, in the min-max multiway cut problem introduced by Svitkina and Tardos [15], each part needs to have exactly one terminal. We prove the following theorem for min-max multicut:

Theorem 2. *Given an edge weighted graph $G = (V, E)$ on n vertices, and a set of source sink pairs $S_G = \{(s_1, t_1), \cdots, (s_T, t_T)\}$, there exists a polynomial time algorithm which outputs a partitioning $\mathcal{P} = \{P_1, \cdots, P_{|\mathcal{P}|}\}$ of G such that all source sink pairs are separated, and $\max_{1 \leq i \leq |\mathcal{P}|} \delta(P_i) \leq \mathcal{O}(\log(n)) \cdot OPT$; where OPT is the value of the optimum solution of min-max multicut.*

We also consider the following variation of min-max multicut called min-max constrained multicut. In this variation, the goal is to partition a graph into a minimum number of parts to separate all source-sink pairs.

Definition 3 *(Min-max Constrained Multicut). An edge weighted graph $G = (V, E)$ and a set of source-sink pairs $\{(s_1, t_1), \cdots, (s_T, t_T)\}$ is given. Given k the minimum number of parts needed to separate all source sink pairs, the goal is to partition G into k parts $\{P_1, \cdots, P_k\}$ which separate all source-sink pairs, and $\max_{1 \leq i \leq k} \delta(P_i)$ is minimized.*

We defer our results for this problem to the full version of this paper. Finally, we get improved approximation ratios for min-max correlation clustering, min-max multicut on graphs excluding a fixed minor.

Theorem 3. *Given an edge weighted graph G excluding $K_{r,r}$ minors, there exist polynomial time $\mathcal{O}(r^2)$-approximation algorithms for min-max correlation clustering and min-max multicut.*

2.1   High Level Ideas

Most algorithms for correlation clustering with the global minimizing disagreement objective use a linear programming relaxation [6,7,10]. The recent work of Charikar, Gupta and Scharwtz also uses a similar linear programming relaxation for the vertex-wise min-max objective [5]. Surprisingly, these relaxations do not work for the min-max correlation clustering problem considered in this paper. Indeed, simply obtaining a linear programming relaxation for the cluster-wise min-max objective looks challenging!

Bansal et al. [4] considered a semidefinite programming (SDP) based approximation algorithm for min-max k balanced partitioning and min-max multiway cut with k terminals. In their approach, instead of finding the entire solution in one shot, they obtain a single part at a time. It is possible to encode the same problem with a linear program albeit with a worse approximation guarantee. They use SDP rounding to obtain a part with low cut capacity, and repeat the process until the parts produce a covering of all the vertices. By properly adjusting the weight of each part, the covering can be obtained efficiently. Finally, they convert the covering to partitioning.

The problem of extracting a single cluster of min-max correlation clustering can be captured by a semidefinite programming formulation. Here it is not over a cut capacity objective, instead we need to simultaneously consider the intra-cluster negative edges as well as inter-cluster positive edges. Indeed, even for the global minimization objective, we are not aware of any good rounding algorithm based on SDP relaxation of correlation clustering. Therefore, rounding the SDP formulation directly looks difficult. To overcome this, we instead consider a new problem of *min-max multicut*. Demaine et al. [10] have shown an approximation preserving reduction between multicut and correlation clustering (for the global objective function). By solving the min-max multicut problem and then using the aforementioned reduction, we solve the min-max correlation clustering problem.

First, the reduction of Demaine et al. [10] is for the global objective, and an equivalence in global objective does not necessarily correspond to equivalency in local min-max objective. Fortunately, we could show indeed such an equivalency can be proven (the details are deferred to the full version). Thus, the "multicut" route seems promising as it optimizes over a cut objective. We consider obtaining each component of the min-max multicut problem, repeat this process to obtain a covering [4], and finally convert the covering to a partitioning.

The major technical challenge comes in rounding the SDP relaxation for the multicut instance where we seek to find a single component with good cut property. In order for the relaxation to be valid, we have to add new constraints so that no source-sink pair (s_i, t_i) appears together. We also need to ensure that the component obtained satisfies a weight lower bound by assigning weights to each vertex. This is important in the next step when we wish to get a covering of all the vertices: we will decrease the weight of the vertices in the component recovered and again recompute the SDP relaxation with the same weight lower bound. This ensures the same component is not repeatedly recovered and a final covering can be obtained. To solve min-max multiway cut, Bansal et al. [4] need to separate k terminals. To do so, they can just guess which of the k terminals if any should appear in the current component with only $k + 1$ guesses. For us, the number of such guesses would be 3^T where T is the number of source sink pairs since for every pair (s_i, t_i), either s_i or t_i or none would be part of the returned component. Since T could be $O(n^2)$ such a guessing is prohibitive. We need to come up with a new approach to address this issue.

We use a SDP relaxation to compute a metric on the graph vertices and add additional constraints to separate source sink pairs along with the spreading constraints from Bansal et al. [4] to recover a component of desired size. Next, we use the SDP separator technique introduced by Bansal et al. [4] to design a rounding algorithm that returns a set $S = \{S_1, S_2, \cdots, S_j\}$, such that for each $S_i \in S$, there are no source-sink pairs in S_i. Bansal et al. [4] need to glue the sets in S and report it as a single component, since they wish to get a solution with specified number of components at the end. However, in min-max multicut problem, the number of components does not matter. Therefore, we do not need to union the sets in S, and as a result no source-sink violations happen.

It is possible to use a linear programming formulation for the detour via multicut and use LP-separators of Bansal et al. [4] in place of orthogonal separators and follow our algorithm. This would achieve a similar bound for min-max multicut and min-max correlation clustering in general graphs, but a much better bound of $\mathcal{O}(r^2 \cdot OPT)$ for graphs that exclude $K_{r,r}$ minors. The details are deferred to the full version.

3   Min-Max Multicut

Given a subset $S \subseteq V$, let $\delta(S)$ denote the number of edges with exactly one end-point in S and let the number of source sink pairs (s_i, t_i) such that both s_i and t_i belong to S be $vio(S)$.

In order to prove Theorem 2, we first wish to find a set $S = \{S_1, \cdots, S_j\}$, such that $\forall S_i \in S, S_i \subseteq V$, and $\delta(S_i) \leq \mathcal{O}(\log(n)) \cdot OPT$, where OPT is the maximum number of cut edges on each part of the optimum partitioning for the min-max multicut problem on graph G. In addition, $\Pr[vio(S_i) \geq 1] \leq 1/n$, where n is the number of vertices in G.

Graph $G = (V, E)$ can have arbitrary edge weights, $w : E \to \mathbb{R}^+$. We assume graph $G = (V, E)$ is also a vertex-weighted graph, and there is a measure η on V such that $\eta(V) = 1$. This measure is used to get a covering of all the vertices. In Sect. 3.4, Theorem 4 is repeatedly applied to generate a family of sets that cover all the vertices. When a vertex is covered its weight is decreased so the uncovered vertices have a higher weight. Constraint $\eta(S) \in \eta(S) = \sum_{i=1}^{j} \eta(S_i) \in [H/4, 12H]$ makes sure the newly computed family of sets S has adequate coverage. Parameter $H \in (0, 1)$ is equal to $1/k$ where k is the number of parts in the optimum partitioning which we guess. Since the maximum number of parts is at most n, $H \geq 1/n$.

After getting a covering of all the vertices, in Sect. 3.4, it is explained how to convert a covering into a partitioning with the properties desired in Theorem 2. In order to prove Theorem 1, in the full version of this paper we show how a $\mathcal{O}(\log n)$-approximation algorithm for min-max multicut implies a $\mathcal{O}(\log n)$-approximation algorithm for min-max correlation clustering.

First we prove the following theorem:

Theorem 4. *We are given an edge-weighted graph $G = (V, w)$, a set of source sink pairs S_G, a measure η on V such that $\eta(V) = 1$, and a parameter $H \in (0, 1)$. Assume there exists a set $T \subseteq V$ such that $\eta(T) \in [H, 2H]$, and $vio(T) = 0$. We design an efficient randomized algorithm to find a set S, where $S = \{S_1, \cdots, S_j\}$ satisfying $\forall S_i \in S, S_i \subseteq V$, $\eta(S) = \sum_{i=1}^{j} \eta(S_i) \in [H/4, 12H]$, and $\forall S_i \in S$, $\Pr[vio(S_i) \geq 1] \leq \frac{1}{n}$, and:*

$$\delta(S_i) \leq \mathcal{O}(\log(n)) \cdot \min\left\{\delta(T) : \eta(T) \in [H, 2H], \forall(s_i, t_i) \in S_G, |\{s_i, t_i\} \cap T| \leq 1\right\}$$

In order to prove this theorem, we use the notion of $m-$orthogonal separators, a distribution over subsets of vectors, introduced by Chlamtac et al. [9] which is explained in the following:

Definition 4. *Let X be an ℓ_2^2 space (i.e a finite collection of vectors satisfying ℓ_2^2 triangle inequalities with the zero vector) and $m > 0$. A distribution over subsets S of X is an $m-$orthogonal separator of X with probability scale $\alpha > 0$, separation threshold $0 < \beta < 1$, and distortion $D > 0$, if the following conditions hold:*

- $\forall u \in X, \Pr(u \in S) = \alpha \|u\|^2$
- $\forall u, v \in X$ *if* $\|u - v\|^2 \geq \beta \min\{\|u\|^2, \|v\|^2\}$ *then* $\Pr(u \in S \text{ and } v \in S) \leq \frac{\min\{\Pr(u \in S), \Pr(v \in S)\}}{m}$
- $\forall u, v \in X$, $\Pr(I_S(u) \neq I_S(v)) \leq \alpha D \cdot \|u - v\|^2$, *where I_S is the indicator function for the set S.*

Operator $\|.\|$ shows the ℓ^2 norm. Chlamtac et al. [9] proposed an algorithm for finding m-orthogonal separators.

Theorem 5 [9]. *There exists a polynomial-time randomized algorithm that given an ℓ_2^2 space X containing 0 and a parameter $m > 0$, and $0 < \beta < 1$, generates an $m-$orthogonal separator with distortion $D = \mathcal{O}_\beta(\sqrt{\log |X| \log m})$ and $\alpha \geq \frac{1}{poly(|X|)}$.*

3.1 SDP Relaxation

In order to prove Theorem 4, we use the following SDP relaxation which is inspired by Bansal et al. [4] except for Constraints 5 and 6. In this relaxation, we assign a vector $\bar{v}$ for each vertex $v \in V$. The objective is to minimize the total weight of cut edges. The set of Constraints 2 are ℓ_2^2 triangle inqualities, and the set of Constraints 3 and 4 are ℓ_2^2 triangle inequalities with the zero vector. The set of Constraints 5 and 6 make sure that for each source-sink pair (s_i, t_i), both s_i and t_i do not belong to S since both vectors $\bar{s}_i$ and $\bar{t}_i$ could not be $\mathbf{1}$ for some fixed unit vector simultaneously. Constraint 7 and the set of Constraints 8 make sure the returned subgraph has the desired size. Suppose now that we have approximately guessed the measure H of the optimal solution $H \leq \eta(S) \leq 2H$. We can ignore all vertices $v \in V$ with $\eta(v) > 2H$ since they do not participate in the optimal solution and thus write the set of Constraints 8. Constraints (9) are spreading constraints introduced by Bansal et al. [4] which ensure size of S is small.

$$\min \sum_{(u,v) \in E} w(u,v) \|\bar{u} - \bar{v}\|^2 \tag{1}$$

$$\|\bar{u} - \bar{w}\|^2 + \|\bar{w} - \bar{v}\|^2 \geq \|\bar{u} - \bar{v}\|^2 \qquad \forall u, v, w \in V \tag{2}$$

$$\|\bar{u} - \bar{w}\|^2 \geq \|\bar{u}\|^2 - \|\bar{w}\|^2 \qquad \forall u, w \in V \tag{3}$$

$$\|\bar{u}\|^2 + \|\bar{v}\|^2 \geq \|\bar{u} - \bar{v}\|^2 \qquad \forall u, v \in V \tag{4}$$

$$\|\bar{s}_i - \bar{t}_i\|^2 \geq \|\bar{s}_i\|^2 \qquad \forall(s_i, t_i) \in S_G \tag{5}$$

$$\|\bar{s}_i - \bar{t}_i\|^2 \geq \|\bar{t}_i\|^2 \qquad \forall(s_i, t_i) \in S_G \tag{6}$$

$$\sum_{v \in V} \|\bar{v}\|^2 \, \eta(v) \geq H \tag{7}$$

$$\|\bar{v}\|^2 = 0 \qquad \text{if } \eta(v) > 2H \tag{8}$$

$$\sum_{v \in V} \eta(v) \cdot \min\{\|\bar{u} - \bar{v}\|^2, \|\bar{u}\|^2\} \geq (1 - 2H) \|\bar{u}\|^2 \qquad \forall u \in V \tag{9}$$

Lemma 1. *Given $S^* = \arg\min \{\delta(T) : \eta(T) \in [H, 2H], \forall(s_i, t_i) \in S_G, |\{s_i, t_i\} \cap T| \leq 1\}$, the optimal value of SDP is at most $\delta(S^*)$.*

Proof. We defer proof to the full version of this paper.

3.2   Approximation Algorithm

In this section, we prove Theorem 4. We propose an approximation algorithm which is inspired by Bansal et al.'s [4] algorithm for small-set expansion (SSE). However, there is a significant difference between our algorithm and theirs. In the SSE problem, one does not need to worry about separating source sink pairs.

First, we solve the SDP relaxation, and then proceed iteratively. In each iteration, we sample an n^3-orthogonal separator S with $\beta = 1/2$ and return it (we repeatedly sample S, until a particular function[1] $f(S)$ has some positive value. Details are deferred to Sect. 3.3). Then, S is removed from graph G and the SDP solution, by zeroing the weight of edges incident on S (i.e discarding these edges), and zeroing the SDP variables corresponding to vertices in S. The algorithm maintains the subsets of vertices removed so far in a set $U \subseteq V$, by initializing $U = \emptyset$, and then at each iteration by updating $U = U \cup \{S\}$. We keep iterating until $\eta(U) = \sum_{S_i \in U} \eta(S_i) \geq H/4$. After the last iteration, if $\eta(U) > H$, we output $F = S$ where S is computed in the last iteration. Otherwise, we put $F = U$. Note that in this case, $U = \{S_1, \cdots, S_{|U|}\}$.

3.3   Analysis

First, let's see what is the effect of algorithm's changes to the SDP solution. By zeroing vectors in S and discarding the edges incident on S, the SDP value may only decrease. Triangle inequalities, and the source-sink constraints still hold. Constraint $\sum_{v \in V} \|\bar{v}\|^2 \eta(v) \geq H$ will be violated due to zeroing some variables. However, since before the last iteration $\eta(U) \leq \frac{H}{4}$, the following constraint still holds:

$$\sum_{v \in V} \|\bar{v}\|^2 \eta(v) \geq \frac{3H}{4} \tag{10}$$

Next, we show the set of spreading constraints (9) will remain satisfied after removing S. Consider the spreading constraint for a fixed vertex u, two cases might happen:

Case 1: If $\exists S \in U$ such that $u \in S$, then u will be removed and $\|\bar{u}\| = 0$, the spreading constraint will be satisfied since RHS is 0.

Case 2: If $\nexists S \in U$ such that $u \in S$, the RHS will not change and we can show that $\min\{\|\bar{u} - \bar{v}\|^2, \|\bar{u}\|^2\}$ does not decrease. If $\nexists S' \in U$ such that $v \in S'$, then the term $\min\{\|\bar{u} - \bar{v}\|^2, \|\bar{u}\|^2\}$ does not change. If $\exists S' \in U$ such that $v \in S'$, then $\min\{\|\bar{u} - \bar{v}\|^2, \|\bar{u}\|^2\} = \|\bar{u}\|^2$ since $\|\bar{v}\| = 0$, and its value does not decrease.

Therefore, in both these cases, the spreading constraints will not be violated.

Lemma 2. *Let S be a sampled n-orthogonal separator. Fix a vertex u. We claim that* $\Pr[\eta(S) \leq 12H \mid u \in S] \geq \frac{7}{8}$.

Proof. We defer proof to Appendix A.1.

---

[1] defined later.

Next, we upper bound $\delta(S)$. By the third property of orthogonal separators:

$$\mathbb{E}[\delta(S)] \leq \alpha D \cdot \sum_{(u,v)\in E} \|\bar{u} - \bar{v}\|^2 \cdot w(u,v) \leq \alpha D \cdot SDP$$

where $D = \mathcal{O}_\beta(\sqrt{\log n \log(n^3)}) = \mathcal{O}(\log n)$. Note that $\beta = 1/2$. Consider the function f:

$$f(S) = \begin{cases} \eta(S) - \delta(S) \cdot \frac{H}{4D \cdot SDP} & \text{if } S \neq \emptyset \text{ and } \eta(S) < 12H \\ 0 & \text{otherwise} \end{cases}$$

We wish to lower bound $\mathbb{E}[f(S)]$. First, we lower bound $\mathbb{E}[\eta(S)]$. As a result of Lemma 2 and the first property of orthogonal separators:

$$\mathbb{E}[\eta(S)] = \sum_{u\in V} \Pr[u \in S \wedge \eta(S) < 12H] \cdot \eta(u)$$

$$= \sum_{u\in V} \Pr[\eta(S) < 12H \mid u \in S] \cdot \Pr[u \in S] \cdot \eta(u) \geq \sum_{u\in V} \frac{7\alpha \|\bar{u}\|^2 \eta(u)}{8}$$

Since $\mathbb{E}[\delta(S)] \leq \alpha D \cdot SDP$ and using Constraint 10:

$$\mathbb{E}[f(S)] \geq \sum_{u\in V} \frac{7\alpha \|\bar{u}\|^2 \eta(u)}{8} - \alpha \cdot D \cdot SDP \cdot \frac{H}{4D \cdot SDP} \geq \frac{7\alpha \frac{3H}{4}}{8} - \frac{\alpha H}{4} = \frac{13}{32}\alpha H$$

We have $f(S) \leq 2nH$ since $\|\bar{u}\| = 0$ whenever $\eta(u) > 2H$. Therefore, $\Pr[f(S) > 0] \geq \frac{\frac{13}{32}\alpha H}{2nH} = \Omega(\frac{\alpha}{n})$. So after $\mathcal{O}(n^2/\alpha)$ samples, with probability exponentially close to 1, the algorithm finds a set S with $f(S) > 0$. $f(S) > 0$ implies $\eta(S) \geq \delta(S) \cdot \frac{H}{4D \cdot SDP}$, therefore $\delta(S) \leq \frac{4D \cdot SDP \cdot \eta(S)}{H}$. Consider the two possible cases for the output F:

Case 1: $F = U = \{S_1, S_2, \cdots, S_{|U|}\}$, and $\eta(F) = \sum_{i=1}^{|U|} \eta(S_i)$. In this case, $\frac{H}{4} \leq \eta(F) \leq H$. The set U is a set of orthogonal separators and each $S_i \in U$ forms a separate part.

Case 2: $F = S$. In this case, let's show the last iteration of step 1 as $U = U_{old} \cup \{S\}$. We know $\eta(U) > H$, and $\eta(U_{old}) < \frac{H}{4}$, therefore $\eta(S) > 3H/4$. Also $f(S) > 0$ implies $\eta(S) \leq 12H$. Therefore, $3H/4 < \eta(S) \leq 12H$.

In both cases, $\frac{H}{4} \leq \eta(F) \leq 12H$.

We showed when a set $S_i \in U$ is sampled, $\delta(S_i) \leq \frac{4D \cdot SDP \cdot \eta(S_i)}{H}$. However, in the LHS of this inequality, edges like (u,v) where $u \in S_j, v \in S_i$ and $j < i$ are not considered. We can show $\sum_{j=1}^{i-1} \delta(S_j, S_i) \leq \sum_{j=1}^{i-1} \frac{4D \cdot SDP \cdot \eta(S_j)}{H} \leq 4D \cdot SDP$ since $\sum_{j=1}^{i-1} \eta(S_j) \leq H$. Therefore, $\delta(S_i) \leq \frac{4D \cdot SDP \cdot \eta(S_i)}{H} + \sum_{j=1}^{i-1} \delta(S_j, S_i) \leq \mathcal{O}(D \cdot SDP)$ since $\eta(S_i) \leq 12H$.

In addition, by the second property of orthogonal separators, for each source-sink pair (s_j, t_j), the probability that both s_j and t_j belong to the orthogonal separator S_i is bounded by $\frac{1}{n^3}$. Therefore, $\Pr[vio(S_i) \geq 1] \leq \frac{T}{n^3} \leq \frac{n^2}{n^3} = \frac{1}{n}$. This completes the proof of Theorem 4.

The following corollary is implied from Theorem 4 and is used in the next section.

Corollary 1. *Given an edge-weighted graph $G = (V, w)$, a set of source sink pairs S_G, a measure η on V such that $\eta(V) = 1$, and a parameter τ, a set $S = \{S_1, \cdots, S_j\}$ could be found satisfying $\forall S_i \in S, S_i \subseteq V, \Pr[vio(S_i) \geq 1] \leq 1/n$, and $\delta(S_i) \leq \mathcal{O}(\log(n)) \cdot OPT$, where $OPT = \arg\min\{\delta(T) : \frac{\eta(T)}{\eta(V)} \geq \tau, vio(T) = 0\}$. In addition, $\eta(S) = \sum_{i=1}^{j} \eta(S_i) \geq \Omega(\tau \cdot \eta(V))$.*

Proof. The algorithm guesses $H \geq \tau$ such that $H \leq \eta(OPT) \leq 2H$. Guessing is feasible since $0 \leq \eta(OPT) \leq n \cdot \eta(u)$, where u is the weight of the heaviest element in OPT, and H can be chosen from the set $\{2^t \eta(u) : u \in V, t = 0, \cdots, \log(n)\}$ of size $\mathcal{O}(n \log(n))$. Theorem 4 is invoked with parameter H. The obtained solution S satisfies the properties of this corollary.

3.4 Covering and Aggregation

Once we find F, we follow the multiplicative update algorithm of [4] with some minor modifications, to get a covering of all the vertices. Then, we use the aggregation step to convert the covering to a partitioning. This step is simpler than [4] since we are not required to maintain any size bound on the subgraphs returned after aggregation.

Theorem 6. *Given graph $G = (V, E)$ and T pairs of source and sink, running Algorithm 1 on this instance outputs a multiset $\mathcal{S}$ that satisfies the following conditions:*

- *for all $S \in \mathcal{S}$: $\delta(S) \leq D \cdot OPT$ where $D = \mathcal{O}(\log(n)), \Pr[vio(S) \geq 1] \leq 1/n$*
- *for all $v \in V$, $\frac{|\{S \in \mathcal{S} : v \in S\}|}{|\mathcal{S}|} \geq \frac{1}{5\gamma kn}$, where $\gamma = \mathcal{O}(1)$ and k is the number of parts in the optimal solution which we guess.*

Proof. We defer proof to Appendix A.2.

Algorithm 1. Covering Procedure for Min-Max Multicut

---

1 Set $t = 1, \mathcal{S} = \emptyset$ and $y^1(v) = 1 \; \forall v \in V$;
2 Guess k, which is the number of parts in the optimal solution;
3 **while** $\sum_{v \in V} y^t(v) > \frac{1}{n}$ **do**
4 $\quad$ Find set $S^t = \{S_1, \cdots, S_j\}$ using Corollary 1, where $\tau = \frac{1}{k}$ and $\forall v \in V, \eta(v) = y^t(v)/\sum_{v \in V} y^t(v)$;
5 $\quad$ $\mathcal{S} = S^t \cup \mathcal{S}$;
6 $\quad$ // Update the weights of the covered vertices;
7 $\quad$ **for** $v \in V$ **do**
8 $\quad\quad$ Set $y^{t+1}(v) = \frac{1}{2} \cdot y^t(v)$ if $\exists S_i \in S^t$ such that $v \in S_i$, and $y^{t+1}(v) = y^t(v)$ otherwise.;
9 $\quad$ Set $t = t + 1$;
10 return $\mathcal{S}$;

Now the covering of G is converted into a partitioning of G without violating min-max objective by much.

Theorem 7. *Given a weighted graph* $G = (V, E)$, *a set of source-sink pairs* $(s_1, t_1), \cdots, (s_T, t_T)$, *and a cover* $\mathcal{S}$ *consisting of subsets of* V *such that:*

- *$\forall v \in V$, v is covered by at least a fraction $\frac{c}{nk}$ of sets $S \in \mathcal{S}$, where k is the number of partitions of the optimum solution which we guessed in the covering section, and $c \in (0, 1]$.*
- *$\forall S \in \mathcal{S}$, $\delta(S) \leq B$, $\Pr[vio(S) \geq 1] \leq 1/n$.*

We propose a randomized polynomial time algorithm which outputs a partition $\mathcal{P}$ of V such that $\forall P_i \in \mathcal{P}$, $\delta(P_i) \leq 2B$, and $\Pr[vio(P_i) \geq 1] \leq 1/n$.

Algorithm 2. Aggregation Procedure For Min-Max Multicut

---

1   Step 1: Sort sets in $\mathcal{S}$ in a random order: $S_1, S_2, \cdots, S_{|\mathcal{S}|}$. Let
     $P_i = S_i \setminus (\cup_{j<i} S_j)$.
2   Step 2: **while** *There is a set P_i such that* $\delta(P_i) > 2B$ **do**
3   $\quad$ Set $P_i = S_i$ and for all $j \neq i$, set $P_j = P_j \setminus S_i$;

Proof. We defer proof to Appendix A.3.

Acknowledgements. We are grateful to Nikhil Bansal for useful discussions during a Dagstuhl workshop on scheduling (18101).

A   Missing Proofs

A.1   Proof of Lemma 2

Proof. Consider a vertex u and let $A_u = \{v : \|\bar{u} - \bar{v}\|^2 \geq \beta \|\bar{u}\|^2\}$ and $B_u = \{v : \|\bar{u} - \bar{v}\|^2 < \beta \|\bar{u}\|^2\}$. Assume for now that $u \in S$. We show with high probability $\eta(A_u \cap S)$ is small, and $\eta(B_u)$ is also small. Vertex u satisfies the spreading constraint. It is easy to see that:

$$(1 - 2H) \|u\|^2 \leq \sum_{v \in V} \eta(v) \cdot \min\{\|\bar{u} - \bar{v}\|^2, \|\bar{u}\|^2\} \leq \beta \|\bar{u}\|^2 \eta(B_u) + \|\bar{u}\|^2 \eta(A_u)$$

Since $\eta(V) = 1$ and $A_u \cup B_u = V$, $\eta(A_u) + \eta(B_u) = 1$, and $\beta = 1/2$ therefore:

$$(1 - 2H) \leq \beta \eta(B_u) + (1 - \eta(B_u)) \tag{11}$$

$$\therefore \eta(B_u) \leq \frac{2H}{1 - \beta} = 4H \tag{12}$$

Consider an arbitrary vertex $v \in A_u$ where $\|\bar{v}\| \neq 0$. By definition of A_u, $\|\bar{u} - \bar{v}\|^2 \geq \beta \|\bar{u}\|^2 \geq \beta \min\{\|\bar{u}\|^2, \|\bar{v}\|^2\}$. Therefore, by the second property of orthogonal separators and since we assumed $u \in S$, then $\Pr[v \in S \mid u \in S] \leq \frac{1}{n^3} \leq H$. The second inequality holds since $H \geq 1/n$.

Now we show a bound for $\mathbb{E}[\eta(A_u \cap S) \mid u \in S]$:

$$\mathbb{E}[\eta(A_u \cap S) \mid u \in S] = \sum_{v \in A_u} \eta(v) \Pr[v \in S \mid u \in S] \leq H$$

Now, we want to bound $\Pr[\eta(S) \geq 12H \mid u \in S]$. The event $\{\eta(S) \geq 12H \mid u \in S\}$ implies the event $\{\eta(A_u \cap S) \geq 8H \mid u \in S\}$ since $\eta(B_u \cap S) \leq \eta(B_u) \leq 4H$. (The second inequality holds by (12)). Now we are ready to complete the proof.

$$\Pr[\eta(S) \geq 12H \mid u \in S] \leq \Pr[\eta(A_u \cap S) \geq 8H \mid u \in S] \leq \frac{\mathbb{E}[\eta(A_u \cap S) \mid u \in S]}{8H} \leq \frac{H}{8H} = 1/8$$

We showed $\Pr[\eta(S) \geq 12H \mid u \in S] \leq 1/8$, therefore $\Pr[\eta(S) \leq 12H \mid u \in S] \geq 7/8$ and the proof is complete.

A.2   Proof of Theorem 6

Proof. For an iteration t, let $Y^t = \sum_{v \in V} y^t(v)$. Consider the optimal solution $\{S_i^*\}_{i=1}^k$ to the min-max multicut problem. There exists at least a $S_j^* \in \{S_i^*\}_{i=1}^k$ with weight greater than or equal to the average $(y_t(S_j^*) \geq \frac{Y^t}{k})$, $vio(S_j^*) = 0$, and $\delta(S_j^*) \leq OPT$. Therefore by Corollary 1 where $H = \frac{1}{k}$, a set $S_t = \{S_1, S_2, \cdots, S_j\}$ could be found where $\forall S_i \in S_t$, $\delta(S_i) \leq \mathcal{O}(\log n) \cdot OPT$, $\Pr[vio(S_i) \geq 1] \leq 1/n$.

Now we show the second property of the theorem holds. Let ℓ denote the number of iterations in the while loop. Let $|\{S \in \mathcal{S} : v \in S\}| = N_v$. By the updating rules $y^{\ell+1}(v) = 1/2^{N_v}$. Therefore $\frac{1}{2^{N_v}} = y^{\ell+1}(v) \leq 1/n$, which implies $N_v \geq \log(n)$. By Corollary 1, $y^t(S^t) \geq \frac{1}{\gamma k} Y^t$ where $\gamma = \mathcal{O}(1)$. Therefore:

$$Y^{t+1} = Y^t - \frac{1}{2} y^t(S^t) \leq (1 - \frac{1}{2\gamma k}) Y^t$$

Which implies $Y^\ell \leq (1 - \frac{1}{2\gamma k})^{\ell-1} Y^1 = (1 - \frac{1}{2\gamma k})^{\ell-1} n$. Also $Y^\ell > 1/n$ therefore, $\ell \leq 1 + 4\gamma k \ln(n) \leq 5\gamma k \log(n)$. In each iteration t, the number of sets in S_t is at most n (since all the sets in S_t are disjoint), therefore $|\mathcal{S}| \leq 5\gamma k n \log(n)$, and the second property is proved.

A.3   Proof of Theorem 7

Proof. A similar proof to the one given by Bansal et al. [4] shows after step 2, for each $P_i \in \mathcal{P}$, $\delta(P_i) \leq 2B$. We start by analyzing Step 1. Observe that after Step 1, the collection of sets $\{P_i\}$ is a partition of V and $P_i \subseteq S_i$ for every i. Particularly, $vio(P_i) \leq vio(S_i)$. Note, however, that the bound $\delta(P_i) \leq B$ may be violated for some i since P_i might be a strict subset of S_i.

We finish the analysis of Step 1 by proving that $\mathbb{E}[\sum_i \delta(P_i)] \leq 2knB/c$. Fix an $i \leq |\mathcal{S}|$ and estimate the expected weight of edges $E(P_i, \cup_{j>i} P_j)$, given that the ith set in the random ordering is S. If an edge (u, v) belongs to $E(P_i, \cup_{j>i} P_j)$, then $(u, v) \in E(S_i, V \setminus S_i) = E(S, V \setminus S)$ and both $u, v \notin \cup_{j<i} S_j$. For any edge $(u, v) \in \delta(S)$ (with $u \in S, v \notin S$), $\Pr[(u, v) \in E(P_i, \cup_{j>i} P_j) \mid S_i = S] \leq \Pr(v \notin \cup_{j<i} S_j \mid S_i = S) \leq (1 - \frac{c}{nk})^{i-1}$, since v is covered by at least $\frac{c}{nk}$ fraction of sets in $\mathcal{S}$ and is not covered by $S_i = S$. Hence,

$$\mathbb{E}[w(E(P_i, \cup_{j>i} P_j)) \mid S_i = S] \leq (1 - \frac{c}{nk})^{i-1} \delta(S) \leq (1 - \frac{c}{nk})^{i-1} B$$

and $\mathbb{E}[w(E(P_i, \cup_{j>i} P_j)) \leq (1 - \frac{c}{nk})^{i-1} B$. Therefore:

$$\mathbb{E}\left[\sum_i \delta(P_i)\right] = 2 \cdot \mathbb{E}\left[\sum_i w(E(P_i, \cup_{j>i} P_j))\right] \leq 2 \sum_{i=0}^{\infty} (1 - \frac{c}{nk})^i B = 2knB/c$$

Now we want to analyze step 2. Consider potential function $\sum_i \delta(P_i)$, we showed after step 1, $\mathbb{E}\left[\sum_i \delta(P_i)\right] \leq 2knB/c$. We prove that this potential function reduces quickly over the iterations of Step 2, thus, Step 2 terminates after a small number of steps. After each iteration of Step 2, the following invariant holds: the collection of sets $\{P_i\}$ is a partition of V and $P_i \subseteq S_i$ for all i. Particularly, $vio(P_i) \leq vio(S_i)$. Using an uncrossing argument, we show at every iteration of the while loop in step 2, $\sum_i \delta(P_i)$ decreases by at least $2B$.

$$\delta(S_i) + \sum_{j \neq i} \delta(P_j \setminus S_i) \leq \delta(S_i) + \sum_{j \neq i} \left(\delta(P_j) + w(E(P_j \setminus S_i, S_i)) - w(E(S_i \setminus P_j, P_j)) \right)$$

$$\leq \delta(S_i) + \sum_{j \neq i} \left(\delta(P_j) \right) + w(E(V \setminus S_i, S_i)) - w(E(P_i, V \setminus P_i))$$

$$= \sum_j \left(\delta(P_j) \right) + 2\delta(S_i) - 2\delta(P_i) \leq \sum_j \left(\delta(P_j) \right) - 2B$$

The above inequalities use the facts that $P_i \subseteq S_i$ for all i and that all the P_j's are disjoint. The second inequality uses the facts that $\sum_{j \neq i} w(E(P_j \setminus S_i, S_i)) = w(E(V \setminus S_i, S_i))$, and $\sum_{j \neq i} w(E(S_i \setminus P_j, P_j)) \geq w(E(P_i, V \setminus P_i))$, which hold since the collection of sets $\{P_i\}$ is a partition of V, and $P_i \subseteq S_i$. In particular, $\sum_{j \neq i} w(E(S_i \setminus P_j, P_j)) \geq w(E(P_i, V \setminus P_i))$ holds since for each edge e if $e \in E(P_i, P_j)$ then $e \in E(S_i \setminus P_j, P_j)$. The last inequality holds since $\delta(S_i) \leq B$ and $\delta(P_i) > 2B$.

This proves that the number of iterations of the while loop is polynomially bounded and after step 2, $\delta(P_i) \leq 2B$ for each P_i.

In addition, since each P_i is a subset of S_i, $vio(P_i) \leq vio(S_i)$. Therefore $\Pr[vio(P_i) \geq 1] \leq 1/n$.

References

1. Ailon, N., Avigdor-Elgrabli, N., Liberty, E., Van Zuylen, A.: Improved approximation algorithms for bipartite correlation clustering. SIAM J. Comput. **41**(5), 1110–1121 (2012)
2. Ailon, N., Charikar, M., Newman, A.: Aggregating inconsistent information: ranking and clustering. J. ACM (JACM) **55**(5), 23 (2008)
3. Bansal, N., Blum, A., Chawla, S.: Correlation clustering. Mach. Learn. **56**(1–3), 89–113 (2004)
4. Bansal, N., et al.: Min-max graph partitioning and small set expansion. SIAM J. Comput. **43**(2), 872–904 (2014)
5. Charikar, M., Gupta, N., Schwartz, R.: Local guarantees in graph cuts and clustering. In: Eisenbrand, F., Koenemann, J. (eds.) IPCO 2017. LNCS, vol. 10328, pp. 136–147. Springer, Cham (2017). https://doi.org/10.1007/978-3-319-59250-3_12
6. Charikar, M., Guruswami, V., Wirth, A.: Clustering with qualitative information. In: 44th Annual IEEE Symposium on Foundations of Computer Science, Proceedings, pp. 524–533. IEEE (2003)

7. Chawla, S., Makarychev, K., Schramm, T., Yaroslavtsev, G.: Near optimal LP rounding algorithm for correlation clustering on complete and complete k-partite graphs. In: Proceedings of the Forty-Seventh Annual ACM Symposium on Theory of Computing, pp. 219–228. ACM (2015)

8. Cheng, Y., Church, G.M.: Biclustering of expression data. In: Ismb, vol. 8, pp. 93–103 (2000)

9. Chlamtac, E., Makarychev, K., Makarychev, Y.: How to play unique games using embeddings. In: 47th Annual IEEE Symposium on Foundations of Computer Science, FOCS 2006, pp. 687–696. IEEE (2006)

10. Demaine, E.D., Emanuel, D., Fiat, A., Immorlica, N.: Correlation clustering in general weighted graphs. Theor. Comput. Sci. **361**(2–3), 172–187 (2006)

11. Kim, S., Nowozin, S., Kohli, P., Yoo, C.D.: Higher-order correlation clustering for image segmentation. In Shawe-Taylor, J., Zemel, R.S., Bartlett, P.L., Pereira, F., Weinberger, K.Q. (eds.) Advances in Neural Information Processing Systems, vol. 24, pp. 1530–1538. Curran Associates, Inc. (2011)

12. Kirillov, A., Levinkov, E., Andres, B., Savchynskyy, B., Rother, C.: Instancecut: from edges to instances with multicut. In: 2017 IEEE Conference on Computer Vision and Pattern Recognition (CVPR), pp. 7322–7331. IEEE (2017)

13. Kriegel, H.-P., Kröger, P., Zimek, A.: Clustering high-dimensional data: a survey on subspace clustering, pattern-based clustering, and correlation clustering. ACM Trans. Knowl. Discov. Data (TKDD) **3**(1), 1 (2009)

14. Puleo, G., Milenkovic, O.: Correlation clustering and biclustering with locally bounded errors. In: International Conference on Machine Learning, pp. 869–877 (2016)

15. Svitkina, Z., Tardos, É.: Min-Max multiway cut. In: Jansen, K., Khanna, S., Rolim, J.D.P., Ron, D. (eds.) APPROX/RANDOM -2004. LNCS, vol. 3122, pp. 207–218. Springer, Heidelberg (2004). https://doi.org/10.1007/978-3-540-27821-4_19

16. Symeonidis, P., Nanopoulos, A., Papadopoulos, A., Manolopoulos, Y.: Nearest-biclusters collaborative filtering with constant values. In: Nasraoui, O., Spiliopoulou, M., Srivastava, J., Mobasher, B., Masand, B. (eds.) WebKDD 2006. LNCS (LNAI), vol. 4811, pp. 36–55. Springer, Heidelberg (2007). https://doi.org/10.1007/978-3-540-77485-3_3

Strong Mixed-Integer Programming Formulations for Trained Neural Networks

Ross Anderson[1], Joey Huchette[1(✉)], Christian Tjandraatmadja[1], and Juan Pablo Vielma[2]

[1] Google Research, Cambridge, USA
{rander,jhuchette,ctjandra}@google.com
[2] MIT, Cambridge, USA
jvielma@mit.edu

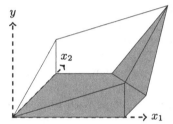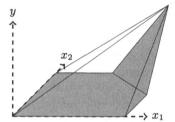

Fig. 1. The convex relaxation for a ReLU neuron using: (**Left**) existing MIP formulations, and (**Right**) the formulations presented in this paper.

Abstract. We present an ideal mixed-integer programming (MIP) formulation for a rectified linear unit (ReLU) appearing in a trained neural network. Our formulation requires a single binary variable and no additional continuous variables beyond the input and output variables of the ReLU. We contrast it with an ideal "extended" formulation with a linear number of additional continuous variables, derived through standard techniques. An apparent drawback of our formulation is that it requires an exponential number of inequality constraints, but we provide a routine to separate the inequalities in linear time. We also prove that these exponentially-many constraints are facet-defining under mild conditions. Finally, we study network verification problems and observe that dynamically separating from the exponential inequalities (1) is much more computationally efficient and scalable than the extended formulation, (2) decreases the solve time of a state-of-the-art MIP solver by a factor of 7 on smaller instances, and (3) nearly matches the dual bounds of a state-of-the-art MIP solver on harder instances, after just a few rounds of separation and in orders of magnitude less time.

Keywords: Mixed-integer programming · Formulations · Deep learning

A. Lodi and V. Nagarajan (Eds.): IPCO 2019, LNCS 11480, pp. 27–42, 2019.
https://doi.org/10.1007/978-3-030-17953-3_3

1   Introduction

Deep learning has proven immensely powerful at solving a number of important predictive tasks arising in areas such as image classification, speech recognition, machine translation, and robotics and control [27,35]. The workhorse model in deep learning is the feedforward network $\mathtt{NN} : \mathbb{R}^{m_0} \to \mathbb{R}^{m_s}$ with rectified linear unit (ReLU) activation functions, for which $\mathtt{NN}(x^0) = x^s$ is defined through

$$x_j^i = \mathtt{ReLU}(w^{i,j} \cdot x^{i-1} + b^{i,j}) \tag{1}$$

for each layer $i \in [\![s]\!] \overset{\text{def}}{=} \{1, \ldots, s\}$ and $j \in [\![m_i]\!]$. Note that the input $x^0 \in \mathbb{R}^{m_0}$ might be high-dimensional, and that the output $x^s \in \mathbb{R}^{m_s}$ may be multivariate. In this recursive description, $\mathtt{ReLU}(v) \overset{\text{def}}{=} \max\{0, v\}$ is the ReLU activation function, and $w^{i,j}$ and $b^{i,j}$ are the weights and bias of an affine function which are learned during the training procedure. Each equation in (1) corresponds to a single *neuron* in the network. Networks with any specialized linear transformations such as convolutional layers can be reduced to this model after training, without loss of generality.

There are numerous contexts in which one may want to solve an optimization problem containing a trained neural network such as $\mathtt{NN}$. For example, such problems arise in deep reinforcement learning problems with high dimensional action spaces and where any of the cost-to-go function, immediate cost, or the state transition functions are learned by a neural network [3,19,40,44,55]. Alternatively, there has been significant recent interest in verifying the robustness of trained neural networks deployed in systems like self-driving cars that are incredibly sensitive to unexpected behavior from the machine learning model [15,43,48]. Relatedly, a string of recent work has used optimization over neural networks trained for visual perception tasks to *generate* new images which are "most representative" for a given class [42], are "dreamlike" [41], or adhere to a particular artistic style via neural style transfer [26].

1.1   MIP Formulation Preliminaries

In this work, we study mixed-integer programming (MIP) approaches for optimization problems containing trained neural networks. In contrast to heuristic or local search methods often deployed for the applications mentioned above, MIP offers a framework for producing provably optimal solutions. This is of particular interest in the verification problem, where rigorous dual bounds can guarantee robustness in a way that purely primal methods cannot.

We focus on constructing MIP formulations for the *graph* of ReLU neurons:

$$\mathrm{gr}(\mathtt{ReLU} \circ f; [L, U]) \overset{\text{def}}{=} \{ (x, (\mathtt{ReLU} \circ f)(x)) \mid L \leqslant x \leqslant U \}, \tag{2}$$

where $\circ$ is the standard function composition operator $(g \circ f)(x) = g(f(x))$. This substructure consists of a single ReLU activation function, taking as input an affine function $f(x) = w \cdot x + b$ over a η-dimensional box-constrained input domain. The nonlinearity is handled by introducing an auxiliary binary variable z to indicate whether $(\text{ReLU} \circ f)(x) = 0$ or $(\text{ReLU} \circ f)(x) = f(x)$ for a given value of x. We focus on these particular substructures because we can readily produce a MIP formulation for the entire network as the composition of formulations for each individual neuron.[1]

A MIP formulation is *ideal* if the extreme points of its linear programming (LP) relaxation are integral. Ideal formulations are highly desirable from a computational perspective, and offer the strongest possible convex relaxation for the set being formulated [50].

Our main contribution is an ideal formulation for a single ReLU neuron with no auxiliary continuous variables and an exponential number of inequality constraints. We show that each of these exponentially-many constraints is facet-defining under very mild conditions. We also provide a simple linear-time separation routine to generate the most violated inequality from the exponential family. This formulation is derived by constructing an ideal extended formulation that uses η auxiliary continuous variables and projecting them out. We evaluate our methods computationally on verification problems for image classification networks trained on the MNIST digit dataset, where we observe that separating over these exponentially-many inequalities solves smaller instances faster than using Gurobi's default cut generation by a factor of 7, and (nearly) matches the dual bounds on larger instances in orders of magnitude less time.

1.2   Relevant Prior Work

In recent years a number of authors have used MIP formulations to model trained neural networks [14,16,20,25,32,38,44,46,47,49,55,56], mostly applying big-M formulation techniques to ReLU-based networks. When applied to a single neuron of the form (2), these big-M formulations will not be ideal or offer an exact convex relaxation; see Example 1 for an illustration. Additionally, a stream of literature in the deep learning community has studied convex relaxations in the original space of input/output variables x and y (or a dual representation thereof), primarily for verification tasks [9,22,23]. It has been shown that these convex relaxations are equivalent to those provided by the standard big-M MIP formulation, after projecting out the auxiliary binary variables (e.g. [46]). Moreover, some authors have investigated how to use convex relaxations within the training procedure in the hopes of producing neural networks with a priori robustness guarantees [21,53,54].

Beyond MIP and convex relaxations, a number of authors have investigated other algorithmic techniques for modeling trained neural networks in optimization problems, drawing primarily from the satisfiability, constraint programming, and global optimization communities [7,8,33,37,45]. Another intriguing direction studies restrictions to the space of models that may make the optimization problem over the network inputs simpler: for example, the classes of binarized [34] or input convex [1] neural networks.

---

[1] Further analysis of the interactions between neurons can be found in the full-length version of this extended abstract [2].

Broadly, our work fits into a growing body of research in prescriptive analytics and specifically the "predict, then optimize" framework, which considers how to embed trained machine learning models into optimization problems [11,12,17,18, 24,28,39]. Additionally, the formulations presented below have connections with existing structures studied in the MIP and constraint programming community like indicator variables and on/off constraints [4,10,13,29,30].

1.3   Starting Assumptions and Notation

We will assume that $-\infty < L_i < U_i < \infty$ for each input component i. While a bounded input domain will make the formulations and analysis considerably more difficult than the unbounded setting (see [4] for a similar phenomenon), it ensures that standard MIP representability conditions are satisfied (e.g. [50, Sect. 11]). Furthermore, variable bounds are natural for many applications (for example in verification problems), and are absolutely essential for ensuring reasonable dual bounds.

Define $\check{L}, \check{U} \in \mathbb{R}^{\eta}$ such that, for each $i \in [\![\eta]\!]$,

$$\check{L}_i = \begin{cases} L_i & \text{if } w_i \geqslant 0 \\ U_i & \text{if } w_i < 0 \end{cases} \quad \text{and} \quad \check{U}_i = \begin{cases} U_i & \text{if } w_i \geqslant 0 \\ L_i & \text{if } w_i < 0 \end{cases}.$$

This definition implies that $w_i \check{L}_i \leqslant w_i \check{U}_i$ for each i, which simplifies the handling of negative weights $w_i < 0$. Take the values $M^+(f) \stackrel{\text{def}}{=} \max_{\tilde{x} \in [L,U]} f(\tilde{x}) \equiv w \cdot \check{U} + b$ and $M^-(f) \stackrel{\text{def}}{=} \min_{\tilde{x} \in [L,U]} f(\tilde{x}) \equiv w \cdot \check{L} + b$. Define $\text{supp}(w) \stackrel{\text{def}}{=} \{\, i \in [\![\eta]\!] \mid w_i \neq 0 \,\}$. Finally, take $\mathbb{R}_{\geqslant 0} \stackrel{\text{def}}{=} \{\, x \in \mathbb{R} \mid x \geqslant 0 \,\}$ as the nonnegative orthant.

We say that *strict activity* holds for a given ReLU neuron $\text{gr}(\texttt{ReLU} \circ f; [L, U])$ if $M^-(f) < 0 < M^+(f)$, or in other words, if $\text{gr}(\texttt{ReLU} \circ f; [L, U])$ is not equal to either $\text{gr}(0; [L, U])$ or $\text{gr}(f; [L, U])$. We assume for the remainder that strict activity holds for each ReLU neuron. This assumption is not onerous, as otherwise, the nonlinearity can be replaced by an affine function (either 0 or $w \cdot x + b$). Moreover, strict activity can be verified or disproven in time linear in η.

2   The ReLU Neuron

The ReLU is the workhorse of deep learning models: it is easy to reason about, introduces little computational overhead, and despite its simple structure is nonetheless capable of articulating complex nonlinear relationships.

2.1   A Big-M Formulation

A standard big-M formulation for $\text{gr}(\texttt{ReLU} \circ f; [L, U])$ is:

$$y \geqslant f(x) \tag{3a}$$

$$y \leqslant f(x) - M^-(f) \cdot (1 - z) \tag{3b}$$

$$y \leqslant M^+(f) \cdot z \tag{3c}$$

$$(x, y, z) \in [L, U] \times \mathbb{R}_{\geqslant 0} \times \{0, 1\}. \tag{3d}$$

This is the formulation used recently in the bevy of papers referenced in Sect. 1.2. Unfortunately, this formulation is not necessarily ideal, as illustrated by the following example.

Example 1. If $f(x) = x_1 + x_2 - 1.5$, formulation (3a–3d) for $gr(\texttt{ReLU} \circ f; [0,1]^2)$ is

$$y \geqslant x_1 + x_2 - 1.5 \tag{4a}$$
$$y \leqslant x_1 + x_2 - 1.5 + 1.5(1 - z) \tag{4b}$$
$$y \leqslant 0.5z \tag{4c}$$
$$(x, y, z) \in [0, 1]^2 \times \mathbb{R}_{\geqslant 0} \times [0, 1] \tag{4d}$$
$$z \in \{0, 1\}. \tag{4e}$$

The point $(\hat{x}, \hat{y}, \hat{z}) = ((1,0), 0.25, 0.5)$ is feasible for the LP relaxation (4a–4d); however, $(\hat{x}, \hat{y}) \equiv ((1,0), 0.25)$ is not in $\text{Conv}(gr(\texttt{ReLU} \circ f; [0,1]^2))$, and so the formulation does not offer an exact convex relaxation (and, hence, is not ideal). See Fig. 1 for an illustration: on the left, of the big-M formulation projected to (x, y)-space, and on the right, the tightest possible convex relaxation.

The integrality gap of (3a–3d) can be arbitrarily bad, even in fixed dimension η.

Example 2. Fix $\gamma \in \mathbb{R}_{\geqslant 0}$ and even $\eta \in \mathbb{N}$. Take the affine function $f(x) = \sum_{i=1}^{\eta} x_i$, the input domain $[L, U] = [-\gamma, \gamma]^{\eta}$, and the point $\hat{x} = \gamma \cdot (1, -1, \cdots, 1, -1)$ as a scaled vector of alternating ones and negative ones. We can check that $(\hat{x}, \hat{y}, \hat{z}) = (\hat{x}, \frac{1}{2}\gamma\eta, \frac{1}{2})$ is feasible for the LP relaxation of the big-M formulation (3a–3d). Additionally, $f(\hat{x}) = 0$, and for any $\tilde{y}$ such that $(\hat{x}, \tilde{y}) \in \text{Conv}(gr(\texttt{ReLU} \circ f; [L, U]))$, then $\tilde{y} = 0$ necessarily. Therefore, there exists a fixed point $\hat{x}$ in the input domain where the tightest possible convex relaxation (for example, from an ideal formulation) is exact, but the big-M formulation deviates from this value by at least $\frac{1}{2}\gamma\eta$.

Intuitively, this example suggests that the big-M formulation is particularly weak around the boundary of the input domain, as it cares only about the value $f(x)$ of the affine function, and not the particular input value x.

2.2   An Ideal Extended Formulation

It is possible to produce an ideal *extended* formulation for the ReLU neuron by introducing auxiliary continuous variables. The "multiple choice" formulation is

$$(x, y) = (x^0, y^0) + (x^1, y^1) \tag{5a}$$

$$y^0 = 0 \geqslant w \cdot x^0 + b(1 - z) \tag{5b}$$

$$y^1 = w \cdot x^1 + bz \geqslant 0 \tag{5c}$$

$$L(1 - z) \leqslant x^0 \leqslant U(1 - z) \tag{5d}$$

$$Lz \leqslant x^1 \leqslant Uz \tag{5e}$$

$$z \in \{0, 1\}, \tag{5f}$$

is an ideal extended formulation for piecewise linear functions [52]. It can alternatively be derived from techniques introduced by Balas [5,6]. Although the multiple choice formulation offers the tightest possible convex relaxation for a single neuron, it requires a copy x^0 of the input variables (note that it is straightforward to use Eq. (5a) to eliminate the second copy x^1). This means that when the multiple choice formulation is applied to every neuron in the network to formulate NN, the total number of continuous variables required is $m_0 + \sum_{i=1}^{r}(m_{i-1}+1)m_i$ (using the notation of (1), where m_i is the number of neurons in layer i). In contrast, the big-M formulation requires only $m_0 + \sum_{i=1}^{r} m_i$ continuous variables to formulate the entire network. As we will see in Sect. 3.2, the quadratic growth in size of the extended formulation can quickly become burdensome. Additionally, a folklore observation in the MIP community is that multiple choice formulations tend to not perform as well as expected in simplex-based branch-and-bound algorithms, likely due to degeneracy introduced by the block structure [51].

2.3   An Ideal Non-extended Formulation

We now present a non-extended ideal formulation for the ReLU neuron, stated only in terms of the original variables (x, y) and the single binary variable z. Put another way, it is the strongest possible tightening that can be applied to the big-M formulation (3a–3d) and so matches the strength of the multiple choice formulation without the additional continuous variables.

Proposition 1. *Take some affine function $f(x) = w \cdot x + b$ over input domain $[L, U]$. The following is an ideal formulation for $\mathrm{gr}(\mathtt{ReLU} \circ f; [L, U])$:*

$$y \geqslant w \cdot x + b \tag{6a}$$

$$y \leqslant \sum_{i \in I} w_i(x_i - \check{L}_i(1 - z)) + \left(b + \sum_{i \notin I} w_i \check{U}_i\right) z \quad \forall I \subseteq \mathrm{supp}(w) \tag{6b}$$

$$(x, y, z) \in [L, U] \times \mathbb{R}_{\geqslant 0} \times \{0, 1\} \tag{6c}$$

Proof. See Appendix A.1.   □

Furthermore, each of the exponentially-many inequalities in (6b) is necessary.

Proposition 2. *Each inequality in* (6b) *is facet-defining.*

Proof. See Appendix A.2.   □

We require the assumption of strict activity above, as introduced in Sect. 1.3. Under the same condition, it is also possible to show that (6a) is facet-defining, but we omit it in this extended abstract for brevity. As a result of this and Proposition 2, the formulation (6a–6c) is minimal (modulo variable bounds).

The proof of Proposition 2 offers a geometric interpretation of the facets induced by (6b). Each facet is a convex combination of two faces: an $(\eta - |I|)$-dimensional face consisting of all feasible points with $z = 0$ and $x_i = \check{L}_i$ for all $i \in [\![\eta]\!] \setminus I$, and an $|I|$-dimensional face consisting of all feasible points with $z = 1$ and $x_i = \check{U}_i$ for all $i \in I$.

It is also possible to separate from the family (6b) in time linear in η.

Proposition 3. *Take a point* $(\hat{x}, \hat{y}, \hat{z}) \in [L, U] \times \mathbb{R}_{\geqslant 0} \times [0, 1]$, *along with the set*

$$\hat{I} = \left\{ i \in \operatorname{supp}(w) \;\middle|\; w_i \hat{x}_i < w_i \left(\check{L}_i (1 - \hat{z}) + \check{U}_i \hat{z} \right) \right\}.$$

If any constraint in the family (6b) *is violated at* $(\hat{x}, \hat{y}, \hat{z})$, *then the one corresponding to* $\hat{I}$ *is the most violated.*

Proof. Follows from inspecting the family (6b): each has the same left-hand-side, and so to maximize violation, it suffices to select the subset I that minimizes the right-hand-side. This can be performed in a separable manner, independently for each component $i \in \operatorname{supp}(w)$, giving the result.   □

Observe that the inequalities (3b) and (3c) are equivalent to those in (6b) with $I = \operatorname{supp}(w)$ and $I = \varnothing$, respectively (modulo components i with $w_i = 0$). This suggests an iterative scheme to produce strong relaxations for ReLU neurons: start with the big-M formulation (3a–3d), and use Proposition 3 to separate strengthening inequalities from the exponential family (6b) as they are needed. We evaluate this approach in the following computational study.

3   Computational Experiments

To conclude the work, we study the strength of the ideal formulations presented in Sect. 2 for individual ReLU neurons. We study the verification problem on image classification networks trained on the canonical MNIST digit dataset [36]. We train a neural network $f : [0, 1]^{28 \times 28} \to \mathbb{R}^{10}$, where the 10 outputs correspond to the logits for each of the digits from 0 to 9. Given a labeled image $\tilde{x} \in [0, 1]^{28 \times 28}$, our goal is to prove or disprove the existence of a perturbation of $\tilde{x}$ such that the neural network f produces a wildly different classification result. If $f(\tilde{x})_i = \max_{j=1}^{10} f(\tilde{x})_j$, then image $\tilde{x}$ is placed in class i. To evaluate robustness around $\tilde{x}$ with respect to class j, we can solve the following optimization problem for some small constant $\epsilon > 0$:

$$\max_{a: ||a||_\infty \leqslant \epsilon} \; f(\tilde{x} + a)_j - f(\tilde{x} + a)_i.$$

If the optimal solution (or a valid dual bound thereof) is less than zero, this verifies that our network is robust around $\tilde{x}$ in the sense that we cannot produce a small perturbation that will flip the classification from i to j.

We train a smaller and a larger model, each with two convolutional layers with ReLU activation functions, feeding into a dense layer of ReLU neurons, and then a final dense linear layer. TensorFlow pseudocode specifying the two network architectures is included in Fig. 2. We generate 100 instances for each network by randomly selecting images $\tilde{x}$ with true label i from the test data, along with a random target adversarial class $j \neq i$. Note that we make no attempts to utilize recent techniques that train the networks to be verifiable [21,53,54,56].

```
input = placeholder(float32, shape=(28,28))
conv1 = conv2d(input, filters=4, kernel_size=4,
               strides=(2,2), activation=relu, use_bias=True)
conv2 = conv2d(conv1, filters=4, kernel_size=4,
               strides=(2,2), use_bias=True)
flatten = reshape(conv2, [5*5*4])
dense = dense(flatten, 16, activation=relu, use_bias=True)
logits = dense(dense, 10, use_bias=True)
```

(a) Smaller ReLU network.

```
input = placeholder(float32, shape=(28,28))
conv1 = conv2d(input, filters=16, kernel_size=4,
               strides=(2,2), activation=relu, use_bias=True)
conv2 = conv2d(conv1, filters=32, kernel_size=4,
               strides=(2,2), activation=relu, use_bias=True)
flatten = reshape(conv2, [5*5*32])
dense = dense(flatten, 100, activation=relu, use_bias=True)
logits = dense(dense, 10, use_bias=True)
```

(b) Larger ReLU network.

Fig. 2. TensorFlow pseudocode specifying the two network architectures used.

For all experiments, we use the Gurobi v7.5.2 solver, running with a single thread on a machine with 128 GB of RAM and 32 CPUs at 2.30 GHz. We use a time limit of 30 min (1800 s) for each run. We perform our experiments using the tf.opt package for optimization over trained neural networks; tf.opt is under active development at Google, with the intention to open source the project in the future. Below, the *big-M* + (6b) method is the big-M formulation (3a–3d) paired with separation[2] over the exponential family (6b), and with Gurobi's cutting plane generation turned off. Similarly, the *big-M* and the *extended* methods are the big-M formulation (3a–3d) and the extended formulation (5a– 5f) respectively, with default Gurobi settings. Finally, the *big-M* + *no cuts* method turns off Gurobi's cutting plane generation without adding separation over (6b).

---

[2] We use cut callbacks in Gurobi to inject separated inequalities into the cut loop. While this offers little control over when the separation procedure is run, it allows us to take advantage of Gurobi's sophisticated cut management implementation.

3.1 Small ReLU Network

We start with a smaller ReLU network whose architecture is depicted in Tensor-Flow pseudocode in Fig. 2a. The model attains 97.2% test accuracy. We select a perturbation ball radius of $\epsilon = 0.1$. We report the results in Table 1 and in Fig. 3. The big-M + (6b) method solves 7 times faster on average than the big-M formulation. Indeed, for 79 out of 100 instances the big-M method does not prove optimality after 30 min, and it is never the fastest choice (the "win" column). Moreover, the big-M + no cuts times out on every instance, implying that using *some* cuts is important. The extended method is roughly 5 times slower than the big-M + (6b) method, but only exceeds the time limit on 19 instances, and so is substantially more reliable than the big-M method for a network of this size. From this, we conclude that the additional strength offered by the ideal formulations (5a–5f) and (6a–6c) can offer substantial computational improvement over the big-M formulation (3a–3d).

3.2 Larger ReLU Network

Now we turn to the larger ReLU network described in Fig. 2b. The trained model attains 98.5% test accuracy. We select a perturbation ball radius of $\epsilon = 10/256$.

Table 1. Results for smaller network. Shifted geometric mean for time and optimality gap taken over 100 instances (shift of 10 and 1, respectively). The "win" column is the number of (solved) instances on which the method is the fastest.

Method	Time (s)	Optimality gap	Win
big-M + (6b)	174.49	0.53%	81
big-M	1233.49	6.03%	0
big-M + no cuts	1800.00	125.6%	0
Extended	890.21	1.26%	6

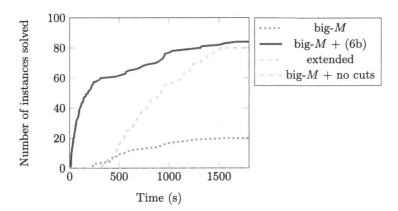

Fig. 3. Number of small network instances solved within a given amount of time. Curves to the upper left are better, with more instances solved in less time.

For these larger networks, we eschew solving the problems to optimality and focus on the quality of the dual bound available at the root node. As Gurobi does not reliably produce feasible primal solutions for these larger instances, we turn off primal heuristics and compare the approaches based on the "verification gap", which measures how far the dual bound is from proving robustness (i.e. an objective value of 0). To evaluate the quality of a dual bound, we measure the "improvement percentage" $\frac{\text{big\_M\_bound} - \text{other\_bound}}{\text{big\_M\_bound}}$, where our baseline for comparison, big_M_bound, is the bound from the big-M + no cuts method, and other_bound is the dual bound being compared.

Table 2. Results at the root node for larger network. Shifted geometric mean of bound, time, and improvement over 100 instances (shift of 10).

Method	Bound	Time (s)	Improvement
big-M + no cuts	302.03	3.08	–
big-M + (6b)	254.95	8.13	15.44%
big-M	246.87	612.65	18.08%
big-M + 15 s timeout	290.21	15.00	3.75%
Extended	–	1800.00	–

We report aggregated results over 100 instances in Table 2. First, we are unable to solve even the LP relaxation of the extended method on any of the instances in the allotted 30 min, due to the quadratic growth in size. In contrast, the LP relaxation of the big-M + no cuts method can be solved very quickly. The big-M + (6b) method strengthens this LP bound by more than 15% on average, and only takes roughly 2.5× as long. This is not only because the separation runs very quickly, but also for a technical reason: when Gurobi's cutting planes are disabled, the callback separating over (6b) is only called a small number of times, as determined by Gurobi's internal cut selection procedure. Therefore, this 15% improvement is the result of only a small number of separation rounds, not an exhaustive iterative procedure (i.e. Gurobi terminates the cut loop well before all violated inequalities have been separated).

We may compare these results against the big-M method, which is able to provide a modestly better bound (roughly 18% improvement), but requires almost two orders of magnitude more time to produce the bound. For another comparison, *big-M + 15 s timeout*, we set a smaller time limit of 15 s on Gurobi, which is a tighter upper bound on the maximum time used by the big-M + (6b) method. In this short amount of time, Gurobi is not able to improve the bound substantially, with less than 4% improvement. This suggests that the inequalities (6b) are not trivial to infer by generic cutting plane methods, and that it takes Gurobi many rounds of cut generation to achieve the same level of bound improvement we derive from restricting ourselves to those cuts in (6b).

Acknowledgement. The authors gratefully acknowledge Yeesian Ng and Ondřej Sýkora for many discussions on the topic of this paper, and for their work on the development of the `tf.opt` package used in the computational experiments.

A   Deferred Proofs

A.1   Proof of Proposition 1

Proof. The result follows from applying Fourier–Motzkin elimination to (5a–5f) to project out the x^0, x^1, y^0, and y^1 variables; see [31, Chap. 13] for an explanation of the approach. We start by eliminating the x^1, y^0, and y^1 using the equations in (5a), (5b), and (5c), respectively, leaving only x^0.

First, if there is some input component i with $w_i = 0$, then x_i^0 only appears in the constraints (5d–5e), and so the elimination step produces $L_i \leqslant x_i \leqslant U_i$.

Second, if there is some i with $w_i < 0$, then we introduce an auxiliary variable $\tilde{x}_i$ with the equation $\tilde{x}_i = -x_i$. We then replace $w_i \leftarrow |w_i|$, $L_i \leftarrow -U_i$, and $U_i \leftarrow -L_i$, and proceed as follows under the assumption that $w > 0$.

Applying the Fourier-Motzkin procedure to eliminate x_1^0 gives the inequalities

$$y \geqslant w \cdot x + b$$

$$y \leqslant w_1 x_1 - w_1 L_1 (1 - z) + \sum_{i>1} w_i x_i^0 + bz$$

$$y \leqslant w_1 U_1 z + \sum_{i>1} w_i x_i^0 + bz$$

$$y \geqslant w_1 x_1 - w_1 U_1 (1 - z) + \sum_{i>1} w_i x_i^0 + bz$$

$$y \geqslant w_1 L_1 z + \sum_{i>1} w_i x_i^0 + bz$$

$$L_1 \leqslant x_1 \leqslant U_1,$$

along with the existing inequalities in (5a–5f) where the x_1^0 coefficient is zero. Repeating this procedure for each remaining component of x^0 yields the linear system

$$y \geqslant w \cdot x + b \tag{7a}$$

$$y \leqslant \sum_{i \in I} w_i x_i - \sum_{i \in I} w_i L_i (1 - z) + \left(b + \sum_{i \notin I} w_i U_i \right) z \quad \forall I \subseteq \text{supp}(w) \tag{7b}$$

$$y \geqslant \sum_{i \in I} w_i x_i - \sum_{i \in I} w_i U_i (1 - z) + \left(b + \sum_{i \notin I} w_i L_i \right) z \quad \forall I \subseteq \text{supp}(w) \tag{7c}$$

$$(x, y, z) \in [L, U] \times \mathbb{R}_{\geqslant 0} \times [0, 1]. \tag{7d}$$

Moreover, we can show that the family of inequalities (7c) is redundant, and can therefore be removed. Fix some $I \subseteq \text{supp}(w)$, and take $h(I) \stackrel{\text{def}}{=} \sum_{i \in I} w_i \check{L}_i + \sum_{i \notin I} w_i \check{U}_i + b$. If $h(\llbracket \eta \rrbracket \setminus I) \geqslant 0$, we can express the inequality in (7c) corresponding to the set I as a conic combination of the remaining constraints as:

$$
\begin{array}{lll}
y \geqslant w \cdot x + b & \times & 1 \\
0 \geqslant L_i - x_i & \times & w_i \qquad\qquad \forall i \notin I \\
0 \geqslant z - 1 & \times & h(\llbracket \eta \rrbracket \setminus I)
\end{array}
$$

Alternatively, if $h(\llbracket \eta \rrbracket \setminus I) < 0$, we can express the inequality in (7c) corresponding to the set I as a conic combination of the remaining constraints as:

$$
\begin{array}{lll}
y \geqslant 0 & \times & 1 \\
0 \geqslant x_i - U_i & \times & w_i \qquad\qquad \forall i \in I \\
0 \geqslant -z & \times & -h(\llbracket \eta \rrbracket \setminus I)
\end{array}
$$

To complete the proof, for any components i where we introduced an auxiliary variable $\tilde{x}_i$, we use the corresponding equation $\tilde{x}_i = -x_i$ to eliminate x_i and replace it $\tilde{x}_i$, giving the result. $\qquad\qquad\qquad\qquad\qquad\qquad\qquad\square$

A.2   Proof of Proposition 2

Proof. We fix $I = \{\kappa+1, \ldots, \eta\}$ for some κ; this is without loss of generality by permuting the rows of the matrices presented below. Additionally, we presume that $w \geqslant 0$, which allows us to infer that $\check{L} = L$ and $\check{U} = U$. This is also without loss of generality by appropriately interchanging $+$ and $-$ in the definition of the $\tilde{p}^k$ below. In the following, references to (6b) are taken to be references to the inequality in (6b) corresponding to the subset I.

Take the two points $p^0 = (x, y, z) = (L, 0, 0)$ and $p^1 = (U, f(U), 1)$. Each point is feasible with respect to (6a–6c) and satisfies (6b) at equality. Then for some $\epsilon > 0$ and for each $i \in \llbracket \eta \rrbracket \setminus I$, take $\tilde{p}^i = (x, y, z) = (L + \epsilon e^i, 0, 0)$. Similarly, for each $i \in I$, take $\tilde{p}^i = (x, y, z) = (U - \epsilon e^i, f(U - \epsilon e^i), 1)$. From the strict activity assumption, there exists some $\epsilon > 0$ sufficiently small such that each $\tilde{p}^k$ is feasible with respect to (6a–6c) and satisfies (6b) at equality.

This leaves us with $\eta + 2$ feasible points satisfying (6b) at equality; the result then follows by showing that the points are affinely independent. Take the matrix

$$
\begin{pmatrix}
p^1 - p^0 \\
\tilde{p}^1 - p^0 \\
\vdots \\
\tilde{p}^\kappa - p^0 \\
\tilde{p}^{\kappa+1} - p^0 \\
\vdots \\
\tilde{p}^\eta - p^0
\end{pmatrix}
=
\begin{pmatrix}
U - L & f(U) & 1 \\
\epsilon e^1 & 0 & 0 \\
\vdots & \vdots & \vdots \\
\epsilon e^\kappa & 0 & 0 \\
U - L - \epsilon e^{\kappa+1} & f(U - \epsilon e^{\kappa+1}) & 1 \\
\vdots & \vdots & \vdots \\
U - L - \epsilon e^\eta & f(U - \epsilon e^\eta) & 1
\end{pmatrix}
\cong
\begin{pmatrix}
U - L & f(U) & 1 \\
\epsilon e^1 & 0 & 0 \\
\vdots & \vdots & \vdots \\
\epsilon e^\kappa & 0 & 0 \\
-\epsilon e^{\kappa+1} & -w_{\kappa+1}\epsilon & 0 \\
\vdots & \vdots & \vdots \\
-\epsilon e^\eta & -w_\eta\epsilon & 0
\end{pmatrix},
$$

where the third matrix is constructed by subtracting the first row to each of row $\kappa + 2$ to $\eta + 1$ (i.e. those corresponding to $\tilde{p}^i - p^0$ for $i > \kappa$), and is taken to mean congruency with respect to elementary row operations. If we permute the last column (corresponding to the z variable) to the first column, we observe that the resulting matrix is upper triangular with a nonzero diagonal, and so has full row rank. Therefore, the starting matrix also has full row rank, as we only applied elementary row operations, and therefore the $\eta + 2$ points are affinely independent, giving the result.   □

References

1. Amos, B., Xu, L., Kolter, J.Z.: Input convex neural networks. In: Precup, D., Teh, Y.W. (eds.) Proceedings of the 34th International Conference on Machine Learning. Proceedings of Machine Learning Research, vol. 70, pp. 146–155. PMLR, International Convention Centre, Sydney, Australia, 06–11 August 2017
2. Anderson, R., Huchette, J., Tjandraatmadja, C., Vielma, J.P.: Strong convex relaxations and mixed-integer programming formulations for trained neural networks (2018). https://arxiv.org/abs/1811.01988
3. Arulkumaran, K., Deisenroth, M.P., Brundage, M., Bharath, A.A.: Deep reinforcement learning: a brief survey. IEEE Signal Process. Mag. **34**(6), 26–38 (2017)
4. Atamtürk, A., Gómez, A.: Strong formulations for quadratic optimization with M-matrices and indicator variables. Math. Program. **170**, 141–176 (2018)
5. Balas, E.: Disjunctive programming and a hierarchy of relaxations for discrete optimization problems. SIAM J. Algorithmic Discret. Methods **6**(3), 466–486 (1985)
6. Balas, E.: Disjunctive programming: properties of the convex hull of feasible points. Discret. Appl. Math. **89**, 3–44 (1998)
7. Bartolini, A., Lombardi, M., Milano, M., Benini, L.: Neuron constraints to model complex real-world problems. In: Lee, J. (ed.) CP 2011. LNCS, vol. 6876, pp. 115–129. Springer, Heidelberg (2011). https://doi.org/10.1007/978-3-642-23786-7_11
8. Bartolini, A., Lombardi, M., Milano, M., Benini, L.: Optimization and controlled systems: a case study on thermal aware workload dispatching. In: Proceedings of the Twenty-Sixth AAAI Conference on Artificial Intelligence, pp. 427–433 (2012)
9. Bastani, O., Ioannou, Y., Lampropoulos, L., Vytiniotis, D., Nori, A.V., Criminisi, A.: Measuring neural net robustness with constraints. In: Advances in Neural Information Processing Systems, pp. 2613–2621 (2016)
10. Belotti, P., et al.: On handling indicator constraints in mixed integer programming. Comput. Optim. Appl. **65**(3), 545–566 (2016)
11. Bertsimas, D., Kallus, N.: From predictive to prescriptive analytics. Management Science (2018). https://arxiv.org/abs/1402.5481
12. Biggs, M., Hariss, R.: Optimizing objective functions determined from random forests (2017). https://papers.ssrn.com/sol3/papers.cfm?abstract_id=2986630
13. Bonami, P., Lodi, A., Tramontani, A., Wiese, S.: On mathematical programming with indicator constraints. Math. Program. **151**(1), 191–223 (2015)
14. Bunel, R., Turkaslan, I., Torr, P.H., Kohli, P., Kumar, M.P.: A unified view of piecewise linear neural network verification. In: Advances in Neural Information Processing Systems (2018)
15. Carlini, N., Wagner, D.: Towards evaluating the robustness of neural networks. In: 2017 IEEE Symposium on Security and Privacy (SP), pp. 39–57 (2017)
16. Cheng, C.-H., Nührenberg, G., Ruess, H.: Maximum resilience of artificial neural networks. In: D'Souza, D., Narayan Kumar, K. (eds.) ATVA 2017. LNCS, vol. 10482, pp. 251–268. Springer, Cham (2017). https://doi.org/10.1007/978-3-319-68167-2_18

17. Deng, Y., Liu, J., Sen, S.: Coalescing data and decision sciences for analytics. In: Recent Advances in Optimization and Modeling of Contemporary Problems. INFORMS (2018)
18. Donti, P., Amos, B., Kolter, J.Z.: Task-based end-to-end model learning in stochastic optimization. In: Guyon, I., et al. (eds.) Advances in Neural Information Processing Systems, vol. 30, pp. 5484–5494. Curran Associates, Inc. (2017)
19. Dulac-Arnold, G., et al.: Deep reinforcement learning in large discrete action spaces (2015). https://arxiv.org/abs/1512.07679
20. Dutta, S., Jha, S., Sankaranarayanan, S., Tiwari, A.: Output range analysis for deep feedforward neural networks. In: Dutle, A., Muñoz, C., Narkawicz, A. (eds.) NFM 2018. LNCS, vol. 10811, pp. 121–138. Springer, Cham (2018). https://doi.org/10.1007/978-3-319-77935-5_9
21. Dvijotham, K., et al.:: Training verified learners with learned verifiers (2018). https://arxiv.org/abs/1805.10265
22. Dvijotham, K., Stanforth, R., Gowal, S., Mann, T., Kohli, P.: A dual approach to scalable verification of deep networks. In: Thirty-Fourth Conference Annual Conference on Uncertainty in Artificial Intelligence (2018)
23. Ehlers, R.: Formal verification of piece-wise linear feed-forward neural networks. In: D'Souza, D., Narayan Kumar, K. (eds.) ATVA 2017. LNCS, vol. 10482, pp. 269–286. Springer, Cham (2017). https://doi.org/10.1007/978-3-319-68167-2_19
24. Elmachtoub, A.N., Grigas, P.: Smart "Predict, then Optimize" (2017). https://arxiv.org/abs/1710.08005
25. Fischetti, M., Jo, J.: Deep neural networks and mixed integer linear optimization. Constraints **23**, 296–309 (2018)
26. Gatys, L.A., Ecker, A.S., Bethge, M.: A neural algorithm of artistic style (2015). https://arxiv.org/abs/1508.06576
27. Goodfellow, I., Bengio, Y., Courville, A.: Deep Learning, vol. 1. MIT Press, Cambridge (2016)
28. den Hertog, D., Postek, K.: Bridging the gap between predictive and prescriptive analytics - new optimization methodology needed (2016). http://www.optimization-online.org/DB_HTML/2016/12/5779.html
29. Hijazi, H., Bonami, P., Cornuéjols, G., Ouorou, A.: Mixed-integer nonlinear programs featuring "on/off" constraints. Comput. Optim. Appl. **52**(2), 537–558 (2012)
30. Hijazi, H., Bonami, P., Ouorou, A.: A note on linear on/off constraints (2014). http://www.optimization-online.org/DB_FILE/2014/04/4309.pdf
31. Hooker, J.: Logic-Based Methods for Optimization: Combining Optimization and Constraint Satisfaction. Wiley, Hoboken (2011)
32. Huchette, J.: Advanced mixed-integer programming formulations: methodology, computation, and application. Ph.D. thesis, Massachusetts Institute of Technology (June 2018)
33. Katz, G., Barrett, C., Dill, D.L., Julian, K., Kochenderfer, M.J.: Reluplex: an efficient SMT solver for verifying deep neural networks. In: Majumdar, R., Kunčak, V. (eds.) CAV 2017. LNCS, vol. 10426, pp. 97–117. Springer, Cham (2017). https://doi.org/10.1007/978-3-319-63387-9_5
34. Khalil, E.B., Gupta, A., Dilkina, B.: Combinatorial attacks on binarized neural networks. In: International Conference on Learning Representations (2019)
35. LeCun, Y., Bengio, Y., Hinton, G.: Deep learning. Nature **521**(7553), 436–444 (2015)

36. LeCun, Y., Bottou, L., Bengio, Y., Haffner, P.: Gradient-based learning applied to document recognition. Proc. IEEE **86**, 2278–2324 (1998)
37. Lombardi, M., Gualandi, S.: A lagrangian propagator for artificial neural networks in constraint programming. Constraints **21**(4), 435–462 (2016)
38. Lomuscio, A., Maganti, L.: An approach to reachability analysis for feed-forward ReLU neural networks (2017). https://arxiv.org/abs/1706.07351
39. Mišić, V.V.: Optimization of tree ensembles (2017). https://arxiv.org/abs/1705.10883
40. Mladenov, M., Boutilier, C., Schuurmans, D., Elidan, G., Meshi, O., Lu, T.: Approximate linear programming for logistic Markov decision processes. In: Proceedings of the Twenty-sixth International Joint Conference on Artificial Intelligence (IJCAI 2017), pp. 2486–2493, Melbourne, Australia (2017)
41. Mordvintsev, A., Olah, C., Tyka, M.: Inceptionism: going deeper into neural networks (2015). https://ai.googleblog.com/2015/06/inceptionism-going-deeper-into-neural.html
42. Olah, C., Mordvintsev, A., Schubert, L.: Feature Visualization. Distill (2017). https://distill.pub/2017/feature-visualization
43. Papernot, N., McDaniel, P., Jha, S., Fredrikson, M., Celik, Z.B., Swami, A.: The limitations of deep learning in adversarial settings. In: IEEE European Symposium on Security and Privacy, pp. 372–387, March 2016
44. Say, B., Wu, G., Zhou, Y.Q., Sanner, S.: Nonlinear hybrid planning with deep net learned transition models and mixed-integer linear programming. In: Proceedings of the Twenty-Sixth International Joint Conference on Artificial Intelligence, IJCAI 2017, pp. 750–756 (2017)
45. Schweidtmann, A.M., Mitsos, A.: Global deterministic optimization with artificial neural networks embedded. J. Optim. Theory Appl. **180**(3), 925–948 (2019)
46. Serra, T., Ramalingam, S.: Empirical bounds on linear regions of deep rectifier networks (2018). https://arxiv.org/abs/1810.03370
47. Serra, T., Tjandraatmadja, C., Ramalingam, S.: Bounding and counting linear regions of deep neural networks. In: Thirty-Fifth International Conference on Machine Learning (2018)
48. Szegedy, C., et al.: Intriguing properties of neural networks. In: International Conference on Learning Representations (2014)
49. Tjeng, V., Xiao, K., Tedrake, R.: Verifying neural networks with mixed integer programming. In: International Conference on Learning Representations (2019)
50. Vielma, J.P.: Mixed integer linear programming formulation techniques. SIAM Rev. **57**(1), 3–57 (2015)
51. Vielma, J.P.: Small and strong formulations for unions of convex sets from the Cayley embedding. Math. Program. (2018)
52. Vielma, J.P., Nemhauser, G.: Modeling disjunctive constraints with a logarithmic number of binary variables and constraints. Math. Program. **128**(1–2), 49–72 (2011)
53. Wong, E., Kolter, J.Z.: Provable defenses against adversarial examples via the convex outer adversarial polytope. In: International Conference on Machine Learning (2018)
54. Wong, E., Schmidt, F., Metzen, J.H., Kolter, J.Z.: Scaling provable adversarial defenses. In: 32nd Conference on Neural Information Processing Systems (2018)

55. Wu, G., Say, B., Sanner, S.: Scalable planning with Tensorflow for hybrid nonlinear domains. In: Advances in Neural Information Processing Systems, pp. 6276–6286 (2017)
56. Xiao, K.Y., Tjeng, V., Shafiullah, N.M., Madry, A.: Training for faster adversarial robustness verification via inducing ReLU stability. In: International Conference on Learning Representations (2019)

Extended Formulations from Communication Protocols in Output-Efficient Time

Manuel Aprile[1]($\boxtimes$) and Yuri Faenza[2]

[1] DISOPT, EPFL, Lausanne, Switzerland
manuelf.aprile@gmail.com
[2] IEOR, Columbia University, New York, USA
yf2414@columbia.edu

Abstract. Deterministic protocols are well-known tools to obtain extended formulations, with many applications to polytopes arising in combinatorial optimization. Although constructive, those tools are not output-efficient, since the time needed to produce the extended formulation also depends on the number of rows of the slack matrix (hence, of the exact description in the original space). We give general sufficient conditions under which those tools can be implemented as to be output-efficient, showing applications to e.g. Yannakakis' extended formulation for the stable set polytope of perfect graphs, for which, to the best of our knowledge, an efficient construction was previously not known. For specific classes of polytopes, we give also a direct, efficient construction of those extended formulations. Finally, we deal with extended formulations coming from certain unambiguous non-deterministic protocols.

Keywords: Communication protocols · Extended formulations · Perfect graphs

1 Introduction

Linear extended formulations are a fundamental tool in integer programming and combinatorial optimization, as they allow to reduce an optimization problem over a polyhedron P to an analogous one over a polyhedron Q that linearly projects to P. When Q can be described with much fewer inequalities than P (typically, polynomial vs. exponential in the dimension of P), this leads to a computational speedup. Q as above is called an *extension* of P, any set of linear inequalities describing Q is an *extended formulation*, and the minimum number of inequalities in an extended formulation for P is called the *extension complexity* of P, and denoted by $\mathrm{xc}(P)$. Computing or bounding the extension complexity of polytopes has been an important topic in recent years, see e.g. [6,12,27].

Lower bounds on extension complexity are usually unconditional: neither they rely on any complexity theory assumptions, nor they take into account the

© Springer Nature Switzerland AG 2019
A. Lodi and V. Nagarajan (Eds.): IPCO 2019, LNCS 11480, pp. 43–56, 2019.
https://doi.org/10.1007/978-3-030-17953-3_4

time needed to produce the extension or the encoding length of coefficients in the inequalities. Upper bounds are often constructive and produce an extended formulation in time polynomial (often linear) in its size. Examples of the latter include Balas' union of polytopes, reflection relations, and branched polyhedral branching systems (see e.g. [9, 18]).

The fact that we can construct extended formulations *efficiently* is crucial, since their final goal is to make certain optimization problems (more) tractable. It is interesting to observe that there is indeed a gap between the *existence* of certain extended formulations, and the fact that we can construct them efficiently: in [5], it is shown that there is a small extended formulation for the stable set polytope that is $O(\sqrt{n})$-approximated (n being the number of nodes of the graph), but we do not expect to obtain it efficiently because of known hardness results [17]. In another case, a proof of the existence of a formulation of subexponential size with integrality gap $2 + \epsilon$ for min-knapsack [4] predated the efficient construction of a formulation with these properties [11].

In this paper, we investigate the efficiency of an important tool for producing extended formulation: communication protocols. In a striking result, Yannakakis [30] showed that a deterministic communication protocol computing the slack matrix of a polytope $P = \{x : Ax \leq b\} \subseteq \mathbb{R}^n$ can be used to produce an extended formulation for P. The number of inequalities of the latter is at most 2^c, where c is the *complexity* of the protocol (see Sect. 2 for definitions). Hence, deterministic protocols can be used to provide upper bounds on extension complexity of polytopes. This reduction is constructive, but not efficient. Indeed, it produces an extended formulation with $n + 2^c$ variables, 2^c inequalities, and an equation per row of A. Basic linear algebra implies that most equations are redundant, but in order to obtain a basis we may have to go through the full (possibly exponential-sized) list. The main application of Yannakakis' technique is arguably given in his original paper, and deals with the stable set polytope of perfect graphs. This is a class of polytopes that has received much attention in the literature [8, 16]. They also play an important role in extension complexity: while many open problems in the area were settled one after the other [6, 12, 19, 27], we still do not know if the stable set polytope of perfect graphs has polynomial-size extension complexity. Yannakakis' protocol gives an upper bound of $n^{O(n \log n)}$, while the best lower bound is as small as $\Omega(n \log n)$ [2]. On the other hand, a maximum stable set in a perfect graph can be computed efficiently via a poly-size *semidefinite* extension known as Lovasz' Theta body [23]. This can also be used, together with the ellipsoid method, to efficiently find a coloring of a perfect graph, see e.g. [28, Section 67.1]. We remark that designing a combinatorial (or at least SDP-free) polynomial-time algorithm to find a maximum stable set in perfect graphs, or to color them, is a main open problem [7].

Our Results. In this paper, we investigate conditions under which we can explicitly obtain an extended formulation from a communication protocol in time polynomial in the size of the formulation itself. We first show a general algorithm that achieves this for any deterministic protocol, given a compact representation of the protocol as a labelled tree and of certain extended formulations associated to leaves of the tree. The algorithm runs in linear time in the input size and is flexible, in that it also handles non-exact extended formulations. We then show that in some cases one can obtain those extended formulations directly, without relying on this general algorithm. This may be more interesting computationally. We show applications of our techniques in the context of (not only) perfect graphs. Our most interesting application is to Yannakakis' original protocol, whose associated extended formulation we construct in time linear in the size of the formulation itself, which is $n^{O(\log n)}$. For perfect graphs, this gives a subexponential SDP-free algorithm that computes a maximum stable set (resp. an optimal coloring). For general graphs, this gives a new relaxation of the stable set polytope which is (strictly) contained in the clique relaxation. Finally, we discuss extended formulations from certain unambiguous non-deterministic protocols.

Note: Extended discussions and several proofs are deferred to the journal version.

2   Preliminaries

Deterministic Communication Protocols. We start by describing the general setting of communication protocols, referring to [20] for details. Let M be a non-negative matrix with row (resp. column) set X (resp. Y), and two agents Alice and Bob. Alice is given as input a row index $i \in X$, Bob a column index $j \in Y$, and they aim at determining M_{ij} by exchanging information according to some pre-specified mechanism, that goes under the name of *deterministic protocol*. Such a protocol can be modelled as a rooted tree, with each vertex modelling a step where one of Alice or Bob sends a bit (hence labelled with A or B), and its children representing subsequent steps of the protocol. The tree is therefore binary, with each edge representing a 0 or a 1 sent. The *leaves* of the tree indicate the termination of the protocol and are labelled with the corresponding output. At each step of the protocol, the actions of Alice (resp. Bob) only depend on her (resp. his) input and on what they exchanged so far. The protocol is said to *compute* M if, for any input i of Alice and j of Bob, it returns M_{ij}. Hence, a deterministic protocol can be identified by the following parameters: a rooted binary tree τ with node set $\mathcal{V}$; a function $\ell : \mathcal{V} \to \{A, B\}$ ("Alice","Bob") associating each vertex to its type; for each leaf $v \in \mathcal{V}$, a non-negative number Λ_v corresponding to the value output at v; for each $v \in \mathcal{V}$, the set S_v of pairs $(i, j) \subseteq X \times Y$ such that, on input (i, j), the step corresponding to node v is executed during the protocol. We represent this compactly by $(\tau, \ell, \Lambda, \{S_v\}_{v \in \mathcal{V}})$. It can be shown that each S_v is a *rectangle*, i.e. a submatrix of M. For a leaf v of τ, all entries of S_v have the same value Λ_v, i.e. they form a submatrix of M with constant values. Such submatrices are called *monochromatic rectangles*. We assume that $S_v \neq \emptyset$ for each v.

   An *execution* of the protocol is a path of τ from the root to a leaf, whose edges correspond to the bits sent during the execution. The *complexity* of the protocol is given by the height h of the tree τ. A deterministic protocol computing M gives a partition of M in at most 2^h monochromatic rectangles. We remark that one can obtain a protocol (and a partition in rectangles) for M^T given a protocol for M by just exchanging the roles of Alice and Bob.

Extended Formulations and How to Find Them. We follow here the framework introduced in [26], that extends [30]. Consider a pair of polytopes (P, Q) with $P = \text{conv}(v_1, \dots, v_n) \subseteq Q = \{x \in \mathbb{R}^d : Ax \leq b\} \subseteq \mathbb{R}^d$, where A has m rows. A polyhedron $Q \in \mathbb{R}^{d'}$ is an *extension* for the pair (P, Q) if there is a projection $\pi : \mathbb{R}^{d'} \to \mathbb{R}^d$ such that $P \subseteq \pi(R) \subseteq Q$. An *extended formulation* for (P, Q) is a set of linear inequalities describing R as above, and the minimum number of inequalities in an extended formulation for (P, Q) is its *extension complexity*. The *slack matrix* $M(P, Q)$ of the pair (P, Q) is the non-negative $m \times n$ matrix with $M(P, Q)_{i,j} = b_i - a_i^\top v_j$, where a_i is the i-th row of A. A *non-negative factorization* of M is a pair of non-negative matrices (T, U) such that $M = TU$. The *non-negative rank* of M is the smallest intermediate dimension in a non-negative factorization of M.

Theorem 1 [26] *[Yannakakis' Theorem for pairs of polytopes]. Given a slack matrix M of a pair of polytopes (P, Q) of dimension at least 1, the extension complexity of (P, Q) is equal to the non-negative rank of M. In particular, if $M = TU$ is a non-negative factorization of M, then $P \subseteq \{x : \exists \, y \geq 0 : Ax + Ty = b\} \subseteq Q$.*

Hence, a factorization of the slack matrix of intermediate dimension N gives an extended formulation of size N (i.e., with N inequalities). However such formulation has as many equations as the number of rows of A.

Now assume we have a deterministic protocol of complexity c for computing $M = M(P, Q)$. The protocol gives a partition of M into at most 2^c monochromatic rectangles. This implies that $M = R_1 + \dots + R_N$, where $N \leq 2^c$ and each R_i is a rank 1 matrix corresponding to a monochromatic rectangle of non-zero value. Hence M can be written as a product of two non-negative matrices U, T of intermediate dimension N, where $T_{i,j} = 1$ if the (monochromatic) rectangle R_j contains row index i and 0 otherwise, and $U_{i,j}$ is equal to the value of R_i if R_i contains column index j, and 0 otherwise. As a consequence of Theorem 1, this yields an extended formulation for (P, Q). In particular, let $\mathcal{R}^1$ be the set of monochromatic, non-zero rectangles of M produced by the protocol and, for $i = 1, \dots, m$, let $\mathcal{R}^1_i \subset \mathcal{R}^1$ be the set of rectangles whose row index set includes i. Then the following is an extended formulation for (P, Q):

$$a_i x + \sum_{R \in \mathcal{R}^1_i} y_R = b_i \quad \forall \, i = 1, \dots, m \tag{1}$$

$$y \geq 0$$

Again, the formulation has as many equations as the number of rows of A, and it is not clear how to get rid of non-redundant equations efficiently. Note that all definitions and facts from this section specialize to those from [30] for a single polytope when $P = Q$.

Stable Set Polytope and QSTAB(G). The stable set polytope STAB(G) is the convex hull of the characteristic vectors of stable (also, independent) sets of a graph G. It has exponential extension complexity [12,15]. The *clique relaxation* of STAB(G) is:

$$\text{QSTAB}(G) = \left\{ x \in \mathbb{R}_+^d : \sum_{v \in C} x_v \le 1 \text{ for all cliques } C \text{ of } G \right\}. \qquad (2)$$

Note that in (2) one could restrict to maximal cliques, even though in the following we will consider all cliques when convenient. As a consequence of the equivalence between separation and optimization, optimizing over QSTAB(G) is NP-hard for general graphs, see e.g. [28]. However, the clique relaxation is exact for perfect graphs, for which the optimization problem is polynomial-time solvable via semidefinite programming (see Sect. 1).

Theorem 2 ([8]). *A graph G is perfect if and only if* STAB(G) = QSTAB(G).

The following result from [30] is crucial for this paper.

Theorem 3. *Let G be a graph with n vertices. There is a deterministic protocol of complexity $O(\log^2 n)$ computing the slack matrix of the pair* (STAB(G), QSTAB(G)). *Hence, there is an extended formulation of size $n^{O(\log(n))}$ for* (STAB(G), QSTAB(G)).

We remark that, when G is perfect, Theorem 3 gives a quasipolynomial size extended formulation for $STAB(G)$. However, as discussed above, it is not clear how to obtain such formulation in subexponential time.

3   A General Approach

We present here a general technique to explicitly and efficiently produce extended formulations from deterministic protocols, starting with an informal discussion.

The natural approach to reduce the size of (1) is to eliminate redundant equations. However, the structure of the coefficient matrix depends both on A and on rectangles $\mathcal{R}_i$'s of the factorization, which can have a complex behaviour. The reader is encouraged to try e.g. on the extended formulations obtained via Yannakakis' protocol for STAB(G), G perfect: the sets $\mathcal{R}_i$'s have very non-trivial relations with each other that depend heavily on the graph, and we did not manage to directly reduce the system (1) for general perfect graphs. Theorem 5 shows how to bypass this problem: in order to reconstruct the original extended formulation, all we need are (hopefully simpler) extended formulations for certain polytopes associated to all leaves of the protocol. We first recall a well-known fact [3], in the version given in [29, Section 3.1.1].

Theorem 4. *Let $P_1, P_2 \subset \mathbb{R}^d$ be non-empty polytopes, with $P_i = \pi_i\{y \in \mathbb{R}^{m_i} : A^i y \leq b^i\}$, where $\pi_i : \mathbb{R}^{m_i} \to d$ is a linear map, for $i = 1, 2$. Let $P = \mathrm{conv}(P_1 \cup P_2)$. Then we have:*

$$P = \{x \in \mathbb{R}^d : \exists\, y^1 \in \mathbb{R}^{m_1}, y^2 \in \mathbb{R}^{m_2}, \lambda \in \mathbb{R} : x = \pi_1(y^1) + \pi_2(y^2),$$
$$A^1 y^1 \leq \lambda b^1, A^1 y^2 \leq (1 - \lambda) b^2, 0 \leq \lambda \leq 1\}.$$

Moreover, the inequality $\lambda \geq 0$ ($\lambda \leq 1$ respectively) is redundant if P_1 (P_2) has dimension at least 1. Hence $\mathrm{xc}(P) \leq \mathrm{xc}(P_1) + \mathrm{xc}(P_2) + |\{i : \dim(P_i) = 0\}|$.

We now give the main theorem of this section. Note that, while the result relies on the existence of a deterministic protocol $(\tau, \ell, \Lambda, \{S_v\}_{v \in \mathcal{V}})$, its complexity does not depend on the encoding of Λ and $\{S_v\}_{v \in \mathcal{V}}$.

Theorem 5. *Let S be a slack matrix for a pair (P, Q), where*

$$P = \mathrm{conv}\{x_1^*, \ldots, x_n^*\} \subseteq Q = \{x \in \mathbb{R}^d : a_i x \leq b_i \text{ for } i = 1, \ldots, m\} \subseteq \mathbb{R}^d$$

are polytopes and for $j \in [d]$, let ℓ_j (resp. u_j) be a valid upper bound (resp. lower bound) on the variable x_j in Q. Assume there exists a deterministic protocol $(\tau, \ell, \Lambda, \{S_v\}_{v \in \mathcal{V}})$ with complexity c computing S, and let $\mathcal{R}$ be the set of monochromatic rectangles in which it partitions S (hence $c \leq \lceil \log_2 |\mathcal{R}| \rceil$). For $R \in \mathcal{R}$, let $P_R = \mathrm{conv}\{x_j^* : j \text{ is a column of } R\}$ and $Q_R = \{x \in \mathbb{R}^d : a_i x \leq b_i \,\forall\, i \text{ row of } R; \ell_j \leq x_j \leq u_j \text{ for all } j \in [d]\}$.

Suppose we are given τ, ℓ and, for each $R \in \mathcal{R}$, an extended formulation T_R for (P_R, Q_R). Let $\sigma(T_R)$ be the size (number of inequalities) of T_R, and $\sigma_+(T_R)$ be the total encoding length of the description of T_R (including the number of inequalities, variables and equations). Then we can construct an extended formulation for (P, Q) of size linear in $\sum_{R \in \mathcal{R}} \sigma(T_R)$ in time linear in $\sum_{R \in \mathcal{R}} \sigma_+(T_R)$.

Proof. We can assume without loss of generality that τ is a complete binary tree, i.e. each node of the protocol other than the leaves has exactly two children. Let $\mathcal{V}$ be the set of nodes of τ and $v \in \mathcal{V}$. Recall that S_v is the (non-necessarily monochromatic) rectangle given by all pairs (i, j) such that, on input (i, j), the execution of the protocol visits node v. Let us define, for any such S_v, a pair (P_v, Q_v) with $P_v = \mathrm{conv}\{x_j^* : j \text{ is a column of } S_v\}$ and

$$Q_v = \{x \in \mathbb{R}^d : a_i x \leq b_i \,\forall\, i \text{ row of } S_v; \ell_j \leq x_j \leq u_j \text{ for all } j \in [d]\}.$$

Clearly $P_v \subseteq P \subseteq Q \subseteq Q_v$, and Q_v is a polytope. Moreover, $S_\rho = S, P_\rho = P$, and $Q_\rho = Q$ for the root ρ of τ. We now show how to obtain an extended formulation T_v for the pair (P_v, Q_v) given extended formulations T_{v_i}'s for (P_{v_i}, Q_{v_i}), $i = 0, 1$, where v_0 (resp. v_1) are the two children nodes of v in τ.

Assume first that v is labelled A. Then we have $S_v^T = \begin{bmatrix} S_{v_0} & S_{v_1} \end{bmatrix}^T$ (up to permutation of rows), since the bit sent by Alice at v splits S_v in two rectangles by rows – those corresponding to rows where she sends 1 and those corresponding to rows where she sends 0. Therefore $P_v = P_{v_0} = P_{v_1}$ and $Q_v = Q_{v_0} \cap Q_{v_1}$. Hence we have $P_v \subseteq \pi_0(T_{v_0}) \cap \pi_1(T_{v_1}) \subseteq Q_v$, where π_i is a projection from the space of T_{v_i} to $\mathbb{R}^d$. An extended formulation for $T_v := \pi_0(T_{v_0}) \cap \pi_1(T_{v_1})$ can be obtained efficiently by juxtaposing the formulations of T_{v_0}, T_{v_1}.

If conversely v is labelled B, similarly we have $S_v = \begin{bmatrix} S_{v_0} & S_{v_1} \end{bmatrix}$ (up to permutations of columns). Hence, $P_v = \text{conv}\{P_{v_0} \cup P_{v_1}\}$ and $Q_v = Q_{v_0} = Q_{v_1}$, which implies $P_v \subseteq \text{conv}\{\pi_0(T_{v_0}) \cup \pi_1(T_{v_1})\} \subseteq Q_v$. An extended formulation for $T_v := \text{conv}\{\pi_0(T_{v_0}) \cup \pi_1(T_{v_1})\}$ can be obtained efficiently by applying Theorem 4 to the formulations of T_{v_0}, T_{v_1}. Iterating the procedure, in a bottom-up approach we obtain an extended formulation for (P, Q) from extended formulations of (P_v, Q_v), for each leaf v of the protocol.

We now bound the total encoding size of our formulation. If $T_v = \pi_0(T_{v_0}) \cap \pi_1(T_{v_1})$, then $\sigma_+(T_v) \leq \sigma_+(T_{v_0}) + \sigma_+(T_{v_1})$. Consider now $T_v = \text{conv}\{\pi_0(T_{v_0}) \cup \pi_1(T_{v_1})\}$. From Theorem 4 we have $\sigma_+(T_v) \leq \sigma_+(T_{v_0}) + \sigma_+(T_{v_1}) + O(d)$. Since the binary tree associated to the protocol is complete, it has size linear in the number of leaves, hence for the final formulation T_ρ we have

$$\sigma_+(T_\rho) \leq \sum_{R \in \mathcal{R}} (\sigma_+(T_R) + O(d)) = O\left(\sum_{R \in \mathcal{R}} \sigma_+(T_R) \right),$$

where the last equation follows from the fact that we can assume $\sigma_+(T_R) \geq d$ for any $R \in \mathcal{R}$. The bounds on the size of T_ρ and on the time needed to construct the formulation are derived analogously. $\square$

A couple of remarks on Theorem 5 are in order. The reader may recognize that the proof of Theorem 5 has a similar flavour to that of the main result in [11], where a technique is given to construct approximate extended formulations for 0/1 polytopes using Boolean formulas. However, those two results seem incomparable, in the sense that one does not follow from the other. It is unclear whether a more general framework encompassing both those techniques would have interesting applications.

The formulation produced by Theorem 5 may not be *exactly* the one given by the corresponding protocol. Also, even for the special case $P = Q$, the proof relies on the version of Yannakakis' theorem for pairs of polytopes. On the other hand, it does not strictly require that we reach the leaves of the protocol – a similar bottom-up approach would work starting at any node v if we have an extended formulation for (P_v, Q_v).

Using Theorem 5 it is possible to efficiently produce the formulation from Theorem 3, see the appendix. We instead show a direct derivation of a formulation of the same size in the next section. An application of Theorem 5 to min-up/min-down polytopes is also mentioned in Sect. 5.

4   Direct Derivations

Complement Graphs. An extended formulation for $STAB(G)$, G perfect, can be efficiently obtained from an extended formulation of $STAB(\bar{G})$, keeping a similar dimension (including the number of equations). We use the following two known facts.

Lemma 6 ([28]). *G is a perfect graph if and only if* $\text{STAB}(G) = \{x : x \geq 0, x^T y \leq 1 \,\forall\, y \in \text{STAB}(\bar{G})\}$.

Lemma 7 ([25,29]). *Given a non-empty polyhedron Q and $\gamma \in \mathbb{R}$, let $P = \{x : x^T y \leq \gamma \,\forall\, y \in Q\}$. If $Q = \{y : \exists\, z : Ay + Bz \leq b, Cy + Dz = d\}$, then we have that*

$$P = \{x : \exists \lambda \geq 0, \ \mu : A^T \lambda + C^T \mu = x, \quad B^T \lambda + D^T \mu = 0, \ b^T \lambda + d^T \mu \leq \gamma\}.$$

Hence $xc(P) \leq xc(Q) + 1$.

The next fact then follows immediately.

Corollary 8. *Let G be a perfect graph on n vertices such that $\text{STAB}(\bar{G})$ admits an extended formulation Q with r additional variables (i.e. $n + r$ variables in total), m inequalities and k equations. Then $\text{STAB}(G)$ admits an extended formulation with $m + k$ additional variables, $m + 1$ inequalities, $n + r$ equations, which can be written down efficiently given Q.*

Stable Set Polytopes of Perfect Graphs. We now present an algorithm that, given a perfect graph G on n vertices, produces an explicit extended formulation for $\text{STAB}(G)$ of size $n^{O(\log n)}$, in time bounded by $n^{O(\log n)}$. The algorithm is based on a decomposition approach inspired by Yannakakis' protocol [30], even though the formulation obtained has a different form than the one obtained via Yannakakis' protocol as in (1). A key tool is the next Lemma, whose proof can be found in the journal version. For a vertex v of G, $N^+(v) := N(v) \cup \{v\}$ denotes the inclusive neighbourhood of v.

Lemma 9. *Let G be a perfect graph on vertex set $V = \{v_1, \ldots, v_n\}$, and, fix k with $1 \leq k \leq n$. Let G_i be the induced subgraph of G on vertex set $V_i = N^+(v_i) \setminus \{v_1, \ldots, v_{i-1}\}$ for $i = 1, \ldots, k$, and G_0 the induced subgraph of G on vertex set $V_0 = \{v_{k+1}, \ldots, v_n\}$. Then we have $\text{STAB}(G) = (\text{STAB}(G_0) \times \mathbb{R}^{V \setminus V_0}) \cap \cdots \cap (\text{STAB}(G_k) \times \mathbb{R}^{V \setminus V_k})$.*

A simple though important observation for our approach is the following.

Observation 10. *Let $P_1, \ldots, P_k \in \mathbb{R}^n$ be polytopes and $P = P_1 \cap \cdots \cap P_k$, and let Q_i be an extended formulation for P_i for $i = 1, \ldots, k$, i.e. $P_i = \{x \in \mathbb{R}^n : \exists\, y^{(i)} \in \mathbb{R}^{r_i} : (x, y^{(i)}) \in Q_i\}$. Then $P = \{x \in \mathbb{R}^n : \text{for } i = 1, \ldots, k \ \exists\, y^{(i)} \in \mathbb{R}^{r_i} : (x, y^{(i)}) \in Q_i\}$.*

Theorem 11. *Let G be a perfect graph on n vertices. There is an algorithm that, on input G, outputs an extended formulation of $\text{STAB}(G)$ of size $n^{O(\log n)}$ in $n^{O(\log n)}$ time.*

Proof. We argue by induction on n. In this proof, logarithms are in base 2 (note that this was not needed earlier, because of big O notation). The base cases for n bounded by a constant are trivial, as the size of the classical formulation (and the time to obtain it) is constant too. For general n, let $v_1, \ldots, v_k$ be the vertices of G with degree at most $n/2$ and $G_0, \ldots, G_k$ be defined as in Lemma 9. First, assume that $k \geq n/2$, hence $G_0, \ldots, G_k$ have all size at most $n/2 + 1$. By induction, and thanks to Lemma 9 and Observation 10, running the algorithm on $G_0, \ldots, G_k$ and then applying Lemma 9, we produce an extended formulation of STAB(G) from those of STAB(G_0), $\ldots$, STAB(G_k) of size at most $n \cdot \left(\frac{n}{2} + 1\right)^{c \log\left(\frac{n}{2}+1\right)}$ for some constant $c > 0$, but this is at most $n^{c \log n}$ (assuming without loss of generality that $c \geq 2$). The same bound holds for the total running time.

If $k < n/2$, we take the complement graph $\bar{G}$, hence swapping low degree and high degree vertices: we now have $k \geq n/2$, hence by the previous case the algorithm obtains a formulation of STAB($\bar{G}$) of size at most $n \cdot \left(\frac{n}{2} + 1\right)^{c \log\left(\frac{n}{2}+1\right)}$. We can then use Lemma 7 to efficiently obtain a formulation for STAB(G), which by Corollary 8 has size at most $n \cdot \left(\frac{n}{2} + 1\right)^{c \log\left(\frac{n}{2}+1\right)} + 1 \leq n^{c \log n}$. Similar calculations bound the number of variables and equations of the formulation. The same bound holds for the total running time. □

Extension to Non-perfect Graphs and General Decomposition Trees. Although in the previous section we restricted ourselves to perfect graphs for ease of exposition, it is not hard to show that the above algorithm can be used on general graphs, yielding an extended formulation of (STAB(G), QSTAB(G)) of quasipolynomial size. Moreover, one can notice that the correctness of the algorithm does not depend on the decomposition procedure chosen: informally, one can define an arbitrary *decomposition tree* whose root corresponds to G and whose nodes correspond to either decomposing the graph as in Lemma 9 (with any choice of vertices $v_1, \ldots, v_k$) or taking the complement graph. Using such a tree and proceeding similarly as above we can obtain an extended formulation whose size only depends on the size of the tree and on the formulations we have for the subgraphs corresponding to leaves of the tree.

Theorem 12. *Let G be a graph with n vertices. There is an algorithm that, on input G, outputs an extended formulation of (STAB(G), QSTAB(G)) of size $n^{O(\log n)}$ in $n^{O(\log n)}$ time.*

Theorem 12, whose proof can be found in the appendix, generalizes Theorem 11 and, we believe, can have interesting computational consequences: indeed, there is interest in producing relaxations of STAB(G) without explicitly computing the Theta body (see for instance [13]). We conclude by remarking that, as a consequence of the main result of [14], there exists a constant $c > 0$ such that, there is no extended formulation of size $O(n^{\log^c n})$ for (STAB(G), QSTAB(G)): this limits the extent to which Theorem 12 could be improved for general (non-perfect) graphs.

Claw-Free Perfect Graphs and Extensions. Let $t \in \mathbb{N}_{\geq 3}$. A graph G is $K_{1,t}$-free if it does not contain $K_{1,t}$ (the complete bipartite graph with bipartitions of size 1 and t, respectively) as an induced subgraph. When $t = 3$, G is called *claw-free*. The following protocol for $STAB(G)$ when G is perfect and $K_{1,t}$-free appeared in [10]. Let Alice be given a clique C and Bob a stable set S. Alice sends $v \in C$. If $v \in S$, Bob outputs 0. Else, he sends $S' := S \cap N(v)$. Since G is $K_{1,t}$-free, $|S'| \leq t - 1$. Since $S \cap C \subset S \cap N^+(v) = S'$, Alice then outputs $1 - |C \cap S'| = 1 - |C \cap S|$. This gives an extended formulation of size $O(|V(G)|)^t$, with one equation per clique of G. It turns out that a much smaller set of equations is needed.

Theorem 13. *Let G be a $K_{1,t}$-free perfect graph. Let $\mathcal{R}$ be the set of monochromatic rectangles relative to the protocol for G. Following (1), denote by $\mathcal{R}_C$ the set of rectangles including the row index corresponding to the clique C. Then the following is an extended formulation of $STAB(G)$.*

$$x(C) + \sum_{R \in \mathcal{R}_C} y_R = 1 \quad \forall C \text{ clique of } G, |C| \leq 2$$

$$x, y \geq 0.$$

5   Further Applications and Extensions

The techniques developed in this papers can be applied to other extended formulations arising from deterministic protocols. Two more examples are: *Min-up/min-down* polytopes, which have been introduced in [22] to model scheduling problems with machines that have a physical constraint on the frequency of switches between the operating and not operating states; *Threshold-free* graphs, where a threshold graph is a graph for which there is an ordering of the vertices $v_1, \ldots, v_n$, such that for each i v_i is either complete or anticomplete to $v_{i+1}, \ldots, v_n$. We remark that a deterministic protocol for the clique-stable set incidence matrix of threshold-free graphs is known [21]. We defer details to the journal version.

Deterministic protocols are not the only that can be employed to produce extended formulations. In fact, extended formulations are *equivalent* to *randomized* protocols that compute the slack matrix in expectation, as defined in [10]. Even more, every extended formulation is obtained from a simple, one-way randomized protocol, see again [10]. Because of this strong equivalence, it seems hard to obtain a general algorithm as in Theorem 5. On the other hand, *non-deterministic unambiguous* protocols can also be used to show the existence of extended formulations, and seem easier to deal with. For instance, an extended formulation for a comparability graphs from a non-deterministic unambiguous protocol has been given in [30]. Again, this result does not imply that one can build the formulation efficiently. However, the construction can be made efficient by employing the techniques developed in the present paper. In particular, Theorem 13 holds also when G is a comparability graph (we remark that an extended formulation for the stable set polytopes of comparability graphs which does not use the protocol has been given in [24], to which we also refer to for the definition of comparability graphs). We defer details and the extension of those techniques to more general unambiguous non-deterministic protocols to the journal version.

Acknowledgements. We thank Mihalis Yannakakis for inspiring discussions and Samuel Fiorini for useful comments on [1], where many of the results presented here appeared. Manuel Aprile would also like to thank Aurélie Lagoutte and Nicolas Bousquet for useful discussions.

Appendix

Application of Theorem 5 to $(STAB(G), QSTAB(G))$. We now sketch how to apply Theorem 5 to the protocol from Theorem 3 in order to obtain an extended formulation for $(STAB(G), QSTAB(G))$ in time $n^{O(\log(n))}$.

For a detailed description of the protocol from [30] and for full details of the proof, we refer to the journal version. Here we recall that, at the beginning of the protocol, Alice is given a clique C of G as input and Bob a stable set S, and they want to compute the entry of the slack matrix of $STAB(G)$ corresponding to C, S, i.e. to establish whether C, S intersect or not. To do that, they exchange vertices of their respective sets. The number of vertices exchanged in total is at most $\lceil \log_2 n \rceil$. Hence, the protocol partitions the slack matrix of $(STAB(G), QSTAB(G))$ in a collection $\mathcal{R}$ of $n^{O(\log n)}$ monochromatic rectangles. Each rectangle $R \in \mathcal{R}$ is univocally identified by the list of vertices sent by Alice and by Bob during the execution of the protocol, which we denote by C_R and S_R respectively. Notice that $|C_R| + |S_R| \le \lceil \log_2 n \rceil$. For any clique C (resp. stable set) whose corresponding row (resp. column) is in R, we write $C \in R$ (resp. $S \in R$). If $C \in R$, then $C_R \subseteq C$ (in particular, C_R is a clique), and similarly $S \in R$ implies $S_R \subseteq S$ (and S_R is a stable set). We let P_R be the convex hull of stable sets $S \in R$ and Q_R the set of clique inequalities corresponding to cliques $C \in R$, together with the unit cube constraints.

To apply Theorem 5 we need to have a description of τ, ℓ and the extended formulations for (P_R, Q_R), for each $R \in \mathcal{R}$. These are computed as follows.

1. τ, ℓ, and (C_R, S_R) for all $R \in \mathcal{R}$. Enumerate all cliques and stable sets of G of combined size at most $\lceil \log_2 n \rceil$ and run the protocol on each of those input pairs. Each of those inputs gives a path in the tree (with the corresponding ℓ), terminating in a leaf v, corresponding to a rectangle R. τ is given by the union of those paths, hence it has size $n^{O(\log n)}$. Moreover, the inclusion-wise minimal input pairs corresponding to the same rectangle R give C_R, S_R.
2. Extended formulations of (P_R, Q_R), for each $R \in \mathcal{R}$. From the discussion in Sect. 2, it follows that an extended formulation of (P_R, Q_R) is given by

$$T_R := \{x \in \mathbb{R}^d, y_R \in \mathbb{R} : x(C) + y_R = 1 \ \forall \ C \in R, \ y_R \ge 0, 0 \le x \le 1\}. \quad (3)$$

We now claim that the equations in the description above, which can be exponentially many, are implied by a much smaller system (of size at most

n):

$$T_R = \{x \in \mathbb{R}^d, y_R \in \mathbb{R} : x(C_R) + y_R = 1$$
$$x(C_R + v) + y_R = 1 \; \forall \, v \in V \setminus C_R : C_R + v \in R$$
$$y_R \geq 0, \; 0 \leq x \leq 1\}.$$

Let $C \in R$. We only need to show that $x(C) + y_R$ is a linear combination of the left-hand sides of the equations above. For any $v \in C \setminus C_R$, one can show that $C_R + v \in R$. Hence we obtain, as required,

$$\sum_{v \in C \setminus C_R} (x(C_R + v) + y_R) - (|C \setminus C_R| - 1)(x(C_R) + y_R) = x(C) + y_R.$$

Proof of Theorem 12 (Sketch). The decomposition scheme outlined in the proof of Theorem 11 can be associated to a *decomposition tree* $\tau = \tau(G)$ on $n^{O(\log n)}$ nodes as follows: at each step, either decompose the current graph H using Lemma 9 (in which case each children is one of the H_i's), or take the complement (in which case there is a single child, associated to $\bar{H}$). We will abuse notation and identify a node of the decomposition tree and the corresponding subgraph. Note, in particular, that this decomposition tree does not depend on the fact that G is perfect, hence it can be applied here as well.

We can assume that, for each leaf L of τ, we are given an extended formulation T_L of $(\mathrm{STAB}(L), \mathrm{QSTAB}(L))$. Consider the extended formulation, which we call $\eta(G)$, obtained by traversing τ bottom-up and applying the following:

1. if a non-leaf node H of τ has a single child $\bar{H}$, then define $\eta(H)$ to be equal to the extended formulation of $\{x \in \mathbb{R}^{V(H)} : x^T y \leq 1 \; \forall \, y \in \eta(\bar{H})\}$, obtained by applying Lemma 7.
2. otherwise if H has children $H_0, \ldots, H_k$, then we define $\eta(H)$ to be the extended formulation obtained from $\eta(H_0), \ldots, \eta(H_k)$ using Observation 10.

We only need to prove that $\eta(G)$ is an extended formulation for $(\mathrm{STAB}(G), \mathrm{QSTAB}(G))$, as the efficiency aspects have been discussed in Theorems 5 and 11. We proceed by induction on the height of τ, in particular we prove that for a node H of τ, $\eta(H)$ is an extended formulation of $(\mathrm{STAB}(H), \mathrm{QSTAB}(H))$, assuming this is true for the children of H. If H is a leaf of τ, then there is nothing to prove. Otherwise, we need to analyze two cases:

1. H has a single child, labelled $\bar{H}$. Assume that $\mathrm{STAB}(\bar{H}) \subseteq \pi(\eta(\bar{H})) \subseteq \mathrm{QSTAB}(\bar{H})$, where π is the projection on the appropriate space. Let S be a stable set in H, and χ^S the corresponding incidence vector. For any $y \in \pi(\eta(\bar{H})) \subseteq \mathrm{QSTAB}(\bar{H})$, we have $y^T \chi^S = y(S) \leq 1$ as S is a clique in $\bar{H}$, hence since $\eta(H)$ (hence $\pi(\eta(H)))$ is clearly convex it follows that $\mathrm{STAB}(H) \subseteq \pi(\eta(H))$. Now, for a clique C of H and $x \in \pi(\eta(H))$, one has $\chi^C \in \pi(\eta(\bar{H}))$ hence $x(C) = x^T \chi^C \leq 1$, proving $\pi(\eta(H)) \subseteq \mathrm{QSTAB}(H)$.

2. H has children $H_0, \ldots, H_k$. Assume that for $i = 0, \ldots, k$, $\text{STAB}(H_i) \subseteq \pi(\eta(H_i)) \subseteq \text{QSTAB}(H_i)$ for π as above. Let χ^S be the characteristic vector of a stable set S of H, for every $i = 0, \ldots, k$ $S \cap V(H_i)$ is a stable set in H_i, hence $\chi^S \in \pi(\eta(H_i)) \times \mathbb{R}^{V(H) \setminus V(H_i)}$, and by convexity we conclude again that $\text{STAB}(H) \subseteq \pi(\eta(H))$. Finally, let $x \in \pi(\eta(H))$, and let C be a clique of H. It follows from the way we decompose H, that C is contained in H_i for some i: indeed, if $\{v_1, \ldots, v_k\} \cap C \neq \emptyset$, let i be the minimum such that $v_i \in C$, then $C \subseteq N^+(V_i) \setminus \{v_1, \ldots, v_{i-1}\} = V_i$ by definition; if $\{v_1, \ldots, v_k\} \cap C = \emptyset$, then $C \subseteq V_0$. By induction hypothesis, $\pi(\eta(H_i)) \subseteq \text{QSTAB}(H_i)$. But then $x(C) \leq 1$, and since this holds for all the cliques of H, we have $\pi(\eta(H)) \subseteq \text{QSTAB}(H)$. $\square$

References

1. Aprile, M.: On some problems related to 2-level polytopes. Ph.D. thesis, École Polytechnique Fédérale de Lausanne (2018)
2. Aprile, M., Faenza, Y., Fiorini, S., Huynh, T., Macchia, M.: Extension complexity of stable set polytopes of bipartite graphs. In: Bodlaender, H.L., Woeginger, G.J. (eds.) WG 2017. LNCS, vol. 10520, pp. 75–87. Springer, Cham (2017). https://doi.org/10.1007/978-3-319-68705-6_6
3. Balas, E.: Disjunctive programming. Ann. Discrete Math. **5**, 3–51 (1979)
4. Bazzi, A., Fiorini, S., Huang, S., Svensson, O.: Small extended formulation for knapsack cover inequalities from monotone circuits. In: Proceedings of the Twenty-Eighth Annual ACM-SIAM Symposium on Discrete Algorithms, pp. 2326–2341. SIAM (2017)
5. Bazzi, A., Fiorini, S., Pokutta, S., Svensson, O.: No small linear program approximates vertex cover within a factor $2 - \epsilon$. Math. Oper. Res. **44**(1), 147–172 (2018)
6. Chan, S.O., Lee, J.R., Raghavendra, P., Steurer, D.: Approximate constraint satisfaction requires large LP relaxations. J. ACM (JACM) **63**(4), 34 (2016)
7. Chudnovsky, M., Trotignon, N., Trunck, T., Vušković, K.: Coloring perfect graphs with no balanced skew-partitions. J. Comb. Theory Ser. B **115**, 26–65 (2015)
8. Chvátal, V.: On certain polytopes associated with graphs. J. Comb. Theory Ser. B **18**, 138–154 (1975)
9. Conforti, M., Cornuéjols, G., Zambelli, G.: Extended formulations in combinatorial optimization. 4OR **8**(1), 1–48 (2010)
10. Faenza, Y., Fiorini, S., Grappe, R., Tiwary, H.R.: Extended formulations, nonnegative factorizations, and randomized communication protocols. Math. Program. **153**(1), 75–94 (2015)
11. Fiorini, S., Huynh, T., Weltge, S.: Strengthening convex relaxations of 0/1-sets using Boolean formulas. arXiv preprint arXiv:1711.01358 (2017)
12. Fiorini, S., Massar, S., Pokutta, S., Tiwary, H.R., De Wolf, R.: Linear vs. semidefinite extended formulations: exponential separation and strong lower bounds. In: Proceedings of the Forty-Fourth Annual ACM Symposium on Theory of Computing, pp. 95–106. ACM (2012)
13. Giandomenico, M., Letchford, A.N., Rossi, F., Smriglio, S.: Ellipsoidal relaxations of the stable set problem: theory and algorithms. SIAM J. Optim. **25**(3), 1944–1963 (2015)
14. Göös, M.: Lower bounds for clique vs. independent set. In: 2015 IEEE 56th Annual Symposium on Foundations of Computer Science, pp. 1066–1076. IEEE (2015)

15. Göös, M., Jain, R., Watson, T.: Extension complexity of independent set polytopes. SIAM J. Optim. **47**(1), 241–269 (2018)
16. Grötschel, M., Lovász, L., Schrijver, A.: Polynomial algorithms for perfect graphs. N.-Holl. Math. Stud. **88**, 325–356 (1984)
17. Håstad, J.: Some optimal inapproximability results. J. ACM (JACM) **48**(4), 798–859 (2001)
18. Kaibel, V.: Extended formulations in combinatorial optimization. OPTIMA **85**, 2–7 (2011)
19. Kaibel, V., Pashkovich, K.: Constructing extended formulations from reflection relations. In: Jünger, M., Reinelt, G. (eds.) Facets of Combinatorial Optimization, pp. 77–100. Springer, Heidelberg (2013). https://doi.org/10.1007/978-3-642-38189-8_4
20. Kushilevitz, E., Nisan, N.: Communication Complexity. Cambridge University Press, Cambridge (1996)
21. Lagoutte, A.: Personal communication, Cargese, Corsica, 18 October 2018
22. Lee, J., Leung, J., Margot, F.: Min-up/min-down polytopes. Discrete Optim. **1**(1), 77–85 (2004)
23. Lovász, L.: On the Shannon capacity of a graph. IEEE Trans. Inf. Theory **25**(1), 1–7 (1979)
24. Lovász, L.: Stable sets and polynomials. Discrete Math. **124**(1–3), 137–153 (1994)
25. Martin, R.K.: Using separation algorithms to generate mixed integer model reformulations. Oper. Res. Lett. **10**(3), 119–128 (1991)
26. Pashkovich, K.: Extended formulations for combinatorial polytopes. Ph.D. thesis, Otto-von-Guericke-Universität Magdeburg (2012)
27. Rothvoß, T.: The matching polytope has exponential extension complexity. J. ACM (JACM) **64**(6), 41 (2017)
28. Schrijver, A.: Combinatorial Optimization: Polyhedra and Efficiency, vol. 24. Springer, Heidelberg (2002)
29. Weltge, S.: Sizes of linear descriptions in combinatorial optimization. Ph.D. thesis, Otto-von-Guericke-Universität Magdeburg, Fakultät für Mathematik (2015)
30. Yannakakis, M.: Expressing combinatorial optimization problems by linear programs. J. Comput. Syst. Sci. **43**, 441–466 (1991)

Sub-Symmetry-Breaking Inequalities for ILP with Structured Symmetry

Pascale Bendotti[1,2], Pierre Fouilhoux[1(✉)], and Cécile Rottner[1,2]

[1] Sorbonne Université, LIP6 CNRS 7606, Paris, France
pierre.fouilhoux@lip6.fr
[2] EDF R&D, Palaiseau, France
{pascale.bendotti,cecile.rottner}@edf.fr

Abstract. We consider integer linear programs whose solutions are binary matrices and whose (sub-)symmetry groups are symmetric groups acting on (sub-)columns. We propose a framework to build (sub-)symmetry-breaking inequalities for such programs, by introducing one additional variable per sub-symmetry group considered. The proposed framework is applied to derive inequalities breaking both symmetries and sub-symmetries in the graph coloring problem and in the ramp constrained min-up/min-down unit commitment problem.

Keywords: Integer linear programming ·
(Sub)-symmetry breaking inequalities · Graph Coloring Problem ·
Unit Commitment Problem

1 Introduction

It is well known that symmetries arising in integer linear programs can impair the solution process, in particular when symmetric solutions lead to an excessively large Branch-and-Bound (B&B) search tree. Various techniques, so called *Symmetry-Breaking Techniques* (SBT), are available to handle symmetries in integer linear programs of the form (ILP) $\min\{cx \mid x \in \mathcal{X}\}$ with $c \in \mathbb{R}^n$, and $\mathcal{X} \subseteq \mathcal{P}(m, n)$, where $\mathcal{P}(m, n)$ is the set of $m \times n$ binary matrices. A symmetry is defined as a permutation π of the indices $\{(i, j) \mid 1 \leq i \leq m, 1 \leq j \leq n\}$ such that for any solution matrix $x \in \mathcal{X}$, matrix $\pi(x)$ is also solution with the same cost, *i.e.*, $\pi(x) \in \mathcal{X}$ and $c(x) = c(\pi(x))$. The *symmetry group* $\mathcal{G}$ of (ILP) is the set of all such permutations. It partitions the solution set $\mathcal{X}$ into *orbits*, *i.e.*, two matrices are in the same orbit if there exists a permutation in $\mathcal{G}$ sending one to the other. A *subproblem* is problem (ILP) restricted to a subset of $\mathcal{X}$. In [4], symmetries arising in solution subsets of (ILP) are called *sub-symmetries*. Such sub-symmetries may be undetected in $\mathcal{G}$.

In this article, we focus on structured symmetries arising from (sub-)symmetry groups containing all sub-column permutations of a given solution submatrix. These symmetry groups are assumed to be known or previously detected [6,21]. A first idea to break symmetries is to reformulate the problem

© Springer Nature Switzerland AG 2019
A. Lodi and V. Nagarajan (Eds.): IPCO 2019, LNCS 11480, pp. 57–71, 2019.
https://doi.org/10.1007/978-3-030-17953-3_5

using integer variables summing the variables along orbits. Such a reformulation aggregates variables, thus reduces the size of the resulting ILP [20]. However, this can be used only when solutions can be disaggregated, e.g., when the integer decomposition property [2] holds. A more general idea to break symmetries is, in each orbit, to pick one solution, defined as the *representative*, and then restrict the solution set to the set of all representatives. The most common choice of representative is based on the lexicographical order. Column $y \in \{0, 1\}^m$ is said to be *lexicographically*, denoted by lex., *greater than* column $z \in \{0, 1\}^m$ if there exists $i \in \{1, ..., m-1\}$ such that $\forall i' \leq i,\ y_{i'} = z_{i'}$ and $y_{i+1} > z_{i+1}$, *i.e.*, $y_{i+1} = 1$ and $z_{i+1} = 0$. We write $y \succ z$ (resp. $y \succeq z$) if y is lex. greater than z (resp. greater than or equal to z). A technique is said to be *full symmetry-breaking* (resp. *partial symmetry-breaking*) if the solution set is exactly (resp. partially) restricted to the representative set.

Many SBT are based on pruning and fixing rules in the B&B tree [4,11,17, 26,30]. Other SBT rely on full or partial Symmetry-Breaking Inequalities (SBI). Note that the size of the LP solved at each node of the branching tree is generally invariant under pruning and fixing techniques, whereas it is increased using SBI.

SBI can be derived from the linear description of the convex hull of an arbitrary representative set [12]. In most works, each chosen representative x is lex. maximal in its orbit, *i.e.*, $x \succeq g(x)$, for each $g \in \mathcal{G}$. The convex hull of the representative set is called the *symmetry-breaking polytope* [14]. When x is a matrix and symmetry group $\mathcal{G}$ acts on the columns of x, the symmetry-breaking polytope is called *orbitope*. Even if complete linear descriptions for symmetry-breaking polytopes may be hard to reach in general, ILP formulations for these polytopes still yield full SBI [14]. Instead of considering orbits of solutions, [21,22] introduce inequalities enforcing a lexicographical order within orbits.

If symmetry group $\mathcal{G}$ is the *symmetric group* $\mathfrak{S}_n$ acting on the columns, the chosen representative x of an orbit may be such that its columns $x(1)$, ..., $x(n)$ are lex. non-increasing, *i.e.*, for all $j < n$, $x(j) \succeq x(j + 1)$. The convex hull of all $m \times n$ binary matrices with lex. non-increasing columns is called the *full orbitope* [18]. An $O(mn^3)$ extended formulation is given in [16]. To the best of our knowledge, it has never been used in practice to handle symmetries. A complete linear description of the full orbitope in the x variable space seems hard to reach [24]. To ensure that the integer solutions are in the full orbitope, Friedman [12] introduces full SBI with exponentially valued coefficients, causing numerical intractability. Some alternative inequalities featuring binary coefficients can also be used, at the expense of losing the full symmetry-breaking property. Examples are *column inequalities* [18,29], or the partial symmetry-breaking form of Friedman inequalities [15,23].

We propose a general framework (Sect. 2) to build full Sub-Symmetry-Breaking Inequalities (SSBI) in order to handle sub-symmetries arising from solution subsets whose symmetry groups contain the symmetric group acting on some sub-columns. One additional variable per subset Q considered may be needed in these inequalities, depending on whether variables x are sufficient to indicate that "x belongs to subset Q". The proposed framework is applied to derive such inequalities when the symmetry group is the symmetric group $\mathfrak{S}_n$ acting on the columns (Sect. 3). It is also applied to derive efficient SSBI for the Graph Coloring Problem (Sect. 4) and for the Ramp constrained Min-up/min-down Unit Commitment Problem (Sect. 5).

2 Sub-Symmetry-Breaking Inequalities

For a given solution subset Q, the symmetry group $\mathcal{G}_Q$ of the corresponding subproblem is different from $\mathcal{G}$ and may contain symmetries undetected in $\mathcal{G}$. In practice it is too expensive to compute the symmetry group for every subset $Q \subset \mathcal{X}$. However for many problems, symmetries of $\mathcal{G}$ can be deduced from the problem's structure, and so can symmetries of $\mathcal{G}_Q$, for some particular solution subsets Q. In this case, symmetries of $\mathcal{G}_Q$ are a priori known, and thus do not need to be computed. Such symmetries may be handled together with symmetries of $\mathcal{G}$. In this section, we introduce SSBI designed to simultaneously handle symmetries and sub-symmetries in symmetric groups. First, we briefly recall the concepts of sub-symmetry in ILP introduced in [4].

Consider a subset $Q \subset \mathcal{X}$ of solutions of (ILP). The sub-symmetry group $\mathcal{G}_Q$ relative to subset Q is defined as the symmetry group of subproblem $\min\{cx \mid x \in Q\}$. Permutations in sub-symmetry group $\mathcal{G}_Q$ are referred to as *sub-symmetries*.

Let $\{Q_s \subset \mathcal{X}, \ s \in \{1, ..., q\}\}$ be a set of solution subsets. To each Q_s, $s \in \{1, ..., q\}$, there corresponds a sub-symmetry group $\mathcal{G}_{Q_s}$. Let O_k^s, $k \in \{1, ..., o_s\}$, be the orbits defined by $\mathcal{G}_{Q_s}$ on subset Q_s, $s \in \{1, ..., q\}$, and $\mathcal{O} = \{O_k^s, \ k \in \{1, ..., o_s\}, \ s \in \{1, ..., q\}\}$. For given $x \in \mathcal{P}(m, n)$, let us define $\mathcal{G}(x) = \bigcup_{Q_s \ni x} \mathcal{G}_{Q_s}$, the set of all permutations π in $\bigcup_{s=1}^q \mathcal{G}_{Q_s}$ such that π can be applied to x. Matrix x' is said to be in relation with $x \in \mathcal{P}(m, n)$ if there exist $r \in \mathbb{N}$ and permutations $\pi_1, ..., \pi_r$ such that $\pi_k \in \mathcal{G}(\pi_{k-1} \circ ... \circ \pi_1(x))$, $\forall k \in \{1, ..., r\}$, and $x' = \pi_r \circ \pi_{r-1} \circ ... \circ \pi_1(x)$. The *generalized orbit* $\mathbb{O}$ of x w.r.t. $\{Q_s, s \in \{1, ..., q\}\}$ is thus the set of all x' in relation with x. By definition, for any generalized orbit $\mathbb{O}$, there exist orbits $\sigma_1, ..., \sigma_p \in \mathcal{O}$ such that $\mathbb{O} = \cup_{s=1}^p \sigma_s$. To each orbit σ, there corresponds a representative $\rho(\sigma)$. The set of representatives $\{\rho(\sigma), \sigma \in \mathcal{O}\}$ is said to be *orbit-compatible* if for any generalized orbit $\mathbb{O} = \cup_{s=1}^p \sigma_s$, $\sigma_1, ..., \sigma_p \in \mathcal{O}$, there exists j such that $\rho(\sigma_j) = \rho(\sigma_s)$ for all i such that $\rho(\sigma_j) \in \sigma_s$. Such a solution $\rho(\sigma_j)$ is said to be a *generalized representative* of $\mathbb{O}$.

Given $x \in \mathcal{X}$ and sets $R \subset \{1, ..., m\}$ and $C \subset \{1, ..., n\}$, we consider submatrix (R, C) of x, denoted by $x(R, C)$, obtained by considering columns C of x on rows R only. Symmetry group $\mathcal{G}_Q$ is the *sub-symmetric group* w.r.t. (R, C) if it is the set of all permutations of the columns of $x(R, C)$. If $\mathcal{G}_Q$ is the sub-symmetric group w.r.t. (R, C) then subset Q is said to be *sub-symmetric* w.r.t. (R, C).

Consider a set $\mathbb{S}$ of Q_s, $s \in \{1, ..., q\}$, such that each subset $Q_s \subseteq \mathcal{X}$, $s \in \{1, ..., q\}$, is sub-symmetric w.r.t. (R_s, C_s). For each orbit O_k^s, $k \in \{1, ..., o_s\}$ of $\mathcal{G}_{Q_s}$, $s \in \{1, ..., q\}$, its representative $x_k^s \in O_k^s$ is chosen to be such that submatrix $x_k^s(R_s, C_s)$ is lex. maximal, *i.e.*, its columns are lex. non-increasing. Such x_k^s is said to be the *lex-max* of orbit O_k^s w.r.t. (R_s, C_s).

Property 1 ([4]). $\{x_k^s, \ k \in \{1, ..., o_s\}, \ s \in \{1, ..., q\}\}$ is orbit-compatible.

The *full sub-orbitope* $\mathcal{P}_{sub}(\mathbb{S})$ associated to $\mathbb{S}$ is the convex hull of binary matrices x such that for each $s \in \{1, ..., q\}$, if $x \in Q_s$ then the columns of $x(R_s, C_s)$ are lex. non-increasing.

2.1 Definition and Validity of Sub-Symmetry-Breaking Inequalities

Consider a set $\mathbb{S}$ of solution subsets Q_s, $s \in \{1, ..., q\}$, such that each subset Q_s, $s \in \{1, ..., q\}$, is sub-symmetric w.r.t. (R_s, C_s). Consider an integer variable z_s, $s \in \{1, ..., q\}$, such that $z_s = 0$ if variable $x \in Q_s$, and such that $z_s \geq 1$ if $x \notin Q_s$. For any $x \in \mathcal{X}$, function Z associates x to a vector $Z(x)$ such that z_s, $s \in \{1, ..., q\}$, is the sth component of $Z(x)$ denoted by $Z_s(x)$.

Note that in many cases, function Z is linear, *i.e.*, each integer variable z_s is a linear expression of variables x. In such cases, no additional variable z_s is needed, as $z_s = Z(x)$. In some cases where function Z is not linear, variable z_s can be linearly expressed from variables x using only a few additional inequalities or integer variables.

Given c, $c' \in C_s$ such that $c < c'$, the *sub-symmetry-breaking inequality*, denoted by $(Q_s(c, c'))$, is defined as follows.

$$x_{r_1, c'} \leq z_s + x_{r_1, c} \qquad \text{where } r_1 = \min(R_s) \tag{1}$$

For each orbit O_k^s, $k \in \{1, ..., o_s\}$, of $\mathcal{G}_{Q_s}$, $s \in \{1, ..., q\}$, the chosen representative is the lex-max of orbit O_k^s w.r.t. (R_s, C_s). Then by Property 1, this set of representatives is orbit-compatible. In particular, solution set $\mathcal{X}$ can be restricted to the set of representatives by considering its intersection with the full sub-orbitope $\mathcal{P}_{sub}(\mathbb{S})$. If $x \in Q_s$, inequality $(Q_s(c, c'))$ enforces that the first row of submatrix $x(R_s, C_s)$ is lex. non-increasing, hence the following.

Lemma 1 (Validity). *If* $x \in \mathcal{P}_{sub}(\mathbb{S})$*, then* $(x, Z(x))$ *satisfies inequality* $(Q_s(c, c'))$ *for each* $s \in \{1, ..., q\}$ *and* c*,* $c' \in C_s$ *such that* $c < c'$*.*

Note that an inequality similar to (1) applied to a row of R_s distinct from r_1 may not be valid when used alongside with (1), as shown in Example 1.

Example 1. Let $\mathbb{S} = \{Q_1\}$, $q = 1$, where subset $Q_1 = \{x \in \mathcal{P}(4, 3) \cap \mathcal{X} \mid \sum_{c=1}^{3} x_{2,c} = 3\}$ Let us suppose the symmetry group of Q_1 is the sub-symmetric group w.r.t. submatrix $(\{3, 4\}, \{1, 2, 3\})$. Variable z_1 can be defined using equality $z_1 = 3 - \sum_{c=1}^{3} x_{2,c}$. Note that $z_1 = Z_1(x) = 0$ when $\sum_{c=1}^{3} x_{2,c} = 3$, *i.e.*, $x \in Q_1$, and is positive otherwise. Here the first row in R_1 is $r_1 = \min(R_1) = 3$, thus given c, $c' \in \{1, 2, 3\}$, $c < c'$, inequality $(Q_1(c, c'))$ is $x_{3,c'} \leq (3 - \sum_{j=1}^{3} x_{2,j}) + x_{3,c}$. This inequality enforces that row 3 of a solution matrix x is lex. ordered, *i.e.*, $x_{3,1} \geq x_{3,2} \geq x_{3,3}$, whenever $\sum_{c=1}^{3} x_{2,c} = 3$. Now let x^1, $x^2 \in Q_1$:

$$x^1 = \begin{bmatrix} 1 & 0 & 0 \\ 1 & 1 & 1 \\ 1 & 0 & 0 \\ 0 & 1 & 1 \end{bmatrix} \qquad \text{and} \qquad x^2 = \begin{bmatrix} 1 & 0 & 0 \\ 1 & 1 & 1 \\ 0 & 0 & 1 \\ 1 & 1 & 0 \end{bmatrix}$$

Inequality $(Q_1(c, c'))$ cuts off solution x^2 from the feasible set. Inequality (1) applied to row 4 is $x_{4,c'} \leq (3 - \sum_{j=1}^{3} x_{2,j}) + x_{4,c}$ This inequality would cut off x^1. This shows that these two inequalities cannot be used simultaneously.

Note that in the general case, inequalities (1) may only be partial-symmetry-breaking. Indeed, for given $s \in \{1, ..., q\}$ and c, $c' \in C_s$ such that $c < c'$, inequality $(Q_s(c, c'))$ only enforces that the first row of submatrix $x(R_s, C_s)$ is lex. non-increasing when $x \in Q_s$. In the case when $x_{r_1,c'} < x_{r_1,c}$, then sub-columns $x(R_s, \{c'\}) \prec x(R_s, \{c\})$. Otherwise, when $x_{r_1,c'} = x_{r_1,c}$, inequality (1) is not sufficient to select the lexmax representatives.

To enforce a lexicographical order, subsequent rows of submatrix $x(R_s, C_s)$ should be considered until a tie-break row is found. It is shown in the next section that inequalities $(Q_s(c, c'))$ for all $s \in \{1, ..., q\}$ and $c < c' \in C_s$ enforce that $x \in \mathcal{P}_{sub}(\mathbb{S})$ provided a tie-break condition on set $\mathbb{S}$ is fulfilled.

2.2   Full Symmetry-Breaking Sufficient Condition

We introduce a condition for inequalities (1) to be full symmetry-breaking.

For each $s \in \{1, ..., q\}$, consider $R_s = \{r_1^s, ..., r_{|R_s|}^s\}$ and $C_s = \{c_1^s, ..., c_{|C_s|}^s\}$, where $r_1^s < ... < r_{|R_s|}^s$ and $c_1^s < ... < c_{|C_s|}^s$. For given $s \in \{1, ..., q\}$ and any two columns $c_{l-1}^s, c_l^s \in C_s$, if there is a solution $x \in Q_s$ such that columns c_{l-1}^s and c_l^s are equal from row r_1^s to row r_{k-1}^s, it must be ensured that row r_k^s is lex. non increasing, i.e., $x_{r_k^s, c_{l-1}^s} \geq x_{r_k^s, c_l^s}$. The key idea is to exhibit another set $Q_p \in \mathbb{S}$ for quadruple (Q_s, k, l, x), such that Q_p contains x and is sub-symmetric w.r.t. (R_p, C_p), where the first row of R_p is r_k^s and C_p contains columns c_{l-1}^s and c_l^s. Then inequality $(Q_p(c_{l-1}^s, c_l^s))$ will ensure that $x_{r_k^s, c_{l-1}^s} \geq x_{r_k^s, c_l^s}$. For each quartet (Q_s, k, l, x), the existence of such a subset Q_p in $\mathbb{S}$ will be ensured by tie-break condition $(\mathcal{C})$, defined as follows:

$$(\mathcal{C}) \begin{cases} \forall s \in \{1, ..., q\}, \ \forall k \in \{2, ..., |R_s|\}, \ \forall l \in \{2, ..., |C_s|\} \\ \text{If } x \in Q_s \text{ s. t. } x_{r_{k'}^s, c_{l-1}^s} = x_{r_{k'}^s, c_l^s}, \ \forall k' \in \{1, ..., k-1\}, \text{then} \\ \text{there exists } p \in \{1, ..., q\} \text{ s. t. } x \in Q_p, \ C_p \supseteq \{c_{l-1}^s, c_l^s\} \text{ and } r_k^s = \min(R_p) \end{cases}$$

If condition $(\mathcal{C})$ holds, inequalities $(Q_s(c_{l-1}^s, c_l^s))$, $s \in \{1, ..., q\}$, $l \in \{2, ..., |C_s|\}$ exactly restrict the solution set to the representative set $\mathcal{X} \cap \mathcal{P}_{sub}(\mathbb{S})$. They are thus full symmetry-breaking, w.r.t. the sub-symmetries defined by $\mathbb{S}$. Thus:

Theorem 1. *If $(\mathcal{C})$ holds, then $x \in \mathcal{P}_{sub}(\mathbb{S})$ iff $(x, Z(x))$ satisfies $(Q_s(c_{l-1}^s, c_l^s))$, $\forall s \leq q$, $\forall l \in \{2, ..., |C_s|\}$*

For general set $\mathbb{S}$, condition $(\mathcal{C})$ may not hold. Fortunately, we can construct from $\mathbb{S}$ another set $\widetilde{\mathbb{S}}$ satisfying $(\mathcal{C})$ and such that $\mathcal{P}_{sub}(\widetilde{\mathbb{S}}) = \mathcal{P}_{sub}(\mathbb{S})$. The idea is to divide each Q_s, $s \in \{1, ..., q\}$, in smaller subsets such that for each $r_k^s \in R_s$ and each $c_l^s \in C_s$, there is a subset Q sub-symmetric w.r.t. $(R, C) = (\{r_k^s, ..., r_{|R_s|}^s\}, \{c_{l-1}^s, c_l^s\})$.

The set $\widetilde{\mathbb{S}}$ is defined as $\widetilde{\mathbb{S}} = \{\widetilde{Q}_s(k,l) \mid s \in \{1, ..., q\},\ k \in \{1, ..., |R_s|\},\ l \in \{2, ..., |C_s|\}\}$, where for each $s \in \{1, ..., q\},\ l \in \{2, ..., |C_s|\},\ k \in \{1, ..., |R_s|\}$,
$$\widetilde{Q}_s(k,l) = \{x \in Q_s \mid x_{r,c_{l-1}^s} = x_{r,c_l^s},\ \forall r \in \{r_1^s, ..., r_{k-1}^s\}\}$$
Note that for solution $x \in Q_s$ such that columns c_{l-1}^s and c_l^s are equal from row r_1^s to r_{k-1}^s, the set exhibited for quartet (Q_s, k, l, x) is $\widetilde{Q}_s(k,l)$. Note also that $\widetilde{Q}_s(1,l) = Q_s$, $l \in \{2, ..., |C_s|\}$. We thus have the following result:

Lemma 2. *Set $\widetilde{\mathbb{S}}$ satisfies $(\mathcal{C})$ and is such that $\mathcal{P}_{sub}(\widetilde{\mathbb{S}}) = \mathcal{P}_{sub}(\mathbb{S})$.*

It follows, from Theorem 1, that inequalities $(Q(c, c'))$, $c < c' \in C$, $Q \in \widetilde{\mathbb{S}}$ are full symmetry-breaking w.r.t. the sub-symmetries defined by $\mathbb{S}$.

Corollary 1. *If for each $Q \in \widetilde{\mathbb{S}}$, $(x, Z(x))$ satisfies inequality $(Q(c, c'))$, $\forall c < c' \in C$, then $x \in \mathcal{P}_{sub}(\mathbb{S})$.*

Set $\widetilde{\mathbb{S}}$ can be considered instead of $\mathbb{S}$, thus implying that at least one inequality (resp. at most one variable) can be added per subset $Q \in \widetilde{\mathbb{S}}$, i.e., $O(qmn)$ inequalities at least (resp. variables at most).

3   Application to the Symmetric Group Case

In this section, we apply the framework of Sect. 2 to any problem whose symmetry group $\mathcal{G}$ is the symmetric group $\mathfrak{S}_n$ acting on the columns. The collection $\mathbb{S}_{\mathfrak{S}}$ of subsets considered will lead to inequalities restricting any solution $x \in \mathcal{X}$ to be in the full orbitope. These inequalities feature variables z which can be explicitly expressed from x with $O(mn)$ linear inequalities. Here, the sub-symmetries considered are restrictions of symmetries' actions to solution subsets.

A complete linear description of the 2-column full orbitope, featuring additional integer variables, is proposed in [24]. In the general n-column case, we show that these inequalities can also be derived using the framework described in Sect. 2, and can be used as full SBI.

We consider $\mathbb{S}_{\mathfrak{S}} = \{Q_{i,j},\ i \in \{0\} \cup \{1, ..., m-1\},\ j \in \{2, ..., n\}\}$, where $Q_{i,j} = \{x \in \mathcal{X} \mid x_{i',j-1} = x_{i',j}\ \forall i' \in \{1, ..., i\}\}$. Subset $Q_{i,j}$ is the set of feasible solutions such that columns $j-1$ and j are equal from row 1 to row i. Note that $Q_{0,j} = \mathcal{X}$. The symmetry group of $Q_{i,j}$ is then the sub-symmetric group w.r.t. $(R_i, \{j-1, j\})$ where $R_i = \{i+1, ..., m\}$. It can be readily checked that in this case, $\mathbb{S}$ already satisfies condition $(\mathcal{C})$.

Let variable $z_{i,j}$ be such that $z_{i,j} = 0$ if $x \in Q_{i,j}$ and 1 otherwise. Note that for all $j \in \{2, ..., n\}$, $Q_{0,j} = \mathcal{X}$, thus $z_{0,j} = 0$, $\forall x \in \mathcal{X}$. Note also that $\mathcal{X} \cap \mathcal{P}_{sub}(\mathbb{S}_{\mathfrak{S}})$ is a subset of the full orbitope. Thus, given that the columns of any $x \in \mathcal{X} \cap \mathcal{P}_{sub}(\mathbb{S}_{\mathfrak{S}})$ are in a non-increasing lexicographical order, function Z is such that $Z(x) = z$, where z satisfies the following linear inequalities.

$$
\begin{cases}
z_{1,j-1} = x_{1,j-1} - x_{1,j} & \forall j \in \{2, ..., n\} & (2a) \\
z_{i,j-1} \le z_{i-1,j-1} + x_{i,j-1} & \forall i \in \{2, ..., m\},\ j \in \{2, ..., n\} & (2b) \\
z_{i,j-1} + x_{i,j} \le 1 + z_{i-1,j-1} & \forall i \in \{2, ..., m\},\ j \in \{2, ..., n\} & (2c) \\
x_{i,j-1} \le z_{i,j-1} + x_{i,j} & \forall i \in \{2, ..., m\},\ j \in \{2, ..., n\} & (2d) \\
z_{i-1,j-1} \le z_{i,j-1} & \forall i \in \{2, ..., m\},\ j \in \{2, ..., n\} & (2e)
\end{cases}
$$

Constraint (2a) sets variable $z_{1,j-1}$ to 1 whenever columns $j-1$ and j are different and in a non-increasing lexicographical order on row 1, and to 0 when they are equal. Constraint (2b) (resp. (2c)) sets variable $z_{i,j-1}$ to 0 when $z_{i-1,j-1} = 0$ and columns $j-1$ and j are equal to 0 (resp. 1) on row i. Constraint (2d) sets variable $z_{i,j-1}$ to 1 if columns $j-1$ and j are different and in a non-increasing lexicographical order on row i. Constraint (2e) sets $z_{i,j-1}$ to 1 when variable $z_{i-1,j-1} = 1$, i.e., when columns $j-1$ and j are different before row i.

For each $i \in \{0, ..., m-1\}$ and $j \in \{2, ..., n\}$ inequality (1) is inequality $(Q_{i,j}(j-1, j))$ as follows: $x_{i+1,j} \le z_{i,j-1} + x_{i+1,j-1}$, $\forall i \in \{1, ..., m\}$, $\forall j \in \{2, ..., n\}$. It ensures that if columns $j-1$ and j of x are equal from row 1 to i, then row $i+1$ is in a non-increasing lexicographical order.

Note that if $z_{i-1,j-1} - z_{i,j-1} = -1$ then necessarily $x_{i,j} = 0$. Thus inequality $((Q_{i,j}(j-1, j)))$ can be lifted to

$$ x_{i,j} \le (2z_{i-1,j-1} - z_{i,j-1}) + x_{i,j-1} \tag{3} $$

In the special case when $n = 2$, by replacing variable $z_{i,j}$ by $y_{i,j}$ where $z_{i,j} = 1 - \sum_{i'=1}^{i} y_{i',j}$, for each $i \in \{1, ..., m\}$, $j \in \{1, 2\}$, inequalities (2a)–(3) yield the complete linear description of the 2-column full orbitope proposed in [24]. In the general n-column case, inequalities (2a)–(3) are still full symmetry-breaking (by Theorem 1), and then can be used to restrict the feasible set to any full orbitope. In this case, $O(mn)$ additional variables and constraints are needed.

4 Application to the Graph Coloring Problem

Given an undirected graph $G = (V, E)$ with $|V| = n$, a *vertex coloring* of G is an assignment of values $\{1, ..., n\}$, denoted as *colors*, to the nodes so that no two adjacent nodes receive the same color. The minimum number of colors in a vertex coloring of G is called the *chromatic number* $\chi(G)$ of G. The *vertex coloring problem* is to find a vertex coloring with a minimum number of colors.

The column generation based linear program, proposed in [27], provides very good lower bounds. This approach is also used to come up with ILPs [13,25]. On the opposite, to obtain a compact ILP, a classical formulation is proposed as an extension of the stable set formulation [9]. However this formulation is very symmetric as a new vertex coloring can be obtained from another by permuting the color indices. An extension of this formulation has been devised in [7,10] using the notion of representative nodes, i.e., nodes representing a color. Some SBI for the classical graph coloring formulation are deduced from the linear description of the partitioning orbitope [18].

We consider the classical vertex coloring formulation where variable x_k^i indicates that color $k \in \{1, ..., K\}$ is assigned to vertex $i \in \{1, ..., n\}$. The feasible solution set is denoted by $\mathcal{X}_{col}$. Let S_1 and S_2 be two disjoint stable subsets of V and let $W \subseteq V$ be such that S_1 and S_2 have the same neighborhood in W, i.e., $N(S_1) \cap W = N(S_2) \cap W$. Let K be an upper bound on $\chi(G)$.

Consider solution subset

$$Q_{c_1,c_2}^{S_1,S_2,W} = \{x \in \mathcal{X}_{col} \mid x_{c_1}^i = 1 \ \forall i \in S_1, \ x_{c_2}^i = 1 \ \forall i \in S_2,$$
$$x_{c_1}^i = 0 \ \forall i \in U \backslash N(S_1), \ x_{c_2}^i = 0 \ \forall i \in U \backslash N(S_2)\}$$

where $U = V \backslash (W \cup S_1 \cup S_2)$. Subset $Q_{c_1,c_2}^{S_1,S_2,W}$ contains all colorings such that S_1 has color c_1, S_2 has color c_2, and elements outside S_1, S_2 and W are not colored by c_1 or c_2. Note that subset $Q_{c_1,c_2}^{S_1,S_2,W}$ is sub-symmetric w.r.t. $(R, \{c_1, c_2\})$, where $R = W \backslash (N(S_1) \cap N(S_2))$.

Variable z associated to $Q_{c_1,c_2}^{S_1,S_2,W}$ can be linearly expressed in terms of x variables, as follows

$$z = \sum_{s \in S_1}(1 - x_{c_1}^s) + \sum_{s \in S_2}(1 - x_{c_2}^s) + \sum_{r \in U \backslash N(s_1)} x_{c_1}^r + \sum_{r \in U \backslash N(s_2)} x_{c_2}^r$$

As this is a partition problem, there is exactly one 1-entry on each row. Therefore, Hence if $x \in Q_{c_1,c_2}^{S_1,S_2,W}$ and if sub-matrix $(R, \{c_1, c_2\})$ has lex. non-increasing columns, then variable $x_{c_2}^{w_1} = 0$. Thus the corresponding SSBI is $x_{c_2}^{w_1} \leq z$, where $w_1 = \min R$.

For a set $\mathbb{S}$ containing arbitrary $Q_{c_1,c_2}^{S_1,S_2,W}$, condition $(\mathcal{C})$ is not necessarily satisfied. For each $Q_{c_1,c_2}^{S_1,S_2,W} \in \mathbb{S}$, for any $r \in \{2, ..., |W|\}$, set $\widetilde{Q}_{c_1,c_2}^{S_1,S_2,W}(r)$ is defined as $\widetilde{Q}_{c_1,c_2}^{S_1,S_2,W}(r) = Q_{c_1,c_2}^{s_1,s_2,W} \cap \{x \mid x_{c_1}^i = x_{c_2}^i, \ \forall i \in \{w_1, ..., w_{r-1}\}\}$, where $W = \{w_1, ..., w_{|W|}\}$, $w_1 < ... < w_{|W|}$. Let us then consider set $\widetilde{\mathbb{S}}$ containing the sets in $\mathbb{S}$ and sets $\widetilde{Q}_{c_1,c_2}^{S_1,S_2,W}(r)$, for each $r \in \{2, ..., |W|\}$. By Lemma 2, set $\widetilde{\mathbb{S}}$ satisfies condition $(\mathcal{C})$ and therefore the associated SSBI are full symmetry-breaking. Note that variable $\widetilde{z}$ associated to $\widetilde{Q}_{c_1,c_2}^{S_1,S_2,W}(r)$ can be expressed as $\widetilde{z} = z + \sum_{i=1}^{r-1} x_{c_1}^{w_i}$, where z is the variable associated to set $Q_{c_1,c_2}^{S_1,S_2,W}$. Indeed, if sub-symmetries are broken, then $x_{c_2}^{w_i}$ cannot be 1 if $\sum_{j=1}^{i-1} x_{c_1}^{w_j}$ is 0.

Symmetry group $\mathcal{G}$ of the classical vertex coloring formulation contains all column permutations of solution matrices (where a column corresponds to a color). For any partitioning problem, *column inequalities* introduced in [18] are full symmetry-breaking w.r.t. such symmetries. We can show that these inequalities can also be derived using our framework, by considering solution subsets $Q_{k,k+1}^{\varnothing,\varnothing,V}$, $\widetilde{Q}_{k,k+1}^{\varnothing,\varnothing,V}(i)$, $i \in V$ and variables $\widetilde{z} = \sum_{j=k}^{i-1} x_k^j - \sum_{k'=k+2}^{\min(i,K)} x_{k'}^i$.

Promising experimental results are obtained on DIMACS graph coloring benchmark instances [1] classified as NP-m (*i.e.*, solvable within a minute). We compare our inequalities to the classical vertex coloring formulation (denoted by F), and to the latter formulation in which column inequalities are added (denoted by F-Col), each being implemented with Cplex 12.8 C++ API in default setting. This means in particular that Cplex internal SBT are turned on. We denote by F-Sub the classical formulation with column and SSBI, derived for some sets S_1 and S_2 of size 1.

Formulation F-Sub improves F-Col on some instances. Moreover, on some instances sets, *e.g.*, *ash* or *le*, SSBI prove to be extremely efficient as they improve F's CPU time by a factor up to 30 and F-Col's by a factor up to 3.

5   Application to the Unit Commitment Problem

The Unit Commitment Problem (UCP) has demonstrated to be a good candidate to apply SBT [4,20,23,29]. Given a discrete time horizon $\mathcal{T} = \{1,...,T\}$, a demand for electric power D_t is to be met at each time period $t \in \mathcal{T}$. Power is provided by a set $\mathcal{N}$ of n production units. At each time period, unit $j \in \mathcal{N}$ is either down or up, and in the latter case, its production is within $[P_{min}^j, P_{max}^j]$. Each unit must satisfy min-up (resp. min-down) times, *i.e.*, it must remain up (resp. down) during at least L^j (resp. ℓ^j) periods after start up (resp. shut down). Each unit j also features three costs: a fixed cost c_f^j, incurred each time period the unit is up; a start-up cost c_0^j, incurred each time the unit starts up; and a cost c_p^j proportional to its production. The Min-up/min-down UCP (MUCP) is to find a production plan minimizing the total cost while satisfying the demand and the minimum up and down time constraints. The MUCP is strongly NP-hard [5]. In the real-world UCP, some technical constraints have to be taken into account, such as ramp constraints. The MUCP is the combinatorial core structure of the UCP. In this article, we consider the RMUCP, *i.e.* the MUCP featuring *ramp-up* (resp. *ramp-down*) constraints, *i.e.*, the maximum increase (resp. decrease) in generated power from time period t to time period $t+1$ is RU^j (resp. RD^j). Moreover, if unit i starts up at time t (resp. shuts down at time $t+1$), its production at time t cannot be higher than SU^j (resp. SD^j).

The classical formulation of the RMUCP features two sets of binary variables: up variables $x_{t,j}$, indicating if unit j is up at time t, and start-up variables $u_{t,j}$, indicating if unit j starts up at time t [3,29,32]. The feasible set is denoted by $\mathcal{X}_{UCP}$. The RMUCP formulation including ramp constraints can be further strengthened with valid inequalities as proposed in [19,28,31].

In practical instances, there are H sets of n_h identical units, *i.e.*, units with same characteristics. Assuming a solution is expressed as a matrix where column j corresponds to the up/down trajectory of unit j over the time horizon, then any permutation of columns corresponding to identical units leads to another solution with same cost. Moreover, in some subproblems, there exist symmetries not contained in the symmetry group of the original problem, arising from the possibility of permuting some sub-columns. Consider two identical units and suppose at some time period t, these two units are down and ready to start up. Then their plans after t can be permuted, even if they do not have the same up/down plan before t. This still holds when ramp-constraints are considered.

5.1   Sub-Symmetry-Breaking Inequalities for the RMUCP

For each $t \in \mathcal{T}$ and any two units $j < j'$ of type h, consider $\overline{Q}_{j,j'}^{t,h} \subset \mathcal{X}_{UCP}$:

$$\overline{Q}_{j,j'}^{t,h} = \left\{ x \in \mathcal{X}_{UCP} \mid x_{t',i} = 0, \; \forall t' \in \{t - \ell^i, ..., t - 1\}, \; \forall i \in \{j, j'\} \right\}$$

Subset $\overline{Q}_{j,j'}^{t,h}$ is sub-symmetric w.r.t. the submatrix defined by rows and columns $(\{t, ..., T\}, \{j, j'\})$. Most of these sub-symmetries, referred to as *Start-up sub-symmetries*, are not detected in the symmetry group of the RMUCP. Note that $\overline{Q}_{j,j'}^{t,h}$ is different from subsets $Q_{i,j}$ defined in Sect. 3. Applying results from Sect. 2, variables $\overline{z}_{j,j'}^{t,h}$, indicating whether $x \in \overline{Q}_{j,j'}^{t,h}$, are

$$\overline{z}_{j,j'}^{t,h} = x_{t-\ell^j,j'} + \sum_{t'=t-\ell^j+1}^{t-1} u_{t',j'} + x_{t-\ell^j,j} + \sum_{t'=t-\ell^j+1}^{t-1} u_{t',j}$$

Consider $\mathbb{S}_{UCP} = \{\overline{Q}_{j,j'}^{t,h}, \; t \in \mathcal{T}, \; \forall h, \; j < j' \text{ of type } h\}$. Then set $\mathbb{S}$ directly satisfies condition $\mathcal{C}$. For each h, $j < j'$ and $t \in \mathcal{T}$, inequalities $(\overline{Q}_{j,j'}^{t,h})$ are :

$$x_{t,j'} \leq \left[x_{t-\ell^h,j'} + \sum_{t'=t-\ell^h+1}^{t-1} u_{t',j'} \right] + \left[x_{t-\ell^h,j} + \sum_{t'=t-\ell^h+1}^{t-1} u_{t',j} \right] + x_{t,j}.$$

Inequalities $(\overline{Q}_{j,j'}^{t,h})$ can be further strengthened, using the relationship between variables x and u. First note that by definition of variables w: $x_{t,j'} - \left[x_{t-\ell^h,j'} + \sum_{t'=t-\ell^h+1}^{t-1} u_{t',j'} \right] = u_{t,j'} - \sum_{t'=t-\ell^h+1}^{t} w_{t',j'}$ As if $u_{t,j'} = 1$, then $\sum_{t'=t-\ell^h+1}^{t} w_{t',j'} = 0$, the following *Start-Up-Ready* inequalities are valid and stronger than inequalities $(\overline{Q}_{j,j'}^{t,h})$:

$$u_{t,j'} \leq \left[x_{t-\ell^h,j} + \sum_{t'=t-\ell^h+1}^{t-1} u_{t',j} \right] + x_{t,j} \tag{4}$$

5.2   Experimental Results

We compare symmetry-breaking formulations for the RMUCP. Whereas the aggregated formulation proposed in [20] is very efficient for the MUCP, it can no longer be used for the RMUCP. An alternative formulation is the so-called aggregated interval formulation [20]. The corresponding results (not included in this paper) revealed computation times at least one order of magnitude slower than those obtained with any other proposed formulation. Furthermore, for some of the largest instances, the root node cannot be processed at all within the time limit. As shown in [29], neither Friedman inequalities nor column inequalities are competitive w.r.t. the classical UCP formulation when solved by Cplex. On the opposite, partial Friedman SBI has been shown to outperform Cplex in [23]. Hence the following formulations for the ramp-constrained MUCP are compared:

- $F(x, u)$: (x, u)-formulation.
- $W(x, u)$: (x, u)-formulation with partial Friedman SBI.
- $F(x, u, z)$: (x, u)-formulation with variables z, inequalities (2a)-(2e) and (3).
- $LF(x, u)$: (x, u)-formulation with SSBI (4).

Formulation $F(x, u, z)$ is obtained by adding (2a)–(3) to $F(x, u)$. Taking into account sub-symmetries leads to $LF(x, u)$ featuring Start-Up-Ready inequalities (4), in place of (2a)–(3). Note that start-up sub-symmetries are not handled by $F(x, u)$, $W(x, u)$ and $F(x, u, z)$. Each formulation features $O(nT)$ variables.

All experiments are performed using the same settings as in Sect. 4. The RMUCP instances are solved within a 3600 s timelimit. In Appendix A, the generation of RMUCP instances is described, along with some statistics on the instances characteristics in Table 1, and in Appendix B performance indicators in Table 2. In short, formulation $LF(x, u)$ appears to be the most efficient.

6   Perspectives

A perspective is to use the proposed framework to derive new SSBI. For the GCP, more experimental work would be necessary to embed SSBI in an efficient resolution setting. Obviously the proposed framework can be further applied to other problems with structured symmetries, *e.g.*, covering problems.

A   RMUCP Instances

We generate RMUCP instances as follows. For each instance, we generate a "2-peak per day" type demand with a large variation between peak and off-peak values: during one day, the typical demand in energy has two peak periods, one in the morning and one in the evening. The amplitudes between peak and off-peak periods have similar characteristics to those in the dataset from [8].

We consider the parameters $(P_{min}, P_{max}, L, \ell, c_f, c_0, c_p)$ of each unit from the dataset presented in [8]. We draw a correlation matrix between these characteristics and define a possible range for each characteristic. In order to introduce symmetries in our instances, some units are randomly generated based on the parameters correlations and ranges. Each unit generated is duplicated d times, where d is randomly selected in $[1, \frac{n}{F}]$ in order to obtain a total of n units. The parameter F is called symmetry factor, and can vary from 2 to 4 depending on the value of n. Note that these instances are generated along the same lines as literature instances considered in [3], but with different F factors.

In order to determine which symmetry-breaking technique performs best w.r.t. the number of rows and columns of matrices in feasible set $\mathcal{X}$, we consider various instance sizes $n \in \{20, 30, 60\}$ and $T \in \{48, 96\}$, and various symmetry factors $F \in \{2, 3, 4\}$.

For each size (n, T) and symmetry factor F, we generate a set of 20 instances. Symmetry factor $F = 4$ is not considered for instances with a small number n of units ($n = 20$ or 30), as it leads to very small sets of identical units. The ramp characteristics are the following: $RU^j = (P_{max}^j - P_{min}^j)/3$, $RD^j = (P_{max}^j - P_{min}^j)/2$ and $SU^j = SD^j = P_{min}^j$.

Table 1 provides some statistics on the instances characteristics. For each instance, a group is a set of two or more units with the same characteristics. Each unit which has not been duplicated is a singleton. The first and second entries column-wise are the number of singletons and groups. The third entry is the average group size and the fourth entry is the maximum group size. Each entry row-wise corresponds to the average value obtained over 20 instances with same size (n, T) and same symmetry factor F.

Table 1. Instance characteristics

Size (n, T)	Sym. factor	Nb singl.	Nb groups	Av. group size	Group max. size
(20, 48)	$F = 3$	1.25	4.90	3.96	5.75
	$F = 2$	0.75	3.20	6.45	8.75
(20, 96)	$F = 3$	0.90	4.75	4.08	5.60
	$F = 2$	0.75	3.45	5.93	8.65
(30, 48)	$F = 3$	1.10	5.35	5.51	9.45
	$F = 2$	0.25	3.85	8.30	12.60
(30, 96)	$F = 3$	0.40	5.25	5.97	8.65
	$F = 2$	0.55	4.05	7.59	11.40
(60, 48)	$F = 4$	0.80	7.70	7.86	13.20
	$F = 3$	0.55	5.80	10.90	17.80
	$F = 2$	0.20	4.75	13.90	23.80
(60, 96)	$F = 4$	0.60	7.90	7.79	13.20
	$F = 3$	0.30	5.95	10.50	16.60
	$F = 2$	0.20	4.35	14.80	24.90

B   Tables Relative to Results for the RMUCP

Table 2 provides, for each formulation and each group of 20 instances:

    #opt: Number of instances solved to optimality,
    #nodes: Average number of nodes,
    gap: Average optimality gap,
    CPU time: Average CPU time in seconds.

Note that a sign "-" in the column entry corresponding to the CPU time means that no instance could be solved within the time limit.

Table 2. Comparison of formulations for MUCP instances with symmetries.

Instances		Formulation	#opt	#nodes	gap (%)	CPU time
(20, 48)	$F = 2$	$F(x, u)$	9	667 974	0.009 16	2061.6
		$W(x, u)$	10	232 589	0.011 15	1965.2
		$F(x, u, z)$	11	139 493	0.009 91	1840.4
		$LF(x, u)$	16	242 096	0.001 89	980.4
	$F = 3$	$F(x, u)$	13	634 436	0.002 96	1424.7
		$W(x, u)$	16	314 447	0.004 40	1295.9
		$F(x, u, z)$	18	102 717	0.002 26	998.0
		$LF(x, u)$	20	30 014	0	132.8

Table 2. (*continued*)

Instances		Formulation	#opt	#nodes	gap (%)	CPU time
(20, 96)	$F = 2$	$F(x, u)$	5	702 415	0.007 76	2781.9
		$W(x, u)$	4	233 582	0.025 84	3058.1
		$F(x, u, z)$	8	61 384	0.006 81	2556.5
		$LF(x, u)$	6	160 150	0.007 18	2675.6
	$F = 3$	$F(x, u)$	7	989 738	0.006 44	2470.2
		$W(x, u)$	5	198 137	0.014 66	2725.6
		$F(x, u, z)$	12	87 375	0.004 24	1819.7
		$LF(x, u)$	15	186 018	0.005 65	1794.7
(30, 48)	$F = 2$	$F(x, u)$	4	354 029	0.018 03	2924.7
		$W(x, u)$	7	210 032	0.011 00	2535.4
		$F(x, u, z)$	4	71 467	0.025 47	2969.0
		$LF(x, u)$	15	219 655	0.002 04	1341.8
	$F = 3$	$F(x, u)$	6	379 482	0.012 13	2676.9
		$W(x, u)$	10	240 767	0.006 98	1931.4
		$F(x, u, z)$	5	107 609	0.016 23	2736.1
		$LF(x, u)$	16	191 113	0.002 19	965.7
(30, 96)	$F = 2$	$F(x, u)$	3	390 666	0.004 63	3069.8
		$W(x, u)$	4	121 205	0.007 55	3130.1
		$F(x, u, z)$	5	46 869	0.009 18	3107.7
		$LF(x, u)$	9	315 503	0.002 38	2263.5
	$F = 3$	$F(x, u)$	5	460 304	0.003 24	2927.0
		$W(x, u)$	3	211 303	0.004 65	3130.5
		$F(x, u, z)$	4	61 994	0.004 55	3059.7
		$LF(x, u)$	12	183 633	0.000 77	1852.9
(60, 48)	$F = 2$	$F(x, u)$	1	757 017	0.003 09	3437.6
		$W(x, u)$	4	203 485	0.002 85	3046.2
		$F(x, u, z)$	6	66 272	0.037 46	2839.8
		$LF(x, u)$	5	569 546	0.001 26	2710.6
	$F = 3$	$F(x, u)$	1	850 192	0.002 68	3422.5
		$W(x, u)$	6	192 656	0.002 45	2689.3
		$F(x, u, z)$	9	40 680	0.003 97	2527.5
		$LF(x, u)$	14	493 254	0.000 40	1450.2
	$F = 4$	$F(x, u)$	7	870 666	0.002 43	2582.4
		$W(x, u)$	10	295 149	0.000 95	1971.9
		$F(x, u, z)$	14	33 574	0.000 53	1623.1
		$LF(x, u)$	15	459 142	0.000 27	1043.8
(60, 96)	$F = 2$	$F(x, u)$	0	120 125	0.012 62	–
		$W(x, u)$	0	23 851	0.051 90	–
		$F(x, u, z)$	0	3 813	0.528 55	–
		$LF(x, u)$	0	52 226	0.011 25	–
	$F = 3$	$F(x, u)$	0	144 265	0.014 90	–
		$W(x, u)$	0	50 841	0.018 15	–
		$F(x, u, z)$	0	6 404	0.034 76	–
		$LF(x, u)$	0	83 335	0.013 11	–
	$F = 4$	$F(x, u)$	0	230 935	0.009 56	–
		$W(x, u)$	0	92 298	0.010 63	–
		$F(x, u, z)$	0	9 616	0.015 89	–
		$LF(x, u)$	2	150 692	0.006 56	3467.7

References

1. Graph coloring benchmark. https://sites.google.com/site/graphcoloring/vertex-coloring
2. Baum, S., Trotter, L.E.: Integer rounding and polyhedral decomposition for totally unimodular systems. In: Henn, R., Korte, B., Oettli, W. (eds.) Optimization and Operations Research. LNCS, vol. 157, pp. 15–23. Springer, Heidelberg (1978). https://doi.org/10.1007/978-3-642-95322-4_2
3. Bendotti, P., Fouilhoux, P., Rottner, C.: The min-up/min-down unit commitment polytope. J. Comb. Optim. **36**(3), 1024–1058 (2018)
4. Bendotti, P., Fouilhoux, P., Rottner, C.: Orbitopal fixing for the full (sub-)orbitope and application to the unit commitment problem. Optimization Online (2018). http://www.optimization-online.org/DB_HTML/2017/10/6301.html
5. Bendotti, P., Fouilhoux, P., Rottner, C.: On the complexity of the unit commitment problem. Ann. Oper. Res. **274**(1), 119–130 (2019)
6. Berthold, T., Pfetsch, M.E.: Detecting orbitopal symmetries. In: Fleischmann, B., Borgwardt, K.H., Klein, R., Tuma, A. (eds.) Operations Research Proceedings 2008, pp. 433–438. Springer, Heidelberg (2009). https://doi.org/10.1007/978-3-642-00142-0_70
7. Burke, E.K., Mareček, J., Parkes, A.J., Rudová, H.: A supernodal formulation of vertex colouring with applications in course timetabling. Ann. Oper. Res. **79**, 105–130 (2010)
8. Carrion, M., Arroyo, J.M.: A computationally efficient mixed-integer linear formulation for the thermal unit commitment problem. IEEE Trans. Power Syst. **21**, 1371–1378 (2006)
9. Coll, P., Marenco, J., Díaz, I., Zabala, P.: Facets of the graph coloring polytope. Ann. Oper. Res. **116**, 79–90 (2002)
10. Figueiredo, R., Barbosa, V., Maculan, N., Souza, C.: Acyclic orientations with path constraints. RAIRO-Oper. Res. **42**, 455–467 (2008)
11. Fischetti, M., Lodi, A., Salvagnin, D.: Just MIP it!. In: Maniezzo, V., Stützle, T., Voß, S. (eds.) Matheuristics. Annals of Information Systems, vol. 10, pp. 39–70. Springer, Boston (2009). https://doi.org/10.1007/978-1-4419-1306-7_2
12. Friedman, E.J.: Fundamental domains for integer programs with symmetries. In: Dress, A., Xu, Y., Zhu, B. (eds.) COCOA 2007. LNCS, vol. 4616, pp. 146–153. Springer, Heidelberg (2007). https://doi.org/10.1007/978-3-540-73556-4_17
13. Gualandi, S., Malucelli, F.: Exact solution of graph coloring problems via constraint programming and column generation. INFORMS J. Comput. **24**(1), 81–100 (2012)
14. Hojny, C., Pfetsch, M.E.: Polytopes associated with symmetry handling. Optimization Online (2017). http://www.optimization-online.org/DB_HTML/2017/01/5835.html
15. Jans, R.: Solving lot-sizing problems on parallel identical machines using symmetry-breaking constraints. INFORMS J. Comput. **21**(1), 123–136 (2009)
16. Kaibel, V., Loos, A.: Branched polyhedral systems. In: Eisenbrand, F., Shepherd, F.B. (eds.) IPCO 2010. LNCS, vol. 6080, pp. 177–190. Springer, Heidelberg (2010). https://doi.org/10.1007/978-3-642-13036-6_14
17. Kaibel, V., Peinhardt, M., Pfetsch, M.E.: Orbitopal fixing. In: Fischetti, M., Williamson, D.P. (eds.) IPCO 2007. LNCS, vol. 4513, pp. 74–88. Springer, Heidelberg (2007). https://doi.org/10.1007/978-3-540-72792-7_7
18. Kaibel, V., Pfetsch, M.E.: Packing and partitioning orbitopes. Math. Program. **114**(1), 1–36 (2008)

19. Knueven, B., Ostrowski, J., Wang, J.: Generating cuts from the ramping polytope for the unit commitment problem. Optimization Online (2016). http://www.optimization-online.org/DB_HTML/2015/09/5099.html
20. Knueven, B., Ostrowski, J., Watson, J.P.: Exploiting identical generators in unit commitment. IEEE Trans. Power Syst. **33**(4), 4496–4507 (2018)
21. Liberti, L.: Reformulations in mathematical programming: automatic symmetry detection and exploitation. Math. Program. **131**(1), 273–304 (2012)
22. Liberti, L., Ostrowski, J.: Stabilizer-based symmetry breaking constraints for mathematical programs. J. Glob. Optim. **60**(2), 183–194 (2014)
23. Lima, R.M., Novais, A.Q.: Symmetry breaking in MILP formulations for unit commitment problems. Comput. Chem. Eng. **85**, 162–176 (2016)
24. Loos, A.: Describing orbitopes by linear inequalities and projection based tools. Ph.D. thesis, Universität Magdeburg (2011)
25. Malaguti, E., Monaci, M., Toth, P.: An exact approach for the vertex coloring problem. Discrete Optim. **8**(2), 174–190 (2010)
26. Margot, F.: Exploiting orbits in symmetric ILP. Math. Program. **98**(1), 3–21 (2003)
27. Mehrotra, A., Trick, M.: A column generation approach for graph coloring. INFORMS J. Comput. **8**, 344–354 (1996)
28. Ostrowski, J., Anjos, M.F., Vannelli, A.: Tight mixed integer linear programming formulations for the unit commitment problem. IEEE Trans. Power Syst. **27**, 39–46 (2012)
29. Ostrowski, J., Anjos, M.F., Vannelli, A.: Modified orbital branching for structured symmetry with an application to unit commitment. Math. Program. **150**(1), 99–129 (2015)
30. Ostrowski, J., Linderoth, J., Rossi, F., Smriglio, S.: Orbital branching. Math. Program. **126**(1), 147–178 (2011)
31. Pan, K., Guan, Y.: A polyhedral study of the integrated minimum-up/-down time and ramping polytope. Optimization Online (2016). http://www.optimization-online.org/DB_HTML/2015/08/5070.html
32. Rajan, D., Takriti, S.: Minimum up/down polytopes of the unit commitment problem with start-up costs. IBM Research Report (2005)

Intersection Cuts for Polynomial Optimization

Daniel Bienstock[1], Chen Chen[2], and Gonzalo Muñoz[3]([✉])

[1] IEOR, Columbia University, New York, USA
dano@columbia.edu
[2] ISE, The Ohio State University, Columbus, USA
chen.8018@osu.edu
[3] IVADO Fellow, Polytechnique Montréal, Montreal, Canada
gonzalo.munoz@polymtl.ca

Abstract. We consider dynamically generating linear constraints (cutting planes) to tighten relaxations for polynomial optimization problems. Many optimization problems have feasible set of the form $S \cap P$, where S is a closed set and P is a polyhedron. Integer programs are in this class and one can construct intersection cuts using convex "forbidden" regions, or S-free sets. Here, we observe that polynomial optimization problems can also be represented as a problem with linear objective function over such a feasible set, where S is the set of real, symmetric matrices representable as outer-products of the form xx^T. Accordingly, we study outer-product-free sets and develop a thorough characterization of several (inclusion-wise) maximal intersection cut families. In addition, we present a cutting plane approach that guarantees polynomial-time separation of an extreme point in $P \setminus S$ using our outer-product-free sets. Computational experiments demonstrate the promise of our approach from the point of view of strength and speed.

1 Introduction

In this work we focus on polynomial optimization:

$$\min \ p_0(x)$$
$$\textbf{(PO)} \quad \text{s.t.} \ p_i(x) \le 0 \quad i = 1, ..., m,$$

where each p_i is a polynomial function with respect to $x \in \mathbb{R}^n$. We consider the dynamic generation of linear valid inequalities, i.e. cutting planes, to tighten relaxations of **PO**. Cuts for polynomial optimization are typically generated for a single nonlinear term or function (e.g. [7,35,37,41,45,51–53]) over a simple subset of linear constraints such as box constraints. In contrast, we develop general-purpose cuts that have the potential to involve all variables simultaneously. To the best of our knowledge there are two papers (applicable to polynomial optimization) that are similar to our work in this regard. The disjunctive cuts of Saxena, Bonami, and Lee [46,47], and the work of Ghaddar, Vera, and

© Springer Nature Switzerland AG 2019
A. Lodi and V. Nagarajan (Eds.): IPCO 2019, LNCS 11480, pp. 72–87, 2019.
https://doi.org/10.1007/978-3-030-17953-3_6

Anjos [26] who propose a lift-and-project method using moment relaxations. Polynomial-time separation from these procedures is not guaranteed in general.

We adopt the geometric perspective for generating cuts, in which cuts for a region of the form $S \cap P$, with P a polyhedron and S a closed set, are derived from convex forbidden zones, or S-free sets. The S-free approach developed in the context of mixed-integer programming, with S typically considered to be the integer lattice. Practical applications of this technique have so far focused on natural extensions such as conic integer programming (e.g. [4,30,42]) and bilevel mixed-integer linear programming [23]. In contrast, **PO** represents an essentially different domain of application since variables here are continuous.

We work with a representation of **PO** that uses a symmetric matrix of decision variables, and yields an *equivalent* formulation with a linear objective function and a feasible region of the form $S \cap P$, with S the (closed) set of symmetric matrices that can be represented as outer products of the form xx^T— accordingly, we study *outer-product-free* sets. Several families of full-dimensional (inclusion-wise) maximal outer-product-free sets are identified in Theorems 3 and 4 of Sect. 5. Furthermore, we derive an oracle-based outer-product-free set in Sect. 4. With the aforementioned results we develop a cut generation procedure (see Sect. 6) that has (to our knowledge) the following unique properties: any infeasible extreme point of a (lifted) polyhedral relaxation of **PO** can be separated in polynomial time; and variable bounds are not required. In Sect. 7 we demonstrate the practical effectiveness of our approach over a variety of instances using a straightforward pure cutting-plane setup. The speed of our separation routines and the quality of the resulting cuts strongly suggest the viability of our cut families within a full-fledged branch-and-cut solver.

1.1   Notation

Denote the interior of a set $\mathrm{int}(\cdot)$ and its boundary $\mathrm{bd}(\cdot)$. The convex hull of a set is denoted $\mathrm{conv}(\cdot)$, and its closure is $\mathrm{clconv}(\cdot)$; likewise, the conic hull of a set is $\mathrm{cone}(\cdot)$, and its closure $\mathrm{clcone}(\cdot)$. For a point x and nonempty set S in $\mathbb{R}^n$, we define $d(x, S) := \inf_{s \in S}\{\|x - s\|_2\}$; note that for S closed we can replace the infimum with minimum. Denote the ball with center x and radius r to be $\mathcal{B}(x, r)$. $\langle \cdot, \cdot \rangle$ denotes the matrix inner product and $\|\cdot\|_F$ the Frobenius norm. A positive semidefinite matrix may be referred to as a PSD matrix for short, and likewise NSD refers to negative semidefinite.

2 S-free Sets and the Intersection Cut

Definition 1. *A set $C \subset \mathbb{R}^n$ is S-free if $int(C) \cap S = \emptyset$ and C is convex.*

For any S-free set C we have $S \cap P \subseteq \mathrm{clconv}(P \setminus \mathrm{int}(C))$, and so any valid inequalities for $\mathrm{clconv}(P \setminus \mathrm{int}(C))$ are valid for $S \cap P$. Hillestad and Jacobsen [29], and later on Sen and Sherali [48], provide results regarding polyhedrality of $\mathrm{clconv}(P \setminus \mathrm{int}(C))$. Averkov [5] provides theoretical consideration on how one can derive cuts from C. In specific instances, $\text{conv}(P \setminus \text{int}(C))$ can be fully described (see [10,11,30,42]), however, separating over $P \setminus \text{int}(C)$ is NP-hard [25]. A standard workaround is to find a simplicial cone P' containing P and apply Balas' intersection cut [6] for $P' \setminus \text{int}(C)$ (also see Tuy [54]). Larger S-free sets can be useful for generating deeper cuts [17].

Definition 2. *An S-free set C is* maximal *if $V \not\supset C$ for all S-free V.*

Under certain conditions (see [9,17,18,30]), maximal S-free sets are sufficient to generate all nontrivial cuts for a problem. When $S = \mathbb{Z}^n$, C is called a lattice-free set. Maximal lattice-free sets are well-studied in integer programming theory [1,2,8,14,20,27,30,36], and the notion of S-free sets was introduced as a generalization [21].

2.1   The Intersection Cut

Let $P' \supseteq P$ be a simplicial conic relaxation of P: a displaced polyhedral cone with apex $\bar{x}$ and defined by the intersection of n linearly independent halfspaces. P' may be written as follows:

$$P' = \{\bar{x} + \sum_{j=1}^{n} \lambda_j r^j : \lambda \geq 0\}. \tag{1}$$

Each extreme ray of P' is of the form $\{\bar{x} + \lambda_j r^j | \lambda_j \geq 0\}$. Alternatively, the simplicial conic relaxation can be given in inequality form

$$P' = \{x | Ax \leq b\}, \tag{2}$$

where A is an invertible matrix. Note that any basis of P would be suitable to derive P'. The apex $\bar{x} = A^{-1}b$, and the rays r^j in (1) can be obtained directly from A: for each j, one can identify $-r^j$ as the jth column of the inverse of A.

We shall assume $\bar{x} \notin S$, so that $\bar{x}$ is to be separated from S via separation from $P' \setminus \text{int}(C)$, with C an S-free set with $\bar{x}$ in its interior. Since $\bar{x} \in \text{int}(C)$, there must exist $\lambda > 0$ such that $\bar{x} + \lambda_j r^j \in \text{int}(C) \, \forall j$. Also, each extreme ray is either entirely contained in C, i.e. $\bar{x} + \lambda_j r^j \in \text{int}(C) \forall \lambda_j \geq 0$, or else there is an intersection point with the boundary: $\exists \lambda_j^* : \bar{x} + \lambda_j^* r^j \in \text{bd}(C)$. We refer to λ_j^* as the *step length* in the latter case, and for convenience, we define the step length $\lambda_j^* = \infty$ in the former case. The *intersection cut* is the halfspace whose boundary contains each intersection point (given by $\lambda_j^* < \infty$) and that is parallel to all extreme rays contained in C.

Given $\lambda_j^* \in (0, \infty] \, \forall j = 1, \ldots, n$, Balas [6, Theorem 2] provides a closed-form expression for the intersection cut $\pi x \leq \pi_0$:

$$\pi_0 = \sum_{i=1}^{n}(1/\lambda_i^*)b_i - 1, \quad \pi_j = \sum_{i=1}^{n}(1/\lambda_i^*)a_{ij}, \tag{3}$$

where a_{ij} are the entries of A in (2) and $1/\infty := 0$ [6, p. 34].

The validity of the intersection cut and a condition in which the cut gives the convex hull of $P' \setminus \text{int}(C)$ is established by Balas [6, Theorem 1]. A more detailed analysis, as well as a strengthening procedure for infinite step lengths is provided in our full-length paper [12].

3 Moment-Based Reformulation of Polynomial Optimization

Our approach to **PO** leverages the moment/sum-of-squares approach (see [33, 34]) from which a definition of the feasible set as $S \cap P$ is naturally obtained.

Let $m_r = [1, x_1, \ldots, x_n, x_1 x_2, \ldots, x_n^2, \ldots, x_n^r]$ be a vector of all monomials up to degree r. Any polynomial may be written as $p_i(x) = m_r^T A_i m_r$ (provided r is sufficiently large), where A_i is a symmetric matrix derived from p_i. We can apply this transformation to **PO** to obtain a lifted representation **LPO**:

$$\min \langle A_0, X \rangle$$
$$(\textbf{LPO}) \text{ s.t. } \langle A_i, X \rangle \leq b_i, \ i = 1, \ldots, m, \tag{4a}$$
$$X = m_r m_r^T. \tag{4b}$$

Denote $n_r := \binom{n+r}{r}$, i.e. the length of m_r. Here $A_i \in \mathbb{S}^{n_r \times n_r}$ are symmetric matrices of data, and $X \in \mathbb{S}^{n_r \times n_r}$ is a symmetric matrix of decision variables. The problem has linear objective function, linear constraints (4a), and nonlinear constraints (4b). One can replace the moment matrix condition $X = m_r m_r^T$ with the equivalent conditions of $X \succeq 0, \text{rank}(X) \leq 1$ and linear consistency constraints among entries from X representing the same monomial. Dropping the nonconvex rank one constraint yields the standard semidefinite relaxation [50].

On the other hand, the feasible region of **LPO** has a natural description as an intersection of a polyhedron P_{OP}, that corresponds to linear constraints (4a) together with consistency constraints, and the following closed set,

$$S_{OP} := \{ X \in \mathbb{S}^{n_r \times n_r} : X = xx^T, x \in \mathbb{R}^{n_r} \}.$$

Accordingly, we shall study sets that are *outer-product-free* (OPF): closed, convex sets in $\mathbb{S}^{n_r \times n_r}$ with interiors that do not intersect with S_{OP}. In what follows, suppose we have an extreme point $\bar{X} \in P_{OP} \setminus S_{OP}$ with spectral decomposition $\bar{X} := \sum_{i=1}^{n_r} \lambda_i d_i d_i^T$ and ordering $\lambda_1 \geq \ldots \geq \lambda_{n_r}$. We seek to separate $\bar{X}$.

4 Oracle-Based Outer-Product-Free Sets

If one has access to a distance oracle to the set S, one can easily construct an OPF set, namely, an OPF *ball*. In the case of **LPO** this corresponds to the distance to the nearest symmetric outer product. This distance is a special case of the following PSD matrix approximation problem, given an integer $q > 0$:

$$(\textbf{PMA}) \quad \min_Y \left\{ \|\bar{X} - Y\| : \text{rank}(Y) \leq q, \ Y \succeq 0 \right\}.$$

Here $\| \cdot \|$ is a unitarily invariant matrix norm such as the Frobenius norm, $\| \cdot \|_F$. Dax [19] proves the following:

Theorem 1 (Dax's Theorem). *Let k be the number of nonnegative eigenvalues of $\bar{X}$. For $q = 1, \ldots, n - 1$, an optimal solution to **PMA** is given by $Y = \sum_{i=1}^{\min\{k,q\}} \lambda_i d_i d_i^T$.*

When $\bar{X}$ is not NSD, the solution from Dax's theorem coincides with Eckart-Young-Mirsky [22,40] solution to **PMA** without the PSD constraint. The optimal PSD approximant allows us to construct an OPF ball:

$$\mathcal{B}_{\text{oracle}}(\bar{X}) := \begin{cases} \mathcal{B}(\bar{X}, \|\bar{X}\|_F), & \text{if } \bar{X} \text{ is NSD,} \\ \mathcal{B}(\bar{X}, \|\sum_{i=2}^{n} \lambda_i d_i d_i^T\|_F), & \text{otherwise.} \end{cases}$$

Corollary 1. $\mathcal{B}_{oracle}(\bar{X})$ *is OPF.*

Proof. Setting $q = 1$ in Dax's Theorem, we see that the nearest symmetric outer product is either $\lambda_1 d_1 d_1^T$ if $\lambda_1 > 0$, or else the zeros matrix. □

For **LPO** we can use a simple geometric construction involving Theorem 2 to obtain an OPF *cone* from the oracle ball. This extension is detailed in [12].

5   Maximal Outer-Product-Free Sets

5.1   General Properties of Maximal Outer-Product-Free Sets

We now turn to characterizing and finding maximal OPF sets. Our first Theorem is a building block towards maximality.

Theorem 2. *Let $C \subset \mathbb{S}^{n_r \times n_r}$ be a full-dimensional OPF set. Then clcone(C) is OPF. In particular, every full-dimensional maximal OPF set is a convex cone.*

Proof. Suppose clcone(C) is not OPF; since it is closed and convex, then by definition of OPF sets there must exist $d \in \mathbb{R}^{n_r}$ such that dd^T is in its interior. If d is the zeros vector, then int(C) also contains the origin, which contradicts the condition of C being OPF. Otherwise the ray r^0 emanating from the origin with nonzero direction dd^T is entirely contained in and hence is an interior ray of clcone(C). By convexity, the interior of cone(C) is the same as the interior of its closure, so r^0 is also an interior ray of cone(C). From this, it can be proved that r^0 must pass through the interior of C (see [12]). But every point along r^0 is a symmetric outer-product, which again implies that C is not OPF. □

We can also obtain the following properties regarding the geometry of maximal OPF sets via their supporting halfspaces.

Definition 3. *A supporting halfspace of a closed, convex set S contains S and its boundary is a supporting hyperplane of S.*

Lemma 1. *Let C be a full-dimensional maximal OPF set. Every supporting halfspace of C is of the form $\langle A, X \rangle \geq 0$ for some $A \in \mathbb{S}^{n_r \times n_r}$.*

Proof. From Theorem 2 we have that C is a convex cone. From this it follows that a supporting halfspace $\langle A, X \rangle \geq b$ must have $b = 0$. □

From Lemma 1 we may characterize a maximal OPF set as $C = \{X \in \mathbb{S}^{n_r \times n_r} | \langle A_i, X \rangle \geq 0 \; \forall i \in I\}$, with I a potentially infinite index set.

Theorem 3. *The halfspace $\langle A, X \rangle \geq 0$ is maximal OPF iff A is NSD.*

Proof. If A has a strictly positive eigenvalue, then $\langle A, dd^T \rangle > 0$ for some d, and so the halfspace is not OPF. If A is NSD then $\langle A, dd^T \rangle = d^T A d \leq 0 \; \forall d \in \mathbb{R}^{n_r}$, so the halfspace is OPF. For maximality, suppose the halfspace is strictly contained in another OPF set $\bar{C}$. Then there exists some $\bar{X} \in \text{int}(\bar{C})$ such that $\langle A, \bar{X} \rangle < 0$. Thus, $-\bar{X} \in \text{int}(\bar{C})$ and so is the zeros matrix. Thus $\bar{C}$ cannot be OPF. □

5.2 Maximal Outer-Product-Free Sets Derived from 2 × 2 Submatrices

Theorem 3 provides our first explicit family of maximal OPF sets. Another family is suggested by the following result by Kocuk, Dey, and Sun [31]:

Proposition 1 (KDS Proposition). *A nonzero, Hermitian matrix X is PSD and has rank one iff all the 2×2 minors of X are zero and the diagonal elements of X are nonnegative.*

In what follows, denote the entries of a 2×2 submatrix of X from some rows $i_1 < i_2$ and columns $j_1 < j_2$ as $X_{[[i_1,i_2],[j_1,j_2]]} := \begin{bmatrix} a & b \\ c & d \end{bmatrix}$.

Lemma 2. *Let $\lambda \in \mathbb{R}^2$ with $\|\lambda\|_2 = 1$. (5a) and (5b) describe an OPF set:*

$$\lambda_1(a+d)/2 + \lambda_2(b-c)/2 \geq \|(b+c)/2, (a-d)/2\|_2, \tag{5a}$$
$$\lambda_1(b+c)/2 + \lambda_2(a-d)/2 \geq \|(a+d)/2, (b-c)/2\|_2. \tag{5b}$$

Proof. The set defined by $ad \geq bc$ is OPF (Proposition 1). The proof follows from checking that (5a) defines a subset of it. Similarly, (5b) defines a subset of $ad \leq bc$. □

The following Theorem provides an extensive list of *maximal* OPF sets that can be obtained from Lemma 2. We leave the proof in the Appendix.

Theorem 4. *(5a) describes a maximal OPF set if*

(i) $\lambda_1 = 1, \lambda_2 = 0$, and neither b nor c are diagonal entries;
(ii) $\lambda_1 = 0, \lambda_2 = 1$, and b is a diagonal entry;
(iii) $\lambda_1 = 0, \lambda_2 = -1$, and c is a diagonal entry;
(iv) $\lambda_1^2 + \lambda_2^2 = 1$, and none of a, b, c, d are diagonal entries.

Similarly, (5b) describes a maximal *OPF set if*
(v) $\lambda_1 = 1, \lambda_2 = 0$, *and either b or c is a diagonal entry;*
(vi) $\lambda_1 = 0, \lambda_2 = 1$, *and a but not d is a diagonal entry;*
(vii) $\lambda_1 = 0, \lambda_2 = -1$, *and d but not a is a diagonal entry;*
(viii) $\lambda_1^2 + \lambda_2^2 = 1$, *and none of a, b, c, d are diagonal entries.*

The following theorem shows that the maximal OPF sets we have identified characterizes all such sets in the special case where $n_r = 2$.

Theorem 5. *In $\mathbb{S}^{2 \times 2}$ every full-dimensional maximal OPF set is either the cone of PSD matrices or a halfspace of the form $\langle A, X \rangle \geq 0$, where A is NSD.*

Proof. See [12] for details.

6   Implementation of Intersection Cuts

Suppose that we have a simplicial conic relaxation of P_{OP} with apex $\bar{X}$. This section provides a brief overview on how to generate a cutting plane to separate $\bar{X}$ from $P_{OP} \cap S_{OP}$ using results from Sects. 4 and 5.

6.1   Step 1: Selecting an Outer-Product-Free Set

Separation Using the Distance Oracle. As outlined in Sect. 4, $\mathcal{B}_{\text{oracle}}(\bar{X})$, or its conic extension/strengthening can always be used to separate $\bar{X}$.

Separation Using Halfspaces. Theorem 3 shows that certain halfspaces are OPF sets. Moreover, it is not hard to see that the halfspaces of Theorem 3 imply and provide no more than the family of cuts equivalent to the PSD condition:

$$d^T X d \geq 0 \quad \forall d \in \mathbb{R}^n \iff X \succeq 0.$$

Choosing d equal to the eigenvectors of $\bar{X}$ provides polynomial-time separation (given fixed numerical tolerances); this is a well-studied linear outer-approximation procedure for semidefinite programming problems [32,44,47,49], and here we provide a new interpretation of them via the maximal OPF property.

Separation with all 2×2 Submatrices of $\bar{X}$. From Proposition 1 we have that $\bar{X} \notin S_{OP}$ implies a nonzero 2×2 minor or a negative diagonal term. Supposing the nonnegative diagonal constraints are included in P_{OP}, then at least one of the $\mathcal{O}(n^4)$ 2×2 minors will be nonzero. We can show that for any such minor that is nonzero at least one of the sets described in Theorem 4 will strictly contain $\bar{X}$. There is an additional choice of the λ parameters for sets of the form (iv) and (viii), which in our experiments we set to extreme values $\lambda_1 = 1, \lambda_2 = 0$ or $\lambda_1 = 0, \lambda_2 = 1$. Intermediate values for λ are the subject of ongoing research.

Separation with Only Principal 2×2 Submatrices of $\bar{X}$. An alternative characterization to Proposition 1 is given by Chen, Atamtürk and Oren [16]:

Proposition 2 (CAO Proposition). *For $n > 1$ a nonzero Hermitian PSD $n \times n$ matrix X has rank one iff all of its 2×2 principal minors are zero.*

Hence if $\bar{X}$ is not PSD then it is contained in at least one halfspace described in Theorem 3. Otherwise, $\bar{X}$ has at least one of its $\mathcal{O}(n^2)$ principal 2×2 minors strictly positive, and so (5a) is strictly satisfied for case (i) of Theorem 4.

6.2 Step 2: Generating an Intersection Cut

The halfspace OPF sets can generate a cut directly as mentioned above. We only need the eigenvectors of $\bar{X}$. The remaining OPF sets are a ball, for which the step-lengths for the intersection cuts are simply the ball's radius, and second-order cones. We can thus derive (see [12]) computationally trivial closed-form expressions for step lengths from the interior of a second-order cone to its boundary. This is one of the most crucial features of our proposed cutting planes, as they can be generated with little computational effort and thus making them suitable for their incorporation in a branch-and-cut procedure.

7 Numerical Experiments

We present experiments using a pure cutting-plane algorithm using the cuts described in Sect. 6. The experiments are designed to investigate the stand-alone performance of our cuts, particularly separation speed and effectiveness. The cutting plane algorithm solves an LP relaxation and obtains an (extreme point) optimal solution $\bar{X}$, adds cuts separating $\bar{X}$, and repeats until either:

- A time limit of 600 seconds is reached, or
- The objective value does not improve for 10 iterations, or
- The violation of all cuts is not more than 10^{-6}. Here, if $\pi^T x \leq \pi_0$ is the cut and x^* is the candidate solution, we define the violation as $(\pi^T x^* - \pi_0)/\|\pi\|_1$.

For improving stability, we add a maximum of 10 cuts per iteration (selected using violations) and remove non-active cuts every 15 iterations. Computations are run on a 32-core server with an Intel Xeon Gold 6142 2.60 GHz CPU and 512 GB of RAM. Although the machine is powerful, we run the algorithm single-threaded and the experiments do not require a significant amount of memory; we confirmed that similar performance can be obtained with a laptop. The code is written in C++ using the Eigen library for linear algebra [28]. The LP solver is Gurobi 8.0.0 and, for comparisons, we solve SDP relaxations using the C++ Fusion API of Mosek 8 [43]. Our code is available at https://github.com/g-munoz/poly_cuts_cpp.

Test instances are taken from two sources. First, we consider all 27 problem instances from Floudas et al. [24] (available via GLOBALLib [39]) that have quadratic objective and constraints. Our cuts can accommodate arbitrary polynomial terms, however for implementation purposes reading QCQP problems is more convenient. Second, we consider all 99 instances of BoxQP developed by several authors [15,55]. These problems have simple box constraints $x \in [0,1]^n$ and a nonconvex quadratic objective function. In a recent paper by Bonami et al. [13], the authors show how to obtain cutting planes for this particular case.

We choose the initial LP relaxation to be the standard RLT relaxation of QCQP: setting $r = 1$ in **LPO** and including McCormick estimators for bilinear terms (see [3,38]). Problem sizes vary from 21×21 to 126×126 symmetric matrices of decision variables for BoxQP instances and from 6×6 to 63×63 for GLOBALLib instances. To obtain variable bounds for some of the GLOBALLib instances we apply a simple bound tightening procedure: minimize/maximize a given variable subject to the RLT relaxation. Lastly, we use *Gap Closed* as a measure of quality of the bounds generated by each approach. This is defined as follows: let OPT denote the optimal value of an instance, RLT the optimal value of the standard RLT relaxation, and GLB the objective value obtained after applying the cutting plane procedure. Then Gap Closed $= \frac{GLB-RLT}{OPT-RLT}$.

Results. In Table 1, we show a performance comparison in the selected GLOBALLib instances between our cutting plane algorithm versus the relaxation obtained from adding a PSD requirement for the variable X in the RLT relaxation (SDP). We do not show results for 2 of the instances, as the RLT relaxation is tight for these. The results in Table 1 are very encouraging: in only 4 instances we are not able to reach the same gap closed as the SDP. Moreover, our simple cutting plane approach (almost) always runs in just a few seconds.

In Table 2, we compare our results with the V2 setting used by Saxena, Bonami and Lee [46] in the selected GLOBALLib instances. We chose [46] as a comparison as we find it the most similar to our approach. V2 uses an RLT relaxation for QCQPs and applies two types of cuts: an outer-approximation of the PSD cone and disjunctive cuts for which the separation involves a MIP. We emphasize that these families of cuts are complementary and not competitive. It is also important to mention that the running times in Table 2 for V2 correspond to the reports in [46], published in 2010. While new hardware may improve these times, we believe the conclusions we draw from Table 2 would not change.

For comparison purposes, we turned off our simple bound tightening routine in order to obtain the same initial relaxation value as V2 (and thus the gaps are different than the ones in Table 1). Even doing so, for certain instances of GLOBALLib we did not obtain the same initial bound and thus excluded these from comparison. On the comparable GLOBALLib instances our algorithm terminates with smaller gap closed on average, but it does produce higher gap closed on some instances. The advantage of our cuts is that times are substantially shorter. This is expected, as V2 solves a MIP in the cut generation, while our cuts only require finding eigenvalues and roots of single-variable quadratics. Overall we believe these results are promising, as the cutting planes are able to close a significant amount of gap in many cases, in a very short time.

The results on the BoxQP instances are interesting as well. In the interest of space, we limit ourselves to summarizing them here. The interested reader can find a complete log in https://goo.gl/8wPeY6. We compare with V2 as before,

Table 1. Comparison of intersection cuts versus SDP relaxation in non-convex quadratic GLOBALLib instances.

Instance Name	SDP Gap Closed	Time	Intersection Cuts Gap Closed	Time	Instance Name	SDP Gap Closed	Time	Intersection Cuts Gap Closed	Time
Ex2_1_1	0.00%	0.01	57.70%	0.02	Ex5_2_5	0.00%	1.84	0.00%	7.83
Ex2_1_5	0.00%	0.03	99.57%	0.01	Ex5_3_2	0.10%	0.74	0.00%	0.94
Ex2_1_6	0.00%	0.02	79.56%	0.09	Ex5_3_3	3.75%	105.01	0.50%	8.12
Ex2_1_7	0.00%	0.46	22.59%	0.96	Ex5_4_2	0.00%	0.02	0.40%	0.08
Ex2_1_8	0.00%	1.11	51.89%	1.91	Ex8_4_1	98.43%	0.84	59.42%	23.03
Ex2_1_9	0.00%	0.02	31.92%	0.98	Ex9_1_4	0.00%	0.06	97.65%	0.08
Ex3_1_1	0.00%	0.05	0.71%	0.32	Ex9_2_1	6.25%	0.07	26.55%	4.45
Ex3_1_2	22.41%	0.01	100.00%	0.00	Ex9_2_2	16.67%	0.04	62.14%	1.55
Ex3_1_4	0.00%	0.01	32.61%	0.02	Ex9_2_3	0.00%	0.24	0.00%	0.30
Ex5_2_2_case1	0.00%	0.04	12.84%	0.79	Ex9_2_4	99.83%	0.03	33.33%	0.10
Ex5_2_2_case2	0.00%	0.04	30.25%	0.56	Ex9_2_6	99.76%	0.27	56.64%	0.20
Ex5_2_2_case3	0.00%	0.04	19.15%	0.50	Ex9_2_7	6.25%	0.05	26.55%	4.58
Ex5_2_4	0.00%	0.02	27.55%	0.20					

Table 2. Comparison of Intersection Cuts versus V2 approach of [46] in Non-Convex Quadratic GLOBALLib Instances. Entries labelled NR were not reported in [46].

Instance Name	V2 Gap Closed	Time	Intersection Cuts Gap Closed	Time	Instance Name	V2 Gap Closed	Time	Intersection Cuts Gap Closed	Time
Ex2_1_1	72.62%	704.40	57.70%	0.02	Ex5_2_4	79.31%	68.93	27.14%	0.20
Ex2_1_5	99.98%	0.17	99.68%	0.00	Ex5_2_5	6.27%	3793.17	0.00%	7.59
Ex2_1_6	99.95%	3397.65	88.82%	0.09	Ex5_3_2	7.27%	245.82	0.00%	0.91
Ex2_1_8	84.70%	3632.28	3.08%	1.71	Ex5_3_3	0.21%	3693.76	0.50%	16.73
Ex2_1_9	98.79%	1587.94	32.01%	0.89	Ex5_4_2	27.57%	3614.38	0.56%	0.10
Ex3_1_1	15.94%	3600.27	0.89%	0.11	Ex9_1_4	0.00%	0.60	0.00%	0.06
Ex3_1_2	99.99%	0.08	100.00%	0.00	Ex9_2_1	60.04%	2372.64	27.27%	0.10
Ex3_1_4	86.31%	21.26	32.61%	0.02	Ex9_2_2	88.29%	3606.36	69.88%	8.30
Ex5_2_2_case1	0.00%	0.02	0.00%	0.02	Ex9_2_6	87.93%	2619.02	99.70%	1.76
Ex5_2_2_case2	0.00%	0.05	0.00%	0.08	Ex9_2_8	NR	NR	99.68%	0.03
Ex5_2_2_case3	0.36%	0.36	18.89%	0.11					

and we replicate the same initial relaxation used by Saxena, Bonami and Lee [46], namely the weak RLT relaxation (wRLT)[1]. We also compare against the wRLT relaxation with a PSD constraint (wRLT+SDP). On the 42 BoxQP instances reported in [46], our cuts *always* perform better than both V2 and wRLT+SDP. The latter reaches optimality in seconds, but the relaxation is not strong, as there are missing McCormick inequalities. Intersection Cuts, with a time limit of 600 seconds, is able to close 90.49% gap on average in these instances, while V2 closes 65.28% and wRLT+SDP 51.87%. Even though wRLT is a relaxation that is not used in practice, these experiments still evidence the potential of our proposed cuts, as they close a large amount of gap in a short amount of time, even surpassing the impact of including an explicit SDP constraint.

---

[1] This BoxQP relaxation only adds the "diagonal" McCormick estimates $X_{ii} \leq x_i$.

8 Conclusions

We have introduced intersection cuts in the context of polynomial optimization. Accordingly, we have developed an S-free approach for polynomial optimization, where S is the set of real, symmetric outer products. Our results on full-dimensional maximal OPF sets include a full characterization of such sets when $n_r = 2$ as well as extensive families of maximal OPF sets. Computational experiments have demonstrated the potential of our cuts as a fast way to reduce optimality gaps on a variety of problems using computationally simple closed-form procedures. A full implementation is being considered for future empirical work, incorporating the cuts into a branch-and-cut solver and developing a more sophisticated implementation, e.g. stronger initial relaxations with problem-specific valid inequalities, warm-starting the outer-approximation with an SDP, sparsification of the cuts, advanced cut management, improved scalability, among others.

Acknowledgements. We would like to thank the anonymous reviewers for their valuable comments. This research was partly supported by award ONR N00014-16-1-2889, Conicyt Becas Chile 72130388 and The Institute for Data Valorisation (IVADO).

Appendix

Proof (Theorem 4). The OPF property is given by Lemma 2, so maximality remains. Let C be a set described by (5a) or (5b). It suffices to construct, for *every* symmetric matrix $\bar{X} \notin C$, $Z := zz^T$ such that $Z - \bar{X} \in \text{int}(C)$. This implies $Z \in \text{int}(\text{conv}(C \cup \bar{X}))$. Denote the submatrices of $\bar{X}, Z$:

$$\bar{X}_{[[i_1,i_2],[j_1,j_2]]} := \begin{bmatrix} \bar{a} & \bar{b} \\ \bar{c} & \bar{d} \end{bmatrix}, \; Z_{[[i_1,i_2],[j_1,j_2]]} := \begin{bmatrix} a_Z & b_Z \\ c_Z & d_Z \end{bmatrix}.$$

Furthermore, for convenience let us define the following:

$$\bar{p} := (\bar{a} + \bar{d})/2, \; \bar{q} := (\bar{a} - \bar{d})/2, \; \bar{r} := (\bar{b} + \bar{c})/2, \; \bar{s} := (\bar{b} - \bar{c})/2.$$

Construction for (5a): Suppose $\bar{X}$ violates (5a). We propose the following:

$$a_Z = \bar{q} + \lambda_1 \|\bar{q}, \bar{r}\|_2, \; b_Z = \bar{r} + \lambda_2 \|\bar{q}, \bar{r}\|_2, \; c_Z = \bar{r} - \lambda_2 \|\bar{q}, \bar{r}\|_2, \; d_Z = -\bar{q} + \lambda_1 \|\bar{q}, \bar{r}\|_2 \tag{6}$$

$$\implies \lambda_1(\bar{a} + \bar{d})/2 + \lambda_2(\bar{b} - \bar{c})/2 < \|(\bar{b} + \bar{c})/2, (\bar{a} - \bar{d})/2\|_2$$
$$= \lambda_1(a_Z + d_Z)/2 + \lambda_2(b_Z - c_Z)/2$$

where the last equality follows from $\lambda_1^2 + \lambda_2^2 = 1$. This implies

$$\lambda_1((a_Z - \bar{a}) + (d_Z - \bar{d}))/2 + \lambda_2((b_Z - \bar{b}) - (c_Z - \bar{c}))/2 > 0$$

and since $\|((b_Z - \bar{b}) + (c_Z - \bar{c}))/2, ((a_Z - \bar{a}) - (d_Z - \bar{d}))/2\|_2 = 0$, we conclude $Z - \bar{X} \in \text{int}(C)$.

Construction for (5b): If $\bar{X}$ violates (5b), we use the following construction:

$$a_Z = \bar{p} + \lambda_2\|\bar{p}, \bar{s}\|_2, b_Z = \bar{s} + \lambda_1\|\bar{p}, \bar{s}\|_2, c_Z = -\bar{s} + \lambda_1\|\bar{p}, \bar{s}\|_2, d_Z = \bar{p} - \lambda_2\|\bar{p}, \bar{s}\|_2.$$
$$\implies \lambda_1(\bar{b} + \bar{c})/2 + \lambda_2(\bar{a} - \bar{d})/2 < \|(\bar{a} + \bar{d})/2, (\bar{b} - \bar{c})/2\|_2$$
$$= \lambda_1(b_Z + c_Z)/2 + \lambda_2(a_Z - d_Z)/2,$$
$$\implies \lambda_1((b_Z - \bar{b}) + (c_Z - \bar{c}))/2 + \lambda_2((a_Z - \bar{a}) - (d_Z - \bar{d}))/2 > 0.$$

We conclude $Z - \bar{X} \in \text{int}(C)$ as before, since $\|((a_Z - \bar{a}) + (d_Z - \bar{d}))/2, ((b_Z - \bar{b}) - (c_Z - \bar{c}))/2\|_2 = 0$. It remains to set the other entries of Z and to show it is an outer product.

Claim. For each condition (i)–(viii), $a_Z d_Z = b_Z c_Z$ and all diagonal elements among a_Z, b_Z, c_Z, d_Z are nonnegative.

Proof: First consider conditions (i)–(iv). By construction of (6):

$$a_Z d_Z = -\bar{q}^2 + \lambda_1^2\|\bar{q}, \bar{r}\|_2^2 = \bar{r}^2 - \lambda_2^2\|\bar{q}, \bar{r}\|_2^2 = b_Z c_Z.$$

The second equality is derived from the following identity:

$$\|\bar{q}, \bar{r}\|_2^2 = \bar{q}^2 + \bar{r}^2 \iff -\bar{q}^2 + \lambda_1^2\|\bar{q}, \bar{r}\|_2^2 = \bar{r}^2 - \lambda_2^2\|\bar{q}, \bar{r}\|_2^2.$$

Nonnegativity of diagonal elements follows from $\|\bar{q}, \bar{r}\|_2 \geq \max\{|\bar{q}|, |\bar{r}|\}$. In case (i) only a_Z or d_Z can be diagonal elements, and they are both nonnegative. The other cases can be directly verified. Similarly, for conditions (v)–(viii):

$$a_Z d_Z = \bar{p}^2 - \lambda_2^2\|\bar{p}, \bar{s}\|_2^2 = -\bar{s}^2 + \lambda_1^2\|\bar{p}, \bar{s}\|_2^2 = b_Z c_Z.$$

The second equality is derived from the following identity:

$$\|\bar{p}, \bar{s}\|_2^2 = \bar{p}^2 + \bar{s}^2 \iff -\bar{s}^2 + \lambda_1^2\|\bar{p}, \bar{s}\|_2^2 = \bar{p}^2 - \lambda_2^2\|\bar{p}, \bar{s}\|_2^2.$$

Nonnegativity of diagonal elements follows from the same argument as before, by using the fact that $\|\bar{p}, \bar{s}\|_2 \geq \max\{|\bar{p}|, |\bar{s}|\}$. ∎

To maintain symmetry we set $Z_{i_1,j_1} = Z_{j_1,i_1}$, $Z_{i_1,j_2} = Z_{j_2,i_1}$, $Z_{i_2,j_1} = Z_{j_1,i_2}$, $Z_{i_2,j_2} = Z_{j_2,i_2}$. Now denote $\ell = [i_1, i_2, j_1, j_2]$. If $a_Z = b_Z = c_Z = d_Z = 0$, then we simply set all other entries of Z equal to zero and so Z is the outer-product of the vector of zeroes. Otherwise, consider the following cases.

Case 1: ℓ has 4 unique entries. Suppose w.l.o.g we have an upper-triangular entry $(i_1 < i_2 < j_1 < j_2)$ and furthermore suppose that b_Z is nonzero. Then set

$$Z_\ell := \begin{bmatrix} 1 & d_Z/b_Z & a_Z & b_Z \\ d_Z/b_Z & d_Z^2/b_Z^2 & c_Z & d_Z \\ a_Z & c_Z & a_Z^2 & a_Z b_Z \\ b_Z & d_Z & a_Z b_Z & b_Z^2 \end{bmatrix}$$

and set all remaining entries of Z to zero. Other orderings of indices or the use of a different nonzero entry is handled by relabeling/rearranging column/row order.

Case 2: ℓ has three unique entries. Then, exactly one of a_Z, b_Z, c_Z, d_Z is a diagonal entry, and so cases (i)–(iii), (v)–(vii) apply. If in any of these cases a_Z or d_Z is on the diagonal, by construction $|b_Z| = |c_Z|$. As $a_Z d_Z = b_Z c_Z$, we have $b_Z = c_Z = 0$ iff exactly one of a_Z or d_Z is zero. Likewise, if b_Z or c_Z is a diagonal element, then $|a_Z| = |d_Z|$ and so $a_Z = d_Z = 0$ iff exactly one of b_Z or c_Z are zero.

Suppose a_Z is a nonzero diagonal entry. We propose:

$$Z_{\ell'} = \begin{bmatrix} a_Z & b_Z & c_Z \\ b_Z & b_Z^2/a_Z & d_Z \\ c_Z & d_Z & c_Z^2/a_Z \end{bmatrix}$$

where ℓ' are the unique entries of ℓ. If $a_Z = 0$ and on the diagonal, then we replace b_Z^2/a_Z and c_Z^2/a_Z with $|d_Z|$. If b_Z, c_Z or d_Z is on the diagonal, we use the same construction but with relabeling/rearranging column/row order.

Case 3: ℓ has two unique entries. All remaining entries of Z are set to zero.

For all cases, our construction ensures that all diagonal entries of Z are nonnegative, and all 2×2 minors are zero; by Proposition 1, Z is an outer-product. □

References

1. Andersen, K., Louveaux, Q., Weismantel, R.: An analysis of mixed integer linear sets based on lattice point free convex sets. Math. Oper. Res. **35**(1), 233–256 (2010)
2. Andersen, K., Louveaux, Q., Weismantel, R., Wolsey, L.A.: Inequalities from two rows of a simplex tableau. In: Fischetti, M., Williamson, D.P. (eds.) IPCO 2007. LNCS, vol. 4513, pp. 1–15. Springer, Heidelberg (2007). https://doi.org/10.1007/978-3-540-72792-7_1
3. Anstreicher, K.M.: Semidefinite programming versus the reformulation linearization technique for nonconvex quadratically constrained quadratic programming. J. Glob. Optim. **43**, 471–484 (2009)
4. Atamtürk, A., Narayanan, V.: Conic mixed-integer rounding cuts. Math. Program. **122**, 1–20 (2010)
5. Averkov, G.: On finite generation and infinite convergence of generalized closures from the theory of cutting planes. arXiv preprint arXiv:1106.1526 (2011)
6. Balas, E.: Intersection cuts—a new type of cutting planes for integer programming. Oper. Res. **19**(1), 19–39 (1971)
7. Bao, X., Sahinidis, N.V., Tawarmalani, M.: Multiterm polyhedral relaxations for nonconvex, quadratically constrained quadratic programs. Optim. Methods Softw. **24**(4–5), 485–504 (2009)
8. Basu, A., Conforti, M., Cornuéjols, G., Zambelli, G.: Maximal lattice-free convex sets in linear subspaces. Math. Oper. Res. **35**(3), 704–720 (2010)

9. Basu, A., Conforti, M., Cornuéjols, G., Zambelli, G.: Minimal inequalities for an infinite relaxation of integer programs. SIAM J. Discrete Math. **24**(1), 158–168 (2010)

10. Belotti, P., Góez, J.C., Pólik, I., Ralphs, T.K., Terlaky, T.: On families of quadratic surfaces having fixed intersections with two hyperplanes. Discrete Appl. Math. **161**(16–17), 2778–2793 (2013)

11. Bienstock, D., Michalka, A.: Cutting-planes for optimization of convex functions over nonconvex sets. SIAM J. Optim. **24**, 643–677 (2014)

12. Bienstock, D., Chen, C., Muñoz, G.: Outer-product-free sets for polynomial optimization and oracle-based cuts. arXiv preprint arXiv:1610.04604 (2016)

13. Bonami, P., Günlük, O., Linderoth, J.: Globally solving nonconvex quadratic programming problems with box constraints via integer programming methods. Math. Program. Comput. **10**(3), 333–382 (2018)

14. Borozan, V., Cornuéjols, G.: Minimal valid inequalities for integer constraints. Math. Oper. Res. **34**(3), 538–546 (2009)

15. Burer, S.: Optimizing a polyhedral-semidefinite relaxation of completely positive programs. Math. Program. Comput. **2**(1), 1–19 (2010)

16. Chen, C., Atamtürk, A., Oren, S.S.: A spatial branch-and-cut method for non-convex QCQP with bounded complex variables. Math. Program. **165**(2), 549–577 (2017)

17. Conforti, M., Cornuéjols, G., Daniilidis, A., Lemaréchal, C., Malick, J.: Cut-generating functions and S-free sets. Math. Oper. Res. **40**(2), 276–391 (2014)

18. Cornuéjols, G., Wolsey, L., Yıldız, S.: Sufficiency of cut-generating functions. Math. Program. **152**, 1–9 (2013)

19. Dax, A.: Low-rank positive approximants of symmetric matrices. Adv. Linear Algebra Matrix Theory **4**(3), 172–185 (2014)

20. Dey, S.S., Wolsey, L.A.: Lifting integer variables in minimal inequalities corresponding to lattice-free triangles. In: Lodi, A., Panconesi, A., Rinaldi, G. (eds.) IPCO 2008. LNCS, vol. 5035, pp. 463–475. Springer, Heidelberg (2008). https://doi.org/10.1007/978-3-540-68891-4_32

21. Dey, S.S., Wolsey, L.A.: Constrained infinite group relaxations of MIPs. SIAM J. Optim. **20**(6), 2890–2912 (2010)

22. Eckart, C., Young, G.: The approximation of one matrix by another of lower rank. Psychometrika **1**(3), 211–218 (1936)

23. Fischetti, M., Ljubić, I., Monaci, M., Sinnl, M.: A new general-purpose algorithm for mixed-integer bilevel linear programs. Oper. Res. **65**(6), 1615–1637 (2017)

24. Floudas, C.A., et al.: Handbook of Test Problems in Local and Global Optimization, vol. 33. Springer, Boston (2013). https://doi.org/10.1007/978-1-4757-3040-1

25. Freund, R.M., Orlin, J.B.: On the complexity of four polyhedral set containment problems. Math. Program. **33**(2), 139–145 (1985)

26. Ghaddar, B., Vera, J.C., Anjos, M.F.: A dynamic inequality generation scheme for polynomial programming. Math. Program. **156**(1–2), 21–57 (2016)

27. Gomory, R.E., Johnson, E.L.: Some continuous functions related to corner polyhedra. Math. Program. **3**(1), 23–85 (1972)

28. Guennebaud, G., Jacob, B., et al.: Eigen v3 (2010). http://eigen.tuxfamily.org

29. Hillestad, R.J., Jacobsen, S.E.: Reverse convex programming. Appl. Math. Optim. **6**(1), 63–78 (1980)

30. Kılınç-Karzan, F.: On minimal valid inequalities for mixed integer conic programs. Math. Oper. Res. **41**(2), 477–510 (2015)

31. Kocuk, B., Dey, S.S., Sun, X.A.: Matrix minor reformulation and SOCP-based spatial branch-and-cut method for the AC optimal power flow problem. Math. Program. Comput. **10**(4), 557–596 (2018)
32. Krishnan, K., Mitchell, J.E.: A unifying framework for several cutting plane methods for semidefinite programming. Optim. Methods Softw. **21**, 57–74 (2006)
33. Lasserre, J.B.: Global optimization with polynomials and the problem of moments. SIAM J. Optim. **11**, 796–817 (2001)
34. Laurent, M.: Sums of squares, moment matrices and optimization over polynomials. In: Putinar, M., Sullivant, S. (eds.) Emerging Applications of Algebraic Geometry, pp. 157–270. Springer, New York (2009). https://doi.org/10.1007/978-0-387-09686-5_7
35. Locatelli, M., Schoen, F.: On convex envelopes for bivariate functions over polytopes. Math. Program. **144**, 1–27 (2013)
36. Lovász, L.: Geometry of numbers and integer programming. In: Mathematical Programming: Recent Developments and Applications, pp. 177–210 (1989)
37. Luedtke, J., Namazifar, M., Linderoth, J.: Some results on the strength of relaxations of multilinear functions. Math. Program. **136**(2), 325–351 (2012)
38. McCormick, G.P.: Computability of global solutions to factorable nonconvex programs: part I - convex underestimating problems. Math. Program. **10**(1), 147–175 (1976)
39. Meeraus, A.: GLOBALLib. http://www.gamsworld.org/global/globallib.htm
40. Mirsky, L.: Symmetric gauge functions and unitarily invariant norms. Q. J. Math. **11**(1), 50–59 (1960)
41. Misener, R., Floudas, C.A.: Global optimization of mixed-integer quadratically-constrained quadratic programs (MIQCQP) through piecewise-linear and edge-concave relaxations. Math. Program. **136**(1), 155–182 (2012)
42. Modaresi, S., Kılınç, M.R., Vielma, J.P.: Intersection cuts for nonlinear integer programming: convexification techniques for structured sets. Math. Program. **155**, 1–37 (2015)
43. MOSEK ApS: The MOSEK Fusion API for C++ 8.1.0.63 (2018). https://docs.mosek.com/8.1/cxxfusion/index.html
44. Qualizza, A., Belotti, P., Margot, F.: Linear programming relaxations of quadratically constrained quadratic programs. In: Lee, J., Leyffer, S. (eds.) Mixed Integer Nonlinear Programming, pp. 407–426. Springer, New York (2012). https://doi.org/10.1007/978-1-4614-1927-3_14
45. Rikun, A.D.: A convex envelope formula for multilinear functions. J. Glob. Optim. **10**(4), 425–437 (1997)
46. Saxena, A., Bonami, P., Lee, J.: Convex relaxations of non-convex mixed integer quadratically constrained programs: extended formulations. Math. Program. **124**, 383–411 (2010)
47. Saxena, A., Bonami, P., Lee, J.: Convex relaxations of non-convex mixed integer quadratically constrained programs: projected formulations. Math. Program. **130**, 359–413 (2011)
48. Sen, S., Sherali, H.D.: Nondifferentiable reverse convex programs and facetial convexity cuts via a disjunctive characterization. Math. Program. **37**(2), 169–183 (1987)
49. Sherali, H.D., Fraticelli, B.M.P.: Enhancing RLT relaxations via a new class of semidefinite cuts. J. Glob. Optim. **22**, 233–261 (2002)
50. Shor, N.Z.: Quadratic optimization problems. Sov. J. Circ. Syst. Sci. **25**, 6 (1987)
51. Tardella, F.: Existence and sum decomposition of vertex polyhedral convex envelopes. Optim. Lett. **2**, 363–375 (2008)

52. Tawarmalani, M., Richard, J.P.P., Xiong, C.: Explicit convex and concave envelopes through polyhedral subdivisions. Math. Program. **138**, 1–47 (2013)
53. Tawarmalani, M., Sahinidis, N.V.: Convex extensions and envelopes of lower semi-continuous functions. Math. Program. **93**, 247–263 (2002)
54. Tuy, H.: Concave programming under linear constraints. Sov. Math. **5**, 1437–1440 (1964)
55. Vandenbussche, D., Nemhauser, G.: A branch-and-cut algorithm for nonconvex quadratic programs with box constraints. Math. Program. **102**(3), 559–575 (2005)

Fixed-Order Scheduling on Parallel Machines

Thomas Bosman[1](✉), Dario Frascaria[1], Neil Olver[1,2], René Sitters[1,2], and Leen Stougie[1,2]

[1] Department of Econometrics and Operations Research,
Vrije Universiteit Amsterdam, Amsterdam, The Netherlands
thomas.bosman@vu.nl
[2] CWI, Amsterdam, The Netherlands

Abstract. We consider the following natural scheduling problem: Given a sequence of jobs with weights and processing times, one needs to assign each job to one of m identical machines in order to minimize the sum of weighted completion times. The twist is that for machine the jobs assigned to it must obey the order of the input sequence, as is the case in multi-server queuing systems. We establish a constant factor approximation algorithm for this (strongly NP-hard) problem. Our approach is necessarily very different from what has been used for similar scheduling problems without the fixed-order assumption. We also give a QPTAS for the special case of unit processing times.

1 Introduction

We consider an extremely simple, yet challenging, scheduling principle that arises in many logistic and service applications. Given a sequence of jobs and a set of machines, we need to dispatch the jobs one by one over the machines, where for each machine the ordering of the original sequence is preserved. Hence, each machine must handle the jobs in a first-in first-out (FIFO) order. Each job has a processing time p_j and weight w_j and the goal is to minimize the weighted sum of completion times, where the completion time of a job j is the total processing time of the jobs preceding j (including j) on the same machine.

The FIFO-ordering restriction is common in queuing theory, where the *task assignment problem* [5,8,9] is concerned with the same question, except that jobs arrive stochastically over time. Our problem can be seen as asking for the optimal way of dispatching jobs from a single queue over m server queues under complete information of the processing times, essentially *unzipping* a single queue into m queues. The reverse problem of *zipping* m queues into a single queue, is the classic single machine scheduling problem: $1|\text{chains}|\sum w_j C_j$ (in the 3-field

This work was partially supported by the Netherlands Organisation for Scientific Research (NWO) through a VIDI grant (016.Vidi.189.087) and the Gravitation Programme Networks (024.002.003).

A. Lodi and V. Nagarajan (Eds.): IPCO 2019, LNCS 11480, pp. 88–100, 2019.
https://doi.org/10.1007/978-3-030-17953-3_7

notation by Graham et al. [7]) and can be solved efficiently by greedily selecting a prefix of jobs with largest ratio of total weight over total processing time [12].

In the scheduling literature, the fixed-ordering scheduling problem can be seen as a special case of *scheduling problems with sequence dependent setup times*, where for each pair of jobs there is a changeover cost c_{ij} that is paid if job j is the immediate successor of job i on a machine. Such setup times occur naturally in many industrial applications [1]. Our problem is precisely this, in the special case where $c_{ij} = \infty$ if i is later than j in the ordering, and $c_{ij} = 0$ otherwise. While the problem has received substantial attention (see [1–3] for a comprehensive literature review), almost nothing is known from a theoretical perspective (an exception is [10], but this is concerned with a rather unusual objective function). We believe our work may shed light on this more general problem.

We remark that the online version of the problem, where we need to assign a job before we get to know the next job in the sequence, does not admit a constant-competitive algorithm. Consider, for example, two machines and a sequence of three jobs (where job 1 should be completed first and job 3 last on any machine) with $p_1 = k^2, p_2 = k, p_3 = 1$ and $w_1 = 1, w_2 = k$ and w_3 is either zero or k^3, depending on the schedule of the first two jobs. Here, k is an arbitrarily large number. It is easy to see that a good schedule requires knowledge of the weight of job 3 before deciding whether to put jobs 1 and 2 on the same machine.

Our Contribution. Scheduling problems with weighted completion times objective *without* ordering constraints typically admit good approximations algorithms. For identical machines there is a PTAS [19], while a slightly better than $\frac{3}{2}$-approximation [4] for unrelated machines is known. But even simpler approaches yield a constant factor approximation. On identical machines, it is a classic result due to Kawaguchi and Kyan [11] (see [17] for a modern proof) that scheduling greedily according to Smith ratio (see Sect. 2) is a $\frac{1+\sqrt{2}}{2}$-approximation. For unrelated machines, independent rounding of both a natural time-indexed LP [16] and a (nontrivial) convex quadratic program [18] works, and achieve approximation ratios of $\frac{3}{2} + \epsilon$ and $\frac{3}{2}$ respectively. Another approach is α-*point scheduling* [15], where jobs are sorted according to the time by which an α fraction of the job has been processed in some LP relaxation. The jobs are then scheduled greedily in that order. This method has enabled many algorithmic improvements in scheduling, since it can be modified to deal with additional complications, such as precedence constraints and release times [13].

Fixed-order scheduling appears highly resistant to all these techniques. A big obstacle is that moving even a single pair of jobs onto the same machine can have a catastrophic effect on the objective function if the order is fixed: think of a job with large processing time but minuscule weight, followed by a job with large weight and minuscule processing time. Thus in order to have any hope of a non-trivial performance guarantee, jobs must be assigned to machines in a highly dependent way. To achieve this, our approach radically departs from earlier ones.

We define an important partial order $\prec$: essentially, $j \prec k$ means that not only is j earlier than k in the FIFO ordering, but also this ordering is opposite to what would be preferred according to Smith's rule. Thus as alluded to earlier, if $j \prec k$ we should be particularly careful about assigning j and k to the same machine. Order the machines arbitrarily. A key idea is that at only a constant factor loss, we can restrict our attention to a class of schedules we call *Smith-monotone*, meaning that if $j \prec k$, then j is assigned to an earlier machine (or the same machine) as the one that k is assigned to. Next, we relax the problem by computing each job's completion times partially, based only on jobs preceding it in the partial order $\prec$. While this potentially distorts completion times a lot, we can ensure the amortized effect is not too large by appropriately rounding weights and processing times. Finally, we formulate a new LP using these partial completion times and enforcing the strong structure imposed by Smith monotonicity. This LP can be rounded in a way that completely respects the pairwise probabilities of jobs being assigned to the same machine; Smith monotonicity is crucial here. The rounding is very appealing and natural, and can be seen an analog of α-point scheduling with respect to machine index rather than time.

Finally, we remark that for the case of unit processing times, the complexity of the problem is unknown, but it is unlikely to be APX-hard: we present a QPTAS in Sect. A of the appendix.

2   Problem Definition, Notation and NP-Hardness

We have a set of identical machines $M = \{1, \ldots, m\}$, each of which can process one job at a time, and a totally ordered set of jobs $J = 1, \ldots, n$, where each job $j \in J$ has weight $w_j \in \mathbb{N}$ and processing time $p_j \in \mathbb{N}$. Because of its frequent use we also define notation for the so called *Smith ratio* $s_j := \frac{w_j}{p_j}$ of job j. A feasible schedule $\mu : J \to M$ assigns to every job j a machine $\mu(j)$. Each machine processes the jobs assigned to it in order of their number. The cost of the schedule μ is the sum of weighted completion times, i.e.,

$$\Gamma(\mu) = \sum_{k \in J} \sum_{\substack{j \in J : j \leq k, \\ \mu(j) = \mu(k)}} p_j \cdot w_k. \tag{1}$$

The objective is to minimize the cost of the schedule. We denote by OPT the optimal cost, by $\sigma_\mu : J \to \mathbb{N}$ the function that maps every job to its completion time under μ and by $\prec$ a partial order on J such that $j \prec k$ if and only if $j < k$ and $s_j \leq s_k$.

Note that we explicitly disallow $p_j = 0$. This is for convenience, and ensures that the Smith ratio w_j/p_j is always well defined. Our results easily extend to the case where jobs with zero processing time are allowed.

The problem of minimizing the sum of weighted completion times without ordering constraints is a classic problem that has long been known to be strongly NP-hard [6]. This result extends to fixed-order scheduling as well: given an assignment of jobs to a machine, it is always optimal to schedule them in decreasing order of Smith ratio, so the ordering constraints become redundant if $s_1 \geq s_2 \geq \cdots \geq s_n$.

3   Structural Properties of Optimal Solutions

In this section we will provide a characterization of an optimal solution which will help us construct a constant-factor approximation algorithm in Sect. 4.

Unit Processing Times
Suppose all jobs have equal processing time and hence that the relation $j \prec k$ indicates that $j < k$ and $w_j \leq w_k$. W.l.o.g. we may then as well assume that the processing times are 1. An initial simplification is that we will assume that schedules are *staircase shaped*, in the sense that for every prefix of the jobs $1, \ldots, k$, the number of jobs assigned to each machine decreases monotonically with the machine index. We will use the following equivalent definition.

Definition 1. *A schedule μ is **staircase shaped** if for each job k, $\mu(k) = |\{j < k : \sigma_\mu(j) = \sigma_\mu(k)\}| + 1$.*

Given any schedule μ, we can clearly turn it into a staircase shaped schedule without changing the completion time of any job.

Clearly we want jobs with high weights to be completed early, but this may not always be possible because of the ordering on the jobs. Intuitively, it seems like a good idea to 'reserve' some of the good spots early in the schedule, for higher weight jobs later in the sequence. In staircase shaped schedules this means that early and low weight jobs should be put on low index machines as much as possible. Lemma 1 below makes this more precise.

Definition 2. *A schedule μ is **Smith-monotone** if, for every $j \prec k$, it holds that $\mu(j) \leq \mu(k)$.*

Lemma 1. *For unit processing times, there exists an optimal schedule that is Smith-monotone and staircase shaped.*

Proof. Let us define the potential function $\sum_{j \in J} \mu(j)j$. Let μ be a solution maximizing this potential function among those that are optimal and staircase shaped. Suppose μ is not Smith-monotone. We obtain a contradiction by showing there is a staircase shaped schedule with a higher potential but no higher cost.

Since μ is not Smith-monotone, there exists a pair $j \prec k$ that violates Smith monotonicity. Pick j, k so that there is no other violating pair $j' \prec k'$ between it, i.e with $j \leq j'$ and $k' \leq k$ and at least one of the inequalities strict. We call such a pair *tight*. For $g = j, k$, let $S_g = \{h \in \{j + 1, \ldots, k - 1\} : \mu(h) = \mu(g)\}$ be the set of jobs between j and k on the machine of job g. It follows that $h \in S_k \implies w_h \leq w_j$ and that $h \in S_j \implies w_h \geq w_k$ as, otherwise, the pair $j \prec k$ would not be tight. (In fact, $<$ holds.) By assigning j to the machine of k and vice-versa we get a schedule μ' that improves our potential function. We first show is that the new schedule μ' does not incur a higher cost than μ.

Since only the starting times (whence completion times) of j, k and jobs in S_j and S_k may change, all others remaining equal, it follows that

$$\sum_{h \in \{j\} \cup S_k} \sigma_{\mu'}(h) + \sum_{h \in \{k\} \cup S_j} \sigma_{\mu'}(h) = \sum_{h \in \{j\} \cup S_k} \sigma_\mu(h) + \sum_{h \in \{k\} \cup S_j} \sigma_\mu(h),$$

hence

$$\sum_{h \in \{j\} \cup S_k} (\sigma_{\mu'}(h) - \sigma_\mu(h)) = \sum_{h \in \{k\} \cup S_j} (\sigma_\mu(h) - \sigma_{\mu'}(h)). \tag{2}$$

Now note that the completion times of jobs j and S_k increase, while those of k and S_j decrease, since the schedule was staircase shaped. Combining this with the fact that $w_h \leq w_j$ for $h \in S_k$ and $w_h \geq w_k$ for $h \in S_j$, we can bound the increase in the cost as follows:

$$\sum_{h \in \{j,k\} \cup S_j \cup S_k} w_h(\sigma_{\mu'}(h) - \sigma_\mu(h)) =$$

$$\sum_{h \in \{j\} \cup S_k} w_h(\sigma_{\mu'}(h) - \sigma_\mu(h)) - \sum_{h \in \{k\} \cup S_j} w_h(\sigma_\mu(h) - \sigma_{\mu'}(h)) \leq$$

$$w_j \sum_{h \in \{j\} \cup S_k} (\sigma_{\mu'}(h) - \sigma_\mu(h)) - w_k \sum_{h \in \{k\} \cup S_j} (\sigma_\mu(h) - \sigma_{\mu'}(h)) \leq 0 \quad \text{by (2)}.$$

So the new schedule has a higher potential function and no higher cost. The schedule may not be staircase shaped, however. But simply sorting the jobs fixes this without undoing our work. To see this, note that the set of timeslots occupied on each machine did not change when we modified the schedule. So the fact that the old schedule was staircase shaped, implies that the new schedule still has the following weaker property: if exactly k jobs are scheduled at time t, they are scheduled on the first k machines. By sorting all the jobs assigned to one timeslot by number and assigning them to the first available machine in that order, the potential function can only increase further, while completion times stay the same. Hence, we have found an optimal staircase shaped schedule with higher potential function than we started with, contradicting our choice of the original schedule and concluding the proof.

General Processing Times

Unfortunately, for general processing times we cannot hope for an equally nice structural result. Indeed, there may not be an optimal schedule that is Smith-monotone. However, as we will now show, we may impose this structure with the loss of only a constant factor in the objective.

The proof works by reducing a general instance to one with unit processing times, finding an optimal Smith-monotone schedule, and then rounding it back. Although this reduction is not polynomial time, it will suffice to prove the bound on the optimality ratio. Instead, in the next section we will find that we can bypass the reduction, and approximate such a schedule directly.

Given an instance I to the general problem, let I^{unit} be the instance with unit processing times obtained from I by replacing every job $j \in J$ with a set $U(j) = \{j^1, \ldots, j^{p_j}\}$ of p_j consecutive jobs, each having unit processing time and weight w_j/p_j. Let J^{unit} be the set of jobs of I^{unit} and OPT^{unit} be the optimal cost for I^{unit}.

Lemma 2. $\text{OPT}^{unit} \leq \text{OPT} - \sum_{j \in J} \frac{1}{2}(p_j - 1)w_j$

Proof. The statement follows for the schedule μ^{unit} obtained from the optimal schedule for I by putting all the jobs in $U(j)$ on the machine $\mu(j)$, for all $j \in J$. The completion times of the jobs in $U(j)$ run from $\sigma(j) - p_j + 1$ to $\sigma(j)$, and hence the average completion time is $\sigma(j) - \frac{1}{2}(p_j - 1)$. The proof follows from multiplying by the total weight of the jobs in $U(j)$.

Lemma 3. *An optimal Smith-monotone schedule for I has cost at most* $\text{OPT}^{unit} + \sum_{j \in J}(p_j - 1)w_j$.

Proof. Given an optimal schedule μ^{unit} for I^{unit}, we can create a schedule for I by putting job j on machine i with probability $|\{h \in U(j) : \mu^{unit}(h) = i\}|/|U(j)|$, for all $j \in J$. The expected time spent processing job $j \in J$ on machine i is exactly $|\{h \in U(j) : \mu^{unit}(h) = i\}|$. As a consequence, the expected starting time of a job $j \in J$ is at most the average starting time of the jobs in $U(j)$, and thus the expected completion time is at most the expected completion time of jobs in $U(j)$ plus $p_j - 1$ (which is the difference between the processing time of job j and any job in $U(j)$).

What remains to show is that μ is Smith-monotone. By Lemma 1 we can assume, w.l.o.g., that the optimal solution of I^{unit} satisfies Smith monotonicity. Consider an arbitrary pair $j \prec k$. We have that $j' \prec k'$ for all jobs $j' \in U(j), k' \in U(k)$. This implies that, if any job in $U(j)$ is scheduled on machine i, all jobs in $U(k)$ are scheduled on machines with index not smaller than i. Hence it holds that the machine with highest index to which j may be assigned cannot have index larger than any machine to which k may be assigned. Therefore, for every possible realization of the random schedule, Smith monotonicity is satisfied, completing the proof.

Lemma 4. *An optimal Smith-monotone schedule has cost at most* $\frac{3}{2}\text{OPT}$.

Proof. By Lemma 3, we have that an optimal Smith-monotone schedule has cost at most $\text{OPT}^{unit} + \sum_{j \in J}(p_j - 1)w_j$, which in turn, by Lemma 2, is at most $\text{OPT} + \frac{1}{2}\sum_{j \in J}(p_j - 1)w_j \leq \frac{3}{2}\text{OPT}$.

Though the bound in Lemma 4 is unlikely to be tight, it cannot be improved much further. The Kawaguchi-Kyan bound of $\frac{1+\sqrt{2}}{2} \approx 1.207$ is known to be tight [11,17], and this lower bound carries over to fixed-order scheduling: the worst-case example uses jobs with equal Smith ratios, and in that case reordering the jobs assigned to a machine does not change the cost.

4 A Constant Factor Approximation Algorithm

In this section we will describe an algorithm that proves our main result: Theorem 1. Our approach is as follows: first we round the instance such that all Smith ratios are powers of $\frac{1}{3}$ (by rounding up the weights as appropriate). Given that, we show that a certain relaxed objective function is always within a constant of the original objective function. We then use LP rounding to find a Smith-monotone schedule that is optimal with respect to the relaxed objective function.

Theorem 1. *Fixed-order scheduling can be approximated to within a factor $\frac{27}{2} + 9\sqrt{3} < 29.1$.*

From hereon assume that all Smith ratios are powers of some fixed $\gamma \in (0,1)$. Suppose we relax the objective function (1) to disregard the contribution of processing times of jobs with $j < k$ and $w_j > w_k$ (hence, $j \not\preceq k$) in the completion time of job k. We claim this relaxation loses only a factor $(\frac{4}{1-\sqrt{\gamma}} - 3)$.

Definition 3. *The **partial completion time** of a job $k \in J$ under a schedule μ is*

$$\tilde{c}_k = \sum_{j \preceq k : \mu(j) = \mu(k)} p_j \ .$$

*The **partial cost** of a schedule μ is $\tilde{\Gamma}(\mu) = \sum_{k \in J} w_k \tilde{c}_k$.*

It is crucial that the Smith ratios are powers of γ; this ensures that either $j \preceq k$ or w_j is relatively large compared to w_k. Intuitively, in the latter case we don't care too much about the effect of j's processing time on the lighter-weight job k.

Theorem 2. *Take an instance where all Smith ratios are positive powers of $\gamma \in (0,1)$. Consider any schedule μ and denote its cost by $\Gamma(\mu)$. It holds that*

$$\tilde{\Gamma}(\mu) \leq \Gamma(\mu) \leq \left(\frac{4}{1 - \sqrt{\gamma}} - 3 \right) \cdot \tilde{\Gamma}(\mu).$$

Since it is obvious that $\tilde{\Gamma}(\mu) \leq \Gamma(\mu)$ we prove the upper bound.

To simplify notation assume the instance has only one machine; the result can be applied to each machine individually. For $d \in \mathbb{N}$, let N_d be the set of jobs j with $s_j = \gamma^d$; let W_d and P_d be, respectively, the total weight and total processing time of jobs in N_d. We denote by $H_d = \frac{1}{W_d} \sum_{k \in N_d} w_k \tilde{c}_k$ the weighted average of the partial cost in N_d. It follows that

$$\tilde{\Gamma}(\mu) = \sum_{k \in J} w_k \tilde{c}_k = \sum_{d \in \mathbb{N}} W_d H_d = \sum_{d \in \mathbb{N}} \gamma^d P_d H_d. \tag{3}$$

Our goal is to bound $\Gamma(\mu)$ in the same terms. Since H_d only accounts for the contribution of jobs with Smith ratio at most γ^d in the completion time of jobs in N_d, we need to correct for the other jobs (that are in $N_1, \ldots, N_{d-1}$). In the worst case, all these jobs are scheduled first and hence their processing times need to be added. Therefore we get:

$$\Gamma(\mu) \leq \sum_{d \in \mathbb{N}} W_d \left(H_d + \sum_{i=1}^{d-1} P_i \right) \leq \sum_{d \in \mathbb{N}} \gamma^d P_d \left(H_d + \sum_{i=1}^{d-1} P_i \right). \tag{4}$$

We will prove that this value can be bounded by the desired constant times $\tilde{\Gamma}(\mu)$. Globally our strategy is to show that every newly introduced term $\gamma^d P_d P_i$ can be charged to a term in the expression for $\tilde{\Gamma}(\mu)$ such that no term gets charged more than a $\frac{4}{1-\sqrt{\gamma}} - 4$ fraction of its value.

Before we can proceed we will need the inequality in Lemma 5 below. It says that the average weighted completion time in N_d is at least half the total processing time in N_d, which can intuitively be seen as follows: think of all jobs $j \in N_d$ as blocks of equal width, length p_j and mass w_j. So, all blocks have equal density. Now stack the boxes on top of each other: then P_d corresponds to the total length, and H_d approximately to the center of mass, which is in the middle.

Lemma 5. $P_d \leq 2H_d$.

Proof.

$$H_d = \frac{1}{W_d} \sum_{k \in N_d} w_k \sum_{j \leq k \wedge s_j \leq s_k} p_j \geq \frac{1}{W_d} \sum_{k \in N_d} w_k \underbrace{\sum_{j \leq k \wedge j \in N_d} p_j}_{:=Q}$$

Suppose now that $Q < \frac{P_d}{2}$. Since $W_d = \gamma^d P_d$, it follows that:

$$Q = \frac{1}{P_d} \sum_{k \in N_d} p_k \Big(P_d - \sum_{h > k \wedge h \in N_d} p_h \Big) = P_d - \frac{1}{P_d} \sum_{h \in N_d} p_h \sum_{k < h \wedge k \in N_d} p_k \geq P_d - Q,$$

implying that $Q \geq P_d/2$, a contradiction. □

Lemma 6. $\max\{\gamma^d P_d H_d, \gamma^i P_i H_i\} \geq \frac{1}{2}(\gamma^d P_d P_i)\gamma^{\frac{1}{2}(i-d)}$.

Proof. Suppose that $\gamma^d P_d H_d < \frac{1}{2}\gamma^d P_d P_i \gamma^{\frac{1}{2}(i-d)}$. This implies that

$$H_d < \frac{1}{2} P_i \gamma^{\frac{1}{2}(i-d)}. \tag{5}$$

So we obtain

$$\gamma^d P_d P_i \gamma^{\frac{1}{2}(i-d)} \leq \gamma^d 2 H_d P_i \gamma^{\frac{1}{2}(i-d)} \overset{(5)}{<} \gamma^{d-i}\gamma^i P_i \gamma^{\frac{1}{2}(i-d)} P_i \gamma^{\frac{1}{2}(i-d)} \leq 2\gamma^i H_i P_i,$$

where the first and third inequalities follow from Lemma 5. □

We are now ready to prove Theorem 2.

Proof (Theorem 2). We will prove the following inequality, implying the theorem by (4):

$$\sum_{d \in \mathbb{N}} \gamma^d P_d H_d \Big(\frac{4}{1 - \sqrt{\gamma}} - 4 \Big) \geq \sum_{d \in \mathbb{N}} \gamma^d P_d \sum_{i=1}^{d-1} P_i.$$

Applying Lemma 6 and replacing the max by a sum we get:

$$\sum_{d \in \mathbb{N}} \gamma^d P_d \sum_{i=1}^{d-1} P_i \leq \sum_{d \in \mathbb{N}} \sum_{i=1}^{d-1} \gamma^{\frac{1}{2}(d-i)} 2\max(\gamma^d P_d H_d, \gamma^i P_i H_i)$$

$$\leq \sum_{d \in \mathbb{N}} \sum_{i=1}^{d-1} \gamma^{\frac{1}{2}(d-i)} 2(\gamma^d P_d H_d + \gamma^i P_i H_i)$$

$$\leq \sum_{d \in \mathbb{N}} \sum_{i=1}^{\infty} \gamma^{\frac{1}{2}i} 4(\gamma^d P_d H_d) = \sum_{d \in \mathbb{N}} \gamma^d P_d H_d \Big(\frac{4}{1 - \sqrt{\gamma}} - 4 \Big). \quad □$$

Linear Programming Relaxation

Suppose all Smith ratios are positive powers of $\gamma \in (0,1)$. The following mixed-integer program captures the problem of finding a Smith-monotone ordering that minimizes the modified objective $\tilde{\Gamma}(\mu)$.

$$\min \quad \sum_{k \in J} w_k \tilde{c}_k$$

$$\text{s.t.} \quad u_k \geq u_j + z_{jk} \qquad \forall j \prec k \tag{6}$$

$$u_k \leq m \qquad \forall k \in J \tag{7}$$

$$\tilde{c}_k \geq p_k + \sum_{j \prec k}(1 - z_{jk})p_j \qquad \forall k \in J \tag{8}$$

$$u_k \in \mathbb{N}, z_{jk} \in \{0,1\} \qquad \forall k \in J, j \prec k$$

Here, z_{jk} is the indicator variable for the event that j and k are assigned to different machines. The variable u_k indicates which machine job k is assigned to. Finally, $\tilde{c}_k$ represents the partial completion time of job k. The constraint (6) is valid since we require a Smith-monotone ordering.

Now consider the natural LP relaxation of this mixed-integer program, where we drop the integrality requirements and instead require $1 \leq u_k \leq m, 0 \leq z_{jk} \leq 1$. Denote this relaxation by (LP). Let $(z^*, u^*, \tilde{c}^*)$ be an optimal solution to (LP), with cost OPT_{LP}.

Definition 4. *For $\beta \in (0,1)$, the β-**point schedule** associated to u^* is the schedule obtained by assigning job j to machine $\lceil u_j^* - \beta \rceil$.*

From now on, β will be chosen uniformly at random from $(0,1)$, making the β-point schedule a random schedule.

Let N_d be the set of jobs with Smith ratios γ^d and let $\tilde{C}_k$ be the (random) partial completion time of job k under the β-point schedule. The following statements are easy to verify.

Proposition 1. *For any pair of jobs $j \prec k$, the probability that jobs j and k are assigned to the same machine under the β-point schedule is at most $1 - z_{jk}^*$.*

Proof. This follows immediately from the constraint (6). $\qquad\square$

Proposition 2. *For any $k \in J$, $\mathbb{E}[\tilde{C}_k] \leq \tilde{c}_k^*$. Hence*

$$\mathbb{E}\left[\sum_{k \in J} w_k \tilde{C}_k\right] \leq \sum_{k \in J} w_k \tilde{c}_k^* = \text{OPT}_{\text{LP}}.$$

Proof. This follows from Proposition 1 and (8). $\qquad\square$

Note that a β-point schedule can be derandomized in polynomial time: a job j is always assigned to u_j^* when u_j^* is integral and to either $\lfloor u_j^* \rfloor$ or $\lceil u_j^* \rceil$ otherwise. Let b_j be the maximum value of β for which $\lceil u_j^* - \beta \rceil = \lfloor u_j^* \rfloor$. It follows that all possible schedules are the ones obtained by assigning to β the values in $\{b_j | j \in J\} \cup \{1\}$.

Constant Factor Approximation
We are now ready to complete the main proof of this section.

Proof (Theorem 1). Given an instance I where the jobs have arbitrary Smith ratios, let I^γ be the instance obtained from I by rounding up the weights such that the Smith ratios are powers of γ, where γ will be defined later, and then scaled to be in $(0,1)$. Let OPT_I and OPT_{I^γ} be the optimal cost for the instances I and I^γ respectively. Clearly the cost of any schedule μ on I^γ is at most γ^{-1} times the cost of μ on I and $\mathrm{OPT}_I \leq \mathrm{OPT}_{I^\gamma}$ since the cost of a schedule is linear in the weights (see (1)) and weights do not affect feasibility.

We denote by μ and $\mathrm{OPT}_{\mathrm{LP}}$, respectively, the β-point schedule and the optimal cost of (LP) on I^γ. By Lemma 4, Proposition 2 and Theorem 2, the cost of μ on I is at most

$$\frac{3}{2}\left(\frac{4}{1-\sqrt{\gamma}}-3\right)\cdot\mathrm{OPT}_{\mathrm{LP}} \leq \frac{3}{2}\left(\frac{4}{1-\sqrt{\gamma}}-3\right)\cdot\mathrm{OPT}_{I^\gamma} \leq \frac{3}{2}\left(\frac{4}{1-\sqrt{\gamma}}-3\right)\cdot\mathrm{OPT}_I\gamma^{-1}.$$

Minimizing over γ, this yields an approximation ratio of

$$\min_{\gamma\in(0,1)}\frac{3}{2}\left(\frac{4}{1-\sqrt{\gamma}}-3\right)\cdot\gamma^{-1}=\frac{3}{2}(9+6\sqrt{3})=\frac{27}{2}+9\sqrt{3}$$

when $\gamma=1/3$, concluding the proof.   □

5   Epilogue

Our work suggests many further interesting and natural directions. One is to find a PTAS (or even a polynomial time exact algorithm) for unit processing times, perhaps expanding on the QPTAS in the appendix. Good approximation algorithms for all of the following problems remain open questions.

(1) There are k arrival lines that need to be dispatched over m servers, such that the FIFO ordering in each of the arrival lines is obeyed in each of the server queues.
(2) An arbitrary partial order on the jobs is given, and we require that if two jobs are assigned to the same machine, then the partial order is respected. (1) is exactly this problem, where the partial order is described by k disjoint chains.
(3) Instead of requiring that the order is exactly preserved, a natural relaxation is to allow a *reordering buffer* (see, e.g. [14]) of limited size. Jobs enter the buffer in the given FIFO order, but any job in the buffer can be chosen and assigned to one of the machines.

A   A QPTAS for unit processing times

In this section we sketch a simple quasipolynomial time approximation scheme (QPTAS) for the problem under unit processing times. Note that we do not know if this version of the problem remains NP-hard. However, it seems to capture most of the difficulty, so we feel that tackling this case will help substantially in improving the upper bound for the general case. The QPTAS works by solving a relaxed problem by dynamic programming. We round the completion times to geometric intervals and then we consider schedules in which at any time point only one machine per completion time can accept jobs. This sufficiently reduces the solution space to get a quasipolynomial time algorithm.

The first step is to consider only a logarithmic number of distinct completion times. Let $R = \{\lfloor (1+\epsilon)^i \rfloor : i \in \mathbb{N}\}$ be the set of integers found by rounding down a geometric series growing with rate $1 + \epsilon$. Then order the elements $1 = R_1 < R_2 < \ldots$ and take $R_0 = 0$ by convention. Assume that s is the smallest index such that $R_s \geq n$, and note that $s = O(\log_{1+\epsilon}(n))$. Now consider the objective of minimizing the weighted sum of *rounded* completion times, where each completion time is rounded up to the nearest R_i. Call this the *rounded objective*; clearly the rounded objective of any schedule is at most $1 + \epsilon$ times the actual objective. So, if we can find an optimal segmented schedule for the rounded objective, we immediately get a $(1 + \epsilon)$-approximation to the original problem.

Now we define a restricted type of schedule, which we call a *segmented staircase schedule*. A segmented staircase schedule is similar to a staircase shaped schedule, except that the "steps" are now defined in terms of the rounded completion times. When j is assigned to a machine, it is assigned to the leftmost machine that gives it the same rounded completion time. In other words, if a job j gets assigned to $\mu(j)$ and gets completed at time $t \in (R_i, R_{i+1}]$, then $\mu(j)$ is the lowest index machine for which the number of jobs $k < j$ assigned to it does not exceed $R_{i+1} - 1$.

Lemma 7. *There is an optimal solution to the problem of minimizing the rounded objective that is a segmented staircase schedule.*

Proof. Take μ to be a schedule for which $(\mu(1), \mu(2), \ldots, \mu(n))$ is lexicographically minimal amongst all solutions that are optimal for the rounded objective. Notice that μ must be staircase shaped; otherwise, transforming it into a staircase shaped schedule would yield a schedule μ' in which every job has the same completion time, but which is lexicographically smaller than μ.

Suppose for a contradiction that this schedule is not a segmented staircase schedule. Let j be the last (maximum index) job that violates the rule for a segmented staircase schedule: j is assigned to machine h', but $h < h'$ is the smallest index machine that gives it the same rounded completion time, ignoring all jobs after j. Let k be the first job $k > j$ scheduled on machine h. (If there is no such job, then moving j to h' reduces the lexicographical value and does not increase the total rounded completion time.) Note that since j was chosen maximally, no job ℓ with $j < \ell < k$ is scheduled on machine h'. Moreover, $\sigma_\mu(k) > \sigma_\mu(j)$, since μ is staircase shaped, so it must be that j and k are both in the same segment $(R_i, R_{i+1}]$ for some i. So we can simply swap k and j to obtain a lexicographically smaller schedule of the same rounded objective value.

For segmented staircase schedules, we can compactly describe the loads on the machines just prior to assigning job j. Let X_i^j be the number of jobs on machines with load currently in the interval $[R_i, R_{i+1})$ just prior to assigning job j. Since only one of the machines with load in that interval can have strictly more than R_i jobs on it, this number also completely determines how many machines there are with loads exactly R_i. For each j, we have $n^{O(\log_{1+\epsilon} n)}$ options for the values of $X_1^j, X_2^j, \ldots, X_s^j$. Once we know the minimum cost of a schedule attaining each of those options, we can compute the cost of all the schedules up to job $j + 1$. Hence, we get the main result of this section.

Theorem 3. *For any $\epsilon > 0$, there is a $(1 + \epsilon)$-approximation algorithm for fixed-order scheduling with unit processing times with running time $n^{O(\log_{1+\epsilon} n)}$.*

References

1. Allahverdi, A.: The third comprehensive survey on scheduling problems with setup times/costs. Eur. J. Oper. Res. **246**(2), 345–378 (2015)
2. Allahverdi, A., Gupta, J.N., Aldowaisan, T.: A review of scheduling research involving setup considerations. Omega **27**(2), 219–239 (1999)
3. Allahverdi, A., Ng, C., Cheng, T., Kovalyov, M.Y.: A survey of scheduling problems with setup times or costs. Eur. J. Oper. Res. **187**(3), 985–1032 (2008)
4. Bansal, N., Srinivasan, A., Svensson, O.: Lift-and-round to improve weighted completion time on unrelated machines. In: Proceedings of the 48th Annual ACM Symposium on Theory of Computing, pp. 156–167 (2016)
5. Feng, H., Misra, V., Rubenstein, D.: Optimal state-free, size-aware dispatching for heterogeneous M/G/-type systems. Perform. Eval. **62**(1), 475–492 (2005)
6. Garey, M.R., Johnson, D.S.: Computers and Intractability. A Guide to the Theory of NP-Completeness. W. H. Freeman, New York (1979)
7. Graham, R., Lawler, E., Lenstra, J., Kan, A.: Optimization and approximation in deterministic sequencing and scheduling: a survey. In: Discrete Optimization II. Annals of Discrete Mathematics, vol. 5, pp. 287–326. Elsevier (1979)
8. Harchol-Balter, M.: Performance Modeling and Design of Computer Systems: Queueing Theory in Action, 1st edn. Cambridge University Press, New York (2013)
9. Harchol-Balter, M., Crovella, M.E., Murta, C.D.: On choosing a task assignment policy for a distributed server system. IEEE J. Parallel Distrib. Comput. **59**(2), 204–228 (1999)
10. Hiraishi, K., Levner, E., Vlach, M.: Scheduling of parallel identical machines to maximize the weighted number of just-in-time jobs. Comput. Oper. Res. **29**(7), 841–848 (2002)
11. Kawaguchi, T., Kyan, S.: Worst case bound of an LRF schedule for the mean weighted flow-time problem. SIAM J. Comput. **15**(4), 1119–1129 (1986)
12. Lawler, E.L.: Sequencing jobs to minimize total weighted completion time subject to precedence constraints. Ann. Discrete Math. **2**, 7590 (1978)
13. Queyranne, M., Schulz, A.S.: Approximation bounds for a general class of precedence constrained parallel machine scheduling problems. SIAM J. Comput. **35**(5), 1241–1253 (2006)
14. Räcke, H., Sohler, C., Westermann, M.: Online scheduling for sorting buffers. In: Möhring, R., Raman, R. (eds.) Proceedings of 10th Annual European Symposium on Algorithms, pp. 820–832 (2002)

15. Schulz, A.S., Skutella, M.: The power of α-points in preemptive single machine scheduling. J. Sched. **5**(2), 121–133 (2002)
16. Schulz, A.S., Skutella, M.: Scheduling unrelated machines by randomized rounding. SIAM J. Discrete Math. **15**(4), 450–469 (2002)
17. Schwiegelshohn, U.: An alternative proof of the Kawaguchi-Kyan bound for the Largest-Ratio-First rule. Oper. Res. Lett. **39**(4), 255–259 (2011)
18. Skutella, M.: Convex quadratic and semidefinite programming relaxations in scheduling. J. ACM **48**, 206–242 (2001)
19. Skutella, M., Woeginger, G.J.: A PTAS for minimizing the total weighted completion time on identical parallel machines. Math. Oper. Res. **25**(1), 63–75 (2000)

Online Submodular Maximization: Beating 1/2 Made Simple

Niv Buchbinder[1], Moran Feldman[2], Yuval Filmus[3], and Mohit Garg[2(✉)]

[1] Tel Aviv University, Tel Aviv, Israel
niv.buchbinder@gmail.com
[2] The Open University of Israel, Ra'anana, Israel
{moranfe,mohitga}@openu.ac.il
[3] Technion, Haifa, Israel
filmus.yuval@gmail.com

Abstract. The problem of `Submodular Welfare Maximization` (SWM) captures an important subclass of combinatorial auctions and has been studied extensively from both computational and economic perspectives. In particular, it has been studied in a natural online setting in which items arrive one-by-one and should be allocated irrevocably upon arrival. In this setting, it is well known that the greedy algorithm achieves a competitive ratio of $1/2$, and recently Kapralov *et al.* [22] showed that this ratio is optimal for the problem. Surprisingly, despite this impossibility result, Korula *et al.* [25] were able to show that the same algorithm is 0.5052-competitive when the items arrive in a uniformly random order, but unfortunately, their proof is very long and involved. In this work, we present an (arguably) much simpler analysis that provides a slightly better guarantee of 0.5096-competitiveness for the greedy algorithm in the random-arrival model. Moreover, this analysis applies also to a generalization of online SWM in which the sets defining a (simple) partition matroid arrive online in a uniformly random order, and we would like to maximize a monotone submodular function subject to this matroid. Furthermore, for this more general problem, we prove an upper bound of 0.576 on the competitive ratio of the greedy algorithm, ruling out the possibility that the competitiveness of this natural algorithm matches the optimal offline approximation ratio of $1 - 1/e$.

Keywords: Submodular optimization · Online auctions · Greedy algorithms

1 Introduction

The `Submodular Welfare Maximization` problem (SWM) captures an important subclass of combinatorial auctions and has been studied extensively from both computational and economic perspectives. In this problem we are given a set of m items and a set of n bidders, where each bidder has a non-negative

© Springer Nature Switzerland AG 2019
A. Lodi and V. Nagarajan (Eds.): IPCO 2019, LNCS 11480, pp. 101–114, 2019.
https://doi.org/10.1007/978-3-030-17953-3_8

monotone submodular utility function,[1] and the objective is to partition the items among the bidders in a way that maximizes the total utility of the bidders. Interestingly, SWM generalizes other extensively studied problems such as maximum (weighted) matching and budgeted allocation (see [28] for a comprehensive survey).

SWM is usually studied in the value oracle model (see Sect. 2 for definition). In this model the best approximation ratio for SWM is $1 - (1 - 1/n)^n \geq (1 - 1/e)$ [8,16,30]. A different line of work studies SWM in a natural online setting in which items arrive one-by-one and should be allocated irrevocably upon arrival. This setting generalizes, for example, online (weighted) matching and budgeted allocation [1,7,14,21,24,29,32]. It is well known that for this online setting the greedy approach that allocates each item to the bidder with the currently maximal marginal gain for the item is $1/2$-competitive, which is the optimal deterministic competitive ratio [18,22]. While randomization is known to be very helpful for many special cases of online SWM (e.g., matching), Kapralov et al. [22] proved that, unfortunately, this is not the case for online SWM itself— i.e., no (randomized) algorithm can achieve a competitive ratio better than $1/2$ for this problem (unless NP = RP).

A common relaxation of the online setting is to assume that the items arrive in a random order rather than in an adversarial one [9,19]. This model was also studied extensively for special cases of SWM for which improved algorithms were obtained [19,23,26]. Surprisingly, unlike in the adversarial setting, Korula et al. [25] showed that the simple (deterministic) greedy algorithm achieves a competitive ratio of at least 0.5052 in the random arrival model. Unfortunately, the analysis of the greedy algorithm by Korula et al. [25] is very long and involves many tedious calculations, making it very difficult to understand why it works or how to improve it.

1.1   Our Results

In this paper, we study the problem of maximizing a monotone submodular function over a (simple) partition matroid. This problem is a generalization of SWM (see Sect. 2 for exact definitions and a standard reduction between the problems) in which a ground set $\mathcal{N}$ is partitioned into disjoint non-empty sets $P_1, P_2, \ldots, P_m$. The goal is to choose a subset $S \subseteq \mathcal{N}$ that contains at most one element from each set P_i and maximizes a given non-negative monotone submodular function f.[2] We are interested in the performance of the greedy algorithm for this problem when the sets P_i are ordered uniformly at random. A formal description of the algorithm is given as Algorithm 1.

---

[1] A set function $f \colon 2^{\mathcal{N}} \to \mathbb{R}$ is *monotone* if $f(S) \leq f(T)$ for every two sets $S \subseteq T \subseteq \mathcal{N}$ and *submodular* if $f(S \cup \{u\}) - f(S) \geq f(T \cup \{u\}) - f(T)$ for every two such sets and an element $u \in \mathcal{N} \setminus T$.

[2] This constraint on the set of items that can be selected is equivalent to selecting an independent set of the partition matroid $\mathcal{M}$ defined by the partition $\{P_1, P_2, \ldots, P_m\}$.

It is well known that for a fixed (rather than random) permutation π, the greedy algorithm achieves exactly $1/2$-approximation [18]. We prove the following.

Theorem 1. *Algorithm 1 achieves an approximation ratio of at least 0.5096 for the problem of maximizing a non-negative monotone submodular function subject to a partition matroid constraint.*

Algorithm 1. Random Order Greedy$(f, \mathcal{M})$

---
1 Initialize: $A_0 \leftarrow \varnothing$.
2 Let π be a uniformly random permutation of $[m]$.
3 **for** $i = 1$ **to** m **do**
4 $\quad$ Let u_i be the element $u \in P_{\pi(i)}$ maximizing
$\qquad f(u \mid A_{i-1}) \triangleq f(A_{i-1} \cup \{u\}) - f(A_{i-1})$.
5 $\quad A_i \leftarrow A_{i-1} \cup \{u_i\}$.
6 Return A_m.

Through a standard reduction from SWM, this result yields the same guarantee also on the performance of the greedy algorithm for SWM in the random order model. Thus, the result both generalizes and improves over the previously known 0.5052-approximation [25]. Our analysis is also arguably simpler, giving a direct, clean, and short proof that avoids the use of factor revealing LPs.

It should also be mentioned that the result of Korula *et al.* [25] represents the first combinatorial algorithm for offline SWM achieving a better approximation ratio than $1/2$. Analogously, our result is a combinatorial algorithm achieving a better than $1/2$ approximation ratio for the more general problem of maximizing a non-negative monotone submodular function subject to a partition matroid constraint. We remark that in a recent work Buchbinder *et al.* [5] described a (very different) offline combinatorial algorithm which achieves a better than $1/2$ approximation for the even more general problem of maximizing a non-negative monotone submodular function subject to a general matroid constraint. However, the approximation guarantee achieved in [5] is worse, and the algorithm is more complicated and cannot be implemented in an online model.

The greedy algorithm in the random arrival model is known to be $(1 - 1/e)$-competitive for special cases of SWM [19]. For online SWM it is an open question whether the algorithm achieves this (best possible) ratio. However, for the more general problem of maximizing a monotone submodular function over a partition matroid, the following result answers this question negatively. In fact, the result shows that the approximation ratio obtained by the greedy algorithm is quite far from $1 - 1/e \approx 0.632$.

Theorem 2. *There exist a partition matroid $\mathcal{M}$ and a non-negative monotone submodular function f over the same ground set such that the approximation ratio of Algorithm 1 for the problem of maximizing f subject to the constraint defined by $\mathcal{M}$ is at most $19/33 \leq 0.576$.*

Due to space constraints, the proof of Theorem 2 is deferred to the appendix.

1.2   Our Technique

The proof we describe for Theorem 1 consists of two parts. In the first part (Sect. 3.1), we show that when the greedy algorithm considers sets of the partition in a random order, it gains most of the value of its output set during its first iterations (Lemma 3). For example, after viewing 90% of the sets the algorithm already has 49.5% of the value of the optimal solution, which is 99% of its output guarantee according to the standard analysis. Thus, to prove that the greedy algorithm has a better than $1/2$ approximation ratio, it suffices to show that it gets a non-negligible gain from its last iterations.

In the second part of our analysis (Sect. 3.2), we are able to show that this is indeed the case. Intuitively, in this part of the analysis we view the execution of Algorithm 1 as having three stages defined by two integer values $0 < r \leq r' < m$. The first stage consists of the first r iterations of the algorithm, the second stage consists of the next $r' - r$ iterations and the last stage consists of the remaining $m - r'$ iterations. As explained above, by Lemma 3 we get that if r' is large enough, then $f(A_{r'})$ is already very close to $f(OPT)/2$, where OPT is an optimal solution. We use two steps to prove that $f(A_m)$ is significantly larger than $f(A_{r'})$, and thus, achieves a better than $1/2$ approximation ratio. In the first step (Lemma 4), we use symmetry to argue that there are two independent sets of $\mathcal{M}$ that consist only of elements that Algorithm 1 can pick in its second and third phases, and in addition, the value of their union is large. One of these sets consists of the elements of OPT that are available in the final $m - r$ iterations, and the other set (which we denote by C) is obtained by applying an appropriately chosen function to these elements of OPT. In the second step of the analysis, implemented by Lemma 5, we use the fact that the final $m - r'$ elements of C are a random subset of C to argue that they have a large marginal contribution even with respect to the final solution A_m. Combining this with the observation that these elements represent a possible set of elements that Algorithm 1 could pick during its last stage, we get that the algorithm must have made a significant gain during this stage.

1.3   Additional Related Results

The optimal approximation ratio for the problem of maximizing a monotone submodular function subject to a partition matroid constraint (and its special case SWM) is obtained by an algorithm known as (Measured) Continuous Greedy [8,16]. Unfortunately, this algorithm is problematic from a practical point of view since it is based on a continuous relaxation and is quite slow. As discussed above, our first result can be viewed as an alternative simple combinatorial algorithm for this problem, and thus, it is related to a line of work that aims to find better alternatives for Continuous Greedy [3,6,17,31].

While the problem of maximizing a monotone submodular function subject to a partition matroid was studied almost exclusively in the value oracle model, the view of SWM as an auction has motivated its study also in an alternative model known as the demand oracle model. In this model a strictly better than $(1 - 1/e)$-approximation is known for the problem [13].

Another online model, that can be cast as a special case of the random arrival model and was studied extensively, is the i.i.d. stochastic model. In this model input items arrive i.i.d. according to a known or unknown distribution. In the i.i.d. model with a known distribution improved competitive ratios for special cases of SWM are known [2,15,20,27]. Moreover, for the i.i.d. model with an unknown distribution a $(1 - 1/e)$-competitive algorithm is known for SWM as well as for several of its special cases [10,11,22].

2 Preliminaries

For every two sets $S, T \subseteq \mathcal{N}$ we denote the marginal contribution of adding T to S, with respect to a set function f, by $f(T \mid S) \triangleq f(T \cup S) - f(S)$. For an element $u \in \mathcal{N}$ we use $f(u \mid S)$ as shorthands for $f(\{u\} \mid S)$—note that we have already used this notation previously in Algorithm 1.

Following are two well known facts that we use in the analysis of Algorithm 1.

Lemma 1 (Lemma 2.2 of [12]). *Let $f: 2^{\mathcal{N}} \to \mathbb{R}$ be a submodular function, and let T be an arbitrary set $T \subseteq \mathcal{N}$. For every random set $T_p \subseteq T$ which contains every element of T with probability p (not necessarily independently),*

$$\mathbb{E}[f(T_p)] \geq (1 - p) \cdot f(\varnothing) + p \cdot f(T).$$

Observation 1. *For every sets two $S_1 \subseteq S_2 \subseteq \mathcal{N}$ and an additional set $T \subseteq \mathcal{N}$, it holds that $f(S_1 \mid T) \leq f(S_2 \mid T)$ and $f(T \mid S_1) \geq f(T \mid S_2)$.*

The Submodular Welfare Maximization problem (SWM). In this problem we are given a set $\mathcal{N}$ of m items and a set B of n bidders. Each bidder i has a non-negative monotone submodular utility function $f_i: 2^{\mathcal{N}} \to \mathbb{R}_{\geq 0}$; and the goal is to partition the items among the bidders in a way that maximizes $\sum_{i=1}^{m} f_i(S_i)$ where S_i is the set of items allocated to bidder i.

Maximizing a Monotone Submodular Function over a (Simple) Partition Matroid. In this problem we are given a partition matroid $\mathcal{M}$ over a ground set $\mathcal{N}$ and a non-negative monotone submodular function $f: 2^{\mathcal{N}} \to \mathbb{R}_{\geq 0}$. A partition matroid is defined by a partition of its ground set into non-empty disjoint sets $P_1, P_2, \ldots, P_m$. A set $S \subseteq \mathcal{N}$ is independent in $\mathcal{M}$ if $|S \cap P_i| \leq 1$ for every set P_i, and the goal in this problem is to find a set $S \subseteq \mathcal{N}$ that is independent in $\mathcal{M}$ and maximizes f.

In this work we make the standard assumption that the objective function f can be accessed only through a value oracle, *i.e.*, an oracle that given a subset S returns the value $f(S)$.

A Standard Reduction Between the Above Two Problems. Given an instance of SWM, we construct the following equivalent instance of maximizing a monotone submodular function subject to a partition matroid. For each item $u \in \mathcal{N}$ and bidder $i \in B$, we create an element (u, i) which represents the assignment of u to i. Additionally, we define a partition of these elements by constructing for every item u a set $P_u = \{(u, i) \mid i \in B\}$. Finally, for a subset S of the elements, we define

$$f(S) = \sum_{i \in B} f_i(\{u \in \mathcal{N} \mid (u, i) \in S\}).$$

One can verify that for every independent set S the value of f is equal to the total utility of the bidders given the assignment represented by S; and moreover, f is non-negative, monotone and submodular.

It is important to note that running a greedy algorithm that inspects the partitions in a random order after this reduction is the same as running the greedy algorithm on the original SWM instance in the random arrival model.

Additional Technical Reduction. Our analysis of Algorithm 1 uses two integer parameters $0 < r \le r' < m$. A natural way to choose these parameters is to set them to $r = \alpha m$ and $r' = \beta m$, where α and β are rational numbers. Unfortunately, not for every choice of α, β and m these values are integral. The following reduction, which can be proved using standard techniques, allows us to bypass this technical issue (the proof of this reduction and other omitted proofs can be found in the full version of this paper [4]).

Reduction 1. *For any fixed choice of two rational values $\alpha, \beta \in (0,1)$, one may assume that αm and βm are both integral for the purpose of analyzing the approximation ratio of Algorithm 1.*

3   Analysis of the Approximation Ratio

In this section, we analyze Algorithm 1 and lower bound its approximation ratio. The analysis is split between Sects. 3.1 and 3.2. In Sect. 3.1 we present a basic (and quite standard) analysis of Algorithm 1 which only shows that it is a 1/2-approximation algorithm, but proves along the way some useful properties of the algorithm. In Sect. 3.2 we use these properties to present a more advanced analysis of Algorithm 1 which shows that it is a 0.5096-approximation algorithm (and thus proves Theorem 1).

Let us now define some notation that we use in both parts of the analysis. Let OPT be an optimal solution (*i.e.*, an independent set of $\mathcal{M}$ maximizing f). Note that since f is monotone we may assume, without loss of generality, that OPT is a base of $\mathcal{M}$ (*i.e.*, it includes exactly one element of the set P_i for every $1 \le i \le m$). Additionally, for every set $T \subseteq \mathcal{N}$ we denote by $T^{(i)}$ the subset of T that excludes elements appearing in the first i sets out of $P_1, P_2, \ldots, P_m$ when these sets are ordered according to the permutation π. More formally,

$$T^{(i)} = T \setminus \bigcup_{j=1}^{i} P_{\pi(j)} = T \cap \bigcup_{j=i+1}^{m} P_{\pi(j)}.$$

Since π is a uniformly random permutation and OPT contains exactly one element of each set P_i (due to our assumption that it is a base of $\mathcal{M}$), we get the following observation as an immediate consequence.

Observation 2. *For every $0 \le i \le m$, $OPT^{(i)}$ is a uniformly random subset of OPT of size $m - i$.*

3.1 Basic Analysis

In this section, we present a basic analysis of Algorithm 1. Following is the central lemma of this analysis which shows that the expression $f(A_i) + f(S \cup A_i \cup T^{(i)})$ is a non-decreasing function of i for every pair of set $S \subseteq \mathcal{N}$ and base T of $\mathcal{M}$ (recall that A_i is the set constructed by Algorithm 1 during its i-th iteration). It is important to note that this lemma holds deterministically, *i.e.*, it holds for **every** given permutation π.

Lemma 2. *For every subset $S \subseteq \mathcal{N}$, base T of $\mathcal{M}$ and $1 \le i \le m$, $f(A_i) + f(S \cup A_i \cup T^{(i)}) \ge f(A_{i-1}) + f(S \cup A_{i-1} \cup T^{(i-1)})$.*

Proof. Observe that

$$f(A_i) - f(A_{i-1}) = f(u_i \mid A_{i-1}) \ge f(T \cap P_{\pi(i)} \mid A_{i-1})$$
$$\ge f(T \cap P_{\pi(i)} \mid S \cup A_{i-1} \cup T^{(i)}) = f(S \cup A_{i-1} \cup T^{(i-1)}) - f(S \cup A_{i-1} \cup T^{(i)})$$
$$\ge f(S \cup A_{i-1} \cup T^{(i-1)}) - f(S \cup A_i \cup T^{(i)}),$$

where the first inequality follows from the greedy choice of the algorithm, the second inequality holds due to Observation 1 and the final inequality follows from the monotonicity of f. □

The following is an immediate corollary of the last lemma. Note that, like the lemma, it is deterministic and, thus, holds for **every** permutation π.

Corollary 1. *For every subset $S \subseteq \mathcal{N}$, base T of $\mathcal{M}$ and $0 \le i \le m$, $f(A_m) + f(S \cup A_m) \ge f(A_i) + f(S \cup A_i \cup T^{(i)}) \ge f(S \cup T)$.*

Proof. Since $f(A_i) + f(S \cup A_i \cup T^{(i)})$ is a non-decreasing function of i by Lemma 2,

$$f(A_m) + f(S \cup A_m \cup T^{(m)}) \ge f(A_i) + f(S \cup A_i \cup T^{(i)}) \ge f(A_0) + f(S \cup A_0 \cup T^{(0)}).$$

The corollary now follows by recalling that $A_0 = \varnothing$, observing that $f(A_0) \ge 0$ since f is non-negative and noting that by definition $T^{(m)} = \varnothing$ and $T^{(0)} = T$. □

By choosing $S = \varnothing$ and $T = OPT$, the last corollary yields $f(A_m) \ge 1/2 \cdot f(OPT)$, which already proves that Algorithm 1 is a $1/2$-approximation algorithm as promised. The following lemma strengthens this result by showing a lower bound on the value of $f(A_i)$ for every $0 \le i \le m$. Note that this lower bound, unlike the previous one, holds only in expectation over the random choice of the permutation π. Let $g(x) \triangleq x - x^2/2$.

Lemma 3. *For every $0 \le i \le m$, $\mathbb{E}[f(A_i)] \ge g(i/m) \cdot f(OPT)$.*

We omit the proof due to space constraints (it can be found in the full version [4]). Note that Lemma 3 is very similar to known results (see, *e.g.*, [25]).

3.2   Breaking 1/2: An Improved Analysis of Algorithm 1

In this section, we use the properties of Algorithm 1 proved in the previous section to derive a better than $1/2$ lower bound on its approximation ratio and prove Theorem 1. As explained in Sect. 1.2, we view here an execution of Algorithm 1 as consisting of three stages, where the places of transition between the stages are defined by two integer parameters $0 < r \leq r' < m$ whose values are set later in this section to $0.4\,m$ and $0.76\,m$, respectively. The first lemma that we present (Lemma 4) uses a symmetry argument to prove that there are two (not necessarily distinct) independent sets of $\mathcal{M}$ that consist only of elements that Algorithm 1 can pick in its second and third stages (the final $m - r$ iterations), and in addition, the value of their union is large. One of these sets is $OPT^{(r)}$, and the other set is obtained by applying to $OPT^{(r)}$ an appropriately chosen function h. Interestingly, the guarantee of Lemma 4 is particularly strong when the algorithm makes little progress during the second and third stages (*i.e.*, $f(A_m \mid A_{m-r})$ is small), which intuitively is the case in which the basic analysis (from Sect. 3.1) fails to guarantee more than $1/2$-approximation.

Let c be the true (unknown) approximation ratio of Algorithm 1.

Lemma 4. *There exists a function* $h\colon 2^{\mathcal{N}} \to 2^{\mathcal{N}}$ *such that*

(a) for every $1 \leq i \leq m$ *and set* $S \subseteq \mathcal{N}$, $|P_i \cap h(S)| = |P_i \cap S|$.
(b) $\mathbb{E}[f(h(OPT^{(r)}) \cup OPT^{(r)})] \geq f(OPT) - c^{-1} \cdot \mathbb{E}[f(A_m \mid A_{m-r})]$.

Proof. Given Part (b) of the lemma, it is natural to define $h(S)$, for every set $S \subseteq \mathcal{N}$, as the set T maximizing $f(T \cup S)$ among all the sets obeying Part (a) of the lemma (where ties are broken in an arbitrary way). In the rest of the proof we show that this function indeed obeys Part (b).

Observe that

$$
\begin{aligned}
f(A_{m-r} \cup (OPT \setminus OPT^{(m-r)})) &= f(OPT \setminus OPT^{(m-r)} \mid A_{m-r}) + f(A_{m-r}) \\
&\geq f(OPT \setminus OPT^{(m-r)} \mid A_{m-r} \cup OPT^{(m-r)}) + f(A_{m-r}) \\
&= f(A_{m-r} \cup OPT) - f(OPT^{(m-r)} \mid A_{m-r}) \\
&\geq f(OPT) - f(OPT^{(m-r)} \mid A_{m-r}),
\end{aligned}
$$

where first inequality follows from Observation 1 and the second follows by the monotonicity of f. We now note that the last r iterations of Algorithm 1 can be viewed as a standalone execution of this algorithm on the partition matroid defined by the sets $P_{m-r+1}, \ldots, P_m$ and the objective function $f(\cdot \mid A_{m-r})$. Thus, by the definition of c, the expected value of $f(OPT^{(m-r)} \mid A_{m-r})$ is at most $c^{-1} \cdot \mathbb{E}[f(A_m \setminus A_{m-r} \mid A_{m-r})] = c^{-1} \cdot \mathbb{E}[f(A_m \mid A_{m-r})]$. Combining this with the previous inequality, we get

$$
\begin{aligned}
\mathbb{E}[f(h(OPT \setminus OPT^{(m-r)}) &\cup (OPT \setminus OPT^{(m-r)}))] \\
&\geq \mathbb{E}[f(A_{m-r} \cup (OPT \setminus OPT^{(m-r)}))] \geq \mathbb{E}[f(OPT) - f(OPT^{(m-r)} \mid A_{m-r})] \\
&\geq f(OPT) - c^{-1} \cdot \mathbb{E}[f(A_m \mid A_{m-r})],
\end{aligned}
$$

where the first inequality holds due to the definition of h since A_{m-r} obeys Part (a) of the lemma (for $S = OPT \setminus OPT^{(m-r)}$).

To prove the lemma it remains to observe that by Observation 2 the random sets $OPT \setminus OPT^{(m-r)}$ and $OPT^{(r)}$ have the same distribution, which implies that $f(h(OPT \setminus OPT^{(m-r)}) \cup (OPT \setminus OPT^{(m-r)}))$ and $f(h(OPT^{(r)}) \cup OPT^{(r)})$ have the same expectation.   □

Let us denote $C = h(OPT^{(r)})$. Note that C is a random set since $OPT^{(r)}$ is. The following lemma uses the properties of C proved by Lemma 4 to show that Algorithm 1 must make a significant gain during its third stage. Intuitively, the guarantee of this lemma is useful because the basic analysis implies that when Algorithm 1 does not get much more than $1/2$-approximation, its solution gains most of its value early in the execution. Thus, in this case both A_r and A_{m-r} should have a significant fraction of the value of the output set A_m, and therefore, the positive terms on the right hand side of the guarantee of the lemma can counter the negative term.

Lemma 5. *Let* $p = \frac{m-r'}{m-r}$, *then*

$$\mathbb{E}[f(A_m \mid A_{r'})] \geq f(OPT) - (2+p/c) \cdot \mathbb{E}[f(A_m)] + p \cdot \mathbb{E}[f(A_r)] + (p/c) \cdot \mathbb{E}[f(A_{m-r})].$$

Proof. Observe that

$$f(A_m \mid A_{r'}) \geq f(C^{(r')} \mid A_m \cup OPT^{(r)}) = f(A_m \cup C^{(r')} \mid A_r \cup OPT^{(r)}) -$$
$$f(A_m \mid A_r \cup OPT^{(r)}) \geq f(C^{(r')} \mid A_r \cup OPT^{(r)}) - f(A_m \mid A_r \cup OPT^{(r)}),$$

where the first inequality follows by plugging $i = r'$, $T = A_r \cup C$ and $S = A_m \cup OPT^{(r)}$ in the first inequality of Corollary 1, and the second inequality follows by Observation 1.

Similar to what we do in the proof of Lemma 3, let us now denote by π_r an arbitrary injective function from $\{1, \dots, r\}$ to $\{1, \dots, m\}$ and by $\mathcal{E}(\pi_r)$ the event that $\pi(j) = \pi_r(j)$ for every $1 \leq j \leq r$. Observe that conditioned on $\mathcal{E}(\pi_r)$ the set $OPT^{(r)}$ is deterministic, and thus so is the set C which is obtained from $OPT^{(r)}$ by the application of a deterministic function; but $C^{(r')}$ remains a random set that contains every element of C with probability p. Hence, by Lemma 1, conditioned on $\mathcal{E}(\pi_r)$, we get

$$\mathbb{E}[f(C^{(r')} \mid A_r \cup OPT^{(r)})] \geq p \cdot f(C \mid A_r \cup OPT^{(r)}).$$

Taking now expectation over all the possible events $\mathcal{E}(\pi_r)$, and combining with the previous inequality, we get

$$\mathbb{E}[f(A_m \mid A_{r'})] \geq p \cdot \mathbb{E}[f(C \mid A_r \cup OPT^{(r)})] - \mathbb{E}[f(A_m \mid A_r \cup OPT^{(r)})]$$
$$= p \cdot \mathbb{E}[f(C \cup A_r \cup OPT^{(r)})] + (1-p) \cdot \mathbb{E}[f(A_r \cup OPT^{(r)})] - \mathbb{E}[f(A_m \cup OPT^{(r)})]$$
$$\geq p \cdot \mathbb{E}[f(C \cup OPT^{(r)})] + (1-p) \cdot \mathbb{E}[f(A_r \cup OPT^{(r)})] - \mathbb{E}[f(A_m \cup OPT^{(r)})],$$

where the second inequality holds due to the monotonicity of f. We now need to bound all the terms on the right hand side of the last inequality. The first term is lower bounded by $p \cdot f(OPT) - (p/c) \cdot \mathbb{E}[f(A_m \mid A_{m-r})]$ due to Lemma 4. A lower bound of $f(OPT) - f(A_r)$ on the expression $f(A_r \cup OPT^{(r)})$ follows from the second inequality of Corollary 1 by setting $T = OPT$ and $S = \varnothing$. Finally, an upper bound of $2f(A_m) - f(A_r)$ on the expression $f(A_m \cup OPT^{(r)})$ follows from the first inequality of the same corollary by setting $T = OPT$ and $S = A_m$. Plugging all these bounds into the previous inequality yields

$$\mathbb{E}[f(A_m \mid A_{r'})] \geq p \cdot f(OPT) - (p/c) \cdot \mathbb{E}[f(A_m \mid A_{m-r})] +$$
$$(1-p) \cdot \mathbb{E}[f(OPT) - f(A_r)] - \mathbb{E}[2f(A_m) - f(A_r)]$$
$$= f(OPT) - (2+p/c) \cdot \mathbb{E}[f(A_m)] + p \cdot \mathbb{E}[f(A_r)] + (p/c) \cdot \mathbb{E}[f(A_{m-r})]. \quad \square$$

We are now ready to prove Theorem 1.

Proof (Proof of Theorem 1). Let $q = r/m$. By Lemma 5,

$$\mathbb{E}[f(A_m)] = \mathbb{E}[f(A_m \mid A_{r'})] + \mathbb{E}[f(A_{r'})] \geq f(OPT) - (2+p/c) \cdot \mathbb{E}[f(A_m)] +$$
$$p \cdot \mathbb{E}[f(A_r)] + (p/c) \cdot \mathbb{E}[f(A_{m-r})] + \mathbb{E}[f(A_{r'})].$$

Rearranging this inequality, and using the lower bound on $\mathbb{E}[f(A_i)]$ given by Lemma 3, we get

$$(3+p/c) \cdot \mathbb{E}[f(A_m)] \geq [1 + p \cdot g(q) + (p/c) \cdot g(1-q) + g(1-p+pq)] \cdot f(OPT)$$
$$= [1 + pq(1-q/2) + (p/c)(1-q^2)/2 + (1-p^2+2p^2q-p^2q^2)/2] \cdot f(OPT)$$
$$= \frac{1}{2}[3 + pq(2-q) + pc^{-1}(1-q^2) - p^2(1-q)^2] \cdot f(OPT).$$

Thus, the approximation ratio of Algorithm 1 is at least

$$\frac{3 + pq(2-q) + pc^{-1}(1-q^2) - p^2(1-q)^2}{6 + 2pc^{-1}}.$$

Since c is the true approximation ratio of this algorithm by definition, we get

$$c \geq \frac{3 + pq(2-q) + pc^{-1}(1-q^2) - p^2(1-q)^2}{6 + 2pc^{-1}}.$$

We now choose $p = q = 0.4$. Notice that these values for p and q can be achieved by setting $r = 0.4\,m$ and $r' = 0.76\,m$, and moreover, we can assume that this is a valid choice for r and r' by Reduction 1. Plugging these values of p and q into the last inequality and simplifying, we get $6c^2 - 2.3984c - 0.336 \geq 0$. One can verify that all the positive solutions for this inequality are larger than 0.5096, which completes the proof of the theorem. $\quad \square$

Acknowledgment. We thank Nitish Korula, Vahab S. Mirrokni and Morteza Zadimoghaddam for sharing with us the full version of their paper [25]. The research of Niv Buchbinder was supported by ISF grant 1585/15 and BSF grant 2014414. The research of Moran Feldman and Mohit Garg was supported in part by ISF grant 1357/16. Yuval Filmus is a Taub Fellow—supported by the Taub Foundations. His research was funded by ISF grant 1337/16.

A   Upper Bounding the Approximation Ratio

In this section, we prove Theorem 2. We first give in Sect. A.1 a simple proof of a weaker form of the theorem with a bound of $7/12 \approx 0.583$ instead of $19/33 \approx 0.576$. The proof of the theorem as stated appears in Sect. A.2.

A.1   A Simple Weaker Bound

Let us construct a partition matroid over a ground set $\mathcal{N}$ consisting of twelve elements and a non-negative monotone submodular function $f \colon 2^{\mathcal{N}} \to \mathbb{R}_{\geq 0}$ over the same ground set. The partition matroid is defined by a partition of the ground set into three sets: $P_x = \{x_1, x_2, x_3, x_4\}$, $P_y = \{y_1, y_2, y_3, y_4\}$ and $P_z = \{z_1, z_2, z_3, z_4\}$. To define the function f, we view each element of $\mathcal{N}$ as a subset of an underlying universe $\mathcal{U}$ consisting of 12 elements: $\mathcal{U} = \{\alpha_1, \cdots, \alpha_4, \beta_1, \cdots, \beta_4, \gamma_1, \cdots, \gamma_4\}$. The function f is then given as the coverage function $f(S) = |\bigcup_{u \in S} u|$ (coverage functions are known to be non-negative, monotone and submodular). The following table completes the definition of f by specifying the exact subset of $\mathcal{U}$ represented by each element of $\mathcal{N}$:

Elements of P_x	Elements of P_y	Elements of P_z
$x_1 = \{\alpha_1, \alpha_2, \alpha_3, \alpha_4\}$	$y_1 = \{\beta_1, \beta_2, \beta_3, \beta_4\}$	$z_1 = \{\gamma_1, \gamma_2, \gamma_3, \gamma_4\}$
$x_2 = \{\beta_1, \beta_2, \gamma_1, \gamma_2\}$	$y_2 = \{\alpha_1, \alpha_2, \gamma_1, \gamma_2\}$	$z_2 = \{\alpha_1, \alpha_2, \beta_1, \beta_2\}$
$x_3 = \{\beta_1, \gamma_3\}$	$y_3 = \{\alpha_1, \gamma_3\}$	$z_3 = \{\alpha_1, \beta_3\}$
$x_4 = \{\beta_3, \gamma_1\}$	$y_4 = \{\alpha_3, \gamma_1\}$	$z_4 = \{\alpha_3, \beta_1\}$

It is easy to verify that the optimum solution for this instance (i.e., the independent set of $\mathcal{M}$ maximizing f) is the set $\{x_1, y_1, z_1\}$ whose value is 12. To analyze the performance of Algorithm 1 on this instance, we must set a tie breaking rule. Here we assume that the algorithm always breaks ties in favor of the element with the higher index, but it should be noted that a small perturbation of the values of f can be used to make the analysis independent of the tie breaking rule used (at the cost of weakening the impossibility proved by an additive ε term for an arbitrary small constant $\varepsilon > 0$).

Now, consider the case that the set P_x arrives first, followed by P_y and finally P_z. One can check that in this case the greedy algorithm picks the elements x_2, y_3 and z_4 (in this order), and that their marginal contributions upon arrival are $4, 2$ and 1, respectively. Similarly, it can be checked that the exact same marginal contributions also appear in every one of the other five possible arrival orders of the sets P_x, P_y, P_z. Thus, regardless of the arrival order, the approximation ratio achieved by Algorithm 1 for the above instance is only $(4+2+1)/12 = 7/12$.

Remark: It should be noted that by combining multiple independent copies of the above described instance, one can get an arbitrarily large instance for which the approximation ratio of Algorithm 1 is only $7/12$. This rules out the possibility that the approximation ratio of Algorithm 1 approaches $1 - 1/e$ for large enough instances.

A.2   Stronger Upper Bound

We now get to proving Theorem 2 with the stated bound of 19/33. The proof is similar to the one given in Sect. A.1, but the set system we need to use is more complicated.

We construct a partition matroid over a ground set $\mathcal{N}$ consisting of 32 elements and a non-negative monotone submodular function $f: 2^{\mathcal{N}} \to \mathbb{R}_{\geq 0}$ over the same ground set. The ground set of the partition matroid consists of four types of elements:

- o_1, o_2, o_3, o_4
- x_1, x_2, x_3, x_4
- y_{ij} for distinct $i, j \in \{1, 2, 3, 4\}$
- z_{ijk} for distinct $i, j, k \in \{1, 2, 3, 4\}$, where we ignore the order between the first two indices (i.e., z_{ijk} and z_{jik} are two names for the same element)

The partition matroid is defined by a partition of the ground set into four parts P_1, P_2, P_3, P_4, where part P_i is comprised of the 8 elements of the forms $o_i, x_i, y_{ji}, z_{jki}$.

As in Sect. A, the function f is a coverage function, but this time a weighted one. The universe $\mathcal{U}$ used to define this function consists of 28 elements:

$$\mathcal{U} = \{a_i, b_i, c_i, d_i, e_i, f_i, g_i : i \in \{1, 2, 3, 4\}\}.$$

The weights of the elements in this universe are given by the following function $w: \mathcal{U} \to \mathbb{R}_{\geq 0}$:

v	a_i	b_i	c_i	d_i	e_i	f_i	g_i
$w(v)$	14	14	8	5	4	7	14

We extend w to subsets of $\mathcal{U}$ by defining $w(V) = \sum_{v \in V} w(v)$. The function f is then the weighted coverage function given by the formula $f(S) = w(\bigcup_{v \in S} v)$. Like coverage functions, weighted coverage functions (with non-negative weights) are also known to be non-negative, monotone and submodular.

In order to complete the definition of f, we need to specify the sets that the elements of $\mathcal{N}$ correspond to:

$$o_i = \{a_i, b_i, c_i, d_i, e_i, f_i, g_i\}$$
$$x_i = \{b_j, c_j : j \neq i\}$$
$$y_{ij} = \{c_i, e_j\} \cup \{d_k, e_k, f_k : k \neq i, j\}$$
$$z_{ijk} = \{f_i, f_j, g_\ell\}, \text{ where } \ell \text{ is the unique element of } \{1, 2, 3, 4\} \setminus \{i, j, k\}$$

It is easy to verify that the optimal solution for this instance is $\{o_1, o_2, o_3, o_4\}$ and that it achieves a total weight of 264. In contrast, by choosing an appropriate tie breaking rule, we can cause Algorithm 1 to act in the following way when presented with the parts P_i, P_j, P_k, P_ℓ: choose $x_i, y_{ij}, z_{ijk}, o_\ell$ (which has total weight 152, and thus, yields an approximation ratio of $152/264 = 19/33$).

References

1. Aggarwal, G., Goel, G., Karande, C., Mehta, A.: Online vertex-weighted bipartite matching and single-bid budgeted allocations. In: SODA, pp. 1253–1264 (2011)
2. Alaei, S., Hajiaghayi, M., Liaghat, V.: Online prophet-inequality matching with applications to ad allocation. In: EC, pp. 18–35 (2012)
3. Badanidiyuru, A., Vondrák, J.: Fast algorithms for maximizing submodular functions. In: SODA, pp. 1497–1514 (2014)
4. Buchbinder, N., Feldman, M., Filmus, Y., Garg, M.: Online submodular maximization: Beating 1/2 made simple. CoRR abs/1807.05529 (2018). http://arxiv.org/abs/1807.05529
5. Buchbinder, N., Feldman, M., Garg, M.: Deterministic $(1/2 + \varepsilon)$-approximation for submodular maximization over a matroid. In: SODA, pp. 241–254 (2019)
6. Buchbinder, N., Feldman, M., Schwartz, R.: Comparing apples and oranges: query trade-off in submodular maximization. Math. Oper. Res. **42**(2), 308–329 (2017)
7. Buchbinder, N., Jain, K., Naor, J.S.: Online primal-dual algorithms for maximizing ad-auctions revenue. In: Arge, L., Hoffmann, M., Welzl, E. (eds.) ESA 2007. LNCS, vol. 4698, pp. 253–264. Springer, Heidelberg (2007). https://doi.org/10.1007/978-3-540-75520-3_24
8. Călinescu, G., Chekuri, C., Pál, M., Vondrák, J.: Maximizing a monotone submodular function subject to a matroid constraint. SIAM J. Comput. **40**(6), 1740–1766 (2011)
9. Devanur, N.R., Hayes, T.P.: The adwords problem: online keyword matching with budgeted bidders under random permutations. In: EC, pp. 71–78 (2009)
10. Devanur, N.R., Jain, K., Sivan, B., Wilkens, C.A.: Near optimal online algorithms and fast approximation algorithms for resource allocation problems. In: EC, pp. 29–38 (2011)
11. Devanur, N.R., Sivan, B., Azar, Y.: Asymptotically optimal algorithm for stochastic adwords. In: EC, pp. 388–404 (2012)
12. Feige, U., Mirrokni, V.S., Vondrák, J.: Maximizing non-monotone submodular functions. SIAM J. Comput. **40**(4), 1133–1153 (2011)
13. Feige, U., Vondrák, J.: The submodular welfare problem with demand queries. Theory Comput. **6**(1), 247–290 (2010)
14. Feldman, J., Korula, N., Mirrokni, V., Muthukrishnan, S., Pál, M.: Online ad assignment with free disposal. In: Leonardi, S. (ed.) WINE 2009. LNCS, vol. 5929, pp. 374–385. Springer, Heidelberg (2009). https://doi.org/10.1007/978-3-642-10841-9_34
15. Feldman, J., Mehta, A., Mirrokni, V.S., Muthukrishnan, S.: Online stochastic matching: beating 1-1/e. In: FOCS, pp. 117–126 (2009)
16. Feldman, M., Naor, J., Schwartz, R.: A unified continuous greedy algorithm for submodular maximization. In: FOCS, pp. 570–579 (2011)
17. Filmus, Y., Ward, J.: Monotone submodular maximization over a matroid via non-oblivious local search. SIAM J. Comput. **43**(2), 514–542 (2014)
18. Fisher, M.L., Nemhauser, G.L., Wolsey, L.A.: An analysis of approximations for maximizing submodular set functions - II. Math. Program. Study **8**, 73–87 (1978)
19. Goel, G., Mehta, A.: Online budgeted matching in random input models with applications to adwords. In: SODA, pp. 982–991 (2008)
20. Haeupler, B., Mirrokni, V.S., Zadimoghaddam, M.: Online stochastic weighted matching: improved approximation algorithms. In: Chen, N., Elkind, E., Koutsoupias, E. (eds.) WINE 2011. LNCS, vol. 7090, pp. 170–181. Springer, Heidelberg (2011). https://doi.org/10.1007/978-3-642-25510-6_15

21. Kalyanasundaram, B., Pruhs, K.: An optimal deterministic algorithm for online b-matching. Theor. Comput. Sci. **233**(1–2), 319–325 (2000)
22. Kapralov, M., Post, I., Vondrák, J.: Online submodular welfare maximization: greedy is optimal. In: SODA, pp. 1216–1225 (2013)
23. Karande, C., Mehta, A., Tripathi, P.: Online bipartite matching with unknown distributions. In: STOC, pp. 587–596 (2011)
24. Karp, R.M., Vazirani, U.V., Vazirani, V.V.: An optimal algorithm for on-line bipartite matching. In: STOC, pp. 352–358 (1990)
25. Korula, N., Mirrokni, V.S., Zadimoghaddam, M.: Online submodular welfare maximization: greedy beats 1/2 in random order. In: STOC, pp. 889–898 (2015)
26. Mahdian, M., Yan, Q.: Online bipartite matching with random arrivals: an approach based on strongly factor-revealing LPs. In: STOC, pp. 597–606 (2011)
27. Manshadi, V.H., Gharan, S.O., Saberi, A.: Online stochastic matching: online actions based on offline statistics. Math. Oper. Res. **37**(4), 559–573 (2012)
28. Mehta, A.: Online matching and ad allocation. Found. Trends Theor. Comput. Sci. **8**(4), 265–368 (2013)
29. Mehta, A., Saberi, A., Vazirani, U.V., Vazirani, V.V.: Adwords and generalized online matching. J. ACM **54**(5), 22 (2007)
30. Mirrokni, V.S., Schapira, M., Vondrák, J.: Tight information-theoretic lower bounds for welfare maximization in combinatorial auctions. In: EC, pp. 70–77 (2008)
31. Mirzasoleiman, B., Badanidiyuru, A., Karbasi, A., Vondrák, J., Krause, A.: Lazier than lazy greedy. In: AAAI, pp. 1812–1818 (2015)
32. Zadimoghaddam, M.: Online weighted matching: beating the 1/2 barrier. CoRR abs/1704.05384 (2017)

Improving the Integrality Gap
for Multiway Cut

Kristóf Bérczi[1](✉) (ID), Karthekeyan Chandrasekaran[2], Tamás Király[1] (ID), and Vivek Madan[2]

[1] MTA-ELTE Egerváry Research Group, Department of Operations Research, Eötvös Loránd University, Budapest, Hungary
{berkri,tkiraly}@cs.elte.hu
[2] University of Illinois, Urbana-Champaign, USA
{karthe,vmadan2}@illinois.edu

Abstract. In the multiway cut problem, we are given an undirected graph with non-negative edge weights and a collection of k terminal nodes, and the goal is to partition the node set of the graph into k non-empty parts each containing exactly one terminal so that the total weight of the edges crossing the partition is minimized. The multiway cut problem for $k \geq 3$ is APX-hard. For arbitrary k, the best-known approximation factor is 1.2965 due to Sharma and Vondrák [12] while the best known inapproximability result due to Angelidakis, Makarychev and Manurangsi [1] rules out efficient algorithms to achieve an approximation factor that is less than 1.2. In this work, we improve on the lower bound to 1.20016 by constructing an integrality gap instance for the CKR relaxation.

A technical challenge in improving the gap has been the lack of geometric tools to understand higher-dimensional simplices. Our instance is a non-trivial 3-dimensional instance that overcomes this technical challenge. We analyze the gap of the instance by viewing it as a convex combination of 2-dimensional instances and a uniform 3-dimensional instance. We believe that this technique could be exploited further to construct instances with larger integrality gap. One of the ingredients of our proof technique is a generalization of a result on *Sperner admissible labelings* due to Mirzakhani and Vondrák [11] that might be of independent combinatorial interest.

Keywords: Combinatorial optimization · Multiway cut · Integrality gap · Approximation

1 Introduction

In the multiway cut problem, we are given an undirected graph with non-negative edge weights and a collection of k terminal nodes and the goal is to find a minimum weight subset of edges to delete so that the k input terminals cannot reach each other. Equivalently, the goal is to find a partition of the vertex set into

© Springer Nature Switzerland AG 2019
A. Lodi and V. Nagarajan (Eds.): IPCO 2019, LNCS 11480, pp. 115–127, 2019.
https://doi.org/10.1007/978-3-030-17953-3_9

k sets with each set containing exactly one terminal such that the total weight of the edge boundaries of the sets is minimized. For convenience, we will use k-*way cut* to denote this problem when we would like to highlight the dependence on k and *multiway cut* to denote this problem when k grows with the size of the input graph. The 2-way cut problem is the classic minimum $\{s, t\}$-cut problem which is solvable in polynomial time. For $k \geq 3$, Dahlhaus, Johnson, Papadimitriou, Seymour and Yannakakis [7] showed that the k-way cut problem is APX-hard and gave a $(2 - 2/k)$-approximation. Owing to its applications in partitioning and clustering, k-way cut has been an intensely investigated problem in the algorithms literature. Several novel rounding techniques in the approximation literature were discovered to address the approximability of this problem.

The known approximability as well as inapproximability results are based on a linear programming relaxation, popularly known as the CKR relaxation in honor of the authors—Călinescu, Karloff and Rabani—who introduced it [5]. The CKR relaxation takes a geometric perspective of the problem. For a graph $G = (V, E)$ with edge weights $w : E \to \mathbb{R}_+$ and terminals $t_1, \ldots, t_k$, the CKR relaxation is given by

$$\min \quad \frac{1}{2} \sum_{e=\{u,v\}\in E} w(e)\|x^u - x^v\|_1$$

$$\text{s.t. } x^u \in \Delta_k \qquad\qquad \forall\, u \in V,$$
$$x^{t_i} = e^i \qquad\qquad \forall\, i \in [k],$$

where $\Delta_k := \{(x_1, \ldots, x_k) \in [0,1]^k : \sum_{i=1}^k x_i = 1\}$ is the $(k-1)$-dimensional simplex, $e^i \in \{0,1\}^k$ is the extreme point of the simplex along the i-th coordinate axis, i.e., $e^i_j = 1$ if and only if $j = i$, and $[k]$ denotes the set $\{1, \ldots, k\}$.

Călinescu, Karloff and Rabani designed a rounding scheme for the relaxation which led to a $(3/2 - 1/k)$-approximation thus improving on the $(2 - 2/k)$-approximation by Dahlhaus et al. For 3-way cut, Cheung, Cunningham and Tang [6] as well as Karger, Klein, Stein, Thorup and Young [9] designed alternative rounding schemes that led to a $12/11$-approximation factor and also exhibited matching integrality gap instances. We recall that the integrality gap of an instance to the LP is the ratio between the integral optimum value and the LP optimum value. Determining the exact integrality gap of the CKR relaxation for $k \geq 4$ has been an intriguing open question. After the results by Karger et al. and Cunningham et al., a rich variety of rounding techniques were developed to improve the approximation factor of k-way cut for $k \geq 4$ [3,4,12]. The best-known approximation factor for multiway cut is 1.2965 due to Sharma and Vondrák [12].

On the hardness of approximation side, Manokaran, Naor, Raghavendra and Schwartz [10] showed that the hardness of approximation for k-way cut is at least the integrality gap of the CKR relaxation assuming the Unique Games Conjecture (UGC). More precisely, if the integrality gap of the CKR relaxation for k-way cut is τ_k, then it is UGC-hard to approximate k-way cut within a factor of $\tau_k - \epsilon$ for every constant $\epsilon > 0$. As an immediate consequence of this result, we know that the 12/11-approximation factor for 3-way cut is tight. For k-way cut, Freund and Karloff [8] constructed an instance showing an integrality gap of $8/(7 + (1/(k-1)))$. This was the best known integrality gap until last year when Angelidakis, Makarychev and Manurangsi [1] gave a remarkably simple construction showing an integrality gap of $6/(5 + (1/(k-1)))$ for k-way cut. In particular, this gives an integrality gap of 1.2 for multiway cut.

We note that the known upper and lower bounds on the approximation factor for multiway cut match only up to the first decimal digit and thus the approximability of this problem is far from resolved. Indeed Angelidakis, Makarychev and Manurangsi raise the question of whether the lower bound can be improved. In this work, we improve on the lower bound by constructing an instance with integrality gap 1.20016.

Theorem 1. *For every constant $\epsilon > 0$, there exists an instance of multiway cut such that the integrality gap of the CKR relaxation for that instance is at least $1.20016 - \epsilon$.*

The above result in conjunction with the result of Manokaran et al. immediately implies that multiway cut is UGC-hard to approximate within a factor of $1.20016 - \epsilon$ for every constant $\epsilon > 0$.

One of the ingredients of our technique underlying the proof of Theorem 1 is a new generalization of a result on *Sperner admissible labelings* due to Mirzakhani and Vondrák [11] that might be of independent combinatorial interest (see Theorem 6).

2   Background and Result

Before outlining our techniques, we briefly summarize the background literature that we build upon to construct our instance. We rely on two significant results from the literature. In the context of the k-way cut problem, a *cut* is a function $P : \Delta_k \to [k+1]$ such that $P(e^i) = i$ for all $i \in [k]$, where we use the notation $[k] := \{1, 2, \ldots, k\}$. The use of $k+1$ labels as opposed to k labels to describe a cut is slightly non-standard, but is useful for reasons that will become clear later on. The approximation ratio $\tau_k(\mathcal{P})$ of a distribution $\mathcal{P}$ over cuts is given by its *maximum density*:

$$\tau_k(\mathcal{P}) := \sup_{x,y \in \Delta_k, x \neq y} \frac{\Pr_{P \sim \mathcal{P}}(P(x) \neq P(y))}{(1/2)\|x - y\|_1}.$$

Karger et al. [9] defined $\tau_k^* := \inf_{\mathcal{P}} \tau_k(\mathcal{P})$, and moreover showed that there exists $\mathcal{P}$ that achieves the infimum. Hence, $\tau_k^* = \min_{\mathcal{P}} \tau_k(\mathcal{P})$. With this definition of τ_k^*, Karger et al. [9] showed that for every $\epsilon > 0$, there is an instance of multiway cut with k terminals for which the integrality gap of the CKR relaxation is at least $\tau_k^* - \epsilon$. Thus, Karger et al.'s result reduced the problem of constructing an integrality gap instance for multiway cut to proving a lower bound on τ_k^*.

Next, Angelidakis, Makarychev and Manurangsi [1] reduced the problem of lower bounding τ_k^* further by showing that it is sufficient to restrict our attention to *non-opposite cuts* as opposed to all cuts. A cut P is a *non-opposite cut* if $P(x) \in \text{Support}(x) \cup \{k+1\}$ for every $x \in \Delta_k$ where, we use the notation $\text{Support}(x) := \{i \in [k] \mid x_i \neq 0\}$. Let $\Delta_{k,n} := \Delta_k \cap ((1/n)\mathbb{Z})^k$ where $(1/n)\mathbb{Z} := \{i/n \mid i \in \mathbb{Z}\}$. For a distribution $\mathcal{P}$ over cuts, let

$$\tau_{k,n}(\mathcal{P}) := \max_{x,y \in \Delta_{k,n}, x \neq y} \frac{\text{Pr}_{P \sim \mathcal{P}}(P(x) \neq P(y))}{(1/2)\|x - y\|_1}, \text{ and}$$

$$\tilde{\tau}_{k,n}^* := \min\{\tau_{k,n}(\mathcal{P}) : \mathcal{P} \text{ is a distribution over non-opposite cuts}\}.$$

Angelidakis, Makarychev and Manurangsi showed that $\tilde{\tau}_{k,n}^* - \tau_K^* = O(kn/(K - k))$ for all $K > k$. Thus, in order to lower bound τ_K^*, it suffices to lower bound $\tilde{\tau}_{k,n}^*$. That is, it suffices to construct an instance that has *large integrality gap against non-opposite cuts*.

As a central contribution, Angelidakis, Makarychev and Manurangsi constructed an instance showing that $\tilde{\tau}_{3,n}^* \geq 1.2 - O(1/n)$. Now, by setting $n = \Theta(\sqrt{K})$, we see that τ_K^* is at least $1.2 - O(1/\sqrt{K})$. Furthermore, they also showed that their lower bound on $\tilde{\tau}_{3,n}^*$ is almost tight, i.e., $\tilde{\tau}_{3,n}^* \leq 1.2$. The salient feature of this framework is that in order to improve the lower bound on τ_K^*, it suffices to improve $\tilde{\tau}_{k,n}^*$ for some $4 \leq k < K$.

The main technical challenge towards improving $\tilde{\tau}_{4,n}^*$ is that one has to deal with the 3-dimensional simplex Δ_4. Indeed, all known gap instances including that of Angelidakis, Makarychev and Manurangsi are constructed using the 2-dimensional simplex. In the 2-dimensional simplex, the properties of non-opposite cuts are easy to visualize and their cut-values are convenient to characterize using simple geometric observations. However, the values of non-opposite cuts in the 3-dimensional simplex become difficult to characterize. Our main contribution is a simple argument based on properties of lower-dimensional simplices that overcomes this technical challenge. We construct a 3-dimensional instance that has gap larger than 1.2 against non-opposite cuts.

Theorem 2. $\tilde{\tau}_{4,n}^* \geq 1.20016 - O(1/n)$.

Theorem 1 follows from Theorem 2 using the above arguments.

3   Outline of Ideas

Let $G = (V, E)$ be the graph with node set $\Delta_{4,n}$ and edge set $E_{4,n} := \{xy : x, y \in \Delta_{4,n}, \|x - y\|_1 = 2/n\}$, where the terminals are the four unit vectors. In order to lower bound $\tilde{\tau}_{4,n}^*$, we will come up with weights on the edges of G such that every non-opposite cut has cost at least $\alpha = 1.20016$ and moreover the cumulative weight of all edges is $n + O(1)$. This suffices to lower bound $\tilde{\tau}_{4,n}^*$ by the following proposition.

Proposition 1. *Suppose that there exist weights $w : E_{4,n} \to \mathbb{R}_{\geq 0}$ on the edges of G such that every non-opposite cut has cost at least α and the cumulative weight of all edges is $n + O(1)$. Then, $\tilde{\tau}_{4,n}^* \geq \alpha - O(1/n)$.*

Proof. For an arbitrary distribution $\mathcal{P}$ over non-opposite cuts, we have

$$
\tau_{k,n}(\mathcal{P}) = \max_{x,y\in\Delta_{k,n}, x\neq y} \frac{\Pr_{P\sim\mathcal{P}}(P(x)\neq P(y))}{(1/2)\|x-y\|_1} \geq \max_{xy\in E_{4,n}} \frac{\Pr_{P\sim\mathcal{P}}(P(x)\neq P(y))}{(1/2)\|x-y\|_1}
$$

$$
= \max_{xy\in E_{4,n}} \frac{\Pr_{P\sim\mathcal{P}}(P(x)\neq P(y))}{1/n} \geq \sum_{xy\in E_{4,n}} \frac{w(xy)\Pr_{P\sim\mathcal{P}}(P(x)\neq P(y))}{(1/n)(\sum_{e\in E_{4,n}} w(e))}
$$

$$
\geq \frac{\alpha}{1+O(1/n)} = \alpha - O(1/n),
$$

where the last inequality follows from the hypothesis that every non-opposite cut has cost at least α and the cumulative weight of all edges is $n + O(1)$.   □

We obtain our weighted instance from four instances that have large gap against different types of cuts, and then compute the convex combination of these instances that gives the best gap against all non-opposite cuts. All of our four instances are defined as edge-weights on the graph $G = (V, E)$. We identify $\Delta_{3,n}$ with the facet of $\Delta_{4,n}$ defined by $x_4 = 0$. Our first three instances are 2-dimensional instances, i.e. only edges induced by $\Delta_{3,n}$ have positive weight. The fourth instance has uniform weight on $E_{4,n}$.

We first explain the motivation behind Instances 1, 2, and 4, since these are easy to explain. Let $L_{ij} := \{xy \in E_{4,n} : \text{Support}(x), \text{Support}(y) \subseteq \{i,j\}\}$.

- Instance 1 is simply the instance of Angelidakis, Makarychev and Manurangsi [1] on $\Delta_{3,n}$. It has gap $1.2 - \frac{1}{n}$ against all non-opposite cuts, since non-opposite cuts in $\Delta_{4,n}$ induce non-opposite cuts on $\Delta_{3,n}$. Additionally, we show in Lemma 2 that the gap is strictly larger than 1.2 by a constant if the following two conditions hold for the cut:
  - there exist $i, j \in [3]$ such that L_{ij} contains only one edge whose end-nodes have different labels (a cut with this property is called a *non-fragmenting cut*), and
  - $\Delta_{3,n}$ has a lot of nodes with label 5.
- Instance 2 has uniform weight on L_{12}, L_{13} and L_{23}, and 0 on all other edges. Here, a cut in which each L_{ij} contains at least two edges whose end-nodes have different labels (a *fragmenting cut*) has large weight. Consequently, this instance has gap at least 2 against such cuts.
- Instance 4 has uniform weight on all edges in $E_{4,n}$. A beautiful result due to Mirzakhani and Vondrák [11] implies that non-opposite cuts with no node of label 5 have large weight. Consequently, this instance has gap at least $3/2$ against such cuts. We extend their result in Lemma 1 to show that the weight remains large if $\Delta_{3,n}$ has few nodes with label 5.

At first glance, the arguments above seem to suggest that some convex combination of these three instances already has gap strictly larger than 1.2 for all non-opposite cuts. However, there exist two non-opposite cuts such that at least one of them has cost at most 1.2 in every convex combination of these three instances (see full version [2]). One of these two cuts is a fragmenting cut that has almost zero cost in Instance 4 and the best possible cost, namely 1.2, in

Instance 1. Instance 3 is constructed specifically to boost the cost against this non-opposite cut. It has positive uniform weight on 3 equilateral triangles, incident to e^1, e^2 and e^3 on the face $\Delta_{3,n}$. We call the edges of these triangles *red edges*. The side length of these triangles is a parameter, denoted by c, that is optimized at the end of the proof. Essentially, we show that if a non-opposite cut has small cost both on Instance 1 and Instance 4 (i.e., weight 1.2 on Instance 1 and $O(1/n^2)$ weight on Instance 4), then it must contain red edges.

Our lower bound of 1.20016 is obtained by optimizing the coefficients of the convex combination and the parameter c. By Proposition 1 and the results of Angelidakis, Makarychev and Manurangsi, we obtain that $\tau_K^* \geq 1.20016 - O(1/\sqrt{K})$, i.e., the integrality gap of the CKR relaxation for k-way cut is at least $1.20016 - O(1/\sqrt{k})$. We complement our lower bound of 1.20016 by also showing that the best possible gap that can be achieved using convex combinations of our four instances is 1.20067 (see Theorem 4). We refer the reader to the full version of this work [2] for all missing proofs.

4   A 3-dimensional Gap Instance Against Non-opposite Cuts

We will focus on the graph $G = (V, E)$ with the node set $V := \Delta_{4,n}$ being the discretized 3-dimensional simplex and the edge set $E_{4,n} := \{xy : x, y \in \Delta_{4,n}, \|x - y\|_1 = 2/n\}$. The four terminals $s_1, \ldots, s_4$ will be the four extreme points of the simplex, namely $s_i = e^i$ for $i \in [4]$. In this context, a cut is a function $P : V \to [5]$ such that $P(s_i) = i$ for all $i \in [4]$. The *cut-set* corresponding to P is defined as $\delta(P) := \{xy \in E_{4,n} : P(x) \neq P(y)\}$. For a set S of nodes, we will also use $\delta(S)$ to denote the set of edges with exactly one end node in S. Given a weight function $w : E_{4,n} \to \mathbb{R}_+$, the *cost* of a cut P is $\sum_{e \in \delta(P)} w(e)$. Our goal is to come up with weights on the edges so that the resulting 4-way cut instance has gap at least 1.20016 against non-opposite cuts.

We recall that L_{ij} denotes the boundary edges between terminals s_i and s_j. We will denote the boundary nodes between terminals s_i and s_j as V_{ij}, i.e., $V_{ij} := \{x \in \Delta_{4,n} : \text{Support}(x) \subseteq \{i, j\}\}$. Let $c \in (0, 1/2)$ be a constant to be fixed later, such that cn is integral. For each $i \in [3]$, we define $U_i := \{x \in \Delta_{4,n} : x_4 = 0, x_i = 1 - c\}$, $R_i := U_i \cup \{x \in V_{ij} : x_i \geq 1 - c, j \in [3] \setminus \{i\}\}$, $\text{Closure}(R_i) := \{x \in \Delta_{4,n} : x_4 = 0, x_i \geq 1 - c\}$, and $\Gamma_i := \{xy \in E_{4,n} : x, y \in R_i\}$. We will refer to the nodes in R_i as red[1] nodes near terminal s_i and the edges in Γ_i as the red edges near terminal s_i (see Fig. 1b). Let $\text{Face}(s_1, s_2, s_3)$ denote the subgraph of G induced by the nodes whose support is contained in $\{1, 2, 3\}$. We emphasize that red edges and red nodes are present only in $\text{Face}(s_1, s_2, s_3)$ and that the total number of red edges is exactly $9cn$.

---

[1] We use the term "red" as a convenient way for the reader to remember these nodes and edges. The exact color is irrelevant.

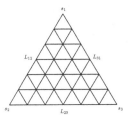

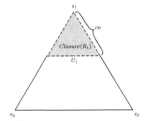

(a) One face of the simplex with edge-sets L_{12}, L_{23} and L_{31}.

(b) Definition of red nodes and edges near terminal s_1. Dashed part corresponds to (R_1, Γ_1).

Fig. 1. Notation on Face(s_1, s_2, s_3).

4.1 Gap Instance as a Convex Combination

Our gap instance is a convex combination of the following four instances.

1. **Instance I_1.** Our first instance constitutes the 3-way cut instance constructed by Angelidakis, Makarychev and Manurangsi [1] that has gap 1.2 against non-opposite cuts. To ensure that the total weight of all the edges in their instance is exactly n, we will scale their instance by $6/5$. Let us denote the resulting instance as J. In I_1, we simply use the instance J on Face(s_1, s_2, s_3) and set the weights of the rest of the edges in $E_{4,n}$ to be zero.
2. **Instance I_2.** In this instance, we set the weights of the edges in L_{12}, L_{23}, L_{13} to be $1/3$ and the weights of the rest of the edges in $E_{4,n}$ to be zero.
3. **Instance I_3.** In this instance, we set the weights of the red edges to be $1/9c$ and the weights of the rest of the edges in $E_{4,n}$ to be zero.
4. **Instance I_4.** In this instance, we set the weight of edges in $E_{4,n}$ to be $1/n^2$.

We note that the total weight of all edges in each of the above instances is $n + O(1)$. For multipliers $\lambda_1, \ldots, \lambda_4 \geq 0$ to be chosen later that will satisfy $\sum_{i=1}^{4} \lambda_i = 1$, let the instance I be the convex combination of the above four instances, i.e., $I = \lambda_1 I_1 + \lambda_2 I_2 + \lambda_3 I_3 + \lambda_4 I_4$. By the properties of the four instances, it immediately follows that the total weight of all edges in the instance I is also $n + O(1)$.

4.2 Gap of the Convex Combination

The following theorem is the main result of this section.

Theorem 3. *For every $n \geq 10$ and $c \in (0, 1/2)$ such that cn is integer, every non-opposite cut on I has cost at least the minimum of the following two terms:*

(i) $\lambda_2 + (1.2 - \frac{1}{n})\lambda_1 + \min_{\alpha \in [0, \frac{1}{2}]} \left\{ 0.4\alpha\lambda_1 + 3\left(\frac{1}{2} - \alpha\right)\lambda_4 \right\}$

(ii) $2\lambda_2 + (1.2 - \frac{5}{2n})\lambda_1 + 3\min\left\{ \frac{2\lambda_3}{9c}, \min_{\alpha \in [0, \frac{c^2}{2}]} \left\{ 0.4\alpha\lambda_1 + 3\left(\frac{c^2}{2} - \alpha\right)\lambda_4 \right\} \right\}$

Before proving Theorem 3, we see its consequence.

Corollary 1. *There exist constants $c \in (0, 1/2)$ and $\lambda_1, \lambda_2, \lambda_3, \lambda_4 \geq 0$ with $\sum_{i=1}^{4} \lambda_i = 1$ such that the cost of every non-opposite cut in the resulting convex combination I is at least $1.20016 - O(1/n)$.*

Proof. The corollary follows from Theorem 3 by setting $\lambda_1 = 0.751652$, $\lambda_2 = 0.147852$, $\lambda_3 = 0.000275$, $\lambda_4 = 0.100221$ and $c = 0.074125$ (this is the optimal setting to achieve the largest lower bound based on Theorem 3). $\square$

Corollary 1 in conjunction with Proposition 1 immediately implies Theorem 2. The following theorem complements Corollary 1 by giving an upper bound on the best possible gap that is achievable using the convex combination of our four instances.

Theorem 4. *For every constant $c \in (0, 1/2)$ and every $\lambda_1, \lambda_2, \lambda_3, \lambda_4 \geq 0$ with $\sum_{i=1}^{4} \lambda_i = 1$, there exists a non-opposite cut whose cost in the resulting convex combination I is at most $1.20067 + O(1/n)$.*

In light of Corollary 1 and Theorem 4, if we believe that the integrality gap of the CKR relaxation is more than 1.20067, then considering convex combination of alternative instances is a reasonable approach towards proving this.

The rest of the section is devoted to proving Theorem 3. We rely on two main ingredients in the proof. The first ingredient is a statement about non-opposite cuts in the 3-dimensional discretized simplex. We prove this in the appendix, where we also give a generalization to higher dimensional simplices, which might be of independent interest.

Lemma 1. *Let P be a non-opposite cut on $\Delta_{4,n}$ with $\alpha(n + 1)(n + 2)$ nodes from $Face(s_1, s_2, s_3)$ labeled as 1, 2, or 3 for some $\alpha \in [0, 1/2]$. Then, $|\delta(P)| \geq 3\alpha n(n + 1)$.*

The constant 3 that appears in the conclusion of Lemma 1 is the best possible for any fixed α (if $n \to \infty$). To see this, consider the non-opposite cut P obtained by labeling s_i to be i for every $i \in [4]$, all nodes at distance at most $\sqrt{2}\alpha n$ from s_1 to be 1, and all remaining nodes to be 5. The number of nodes from $Face(s_1, s_2, s_3)$ labeled as 1, 2, or 3 is $\alpha n^2 + O(n)$. The number of edges in the cut is $3\alpha n^2 + O(n)$.

The second ingredient involves properties of the 3-way cut instance constructed by Angelidakis, Makarychev and Manurangsi [1]. We need two properties that are summarized in Lemma 2 and Corollary 2. We define a cut $Q : \Delta_{3,n} \to [4]$ to be a *fragmenting cut* if $|\delta(Q) \cap L_{ij}| \geq 2$ for every distinct $i, j \in [3]$; otherwise it is a *non-fragmenting cut*.

The first property is that non-opposite non-fragmenting cuts in $\Delta_{3,n}$ that label a large number of nodes with label 4 have cost much larger than 1.2. Recall that J is the instance constructed by Angelidakis, Makarychev and Manurangsi [1] with edge weights scaled by $6/5$ so that the total edge-weight is exactly n.

Lemma 2. *Let $Q : \Delta_{3,n} \to [4]$ be a non-opposite cut with αn^2 nodes labeled as 4. If Q is a non-fragmenting cut and $n \geq 10$, then the cost of Q on J is at least $1.2 + 0.4\alpha - \frac{1}{n}$.*

We show Lemma 2 by modifying Q to obtain a non-opposite cut Q' while reducing its cost by 0.4α. By the main result of [1], the cost of every non-opposite cut Q' on J is at least $1.2 - \frac{1}{n}$. Therefore, it follows that the cost of Q on J is at least $1.2 - \frac{1}{n} + 0.4\alpha$. We emphasize that while it might be possible to improve the constant 0.4 that appears in the conclusion of Lemma 2, it does not lead to much improvement on the overall integrality gap as illustrated by Theorem 4.

The second property is that non-opposite cuts which do not remove any of the red edges, but label a large number of nodes in the red region with label 4 have cost much larger than 1.2.

Corollary 2. *Let* $Q : \Delta_{3,n} \to [4]$ *be a non-opposite cut and* $n \geq 10$. *For each* $i \in [3]$, *let*

$$A_i := \begin{cases} \{v \in Closure(R_i) : Q(v) = 4\} & \text{if } \delta(Q) \cap \Gamma_i = \emptyset, \\ \emptyset & \text{otherwise.} \end{cases}$$

Then, the cost of Q *on* J *is at least* $1.2 + 0.4 \sum_{i=1}^{3} |A_i|/n^2 - \frac{5}{2n}$.

In order to show Corollary 2, we first derive that the cost of the edges $\delta(\cup_{i=1}^{3} A_i)$ in the instance J is at least $0.4 \sum_{i=1}^{3} |A_i|/n^2 - \frac{3}{2n}$ using Lemma 2. Next, we modify Q to obtain a non-opposite cut Q' such that $\delta(Q') = \delta(Q) \backslash \delta(\cup_{i=1}^{3} A_i)$. By the main result of [1], the cost of every non-opposite cut Q' on J is at least $1.2 - \frac{1}{n}$. Therefore, it follows that the cost of Q on J is at least $1.2 + 0.4 \sum_{i=1}^{3} |A_i|/n^2$.

We now have the ingredients to prove Theorem 3.

Proof of Theorem 3. Let $P : \Delta_{4,n} \to [5]$ be a non-opposite cut. Let Q be the cut P restricted to $\text{Face}(s_1, s_2, s_3)$, i.e., for every $v \in \Delta_{4,n}$ with $\text{Support}(v) \subseteq [3]$, let

$$Q(v) := \begin{cases} P(v) & \text{if } P(v) \in \{1, 2, 3\}, \\ 4 & \text{if } P(v) = 5. \end{cases}$$

We consider two cases.

Case 1: Q is a non-fragmenting cut. Let the number of nodes in $\text{Face}(s_1, s_2, s_3)$ that are labeled by Q as 4 (equivalently, labeled by P as 5) be $\alpha(n+1)(n+2)$ for some $\alpha \in [0, \frac{1}{2}]$. Since $|\{x \in \text{Face}(s_1, s_2, s_3) : Q(x) = 4\}| \geq \alpha n^2$, Lemma 2 implies that the cost of Q on J, and hence the cost of P on I_1, is at least $1.2 + 0.4\alpha - \frac{1}{n}$. Moreover, the cost of P on I_2 is at least 1 since at least one edge in L_{ij} should be in $\delta(P)$ for every pair of distinct $i, j \in [3]$. To estimate the cost on I_4, we observe that the number of nodes on $\text{Face}(s_1, s_2, s_3)$ labeled by P as 1, 2, or 3 is $(1/2 - \alpha)(n+1)(n+2)$. By Lemma 1, we have that $|\delta(P)| \geq 3(1/2 - \alpha)n(n+1)$ and thus, the cost of P on I_4 is at least $3(1/2 - \alpha)$. Therefore, the cost of P on the convex combination instance I is at least $\lambda_2 + \left(1.2 - \frac{1}{n}\right)\lambda_1 + \min_{\alpha \in [0, \frac{1}{2}]} \left\{0.4\alpha\lambda_1 + 3\left(\frac{1}{2} - \alpha\right)\lambda_4\right\}$.

Case 2: Q is a fragmenting cut. Then, the cost of P on I_2 is at least 2 as a fragmenting cut contains at least 2 edges from each L_{ij} for distinct $i, j \in [3]$.

We will now compute the cost of P on the other instances. Let $r := |\{i \in [3] : \delta(P) \cap \Gamma_i \neq \emptyset\}|$, i.e., r is the number of red triangles that are intersected by the cut P. We will derive lower bounds on the cost of the cut in each of the three instances I_1, I_3 and I_4 based on the value of $r \in \{0, 1, 2, 3\}$. For each $i \in [3]$, let

$$A_i := \begin{cases} \{v \in \text{Closure}(R_i) : P(v) = 5\} & \text{if } \delta(P) \cap \Gamma_i = \emptyset, \\ \emptyset & \text{otherwise,} \end{cases}$$

and let $\alpha := |A_1 \cup A_2 \cup A_3| / ((n + 1/c)(n + 2/c))$. Since $c < 1/2$, the sets A_i and A_j are disjoint for distinct $i, j \in [3]$. We note that $\alpha \in [0, (3 - r)c^2/2]$ since $|A_i| \leq (cn + 1)(cn + 2)/2$ and $A_i \cap A_j = \emptyset$.

In order to lower bound the cost of P on I_1, we will use Corollary 2. We recall that Q is the cut P restricted to $\text{Face}(s_1, s_2, s_3)$, so the cost of P on I_1 is the same as the cost of Q on J. Moreover, by Corollary 2, the cost of Q on J is at least $1.2 + 0.4\alpha - \frac{5}{2n}$, because $\alpha \leq \sum_{i=1}^{3} |A_i|/n^2$. Hence, the cost of P on I_1 is at least $1.2 + 0.4\alpha - \frac{5}{2n}$.

We now show that the cost of P on I_3 is at least $2r/9c$. For this, we will show that for $i \in [3]$, if $\delta(P) \cap \Gamma_i \neq \emptyset$, then $|\delta(P) \cap \Gamma_i| \geq 2$. Indeed, the subgraph (R_i, Γ_i) is a cycle and therefore, if $P(x) \neq P(y)$ for some $xy \in \Gamma_i$, then the path $\Gamma_i - xy$ must also contain two consecutive nodes labeled differently by P.

Next we compute the cost of P on I_4. If $r = 3$, then the cost of P on I_4 is at least 0. Suppose $r \in \{0, 1, 2\}$. For a red triangle $i \in [3]$ with $\delta(P) \cap \Gamma_i = \emptyset$, we have at least $(cn + 1)(cn + 2)/2 - |A_i|$ nodes from $\text{Closure}(R_i)$ that are labeled as 1, 2, or 3. Moreover, the nodes in $\text{Closure}(R_i)$ and $\text{Closure}(R_j)$ are disjoint for distinct $i, j \in [3]$. Hence, the number of nodes in $\text{Face}(s_1, s_2, s_3)$ that are labeled as 1, 2, or 3 is at least $(3 - r)(cn + 1)(cn + 2)/2 - \alpha(n + 1/c)(n + 2/c) = ((3-r)c^2/2 - \alpha)(n + 1/c)(n + 2/c)$, which is at least $((3-r)c^2/2 - \alpha)(n + 1)(n + 2)$, since $c \leq 1$. Therefore, by Lemma 1, we have $|\delta(P)| \geq 3((3 - r)c^2/2 - \alpha)n^2$ and thus, the cost of P on I_4 is at least $3((3 - r)c^2/2 - \alpha)$.

Thus, the cost of P on the convex combination instance I is at least $2\lambda_2 + (1.2 - \frac{5}{2n})\lambda_1 + \gamma(r, \alpha)$ for some $\alpha \in [0, (3 - r)c^2/2]$, where

$$\gamma(r, \alpha) := \begin{cases} \frac{6\lambda_3}{9c}, & \text{if } r = 3, \\ 0.4\alpha\lambda_1 + \frac{2r}{9c}\lambda_3 + 3\left(\frac{(3-r)c^2}{2} - \alpha\right)\lambda_4, & \text{if } r \in \{0, 1, 2\}. \end{cases}$$

In particular, the cost of P on the convex combination instance I is at least $2\lambda_2 + (1.2 - 5/(2n))\lambda_1 + \gamma^*$, where $\gamma^* := \min_{r \in \{0,1,2,3\}} \min_{\alpha \in [0, \frac{(3-r)c^2}{2}]} \gamma(r, \alpha)$. The following claim completes the proof of the theorem.

Claim. $\gamma^* \geq 3 \min \left\{ \frac{2\lambda_3}{9c}, \min_{\alpha \in [0, \frac{c^2}{2}]} \left\{ 0.4\alpha\lambda_1 + 3\left(\frac{c^2}{2} - \alpha\right)\lambda_4 \right\} \right\}.$ □

Acknowledgements. Kristóf was supported by the ÚNKP-18-4 New National Excellence Program of the Ministry of Human Capacities. Karthekeyan was supported by

NSF grant CCF-1814613. Tamás was supported by the Hungarian National Research, Development and Innovation Office – NKFIH grant K120254. Vivek was supported by NSF grant CCF-1319376.

Appendix: Size of Non-opposite Cuts in $\Delta_{k,n}$

In this section, we prove Lemma 1. In fact, we prove a general result for $\Delta_{k,n}$, that may be useful for constructing instances with larger gap by considering higher dimensional simplices. Our result is an extension of a theorem of Mirzakhani and Vondrák [11] on Sperner-admissible labelings.

A labeling $\ell : \Delta_{k,n} \to [k]$ is *Sperner-admissible* if $\ell(x) \in \text{Support}(x)$ for every $x \in \Delta_{k,n}$. We say that $x \in \Delta_{k,n}$ has an *inadmissible label* if $\ell(x) \notin \text{Support}(x)$. Let $H_{k,n}$ denote the hypergraph whose node set is $\Delta_{k,n}$ and whose hyperedge set is $\mathcal{E} := \left\{ \left\{ \frac{n-1}{n}x + \frac{1}{n}e_1, \frac{n-1}{n}x + \frac{1}{n}e_2, \ldots, \frac{n-1}{n}x + \frac{1}{n}e_k \right\} : x \in \Delta_{k,n-1} \right\}$. Each hyperedge $e \in \mathcal{E}$ has k nodes, and if $x, y \in e$, then there exist distinct $i, j \in [n]$ such that $x - y = \frac{1}{n}e_i - \frac{1}{n}e_j$. We remark that $H_{k,n}$ has $\binom{n+k-1}{k-1}$ nodes and $\binom{n+k-2}{k-1}$ hyperedges. Geometrically, the hyperedges correspond to simplices that are translates of each other and share at most one node. Given a labeling ℓ, a hyperedge of $H_{k,n}$ is *monochromatic* if all of its nodes have the same label. Mirzakhani and Vondrák showed the following.

Theorem 5 (Proposition 2.1 in [11]). *Let ℓ be a Sperner-admissible labeling of $\Delta_{k,n}$. Then, the number of monochromatic hyperedges in $H_{k,n}$ is at most $\binom{n+k-3}{k-1}$, and therefore the number of non-monochromatic hyperedges is at least $\binom{n+k-3}{k-2}$.*

Our main result of this section is an extension of the above result to the case when there are some inadmissible labels on a single face of $\Delta_{k,n}$. We show that a labeling in which all inadmissible labels are on a single face still has a large number of non-monochromatic hyperedges. We will denote the nodes $x \in \Delta_{k,n}$ with $\text{Support}(x) \subseteq [k-1]$ as $\text{Face}(s_1, \ldots, s_{k-1})$.

Theorem 6. *Let ℓ be a labeling of $\Delta_{k,n}$ such that all inadmissible labels are on $\text{Face}(s_1, \ldots, s_{k-1})$ and the number of nodes with inadmissible labels is $\beta \frac{(n+k-2)!}{n!}$ for some β. Then, the number of non-monochromatic hyperedges of $H_{k,n}$ is at least $\left(\frac{1}{(k-2)!} - \beta \right) \frac{(n+k-3)!}{(n-1)!}$.*

Proof. Let $Z := \{x \in \text{Face}(s_1, \ldots, s_{k-1}) : \ell(x) = k\}$, i.e. Z is the set of nodes in $\text{Face}(s_1, \ldots, s_{k-1})$ having an inadmissible label. Let us call a hyperedge of $H_{k,n}$ *inadmissible* if the label of one of its nodes is inadmissible.

Claim. There are at most $\beta \frac{(n+k-3)!}{(n-1)!}$ inadmissible monochromatic hyperedges.

Proof. Let $\mathcal{E}'$ be the set of inadmissible monochromatic hyperedges. Each hyperedge $e \in \mathcal{E}'$ has exactly $k-1$ nodes from $\text{Face}(s_1, \ldots, s_{k-1})$ and they all have the same label as e is monochromatic. Thus, each $e \in \mathcal{E}'$ contains $k-1$ nodes from Z. We define an injective map $\varphi : \mathcal{E}' \to Z$ by letting $\varphi(e)$ to be the node $x \in e \cap Z$ with the largest 1st coordinate. Notice that if $x = \varphi(e)$, then the other nodes of e are $x - (1/n)e_1 + (1/n)e_i$ $(i = 2, \ldots, k)$, and all but the last one are in Z. In particular, x_1 is positive.

Let $Z' \subseteq Z$ be the image of φ. For $x \in Z$ and $i \in \{2, \ldots, k-1\}$, let $Z_x^i := \{y \in Z : y_j = x_j \; \forall j \in [k-1] \setminus \{1, i\}\}$. Since $y_k = 0$ and $\|y\|_1 = 1$ for every $y \in Z$, the nodes of Z_x^i are on a line containing x. It also follows that $Z_x^i \cap Z_x^j = \{x\}$ if $i \neq j$. Let $Z'' := \{x \in Z : \exists i \in \{2, \ldots, k-1\}$ such that $x_i \geq y_i \; \forall y \in Z_x^i\}$. We observe that if $x \in Z'$, then for each $i \in \{2, \ldots, k-1\}$, the node $y = x - (1/n)e_1 + (1/n)e_i$ is in Z and hence, $y \in Z_x^i$ with $y_i > x_i$. In particular, this implies that $Z' \cap Z'' = \emptyset$. We now compute an upper bound on the size of $Z \setminus Z''$, which gives an upper bound on the size of Z' and hence also on the size of $\mathcal{E}'$, as $|Z'| = |\mathcal{E}'|$. For each node $x \in Z \setminus Z''$ and for every $i \in \{2, \ldots, k-1\}$, let z_x^i be the node in $Z'' \cap Z_x^i$ with the largest ith coordinate. Clearly $z_x^i \neq z_x^j$ if $i \neq j$, because $Z_x^i \cap Z_x^j = \{x\}$.

For given $y \in Z''$ and $i \in \{2, \ldots, k-1\}$, we want to bound the size of $S := \{x \in Z \setminus Z'' : z_x^i = y\}$. Consider $a \in S$. Then, $z_a^i = y$ implies that the node in $Z'' \cap Z_a^i$ with the largest i-th coordinate is y. That is, $y_j = a_j$ for all $j \in [k-1] \setminus \{1, i\}$ and moreover $y_i \geq a_i$. If $y_i = a_i$, then $y = a$, so a is in Z'' which contradicts $a \in S$. Thus, $y_i > a_i$ for any $a \in S$, i.e. the nodes in S are on the line Z_y^i and their i-th coordinate is strictly smaller than y_i. This implies that $|S| \leq ny_i$. Consequently, the size of the set $\{x \in Z \setminus Z'' : y = z_x^i$ for some $i \in \{2, \ldots, k-1\}\}$ is at most n, since $\sum_{i=2}^{k-2} y_i \leq \|y\|_1 = 1$.

For each $x \in Z \setminus Z''$, we defined $k-2$ distinct nodes $z_x^2, \ldots, z_x^{k-1} \in Z''$. Moreover, for each $y \in Z''$, we have at most n distinct nodes x in $Z \setminus Z''$ for which there exists $i \in \{2, \ldots, k-1\}$ such that $y = z_x^i$. Hence, $(k-2)|Z \setminus Z''| \leq n|Z''|$, and therefore $|Z \setminus Z''| \leq (n/(n+k-2))|Z|$. This gives $|\mathcal{E}'| = |Z'| \leq |Z \setminus Z''| \leq \frac{n}{n+k-2}|Z| \leq \beta \frac{(n+k-2)!}{n!} \frac{n}{n+k-2} = \beta \frac{(n+k-3)!}{(n-1)!}$, as required. $\qquad\square$

Let ℓ' be a Sperner-admissible labeling obtained from ℓ by changing the label of each node in Z to an arbitrary admissible label. By Theorem 5, the number of monochromatic hyperedges for ℓ' is at most $\binom{n+k-3}{k-1}$. By combining this with the above claim, we get that the number of monochromatic hyperedges for ℓ is at most $\binom{n+k-3}{k-1} + \beta \frac{(n+k-3)!}{(n-1)!}$. Since $H_{k,n}$ has $\binom{n+k-2}{k-1}$ hyperedges, the number of non-monochromatic hyperedges is at least $\binom{n+k-2}{k-1} - \binom{n+k-3}{k-1} - \beta \frac{(n+k-3)!}{(n-1)!} = \left(\frac{1}{(k-2)!} - \beta\right)\frac{(n+k-3)!}{(n-1)!}$. $\qquad\square$

We note that Theorem 6 is tight for the extreme cases where $\beta = 0$ and $\beta = 1/(k-2)!$.

We now derive Lemma 1 from Theorem 6.

Proof of Lemma 1. Let ℓ be the labeling of $\Delta_{4,n}$ obtained from P by setting $\ell(x) = 4$ if $P(x) = 5$, and $\ell(x) = P(x)$ otherwise. This is a labeling with $(\frac{1}{2} - \alpha)(n+1)(n+2)$ nodes having an inadmissible label, all on Face(s_1, s_2, s_3). We apply Theorem 6 with parameters $k = 4$, $\beta = \frac{1}{2} - \alpha$, and the labeling ℓ. By the theorem, the number of non-monochromatic hyperedges in $H_{4,n} = (\Delta_{4,n}, \mathcal{E})$ under labeling ℓ is at least $\alpha n(n + 1)$.

We observe that for each hyperedge $e = \{u_1, u_2, u_3, u_4\} \in \mathcal{E}$, the subgraph $G[e]$ induced by the nodes in e contains 6 edges. Also, for any two hyperedges e_1 and e_2, the edges in the induced subgraphs $G[e_1]$ and $G[e_2]$ are disjoint as e_1 and e_2 can share at most one node. Moreover, for each non-monochromatic hyperedge $e \in \mathcal{E}$, at least 3 edges of $G[e]$ are in $\delta(P)$. Thus, the number of edges of G that are in $\delta(P)$ is at least $3\alpha n(n + 1)$. $\qquad\square$

References

1. Angelidakis, H., Makarychev, Y., Manurangsi, P.: An improved integrality gap for the Călinescu-Karloff-Rabani relaxation for multiway cut. In: Integer Programming and Combinatorial Optimization, IPCO 2017, pp. 39–50 (2017)
2. Bérczi, K., Chandrasekaran, K., Király, T., Madan, V.: Improving the integrality gap for multiway cut (2018). Preprint: https://arxiv.org/abs/1807.09735
3. Buchbinder, N., Naor, J., Schwartz, R.: Simplex partitioning via exponential clocks and the multiway cut problem. In: Proceedings of the Forty-Fifth Annual ACM Symposium on Theory of Computing, STOC 2013, pp. 535–544 (2013)
4. Buchbinder, N., Schwartz, R., Weizman, B.: Simplex transformations and the multiway cut problem. In: Proceedings of the Twenty-Eighth Annual ACM-SIAM Symposium on Discrete Algorithms, SODA 2017, pp. 2400–2410 (2017)
5. Călinescu, G., Karloff, H., Rabani, Y.: An improved approximation algorithm for multiway cut. J. Comput. Syst. Sci. **60**(3), 564–574 (2000)
6. Cheung, K., Cunningham, W., Tang, L.: Optimal 3-terminal cuts and linear programming. Math. Programm. **106**(1), 1–23 (2006)
7. Dahlhaus, E., Johnson, D., Papadimitriou, C., Seymour, P., Yannakakis, M.: The complexity of multiterminal cuts. SIAM J. Comput. **23**(4), 864–894 (1994)
8. Freund, A., Karloff, H.: A lower bound of $8/(7+1/(k-1))$ on the integrality ratio of the Călinescu-Karloff-Rabani relaxation for multiway cut. Inf. Process. Lett. **75**(1), 43–50 (2000)
9. Karger, D., Klein, P., Stein, C., Thorup, M., Young, N.: Rounding algorithms for a geometric embedding of minimum multiway cut. Math. Oper. Res. **29**(3), 436–461 (2004)
10. Manokaran, R., Naor, J., Raghavendra, P., Schwartz, R.: SDP gaps and UGC hardness for multiway cut, 0-extension, and metric labeling. In: Proceedings of the Fortieth Annual ACM Symposium on Theory of Computing, STOC 2008, pp. 11–20 (2008)
11. Mirzakhani, M., Vondrák, J.: Sperner's colorings, hypergraph labeling problems and fair division. In: Proceedings of the twenty-sixth Annual ACM-SIAM Symposium on Discrete Algorithms, SODA 2015, pp. 873–886 (2015)
12. Sharma, A., Vondrák, J.: Multiway cut, pairwise realizable distributions, and descending thresholds. In: Proceedings of the Forty-Sixth Annual ACM Symposium on Theory of Computing, STOC 2014, pp. 724–733 (2014)

ℓ_1-sparsity Approximation Bounds
for Packing Integer Programs

Chandra Chekuri, Kent Quanrud, and Manuel R. Torres[✉]

University of Illinois at Urbana-Champaign, Urbana, IL, USA
{chekuri,quanrud2,manuelt2}@illinois.edu

Abstract. We consider approximation algorithms for packing integer programs (PIPs) of the form $\max\{\langle c, x \rangle : Ax \le b, x \in \{0,1\}^n\}$ where c, A, and b are nonnegative. We let $W = \min_{i,j} b_i / A_{i,j}$ denote the width of A which is at least 1. Previous work by Bansal et al. [1] obtained an $\Omega(\frac{1}{\Delta_0^{1/\lfloor W \rfloor}})$-approximation ratio where Δ_0 is the maximum number of nonzeroes in any column of A (in other words the ℓ_0-column sparsity of A). They raised the question of obtaining approximation ratios based on the ℓ_1-column sparsity of A (denoted by Δ_1) which can be much smaller than Δ_0. Motivated by recent work on covering integer programs (CIPs) [4,7] we show that simple algorithms based on randomized rounding followed by alteration, similar to those of Bansal et al. [1] (but with a twist), yield approximation ratios for PIPs based on Δ_1. First, following an integrality gap example from [1], we observe that the case of $W = 1$ is as hard as maximum independent set even when $\Delta_1 \le 2$. In sharp contrast to this negative result, as soon as width is strictly larger than one, we obtain positive results via the natural LP relaxation. For PIPs with width $W = 1 + \epsilon$ where $\epsilon \in (0, 1]$, we obtain an $\Omega(\epsilon^2/\Delta_1)$-approximation. In the large width regime, when $W \ge 2$, we obtain an $\Omega((\frac{1}{1+\Delta_1/W})^{1/(W-1)})$-approximation. We also obtain a $(1-\epsilon)$-approximation when $W = \Omega(\frac{\log(\Delta_1/\epsilon)}{\epsilon^2})$.

Keywords: Packing integer programs · Approximation algorithms · ℓ_1-column sparsity

1 Introduction

Packing integer programs (abbr. PIPs) are an expressive class of integer programs of the form:

$$\text{maximize } \langle c, x \rangle \text{ over } x \in \{0,1\}^n \text{ s.t. } Ax \le b,$$

C. Chekuri and K. Quanrud supported in part by NSF grant CCF-1526799. M. Torres supported in part by fellowships from NSF and the Sloan Foundation.

A. Lodi and V. Nagarajan (Eds.): IPCO 2019, LNCS 11480, pp. 128–140, 2019.
https://doi.org/10.1007/978-3-030-17953-3_10

where $A \in \mathbb{R}_{\geq 0}^{m \times n}$, $b \in \mathbb{R}_{\geq 0}^m$ and $c \in \mathbb{R}_{\geq 0}^n$ all have nonnegative entries[1]. Many important problems in discrete and combinatorial optimization can be cast as special cases of PIPs. These include the maximum independent set in graphs and hypergraphs, set packing, matchings and b-matchings, knapsack (when $m = 1$), and the multi-dimensional knapsack. The maximum independent set problem (MIS), a special case of PIPs, is NP-hard and unless $P = NP$ there is no $n^{1-\epsilon}$-approximation where n is the number of nodes in the graph [10,18]. For this reason it is meaningful to consider special cases and other parameters that control the difficulty of PIPs. Motivated by the fact that MIS admits a simple $\frac{1}{\Delta(G)}$-approximation where $\Delta(G)$ is the maximum degree of G, previous work considered approximating PIPs based on the maximum number of nonzeroes in any column of A (denoted by Δ_0); note that when MIS is written as a PIP, Δ_0 coincides with $\Delta(G)$. As another example, when maximum weight matching is written as a PIP, $\Delta_0 = 2$. Bansal et al. [1] obtained a simple and clever algorithm that achieved an $\Omega(1/\Delta_0)$-approximation for PIPs via the natural LP relaxation; this improved previous work of Pritchard [13,14] who was the first to obtain an approximation for PIPs only as a function of Δ_0. Moreover, the rounding algorithm in [1] can be viewed as a contention resolution scheme which allows one to get similar approximation ratios even when the objective is submodular [1,6]. It is well-understood that PIPs become easier when the entries in A are small compared to the packing constraints b. To make this quantitative we consider the well-studied notion called the *width* defined as $W := \min_{i,j:A_{i,j}>0} b_i/A_{i,j}$. Bansal et al. obtain an $\Omega((\frac{1}{\Delta_0})^{1/\lfloor W \rfloor})$-approximation which improves as W becomes larger. Although they do not state it explicitly, their approach also yields a $(1 - \epsilon)$-approximation when $W = \Omega(\frac{1}{\epsilon^2} \log(\Delta_0/\epsilon))$.

Δ_0 is a natural measure for combinatorial applications such as MIS and matchings where the underlying matrix A has entries from $\{0,1\}$. However, in some applications of PIPs such as knapsack and its multi-dimensional generalization which are more common in resource-allocation problems, the entries of A are arbitrary rational numbers (which can be assumed to be from the interval $[0,1]$ after scaling). In such applications it is natural to consider another measure of column-sparsity which is based on the ℓ_1 norm. Specifically we consider Δ_1, the maximum column sum of A. Unlike Δ_0, Δ_1 is not scale invariant so one needs to be careful in understanding the parameter and its relationship to the width W. For this purpose we normalize the constraints $Ax \leq b$ as follows. Let $W = \min_{i,j:A_{i,j}>0} b_i/A_{i,j}$ denote the width as before (we can assume without loss of generality that $W \geq 1$ since we are interested in integer solutions). We can then scale each row A_i of A separately such that, after scaling, the i'th constraint reads as $A_i x \leq W$. After scaling all rows in this fashion, entries of A are in the interval $[0,1]$, and the maximum entry of A is equal to 1. Note that this scaling process does not alter the original width. We let Δ_1 denote the maximum column sum of A after this normalization and observe that $1 \leq \Delta_1 \leq \Delta_0$. In many

---

[1] We can allow the variables to have general integer upper bounds instead of restricting them to be boolean. As observed in [1], one can reduce this more general case to the $\{0, 1\}$ case without too much loss in the approximation.

settings of interest $\Delta_1 \ll \Delta_0$. We also observe that Δ_1 is a more robust measure than Δ_0; small perturbations of the entries of A can dramatically change Δ_0 while Δ_1 changes minimally.

Bansal et al. raised the question of obtaining an approximation ratio for PIPs as a function of only Δ_1. They observed that this is not feasible via the natural LP relaxation by describing a simple example where the integrality gap of the LP is $\Omega(n)$ while Δ_1 is a constant. In fact their example essentially shows the existence of a simple approximation preserving reduction from MIS to PIPs such that the resulting instances have $\Delta_1 \leq 2$; thus no approximation ratio that depends only on Δ_1 is feasible for PIPs unless $P = NP$. These negative results seem to suggest that pursuing bounds based on Δ_1 is futile, at least in the worst case. However, the starting point of this paper is the observation that both the integrality gap example and the hardness result are based on instances where the width W of the instance is arbitrarily close to 1. We demonstrate that these examples are rather brittle and obtain several positive results when we consider $W \geq (1 + \epsilon)$ for any fixed $\epsilon > 0$.

1.1   Our Results

Our first result is on the hardness of approximation for PIPs that we already referred to. The hardness result suggests that one should consider instances with $W > 1$. Recall that after normalization we have $\Delta_1 \geq 1$ and $W \geq 1$ and the maximum entry of A is 1. We consider three regimes of W and obtain the following results, all via the natural LP relaxation, which also establish corresponding upper bounds on the integrality gap.

(i) $1 < W \leq 2$. For $W = 1 + \epsilon$ where $\epsilon \in (0, 1]$ we obtain an $\Omega(\frac{\epsilon^2}{\Delta_1})$-approximation.

(ii) $W \geq 2$. We obtain an $\Omega((\frac{1}{1+\frac{\Delta_1}{W}})^{1/(W-1)})$-approximation which can be simplified to $\Omega((\frac{1}{1+\Delta_1})^{1/(W-1)})$ since $W \geq 1$.

(iii) A $(1 - \epsilon)$-approximation when $W = \Omega(\frac{1}{\epsilon^2} \log(\Delta_1/\epsilon))$.

Our results establish approximation bounds based on Δ_1 that are essentially the same as those based on Δ_0 as long as the width is not too close to 1. We describe randomized algorithms which can be derandomized via standard techniques. The algorithms can be viewed as contention resolution schemes, and via known techniques [1,6], the results yield corresponding approximations for submodular objectives; we omit these extensions in this version.

All our algorithms are based on a simple randomized rounding plus alteration framework that has been successful for both packing and covering problems. Our scheme is similar to that of Bansal et al. at a high level but we make a simple but important change in the algorithm and its analysis. This is inspired by recent work on covering integer programs [4] where ℓ_1-sparsity based approximation bounds from [7] were simplified.

1.2   Other Related Work

We note that PIPs are equivalent to the multi-dmensional knapsack problem. When $m = 1$ we have the classical knapsack problem which admits a very efficient FPTAS (see [2]). There is a PTAS for any fixed m [8] but unless $P = NP$ an FPTAS does not exist for $m = 2$.

Approximation algorithms for PIPs in their general form were considered initially by Raghavan and Thompson [15] and refined substantially by Srinivasan [16]. Srinivasan obtained approximation ratios of the form $\Omega(1/n^W)$ when A had entries from $\{0, 1\}$, and a ratio of the form $\Omega(1/n^{1/\lfloor W \rfloor})$ when A had entries from $[0, 1]$. Pritchard [13] was the first to obtain a bound for PIPs based solely on the column sparsity parameter Δ_0. He used iterated rounding and his initial bound was improved in [14] to $\Omega(1/\Delta_0^2)$. The current state of the art is due to Bansal et al. [1]. Previously we ignored constant factors when describing the ratio. In fact [1] obtains a ratio of $(1 - o(1)\frac{e-1}{e^2 \Delta_0})$ by strengthening the basic LP relaxation.

In terms of hardness of approximation, PIPs generalize MIS and hence one cannot obtain a ratio better than $n^{1-\epsilon}$ unless $P = NP$ [10,18]. Building on MIS, [3] shows that PIPs are hard to approximate within a $n^{\Omega(1/W)}$ factor for any constant width W. Hardness of MIS in bounded degree graphs [17] and hardness for k-set-packing [11] imply that PIPs are hard to approximate to within $\Omega(1/\Delta_0^{1-\epsilon})$ and to within $\Omega((\log \Delta_0)/\Delta_0)$ when Δ_0 is a sufficiently large constant. These hardness results are based on $\{0, 1\}$ matrices for which Δ_0 and Δ_1 coincide.

There is a large literature on deterministic and randomized rounding algorithms for packing and covering integer programs and connections to several topics and applications including discrepancy theory. ℓ_1-sparsity guarantees for covering integer programs were first obtained by Chen, Harris and Srinivasan [7] partly inspired by [9].

2   Hardness of Approximating PIPs as a Function of Δ_1

Bansal et al. [1] showed that the integrality gap of the natural LP relaxation for PIPs is $\Omega(n)$ even when Δ_1 is a constant. One can use essentially the same construction to show the following theorem whose proof can be found in the appendix.

Theorem 1. *There is an approximation preserving reduction from MIS to instances of PIPs with $\Delta_1 \leq 2$.*

Unless $P = NP$, MIS does not admit a $n^{1-\epsilon}$-approximation for any fixed $\epsilon > 0$ [10,18]. Hence the preceding theorem implies that unless $P = NP$ one cannot obtain an approximation ratio for PIPs solely as a function of Δ_1.

Round-and-Alter Framework: input A, b, and α

    let x be the optimum fractional solution of the natural LP relaxation

    for $j \in [n]$, set x'_j to be 1 independently with probability αx_j and 0 otherwise

    $x'' \leftarrow x'$

    for $i \in [m]$ **do**

        find $S \subseteq [n]$ such that setting $x'_j = 0$ for all $j \in S$ would satisfy $\langle e_i, Ax' \rangle \leq b_i$

        for all $j \in S$, set $x''_j = 0$

    end for

    return x''

Fig. 1. Randomized rounding with alteration framework.

3 Round and Alter Framework

The algorithms in this paper have the same high-level structure. The algorithms first scale down the fractional solution x by some factor α, and then randomly round each coordinate independently. The rounded solution x' may not be feasible for the constraints. The algorithm alters x' to a feasible x'' by considering each constraint separately in an arbitrary order; if x' is not feasible for constraint i some subset S of variables are chosen to be set to 0. Each constraint corresponds to a knapsack problem and the framework (which is adapted from [1]) views the problem as the intersection of several knapsack constraints. A formal template is given in Fig. 1. To make the framework into a formal algorithm, one must define α and how to choose S in the for loop. These parts will depend on the regime of interest.

For an algorithm that follows the round-and-alter framework, the expected output of the algorithm is $\mathbb{E}\left[\langle c, x'' \rangle\right] = \sum_{j=1}^{n} c_j \cdot \Pr[x''_j = 1]$. Independent of how α is defined or how S is chosen, $\Pr[x''_j = 1] = \Pr[x''_j = 1 | x'_j = 1] \cdot \Pr[x'_j = 1]$ since $x''_j \leq x'_j$. Then we have

$$\mathbb{E}[\langle c, x'' \rangle] = \alpha \sum_{j=1}^{n} c_j x_j \cdot \Pr[x''_j = 1 | x'_j = 1].$$

Let E_{ij} be the event that x''_j is set to 0 when ensuring constraint i is satisfied in the for loop. As x''_j is only set to 0 if at least one constraint sets x''_j to 0, we have

$$\Pr[x''_j = 0 | x'_j = 1] = \Pr\left[\bigcup_{i \in [m]} E_{ij} \Big| x'_j = 1\right] \leq \sum_{i=1}^{m} \Pr[E_{ij} | x'_j = 1].$$

Combining these two observations, we have the following lemma, which applies to all of our subsequent algorithms.

Lemma 1. *Let $\mathcal{A}$ be a randomized rounding algorithm that follows the round-and-alter framework given in Fig. 1. Let x' be the rounded solution obtained with*

```
round-and-alter-by-sorting(A, b, α₁):
    let x be the optimum fractional solution of the natural LP relaxation
    for j ∈ [n], set x'ⱼ to be 1 independently with probability α₁xⱼ and 0 otherwise
    x″ ← x′
    for i ∈ [m] do
        sort and renumber such that A_{i,1} ≤ ··· ≤ A_{i,n}
        s ← max{ℓ ∈ [n] : ∑_{j=1}^{ℓ} A_{i,j}x'ⱼ ≤ bᵢ}
        for each j ∈ [n] such that j > s, set x″ⱼ = 0
    end for
    return x″
```

Fig. 2. Round-and-alter in the large width regime. Each constraint sorts the coordinates in increasing size and greedily picks a feasible set and discards the rest.

scaling factor α. Let E_{ij} be the event that x''_j is set to 0 by constraint i. If for all $j \in [n]$ we have $\sum_{i=1}^{m} \Pr[E_{ij} | x'_j = 1] \leq \gamma$, then $\mathcal{A}$ is an $\alpha(1 - \gamma)$-approximation for PIPs.

We will refer to the quantity $\Pr[E_{ij} | x'_j = 1]$ as the *rejection probability* of item j in constraint i. We will also say that constraint i *rejects* item j if x''_j is set to 0 in constraint i.

4   The Large Width Regime: $W \geq 2$

In this section, we consider PIPs with width $W \geq 2$. Recall that we assume $A \in [0,1]^{m \times n}$ and $b_i = W$ for all $i \in [m]$. Therefore we have $A_{i,j} \leq W/2$ for all i, j and from a knapsack point of view all items are "small". We apply the round-and-alter framework in a simple fashion where in each constraint i the coordinates are sorted by the coefficients in that row and the algorithm chooses the largest prefix of coordinates that fit in the capacity W and the rest are discarded. We emphasize that this sorting step is crucial for the analysis and differs from the scheme in [1]. Figure 2 describes the formal algorithm.

The Key Property for the Analysis: The analysis relies on obtaining a bound on the rejection probability of coordinate j by constraint i. Let X_j be the indicator variable for j being chosen in the first step. We show that $\Pr[E_{ij} | X_j = 1] \leq cA_{ij}$ for some c that depends on the scaling factor α. Thus coordinates with smaller coefficients are less likely to be rejected. The total rejection probability of j, $\sum_{i=1}^{m} \Pr[E_{ij} | X_j = 1]$, is proportional to the column sum of coordinate j which is at most Δ_1.

The analysis relies on the Chernoff bound, and depending on the parameters, one needs to adjust the analysis. In order to highlight the main ideas we provide a detailed proof for the simplest case and include the proofs of some of the other cases in the appendix. The rest of the proofs can be found in the full version [5].

4.1   An $\Omega(1/\Delta_1)$-approximation Algorithm

We show that round-and-alter-by-sorting yields an $\Omega(1/\Delta_1)$-approximation if we set the scaling factor $\alpha_1 = \frac{1}{c_1\Delta_1}$ where $c_1 = 4e^{1+1/e}$.

The rejection probability is captured by the following main lemma.

Lemma 2. *Let $\alpha_1 = \frac{1}{c_1\Delta_1}$ for $c_1 = 4e^{1+1/e}$. Let $i \in [m]$ and $j \in [n]$. Then in the algorithm* round-and-alter-by-sorting(A, b, α_1), *we have* $\Pr[E_{ij}|X_j = 1] \le \frac{A_{i,j}}{2\Delta_1}$.

Proof. At iteration i of round-and-alter-by-sorting, after the set $\{A_{i,1}, \ldots, A_{i,n}\}$ is sorted, the indices are renumbered so that $A_{i,1} \le \cdots \le A_{i,n}$. Note that j may now be a different index j', but for simplicity of notation we will refer to j' as j. Let $\xi_\ell = 1$ if $x'_\ell = 1$ and 0 otherwise. Let $Y_{ij} = \sum_{\ell=1}^{j-1} A_{i,\ell}\xi_\ell$.

If E_{ij} occurs, then $Y_{ij} > W - A_{i,j}$, since x''_j would not have been set to zero by constraint i. That is,

$$\Pr[E_{ij}|X_j = 1] \le \Pr[Y_{ij} > W - A_{i,j}|X_j = 1].$$

The event $Y_{ij} > W - A_{i,j}$ does not depend on x'_j. Therefore,

$$\Pr[Y_{ij} > W - A_{i,j}|X_j = 1] \le \Pr[Y_{ij} \ge W - A_{i,j}].$$

To upper bound $\mathbb{E}[Y_{ij}]$, we have

$$\mathbb{E}[Y_{ij}] = \sum_{\ell=1}^{j-1} A_{i,\ell} \cdot \Pr[X_\ell = 1] \le \alpha_1 \sum_{\ell=1}^{n} A_{i,\ell}x_\ell \le \alpha_1 W.$$

As $A_{i,j} \le 1$, $W \ge 2$, and $\alpha_1 < 1/2$, we have $\frac{(1-\alpha_1)W}{A_{i,j}} > 1$. Using the fact that $A_{i,j}$ is at least as large as all entries $A_{i,j'}$ for $j' < j$, we satisfy the conditions to apply the Chernoff bound in Theorem 7. This implies

$$\Pr[Y_{ij} > W - A_{i,j}] \le \left(\frac{\alpha_1 e^{1-\alpha_1}W}{W - A_{i,j}}\right)^{(W-A_{i,j})/A_{i,j}}.$$

Note that $\frac{W}{W-A_{i,j}} \le 2$ as $W \ge 2$. Because $e^{1-\alpha_1} \le e$ and by the choice of α_1, we have

$$\left(\frac{\alpha_1 e^{1-\alpha_1}W}{W - A_{i,j}}\right)^{(W-A_{i,j})/A_{i,j}} \le (2e\alpha_1)^{(W-A_{i,j})/A_{i,j}} = \left(\frac{1}{2e^{1/e}\Delta_1}\right)^{(W-A_{i,j})/A_{i,j}}.$$

Then we prove the final inequality in two parts. First, we see that $W \ge 2$ and $A_{i,j} \le 1$ imply that $\frac{W-A_{i,j}}{A_{i,j}} \ge 1$. This implies

$$\left(\frac{1}{2\Delta_1}\right)^{(W-1)/A_{i,j}} \le \frac{1}{2\Delta_1}.$$

Second, we see that

$$(1/e^{1/e})^{(W-A_{i,j})/A_{i,j}} \le (1/e^{1/e})^{1/A_{i,j}} \le A_{i,j}$$

for $A_{i,j} \le 1$, where the first inequality holds because $W - A_{i,j} \ge 1$ and the second inequality holds by Lemma 7. This concludes the proof.

Theorem 2. *When setting $\alpha_1 = \frac{1}{c_1 \Delta_1}$ where $c_1 = 4e^{1+1/e}$, for PIPs with width $W \geq 2$, round-and-alter-by-sorting(A, b, α_1) is a randomized $(\alpha_1/2)$-approximation algorithm.*

Proof. Fix $j \in [n]$. By Lemma 2 and the definition of Δ_1, we have

$$\sum_{i=1}^{m} \Pr[E_{ij}|X_j = 1] \leq \sum_{i=1}^{m} \frac{A_{i,j}}{2\Delta_1} \leq \frac{1}{2}.$$

By Lemma 1, which shows that upper bounding the sum of the rejection probabilities by γ for every item leads to an $\alpha_1(1 - \gamma)$-approximation, we get the desired result.

4.2 An $\Omega(\frac{1}{(1+\Delta_1/W)^{1/(W-1)}})$-approximation

We improve the bound from the previous section by setting $\alpha_1 = \frac{1}{c_2(1+\Delta_1/W)^{1/(W-1)}}$ where $c_2 = 4e^{1+2/e}$. Note that the scaling factor becomes larger as W increases. The proof of the following lemma can be found in the appendix.

Lemma 3. *Let $\alpha_1 = \frac{1}{c_2(1+\Delta_1/W)^{1/(W-1)}}$ for $c_2 = 4e^{1+2/e}$. Let $i \in [m]$ and $j \in [n]$. Then in the algorithm round-and-alter-by-sorting(A, b, α_1), we have $\Pr[E_{ij}|X_j = 1] \leq \frac{A_{i,j}}{2\Delta_1}$.*

If we replace Lemma 2 with Lemma 3 in the proof of Theorem 2, we obtain the following stronger guarantee.

Theorem 3. *When setting $\alpha_1 = \frac{1}{c_2(1+\Delta_1/W)^{1/(W-1)}}$ where $c_2 = 4e^{1+2/e}$, for PIPs with width $W \geq 2$, round-and-alter-by-sorting(A, b, α_1) is a randomized $(\alpha_1/2)$-approximation.*

4.3 A $(1 - O(\epsilon))$-approximation When $W \geq \Omega(\frac{1}{\epsilon^2} \ln(\frac{\Delta_1}{\epsilon}))$

In this section, we give a randomized $(1-O(\epsilon))$-approximation for the case when $W \geq \Omega(\frac{1}{\epsilon^2} \ln(\frac{\Delta_1}{\epsilon}))$. We use the algorithm round-and-alter-by-sorting in Fig. 2 with the scaling factor $\alpha_1 = 1 - \epsilon$.

Lemma 4. *Let $0 < \epsilon < \frac{1}{e}$, $\alpha_1 = 1 - \epsilon$, and $W = \frac{2}{\epsilon^2} \ln(\frac{\Delta_1}{\epsilon}) + 1$. Let $i \in [m]$ and $j \in [n]$. Then in round-and-alter-by-sorting(A, b, α_1), we have $\Pr[E_{ij}|X_j = 1] \leq e \cdot \frac{\epsilon A_{i,j}}{\Delta_1}$.*

Lemma 4 implies that we can upper bound the sum of the rejection probabilities for any item j by $e\epsilon$, leading to the following theorem.

Theorem 4. *Let $0 < \epsilon < \frac{1}{e}$ and $W = \frac{2}{\epsilon^2} \ln(\frac{\Delta_1}{\epsilon}) + 1$. When setting $\alpha_1 = 1 - \epsilon$ and $c = e + 1$, round-and-alter-by-sorting(A, b, α_1) is a randomized $(1 - c\epsilon)$-approximation algorithm.*

5 The Small Width Regime: $W = (1 + \epsilon)$

We now consider the regime when the width is small. Let $W = 1 + \epsilon$ for some $\epsilon \in (0, 1]$. We cannot apply the simple sorting based scheme that we used for the large width regime. We borrow the idea from [1] in splitting the coordinates into big and small in each constraint; now the definition is more refined and depends on ϵ. Moreover, the small coordinates and the big coordinates have their own reserved capacity in the constraint. This is crucial for the analysis. We provide more formal details below.

We set α_2 to be $\frac{\epsilon^2}{c_3 \Delta_1}$ where $c_3 = 8e^{1+2/e}$. The alteration step differentiates between "small" and "big" coordinates as follows. For each $i \in [m]$, let $S_i = \{j : A_{i,j} \leq \epsilon/2\}$ and $B_i = \{j : A_{i,j} > \epsilon/2\}$. We say that an index j is *small* for constraint i if $j \in S_i$. Otherwise we say it is *big* for constraint i when $j \in B_i$. For each constraint, the algorithm is allowed to pack a total of $1 + \epsilon$ into that constraint. The algorithm separately packs small indices and big indices. In an ϵ amount of space, small indices that were chosen in the rounding step are sorted in increasing order of size and greedily packed until the constraint is no longer satisfied. The big indices are packed by arbitrarily choosing one and packing it into the remaining space of 1. The rest of the indices are removed to ensure feasibility. Figure 3 gives pseudocode for the randomized algorithm round-alter-small-width which yields an $\Omega(\epsilon^2/\Delta_1)$-approximation.

---

round-alter-small-width$(A, b, \epsilon, \alpha_2)$:
    let x be the optimum fractional solution of the natural LP relaxation
    for $j \in [n]$, set x_j' to be 1 independently with probability $\alpha_2 x_j$ and 0 otherwise
    $x'' \leftarrow x'$
    for $i \in [m]$ do
        if $|S_i| = 0$ then
            $s \leftarrow 0$
        else
            sort and renumber such that $A_{i,1} \leq \cdots \leq A_{i,n}$
            $s \leftarrow \max \left\{ \ell \in S_i : \sum_{j=1}^{\ell} A_{i,j} x_j' \leq \epsilon \right\}$
        end if
        if $|B_i| = 0$, then $t = 0$, otherwise let t be an arbitrary element of B_i
        for each $j \in [n]$ such that $j > s$ and $j \neq t$, set $x_j'' = 0$
    end for
    return x''

Fig. 3. By setting the scaling factor $\alpha_2 = \frac{\epsilon^2}{c \Delta_1}$ for a sufficiently large constant c, round-alter-small-width is a randomized $\Omega(\epsilon^2/\Delta_1)$-approximation for PIPs with width $W = 1 + \epsilon$ for some $\epsilon \in (0, 1]$ (see Theorem 5).

It remains to bound the rejection probabilities. Recall that for $j \in [n]$, we define X_j to be the indicator random variable $\mathbb{1}(x_j' = 1)$ and E_{ij} is the event that j was rejected by constraint i.

We first consider the case when index j is big for constraint i. Note that it is possible that there may not exist any big indices for a given constraint. The same holds true for small indices.

Lemma 5. *Let $\epsilon \in (0,1]$ and $\alpha_2 = \frac{\epsilon^2}{c_3 \Delta_1}$ where $c_3 = 8e^{1+2/e}$. Let $i \in [m]$ and $j \in B_i$. Then in* round-alter-small-width$(A, b, \epsilon, \alpha_2)$, *we have* $\Pr[E_{ij}|X_j = 1] \leq \frac{A_{i,j}}{2\Delta_1}$.

Proof. Let $\mathcal{E}$ be the event that there exists $j' \in B_i$ such that $j' \neq j$ and $X_{j'} = 1$. Observe that if E_{ij} occurs and $X_j = 1$, then it must be the case that at least one other element of B_i was chosen in the rounding step. Thus,

$$\Pr[E_{ij}|X_j = 1] \leq \Pr[\mathcal{E}] \leq \sum_{\substack{\ell \in B_i \\ \ell \neq j}} \Pr[X_\ell = 1] \leq \alpha_2 \sum_{\ell \in B_i} x_\ell,$$

where the second inequality follows by the union bound. Observe that for all $\ell \in B_i$, we have $A_{i,\ell} > \epsilon/2$. By the LP constraints, we have $1 + \epsilon \geq \sum_{\ell \in B_i} A_{i,\ell} x_\ell > \frac{\epsilon}{2} \cdot \sum_{\ell \in B_i} x_\ell$. Thus, $\sum_{\ell \in B_i} x_\ell \leq \frac{1+\epsilon}{\epsilon/2} = 2/\epsilon + 2$.

Using this upper bound for $\sum_{\ell \in B_i} x_\ell$, we have

$$\alpha_2 \sum_{\ell \in B_i} x_\ell \leq \frac{\epsilon^2}{c_3 \Delta_1} \left(\frac{2}{\epsilon} + 2 \right) \leq \frac{4\epsilon}{c_3 \Delta_1} \leq \frac{A_{i,j}}{2\Delta_1},$$

where the second inequality utilizes the fact that $\epsilon \leq 1$ and the third inequality holds because $c_3 \geq 16$ and $A_{i,j} > \epsilon/2$.

Next we consider the case when index j is small for constraint i. The analysis here is similar to that in the preceding section with width at least 2 and thus the proof is deferred to the full version [5].

Lemma 6. *Let $\epsilon \in (0,1]$ and $\alpha_2 = \frac{\epsilon^2}{c_3 \Delta_1}$ where $c_3 = 8e^{1+2/e}$. Let $i \in [m]$ and $j \in S_i$. Then in* round-alter-small-width$(A, b, \epsilon, \alpha_2)$, *we have* $\Pr[E_{ij}|X_j = 1] \leq \frac{A_{i,j}}{2\Delta_1}$.

Theorem 5. *Let $\epsilon \in (0,1]$. When setting $\alpha_2 = \frac{\epsilon^2}{c_3 \Delta_1}$ for $c_3 = 8e^{1+2/e}$, for PIPs with width $W = 1 + \epsilon$,* round-alter-small-width$(A, b, \epsilon, \alpha_2)$ *is a randomized $(\alpha_2/2)$-approximation algorithm.*

Proof. Fix $j \in [n]$. Then by Lemmas 5 and 6 and the definition of Δ_1, we have

$$\sum_{i=1}^{m} \Pr[E_{ij}|X_j = 1] \leq \sum_{i=1}^{m} \frac{A_{i,j}}{2\Delta_1} \leq \frac{1}{2}.$$

Recall that Lemma 1 gives an $\alpha_2(1 - \gamma)$-approximation where γ is an upper bound on the sum of the rejection probabilities for any item. This concludes the proof.

Appendix

A   Chernoff Bounds and Useful Inequalities

The following standard Chernoff bound is used to obtain a more convenient Chernoff bound in Theorem 7. The proof of Theorem 7 follows directly from choosing δ such that $(1 + \delta)\mu = W - \beta$ and applying Theorem 6.

Theorem 6 ([12]). *Let $X_1, \ldots, X_n$ be independent random variables where X_i is defined on $\{0, \beta_i\}$, where $0 < \beta_i \leq \beta \leq 1$ for some β. Let $X = \sum_i X_i$ and denote $\mathbb{E}[X]$ as μ. Then for any $\delta > 0$,*

$$\Pr[X \geq (1 + \delta)\mu] \leq \left(\frac{e^\delta}{(1 + \delta)^{1+\delta}} \right)^{\mu/\beta}$$

Theorem 7. *Let $X_1, \ldots, X_n \in [0, \beta]$ be independent random variables for some $0 < \beta \leq 1$. Suppose $\mu = \mathbb{E}[\sum_i X_i] \leq \alpha W$ for some $0 < \alpha < 1$ and $W \geq 1$ where $(1 - \alpha)W > \beta$. Then*

$$\Pr\left[\sum_i X_i > W - \beta \right] \leq \left(\frac{\alpha e^{1-\alpha} W}{W - \beta} \right)^{(W-\beta)/\beta}.$$

Lemma 7. *Let $x \in (0, 1]$. Then $(1/e^{1/e})^{1/x} \leq x$.*

Lemma 8. *Let $y \geq 2$ and $x \in (0, 1]$. Then $x/y \geq (1/e^{2/e})^{y/2x}$.*

B   Skipped Proofs

B.1   Proof of Theorem 1

Proof. Let $G = (V, E)$ be an undirected graph without self-loops and let $n = |V|$. Let $A \in [0, 1]^{n \times n}$ be indexed by V. For all $v \in V$, let $A_{v,v} = 1$. For all $uv \in E$, let $A_{u,v} = A_{v,u} = 1/n$. For all the remaining entries in A that have not yet been defined, set these entries to 0. Consider the following PIP:

$$\text{maximize } \langle x, 1 \rangle \text{ over } x \in \{0, 1\}^n \text{ s.t. } Ax \leq 1. \tag{1}$$

Let S be the set of all feasible integral solutions of (1) and $\mathcal{I}$ be the set of independent sets of G. Define $g : S \to \mathcal{I}$ where $g(x) = \{v : x_v = 1\}$. To show g is surjective, consider a set $I \in \mathcal{I}$. Let y be the characteristic vector of I. That is, y_v is 1 if $v \in I$ and 0 otherwise. Consider the row in A corresponding to an arbitrary vertex u where $y_u = 1$. For all $v \in V$ such that v is a neighbor to u, $y_v = 0$ as I is an independent set. Thus, as the nonzero entries in A of the row corresponding to u are, by construction, the neighbors of u, it follows that the constraint corresponding to u is satisfied in (1). As u is an arbitrary vertex, it follows that y is a feasible integral solution to (1) and as $I = \{v : y_v = 1\}$, $g(y) = I$.

Define $h : S \to \mathbb{N}_0$ such that $h(x) = |g(x)|$. It is clear that $\max_{x \in S} h(x)$ is equal to the optimal value of (1). Let I_{max} be a maximum independent set of G. As g is surjective, there exists $z \in S$ such that $g(z) = I_{max}$. Thus, $\max_{x \in S} h(x) \geq |I_{max}|$. As $\max_{x \in S} h(x)$ is equal to the optimum value of (1), it follows that a β-approximation for PIPs implies a β-approximation for maximum independent set.

Furthermore, we note that for this PIP, $\Delta_1 \leq 2$, thus concluding the proof.

B.2 Proof of Lemma 3

Proof. The proof proceeds similarly to the proof of Lemma 2. Since $\alpha_1 < 1/2$, everything up to and including the application of the Chernoff bound there applies. This gives that for each $i \in [m]$ and $j \in [n]$,

$$\Pr[E_{ij}|X_j = 1] \leq (2e\alpha_1)^{(W - A_{i,j})/A_{i,j}} .$$

By choice of α_1, we have

$$(2e\alpha_1)^{(W - A_{i,j})/A_{i,j}} = \left(\frac{1}{2e^{2/e}(1 + \Delta_1/W)^{1/(W-1)}} \right)^{(W - A_{i,j})/A_{i,j}}$$

We prove the final inequality in two parts. First, note that $\frac{W - A_{i,j}}{A_{i,j}} \geq W - 1$ since $A_{i,j} \leq 1$. Thus,

$$\left(\frac{1}{2(1 + \Delta_1/W)^{1/(W-1)}} \right)^{(W - A_{i,j})/A_{i,j}} \leq \frac{1}{2^{W-1}(1 + \Delta_1/W)} \leq \frac{W}{2\Delta_1}.$$

Second, we see that

$$\left(\frac{1}{e^{2/e}} \right)^{(W - A_{i,j})/A_{i,j}} \leq \left(\frac{1}{e^{2/e}} \right)^{W/2A_{i,j}} \leq \frac{A_{i,j}}{W}$$

for $A_{i,j} \leq 1$, where the first inequality holds because $W \geq 2$ and the second inequality holds by Lemma 8.

References

1. Bansal, N., Korula, N., Nagarajan, V., Srinivasan, A.: Solving packing integer programs via randomized rounding with alterations. Theory Comput. **8**(24), 533–565 (2012). https://doi.org/10.4086/toc.2012.v008a024
2. Chan, T.M.: Approximation schemes for 0-1 knapsack. In: 1st Symposium on Simplicity in Algorithms (2018)
3. Chekuri, C., Khanna, S.: On multidimensional packing problems. SIAM J. Comput. **33**(4), 837–851 (2004)
4. Chekuri, C., Quanrud, K.: On approximating (sparse) covering integer programs. In: Proceedings of the Thirtieth Annual ACM-SIAM Symposium on Discrete Algorithms, pp. 1596–1615. SIAM (2019)

5. Chekuri, C., Quanrud, K., Torres, M.R.: ℓ_1-sparsity approximation bounds for packing integer programs (2019). arXiv preprint: arXiv:1902.08698
6. Chekuri, C., Vondrák, J., Zenklusen, R.: Submodular function maximization via the multilinear relaxation and contention resolution schemes. SIAM J. Comput. **43**(6), 1831–1879 (2014)
7. Chen, A., Harris, D.G., Srinivasan, A.: Partial resampling to approximate covering integer programs. In: Proceedings of the Twenty-Seventh Annual ACM-SIAM Symposium on Discrete Algorithms, pp. 1984–2003. Society for Industrial and Applied Mathematics (2016)
8. Frieze, A., Clarke, M.: Approximation algorithms for the m-dimensional 0-1 knapsack problem: worst-case and probabilistic analyses. Eur. J. Oper. Res. **15**(1), 100–109 (1984)
9. Harvey, N.J.: A note on the discrepancy of matrices with bounded row and column sums. Discrete Math. **338**(4), 517–521 (2015)
10. Håstad, J.: Clique is hard to approximate within $n^{1-\epsilon}$. Acta Math. **182**(1), 105–142 (1999)
11. Hazan, E., Safra, S., Schwartz, O.: On the complexity of approximating k-set packing. Comput. Complex. **15**(1), 20–39 (2006)
12. Mitzenmacher, M., Upfal, E.: Probability and Computing: Randomized Algorithms and Probabilistic Analysis. Cambridge University Press, Cambridge (2005)
13. Pritchard, D.: Approximability of sparse integer programs. In: Fiat, A., Sanders, P. (eds.) ESA 2009. LNCS, vol. 5757, pp. 83–94. Springer, Heidelberg (2009). https://doi.org/10.1007/978-3-642-04128-0_8
14. Pritchard, D., Chakrabarty, D.: Approximability of sparse integer programs. Algorithmica **61**(1), 75–93 (2011)
15. Raghavan, P., Tompson, C.D.: Randomized rounding: a technique for provably good algorithms and algorithmic proofs. Combinatorica **7**(4), 365–374 (1987)
16. Srinivasan, A.: Improved approximation guarantees for packing and covering integer programs. SIAM J. Comput. **29**(2), 648–670 (1999)
17. Trevisan, L.: Non-approximability results for optimization problems on bounded degree instances. In: Proceedings of the Thirty-Third Annual ACM Symposium on Theory of Computing, pp. 453–461. ACM (2001)
18. Zuckerman, D.: Linear degree extractors and the inapproximability of max clique and chromatic number. In: Proceedings of the Thirty-Eighth Annual ACM Symposium on Theory of Computing, pp. 681–690. ACM (2006)

A General Framework for Handling Commitment in Online Throughput Maximization

Lin Chen[1], Franziska Eberle[2(✉)], Nicole Megow[2], Kevin Schewior[3,4], and Cliff Stein[5]

[1] Department of Computer Science, University of Houston, Houston, TX, USA
chenlin198662@gmail.com
[2] Department for Mathematics/Computer Science,
University of Bremen, Bremen, Germany
{feberle,nicole.megow}@uni-bremen.de
[3] Fakultät für Informatik, Technische Universität München, München, Germany
kschewior@gmail.com
[4] Département d'Informatique, École Normale Supérieure, Paris, France
[5] Department of IEOR, Columbia University, New York, USA
cliff@ieor.columbia.edu

Abstract. We study a fundamental online job admission problem where jobs with deadlines arrive online over time at their release dates, and the task is to determine a preemptive single-server schedule which maximizes the number of jobs that complete on time. To circumvent known impossibility results, we make a standard slackness assumption by which the feasible time window for scheduling a job is at least $1+\varepsilon$ times its processing time, for some $\varepsilon > 0$. We quantify the impact that different provider commitment requirements have on the performance of online algorithms. Our main contribution is one universal algorithmic framework for online job admission both with and without commitments. Without commitment, our algorithm with a competitive ratio of $\mathcal{O}(1/\varepsilon)$ is the best possible (deterministic) for this problem. For commitment models, we give the first non-trivial performance bounds. If the commitment decisions must be made before a job's slack becomes less than a δ-fraction of its size, we prove a competitive ratio of $\mathcal{O}(\varepsilon/((\varepsilon - \delta)\delta^2))$, for $0 < \delta < \varepsilon$. When a scheduler must commit upon starting a job, our bound is $\mathcal{O}(1/\varepsilon^2)$. Finally, we observe that for scheduling with commitment the restriction to the "unweighted" throughput model is essential; if jobs have individual weights, we rule out competitive deterministic algorithms.

N. Megow—Supported by the German Science Foundation (DFG) Grant ME 3825/1.
K. Schewior—Supported by CONICYT Grant PII 20150140 and DAAD PRIME program.
C. Stein—Research partly supported by NSF Grants CCF-1714818 and CCF-1822809.

A. Lodi and V. Nagarajan (Eds.): IPCO 2019, LNCS 11480, pp. 141–154, 2019.
https://doi.org/10.1007/978-3-030-17953-3_11

1 Introduction

Many modern computing environments, such as internal clusters and public clouds, involve a centralized system for managing the resource allocation of a large diverse workload [21] with a heterogeneous mixture of jobs. In this paper, we will study scheduling policies, evaluated by the commonly used notion of *throughput* which is the number of jobs completed, or if jobs have weights, the total weight of jobs completed. Throughput is a "social welfare" objective that tries to maximize total utility. To this end, a solution may abort jobs close to their deadlines in favor of many shorter and more urgent tasks [11]. However, for many industrial applications, service providers have to *commit to complete* admitted jobs since without such a guarantee, some applications will fail or customers may be unhappy and choose another environment.

Formally, we consider a fundamental single-machine scheduling model in which jobs arrive online over time at their *release date* r_j. Each job has a *processing time* $p_j \geq 0$, a *deadline* d_j, and possibly a *weight* $w_j > 0$. In order to complete, a job must receive p_j units of processing time in the interval $[r_j, d_j)$. If a schedule completes a set S of jobs, then the *throughput* is $|S|$ while the weighted throughput is $\sum_{j \in S} w_j$. To measure the quality of an online algorithm, we use standard *competitive analysis* where its performance is compared to that of an optimal offline algorithm with full knowledge of the future.

Deadline-based objectives are typically much harder to optimize than other Quality-of-Service (QoS) metrics such as response time or makespan. Indeed, the problem becomes hopeless when *preemption* (interrupting a job and resuming it later) is not allowed: whenever an algorithm starts a job j without being able to preempt it, it may miss the deadlines of an arbitrary number of jobs. For scheduling with commitment, we provide a similarly strong lower bound for the preemptive version of the problem in the presence of weights. Therefore, we focus on *unweighted preemptive online* throughput maximization.

Hard examples for online algorithms tend to involve jobs that arrive and then *must* immediately be processed since $d_j - r_j \approx p_j$. To bar such jobs from a system, we require that any submitted job contains some *slack*. An instance has ε-slack if every job satisfies $d_j - r_j \geq (1 + \varepsilon)p_j$. We develop algorithms whose competitive ratio depends on ε. This slackness parameter captures certain aspects of QoS provisioning and admission control, see, e.g., [13,19], and it has been considered in previous work, e.g., in [2,4,12,14,21,23]. Other results for scheduling with deadlines use speed scaling, which can be viewed as adding slack to the schedule, e.g., [1,3,15,22]. In this paper we quantify the impact that different commitment requirements have on the performance of online algorithms.

1.1 Our Results and Techniques

Our main contribution is a general algorithmic framework, called the **region algorithm**, for online scheduling with and without commitments. We prove performance guarantees which are either tight or constitute the first non-trivial results. We also answer open questions in previous work. We show strong lower bounds for the weighted case. Thus, our algorithms are all for unit weights $w_j = 1$.

Optimal Algorithm for Scheduling Without Commitment. We show that the region algorithm achieves a competitive ratio of $\mathcal{O}(\frac{1}{\varepsilon})$, and give a matching lower bound (ignoring constants) for any deterministic online algorithm.

Impossibility Results for Commitment Upon Job Arrival. In this most restrictive model, an algorithm must decide immediately at a job's release date if the job will be completed. We show that no (randomized) online algorithm admits a bounded competitive ratio. Such a lower bound has only been shown by exploiting job weights [21,25]. Hence, we do not consider this model further.

Scheduling With Commitment. We distinguish two different models: *(i) commitment upon job admission* and *(ii) δ-commitment*. In the first model, an algorithm may discard a job any time before its start. In the second model, an online algorithm must commit to complete a job when its slack has reduced from the original slack requirement of an ε-fraction of the size to a δ-fraction for $0 < \delta < \varepsilon$, modeling an early-enough commitment for mission-critical jobs. We show that implementations of the region algorithm yield a competitive ratio of $\mathcal{O}(1/\varepsilon^2)$ for commitment upon admission and a competitive ratio of $\mathcal{O}(\varepsilon/((\varepsilon - \delta)\delta^2))$, for $0 < \delta < \varepsilon$, in the δ-commitment model. These are the first rigorous non-trivial upper bounds—for any commitment model (excluding $w_j = p_j$).

Instances with arbitrary weights are hopeless without further restrictions. We show that there is no deterministic online algorithm with bounded competitive ratio, neither for commitment upon admission (also shown in [2]) nor for δ-commitment. Informally, our construction implies that there is no deterministic online algorithm with bounded competitive ratio in *any commitment model* in which a scheduler may have to commit to a job before it has completed. (See Sect. 5 for more details.) We rule out bounded performance guarantees for $\varepsilon \in (0,1)$. For sufficiently large slackness ($\varepsilon > 3$), an online algorithm is provided in [2] that has bounded competitive ratio. Our new lower bound answers affirmatively the open question of whether high slackness is indeed required.

Finally, our impossibility result for weighted jobs and the positive result for instances without weights clearly separate the weighted from the unweighted setting. Hence, we do not consider algorithms for weighted throughput.

Our Techniques. Once a job j is admitted to the system, its slack becomes a scarce resource: To complete the job on time one needs to carefully "spend" the slack on admitting jobs to be processed before the deadline of j. Our general framework for admission control, the **region algorithm**, addresses this issue by the concept of "responsibility": Whenever a job j' is admitted while j could be processed, j' becomes responsible for not admitting similar-length jobs for a certain period, its *region*. The intention is that j' reserves time for j to complete. To balance between reservation (commitment to complete j) and performance (loss of other jobs), the algorithm uses the parameters α and β, which specify the length of a region and the similarity of job lengths.

A major difficulty in the analysis is understanding the complex interval structure formed by feasible time windows, regions, and processing time intervals.

Here, the key ingredient is that regions are defined independently of scheduling decisions. Thus, the analysis can be naturally split into two parts. **In the first part,** we argue that the scheduling routine can handle the admitted jobs sufficiently well for aptly chosen parameters α and β. That means that the respective commitment model is obeyed and, if not implied by that, an adequate number of the admitted jobs is completed. **In the second part,** we can disregard how jobs are actually scheduled and argue that the region algorithm admits sufficiently many jobs to be competitive with an optimum solution. The above notion of "responsibility" suggests a proof strategy mapping jobs that are completed in the optimum to the corresponding job that was "responsible" due to its region. Transforming this idea into a charging scheme is, however, a non-trivial task as there might be many ($\gg \Theta(\frac{1}{\varepsilon^2})$) jobs released within the region of a single job j and completed by the optimum but not admitted by the region algorithm due to many consecutive regions of varying size. We develop a careful charging scheme that avoids such overcharging. We handle the complex interval structure by working on a natural tree structure (*interruption tree*) related to the region construction and independent of the actual schedule. Our charging scheme comprises two central routines for distributing charge: Moving charge along a sequence of consecutive jobs (*Push Forward*) or to children (*Push Down*).

1.2   Previous Results

Preemptive online scheduling and admission control have been studied rigorously, see, e.g., [5,12,14] and references therein. Impossibility results for jobs with hard deadlines and without slack have been known for decades [6,7,17,18,20].

Most research on online scheduling does not address commitment. The only results independent of slack (or other job-dependent parameters) concern weighted throughput for the special case $w_j = p_j$, where a constant competitive ratio is possible [6,17,18,24]. In the unweighted setting, a randomized $\mathcal{O}(1)$-competitive algorithm is known [16]. For instances with ε-slack, an $\mathcal{O}(\frac{1}{\varepsilon^2})$-competitive algorithm in the general weighted setting is given in [21]. To the best of our knowledge, no lower bound was known to date.

Much less is known for scheduling with commitment. In the most restrictive model, *commitment upon job arrival*, Lucier et al. [21] rule out competitive online algorithms for any slack parameter ε when jobs have arbitrary weights. For *commitment upon job admission*, they give a heuristic that empirically performs very well but without a rigorous worst-case bound. Azar et al. [2] show that no bounded competitive ratio is possible for weighted throughput maximization for small ε. For the δ-*commitment model*, [2] design (in the context of truthful mechanisms) an online algorithm that is $\mathcal{O}(\frac{1}{\varepsilon^2})$-competitive for large slack ε. They left open if this latter condition is an inherent property of any committed scheduler in this model which we answer affirmatively. The machine utilization variant ($w_j = p_j$) is better tractable as greedy algorithms achieve the best possible competitive ratio $\Theta(\frac{1}{\varepsilon})$ [10,12] in all mentioned commitment models.

2 Our General Framework

2.1 The Region Algorithm

We now present our general algorithmic framework for scheduling with and without commitment. We assume that the slackness constant $\varepsilon > 0$ and, in the δ-commitment model, $0 < \delta < \varepsilon$ are known to the online algorithm.

Algorithm 1.1. Region algorithm

Scheduling routine: At any time t, run an admitted and not yet completed job with shortest processing time.

Event: Upon release of a new job at time t or Upon ending of a region at time t: Call region **preemption routine**.

Region preemption routine:
$k \leftarrow$ the job whose region contains t
$i \leftarrow$ a shortest available job at t, i.e.,
$\quad i = \arg\min\{p_j \mid r_j \le t \text{ and } d_j - t \ge (1+\delta)p_j\}$
If $p_i < \beta p_k$, then
1. admit job i and reserve region $R(i) = [t, t + \alpha p_i)$,
2. update remaining regions $R(j)$ with $R(j) \cap [t, \infty) \ne \emptyset$ as described below

---

We first describe informally three underlying design principles. The third principle is crucial to improve on existing results that only use the first two [21].

1. A running job can be preempted only by smaller jobs (parameter β).
2. A job cannot start for the first time when its remaining slack is too small (constant δ in the δ-commitment model and otherwise set to $\delta = \frac{\varepsilon}{2}$).
3. If a job preempts other jobs, then it takes "responsibility" for a certain time interval (parameter α) in which the jobs it preempted can be processed.

The region algorithm has two parameters, $\alpha \ge 1$ and $0 < \beta < 1$. A *region*, $R(j)$ for job j, is a union of time intervals associated with j, and the size of the region is the sum of sizes of the intervals. Region $R(j)$ will always have size αp_j, although the particular time intervals composing the region may change over time. Regions are always disjoint. Informally, whenever our algorithm starts a job i (we say i is *admitted*) that arrives during the region of an already admitted job j, then the current interval of j is split into two intervals and the region $R(j)$ and all later regions are delayed.

Formally, at any time t, the region algorithm maintains two sets of jobs: *admitted jobs*, which have been started before or at time t, and *available jobs*. A job j is available if it is released before or at time t, is not yet admitted, and it is not too close to its deadline, i.e., $r_j \le t$ and $d_j - t \ge (1+\delta)p_j$. The intelligence of the region algorithm lies in admitting jobs and (re)allocating regions. The actual scheduling decisions then are independent of the regions: at any point in time, schedule the shortest admitted job that has not completed its processing, i.e., schedule admitted jobs in *Shortest Processing Time (SPT)* order. The algorithm never explicitly considers deadlines except when deciding whether to admit jobs.

The region algorithm starts by admitting job 1 at its release date and creating the region $R(1) := [r_1, r_1 + \alpha p_1)$. Two events – the release of a job and the end of a region– trigger the **region preemption** subroutine. This subroutine compares the processing time of the smallest *available* job i with the processing time of the *admitted* job k whose region contains t. If $p_i < \beta p_k$, job i is admitted and the region algorithm reserves the interval $[t, t + \alpha p_i)$ for processing i. Since regions must be disjoint, the algorithm then modifies all other remaining regions, i.e., the parts of regions that belong to $[t, \infty)$ of other jobs j. We refer to the set of such jobs j whose regions have not yet completed by time t as $J(t)$. Intuitively, we preempt the interval of the region containing t and delay its remaining part as well as the remaining regions of all other jobs. Formally, this **update of all remaining regions** is defined as follows. Let k be the one job whose region is interrupted at time t, and let $[a'_k, b'_k)$ be the interval of $R(k)$ containing t. Interval $[a'_k, b'_k)$ is replaced by $[a'_k, t) \cup [t + \alpha p_i, b'_k + \alpha p_i)$. For all other jobs $j \in J(t) \setminus \{k\}$, the remaining region $[a'_j, b'_j)$ of j is replaced by $[a'_j + \alpha p_i, b'_j + \alpha p_i)$. Observe that, although the region of a job may change throughout the algorithm, the starting point of a region for a job will never be changed. See the summary Algorithm 1.1.

We apply the region algorithm in different commitment models with different choices of parameters α and β, which we derive in the following sections. In the δ-commitment model, δ is given as part of the input. In the other models, i.e., without commitment or with commitment upon admission, we simply set $\delta = \frac{\varepsilon}{2}$.

If the region algorithm commits to a job, it does so upon admission, which is, for our algorithm, the same as its start time. The parameter δ determines the latest possible start time of a job, which is then for our algorithm also the latest time the job can be admitted. Thus, for the analysis, the algorithm's execution for commitment upon admission ($\delta = \frac{\varepsilon}{2}$) is a special case of δ-commitment. This is true only for our algorithm, not in general.

2.2   Main Results on the Region Algorithm

In the analysis we focus on instances with small slack ($0 < \varepsilon \leq 1$) as for $\varepsilon > 1$ we run our algorithm simply by setting $\varepsilon = 1$ and obtain constant competitive ratios.

Without commitment, we give an optimal online algorithm which is an exponential improvement upon a previous result [21] (given for weighted throughput). For scheduling with commitment, we give the first rigorous upper bound.

Theorem 1 (Scheduling Without Commitment). *Let $0 < \varepsilon \leq 1$. Choosing $\alpha = 1$, $\beta = \frac{\varepsilon}{4}$, $\delta = \frac{\varepsilon}{2}$, the region algorithm is $\Theta(\frac{1}{\varepsilon})$-competitive for scheduling without commitment.*

Theorem 2 (Scheduling With Commitment). *Let $0 < \delta < \varepsilon \leq 1$. Choosing $\alpha = \frac{8}{\delta}$, $\beta = \frac{\delta}{4}$, the region algorithm is $\mathcal{O}(\frac{\varepsilon}{(\varepsilon - \delta)\delta^2})$-competitive in the δ-commitment model. When the scheduler has to commit upon admission, the region algorithm has a competitive ratio $\mathcal{O}(\frac{1}{\varepsilon^2})$ for $\alpha = \frac{4}{\varepsilon}$ and $\beta = \frac{\varepsilon}{8}$.*

2.3 Interruption Trees

To analyze the performance of the region algorithm, we retrospectively consider the final schedule and the final regions. Let a_j be the admission date of job j which was not changed while executing the algorithm. Let b_j denote the end point of j's region. Then, the convex hull of $R(j)$ is given by $\mathrm{conv}(R(j)) = [a_j, b_j)$.

Fig. 1. Gantt chart of the regions (left) and the interruption tree (right)

Our analysis crucially relies on understanding the interleaving structure of the regions that the algorithm constructs. We use a tree or forest in which each job is represented by one vertex. A job vertex is the child of another vertex if and only if the region of the latter is interrupted by the first one. The leaves correspond to jobs with non-interrupted regions. By adding a machine job M with $p_M := \infty$ and $a_M = -\infty$, we can assume that the instance is represented by a tree which we call *interruption tree*. This idea is visualized in Fig. 1, where the vertical arrows indicate the interruption of a region by another job.

Let $\pi(j)$ denote the *parent* of j. Let T_j be the subtree of the interruption tree rooted in job j and let the forest T_{-j} be T_j without its root j. By abusing notation, we denote the tree/forest as well as its jobs by T_*. A key property of this tree is that the processing times on a path are geometrically decreasing.

Lemma 1. *Let $j_1, \ldots, j_\ell$ be ℓ jobs on a path in the interruption (sub)tree T_j rooted in j such that $\pi(j_{i+1}) = j_i$. Then, $p_{j_\ell} \leq \beta p_{j_{\ell-1}} \cdots \leq \beta^{\ell-1} p_{j_1} \leq \beta^\ell p_j$ and the total processing volume is $\sum_{i=1}^\ell p_{j_i} \leq \sum_{i=1}^\ell \beta^i p_j \leq \frac{\beta}{1-\beta} \cdot p_j$.*

3 Successfully Completing Sufficiently Many Jobs

We show that the region algorithm completes sufficiently many jobs among the admitted jobs on time, when the parameters α, β, and δ are chosen properly.
Scheduling Without Commitment. Let $\delta = \frac{\varepsilon}{2}$ for $0 < \varepsilon \leq 1$.

Theorem 3. *Let $\alpha = 1$ and $\beta = \frac{\varepsilon}{4}$. Then the region algorithm completes at least half of all admitted jobs before their deadline.*

The intuition for setting $\alpha = 1$ and thus reserving regions of minimum size $|R(j)| = p_j$, for any j, is that, due to the scheduling order SPT, a job is always prioritized within its own region and, in the model without commitment, a job does not need to block extra time in the future to ensure the completion of preempted jobs. In order to prove Theorem 3, we show that a late job j implies that the subtree T_j rooted in j contains more finished than unfinished jobs.

Scheduling With Commitment. For both models, commitment at admission and δ-commitment, we give conditions on the choice of α, β, and δ such that every admitted job will complete before its deadline. We restrict in the analysis to the δ-commitment model since the algorithm otherwise runs with $\delta = \frac{\varepsilon}{2}$.

Theorem 4. *Let $\varepsilon, \delta > 0$ be fixed with $\delta < \varepsilon$. If $\alpha \geq 1$ and $0 < \beta < 1$ satisfy*

$$\frac{\alpha - 1}{\alpha} \cdot \left(1 + \delta - \frac{\beta}{1 - \beta}\right) \geq 1, \tag{1}$$

any job j admitted by the algorithm at time $a_j \leq d_j - (1 + \delta)p_j$ finishes by d_j.

For any admitted job j, we consider two types of descendants in the interruption subtree T_j whose regions intersect $[a_j, d_j)$: *(i)* jobs k with $d_j \in conv(R(k))$ form a path in T_j and, thus, Lemma 1 bounds their total processing volume from above by $\frac{\beta}{1-\beta}p_j$, *(ii)* jobs k with $R(k) \subset [a_j, d_j)$ reserve an $(\frac{\alpha-1}{\alpha})$-fraction of $R(k)$ for processing j. Thus, a straightforward calculation implies Theorem 4.

4   Competitiveness: Admission of Sufficiently Many Jobs

Theorem 5. *The number of jobs that an optimal (offline) algorithm can complete on time is by at most a multiplicative factor $\lambda + 1$ larger than the number of jobs admitted by the region algorithm, where $\lambda := \frac{\varepsilon}{\varepsilon-\delta}\frac{\alpha}{\beta}$, for $0 < \delta < \varepsilon \leq 1$.*

To prove the theorem, we fix an instance and an optimal offline algorithm OPT. Let X be the set of jobs that OPT scheduled and the region algorithm did not admit. We can assume that OPT completes all jobs in X on time. Let J denote the jobs that the region algorithm admitted. Then, $X \cup J$ is a superset of the jobs scheduled by OPT. Thus, showing $|X| \leq \lambda|J|$ implies Theorem 5.

To this end, we develop a charging procedure that assigns each job in X to a unique job in J such that each job $j \in J$ is assigned at most $\lambda = \frac{\varepsilon}{\varepsilon-\delta}\frac{\alpha}{\beta}$ jobs. For a job $j \in J$ admitted by the region algorithm we define the subset $X_j \subset X$ based on release dates. Then, we inductively transform the laminar family $(X_j)_{j\in J}$ into a partition $(Y_j)_{j\in J}$ of X with $|Y_j| \leq \lambda$ for all $j \in J$ in the proof of Lemma 2, starting with the leaves in the interruption tree as base case (Appendix, Lemma 4). For the construction of $(Y_j)_{j\in J}$, we heavily rely on the key property (Volume Lemma 3) and Corollary 1.

More precisely, for a job $j \in J$ let X_j be the set of jobs $x \in X$ that were released in the interval $[a_j, b_j)$ and satisfy $p_x < \beta p_{\pi(j)}$. Let $X_j^S := \{x \in X_j : p_x < \beta p_j\}$ and $X_j^B := X_j \setminus X_j^S$ denote the *small* and the *big* jobs, respectively, in X_j. Recall that $[a_j, b_j)$ is the convex hull of the region $R(j)$ of job j and that it includes the convex hulls of the regions of all descendants of j in the interruption tree, i.e., jobs in T_j. In particular, $X_k \subset X_j$ if $k \in T_j$.

Observation 1

1. *Any job $x \in X$ that is scheduled by* OPT *and that is not admitted by the region algorithm is released within the region of some job $j \in J$, i.e., $\bigcup_{j \in J} X_j = X$.*
2. *As the region algorithm admits any job that is small w.r.t. j and released in $R(j)$, it holds that $X_j^S = \bigcup_{k:\pi(k)=j} X_k$.*

Recall that M denotes the machine job. By Observation 1, $X = X_M^S$ and, thus, it suffices to show that $|X_M^S| \le \lambda |J|$. In fact, we show a stronger statement for each job $j \in J$. The number of small jobs in X_j is bounded by $\lambda \tau_j$ where τ_j is the number of descendants of j in the interruption tree, i.e., $\tau_j := |T_{-j}|$.

Lemma 2. *For all $j \in J \cup \{M\}$, $|X_j^S| \le \lambda \tau_j$.*

A proof sketch can be found in the appendix. We highlight the main steps here. The fine-grained definition of the sets X_j in terms of the release dates and the processing times allows us to show that any job j with $|X_j| > (\tau_j + 1)\lambda$ has *siblings* $j_1, \ldots, j_k$ such that $|X_j| + \sum_{i=1}^{k} |X_{j_i}| \le \lambda(\tau_j + 1 + \sum_{i=1}^{k}(\tau_{j_i} + 1))$. We call i and j siblings if they have the same parent in the interruption tree. Simultaneously applying this charging idea to *all* descendants of a job h already proves $|X_h^S| \le \lambda \tau_h$ as $X_h^S = \bigcup_{j:\pi(j)=h} X_j$ by Observation 1.

We prove that this "balancing" of X_j between jobs only happens between siblings $j_1, \ldots, j_k$ with the property that $b_{j_i} = a_{j_{i+1}}$ for $1 \le i < k$. We call such a set of jobs a *string* of jobs. The ellipses in Fig. 1 visualize the maximal strings of jobs. A job j is *isolated* if $b_i \ne a_j$ and $b_j \ne a_i$ for all children $i \ne j$ of $\pi(j)$.

The next (technical) lemma is a key ingredient for the "balancing" of X_j between a string of jobs. For any subset of J, we index the jobs in order of increasing admission points a_j. Conversely, for a subset of X, we order the jobs in increasing order of completion times, C_x^*, in the optimal schedule.

Lemma 3 (Volume Lemma). *Let $f, \ldots, g \in J$ be jobs with a common parent in the interruption tree. Let $x \in \bigcup_{j=f}^{g} X_j$ such that*

$$\sum_{j=f}^{g} \sum_{y \in X_j : C_y^* \le C_x^*} p_y \ge \frac{\varepsilon}{\varepsilon - \delta}(b_g - a_f) + p_x. \tag{V}$$

Then, $p_x \ge \beta p_{j^*}$, where $j^* \in J \cup \{M\}$ is the job whose region contains b_g.

The next corollary follows directly from the Volume Lemma applied to a string of jobs or to a single job $j \in J$ (let $f = j = g$). To see this, recall that X_j contains only jobs that are small w.r.t. $\pi(j)$, i.e., all $x \in X_j$ satisfy $p_x < \beta p_{\pi(j)}$.

Corollary 1. *Let $\{f, \ldots, g\} \subset J$ be a string of jobs and let $x \in \bigcup_{j=f}^{g} X_j$ satisfy (V). Then, the interruption tree contains a sibling j^* of g with $b_g = a_{j^*}$.*

The main part of the proof of Lemma 2 is to show (V) for a string of jobs only relying on $\sum_{j=f}^{g} |X_j| > \lambda \sum_{j=f}^{g}(\tau_j + 1)$. Then, Corollary 1 allows us to charge the "excess" jobs to a subsequent sibling $g + 1$. The relation between processing volume and size of job sets is possible due to the definition of X_j based on T_j.

Proof of Theorem 5. The job set scheduled by OPT clearly is a subset of $X \cup J$, the union of jobs only scheduled by OPT and the jobs admitted by the region algorithm. Thus, it suffices to prove that $|X| \leq \lambda|J|$. By Observation 1, $|X_M^S| \leq \lambda|J|$ implies $|X| \leq \lambda|J|$. This holds by applying Lemma 2 to the machine job M. □

Finalizing the Proofs of Theorems 1 and 2

Proof of Theorem 1. Set $\alpha = 1$ and $\beta = \frac{\varepsilon}{4}$. By Theorem 3 at least half of all admitted jobs complete on time. Theorem 5 implies the competitive ratio $16/\varepsilon$. □
Proof of Theorem 2. Theorem 4, $\alpha = \frac{8}{\delta}$ and $\beta = \frac{\delta}{4}$ imply that the algorithm completes all admitted jobs. Theorem 5 implies the competitive ratio $32/((\varepsilon - \delta)\delta^2 + 1)$. □

5   Lower Bounds on the Competitive Ratio

Theorem 6 (Scheduling Without Commitment). *Every deterministic online algorithm has a competitive ratio* $\Omega(\frac{1}{\varepsilon})$.

Theorem 7 (Commitment Upon Arrival). *No randomized online algorithm has a bounded competitive ratio for commitment upon arrival.*

Theorem 8 (δ-Commitment). *Consider weighted jobs in the δ-commitment model. For any $\delta > 0$ and ε with $\delta \leq \varepsilon < 1+\delta$, no deterministic online algorithm has a bounded competitive ratio.*

In particular, there is no bounded competitive ratio possible for $\varepsilon \in (0, 1)$. A restriction for ε appears to be necessary as Azar et al. [2] provide an upper bound for sufficiently large slackness, i.e., $\varepsilon > 3$. We answer affirmatively the open question in [2] if high slackness is indeed required. Again, this strong impossibility result clearly separates the weighted and the unweighted problem as we show in the unweighted setting a bounded competitive ratio for any $\varepsilon > 0$ (Theorem 2).

6   Concluding Remarks

We provide a general framework for online scheduling of deadline-sensitive jobs with and without commitment. This is the first unifying approach and we believe that it captures well (using parameters) the key design principles needed when scheduling *online*, *deadline-sensitive*, and *with commitment*. Some gaps between upper and lower bounds remain and, clearly, it would be interesting to close them. In fact, the lower bound comes from scheduling without commitment and it is unclear, if scheduling with commitment is truly harder than without. It is somewhat surprising that essentially the same algorithm performs well for both commitment models, commitment upon admission and δ-commitment, whereas a close relation between the models does not seem immediate. It remains open if an algorithm can exploit the seemingly greater flexibility of δ-commitment.

Our focus on unit-weight jobs is justified by strong impossibility results (Theorem 7, 8, [2,21,25]). Thus, for weighted throughput a rethinking of the model is needed. A major difficulty seems to be the interleaving structure of time intervals as special structures (laminar or agreeable intervals) have been proven to be substantially better tractable in related research [8,9].

Finally, while we close the problem of scheduling unweighted jobs without commitment with a best-achievable competitive ratio $\Theta(\frac{1}{\varepsilon})$, it remains open if the weighted setting is indeed harder than the unweighted setting or if the upper bound $\mathcal{O}(\frac{1}{\varepsilon^2})$ in [21] can be improved. Future research on generalizations to multi-processors seems highly relevant. We believe that our general framework is a promising starting point.

A   Appendix

Lemma 4. *Let $\{f, \dots, g\} \subset J$ be jobs at maximal distance from M such that $\sum_{j=f}^{i} |X_j| > \lambda(i + 1 - f)$ holds for all $f \leq i \leq g$. If g is the last such job, there is a sibling j^* of g with $b_g = a_{j^*}$ and $\sum_{j=f}^{j^*} |X_j| \leq \lambda(j^* + 1 - f)$.*

Proof (Sketch). Observe that $[a_f, b_g) = \bigcup_{j=f}^{k} R(g)$ because the leaves $f, \dots, g$ form a string of jobs. Thus, by showing that there is a job $x \in X_f^g := \bigcup_{j=f}^{g} X_j$ satisfying (V), we prove the lemma with the Volume Lemma. We show that for every job $f \leq j \leq g$ there is a set Y_j such that the processing volume of Y_j covers the interval $[a_j, b_j)$ at least $\frac{\varepsilon}{\varepsilon-\delta}$ times. More precisely, $Y_f, \dots, Y_g$ satisfy

$$(i) \bigcup_{j=f}^{g} Y_j \subset X_f^g, \ (ii) |Y_j| = \lambda, \ (iii) Y_j \subset \{x \in X_f^g : p_x \geq \beta p_j\} \text{ for } f \leq j \leq g.$$

Then, (ii) and (iii) imply $\sum_{y \in Y_j} p_y \geq \lambda\beta p_j = \frac{\varepsilon}{\varepsilon-\delta}(b_j - a_j)$. Thus, if $x \notin \bigcup_{j=f}^{g} Y_j$ and x is among those jobs in X_f^g that OPT completes last, (V) is satisfied. We first describe how to find $Y_f, \dots, Y_g$ before we show that these sets satisfy (i) to (iii).

By assumption, $|X_f| > \lambda$. Index the jobs in $X_f = \{x_1, \dots, x_\lambda, x_{\lambda+1}, \dots\}$ in increasing completion times C_x^*. Define $Y_f := \{x_1, \dots, x_\lambda\}$ and $L_f := X_f \backslash Y_f$. Let $Y_f, \dots, Y_j$ and L_j be defined for $f < j + 1 \leq g$. By assumption, $|X_{j+1} \cup L_j| > \lambda$ since $|Y_i| = \lambda$ for $f \leq i \leq j$. We again index the jobs in $X_{j+1} \cup L_j = \{x_1, \dots, x_\lambda, x_{\lambda+1}, \dots\}$ in increasing optimal completion times. Then, $Y_{j+1} := \{x_1, \dots, x_\lambda\}$ and $L_{j+1} := \{x_{\lambda+1}, \dots\}$. Since we move jobs only horizontally to later siblings, we call this procedure **Push Forward**.

By definition, (i) and (ii) are satisfied. Since $f, \dots, g$ are leaves, the jobs in $Y_j \cap X_j$ are big w.r.t. j. Thus, it remains to show that the jobs in L_j are big w.r.t. the next job $j + 1$. To this end, we assume that the jobs in $Y_f, \dots, Y_j$ are big w.r.t. $f, \dots, j$, respectively. If we find an index $f \leq i(x) \leq j$ such that x as well as the jobs in $\bigcup_{i=i(x)}^{j} Y_i$ are released after $a_{i(x)}$ and x completes after every $y \in \bigcup_{i=i(x)}^{j} Y_i$, then the Volume Lemma 3 implies that $x \in L_j$ is big w.r.t. $j+1$. Indeed, then $\sum_{i=i(x)}^{j} \sum_{y \in X_i : C_y^* \leq C_x^*} p_y \geq p_x + \sum_{i=i(x)}^{j} \sum_{y \in Y_i} p_y \geq \frac{\varepsilon}{\varepsilon - \delta}(b_j - a_{i(x)}) + p_x$. By induction, we show the existence of such an index $i(x)$.

By the same argumentation for $j = g$, Corollary 1 implies the lemma. □

Lemma 2. *For all $j \in J \cup \{M\}$, $|X_j^S| \leq \lambda \tau_j$.*

Proof (Sketch). Recall that T_j is the subtree of the interruption tree rooted in $j \in J$ while the forest T_{-j} is T_j without its root j. We show that for all $j \in J \cup \{M\}$ there exists a partition $(Y_k)_{k \in T_{-j}}$ with

(i) $\bigcup_{k \in T_{-j}} Y_k = X_j^S$, (ii) $Y_k \subset \{x \in X_j : p_x \geq \beta p_k\}$, (iii) $|Y_k| \leq \lambda$ for $k \in T_{-j}$.

Then, $|X_j^S| = |\bigcup_{k \in T_{-j}} Y_k| = \sum_{k \in T_{-j}} |Y_k| \leq \tau_j \lambda$ and, thus, the lemma follows.

The proof consists of an outer and an inner induction. The outer induction is on the distance $\varphi(j)$ of a job j from machine job M, i.e., $\varphi(M) := 0$ and $\varphi(j) := \varphi(\pi(j)) + 1$ for $j \in J$. Let $\varphi_{\max} := \max\{\varphi(i) : i \in J\}$. The inner induction uses the idea about pushing jobs $x \in X_j$ to some later sibling of j in the same string of jobs (see proof of Lemma 4).

Let $j \in J$ with $\varphi(j) = \varphi_{\max} - 1$. By Observation 1, $X_j^S = \bigcup_{k:\pi(k)=j} X_k$, where all $k \in T_{-j}$ are leaves at distance $\varphi_{\max}$ from M. To define Y_k for $k \in T_{-j}$ satisfying (i) to (iii), we distinguish three cases:

Case 1. If $k \in T_{-j}$ is isolated, $|X_k| \leq \lambda$ follows directly from the Volume Lemma as otherwise $\sum_{x \in X_k} p_x \geq \lambda \beta p_k + p_x = \frac{\varepsilon}{\varepsilon - \delta}(b_k - a_k) + p_x$ contradicts Corollary 1, where $x \in X_k$ is the last job that OPT completes from the set X_k. Since all jobs in X_k are big w.r.t. k, we set $Y_k := X_k$.

Case 2. For $k \in T_{-j}$ with $|X_k| > \lambda$, we find $Y_f, \ldots, Y_g$ with Lemma 4 and set $Y_{g+1} := X_{g+1} \cup L_g$ where $f \leq k \leq g$ (maximal) satisfy Lemma 2.

Case 3. Consider jobs k in a string with $|X_k| \leq \lambda$ without siblings $f, \ldots, g$ in the same string with $b_g = a_k$ and $\sum_{i=f}^{g} |X_j| > (g - f)\lambda$. This means that such jobs do not receive jobs $x \in X_i$ for $i \neq k$ by the **Push Forward** procedure in Case 2. For such $k \in T_{-j}$ we define $Y_k := X_k$ as in Case 1.

Then, $X_j^S = \bigcup_{k \in T_{-j}} X_k = \bigcup_{k \in T_{-j}} Y_k$ and, thus, (i) to (iii) are satisfied.

We use induction to extend the claim for $\varphi = \varphi_{\max}$ to all $0 \leq \varphi \leq \varphi_{\max}$. Let $\varphi < \varphi_{\max}$ such that $(Y_k)_{k \in T_{-j}}$ satisfying (i) to (iii) exists for all $j \in J$ with $\varphi(j) \geq \varphi$. Fix $j \in J$ with $\varphi(j) = \varphi - 1$. By induction and Observation 1, it holds that $X_j^S = \bigcup_{k:\pi(k)=j} \left(X_k^B \cup \bigcup_{i \in T_{-k}} Y_i \right)$. Now, we use the partitions $(Y_i)_{i \in T_{-k}}$ for k with $\pi(k) = j$ as starting point to find the partition $(Y_k)_{k \in T_{-j}}$. We fix k with $\pi(k) = j$ and distinguish similar three cases as in the base case:

Case 1. If k is isolated, we show that $|X_k| \leq (\tau_k + 1)\lambda$ and develop a procedure to find $(Y_i)_{i \in T_k}$.

By induction, $|X_k^S| \leq \tau_k \lambda$. In the full version of the paper, we prove that $|X_k^B| \leq \lambda + (\tau_k \lambda - |X_k^S|)$. To construct $(Y_i)_{i \in T_k}$, we assign $\min\{\lambda, |X_k^B|\}$ jobs from X_k^B to Y_k. If $|X_k^B| > \lambda$, distribute the remaining jobs according to $\lambda - |Y_i|$ among the descendants of k. Then, $X_k = \bigcup_{i \in T_k} Y_i$. Because a job that is big w.r.t job k is also big w.r.t. all descendants of k, every (new) set Y_i satisfies

(ii) and (iii). We refer to this procedure as **Push Down** since jobs are shifted vertically to descendants.

Case 2. If $|X_k| > (\tau_k + 1)\lambda$, k must belong to a string with similar properties as in Lemma 4, i.e., there is a maximal string of jobs $f, \ldots, g$ containing k such that $\sum_{j=f}^{i} |X_j| > \lambda \sum_{j=f}^{i} \tau_j$ for $f \leq i \leq g$ and $b_j = a_{j+1}$ for $f \leq j < g$.

If the Volume Condition (V) is satisfied, there exists another sibling $g + 1$ that balances the sets $X_f, \ldots, X_g, X_{g+1}$ due to Corollary 1. This is shown by using **Push Down** within a generalization of the **Push Forward** procedure. As the jobs $f, \ldots, g$ may have descendants, we use **Push Forward** to construct the sets $Z_f, \ldots, Z_g$ and $L_f, \ldots, L_g$ with $|Z_k| = \lambda(\tau_k + 1)$. Then, we apply **Push Down** to Z_k and $(Y_i)_{i \in T_{-k}}$ in order to obtain $(Y_i)_{i \in T_k}$ such that they will satisfy $Z_k = \bigcup_{i \in T_k} Y_i$, $Y_i \subset \{x \in X_j : p_x \geq \beta p_i\}$, and $|Y_i| = \lambda$ for $i \in T_k$. Thus, the sets $X_f, \ldots, X_g$ satisfy (V) and we can apply Corollary 1.

Case 3. Any job k with $\pi(k) = j$ that was not yet considered as part of a string must satisfy $|X_k| \leq (\tau_k + 1)\lambda$. We use **Push Down** of Case 1 to get $(Y_i)_{i \in T_k}$. Hence, we have found $(Y_k)_{k \in T_{-j}}$ with the properties (i) to (iii). $\square$

References

1. Agrawal, K., Li, J., Lu, K., Moseley, B.: Scheduling parallelizable jobs online to maximize throughput. In: Bender, M.A., Farach-Colton, M., Mosteiro, M.A. (eds.) LATIN 2018. LNCS, vol. 10807, pp. 755–776. Springer, Cham (2018). https://doi.org/10.1007/978-3-319-77404-6_55
2. Azar, Y., Kalp-Shaltiel, I., Lucier, B., Menache, I., Naor, J., Yaniv, J.: Truthful online scheduling with commitments. In: Proceedings of the ACM Symposium on Economics and Computations (EC), pp. 715–732 (2015)
3. Bansal, N., Chan, H.-L., Pruhs, K.: Competitive algorithms for due date scheduling. In: Arge, L., Cachin, C., Jurdziński, T., Tarlecki, A. (eds.) ICALP 2007. LNCS, vol. 4596, pp. 28–39. Springer, Heidelberg (2007). https://doi.org/10.1007/978-3-540-73420-8_5
4. Baruah, S.K., Haritsa, J.R.: Scheduling for overload in real-time systems. IEEE Trans. Comput. **46**(9), 1034–1039 (1997)
5. Baruah, S.K., Haritsa, J.R., Sharma, N.: On-line scheduling to maximize task completions. In: Proceedings of the IEEE Real-Time Systems Symposium (RTSS), pp. 228–236 (1994)
6. Baruah, S.K., et al.: On the competitiveness of on-line real-time task scheduling. Real-Time Syst. **4**(2), 125–144 (1992)
7. Canetti, R., Irani, S.: Bounding the power of preemption in randomized scheduling. SIAM J. Comput. **27**(4), 993–1015 (1998)
8. Chen, L., Megow, N., Schewior, K.: An $\mathcal{O}(\log m)$-competitive algorithm for online machine minimization. In: Proceedings of the ACM-SIAM Symposium on Discrete Algorithms (SODA), pp. 155–163 (2016)
9. Chen, L., Megow, N., Schewior, K.: The power of migration in online machine minimization. In: Proceedings of the ACM Symposium on Parallelism in Algorithms and Architectures (SPAA), pp. 175–184 (2016)
10. DasGupta, B., Palis, M.A.: Online real-time preemptive scheduling of jobs with deadlines. In: Jansen, K., Khuller, S. (eds.) APPROX 2000. LNCS, vol. 1913, pp. 96–107. Springer, Heidelberg (2000). https://doi.org/10.1007/3-540-44436-X_11

11. Ferguson, A.D., Bodík, P., Kandula, S., Boutin, E., Fonseca, R.: Jockey: guaranteed job latency in data parallel clusters. In: Proceedings of the European Conference on Computer Systems (EuroSys), pp. 99–112 (2012)
12. Garay, J.A., Naor, J., Yener, B., Zhao, P.: On-line admission control and packet scheduling with interleaving. In: Proceedings of the IEEE International Conference on Computer Communications (INFOCOM), pp. 94–103 (2002)
13. Georgiadis, L., Guérin, R., Parekh, A.K.: Optimal multiplexing on a single link: delay and buffer requirements. IEEE Trans. Inf. Theory **43**(5), 1518–1535 (1997)
14. Goldwasser, M.H.: Patience is a virtue: the effect of slack on competitiveness for admission control. In: Proceedings of the ACM-SIAM Symposium on Discrete Algorithms (SODA), pp. 396–405 (1999)
15. Im, S., Moseley, B.: General profit scheduling and the power of migration on heterogeneous machines. In: Proceedings of the ACM Symposium on Parallelism in Algorithms and Architectures (SPAA), pp. 165–173 (2016)
16. Kalyanasundaram, B., Pruhs, K.: Maximizing job completions online. J. Algorithms **49**(1), 63–85 (2003)
17. Koren, G., Shasha, D.E.: MOCA: a multiprocessor on-line competitive algorithm for real-time system scheduling. Theor. Comput. Sci. **128**(1–2), 75–97 (1994)
18. Koren, G., Shasha, D.E.: D^{over}: an optimal on-line scheduling algorithm for overloaded uniprocessor real-time systems. SIAM J. Comput. **24**(2), 318–339 (1995)
19. Liebeherr, J., Wrege, D.E., Ferrari, D.: Exact admission control for networks with a bounded delay service. IEEE/ACM Trans. Netw. **4**(6), 885–901 (1996)
20. Lipton, R.: Online interval scheduling. In: Proceedings of the ACM-SIAM Symposium on Discrete Algorithms (SODA), pp. 302–311 (1994)
21. Lucier, B., Menache, I., Naor, J., Yaniv, J.: Efficient online scheduling for deadline-sensitive jobs: extended abstract. In: Proceedings of the ACM Symposium on Parallelism in Algorithms and Architectures (SPAA), pp. 305–314 (2013)
22. Pruhs, K., Stein, C.: How to schedule when you have to buy your energy. In: Proceedings of the International Conference on Approximation Algorithms for Combinatorial Optimization Problems (APROX), pp. 352–365 (2010)
23. Schwiegelshohn, C., Schwiegelshohn, U.: The power of migration for online slack scheduling. In: Proceedings of the European Symposium of Algorithms (ESA), vol. 57, pp. 75:1–75:17 (2016)
24. Woeginger, G.J.: On-line scheduling of jobs with fixed start and end times. Theor. Comput. Sci. **130**(1), 5–16 (1994)
25. Yaniv, J.: Job scheduling mechanisms for cloud computing. Ph.D. thesis, Technion, Israel (2017)

Lower Bounds and a New Exact Approach for the Bilevel Knapsack with Interdiction Constraints

Federico Della Croce[1,2(✉)] and Rosario Scatamacchia[1]

[1] Dipartimento di Ingegneria Gestionale e della Produzione, Politecnico di Torino,
Corso Duca degli Abruzzi 24, 10129 Torino, Italy
{federico.dellacroce,rosario.scatamacchia}@polito.it
[2] CNR, IEIIT, Torino, Italy

Abstract. We consider the Bilevel Knapsack with Interdiction Constraints, an extension of the classic 0-1 knapsack problem formulated as a Stackelberg game with two agents, a leader and a follower, that choose items from a common set and hold their own private knapsacks. First, the leader selects some items to be interdicted for the follower while satisfying a capacity constraint. Then the follower packs a set of the remaining items according to his knapsack constraint in order to maximize the profits. The goal of the leader is to minimize the follower's profits. The presence of two decision levels makes this problem very difficult to solve in practice: the current state-of-the-art algorithms can solve to optimality instances with 50–55 items at most. We derive effective lower bounds and present a new exact approach that exploits the structure of the induced follower's problem. The approach successfully solves all benchmark instances within one second in the worst case and larger instances with up to 500 items within 60 s.

1 Introduction

Recently, a growing attention has been centered to multilevel programming. Here we focus on bilevel optimization where two agents, denoted as a leader and a follower, play a Stackelberg game [11]. In this game, the leader takes the first decision and then the follower reacts taking into account the leader's strategy. Two standard assumptions hold in a Stackelberg game: *complete information*, that is each agent knows the problem solved by the other agent; *rationale behavior*, namely each agent has no interest in deviating from his own objective.

In this paper, we consider the Bilevel Knapsack with Interdiction Constraints (BKP), as introduced in [6]. The problem is an extension of the classic 0-1 Knapsack Problem (KP) to a Stackelberg game where the leader and the follower choose items from a common set and hold their own private knapsacks. First, the leader selects some items to be interdicted for the follower while satisfying a capacity constraint. Then the follower packs a set of the remaining items according to his knapsack constraint in order to maximize the profits. The goal

© Springer Nature Switzerland AG 2019
A. Lodi and V. Nagarajan (Eds.): IPCO 2019, LNCS 11480, pp. 155–167, 2019.
https://doi.org/10.1007/978-3-030-17953-3_12

of the leader is to minimize the follower's profits. One of the best performing algorithms for BKP is given in [2]. The algorithm, denoted as CCLW, solves to optimality instances with 50 items within a CPU time limit of 3600 s, running out of time in instances with 55 items only. Very recently, an improved branch-and-cut algorithm was given in [7]. The proposed approach manages to solve to optimality all benchmark instances in [2], requiring at most a computation time of about 85 s in an instance with 55 items. We also mention the work of [8] where a heuristic approach is proposed for BKP and other interdiction games.

Other bilevel knapsack problems have been tackled in the literature. In [1], the leader cannot interdict items but modifies the follower's capacity. In [4], the leader can modify the follower's objective function only. As discussed in [2], both problems are easier to handle than BKP. Recently, a polynomial algorithm has been provided in [3] for the continuous BKP.

Our contribution for BKP is twofold. First, we derive effective lower bounds based on mathematical programming. Second, we present a new exact approach that exploits the induced follower's problem and the lower bounds. The proposed approach shows up to be very effective successfully solving all benchmark literature instances provided in [2] within few seconds of computation. Moreover, our algorithm manages to solve to optimality all instances with up to 500 items within a CPU time limit of 60 s. Further details are available in [5].

2   Notation and Problem Formulation

In BKP a set of n items and two knapsacks are given. Each item i ($= 1, \ldots, n$) has associated a profit $p_i > 0$ and a weight $w_i > 0$ for the follower's knapsack and a weight $v_i > 0$ for the leader's knapsack. Leader and follower have different knapsack capacities denoted by C_u and C_l, respectively. Quantities p_i, v_i, w_i ($i = 1, \ldots, n$), C_u, C_l are assumed to be integer, with $v_i \leq C_u$ and $w_i \leq C_l$ for all i. To avoid trivial instances, it is also assumed that $\sum_{i=1}^{n} v_i > C_u$ and $\sum_{i=1}^{n} w_i > C_l$. We introduce 0/1 variables x_i ($i = 1, \ldots, n$) equal to one if the leader selects item i and 0/1 variables y_i equal to one if item i is chosen by the follower. BKP can be modeled as follows:

$$\min \quad \sum_{i=1}^{n} p_i y_i \tag{1}$$

$$\text{subject to} \quad \sum_{i=1}^{n} v_i x_i \leq C_u \tag{2}$$

$$x_i \in \{0, 1\} \qquad i = 1, \ldots, n \tag{3}$$

where $y_1, \ldots, y_n$ solve

$$\text{the follower's problem:} \quad \max \quad \sum_{i=1}^{n} p_i y_i \tag{4}$$

$$\text{subject to} \quad \sum_{i=1}^{n} w_i y_i \leq C_l \tag{5}$$

$$y_i \leq 1 - x_i \qquad i = 1, \ldots, n \tag{6}$$

$$y_i \in \{0, 1\} \qquad i = 1, \ldots, n \tag{7}$$

The leader's objective function (1) minimizes the profits of the follower through the interdiction constraints (6). These constraints ensure that each item i can be selected by the follower, i.e. $y_i \leq 1$, only if the item is not interdicted by the leader, i.e. $x_i = 0$. Constraint (2) represents the leader's capacity constraint. The objective function (4) maximizes the follower's profits and constraint (5) represents the follower's capacity constraint. Constraints (3) and (7) define the domain of the variables.

The optimal solution value of model (1)–(7) is denoted by z^*. The optimal solution vectors of variables x_i and y_i are respectively denoted by x^* and y^*. Notice that in model (1)–(7) there always exists an optimal solution for the leader which is maximal, namely where items are included in the leader's knapsack until there is enough capacity left.

Let us now recall the optimal solution of the continuous relaxation of a standard KP, namely the follower's model (4)–(7) without constraints (6) and constraints (7) replaced by inclusion in $[0, 1]$. Under the assumption $\sum_{i=1}^{n} w_i > C_l$, this solution has the following structure. Consider the sorting of the items by non-increasing ratios of profits over weights:

$$\frac{p_1}{w_1} \geq \frac{p_2}{w_2} \geq \cdots \geq \frac{p_n}{w_n}. \tag{8}$$

According to this order, items $j = 1, 2, \ldots$ are inserted into the knapsack as long as $\sum_{k=1}^{j} w_k \leq C_l$. The first item s which cannot be fully packed is commonly denoted in the knapsack literature as the *split* item (or *break/critical* item). The optimal solution of the KP linear relaxation is given by setting $y_j = 1$ for $j = 1, \ldots, s - 1$, $y_j = 0$ for $j = s + 1, \ldots, n$ and $y_s = (C_l - \sum_{j=1}^{s-1} w_j)/w_s$. The solution with items $1, \ldots, (s - 1)$ is a feasible solution for KP and is commonly denoted as the *split solution*.

In the remainder of the paper, we assume the ordering of the items (8). We denote by $KP(x)$ the follower's knapsack problem induced by a leader's strategy encoded in vector x, i.e. a knapsack problem with item set

$$S := \{i : x_i = 0, x_i \in x\}.$$

We also denote by $KP^{LP}(x)$ the corresponding Linear Programming (LP) relaxation. If $\sum_{i \in S} w_i > C_l$, we define the *critical* item c of $KP^{LP}(x)$ as the last item with a strictly positive value in its optimal solution. Thus, we have $y_c \in (0, 1]$ and a corresponding split solution with profit

$$\sum_{i \in S: i < c} p_i = \sum_{i=1}^{c-1} p_i(1 - x_i) \tag{9}$$

which constitutes a feasible solution for $KP(x)$. Notice that we denote by $z(M)$ the optimal solution value of any given mathematical model M.

3 Computing Lower Bounds on BKP

Consider the optimal solution vector x^*. In the induced follower's knapsack problem $KP(x^*)$ with item set S, two cases can occur: either there is no critical item in $KP^{LP}(x^*)$, namely $\sum_{i \in S} w_i \leq C_l$, or one critical item exists, namely $\sum_{i \in S} w_i > C_l$. The first case can be easily handled by considering that the follower will pack all items not interdicted by the leader. This case is discussed in Sect. 4.2.

In the second case, we derive effective lower bounds on BKP by guessing the critical item of $KP^{LP}(x^*)$ and correspondingly computing the related split solution of the follower's problem. These bounds constitute the main ingredient of the exact approach presented in Sect. 4. Since we do not know a priori the leader's optimal solution x^*, we formulate an Integer Linear Programming (ILP) model where we impose that a given item c must be critical and evaluate the profit of the corresponding split solution in the objective function. We consider binary variables k_j $(j = 1, \ldots, w_c)$ associated with the weight contribution of the critical item and introduce the following model (denoted as $CRIT_1(c)$).

$$CRIT_1(c): \quad \min \quad \sum_{i=1}^{c-1} p_i(1 - x_i) \tag{10}$$

$$\text{subject to} \quad \sum_{i=1}^{n} v_i x_i \leq C_u \tag{11}$$

$$\sum_{i=1}^{c-1} w_i(1 - x_i) + \sum_{j=1}^{w_c} j k_j = C_l \tag{12}$$

$$\sum_{j=1}^{w_c} k_j = 1 \tag{13}$$

$$x_c = 0 \tag{14}$$

$$x_i \in \{0, 1\} \quad i = 1, \ldots, n \tag{15}$$

$$k_j \in \{0, 1\} \quad j = 1, \ldots, w_c \tag{16}$$

The objective function (10) minimizes the value of the split solution. Constraint (11) represents the leader's capacity constraint. Constraints (12) and (13) ensure that item c is critical as it is the last item packed, with a weight in the interval $[1, w_c]$. Constraint (14) indicates that item c can be critical only if it is not interdicted by the leader. Constraints (15) and (16) indicate that all variables are binary. We can state the following proposition.

Proposition 1. *If there exists a critical item c in $KP^{LP}(x^*)$, then $z(CRIT_1(c))$ is a valid lower bound on z^*.*

Proof. Under the assumption that item c is critical in $KP^{LP}(x^*)$, the optimal BKP solution x^* constitutes a feasible solution for model $CRIT_1(c)$. Let denote by z_1 the corresponding solution value that coincides with the value of the split solution in $KP(x^*)$. Since the follower maximizes the profits in $KP(x^*)$ obtaining a solution with a value greater than (or equal to) the one of the split solution, we have $z_1 \leq z^*$. But this means that there exists an optimal solution of model $CRIT_1(c)$ such that $z(CRIT_1(c)) \leq z_1$ which implies a lower bound on z^*.   □

The previous proposition already provides a first significant lower bound for the problem. However, following the reasoning in the proof of Proposition 1, we remark that improved bounds on z^* can be derived by considering any feasible solution for $KP(x^*)$ that might be obtained by removing (adding) items that were not interdicted by the leader and that were selected (not selected) by the split solution, provided that the follower's capacity is not exceeded. Indeed, this corresponds to removing tuples of items $i \in [1, c-1] : x_i = 0$ and/or to adding tuples of items $i \in [c, n] : x_i = 0$ from the split solution without exceeding the follower's capacity.

Notice that, the state-of-the-art algorithms for KP, *Minknap* [10] and *Combo* [9] consider that in general only few items with ratio p_i/w_i close to that of the critical item change their values in an optimal solution with respect to the values taken in the split solution. These items constitute the so-called *core* of the knapsack. *Minknap* and *Combo* start with the computation of the split solution and an expanding core initialized with the critical item only. Then, the algorithms iteratively enlarge the core by evaluating both the removal of items from the split solution and the addition of items after the critical item. The empirical evidence illustrates that an optimal (or close to be optimal) KP solution is typically found after few iterations.

We cannot precisely characterize the features of these exact algorithms by a set of constraints within an ILP model, but we can mimic the same algorithmic reasoning by considering subsets of the items set $c - \delta, ..., c + \delta$ including the critical item c for any given core size $2\delta + 1$. In each subset, the items $i : i \leq c-1$ are removed from the split solution, while the items $j : j \geq c$ are added to the solution. Correspondingly, the initial profit and weight of the split solution are modified by subtracting the profits and the weights of the removed items and by summing up the profits and the weights of the added items.

Then, for any given subset τ of the items set $c - \delta, ..., c + \delta$, let p^τ and w^τ be the overall profit (namely the value of the improvement upon the split solution) and weight contributions of the items in τ, namely:

$$p^\tau = - \sum_{i \in \tau : i < c} p_i + \sum_{j \in \tau : j \geq c} p_j; \tag{17}$$

$$w^\tau = - \sum_{i \in \tau : i < c} w_i + \sum_{j \in \tau : j \geq c} w_j. \tag{18}$$

A subset τ with $p^\tau \leq 0$ is not considered since it does not improve upon the split solution. Instead, an improving subset with $p^\tau > 0$ is feasible only if $w^\tau \leq w_c$ and all items in τ are not interdicted by the leader. In that case, by keeping the notation of model $CRIT_1(c)$, an improvement π can be determined if the following constraint is added:

$$\pi \geq p^\tau \left(\sum_{j=\max\{1; w^\tau\}}^{w_c} k_j - \sum_{i \in \tau} x_i \right). \tag{19}$$

Correspondingly, a new model can be generated by introducing a non-negative variable π that carries the maximum additional profit to the split solution value provided by any of the additional constraints (19) indicated above.

These constraints, denoted as $\mathcal{F}(\pi, x, k)$, link variable π to variables x_i and k_j. The model (denoted as $CRIT_2(c)$) is as follows.

$$CRIT_2(c): \quad \min \quad \sum_{i=1}^{c-1} p_i(1 - x_i) + \pi \tag{20}$$

$$\text{subject to} \quad \mathcal{F}(\pi, x, k) \tag{21}$$

$$(11), (16)$$

$$\pi \geq 0 \tag{22}$$

Clearly, due to the addition of constraints in $\mathcal{F}(\pi, x, k)$, we have $z(CRIT_1(c)) \leq z(CRIT_2(c))$ for any c. Notice that, in all these additional constraints, only items which will not be interdicted by the leader can be packed and the follower's capacity constraint is not violated. We denote as *proper* any set $\mathcal{F}(\pi, x, k)$ that satisfies both conditions. After the set $\mathcal{F}(\pi, x, k)$ is built, variable π will carry the maximum profit obtainable in addition to the profit of the split solution.

Proposition 2. *If $KP^{LP}(x^*)$ admits a critical item c and model $CRIT_2(c)$ has a proper set $\mathcal{F}(\pi, x, k)$, then $z(CRIT_2(c)) \leq z^*$.*

Proof. Since model $CRIT_2(c)$ considers feasible solutions for $KP(x^*)$, the inequality holds by applying the same argument of Proposition 1.   □

We remark that models $CRIT_1(c)$ and $CRIT_2(c)$ contain a pseudo polynomial number of binary variables k_j depending on the magnitude of the follower's weights. Hence, the hardness of these ILP models may increase with the size increase of such input entries.

4   A New Exact Approach for BKP

4.1   Overview

We propose an exact algorithm for BKP that considers the possible existence of a critical item in $KP^{LP}(x^*)$ and exploits the bounds provided by model $CRIT_2(c)$. The key idea of the algorithm is to compute appropriate leader's solutions by exploring the most promising subproblems in terms of lower bounds. This strategy considerably speeds up in practice both the identification and certification of an optimal interdiction structure.

The approach involves two main steps. In the first step, the possible non-existence of a critical item is first evaluated. Then, the approach assumes the existence of a critical item and identifies a set of possible candidate items. For each candidate item c and a parameter δ to identify the core size, model $CRIT_2(c)$ is built by considering several subsets of additional constraints (19). Then the linear relaxation $CRIT_2^{LP}(c)$ is solved, where the integrality constraints (15) and (16) are replaced by inclusion in $[0, 1]$. The feasible problems $CRIT_2^{LP}(c)$ are sorted by increasing optimal value so as to identify an order of the most promising subproblems to explore. A limited number of feasible BKP solutions is also computed in this step.

In the second step, each relevant subproblem is explored by constraint generation until the subproblem can be pruned. An optimal BKP solution is eventually
returned. The approach takes as input five parameters α, β, δ, μ, γ and relies
on an ILP solver along its steps. We discuss the steps of the algorithm in the
following. The corresponding pseudo code is provided in Appendix.

4.2   Step 1

Handling the Possible Non-existence of a Critical Item. We first consider
the case where there does not exist a critical item in $KP^{LP}(x^*)$. Thus, the
follower will select all available items which are not interdicted by the leader and
an optimal solution of BKP is found by solving the following problem NCR.

$$NCR: \quad \min \sum_{i=1}^{n} p_i(1 - x_i) \tag{23}$$

$$\text{subject to} \quad \sum_{i=1}^{n} v_i x_i \leq C_u \tag{24}$$

$$\sum_{i=1}^{n} w_i(1 - x_i) \leq C_l \tag{25}$$

$$x_i \in \{0, 1\} \qquad i = 1, \ldots, n \tag{26}$$

If problem NCR is feasible, let denote by x' the related optimal solution
representing the leader's strategy. The corresponding follower's solution is
denoted by y', with $y'_i = 1 - x'_i$ ($i = 1, \ldots, n$). The current best solution (x^*, y^*)
with value z^* (which will be optimal at the end of the algorithm) is initialized
accordingly (Lines 3–4 of the pseudo code).

Identifying the Relevant Critical Items. We now assume that there exists a
critical item c in $KP^{LP}(x^*)$ (Lines 5–13) and estimate the first and last possible
items l and r that can be critical according to ordering (8). For item l we have

$$l := \min\{j : \sum_{i=1}^{j} w_i \geq C_l\}. \tag{27}$$

All items $1, \ldots, (l-1)$ cannot in fact be critical even without the leader's interdiction. For the last item r, we first compute the maximum weight of the follower
that can be interdicted by the leader (similarly as in [2]) by solving the following
problem (denoted by LW).

$$LW: \quad \max \sum_{i=1}^{n} w_i x_i \tag{28}$$

$$\text{subject to} \quad \sum_{i=1}^{n} v_i x_i \leq C_u \tag{29}$$

$$x_i \in \{0, 1\} \qquad i = 1, \ldots, n \tag{30}$$

Item r is defined as

$$r := \min\{j : \sum_{i=1}^{j} w_i \geq C_l + z(LW)\}. \tag{31}$$

Since from (31) we have $\sum_{i=1}^{r} w_i(1-x_i) \geq C_l$ for any leader's strategy, all items from $(r+1)$ to n cannot be critical.

Building Models $CRIT_2(c)$. For each candidate critical item $c \in [l, r]$, we formulate model $CRIT_2(c)$ by constructing a proper set $\mathcal{F}(\pi, x, k)$ as follows. Consider the subsets involving items in the interval $[c - \delta, c + \delta]$. Even for small value of δ, the number of subsets can be very large. Hence, in order to limit the number of constraints in $\mathcal{F}(\pi, x, k)$, we propose a different strategy that greedily selects the subsets according to the procedure denoted as *ComputeTuples* and sketched in Appendix.

For a given value of δ, we consider the interval of items $[a, b]$, with $a = \max\{1; c - \delta\}$ and $b = \min\{c + \delta; n\}$. Starting by the empty set, we enumerate at most α "backward" sets with items $(c - 1), \ldots, a$ in increasing order of size. Each set has a profit and weight equal to the sum of profits and weights of the included items. We also compute at most β "forward" sets with items $c, \ldots, b$ in increasing order of size and with a weight not superior to the maximum weight of a backward set. This in order to exclude forward sets having less chance to be combined with a backward set.

Then the backward (resp. forward) sets are ordered by increasing (resp. decreasing) profit. We combine each backward set with a forward set and generate a tuple τ. If $p^\tau > 0$ and $w^\tau \leq w_c$, we add constraint (19) to $\mathcal{F}(\pi, x, k)$. We continue adding constraints to $\mathcal{F}(\pi, x, k)$ until their number is superior to an input parameter μ. If not previously included, we also add to set $\mathcal{F}(\pi, x, k)$ the constraint $\pi \geq p_c k_{w_c}$ which handles the possible adding of the critical item to the split solution if the residual capacity is equal to w_c.

Then we solve models $CRIT_2^{LP}(c)$ for each $c \in [l, r]$ and order the models by increasing optimal value so as to have an order of most promising subproblems to explore. If for the first subproblem we have $z(CRIT_2^{LP}(c)) \geq z^*$, an optimal BKP solution is already certified (Line 13 of the pseudo code).

Computing Feasible BKP Solutions. According to the previous order of subproblems, we compute BKP feasible solutions by considering the first γ subproblems (Lines 15–21). For a given item c, we solve model $CRIT_2(c)$ obtaining a solution $\hat{x}$. If $z(CRIT_2(c)) < z^*$, we solve the induced follower's problem $KP(\hat{x})$ with optimal solution $\hat{y}$ and update the current best solution if $z(KP(\hat{x})) < z^*$.

4.3   Step 2

This step considers all relevant (ordered) suproblems $CRIT_2(c)$. For each subproblem, we first test for standard variables fixing and then each subproblem is explored by means of a constraint generation approach (Lines 23–33).

Fixing Variables in Subproblems. For a given problem $CRIT_2^{LP}(c)$, denote the optimal values of variables x_i and k_j by x_i^{LP} and k_j^{LP} respectively. Let r_{x_i} and r_{k_j} be the reduced costs of non basic variables in the optimal solution of

$CRIT_2^{LP}(c)$. We apply then standard variable-fixing techniques from Integer Linear Programming: if the gap between the best feasible solution available and the optimal solution value of the continuous relaxation solution is not greater than the absolute value of a non basic variable reduced cost, then the related variable can be fixed to its value in the continuous relaxation solution. Thus, the following constraints are added to $CRIT_2(c)$:

$$\forall i : |r_{x_i}| \geq z^* - z(CRIT_2^{LP}(c)), \quad x_i = x_i^{LP}; \tag{32}$$

$$\forall j : |r_{k_j}| \geq z^* - z(CRIT_2^{LP}(c)), \quad k_j = k_j^{LP}. \tag{33}$$

Solving Subproblems. For each open subproblem $CRIT_2(c)$, we first solve $CRIT_2(c)$ obtaining a solution $\bar{x}$. If the corresponding objective value is lower than the current best feasible solution value, we solve $KP(\bar{x})$ with solution $\bar{y}$ and if an improving solution is found, the current best solution is updated, as in Sect. 4.2. Then, we add to $CRIT_2(c)$ the constraint

$$\sum_{i:\bar{y}_i=1}^{n} x_i \geq 1. \tag{34}$$

Constraint (34) is a cut imposing that at least one item selected by the follower in solution $\bar{y}$ must be interdicted. We solve $CRIT_2(c)$ with one more constraint and apply the same procedure until $z(CRIT_2(c)) \geq z^*$ or the problem becomes infeasible. At the end of Step 2, the optimal BKP solution (x^*, y^*) is returned (Line 34).

5   Computational Results

All tests were performed on an Intel i7 CPU @ 2.4 GHz with 8 GB of RAM. The code was implemented in the C++ programming language. The ILP solver used along the steps of the algorithm is CPLEX 12.6.2. The parameters of the ILP solver were set to their default values. The BKP instances with $n = 35, 40, 45, 50, 55$ are generated in [2] as follows. Profits p_i and weights w_i of the follower and weights v_i of the leader are integers randomly distributed in $[1, 100]$: 10 instances are generated for each value of n. The follower's capacity C_l is set to $\lceil (INS/11) \sum_i^n w_i \rceil$ where $INS (= 1, \ldots, 10)$ denotes the instance identifier. The leader's capacity is randomly selected in the interval $[C_l - 10; C_l + 10]$.

We first tested our approach on these 50 benchmark instances. After some preliminary computational tests, we chose the following parameter entries for our approach: $\alpha = 100$, $\beta = 100$, $\delta = 10$, $\mu = 150$, $\gamma = 2$. Algorithm CCLW in [2] solves all instances with 50 items within a CPU time limit of 3600 s but runs out of time limit in two instances with 55 items. Algorithm in [7] solves all benchmark instances, requiring at most a computation time of about 85 s for solving an instance with 55 items. The proposed exact approach outperforms the competing algorithms, successfully solving to optimality each instance in at most 1.1 s (the maximum CPU time is reached in an instance with 55 items) with an average of 0.2 s. Notice that the tests in [2] and in [7] were carried out on different but comparable machines in terms of hardware specifications. Furthermore, the computational tests in both [2] and [7] are limited to instances with 55 items. We then tested larger instances with $n = 100, 200, 300, 400, 500$ according to the generation scheme in [2]. For each value of n and INS, we generated 10 instances for a total of 500 instances. For these large instances, we set the parameters of our algorithm to the following values: $\alpha = 500$, $\beta = 500$, $\delta = 20$, $\mu = 1000$, $\gamma = 5$. It is pointed out in [2] that in instances with $INS \geq 5$ the follower's capacity constraint is expected to be inactive and this makes the instances easy to solve. Our computational experiments confirm this trend: the proposed algorithm solves each instance with n from 100 to 500 and $INS \geq 5$ in at most 8 s never invoking Step 2. In the light of this consideration, we report in the following Table 1 only the results for instances with $INS \leq 4$.

Table 1. BKP instances with $n = 100, 200, 300, 400, 500$ and $INS \leq 4$.

n	INS	#Opt	CPU time		# Subproblems in Step 2		# $CRIT_2(\cdot)$ solved	
			Average	Max	Average	Max	Average	Max
100	1	10	2.1	3.0	0.7	2.0	4.8	7.0
	2	10	5.6	9.9	3.8	9.0	8.9	16.0
	3	10	4.3	6.4	2.5	7.0	7.5	12.0
	4	10	2.3	4.5	0.7	4.0	5.2	9.0
200	1	10	5.3	10.7	3.4	7.0	8.9	17.0
	2	10	7.8	12.2	5.0	9.0	10.1	14.0
	3	10	9.1	13.6	6.4	12.0	12.3	19.0
	4	10	6.0	8.6	3.5	8.0	8.3	13.0
300	1	10	6.4	8.3	3.9	8.0	9.0	13.0
	2	10	15.5	37.4	7.2	14.0	13.5	23.0
	3	10	14.0	17.7	10.9	15.0	16.8	24.0
	4	10	8.7	13.2	4.9	11.0	9.9	16.0
400	1	10	8.8	12.3	6.7	10.0	12.8	17.0
	2	10	15.2	18.7	9.1	12.0	15.1	20.0
	3	10	19.0	30.5	12.0	17.0	18.8	32.0
	4	10	12.6	16.5	8.4	23.0	13.8	30.0
500	1	10	11.9	18.2	7.6	13.0	13.1	20.0
	2	10	20.6	26.6	11.0	20.0	17.0	25.0
	3	10	21.2	25.8	12.7	17.0	17.8	22.0
	4	10	15.1	17.1	4.7	8.0	9.8	13.0

The results in the table are summarized in terms of average, maximum CPU time and number of optimal solutions obtained with a time limit of 60 s. We also report average and maximum number of subproblems explored in Step 2. The last column reports average and maximum number of times model $CRIT_2(c)$ is solved along the two steps. The results illustrate the effectiveness of our approach. All instances are solved to optimality requiring 37.4 s at most for an instance with 300 items. The number of subproblems handled by Step 2 is in general limited, reaching a maximum value of 23 (in an instance with 400 items). Also, the number of models $CRIT_2(c)$ to be solved is generally limited and never superior to 32. We finally point out that the number of times constraint (34) is added to each subproblem is limited: in the tested instances, the while–loop of Step 2 executed 8 iterations at most.

Appendix

Compute Tuples(c, α, β, δ, μ)

1: Consider items in the interval $[a, b]$ with $a := \max\{c-\delta; 1\}$, $b := \min\{c+\delta; n\}$.

2: Starting from the empty set and in increasing order of size, enumerate α backward sets with items $(c-1), \ldots, a$. Denote by w_{max} the maximum weight of a backward set. Order the sets by increasing profits.

3: Enumerate β forward sets with items $c, \ldots, b$ in increasing order of size and with a weight not superior to w_{max}. Order the sets by decreasing profits.

4: Take the first available backward set. Merge the set with a forward set and generate tuple τ.

5: If $p^\tau > 0$ and $w^\tau \leq w_c$, add constraint $\pi \geq p^\tau \left(\sum_{j=\max\{1;w^\tau\}}^{w_c} k_j - \sum_{i \in \tau} x_i \right)$ to $\mathcal{F}(\pi, x, k)$.

6: Iterate Steps 4-5 as long as $|\mathcal{F}(\pi, x, k)| \leq \mu$.

7: If not already included, add to $\mathcal{F}(\pi, x, k)$ constraint $\pi \geq p_c k_{w_c}$.

Exact solution approach

1: **Input:** BKP instance, parameters α, β, δ, μ, γ.

$\triangleright$ Step 1

2: *Handle the absence of a critical item:*

3: solve NCR; $z^* \leftarrow +\infty$;

4: **if** NCR *has a feasible solution* **then** $x^* = x'$, $y^* = y'$, $z^* = z(NCR)$; **end if**

5: *Identify the candidate critical items and build models $CRIT_2(c)$:*

6: Compute the interval of critical items $[l, r]$: $l \leftarrow$ apply (27), $r \leftarrow$ apply (31);

7: **for** all c in $[l, r]$ **do**

8:      Build model $CRIT_2(c)$ by procedure *ComputeTuples(c, α, β, δ, μ)*;

9:      Solve model $CRIT_2^{LP}(c)$;

10: **end for**

11: Sort models $CRIT_2(c)$ by increasing $z(CRIT_2^{LP}(c))$.

12: $\Longrightarrow$ Create a list of ordered critical items $L = \{c_1, c_2, \dots\}$;

13: **if** $z(CRIT_2^{LP}(c_1)) \geq z^*$ **then return** (x^*, y^*); **end if**

14: *Compute feasible BKP solutions:*

15: **for** $i = 1, \dots, \gamma$ **do**

16:      **if** $z(CRIT_2^{LP}(c_i)) < z^*$ **then** $\hat{x} \leftarrow$ solve $CRIT_2(c_i)$;

17:          **if** $z(CRIT_2(c_i)) < z^*$ **then** $\hat{y} \leftarrow$ solve $KP(\hat{x})$;

18:              **if** $z(KP(\hat{x})) < z^*$ **then** $x^* = \hat{x}$, $y^* = \hat{y}$, $z^* = z(KP(\hat{x}))$; **end if**

19:          **end if**

20:      **end if**

21: **end for**

$\triangleright$ Step 2

22: *Solve subproblems:*

23: **for** all c in list L **do**

24:      **if** $z(CRIT_2^{LP}(c)) \geq z^*$ **then return** (x^*, y^*); **end if**

25:      Apply (32), (33) and fix variables in $CRIT_2(c)$;

26:      $\bar{x} \leftarrow$ solve $CRIT_2(c)$;

27:      **while** $z(CRIT_2(c)) < z^*$ **do**

28:          $\bar{y} \leftarrow$ solve $KP(\bar{x})$;

29:          **if** $z(KP(\bar{x})) < z^*$ **then** $x^* = \bar{x}$, $y^* = \bar{y}$, $z^* = z(KP(\bar{x}))$; **end if**

30:          Add constraint (34) to $CRIT_2(c)$;

31:          $\bar{x} \leftarrow$ solve $CRIT_2(c)$;

32:      **end while**

33: **end for**

34: **return** (x^*, y^*).

References

1. Brotcorne, L., Hanafi, S., Mansi, R.: One-level reformulation of the bilevel knapsack problem using dynamic programming. Discrete Optim. **10**, 1–10 (2013)
2. Caprara, A., Carvalho, M., Lodi, A., Woeginger, G.: Bilevel knapsack with interdiction constraints. INFORMS J. Comput. **28**, 319–333 (2016)
3. Carvalho, M., Lodi, A., Marcotte, P.: A polynomial algorithm for a continuous bilevel knapsack problem. Oper. Res. Lett. **46**, 185–188 (2018)
4. Chen, L., Zhang, G.: Approximation algorithms for a bi-level knapsack problem. Theor. Comput. Sci. **497**, 1–12 (2013)
5. Della Croce, F., Scatamacchia, R.: A new exact approach for the bilevel knapsack with interdiction constraints (2018). http://arxiv.org/abs/1811.02822
6. DeNegre, S.: Interdiction and discrete bilevel linear programming. Ph.D. thesis. Lehigh University (2011)
7. Fischetti, M., Ljubić, I., Monaci, M., Sinnl, M.: Interdiction games and monotonicity, with application to knapsack problems. INFORMS J. Comput. (2018, to appear), technical report available at: https://homepage.univie.ac.at/ivana.ljubic/research/publications/interdiction_games_and_monotonicity.pdf
8. Fischetti, M., Monaci, M., Sinnl, M.: A dynamic reformulation heuristic for generalized interdiction problems. Eur. J. Oper. Res. **267**, 40–51 (2018)
9. Martello, S., Pisinger, D., Toth, P.: Dynamic programming and strong bounds for the 0-1 knapsack problem. Manag. Sci. **45**, 414–424 (1999)
10. Pisinger, D.: A minimal algorithm for the 0-1 knapsack problem. Oper. Res. **45**, 758–767 (1997)
11. Stackelberg, H.V.: The Theory of the Market Economy. Oxford University Press, Oxford (1952)

On Friedmann's Subexponential Lower Bound for Zadeh's Pivot Rule

Yann Disser[1,2] and Alexander V. Hopp[1,2(✉)]

[1] Graduate School of Computational Engineering, TU Darmstadt,
Darmstadt, Germany
hopp@gsc.tu-darmstadt.de
[2] Department of Mathematics, TU Darmstadt, Darmstadt, Germany
disser@mathematik.tu-darmstadt.de

Abstract. The question whether the Simplex method admits a polynomial time pivot rule remains one of the most important open questions in discrete optimization. Zadeh's pivot rule had long been a promising candidate, before Friedmann (IPCO, 2011) presented a subexponential instance, based on a close relation to policy iteration algorithms for Markov decision processes (MDPs). We investigate Friedmann's lower bound construction and exhibit three flaws in his analysis: We show that (a) the initial policy for the policy iteration does not produce the required occurrence records and improving switches, (b) the specification of occurrence records is not entirely accurate, and (c) the sequence of improving switches described by Friedmann does not consistently follow Zadeh's pivot rule. In this paper, we resolve each of these issues. While the first two issues require only minor changes to the specifications of the initial policy and the occurrence records, the third issue requires a significantly more sophisticated ordering and associated tie-breaking rule that are in accordance with the LEAST-ENTERED pivot rule. Most importantly, our changes do not affect the macroscopic structure of Friedmann's MDP, and thus we are able to retain his original result.

1 Introduction

The Simplex method, originally proposed by Dantzig in 1947 [2], is one of the most important algorithms to solve linear programs in practice. At its core, it operates by maintaining a subset of basis variables while restricting non-basis variables to trivial values, and repeatedly replacing a basis variable according to a fixed *pivot rule* until the objective function value can no longer be improved. Exponential worst-case instances have been devised for many natural pivot rules (see, for example, [1,5,7,8]), and the question whether a polynomial time pivot rule exists remains one of the most important open problems in optimization.

This work is supported by the 'Excellence Initiative' of the German Federal and State Governments and the Graduate School CE at TU Darmstadt.

Full version digitally published at the University and State Library Darmstadt, available at http://tuprints.ulb.tu-darmstadt.de/id/eprint/7557.

© Springer Nature Switzerland AG 2019
A. Lodi and V. Nagarajan (Eds.): IPCO 2019, LNCS 11480, pp. 168–180, 2019.
https://doi.org/10.1007/978-3-030-17953-3_13

Zadeh's LEAST-ENTERED pivot rule [11] was designed to avoid the exponential behavior on known worst-case instances for other pivot rules. It is *memorizing* in that it selects a variable to enter the basis that improves the objective function and has previously been selected least often among all improving variables. For more than thirty years, Zadeh's rule defied all attempts to construct superpolynomial instances, and seemed like a promising candidate for a polynomial pivot rule.

It was a breakthrough when Friedmann eventually presented the first superpolynomial lower bound for Zadeh's pivot rule [4]. His construction uses a connection between the Simplex Algorithm and Howard's Policy Iteration Algorithm [6] for computing optimal policies in Markov decision processes (MDPs). For a given $n \in \mathbb{N}$, Friedmann's construction consists of an MDP of size $\mathcal{O}(n^2)$, an initial policy, and an ordering of the improving switches obeying the LEAST-ENTERED pivot rule. This ordering results in an exponential number of iterations when beginning with the initial policy and repeatedly making improving switches in the specified order. The construction translates into a linear program of the same asymptotic size for which the Simplex Algorithm with Zadeh's pivot rule needs $\Omega(2^n)$ steps. Since the input size is $\mathcal{O}(n^2)$, this in turn results in a superpolynomial lower bound. Recently, an exponential lower bound for Zadeh's pivot rule was found for AUSOs [10], but it is not clear whether this construction can be realized as a linear program. However, the construction is simpler than Friedmann's construction with more natural tie-breaking, and thus presents an alternative approach to devising lower bounds for memorizing pivot rules.

Our Contribution. In this paper, we expose different flaws in Friedmann's construction and present adaptations to eliminate them. We first show that the chosen initial policy does not produce the claimed set of improving switches, and propose a modified initial policy that leads to the desired behavior. Second, we observe that the given formula describing the occurrence records (that count the number of times an improving switch was made) is inaccurate, and provide a (small) correction that does not disturb the overall argument.

Most importantly, we exhibit a significant problem with the order in which the improving switches are applied in [4]. More precisely, we show that this order does not consistently obey Zadeh's pivot rule, and, in fact, that no consistent ordering exists that updates the MDP "level by level" in each phase according to a fixed order. This not only rules out Friedmann's ordering, but shows that a fundamentally different approach to ordering improving switches is needed. To amend this issue, we show the existence of an ordering and a tie-breaking rule compatible with the LEAST-ENTERED rule, such that applying improving switches according to the ordering still proceeds along the same macroscopic phases as intended by Friedmann. In this way, we are able to quantitatively retain Friedmann's superpolynomial lower bound.

Outline. Throughout this paper, we assume some basic familiarity with the construction given in [4] and Markov decision processes in general. We review the most important aspects and notation of [4] in Sect. 2. Section 3 treats issues with the initial policy and the description of the occurrence records and our adaptations to address them. The main part of this paper is Sect. 4, where we show that the sequence of improving switches can be reordered such that the LEAST-ENTERED rule is obeyed.

2 Friedmann's Lower Bound Construction

In [4], Friedmann uses the connection between the Simplex Algorithm and the Policy Iteration Algorithm for obtaining optimal policies in MDPs. Similarly, we also restrict our discussion to policy iteration for MDPs, with the understanding that results carry over to the Simplex Algorithm. We assume knowledge of MDPs and the connection to the Simplex Algorithm and refer to [9] for more information. For convenience, we refer to improving switches simply as switches.

Let $n \in \mathbb{N}$. Friedmann's construction emulates an n-bit binary counter by a Markov decision process G_n. For every n-digit binary number b, there is a policy σ_b for G_n representing b. We denote the i-th bit of b by b_i, so $b = (b_n, \ldots, b_1)$. The MDP G_n is constructed such that applying the Policy Iteration Algorithm using the LEAST-ENTERED rule enumerates the policies σ_0 to σ_{2^n-1}. According to the pivot rule, the algorithm always chooses a switch chosen least often in the past. More specifically, an *occurrence record* ϕ is maintained, and, in every step, a switch minimizing ϕ is chosen. The rule does however not determine *which* switch minimizing ϕ should be chosen, so a tie-breaking rule is needed. For an edge e and a policy σ, we denote the occurrence record of e once σ is reached by $\phi^\sigma(e)$. We denote the set of improving switches with respect to σ by I_σ.

We fix the following notation. The set of n-digit binary numbers is denoted by $\mathbb{B}_n$. For $b \in \mathbb{B}_n, b \neq 0$, $\ell(b)$ denotes the least significant bit of b equal to 1. The unique policy representing $b \in \mathbb{B}_n$ constructed in [4] is denoted by σ_b.

The process G_n can be interpreted as a "fair alternating binary counter" as follows. Usually, when counting from 0 to $2^n - 1$ in binary, less significant bits are switched more often than more significant bits. As the LEAST-ENTERED pivot rule forces the algorithm to switch all bits equally often, the construction must ensure to operate correctly when all bits are switched equally often. This is achieved by representing every bit by *two* gadgets where only one *actively* represents the bit. The gadgets alternate in actively representing the bit.

The construction consists of n structurally identical levels, where level i represents the i-th bit. A large number $N \in \mathbb{N}$ is used for defining the rewards and a small number $\epsilon \geq 0$ is used for defining the probabilities. The i-th level is shown in Fig. 1(a), the coarse structure of the whole MDP in Fig. 1(b).

A number n_v below or next to a vertex v in Fig. 1(a) denotes a reward of $(-N)^{n_v}$ associated with every edge leaving v. Other edges have a reward of 0.

Each level i contains two gadgets attached to the *entry vertex* k_i, called *lanes*. We refer to the left lane as *lane 0* and to the right lane as *lane 1*. Lane j of level i contains a randomization vertex A_i^j and two attached cycles with vertices $b_{i,0}^j$ and $b_{i,1}^j$. These gadgets are called *bicycles*, and we identify the bicycle containing A_i^j with that vertex. For a bicycle A_i^j, the edges $(b_{i,0}^j, A_i^j), (b_{i,1}^j, A_i^j)$ are called *edges of the bicycle*. For a policy σ, the bicycle A_i^j is *closed (w.r.t. σ)* if and only if $\sigma(b_{i,0}^j) = \sigma(b_{i,1}^j) = A_i^j$. A bicycle that is not closed is called *open*.

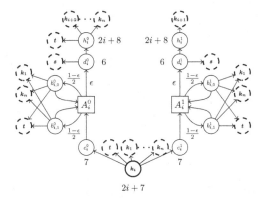

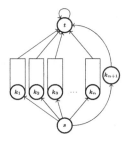

(b) The entry vertices are all connected to s and t. Connections between levels and from the levels to s are not shown here. The vertex k_{n+1} is needed for technical reasons.

(a) Circular vertices are player-controlled, squares are randomization vertices. Bold vertices can be reached from other levels, dashed vertices do not belong to level i.

Fig. 1. Level i of G_n (left) and coarse structure of G_n (right)

The i-th level of G_n corresponds to the i-th bit of the counter. Which of the bicycles of level i is actively representing the i-th bit depends on b_{i+1}. When this bit is equal to 1, A_i^1 is considered active. Otherwise, A_i^0 is considered active. The i-th bit is equal to 1 if and only if the active bicycle in level i is closed.

As initial policy, the MDP is provided the policy σ^* representing 0. Then, a sequence of policies $\sigma_1, \ldots, \sigma_{2^n-1}$ is enumerated by the Policy Iteration Algorithm using the LEAST-ENTERED pivot rule and an (implicit) tie-breaking rule. For $b \in \mathbb{B}_n, b \neq 0$, the policy σ_b should fulfill the following invariants.

1. Exactly the bicycles $A_i^{b_{i+1}}$ corresponding to bits $b_i = 1$ are closed.
2. For all other bicycles A_i^j, it holds that $\sigma_b(b_{i,0}^j) = \sigma_b(b_{i,1}^j) = k_{\ell(b)}$.
3. All entry vertices point to the lane containing the active bicycle if $b_i = 1$ and to $k_{\ell(b)}$ otherwise.
4. The vertex s points to the entry vertex of the least significant set bit.
5. All vertices h_i^0 point to the entry vertex of the first level after level $i+1$ corresponding to a set bit, or to t if no such level exists.
6. The vertex d_i^j points to h_i^j if and only if $b_{i+1} = j$ and to s otherwise.

The Policy Iteration Algorithm is only allowed to switch one edge per iteration. However, the policy σ_{b+1} cannot be reached from σ_b by performing a single switch. Therefore, intermediate policies need to be introduced for the transition from σ_b to σ_{b+1}. These intermediate policies are divided into six *phases*. In each phase, a different "task" is performed within the construction. Let $\ell' := \ell(b+1)$.

1. In phase 1, switches inside of some bicycles are performed to keep the occurrence records of the bicycle edges as balanced as possible. For every open bicycle A_i^j, at least one of the edges $(b_{i,0}^j, A_i^j), (b_{i,1}^j, A_i^j)$ is switched. Some

bicycles are allowed to switch both edges such that their occurrence record can "catch up" with the other edges. In the active bicycle of level ℓ', we also switch both edges, as this bicycle needs to be closed with respect to σ_{b+1}.

2. In phase 2, the new least significant set bit is "made accessible". Thus, $k_{\ell'}$ is switched to $c_{\ell'}^j$, where j is the lane containing the active bicycle.
3. In phase 3, we perform the "resetting process". The entry vertices of all levels i corresponding to bits with $(b+1)_i = 0$ switch to $k_{\ell'}$. The same is done for all vertices $b_{i,l}^j$ contained in inactive bicycles and all vertices $b_{i,l}^j$ corresponding to levels i with $(b+1)_i = 0$.
4. In phase 4, the vertices h_i^0 are updated according to $\ell(b+1)$.
5. In phase 5, we switch s to the entry vertex corresponding to $\ell(b+1)$.
6. In phase 6, we update the vertices d_i^j such that h_i^0 is the target of d_i^0 if and only if $(b+1)_{i+1} = 0$ and h_i^1 is the target of d_i^1 if and only if $(b+1)_{i+1} = 1$.

Before discussing our findings, we need to introduce notation related to binary counting. We further briefly describe the tables contained in [4].

Let $b \in \mathbb{B}_n$. By binary counting, we refer to the process of enumerating the binary representations of all numbers $\tilde{b} \in \{0, 1, \ldots, b\}$. These numbers are used to determine how often and when edges of G_n are improving and will be applied.

Intuitively, we are interested in *schemes* that we observe when counting from 0 to b, or, more precisely, in the set of numbers that *match a scheme*. A *scheme* is a set $S \subseteq \mathbb{N} \times \{0, 1\}$ and b *matches* S if $b_i = q$ for all $(i, q) \in S$. Since the occurrence records of the edges depend on how often a specific scheme occurred when counting from 0 to b, we introduce the following terms.

Definition 1 ([4]). *Let $b \in \mathbb{B}_n, i \in \{1, \ldots, n\}$ and let S be a scheme. The* flip set $F(b, i, S)$ *is the set of all numbers between 0 and b matching S whose least significant bit is the i-th bit. The* flip number *is defined as $f(b, i, S) := |F(b, i, S)|$ and we set $f(b, i) := f(b, i, \emptyset)$.*

We now briefly describe the tables of [4]. For $p \in \{1, \ldots, 6\}$, [4, Table 2] defines the term phase p policy. As we prove later, there is an issue concerning the side conditions of phase 3. For a phase p policy σ, [4, Table 3] contains subsets L_σ^p and supersets U_σ^p of the set of improving switches. The last table is [4, Table 4]. For $b \in \mathbb{B}_n$, it contains the occurrence records $\phi^{\sigma b}$ of the edges with respect to the unique policy representing b. We discuss an issue regarding this table in Sect. 3. Other than correcting these issues, we rely on [4, Tables 2, 3, 4].

3   Initial Policy and Occurrence Records

In this section, we discuss the initial policy σ^* used in [4] and the description of the occurrence records given in [4, Table 4]. We show that σ^* contradicts several aspects of [4], and provide an alternative initial policy resolving these issues. Then, we discuss why the description of the occurrence records given in [4, Table 4] is not entirely accurate and provide a correction of this inaccuracy.

On [4, page 11], the following is stated regarding σ^*: "*As designated initial policy σ^*, we use $\sigma^*(d_i^j) = h_i^j$ and $\sigma^*(\_) = t$ for all other player 0 nodes with non-singular out-degree.*" This initial policy, however, is inconsistent with the sub- and supersets of improving switches given in [4, Table 3] and [4, Lemma 4].[1]

Issue 1. The initial policy σ^* described in [4, page 10] contradicts [4, Table 3] since $I_{\sigma^*} \neq \{(b_{i,r}^j, A_i^j) \colon \sigma^*(b_{i,r}^j) \neq A_i^j\}$. In addition, when the Policy Iteration Algorithm is started with σ^*, at least one of [4, Tables 3, 4] is incorrect for $b = 1$.

Thus, the initial policy needs to be changed. We propose the following policy that resolves both issues. Note that it also fulfills [4, Lemma 1] and can thus indeed be used as initial policy for G_n.

Theorem 1. *Define the policy σ^* via $\sigma^*(d_i^0) := h_i^0$ and $\sigma^*(d_i^1) := s$ for all $i \in \{1, \ldots, n\}$ and $\sigma^*(\cdot) := t$ for all other player-controlled vertices with non-singular out-degree. Then $I_{\sigma^*} = \{(b_{i,r}^j, A_i^j) \colon \sigma^*(b_{i,r}^j) \neq A_i^j\}$ and starting the Policy Iteration Algorithm with σ^* does not contradict [4, Tables 3, 4] for $b = 1$.*

We next prove an issue related to the occurrence records of the bicycles as specified in [4, Table 4].

Issue 2. Let $b < 2^{n-k-1} - 1$ for some $k \in \mathbb{N}$. Assume that the occurrence records of the edges are given by [4, Table 4]. Then, there is a pair $(b_{i,0}^j, A_i^j), (b_{i,1}^j, A_i^j)$ such that at least one of them has a negative occurrence record.

The problem is that the given description does not properly distinguish between inactive bicycles that need to catch up with the counter and inactive bicycles that do not need to do so. Informally, for $b \in \mathbb{B}_n$ the occurrence records once the policy σ_b is reached can be described as follows: (a) Every closed and active bicycle has an occurrence record corresponding to the last time it was closed, (b) every open and active bicycle has an occurrence record of b and (c) inactive bicycles are either "catching up" with other bicycles and thus have an occurrence record less than b or already finished catching up and have an occurrence record of b.

To resolve Issue 2, we formulate an additional condition. It is used to distinguish inactive bicycles that might need to catch up with the counter because they have already been closed once (if $b \geq 2^{i-1} + j \cdot 2^i$), and inactive bicycle that do not need to catch up because they have not been closed before.

To formulate this condition, we need more notation. Let $b \in \mathbb{B}_n$ and A_i^j be a fixed bicycle. We define g as the largest number smaller than b such that the least significant set bit of g has index i and the $(i + 1)$-th bit is equal to j. In addition, we define $z := b - g - 2^{i-1}$ and $\phi^{\sigma_b}(A_i^j) := \phi^{\sigma_b}(b_{i,0}^j, A_i^j) + \phi^{\sigma_b}(b_{i,1}^j, A_i^j)$.

According to the proof of [4, Lemma 5], the switches inside a cycle center A_i^j should then be applied according to the following rules.

---

[1] Proofs for all statements can be found in the full paper [3].

1. If A_i^j is open and active, we switch one edge of the bicycle.
2. Let $j := b_{\ell(b+1)+1}$. In addition to 1., the second edge of $A_{\ell(b+1)}^j$ is switched.
3. If A_i^j is inactive and $b < 2^{i-1} + j \cdot 2^i$, one edge of the bicycle is switched.
4. If A_i^j is inactive, $b \geq 2^{i-1} + j \cdot 2^i$ and $z < \frac{1}{2}(b-1-g)$, both edges of A_i^j are switched. If $z \geq \frac{1}{2}(b-1-g)$, only one edge is switched.

The following theorem gives a correct description of the occurrence records.

Theorem 2. *Suppose that improving switches within bicycles are applied as described by rules 1 to 4. Let $b \in \mathbb{B}_n$ and A_i^j be a bicycle. Then, the occurrence records of $(b_{i,0}^j, A_i^j)$ and $(b_{i,1}^j, A_i^j)$ are correctly specified by the system*

$$|\phi^{\sigma_b}(b_{i,0}^j, A_i^j) - \phi^{\sigma_b}(b_{i,1}^j, A_i^j)| \leq 1 \tag{3.1}$$

$$\phi^{\sigma_b}(A_i^j) = \begin{cases} g+1, & A_i^j \text{ is closed and active} \\ b, & A_i^j \text{ is open and active} \\ b, & A_i^j \text{ is inactive and } b < 2^{i-1} + j \cdot 2^i \\ g+1+2z, & A_i^j \text{ is inactive and } b \geq 2^{i-1} + j \cdot 2^i \end{cases} \tag{3.2}$$

4 Improving Switches of Phase 3

We next discuss the application of improving switches in phase 3. There are two contradictory descriptions in [4] how to apply these. We prove that neither of the given orderings obeys the LEAST-ENTERED rule. We additionally show that a natural adaptation of Friedmann's scheme still does not obey the LEAST-ENTERED rule. We then go on to prove the existence of an ordering and an associated tie-breaking rule that obey the LEAST-ENTERED rule while still producing the intended behavior of Friedmann's construction.

Throughout this section, for a fixed $b \in \mathbb{B}_n$, we use $\ell := \ell(b)$ and $\ell' := \ell(b+1)$.

4.1 Issues with Friedmann's Switching Order

In Sect. 2, we stated that during phase 3, improving switches need to be applied for every entry vertex k_i contained in a level i with $(b+1)_i = 0$. In addition, several bicycles need to be opened. However, according to the description given in [4, pages 9–10], both of these updates should not be performed for all levels but only those with an index smaller than ℓ'. To be precise, the following is stated:[2] *"In the third phase, we perform the major part of the resetting process. By resetting, we mean to unset lower bits again, which corresponds to reopening the respective bicycles. Also, we want to update all other inactive or active but not set bicycles again to move to the entry point $k_{\ell'}$. In other words, we need to update the lower entry points k_z with $z < \ell'$ to move to $k_{\ell'}$, and the bicycle nodes $b_{z,l}^j$ to move to $k_{\ell'}$. We apply these switches by first switching the entry node k_z for some $z < \ell'$ and then the respective bicycle nodes $b_{z,r}^j$."*

However, there is an issue regarding this informal description.

---

[2] The notation in the quote was adapted from [4] to be in line with our paper.

Issue 3. For every $b \in \{1, \ldots, 2^{n-2} - 1\}$, the informal description of phase 3 contradicts [4, Tables 2, 4]. It additionally violates the LEAST-ENTERED pivot rule during the transition from σ_b to σ_{b+1} for every $b \in \{3, \ldots, 2^{n-2} - 2\}$.

In other parts of the construction, Friedmann seems to apply the switches differently, by not only applying them for levels with a lower index than the least significant set bit but for all levels. Especially, the side conditions of [4, Table 2] for defining a phase p policy rely on the fact that these switches are applied for all levels i with $(b + 1)_i = 0$. According to the proof of [4, Lemma 5], the switches need to be applied as follows: (See Footnote 2) *"In order to fulfill all side conditions for phase 3, we need to perform all switches from higher indices to smaller indices, and k_i to $k_{\ell'}$ before $b_{i,r}^j$ with $(b + 1)_{i+1} \neq j$ or $(b + 1)_i = 0$ to $k_{\ell'}$".* However, applying improving switches in this way results in another issue.[3]

Issue 4. Applying the improving switches as described in [4, Lemma 5] does not obey the LEAST-ENTERED pivot rule.

We can show an even stronger statement. (See Footnote 3) Friedmann applies improving switches of phase 3 as follows: During the transition from σ_b to σ_{b+1}, switches are applied "one level after another" where the order of the levels depends on $\ell(b + 1)$. That is, depending on $\ell(b + 1)$, an ordering $S^{\ell(b+1)}$ of the levels is considered and when i_1 appears before i_2 in $S^{\ell(b+1)}$, all switches in level i_1 need to be applied before any switch of level i_2. We prove that applying improving switches in this way violates the LEAST-ENTERED pivot rule at least once, independently of how $S^{\ell(b+1)}$ is chosen.

Issue 5. Consider some $b \in \mathbb{B}_n$ and the transition from σ_b to σ_{b+1}. Suppose that the switches of phase 3 are applied "level by level" according to any fixed ordering of the levels as described above. Further suppose that this ordering only depends on $\ell(b + 1)$. Then, the LEAST-ENTERED pivot rule is violated.

Observe that Issue 5 rules out a broader class of orderings. In some sense, this shows that Friedmann's ordering needs to be changed fundamentally, and cannot be fixed by slight adaptation.

4.2   Fixing the Ordering of the Improving Switches

We now prove the existence of an ordering and an associated tie-breaking rule for the application of the switches of phase 3 that obey the LEAST-ENTERED rule. We then show that these can be used to prove the existence of an ordering and an associated tie-breaking rule that obey the LEAST-ENTERED rule for all phases that produces the intended behavior.

For every phase p policy σ, [4, Table 3] gives a subset L_σ^p and a superset U_σ^p of the improving switches I_σ for σ, see [4, Lemma 4]. The improving switch that

---

[3] The proof can be found in Appendix A.

is then applied in σ is always contained in L_σ^p, and U_σ^p is analyzed instead of I_σ to show that the intended switch can indeed be applied. Now, let σ be a phase 3 policy. We need to compare L_σ^3 and U_σ^3 since all switches that can possibly be applied during phase 3 are contained in U_σ^3. This is done via partitioning U_σ^3. The comparison then enables us to show that there is always a switch contained in L_σ^3 minimizing the occurrence record. This justifies that *"we will only use switches from L_σ^p"* [4, page 12] (at least for phase $p = 3$). We then show the following: All improving switches that should be applied during phase 3 according to [4] can be applied (in a different order) during phase 3, without violating the LEAST-ENTERED pivot rule.

As outlined in Sect. 2, the transition from σ_b to σ_{b+1} is partitioned into six phases. During the third phase, the MDP is reset, that is, some bicycles are opened and the targets of some entry vertices are changed. Therefore, a phase 3 policy σ is always associated with such a transition and we always implicitly consider the underlying transition from σ_b to σ_{b+1}.

We begin by further investigating the occurrence records of switches that should be applied during phase 3, i.e., we analyze the set L_σ^3. First, the occurrence record of these switches is bounded from above by the flip number $f(b, \ell')$.

Lemma 1. *Let σ be a phase 3 policy. Then $\max_{e \in L_\sigma^3} \phi^\sigma(e) \leq f(b, \ell')$.*

The following lemma gives a matching lower bound of $f(b, \ell')$ on all improving switches that should be applied *after* phase 3. It will also be used to estimate the occurrence records of possible improving switches contained in U_σ^3.

Lemma 2. *Let σ be a phase 3 policy. Assume that the Policy Iteration Algorithm is started with the policy σ^*. Then $\min_{e \in L_\sigma^4 \cup L_\sigma^5 \cup L_\sigma^6} \phi^\sigma(e) \geq f(b, \ell')$.*

We now partition U_σ^3 as follows (note that $U_\sigma^4 \subseteq U_\sigma^3$), cf. [4, Table 3]:

$$U_\sigma^{3,1} := \{(k_i, k_z) : \sigma(k_i) \notin \{k_z, k_{\ell'}\}, z \leq \ell' \wedge (b+1)_i = 0\}$$
$$U_\sigma^{3,2} := \{(b_{i,r}^j, k_z) : \sigma(b_{i,r}^j) \notin \{k_z, k_{\ell'}\}, z \leq \ell' \wedge (b+1)_i = 0\}\}$$
$$U_\sigma^{3,3} := \{(b_{i,r}^j, k_z) : \sigma(b_{i,r}^j) \notin \{k_z, k_{\ell'}\}, z \leq \ell' \wedge (b+1)_{i+1} \neq j\}$$
$$U_\sigma^{3,4} := U_\sigma^4$$

Lemma 2 can be used to show that the occurrence records of edges contained in $U^{3,4}$ are too large. To be precise, no switch contained in one of this sets will be applied during phase 3 when following the LEAST-ENTERED rule.

Lemma 3. *Let σ be a phase 3 policy. Then, for all $e \in L_\sigma^3$ and $\tilde{e} \in I_\sigma \cap U_\sigma^{3,4}$, it holds that $\phi^\sigma(e) \leq \phi^\sigma(\tilde{e})$.*

It remains to analyze the sets $U^{3,1}, U^{3,2}$ and $U^{3,3}$. We show that applying certain switches contained in L_σ^3 prevent other switches contained in these sets from being applied. To do so, we introduce subsets of $U_\sigma^{3,1}, U_\sigma^{3,2}$ and $U_\sigma^{3,3}$. The idea is to "slice" these sets such that for each slice, one improving switch prevents the whole slice from being applied. We thus define the following sets:

$$S_{i,\sigma}^{3,1} := \{(k_i, k_z) : \sigma(k_i) \notin \{k_z, k_{\ell'}\}, z \le \ell' \wedge (b+1)_i = 0\} \subseteq U_\sigma^{3,1}$$

$$S_{i,j,r,\sigma}^{3,2} := \{(b_{i,r}^j, k_z) : \sigma(k_i) \notin \{k_z, k_{\ell'}\}, z \le \ell' \wedge (b+1)_i = 0\} \subseteq U_\sigma^{3,2}$$

$$S_{i,j,r,\sigma}^{3,3} := \{(b_{i,r}^j, k_z) : \sigma(k_i) \notin \{k_z, k_{\ell'}\}, z \le \ell' \wedge (b+1)_i \ne j\} \subseteq U_\sigma^{3,3}$$

The informal idea discussed previously is formalized by the following lemma.

Lemma 4. *The following statements hold.*

1. *Let σ be the phase 3 policy in which the switch $(k_i, k_{\ell'})$ is applied. Let σ' be a phase 3 policy of the same transition reached after σ. Then $I_{\sigma'} \cap S_{i,\sigma'}^{3,1} = \emptyset$.*
2. *Let σ be the phase 3 policy in which the improving switch $(b_{i,l}^j, k_{\ell'})$ with $\sigma(b_{i,l}^j) \ne k_{\ell'}$ and $(b+1)_i = 0$ is applied. Let σ' be a phase 3 policy of the same transition reached after σ. Then $I_{\sigma'} \cap S_{i,j,l,\sigma'}^{3,2} = \emptyset$.*
3. *Let σ be the phase 3 policy in which the improving switch $(b_{i,l}^j, k_{\ell'})$ with $\sigma(b_{i,l}^j) \ne k_{\ell'}$ and $(b+1)_{i+1} \ne j$ is applied. Let σ' be a phase 3 policy of the same transition reached after σ. Then $I_{\sigma'} \cap S_{i,j,l,\sigma'}^{3,3} = \emptyset$.*

This statement can then be used to prove the following lemma.

Lemma 5. *Let σ be a phase 3 policy. Then there is an edge $e \in L_\sigma^3$ minimizing the occurrence record among all improving switches.*

This lemma does not yet imply that all switches of phase 3 can be applied since it is not clear why it cannot happen that phase 4 is reached although not all switches of phase 3 were applied yet. However, the following theorem proves that this is impossible (See footnote 3).

Theorem 3. *There is an ordering of the improving switches and an associated tie-breaking rule compatible with the LEAST-ENTERED pivot rule such that all improving switches contained in $L_{\sigma_b}^3$ are applied and the LEAST-ENTERED pivot rule is obeyed during phase 3.*

Although Theorem 3 shows that the improving switches of phase 3 can be applied such that the LEAST-ENTERED rule is obeyed, it does not imply that the transition from σ_b to σ_{b+1} can be executed as intended in [4]. That is, it does not imply that the improving switches of the other phases can be applied as intended. This however can be shown using Theorem 3, yielding the following result.

Theorem 4. *Fix the transition from σ_b to σ_{b+1} for some $\sigma \in \mathbb{B}_n$. There is an order in which to apply improving switches during this transition such that the LEAST-ENTERED rule is obeyed, and the switches of phase p are applied before any switches of phase $p+1$, for every $p \in \{1, \ldots, 5\}$.*

Acknowledgments. The authors are very grateful to Oliver Friedmann for helpful comments and discussions, as well as support in using his implementation of the original construction to verify our findings.

A   Proofs of Selected Statements

This section contains the proofs of the main statements. The proofs use the following two statements whose proofs can be found in [3].

Lemma A.1. *Let $i \in \{2, \ldots, n-2\}$ and $l < i$. Then, there is a number $b \in \mathbb{B}_n$ with $\ell(b+1) = l$ such that for all $j \in \{i+2, \ldots, n\}$, $\phi^{\sigma_b}(k_i, k_{\ell'}) < \phi^{\sigma_b}(k_j, k_{\ell'})$ and $(k_i, k_{\ell'}), (k_j, k_{\ell'}) \in L^3_{\sigma_b}$.*

Lemma A.2. *Assume that for any transition, the switches that should be applied during phase 3 were applied in some order. Let $i \in \{2, \ldots, n-2\}$ and $l < i$. Then there is a $b \in \mathbb{B}_n$ with $\ell(b+1) = l$ such that $\phi^{\sigma_b}(k_{i+1}, k_{\ell'}) < \phi^{\sigma_b}(b^1_{i,r}, k_{\ell'})$, where $r \in \{0, 1\}$ is arbitrary and $(k_{i+1}, k_{\ell'}), (b^1_{i,r}, k_{\ell'}) \in L^3_{\sigma_b}$.*

We now prove the main statements of this paper.

Issue 4. Applying the improving switches as described in [4, Lemma 5] does not obey the LEAST-ENTERED pivot rule.

Proof. According to [4, Lemma 5], the switches of phase 3 should be applied as follows (See footnote 2): *"[...] we need to perform all switches from higher indices to smaller indices, and k_i to $k_{\ell'}$ before $b^j_{i,l}$ with $(b+1)_{i+1} \neq j$ or $(b+1)_i = 0$ to $k_{\ell'}$".*

Let $i \in \{2, \ldots, n-2\}, l < i$ and $j \in \{i+2, \ldots, n-2\}$. By Lemma A.1, there is a number $b \in \mathbb{B}_n$ such that $l = \ell(b+1)$ and $\phi^{\sigma_b}(k_i, k_{\ell'}) < \phi^{\sigma_b}(k_j, k_{\ell'})$. In addition, $(k_i, k_{\ell'}), (k_j, k_{\ell'}) \in L^3_{\sigma_b}$. Therefore, the switch $(k_j, k_{\ell'})$ should be applied before the switch $(k_i, k_{\ell'})$ during the transition from σ_b to σ_{b+1} when following the description of [4].

Consider the phase 3 policy σ of this transition in which the switch $(k_j, k_{\ell'})$ should be applied. Then, since $j > i$ and we *"perform all switches from higher indices to smaller indices"*, the switch $(k_i, k_{\ell'})$ was not applied yet. However, it still is an improving switch for the policy σ. This implies $\phi^{\sigma_b}(k_j, k_{\ell'}) = \phi^{\sigma}(k_j, k_{\ell'})$ and $\phi^{\sigma_b}(k_i, k_{\ell'}) = \phi^{\sigma}(k_i, k_{\ell'})$. Consequently, $\phi^{\sigma_b}(k_i, k_{\ell'}) < \phi^{\sigma_b}(k_j, k_{\ell'})$ implies that $\phi^{\sigma}(k_i, k_{\ell'}) < \phi^{\sigma}(k_j, k_{\ell'})$. Thus, since the edge $(k_i, k_{\ell'})$ is an improving switch for σ having a lower occurrence record than $(k_j, k_{\ell'})$ and σ was chosen as the policy in which $(k_j, k_{\ell'})$ should be applied, the LEAST-ENTERED rule is violated.   □

Issue 5. Consider some $b \in \mathbb{B}_n$ and the transition from σ_b to σ_{b+1}. Suppose that the switches of phase 3 are applied "level by level" according to any fixed ordering of the levels as described above. Further suppose that this ordering only depends on $\ell(b+1)$. Then, the LEAST-ENTERED pivot rule is violated.

Proof. To prove Issue 5, we show that applying the improving switches as discussed before violates Zadeh's LEAST-ENTERED rule several times by showing the following statement: Let S^i be an ordering of $\{1, \ldots, n\}$ for $i \in \{1, \ldots, n\}$. Suppose that the improving switches of phase 3 of the transition from σ_b to σ_{b+1} are applied in the order defined by $S^{\ell(b+1)}$ for all $b \in \mathbb{B}_n$. Then, for every possible least significant bit $l \in \{1, \ldots, n-4\}$, assuming that the ordering S^l obeys the LEAST-ENTERED rule results in a contradiction.

Fix some $l \in \{1, \ldots, n-4\}$. Consider the ordering $S^l = (s_1, \ldots, s_n)$. For any $k \in \{1, \ldots, n\}$, we denote the position of k within S^l by $k^\star$. Towards a contradiction, assume that applying the improving switches level by level according to the ordering S^l obeys the LEAST-ENTERED rule. We show that this assumption yields $(l+1)^\star < (n-1)^\star$ and $(n-1)^\star < (l+1)^\star$.

Let $i \in \{l+1, \ldots, n-2\}$. Then, $i > l$ and therefore, by Lemma A.2, there is a number $b \in \mathbb{B}_n$ with $\ell(b+1) = \ell' = l$ and $\phi^{\sigma_b}(k_{i+1}, k_{\ell'}) < \phi^{\sigma_b}(b^1_{i,r}, k_{\ell'})$ such that $(k_{i+1}, k_{\ell'}), (b^0_{i,r}, k_{\ell'}) \in L^3_{\sigma_b}$. Thus, both switches need to be applied during the transition from σ_b to σ_{b+1}. Because of $\phi^{\sigma_b}(k_{i+1}, k_{\ell'}) < \phi^{\sigma_b}(b^1_{i,r}, k_{\ell'})$, level $i+1$ needs to appear before level i within the ordering S^l. Since this argument can be applied for all $i \in \{l+1, \ldots, n-2\}$, the sequence

$$(n-1, n-2, \ldots, l+1)$$

needs to be a (not necessarily consecutive) subsequence of S^l. In particular, $(n-1)^\star < (l+1)^\star$ since $l+1 \neq n-1$ by assumption.

Let $i = l+1$ and $j \in \{i+2, \ldots, n\}$. Then, by Lemma A.1, there is a number $b \in \mathbb{B}_n$ with $\ell(b+1) = l$ such that $\phi^{\sigma_b}(k_i, k_{\ell'}) < \phi^{\sigma_b}(k_{i+2}, k_{\ell'})$ and $(k_i, k_{\ell'}), (k_{i+2}, k_{\ell'}) \in L^3_{\sigma_b}$. Now, both switches need to be applied during the transition from σ_b to σ_{b+1}. Therefore, for all $i \in \{l+1, \ldots, n-2\}$, level i needs to appear before any of the levels level $j \in \{i+2, \ldots, n\}$ within S^l. But this implies that the sequence

$$(l+1, l+3, l+4, \ldots, n-1, n)$$

needs to be a (not necessarily consecutive) subsequence of S^l. In particular, $(l+1)^\star < (n-1)^\star$ since $n-1 \geq l+3$ as we have $l \leq n-4$ by assumption. This however contradicts $(n-1)^\star < (l+1)^\star$. □

Theorem 3. *There is an ordering of the improving switches and an associated tie-breaking rule compatible with the LEAST-ENTERED pivot rule such that all improving switches contained in $L^3_{\sigma_b}$ are applied and the LEAST-ENTERED pivot rule is obeyed during phase 3.*

Proof. Let σ denote the first phase 3 policy of the transition from σ_b to σ_{b+1}. Then, $L^3_\sigma = L^3_{\sigma_b}$. By Lemma 5, there is an edge $e_1 \in L^3_\sigma$ minimizing the occurrence record I_σ. Applying this switch results in a new phase 3 policy $\sigma[e_1]$ such that $L^3_{\sigma[e_1]} = L^3_\sigma \setminus \{e_1\}$. Now, again by Lemma 5, there is an edge $e_2 \in L^3_{\sigma[e_1]}$ minimizing the occurrence record $I_{\sigma[e_1]}$.

We can now apply the same argument iteratively until we reach a phase 3 policy $\hat{\sigma}$ such that $|L^3_{\hat{\sigma}}| = 1$ while only applying switches contained in $L^3_{\sigma_b}$. Then, by construction and by Lemma 5, $(e_1, e_2, \ldots)$ defines an ordering of the edges of $L^3_{\sigma_b}$ and an associated tie-breaking rule that always follow the LEAST-ENTERED rule. When the policy $\hat{\sigma}$ with $|L^3_{\hat{\sigma}}| = 1$ is reached, applying the remaining improving switch results in a phase 4 policy. Then, all improving switches contained in $L^3_{\sigma_b}$ were applied and the LEAST-ENTERED pivot rule was obeyed. □

References

1. Avis, D., Chvátal, V.: Notes on Bland's pivoting rule. In: Balinski, M.L., Hoffman, A.J. (eds.) Polyhedral Combinatorics. Mathematical Programming Studies, vol. 8, pp. 24–34. (1978). https://doi.org/10.1007/BFb0121192
2. Dantzig, G.B.: Linear Programming and Extensions. Princeton University Press, Princeton (1963)
3. Disser, Y., Hopp, A.V.: On Friedmann's subexponential lower bound for Zadeh's pivot rule. Technical report, Technische Universität Darmstadt (2018). http://tuprints.ulb.tu-darmstadt.de/id/eprint/7557
4. Friedmann, O.: A subexponential lower bound for Zadeh's pivoting rule for solving linear programs and games. In: Proceedings of the 15th International Conference on Integer Programming and Combinatoral Optimization (IPCO), pp. 192–206 (2011)
5. Goldfarb, D., Sit, W.Y.: Worst case behavior of the steepest edge simplex method. Discrete Appl. Math. 1(4), 277–285 (1979)
6. Howard, R.A.: Dynamic Programming and Markov Processes. MIT Press, Cambridge (1960)
7. Jeroslow, R.G.: The simplex algorithm with the pivot rule of maximizing criterion improvement. Discrete Math. 4(4), 367–377 (1973)
8. Klee, V., Minty, G.J.: How good is the simplex algorithm? In: Inequalities III, pp. 159–175. Academic Press, New York (1972)
9. Puterman, M.L.: Markov Decision Processes: Discrete Stochastic Dynamic Programming. Wiley, New York (2005)
10. Thomas, A.: Unique sink orientations: complexity, structure and algorithms. Ph.D. thesis, ETH Zurich (2017)
11. Zadeh, N.: What is the worst case behavior of the simplex algorithm? Technical report 27, Departments of Operations Research, Stanford (1980)

Tight Approximation Ratio for Minimum Maximal Matching

Szymon Dudycz$^{(\boxtimes)}$, Mateusz Lewandowski, and Jan Marcinkowski

Institute of Computer Science, University of Wrocław, Wrocław, Poland
{szymon.dudycz,mateusz.lewandowski,jan.marcinkowski}@cs.uni.wroc.pl

Abstract. We study a combinatorial problem called *Minimum Maximal Matching*, where we are asked to find in a general graph the smallest matching that can not be extended. We show that this problem is hard to approximate with a constant smaller than 2, assuming the Unique Games Conjecture.

As a corollary we show, that Minimum Maximal Matching in bipartite graphs is hard to approximate with constant smaller than $\frac{4}{3}$, with the same assumption. With a stronger variant of the Unique Games Conjecture—that is Small Set Expansion Hypothesis—we are able to improve the hardness result up to the factor of $\frac{3}{2}$.

1 Introduction

Matchings are some of the most central combinatorial structures in theory of algorithms. A routine computing them is a basic puzzle used in numerous results in Computer Science (like Christofides algorithm). Various variants of matchings are studied extensively. Their computation complexity status is usually well-known and some techniques discovered when studying matchings are afterwards employed in other problems.

As we know since 1961, many natural variants of perfect matchings and maximum matchings can be found in polynomial time, even in general graphs. Here we study a different problem—*Minimum Maximal Matching* (MMM). The task is—given graph G, to find an inclusion-wise maximal matching M with the smallest cardinality (or weight in the weighted version).

1.1 Related Work

The MMM problem was studied as early as 1980, when Yannakakis and Gavril showed that it is NP-hard even in some restricted cases [20]. Their paper also presents an equivalence of MMM and *Minimum Edge Dominating Set* (EDS) problem, where the goal is to find minimum cardinality subset of edges F, such that

S. Dudycz—Supported by the Polish National Science Centre grant 2013/11/B/ST6/01748.

M. Lewandowski and J. Marcinkowski—Supported by the Polish National Science Centre grant 2015/18/E/ST6/00456.

A. Lodi and V. Nagarajan (Eds.): IPCO 2019, LNCS 11480, pp. 181–193, 2019.
https://doi.org/10.1007/978-3-030-17953-3_14

every edge in the graph is adjacent to some edge in F. Every maximal matching is already an edge dominating set, and any edge dominating set can be easily transformed to a maximal matching of no larger size. This equivalence does not hold for the weighted variants of the problem.

It is a well known, simple combinatorial fact, that one maximal matching in any graph can not be more than twice as large as another maximal matching. This immediately gives a trivial 2-approximation algorithm for MMM. Coming up with 2-approximation in the weighted variant of either of the problems is more challenging. In 2003, Carr, Fujito, Konjevod and Parekh presented a $2\frac{1}{10}$-approximation algorithm for a weighted EDS problem [3]. Later the approximation was improved to 2 by Fujito and Nagamochi [8].

Some algorithms aiming at approximation ratio better then 2 were also developed for the unweighted problem. Gotthilf, Lewenstein and Rainschmidt came up with a $2 - c\frac{\log n}{n}$-approximation for the general case [9]. Schmied and Viehmann have a better-than-two constant ratio for dense graphs [19].

Finally, hardness results need to be mentioned. In 2006 Chlebík and Chlebíková proved, that it is NP-hard to approximate the problem within factor better than $\frac{7}{6}$ [4]. The result was later improved to 1.207[1] by Escoffier, Monnot, Paschos, and Xiao [7]. $\frac{3}{2}$-hardness results depending on UGC were also obtained [7,19].

1.2   Unique Games Conjecture

Unique Games Conjecture, since being formulated by Khot in 2002 [11], has been used to prove hardness of approximation of many problems. For the survey on UGC results see [12].

Many hardness results obtained from Unique Games Conjecture match previously known algorithms, as is the case, for example, of *Vertex Cover*, *Max Cut* or *Maximum Acyclic Subgraph*. Therefore, it is appealing to use it to obtain new results. While UGC is still open, recently a related 2–2-Games Conjecture has been proved [14], in consequence proving Unique Games Conjecture with partial completeness. This result provides some evidence towards validity of Unique Games Conjecture.

Basing on Unique Games Conjecture we are able to prove the main result of our paper.

Theorem 1. *Assuming Unique Games Conjecture, it is NP-hard to approximate Minimum Maximal Matching with constant better than 2.*

The proof of this theorem relies on the UGC-hardness proof for Vertex Cover of Khot and Regev [15]. In essence, we endeavour to build a matching over the vertices of Vertex Cover.

---

[1]   In their paper they claim 1.18-hardness, which is achieved by approximation preserving reduction from vertex cover problem. Using recent $\sqrt{2}$-hardness for Vertex Cover [14] gives 1.207-hardness for Minimum Maximal Matching.

The Minimum Maximal Matching problem does not seem to be easier on bipartite graphs. All the algorithms mentioned above are defined for general graphs and we are not aware of any ways to leverage the bipartition of the input graph. At the same time, our hardness proof only works for general graphs. With some observations we are able to achieve a hardness result for bipartite graphs, which, however, is not tight.

Theorem 2. *Assuming Unique Games Conjecture, it is NP-hard to approximate bipartite Minimum Maximal Matching with constant better than $\frac{4}{3}$.*

1.3   Obtaining a Stronger Result

The studies on Unique Games Conjecture and hardness of approximation of different problems have led to formulating different hypotheses strengthening upon UGC—among them the *Small Set Expansion Hypothesis* proposed by Raghavendra and Steurer [18], and another conjecture—whose name is not yet established and so far the name *Strong UGC* is used—formulated by Bansal and Khot [1]. A competent discussion on differences between the two conjectures can be found in [16, Appendix C].

To improve our result on bipartite graphs, we construct a reduction from a problem called *Maximum Balanced Biclique* (MBB), where—given a bipartite graph—the goal is to find a maximum biclique with the same number of vertices on each side of the graph. Hardness of approximation results suitable for our reduction have been found starting from both the Small Set Expansion Hypothesis [16] and Strong UGC [2].

Theorem 3. *Assuming Small Set Expansion Hypothesis (or Strong Unique Games Conjecture), it is NP-hard to approximate Bipartite Minimum Maximal Matching with a constant better than $\frac{3}{2}$.*

Due to space limitations, this result is only presented in the full version of our paper (published on arXiv [6]).

2   Revisiting the Khot-Regev Reduction

In their paper [15] Khot and Regev prove the UGC-hardness of approximating Minimum Vertex Cover within a factor smaller than 2. In this section we look at parts of their proof more closely.

Their reduction starts off with an alternative formulation of UGC[2], which, they show, is a consequence of the standard variant.

---

[2] In their paper, Khot and Regev call this formulation "Strong Unique Games Conjecture". Since then, however, the same name has been used to refer another formulation, as in [1], we decided to minimise confusion by not recalling this name.

2.1   Khot-Regev Formulation of Unique Games Conjecture

This formulation talks about a variant of Unique Label Cover problem described by a tuple $\Phi = (X, R, \Psi, E)$. X is a set of variables, E are the edges and Ψ_{x_1, x_2} defines a constraint for every pair of variables connected by an edge. A constraint is a permutation $\Psi_{x_1, x_2} \in R \leftrightarrow R$ meaning that if x_1 is labelled with a colour $r \in R$, x_2 must be labelled with $\Psi_{x_1, x_2}(r)$.

A t-*labelling* is an assignment of subsets $L(x)$ of size $|L(x)| = t$ to the variables. A constraint Ψ_{x_1, x_2} is satisfied by the t-labelling L if there exists a colour $r \in L(x_1)$ such that $\Psi_{x_1, x_2}(r) \in L(x_2)$.

Conjecture 1 (Unique Games Conjecture).

For any $\xi, \gamma > 0$ and $t \in \mathbb{N}$ there exists some $|R|$ such that it is NP-hard to distinguish, given an instance $\Phi = (X, R, \Psi, E)$ which category it falls into:

- (YES instance): There exists a labelling (*1-labelling*) L and a set $X_0 \subseteq X$, $|X_0| \geqslant (1 - \xi)|X|$, such that L satisfies all constraints between vertices of X_0.
- (NO instance): For any t-labelling L and any set $X_0 \subseteq X$, $|X_0| \geqslant \gamma|X|$, not all constraints between variables of X_0 are satisfied by L.

2.2   Weighted Vertex Cover

The next step is a reduction from the UGC to the Minimum Vertex Cover problem. Given an instance $\Phi = (X, R, \Psi, E)$ of Unique Label Cover problem, as described above, we build a graph G_Φ.

For every variable in $x \in X$ we create a *cloud* $\mathscr{C}_x$ of $2^{|R|}$ vertices. Each vertex corresponds to a subset of labels and is denoted by $(x, S) \in |X| \times \mathcal{P}(R)^3$. The weight of a new vertex (x, S), denoted as $w(x, S)$, is equal to

$$\mu(|S|) = \frac{1}{|X|} \cdot p^{|S|}(1 - p)^{|R \setminus S|}$$

where $p = \frac{1}{2} - \varepsilon$ (there is a bias towards smaller sets). The total weight of G_Φ is thus equal to 1.

Next, we connect the vertices (x_1, S_1) and (x_2, S_2) if there is no pair of labels $s_1 \in S_1$ and $s_2 \in S_2$ satisfying the constraint Ψ_{x_1, x_2}. Two lemmas are proved.

Lemma 1 ([15, Sec. 4.2]). *If Φ was a* YES *instance, the graph G_Φ has an independent set of weight at least $\frac{1}{2} - 2\varepsilon$.*

Proof. The instance Φ, being a YES instance, has a labelling L assigning one colour r_x to each variable x. We know, that there is a large set X_0 of variables ($|X_0| \geqslant (1 - \xi)|X|$), such that all constraints between variables of X_0 are satisfied by L.

We now define

$$\mathcal{IS} = \big\{ (x, S) \mid x \in X_0, \, r_x \in S \big\}$$

---
[3] $\mathcal{P}(R)$ denotes a power set of R, that is set of all subsets of R.

and claim, that $\mathcal{IS}$ is an independent set in G_Φ. For any two variables x_1 and x_2 of X_0 we know, that

$$\Psi_{x_1,x_2}(r_{x_1}) = r_{x_2}.$$

Indeed, if we then take the sets of labels $S_1 \ni r_1$ and $S_2 \ni r_2$, they do satisfy the constraint for the variables x_1, x_2. Hence, there is no edge between (x_1, S_1) and (x_2, S_2).

Finally, the weight of $\mathcal{IS}$ is equal to

$$
\begin{aligned}
w\,(\mathcal{IS}) &= \sum_{x \in X_0} \left(\sum_{S \subseteq R, S \ni r_x} w(x,S) \right) = \sum_{x \in X_0} \left(\frac{1}{|X|} \cdot \sum_{k=1}^{|R|} \binom{|R|-1}{k-1} \cdot p^k \cdot (1-p)^{|R|-k} \right) \\
&= \sum_{x \in X_0} \left(p \cdot \frac{1}{|X|} \cdot \sum_{k=0}^{|R|-1} \binom{|R|-1}{k} \cdot p^k \cdot (1-p)^{|R|-1-k} \right) \\
&= \sum_{x \in X_0} \left(p \cdot \frac{1}{|X|} \cdot (p + (1-p))^{|R|-1} \right) \\
&= \frac{|X_0|}{|X|} \cdot p \geqslant (1-\xi)(\tfrac{1}{2} - \varepsilon) > \frac{1}{2} - 2\varepsilon.
\end{aligned}
$$

$\square$

The most of their paper is dedicated to proving the following key lemma.

Lemma 2 ([15, Sec. 4.3]). *If Φ is a NO instance, G_Φ does not have an independent set of weight larger than 2γ.*

Since the Minimum Vertex Cover is a complement of the Maximum Independent Set, we see that it is hard to distinguish between graphs with Minimum Vertex Cover of the weight $\frac{1}{2} + 2\varepsilon$ and those, where Minimum Vertex Cover weights $1 - 2\gamma$.

2.3   Notation

Throughout this paper we are going to use Φ as an instance of Unique Label Cover problem that we are translating to G_Φ. The weight function w on vertices and bias function μ is going to be referred to, as well as the constants ε and γ. When Φ is a YES instance, we are going to refer to the set X_0 as in Conjecture 1, and use the independent set $\mathcal{IS}$ from Lemma 1.

3   Weighted Minimum Maximal Matching

Let us now modify their reduction. The graph G'_Φ gets additional edges between vertices (x, S_1), (x, S_2) if $S_1 \cap S_2 = \varnothing$—they do not assign the same colour to the variable x. Clearly, the Lemmas 1 and 2 still hold for G'_Φ.

Moreover, we introduce the weight function on the edges.

$$w_+\,((x_1, S_1), (x_2, S_2)) \overset{\text{def}}{=\!=} w\,(x_1, S_1) + w\,(x_2, S_2)$$

This weight function is such that for any matching, the weight of matching edges is equal to the weight of matched vertices.

We will now show the similar statements are true for the Minimum Maximal Matching as for the independent set.

Lemma 3. *If Φ was a* YES *instance, the Minimum Maximal Matching in (G'_Φ, w_+) weights at most $\frac{1}{2} + 2\varepsilon$.*

Lemma 4. *If Φ was a* NO *instance, the Minimum Maximal Matching in (G'_Φ, w_+) weights at least $1 - 2\gamma$.*

These lemmas altogether will give us the theorem.

Theorem 4. *Assuming the Unique Games Conjecture, for any $\epsilon > 0$ it is NP-hard to distinguish between graphs with Maximal Matching of weight $\frac{1}{2} + \epsilon$ and those where every Maximal Matching weights at least $1 - \epsilon$.*

This in turn means that—assuming UGC—a polynomial-time approximation algorithm with a factor better than 2 can not be constructed.

Proof (Proof of Lemma 3).

Let us construct a matching M in G'_Φ. The matching will only consist of the edges between vertices corresponding to the same variable in Φ. First we define the part of M restricted to X_0[4].

$$M_0 = \left\{ (x, S_1) \sim (x, S_2) \mid x \in X_0 \wedge S_1 \uplus S_2 = R \setminus \{r_x\} \right\}$$

For vertices in clouds corresponding to variables outside of X_0 we define

$$M_1 = \left\{ (x, S_1) \sim (x, S_2) \mid x \in X_0 \wedge S_1 \uplus S_2 = R \right\}$$

The matching M will be the union of M_0 and M_1.

We can observe, that the vertices matched by M are exactly those, that do not belong to $\mathcal{IS}$. Hence,

$$w_+(M) \leqslant w(V(G'_\Phi)) - w(\mathcal{IS}) \leqslant 1 - \left(\frac{1}{2} - 2\varepsilon\right) = \frac{1}{2} + 2\varepsilon$$

Moreover, since the vertices of M compose a vertex cover, M is a maximal matching. □

Proof (Proof of Lemma 4).

Let M be any maximal matching. The vertices matched by M, $V(M)$ form a vertex cover. Hence, the weight of M is going to be at least as large as the weight of the Minimum Vertex Cover. From Lemma 2 we know, that if Φ was a NO instance, G'_Φ's Minimum Vertex Cover weights at least $1 - 2\gamma$. □

---

[4] $\uplus$ is a disjoint union symbol.

4 Towards the Unweighted MMM: Fractional Matchings

A natural way to reduce a weighted variant of a problem to the unweighted would often be to assume that the weights are integral (that can be achieved by rounding them first at a negligible cost) and copying every vertex as many times, as its weight would suggest. This simple strategy will not however work with instances from previous section, where we were matching pairs of vertices of different weights. Such a matching does not easily translate to the graph with vertex copies. Thus, we want to create different, fractional matching, in which every vertex is matched proportionally to its weight. Then, we can use such matching after copying each vertex.

In order to extend our approximation hardness proof to Minimum Maximal Matching problem in unweighted graphs, we thus need first to modify our weighted reduction a bit. The structure remains the same, but the weight of each edge is now defined to be the minimum of the weights of its endpoints.

$$w_{\min}\left((x_1, S_1), (x_2, S_2)\right) \overset{\text{def}}{=\!=} \min\left\{w(x_1, S_1), w(x_2, S_2)\right\}$$

Similarly to the reasoning presented in the previous section, when G'_Φ is a YES instance, we will want to construct a matching and argue that it is maximal using a known vertex cover.

Definition 1. *A fractional matching is an assignment of values to variables x_e corresponding to edges, such that for every edge e $0 \leqslant x_e \leqslant w_{\min}(e)$ and for every vertex v, the sum $\sum_{(v,w) \in E} x_{(v,w)} \leqslant w(v)$.*

Definition 2. *A fractional matching x saturates a vertex v if $\sum_{(v,w) \in E} x_{(v,w)} = w(v)$. A vertex v is* unmatched *if $\sum_{(v,w) \in E} x_{(v,w)} = 0$.*

As we know already, when Φ is a YES instance, there is a vertex cover in G'_Φ composed of all vertices except those in $\mathcal{IS}$.

Lemma 5. *If Φ was a YES instance, a fractional matching exists that leaves all vertices in $\mathcal{IS}$ unmatched and saturates all the other vertices.*

4.1 Proving Lemma 5

Our matching will again only match vertices in the same clouds. Let us first concentrate on vertices in the cloud $\mathscr{C}_x$ corresponding to a variable $x \notin X_0$. The matching needs to saturate every vertex in $\mathscr{C}_x$.

The fractional matching F can be viewed as a real-valued vector and will be a sum of three matchings. The first one is defined similarly to M_1 in Lemma 3.

$$F^0\left((x, S_1), (x, S_2)\right) = \begin{cases} w_{\min}\left((x, S_1), (x, S_2)\right), & \text{if } S_1 \uplus S_2 = R \\ 0, & \text{otherwise} \end{cases}$$

Recalling, that the weight function w, defined on vertices, has a bias towards smaller sets, we can state the following.

Observation 5. F^0 *saturates all vertices* $(x, S) \in \mathscr{C}_x$ *if* $|S| \geqslant \frac{|R|}{2}$.

Let us now pick $0 < k < \frac{|R|}{2}$ and look at the layer $\mathscr{C}_x^k = \{(x, S) \mid |S| = k\}$. The graph is symmetric, and F^0 matches every vertex with the same weight— $\mu(|R| - k) = \frac{1}{|X|} p^{|R|-k} (1 - p)^k$. In order to build a matching F^1, that saturates all vertices in the layer we build a bipartite graph $\mathcal{B}^k$ out of $\mathscr{C}_x^{k}$[5].

Definition 3. *For every set* S *of size* k, $\mathcal{B}^k$ *has two vertices,* S^L *and* S^R. S_1^L *is connected with* S_2^R *if* $S_1 \cap S_2 = \varnothing$.

The graph $\mathcal{B}_k$ is in fact a *Bipartite Kneser Graph*. As proved in [17], it has a Hamiltonian cycle $\mathcal{H}_k$. We are using this cycle to define F^1—for every edge connecting the sets S_1 and S_2 in $\mathcal{H}_k$ we lay the weight of

$$F^1 \left((x, S_1), (x, S_2) \right) = \frac{1}{4} \left(\mu(k) - \mu(|R| - k) \right)$$

on the edge connecting them in $\mathscr{C}_x^k$.

To saturate the vertices $(x, \varnothing)$ (for $x \notin X_0$), we must realize that all these vertices form a clique in which we can find a Hamiltonian Cycle $\mathcal{H}_\varnothing$. Let us define F^2

$$F^2 \left((x_1, \varnothing), (x_2, \varnothing) \right) = \begin{cases} \frac{\mu(0) - \mu(|R|)}{2}, & \text{for } \{x_1, x_2\} \in \mathcal{H}_\varnothing \\ 0, & \text{otherwise} \end{cases}$$

Lemma 6. $F^0 + F^1 + F^2$ *saturates all vertices in* $\mathscr{C}_x^k$.

Proof. We look at the vertex (x, S). For $0 < |S| < \frac{|R|}{2}$, the Hamiltonian Cycle $\mathcal{H}_k$ visits every set exactly twice (once S^L and once S^R), using four edges incident to it. Hence, the total contribution of F^0 and F^1 is equal to

$$\mu(|R| - k) + 4 \cdot \frac{1}{4} \left(\mu(k) - \mu(|R| - k) \right) = \mu(k) = w(x, S).$$

F^0 contributes $\mu(|R|)$ to the vertex $(x, \varnothing)$, while F^2 contributes $2 \cdot \frac{\mu(0) - \mu(|R|)}{2}$, hence that vertex is also saturated.

Finally, vertices with $S = \varnothing$ are saturated by $F^0 + F^2$. $\square$

When $x \in X_0$. We proceed similarly as for vertices not in X_0. For the cloud $\mathscr{C}_x$ when $x \in X_0$, our first matching F^0 is taking the labeling of the variable x into account. Similarly to Lemma 3, we match (x, S_1) and (x, S_2) if $S_1 \uplus S_2 = R \backslash \{r_x\}$, thus saturating the larger of the sets.

Again, the layer $\mathscr{C}_x^k$ for $k < \frac{|R|-1}{2}$, composed of sets not containing r_x, is a Bipartite Kneser Graph, and we use its Hamiltonian cycle to define F^1.

Also the vertices $(x, \varnothing)$ for $x \in X_0$ form a clique. Once again, we can use the Hamiltonian Cycle in that clique to define F^2.

---

[5] A significantly more crude approach is possible, that just uses every edge equally.

5   Unweighted MMM

Starting with a graph G'_Φ with the weight function w on the vertices, and any precision parameter $\rho > 0$, we are going to construct an unweighted graph $G^\rho_\Phi = (V^\rho, E^\rho)$. The resulting graph size is polynomial in $|\Phi|$ and $\frac{1}{\rho}$.

Definition 4. *Let* $n = |V(G'_\Phi)| \cdot \frac{1}{\rho}$. *For every* $v \in V(G'_\Phi)$ *we set* $n_v = \lceil n \cdot w(v) \rfloor$[6]. *The new set of vertices is going to consist of multiple copies of original vertices; for each vertex* v, *we add* $4 \cdot n_v$ *copies.*

$$V^\rho = \{ \langle v, i \rangle \mid v \in V(G'_\Phi), i \in \{1, \ldots, 4 \cdot n_v\} \}.$$

The edges are going to connect each pair of copies of vertices connected in G'_Φ.

$$E^\rho = \{\{\langle v_1, i_1 \rangle, \langle v_2, i_2 \rangle\} \mid \{v_1, v_2\} \in E(G_\Phi),$$
$$i_1 \in [4 \cdot n_{v_1}], i_2 \in [4 \cdot n_{v_2}]\}.$$

This construction has been presented in [5]. It is shown that any vertex cover $C \subset G'_\Phi$ yields a *product vertex cover* $C^\rho = \bigcup_{v \in C}\{v\} \times [4 \cdot n_v]$ with $\left| w(C) - \frac{|C^\rho|}{|V(G^\rho_\Phi)|} \right| < \rho$ (precision). Moreover, every minimal vertex cover in G^ρ_Φ is a product vertex cover [5, Proposition 8.1].

As before, we are now going to prove two lemmas witnessing the completeness and soundness of our reduction.

Lemma 7 (Soundness). *If* Φ *was a* NO *instance, for every maximal matching* M *in* G^ρ_Φ
$$2 \cdot |M| > |V(G^\rho_\Phi)|(1 - 2\gamma - \rho).$$

Proof. Take any maximal matching M. The $2 \cdot |M|$ vertices matched by it form a vertex cover C. Let C_- be a minimal vertex cover obtained by removing unneeded vertices from C. As presented in [5], C_- is a product vertex cover, which means, there is a vertex cover C_w in G'_Φ with weight

$$w(C_w) < \frac{|C_-|}{V(G^\rho_\Phi)} + \rho \leqslant \frac{|C|}{V(G^\rho_\Phi)} + \rho.$$

On the other hand, from Lemma 2 we have, that $w(C_w) > 1 - 2\gamma$.   □

Lemma 8 (Completeness). *If* Φ *was a* YES *instance, a maximal matching* M *exists in* G^ρ_Φ *with*
$$2 \cdot |M| < |V(G^\rho_\Phi)| \left(\frac{1}{2} + 2\varepsilon + \rho \right).$$

---

[6] $\lceil x \rfloor$ is an integer nearest to x.

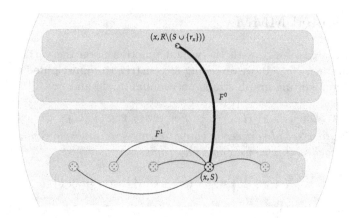

$(x, R\backslash(S \cup \{r_x\}))$

F^0

F^1

(x, S)

Fig. 1. A close-up look at the resulting matching in a cloud of vertices corresponding to the variable $x \in X_0$. The fractional matchings F^0 and F^1 constructed in Sect. 4 can be discretised to match all the copies of respective vertices.

Proof. Take F, a fractional matching on $(G'_\Phi, w_{\min})$ constructed in Lemma 5. When F^0 matches vertices $u = (x, S_1)$ and $v = (x, S_2)$ with some weight $F^0(u, v)$, we are going to match $4 \cdot n \cdot \lceil F^0(u, v) \rceil$ copies of u and v using parallel edges.

Let us focus on a vertex $u = (x, S) \notin \mathcal{IS}$ belonging to a vertex cover of G'_Φ, with $0 < |S| < \frac{|R|}{2}$. It is matched by F^0 to (x, S'), which leaves $4(\lceil w(x, S) \rceil - \lceil w(x, S') \rceil)$ vertices in G^ρ_Φ unmatched. This number is divisible by 4, which allows us to match all the copies of vertices in the Bipartite Kneser Graph according to F^1 (see Fig. 1).

Finally, the number of unmatched copies of the $(x, \varnothing)$ vertices is divisible by 2. We can thus replicate F^2 to match all the remaining copies of these vertices.

Since we are matching every node in a vertex cover of the graph G^ρ_Φ, our matching is maximal and its cardinality is half of the cardinality of the vertex cover.

$$|M| = \frac{1}{2}\left(V(G^\rho_\Phi) - |IS|^\rho\right) < \frac{1}{2}V(G^\rho_\Phi)\left(1 - \left(\frac{1}{2} - 2\varepsilon - \rho\right)\right) \qquad \square$$

6   Conclusion

We would like to finish by discussing potentially interesting open problems. Natural question following our result on MMM is whether other hardness results for Vertex Cover also hold for MMM. In particular, it is known that Vertex Cover on k-hypergraphs is hard to approximate with a constant better than k [15]. Also, the best known NP-hardness of Vertex Cover is $\sqrt{2}$, following the reduction from 2–2 Games Conjecture [13], which has been recently proven [14].

Both of these reductions are very similar to Khot and Regev's UGC-hardness of Vertex Cover. As such they can be used to prove corresponding hardnesses of weighted MMM, by following similar approach as in Sect. 3. They differ, however, in the choice of the weight function of vertices, which turns out to be crucial in terms of unweighted MMM. These weight functions have bias towards bigger sets, so construction described in Sect. 4 can not be used for these problems.

Still the best known NP-hardness of unweighted MMM is 1.207 by Escoffier, Monnot, Paschos, and Xiao [7] and it is an open problem, whether it can be improved to match hardness of Vertex Cover using 2–2 Games Conjecture.

In case of bipartite MMM, there remains a gap between our $\frac{3}{2}$-hardness and best known constant approximation algorithm, which has ratio 2. Showing that bipartite MMM is hard to approximate with a constant better than 2 would immediately imply tight hardness of *Maximum Stable Matching with Ties* [10]. On the other hand, there are no results for MMM leveraging restriction to bipartite graphs. Thus, a potential better than 2 approximation algorithm for bipartite graphs would be interesting for showing structural difference between MMM in bipartite and general graphs.

A Hardness of Bipartite MMM

In this section we will perform a natural reduction to prove the following theorem.

Theorem 6. *Assuming the Unique Games Conjecture, for any $\epsilon > 0$ it is NP-hard to distinguish between balanced bipartite graphs of $2n$ vertices:*

- *(YES instance) with a Maximal Matching of size smaller than $n\left(\frac{1}{2} + \epsilon\right)$.*
- *(NO instance) with no Maximal Matching of size smaller than $n\left(\frac{2}{3} - \epsilon\right)$.*

We will start with the graph G_Φ^ρ defined in Sect. 5. The bipartite graph H_Φ has two copies v^l and v^r of every vertex $v \in G_\Phi^\rho$. The vertices u^l and v^r are connected with an edge if there is an edge (u, v) in G_Φ^ρ. n is going to be equal to $|V(G_\Phi^\rho)|$. We will call this construction *bipartisation* of an undirected graph.

It is easy to see, that if Φ is a YES instance of the Unique Label Cover problem, we can use the matching from Lemma 8 (M in G_Φ^ρ) to produce a maximal matching in H_Φ. For every edge $(u, v) \in M$ we will put its two copies, (u^l, v^r) and (v^l, u^r) into the matching. The resulting matching size is thus equal to $2 \cdot |M| < n(\frac{1}{2} + \epsilon)$.

A.1 Covering with Paths

In order to analyse the NO case, we need to look at the bipartite instance and its matchings from another angle. For any matching in H_Φ, we will view its edges as directed edges in G_Φ^ρ—the vertices on the left will be viewed as *out* vertices, and those on the right as *in* vertices. The graph G_Φ^ρ will thus be covered with directed edges. Every vertex will be incident to at most one outgoing and one incoming edge, which means that the edges will form a structure of directed paths and cycles. The set of these paths and cycles will be called $\mathscr{P}(M)$ for a matching M.

Observation 7. *If M is a maximal matching, every path $P \in \mathscr{P}(M)$ has a length $|P| \geqslant 2$.*

Proof. Assume, that for a maximal matching M in H_Φ there is a length-one path $P = (u,v) \in \mathscr{P}(M)$. This means, that the vertices v^l and u^r are unmatched in M—yet, they are connected with an edge, that can be added to the matching (that would form a length-2 cycle in $\mathscr{P}(M)$). □

We will now use this observation to prove the relation between maximal matchings in H_Φ and vertex covers in G_Φ^ρ.

Lemma 9. *For any maximal matching M in H_Φ, there exists a vertex cover C in G_Φ^ρ of size $|C| \leqslant \frac{3}{2}|M|$.*

Proof. We will construct the vertex cover using paths and cycles of $\mathscr{P}(M)$. For every $P \in \mathscr{P}(M)$ we add all the vertices of P into C. When P is a cycle, it contains as many vertices as edges. A path has at most $\frac{3}{2}$ as many vertices as edges, since its length is at least 2. □

As shown in Lemma 7, when Φ is a NO instance, the Minimum Vertex Cover in G_Φ^ρ has at least $n(1 - \epsilon)$ vertices. The Minimum Maximal Matching in H_Φ must in this case have at least $\frac{2}{3}n(1 - \epsilon) > n(\frac{2}{3} - \epsilon)$ edges.

The hardness coming from Theorem 6 is, that assuming UGC, no polynomial-time algorithm will provide approximation for Minimum Maximal Matching with a factor $\frac{4}{3} - \epsilon$ for any $\epsilon > 0$.

References

1. Bansal, N., Khot, S.: Optimal long code test with one free bit. In: 50th Annual IEEE Symposium on Foundations of Computer Science, FOCS 2009, October 25–27, 2009, Atlanta, Georgia, USA, pp. 453–462. (2009). https://doi.org/10.1109/FOCS.2009.23
2. Bhangale, A., et al.: Bi-covering: covering edges with two small subsets of vertices. SIAM J. Discrete Math. **31**(4), 2626–2646 (2017). https://doi.org/10.1137/16M1082421
3. Carr, R.D., et al.: A $2\frac{1}{10}$-approximation algorithm for a generalizationof the weighted edge-dominating set problem. In: Proceedings of the Algorithms - ESA2000, 8th Annual European Symposium, Saarbrücken, Germany, 5–8 September, 2000, pp. 132–142 (2000). https://doi.org/10.1007/3-540-45253-2_13
4. Chlebík, M., Chlebíková, J.: Approximation hardness of edge dominating set problems. J. Comb. Optim. **11**(3), 279–290 (2006). https://doi.org/10.1007/s10878-006-7908-0
5. Dinur, I., Safra, S.: The importance of being biased. In: Proceedings on 34th Annual ACM Symposium on Theory of Computing, May 19–21, 2002, Montréal, Québec, Canada, pp. 33–42 (2002). https://doi.org/10.1145/509907.509915
6. Dudycz, S., Lewandowski, M., Marcinkowski, J.: Tight Approximation Ratio for Minimum Maximal Matching. In: CoRR abs/1811.08506 (2018). arXiv: 1811.08506

7. Escoffier, B., et al.: New results on polynomial inapproximabilityand fixed parameter approximability of edge dominating set. Theory Comput. Syst. **56**(2), 330–346 (2015). https://doi.org/10.1007/s00224-014-9549-5

8. Fujito, T., Nagamochi, H.: A 2-approximation algorithm for the minimum weight edge dominating set problem. Discrete Appl. Math. **118**(3), 199–207 (2002). https://doi.org/10.1016/S0166-218X(00)00383-8

9. Gotthilf, Z., Lewenstein, M., Rainshmidt, E.: A approximation algorithm for the minimum maximal matching problem. In: Approximation and Online Algorithms, 6th International Workshop, WAOA 2008, Karlsruhe, Germany, September 18–19, 2008. Revised Papers, pp. 267–278 (2008). https://doi.org/10.1007/978-3-540-93980-1_21

10. Huang, C.-C., et al.: A tight approximation bound for the stable marriage problem with restricted ties. In: Approximation, Randomization, and Combinatorial Optimization. Algorithms and Techniques, APPROX/ RANDOM 2015, August 24–26, 2015, Princeton, NJ, USA, pp. 361–380 (2015). https://doi.org/10.4230/LIPIcs.APPROX-RANDOM.2015.361

11. Khot, S.: On the power of unique 2-prover 1-round games. In: Proceedings on 34th Annual ACM Symposium on Theory of Computing, May 19–21, 2002, Montréal, Québec, Canada, pp. 767–775 (2002). https://doi.org/10.1145/509907.510017

12. Khot, S.: On the unique games conjecture (Invited Survey). In: Proceedings of the 25th Annual IEEE Conference on Computational Complexity, CCC 2010, Cambridge, Massachusetts, USA, 9–12 June 2010, pp. 99–121 (2010). https://doi.org/10.1109/CCC.2010.19

13. Khot, S., Minzer, D., Safra, M.: On independent sets, 2-to-2 games, and Grassmann graphs. In: Proceedings of the 49th Annual ACM SIGACT Symposium on Theory of Computing, STOC 2017, Montreal, QC, Canada, 19–23 June 2017, pp. 576–589 (2017). https://doi.org/10.1145/3055399.3055432

14. Khot, S., Minzer, D., Safra, M.: Pseudorandom sets in grassmann graph have near-perfect expansion. In: Electronic Colloquium on Computational Complexity (ECCC), vol. 25, p. 6 (2018). https://eccc.weizmann.ac.il/report/2018/006

15. Khot, S., Regev, O.: Vertex cover might be hard to approximate to within $2-\varepsilon$. J. Comput. Syst. Sci. **74**(3), 335–349 (2008)

16. Manurangsi, P.: Inapproximability of maximum biclique problems, minimum k-cut and densest at-least-k-subgraph from the small set expansion hypothesis. Algorithms **11**(1), 10 (2018). https://doi.org/10.3390/a11010010

17. Mütze, T., Su, P.: Bipartite kneser graphs are hamiltonian. Combinatorica **37**(6), 1207–1219 (2017). https://doi.org/10.1007/s00493-016-3434-6

18. Raghavendra, P., Steurer, D.: Graph expansion and the unique games conjecture. In: Proceedings of the 42nd ACM Symposium on Theory of Computing, STOC 2010, Cambridge, Massachusetts, USA, 5–8 June 2010, pp. 755–764 (2010). https://doi.org/10.1145/1806689.1806788

19. Schmied, R., Viehmann, C.: Approximating edge dominating set in dense graphs. Theor. Comput. Sci. **414**(1), 92–99 (2012). https://doi.org/10.1016/j.tcs.2011.10.001

20. Yannakakis, M., Gavril, F.: Edge dominating sets in graphs. SIAM J. Appl. Math. **38**(3), 364–372 (1980). https://doi.org/10.1137/0138030

Integer Programming and Incidence Treedepth

Eduard Eiben[1], Robert Ganian[2], Dušan Knop[3,4], Sebastian Ordyniak[5(✉)], Michał Pilipczuk[6], and Marcin Wrochna[6,7]

[1] Department of Informatics, University of Bergen, Bergen, Norway
eduard.eiben@uib.no
[2] Algorithms and Complexity Group, Technische Universität Wien, Vienna, Austria
rganian@ac.tuwien.ac.at
[3] Algorithmics and Computational Complexity, Faculty IV, TU Berlin, Berlin, Germany
[4] Department of Theoretical Computer Science, Faculty of Information Technology, Czech Technical University in Prague, Prague, Czech Republic
dusan.knop@fit.cvut.cz
[5] Algorithms Group, Department of Computer Science, University of Sheffield, Sheffield, UK
s.ordyniak@sheffield.ac.uk
[6] Institute of Informatics, University of Warsaw, Warsaw, Poland
{michal.pilipczuk,marcin.wrochna}@mimuw.edu.pl
[7] University of Oxford, Oxford, UK

Abstract. Recently a strong connection has been shown between the tractability of integer programming (IP) with bounded coefficients on the one side and the structure of its constraint matrix on the other side. To that end, integer linear programming is fixed-parameter tractable with respect to the primal (or dual) treedepth of the Gaifman graph of its constraint matrix and the largest coefficient (in absolute value). Motivated by this, Koutecký, Levin, and Onn [ICALP 2018] asked whether it is possible to extend these result to a more broader class of integer linear programs. More formally, is integer linear programming fixed-parameter tractable with respect to the incidence treedepth of its constraint matrix and the largest coefficient (in absolute value)?

We answer this question in negative. We prove that deciding the feasibility of a system in the standard form, $A\mathbf{x} = \mathbf{b}, \mathbf{l} \leq \mathbf{x} \leq \mathbf{u}$, is NP-hard

Eduard Eiben was supported by Pareto-Optimal Parameterized Algorithms (ERC Starting Grant 715744). This work is a part of projects CUTACOMBS, PowAlgDO (M. Wrochna) and TOTAL (M. Pilipczuk) that have received funding from the European Research Council (ERC) under the European Union's Horizon 2020 research and innovation programme (grant agreements No. 714704, No. 714532, and No. 677651). Dušan Knop is supported by DFG, project "MaMu", NI 369/19. Marcin Wrochna is supported by Foundation for Polish Science (FNP) via the START stipend. Robert Ganian acknowledges support from the FWF Austrian Science Fund (Project P31336: NFPC) and is also affiliated with FI MU, Brno, Czech Republic.

© Springer Nature Switzerland AG 2019
A. Lodi and V. Nagarajan (Eds.): IPCO 2019, LNCS 11480, pp. 194–204, 2019.
https://doi.org/10.1007/978-3-030-17953-3_15

even when the absolute value of any coefficient in A is 1 and the incidence treedepth of A is 5. Consequently, it is not possible to decide feasibility in polynomial time even if both the assumed parameters are constant, unless $P = NP$.

Keywords: Integer programming · Incidence treedepth · Gaifman graph · Computational complexity

1   Introduction

In this paper we consider the decision version of Integer Linear Program (ILP) in *standard form*. Here, given a matrix $A \in \mathbb{Z}^{m \times n}$ with m rows (constraints) and n columns and vectors $\mathbf{b} \in \mathbb{Z}^m$ and $\mathbf{l}, \mathbf{u} \in \mathbb{Z}^n$ the task is to decide whether the set

$$\{\mathbf{x} \in \mathbb{Z}^n \mid A\mathbf{x} = \mathbf{b}, \mathbf{l} \leq \mathbf{x} \leq \mathbf{u}\} \tag{SSol}$$

is non-empty. We are going to study structural properties of the incidence graph of the matrix A. An integer program (IP) is a *standard IP* (SIP) if its set of solutions is described by (SSol), that is, if it is of the form

$$\min\{f(\mathbf{x}) \mid A\mathbf{x} = \mathbf{b}, \mathbf{l} \leq \mathbf{x} \leq \mathbf{u}, \mathbf{x} \in \mathbb{Z}^n\}, \tag{SIP}$$

where $f \colon \mathbb{N}^n \to \mathbb{N}$ is the *objective function*; in case f is a linear function the above SIP is said to be a linear SIP. Before we go into more details we first review some recent development concerning algorithms for solving (linear) SIPs in variable dimension with the matrix A admitting a certain decomposition.

Let E be a 2×2 block matrix, that is, $E = \left(\begin{smallmatrix} A_1 & A_2 \\ A_3 & A_4 \end{smallmatrix}\right)$, where $A_1, \ldots, A_4$ are integral matrices. We define an *n-fold 4-block product* of E for a positive integer n as the following block matrix

$$E^{(n)} = \begin{pmatrix} A_1 & A_2 & A_2 & \cdots & A_2 \\ A_3 & A_4 & 0 & \cdots & 0 \\ A_3 & 0 & A_4 & \cdots & 0 \\ \vdots & & & \ddots & \\ A_3 & 0 & 0 & \cdots & A_4 \end{pmatrix},$$

where 0 is a matrix containing only zeros (of appropriate size). One can ask whether replacing A in the definition of the set of feasible solutions (SSol) can give us an algorithmic advantage leading to an efficient algorithm for solving such SIPs. We call such an SIP an *n-fold 4-block IP*. We derive two special cases of the n-fold 4-block IP with respect to special cases for the matrix E (see monographs [4,17] for more information). If both A_1 and A_3 are void (not present at all), then the result of replacing A with $E^{(n)}$ in (SIP) yields the *n-fold IP*. Similarly, if A_1 and A_2 are void, we obtain the *2-stage stochastic IP*.

The first, up to our knowledge, pioneering algorithmic work on n-fold
4-block IPs is due to Hemmecke et al. [9]. They gave an algorithm that given n,
the 2×2 block matrix E, and vectors $\mathbf{w}, \mathbf{b}, \mathbf{l}, \mathbf{u}$ finds an integral vector $\mathbf{x}$ with
$E^{(n)}\mathbf{x} = \mathbf{b}, \mathbf{l} \leq \mathbf{x} \leq \mathbf{u}$ minimizing $\mathbf{wx}$. The algorithm of Hemmecke et al. [9]
runs in time $n^{g(r,s,\|E\|_\infty)}L$, where r is the number of rows of E, s is the number
of columns of E, L is the size of the input, and $g \colon \mathbb{N} \to \mathbb{N}$ is a computable
function. Thus, from the parameterized complexity viewpoint this is an XP algo-
rithm for parameters $r, s, \|E\|_\infty$. This algorithm has been recently improved by
Chen et al. [3] who give better bounds on the function g; it is worth noting that
Chen et al. [3] study also the special case where A_1 is a zero matrix and even
in that case present an XP algorithm. Since the work of Hemmecke et al. [9]
the question of whether it is possible to improve the algorithm to run in time
$g'(r, s, \|E\|_\infty) \cdot n^{O(1)}L$ or not has become a major open question in the area of
mathematical programming.

Of course, the complexity of the two aforementioned special cases of n-fold
4-block IP are extensively studied as well. The first FPT algorithm[1] for the n-fold
IPs (for parameters $r, s, \|E\|_\infty$) is due to Hemmecke et al. [10]. Their algorithm
has been subsequently improved [7,14]. Altmanová et al. [1] implemented the
algorithm of Hemmecke et al. [10] and improved the polynomial factor (achieving
the same running time as Eisenbrand et al. [7]) the above algorithms (from cubic
dependence to $n^2 \log n$). The best running time of an algorithm solving n-fold
IP is due to Jansen et al. [12] and runs in nearly linear time in terms of n.

Last but not least, there is an FPT algorithm for solving the 2-stage stochastic
IP due to Hemmecke and Schultz [11]. This algorithm is, however, based on a
well quasi ordering argument yielding a bound on the size of the Graver basis
for these IPs. Very recently Klein [13] presented a constructive approach using
Steinitz lemma and give the first explicit (and seemingly optimal) bound on the
size of the Graver basis for 2-stage (and multistage) IPs. It is worth noting
that possible applications of 2-stage stochastic IP are much less understood than
those of its counterpart n-fold IP.

In the past few years, algorithmic research in this area has been mainly
application-driven. Substantial effort has been taken in order to find the right
formalism that is easier to understand and yields algorithms having the best
possible ratio between their generality and the achieved running time. It turned
out that the right formalism is connected with variants of the Gaifman graph
(see e.g. [5]) of the matrix A (for the definitions see the Preliminaries section).

Our Contribution. In this paper we focus on the incidence (Gaifman) graph. We
investigate the (negative) effect of the treedepth of the incidence Gaifman graph
on tractability of ILP feasibility.

Theorem 1. *Given a matrix $A \in \{-1, 0, 1\}^{m \times n}$ and vectors $\mathbf{l}, \mathbf{u} \in \mathbb{Z}_\infty^n$. Deciding whether the set defined by (SSol) is non-empty is NP-hard even if $\mathbf{b} = \mathbf{0}$ and
$\mathrm{td}_I(A) \leq 5$.*

---

[1] That is, an algorithm running in time $f(r, s, \|E\|_\infty) \cdot n^{O(1)}L$.

Preliminaries

For integers $m < n$ by $[m : n]$ we denote the set $\{m, m+1, \ldots, n\}$ and $[n]$ is a shorthand for $[1 : n]$. We use bold face letters for vectors and normal font when referring to their components, that is, $\mathbf{x}$ is a vector and x_3 is its third component. For vectors of vectors we first use superscripts to access the "inner vectors", that is, $\mathbf{x} = (\mathbf{x}^1, \ldots, \mathbf{x}^n)$ is a vector of vectors and $\mathbf{x}^3$ is the third vector in this collection.

From Matrices to Graphs. Let A be an $m \times n$ integer matrix. The *incidence Gaifman graph* of A is the bipartite graph $G_I = (R \cup C, E)$, where $R = \{r_1, \ldots, r_m\}$ contains one vertex for each row of A and $C = \{c_1, \ldots, c_n\}$ contains one vertex for each column of A. There is an edge $\{r, c\}$ between the vertex $r \in R$ and $c \in C$ if $A(r, c) \neq 0$, that is, if row r contains a nonzero coefficient in column c. The *primal Gaifman graph* of A is the graph $G_P = (C, E)$, where C is the set of columns of A and $\{c, c'\} \in E$ whenever there exists a row of A with a nonzero coefficient in both columns c and c'. The *dual Gaifman graph* of A is the graph $G_D = (R, E)$, where R is the set of rows of A and $\{r, r'\} \in E$ whenever there exists a column of A with a nonzero coefficient in both rows r and r'.

Treedepth. Undoubtedly, the most celebrated structural parameter for graphs is treewidth, however, in the case of ILPs bounding treewidth of any of the graphs defined above does not lead to tractability (even if the largest coefficient in A is bounded as well see e.g. [14, Lemma 18]). Treedepth is a structural parameter which is useful in the theory of so-called sparse graph classes, see e.g. [16]. Let $G = (V, E)$ be a graph. The treedepth of G, denoted $\operatorname{td}(G)$, is defined by the following recursive formula:

$$\operatorname{td}(G) = \begin{cases} 1 & \text{if } |V(G)| = 1, \\ 1 + \min_{v \in V(G)} \operatorname{td}(G - v) & \text{if } G \text{ is connected with } |V(G)| > 1, \\ \max_{i \in [k]} \operatorname{td}(G_i) & \text{if } G_1, \ldots, G_k \text{ are connected components of } G. \end{cases}$$

Let A be an $m \times n$ integer matrix. The *incidence treedepth* of A, denoted $\operatorname{td}_I(A)$, is the treedepth of its incidence Gaifman graph G_I. The *dual treedepth* of A, denoted $\operatorname{td}_D(A)$, is the treedepth of its dual Gaifman graph G_D. The *primal treedepth* is defined similarly.

The following two well-known theorems will be used in the proof of Theorem 1.

Theorem 2 (Chinese Remainder Theorem). *Let $p_1, \ldots, p_n$ be pairwise co-prime integers greater than 1 and let $a_1, \ldots, a_n$ be integers such that for all $i \in [n]$ it holds $0 \leq a_i < p_i$. Then there exists exactly one integer x such that*

1. $0 \leq x < \prod_{i=1}^{n} p_i$ and
2. $\forall i \in [n] : x \equiv a_i \mod p_i$.

Theorem 3 (Prime Number Theorem). *Let $\pi(n)$ denote the number of primes in $[n]$, then $\pi(n) \in \Theta(\frac{n}{\log n})$.*

It is worth pointing out that, given a positive integer n encoded in unary, it is possible to the n-th prime in polynomial time.

2   Proof of Theorem 1

Before we proceed to the proof of Theorem 1 we include a brief sketch of its idea. To prove NP-hardness, we will give a polynomial time reduction from 3-SAT which is well known to be NP-complete [8]. The proof is inspired by the NP-hardness proof for ILPs given by a set of inequalities, where the primal graph is a star, of Eiben et al. [6].

Proof Idea. Let φ be a 3-CNF formula. We encode an assignment into a variable y. With every variable v_i of the formula φ we associate a prime number p_i. We make $y \bmod p_i$ be the boolean value of the variable v_i; i.e., using auxiliary gadgets we force $y \bmod p_i$ to always be in $\{0, 1\}$. Further, if for a clause $C \in \varphi$ by $\|C\|$ we denote the product of all of the primes associated with the variables occurring in C, then, by Chinese Remainder Theorem, there is a single value in $[\|C\|]$, associated with the assignment that falsifies C, which we have to forbid for $y \bmod \|C\|$. We use the box constraints, i.e., the vectors $\mathbf{l}, \mathbf{u}$, for an auxiliary variable taking the value $y \bmod \|C\|$ to achieve this. For example let $\varphi = (v_1 \vee \neg v_2 \vee v_3)$ and let the primes associated with the three variables be $2, 3$, and 5, respectively. Then we have $\|(v_1 \vee \neg v_2 \vee v_3)\| = 30$ and, since $v_1 = v_3 = \texttt{false}$ and $v_2 = \texttt{true}$ is the only assignment falsifying this clause, we have that 21 is the forbidden value for $y \bmod 30$. Finally, the (SIP) constructed from φ is feasible if and only if there is a satisfying assignment for φ.

Proof (of Theorem 1). Let φ be a 3-CNF formula with n' variables $v_1, \ldots, v_{n'}$ and m' clauses $C_1, \ldots, C_{m'}$ (an instance of 3-SAT). Note that we can assume that none of the clauses in φ contains a variable along with its negation. We will define an SIP, that is, vectors $\mathbf{b}, \mathbf{l}, \mathbf{u}$, and a matrix A with $\mathcal{O}((n' + m')^5)$ rows and columns, whose solution set is non-empty if and only if a satisfying assignment exists for φ. Furthermore, we present a decomposition of the incidence graph of the constructed SIP proving that its treedepth is at most 5. We naturally split the vector $\mathbf{x}$ of the SIP into subvectors associated with the sought satisfying assignment, variables, and clauses of φ, that is, we have $\mathbf{x} = \left(y, \mathbf{x}^1, \ldots, \mathbf{x}^{n'}, \mathbf{z}^1, \ldots, \mathbf{z}^{m'} \right)$. Throughout the proof p_i denotes the i-th prime number.

Variable Gadget. We associate the $\mathbf{x}^i = \left(x_0^i, \ldots, x_{p_i}^i \right)$ part of $\mathbf{x}$ with the variable v_i and bind the assignment of v_i to y. We add the following constraints

$$x_1^i = x_\ell^i \qquad\qquad \forall \ell \in [2 : p_i] \qquad (1)$$

$$x_0^i = y + \sum_{\ell=1}^{p_i} x_\ell^i \qquad\qquad (2)$$

and box constraints

$$-\infty \leq x_\ell^i \leq \infty \qquad\qquad \forall \ell \in [p_i] \qquad (3)$$

$$0 \leq x_0^i \leq 1 \qquad (4)$$

to the SIP constructed so far.

Claim. For given values of x_0^i and y, one may choose the values of x_ℓ^i for $\ell \in [p_i]$ so that (1) and (2) are satisfied if and only if $x_0^i \equiv y \mod p_i$.

Proof. By (1) we know $x_1^i = \cdots = x_{p_i}^i$ and thus by substitution we get the following equivalent form of (2)

$$x_0^i = y + p_i \cdot x_1^i . \qquad (5)$$

But this form is equivalent to $x_0^i \equiv y \mod p_i$. ⌟

Note that by (the proof of) the above claim the conditions (1) and (2) essentially replace the large coefficient (p_i) used in the condition (5). This is an efficient trade-off between large coefficients and incidence treedepth which we are going to exploit once more when designing the clause gadget.

By the above claim we get an immediate correspondence between y and truth assignments for $v_1, \ldots, v_{n'}$. For an integer w and a variable v_i we define the following mapping

$$\text{assignment}(w, v_i) = \begin{cases} \text{true} & \text{if } w \equiv 1 \mod p_i \\ \text{false} & \text{if } w \equiv 0 \mod p_i \\ \text{undefined} & \text{otherwise.} \end{cases}$$

Notice that (4) implies that the the mapping $\text{assignment}(y, v_i) \in \{\text{true}, \text{false}\}$ for $i \in [n']$. We straightforwardly extend the mapping $\text{assignment}(\cdot, \cdot)$ for tuples of variables as follows. For a tuple $\mathbf{a}$ of length ℓ, the value of $\text{assignment}(w, \mathbf{a})$ is $(\text{assignment}(w, a_1), \ldots, \text{assignment}(w, a_\ell))$ and we say that $\text{assignment}(w, \mathbf{a})$ is defined if all of its components are defined.

Clause Gadget. Let C_j be a clause with variables v_e, v_f, v_g. We define $\|C_j\|$ as the product of the primes associated with the variables occurring in C_j, that is, $\|C_j\| = p_e \cdot p_f \cdot p_g$. We associate the $\mathbf{z}^j = \left(z_0^j, \ldots, z_{\|C_j\|}^j \right)$ part of $\mathbf{x}$ with the clause C_j. Let d_j be the unique integer in $[\|C_j\|]$ for which $\text{assignment}(d_j, (v_e, v_f, v_g))$ is defined and gives the falsifying assignment for C_j. The existence and uniqueness of d_j follows directly from the Chinese Remainder Theorem. We add the following constraints

$$z_1^j = z_\ell^j \qquad\qquad \forall \ell \in [2 : \|C_j\|] \qquad (6)$$

$$z_0^j = y + \sum_{1 \leq \ell \leq \|C_j\|} z_\ell^j \qquad (7)$$

and box constraints

$$-\infty \le z_\ell^j \le \infty \qquad\qquad\qquad \forall \ell \in [\|C_j\|] \qquad\qquad (8)$$

$$d_j + 1 \le z_0^j \le \|C_j\| + d_j - 1 \qquad\qquad\qquad\qquad (9)$$

to the SIP constructed so far.

Claim. Let C_j be a clause in φ with variables v_e, v_f, v_g. For given values of y and z_0^j such that assignment$(y, (v_e, v_f, v_g))$ is defined, one may choose the values of z_ℓ^j for $\ell \in [\|C_j\|]$ so that (6), (7), (8) and (9) are satisfied if and only if assignment$(y, (v_e, v_f, v_g))$ satisfies C_j.

Proof. Similarly to the proof of the first claim, (6) and (7) together are equivalent to $z_0^j \equiv y \mod \|C_j\|$. Finally, by (9) we obtain that $z_0^j \ne d_j$ which holds if and only if assignment$(y, (v_e, v_f, v_g))$ satisfies C_j. ⌟

Let $A\mathbf{x} = \mathbf{0}$ be the SIP with constraints (1), (2), (6), and (7) and box constraints $\mathbf{l} \le \mathbf{x} \le \mathbf{u}$ given by (3), (4), (8), (9), and $-\infty \le y \le \infty$. By the first claim, constraints (1), (2), (3), (4), are equivalent to the assertion that assignment$(y, (v_1, \dots, v_{n'}))$ is defined. Then by the second claim, constraints (6), (7), (8), (9) are equivalent to checking that every clause in φ is satisfied by assignment$(y, (v_1, \dots, v_{n'}))$. This finishes the reduction and the proof of its correctness.

In order to finish the proof we have to bound the number of variables and constraints in the presented SIP and to bound the incidence treedepth of A. It follows from the Prime Number Theorem that $p_i = \mathcal{O}(i \log i)$. Hence, the number of rows and columns of A is at most $(n' + m')p_{n'}^3 = \mathcal{O}((n' + m')^5)$.

Claim. It holds that $\text{td}_I(A) \le 5$.

Proof. Let G be the incidence graph of the matrix A. It is easy to verify that y is a cut-vertex in G. Observe that each component of $G - y$ is now either a variable gadget for v_i with $i \in [n']$ (we call such a component a *variable component*) or a clause gadget for C_j with $j \in [m']$ (we call such a component a *clause component*). Let G_v^i be the variable component (of $G - y$) containing variables $\mathbf{x}^i$ and G_c^j be the clause component containing variables $\mathbf{z}^j$. Let $t_v = \max_{\ell \in [n']} \text{td}(G_v^\ell)$ and $t_c = \max_{\ell \in [m']} \text{td}(G_c^\ell)$. It follows that $\text{td}(G) \le 1 + \max(t_v, t_c)$.

Refer to Fig. 1. Observe that if we delete the variable x_1^i together with the constraint (2) from G_v^i, then each component in the resulting graph contains at most two vertices. Each of these components contains either

- a variable x_ℓ^i and an appropriate constraint (1) (the one containing x_ℓ^i and x_0^i) for some $\ell \in [2 : p_i]$ or
- the variable x_0^i.

Since treedepth of an edge is 2 and treedepth of the one vertex graph is 1, we have that $t_v \le 4$.

The bound on t_c follows the same lines as for t_v, since indeed the two gadgets have the same structure. Now, after deleting z_1^j and (7) in G_c^j we arrive to a graph with treedepth of all of its components again bounded by two (in fact, none of its components contain more than two vertices). Thus, $t_v \leq 4$ and the claim follows.

The theorem follows by combining the three above claims.   □

3 Incidence Treedepth of Restricted ILPs

It is worth noting that the proof of Theorem 1 crucially relies on having variables as well as constraints which have high degree in the incidence graph. Thus, it is natural to ask whether this is necessary or, equivalently, whether bounding the degree of variables, constraints, or both leads to tractability. It is well known that if a graph G has bounded degree and treedepth, then it is of bounded size, since indeed the underlying decomposition tree has bounded height and degree and thus bounded number of vertices. Let (SIP) with n variables be given. Let $\mathrm{maxdeg}_C(A)$ denote the maximum arity of a constraint in its constraint matrix A and let $\mathrm{maxdeg}_V(A)$ denote the maximum occurrence of a variable in constraints of A. In other words, $\mathrm{maxdeg}_C(A)$ denotes the maximum number of nonzeros in a row of A and $\mathrm{maxdeg}_V(A)$ denotes the maximum number of nonzeros in a column of A. Now, we get that ILP can be solved in time $f(\mathrm{maxdeg}_C(A), \mathrm{maxdeg}_V(A), \mathrm{td}_I(A))L^{O(1)}$, where f is some computable function and L is the length of the encoding of the given ILP thanks to Lenstra's algorithm [15].

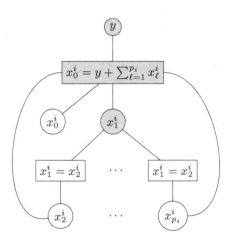

Fig. 1. The variable gadget for u_i of 3-SAT instance together with the global variable y. Variables (of the IP) are in circular nodes while equations are in rectangular ones. The nodes deleted in the proof of the third claim in the proof of Theorem 1 have light gray background.

The above observation can in fact be strengthened—namely, if the arity of all the constraints or the number of occurrences of all the variables in the given SIP is bounded, then we obtain a bound on either primal or dual treedepth. This is formalized by the following lemma.

Lemma 4. *For every (SIP) we have*

$$\mathrm{td}_P(A) \le \mathrm{maxdeg}_C(A) \cdot \mathrm{td}_I(A) \qquad and \qquad \mathrm{td}_D(A) \le \mathrm{maxdeg}_V(A) \cdot \mathrm{td}_I(A).$$

The proof idea is to investigate the definition of the incidence treedepth of A, which essentially boils down to recursively eliminating either a row, or a column, or decomposing a block-decomposable matrix into its blocks. Then, say for the second inequality above, eliminating a column can be replaced by eliminating all the at most $\mathrm{maxdeg}_V(A)$ rows that contain non-zero entries in this column.

It follows that if we bound either $\mathrm{maxdeg}_V(A)$ or $\mathrm{maxdeg}_C(A)$, that is, formally set $\mathrm{maxdeg}(A) = \min\{\mathrm{maxdeg}_V(A), \mathrm{maxdeg}_C(A)\}$, then the linear IP with such a solution set is solvable in time $f(\mathrm{maxdeg}(A), \|A\|_\infty) \cdot n^{O(1)} \cdot L$ thanks to results of Koutecký et al. [14]. Consequently, the use of high-degree constraints and variables in the proof of Theorem 1 is unavoidable.

4 Conclusions

We have shown that, unlike the primal and the dual treedepth, the incidence treedepth of a constraint matrix of (SIP) does not (together with the largest coefficient) provide a way to tractability. This shows our current understanding of the structure of the incidence Gaifman graph is not sufficient. Thus, the effect on tractability of some other "classical" graph parameters shall be investigated. For example we have some preliminary evidences that

– the vertex cover number of the incidence Gaifman graph together with the largest coefficient yields a tractable case and
– the graph in our reduction (Theorem 1) may admit a treecut decomposition of constant width.

We are going to investigate the two above claims in detail in the full version of this paper. Last but not least, all of the above suggest some open questions. Namely, whether ILP parameterized by the largest coefficient and treewidth and the maximum degree of the incidence Gaifman graph is in FPT or not. Furthermore, one may also ask about parameterization by the largest coefficient and the feedback vertex number of the incidence Gaifman graph.

Appendix

Proof of Lemma 4. We prove only the second inequality, as the first one is symmetric. The proof is by induction with respect to the total number of rows and columns of the matrix A. The base of the induction, when A has one row and one column, is trivial, so we proceed to the induction step.

Observe that $G_I(A)$ is disconnected if and only if $G_D(A)$ is disconnected if and only if A is a block-decomposable matrix. Moreover, the incidence treedepth of A is the maximum incidence treedepth among the blocks of A, and the same also holds for the dual treedepth. Hence, in this case we may apply the induction hypothesis to every block of A and combine the results in a straightforward manner.

Assume then that $G_I(A)$ is connected. Then

$$\operatorname{td}(G_I(A)) = 1 + \min_{v \in V(G_I(A))} \operatorname{td}(G_I(A) - v).$$

Let v be the vertex for which the minimum on the right hand side is attained. We consider two cases: either v is a row of A or a column of A.

Suppose first that v is a row of A. Then we have

$$\begin{aligned}
\operatorname{td}(G_D(A)) &\leq 1 + \operatorname{td}(G_D(A) - v) \\
&\leq 1 + \operatorname{maxdeg}_V(A) \cdot \operatorname{td}(G_I(A) - v) \\
&= 1 + \operatorname{maxdeg}_V(A) \cdot (\operatorname{td}(G_I(A)) - 1) \\
&\leq \operatorname{maxdeg}_V(A) \cdot \operatorname{td}(G_I(A))
\end{aligned}$$

as required, where the second inequality follows from applying the induction assumption to A with the row v removed.

Finally, suppose that v is a column of A. Let X be the set of rows of A that contain non-zero entries in column v; then $|X| \leq \operatorname{maxdeg}_V(A)$ and X is non-empty, because $G_I(A)$ is connected. If we denote by $A - v$ the matrix obtained from A by removing column v, then we have

$$\begin{aligned}
\operatorname{td}(G_D(A)) &\leq |X| + \operatorname{td}(G_D(A) - X) \\
&\leq \operatorname{maxdeg}_V(A) + \operatorname{td}(G_D(A - v)) \\
&\leq \operatorname{maxdeg}_V(A) + \operatorname{maxdeg}_V(A) \cdot \operatorname{td}(G_I(A - v)) \\
&\leq \operatorname{maxdeg}_V(A) \cdot \operatorname{td}(G_I(A)),
\end{aligned}$$

as required. Here, in the second inequality we used the fact that $G_D(A) - X$ is a subgraph of $G_D(A - v)$, while in the third inequality we used the induction assumption for the matrix $A - v$. □

References

1. Altmanová, K., Knop, D., Koutecký, M.: Evaluating and tuning n-fold integer programming. In: D'Angelo, G. (ed.) 17th International Symposium on Experimental Algorithms, SEA 2018, L'Aquila, Italy, 27–29 June 2018. LIPIcs, vol. 103, pp. 10:1–10:14. Schloss Dagstuhl - Leibniz-Zentrum fuer Informatik (2018). https://doi.org/10.4230/LIPIcs.SEA.2018.10
2. Chatzigiannakis, I., Kaklamanis, C., Marx, D., Sannella, D. (eds.): 45th International Colloquium on Automata, Languages, and Programming, ICALP 2018, Prague, Czech Republic, 9–13 July 2018. LIPIcs, vol. 107. Schloss Dagstuhl - Leibniz-Zentrum fuer Informatik (2018). http://www.dagstuhl.de/dagpub/978-3-95977-076-7

3. Chen, L., Xu, L., Shi, W.: On the graver basis of block-structured integer programming. CoRR abs/1805.03741 (2018). http://arxiv.org/abs/1805.03741
4. De Loera, J.A., Hemmecke, R., Köppe, M.: Algebraic and Geometric Ideas in the Theory of Discrete Optimization. MOS-SIAM Series on Optimization, vol. 14. SIAM (2013). https://doi.org/10.1137/1.9781611972443
5. Dechter, R.: Chapter 7 - tractable structures for constraint satisfaction problems. In: Rossi, F., van Beek, P., Walsh, T. (eds.) Handbook of Constraint Programming, Foundations of Artificial Intelligence, vol. 2, pp. 209–244. Elsevier (2006). https://doi.org/10.1016/S1574-6526(06)80011-8
6. Eiben, E., Ganian, R., Knop, D., Ordyniak, S.: Unary integer linear programming with structural restrictions. In: Lang, J. (ed.) Proceedings of the Twenty-Seventh International Joint Conference on Artificial Intelligence, IJCAI 2018, Stockholm, Sweden, 13–19 July 2018, pp. 1284–1290. ijcai.org (2018). https://doi.org/10.24963/ijcai.2018/179
7. Eisenbrand, F., Hunkenschröder, C., Klein, K.: Faster algorithms for integer programs with block structure. In: Chatzigiannakis et al. [2], pp. 49:1–49:13. https://doi.org/10.4230/LIPIcs.ICALP.2018.49
8. Garey, M.R., Johnson, D.S.: Computers and Intractability: A Guide to the Theory of NP-Completeness. W. H. Freeman, New York (1979)
9. Hemmecke, R., Köppe, M., Weismantel, R.: A polynomial-time algorithm for optimizing over N-fold 4-block decomposable integer programs. In: Eisenbrand, F., Shepherd, F.B. (eds.) IPCO 2010. LNCS, vol. 6080, pp. 219–229. Springer, Heidelberg (2010). https://doi.org/10.1007/978-3-642-13036-6_17
10. Hemmecke, R., Onn, S., Romanchuk, L.: N-fold integer programming in cubic time. Math. Program. **137**(1–2), 325–341 (2013). https://doi.org/10.1007/978-3-642-13036-6_17
11. Hemmecke, R., Schultz, R.: Decomposition of test sets in stochastic integer programming. Math. Program. **94**(2), 323–341 (2003). https://doi.org/10.1007/s10107-002-0322-1
12. Jansen, K., Lassota, A., Rohwedder, L.: Near-linear time algorithm for n-fold ILPs via color coding. CoRR abs/1811.00950 (2018)
13. Klein, K.: About the complexity of two-stage stochastic IPs. CoRR abs/1901.01135 (2019). http://arxiv.org/abs/1901.01135
14. Koutecký, M., Levin, A., Onn, S.: A parameterized strongly polynomial algorithm for block structured integer programs. In: Chatzigiannakis et al. [2], pp. 85:1–85:14. https://doi.org/10.4230/LIPIcs.ICALP.2018.85
15. Lenstra Jr., H.W.: Integer programming with a fixed number of variables. Math. Oper. Res. **8**(4), 538–548 (1983). https://doi.org/10.1287/moor.8.4.538
16. Nešetřil, J., Ossona de Mendez, P.: Sparsity - Graphs, Structures, and Algorithms. AC, vol. 28. Springer, Heidelberg (2012). https://doi.org/10.1007/978-3-642-27875-4
17. Onn, S.: Nonlinear Discrete Optimization: An Algorithmic Theory (Zurich Lectures in Advanced Mathematics). European Mathematical Society Publishing House (2010)

A Bundle Approach for SDPs with Exact Subgraph Constraints

Elisabeth Gaar$^{(\boxtimes)}$ (iD) and Franz Rendl (iD)

Alpen-Adria-Universität Klagenfurt, Institut für Mathematik,
Universitätsstr. 65-67, 9020 Klagenfurt, Austria
{elisabeth.gaar,franz.rendl}@aau.at

Abstract. The 'exact subgraph' approach was recently introduced as a hierarchical scheme to get increasingly tight semidefinite programming relaxations of several NP-hard graph optimization problems. Solving these relaxations is a computational challenge because of the potentially large number of violated subgraph constraints. We introduce a computational framework for these relaxations designed to cope with these difficulties. We suggest a partial Lagrangian dual, and exploit the fact that its evaluation decomposes into two independent subproblems. This opens the way to use the bundle method from non-smooth optimization to minimize the dual function. Computational experiments on the Max-Cut, stable set and coloring problem show the efficiency of this approach.

Keywords: Semidefinite programming · Relaxation hierarchy ·
Max-Cut · Stable set · Coloring

1 Introduction

The study of NP-hard problems has led to the introduction of various hierarchies of relaxations, which typically involve several levels. Moving from one level to the next the relaxations get increasingly tighter and ultimately the exact optimum may be reached, but the computational effort grows accordingly.

Among the most prominent hierarchies are the polyhedral ones from Boros, Crama and Hammer [3] as well as the ones from Sherali and Adams [20], Lovász and Schrijver [15] and Lasserre [13] which are based on semidefinite programming (SDP). Even though on the starting level they have a simple SDP relaxation, already the first nontrivial level in the hierarchy requires the solution of SDPs in matrices of order $\binom{n}{2}$ and on level k the matrix order is $n^{O(k)}$. Hence they are considered mainly as theoretical tools and from a practical point of view these hierarchies are of limited use.

This project has received funding from the European Union's Horizon 2020 research and innovation programme under the Marie Skłodowska-Curie grant agreement No 764759 and the Austrian Science Fund (FWF): I 3199-N31 and P 28008-N35. We thank three anonymous referees for their constructive comments which substantially helped to improve the presentation of our material.

© Springer Nature Switzerland AG 2019
A. Lodi and V. Nagarajan (Eds.): IPCO 2019, LNCS 11480, pp. 205–218, 2019.
https://doi.org/10.1007/978-3-030-17953-3_16

Not all hierarchies are of this type. In [3], a polyhedral hierarchy for the Max-Cut problem is introduced which maintains $\binom{n}{2}$ variables in all levels, with a growing number of constraints. More recently, Adams, Anjos, Rendl and Wiegele [1] introduced a hierarchy of SDP relaxations which act in the space of symmetric $n \times n$ matrices and at level k of the hierarchy all submatrices of order k have to be 'exact' in a well-defined sense, i.e. they have to fulfill an *exact subgraph constraint* (ESC).

It is the main purpose of this paper to describe an efficient way to optimize over level k of this hierarchy for small values of k, e.g. $k \leqslant 6$, and demonstrate the efficiency of our approach for the Max-Cut, stable set and coloring problem.

Maintaining $\binom{n}{k}$ possible ESCs in an SDP in matrices of order n is computationally infeasible even for $k = 2$ or $k = 3$, because each ESC creates roughly $\binom{k}{2}$ additional equality constraints and at most 2^k additional linear variables.

We suggest the following ideas to overcome this difficulty. First we proceed iteratively, and in each iteration we include only (a few hundred of) the most violated ESCs. More importantly, we propose to solve the dual of the resulting SDP. The structure of this SDP with ESCs admits a reformulation of the dual in the form of a non-smooth convex minimization problem with attractive features. First, any dual solution yields a valid bound for our relaxations, so it is not necessary to carry out the minimization to optimality. Secondly, the dual function evaluation decomposes into two independent problems. The first one is simply a sum of max-terms (one for each subgraph constraint), and the second one consists in solving a 'basic' SDP, independent of the ESCs. The optimizer for this second problem also yields a subgradient of the objective function. With this information at hand we suggest to use the bundle method from non-smooth convex optimization. It provides an effective machinery to get close to a minimizer in few iterations.

As a result we are able to get near optimal solutions where all ESCs for small values of k ($k \leqslant 6$) are satisfied up to a small error tolerance. Our computational results demonstrate the practical potential of this approach.

We finish this introductory section with some notation. We denote the vector of all-ones of size n with $\mathbb{1}_n$ and $\Delta_n = \{x \in \mathbb{R}_+^n : \sum_{i=1}^n x_i = 1\}$. If the dimension is clear from the context we may omit the index and write $\mathbb{1}$ and Δ. Furthermore let $N = \{1, 2, \ldots, n\}$. A graph G on n vertices has vertex set N and edge set E and $\overline{G}$ is its complement graph. $\mathcal{S}_n$ is the set of n-dimensional symmetric matrices.

2   The Problems and Their Semidefinite Relaxations

In the Max-Cut problem a symmetric matrix $L \in \mathcal{S}_n$ is given and $c \in \{-1, 1\}^n$ which maximizes $c^T L c$ should be determined. If the matrix L corresponds to the Laplacian matrix of a (edge-weighted undirected) graph G, this is equivalent to finding a bisection of the vertices of G such that the total weight of the edges joining the two bisection blocks is maximized. Such an edge set is also called a *cut* in G.

Bisections of N can be expressed as $c \in \{-1,1\}^n$ where the two bisection blocks correspond to the entries in c of the same sign. Given $c \in \{-1,1\}^n$ we call $C = cc^T$ a *cut matrix*. The convex hull of all cut matrices (of order n) is denoted by CUT_n or simply CUT if the dimension is clear. Since $c^T L c = \langle L, cc^T \rangle$ Max-Cut can also be written as the following (intractable) linear program

$$z_{mc} = \max\{\langle L, X \rangle : X \in \text{CUT}\}.$$

CUT is contained in the spectrahedron $\mathcal{X}^E = \{X \in \mathcal{S}_n : \text{diag}(X) = \mathbb{1}_n, X \succeq 0\}$, hence

$$\max \left\{ \langle L, X \rangle : X \in \mathcal{X}^E \right\} \tag{1}$$

is a basic semidefinite relaxation for Max-Cut. This model is well-known, attributed to Schrijver and was introduced in a dual form by Delorme and Poljak [4]. It can be solved in polynomial time to a fixed prescribed precision and solving this relaxation for $n = 1000$ takes only a few seconds.

It is well-known that the Max-Cut problem is NP-hard. On the positive side, Goemans and Williamson [8] show that one can find a cut in a graph with nonnegative edge weights of value at least $0.878 z_{mc}$ in polynomial time.

In the stable set problem the input is an unweighted graph G. We call a set of vertices *stable*, if no two vertices are adjacent. Moreover we call a vector $s \in \{0,1\}^n$ a *stable set vector* if it is the incidence vector of a stable set. The convex hull of all stable set vectors of G is denoted with $\text{STAB}(G)$. In the stable set problem we want to determine the *stability number* $\alpha(G)$, which denotes the cardinality of a largest stable set in G, hence $\alpha(G) = \max\left\{ \mathbb{1}^T s : s \in \text{STAB}(G) \right\}$. Furthermore we denote with $\text{STAB}^2(G) = \text{conv}\left\{ ss^T : s \in \text{STAB}(G) \right\}$ the convex hull of all *stable set matrices* ss^T. Then with the arguments of Gaar [7] it is easy to check that $\alpha(G) = \max\{\text{trace}(X) : X \in \text{STAB}^2(G)\}$. Furthermore $\text{STAB}^2(G)$ is contained in the following spectrahedron

$$\mathcal{X}^S = \left\{ X \in \mathcal{S}_n : X_{ij} = 0 \quad \forall \{i,j\} \in E, \ x = \text{diag}(X), \ \begin{pmatrix} 1 & x^T \\ x & X \end{pmatrix} \succeq 0 \right\},$$

which is known as the *theta body* in the literature. Therefore

$$\vartheta(G) = \max \left\{ \text{trace}(X) : X \in \mathcal{X}^S \right\} \tag{2}$$

is a relaxation of the stable set problem. The Lovász theta function $\vartheta(G)$ was introduced in a seminal paper by Lovász [14]. We refer to Grötschel, Lovász and Schrijver [9] for a comprehensive analysis of $\vartheta(G)$.

Determining $\alpha(G)$ is again NP-hard. Contrary to Max-Cut, which has a polynomial time .878-approximation, for every $\varepsilon > 0$ there can be no polynomial time algorithm that approximates $\alpha(G)$ within a factor better than $O(n^{1-\varepsilon})$ unless $P = NP$, see Håstad [11].

The coloring problem for a given graph G consists in determining the *chromatic number* $\chi(G)$, which is the smallest t such that N can be partitioned into t stable sets. Let $S = (s_1, \ldots, s_k)$ be a matrix where each column is a stable set vector and these stable sets partition V into k sets. Let us call such matrices S

stable-set partition matrices (SSPM). The $n \times n$ matrix $X = SS^T$ is called *coloring matrix*. The convex hull of the set of all coloring matrices of G is denoted by $\mathrm{COL}(G)$. We also need the *extended coloring polytope*

$$\mathrm{COL}^\varepsilon(G) = \mathrm{conv}\left\{ \begin{pmatrix} k & \mathbb{1}^T \\ \mathbb{1} & X \end{pmatrix} = \sum_{i=1}^k \begin{pmatrix} 1 \\ s_i \end{pmatrix} \begin{pmatrix} 1 \\ s_i \end{pmatrix}^T : \begin{array}{l} S = (s_1, \ldots, s_k) \text{ is a} \\ \text{SSPM of } G, \ X = SS^T \end{array} \right\}.$$

The difficult set COL^ε can be relaxed to the easier spectrahedron $\mathcal{X}^C$

$$\mathcal{X}^C = \left\{ \begin{pmatrix} t & \mathbb{1}^T \\ \mathbb{1} & X \end{pmatrix} \succeq 0 : \ \mathrm{diag}(X) = \mathbb{1}_n, \ X_{ij} = 0 \ \forall \{i,j\} \in E \right\}$$

and we can consider the semidefinite program

$$t^*(G) = \min\left\{ t : \ \begin{pmatrix} t & \mathbb{1}^T \\ \mathbb{1} & X \end{pmatrix} \in \mathcal{X}^C \right\}. \tag{3}$$

Obviously $t^*(G) \leq \chi(G)$ holds because the SSPM S consisting of $\chi(G)$ stable sets yields a feasible coloring matrix $X = SS^T$ with objective function value $\chi(G)$. It is in fact a consequence of conic duality that $t^*(G) = \vartheta(\overline{G})$ holds.

It is NP-hard to find $\chi(G)$, to find a 4-coloring of a 3-colorable graph [10] and to color a k-colorable graph with $O(k^{\frac{\log k}{25}})$ colors for sufficiently large k, [12].

3   Exact Subgraph Hierarchy

In this section we will discuss how to systematically tighten the relaxations (1), (2) and (3) with 'exactness conditions' imposed on small subgraphs. We obtained these relaxations by relaxing the feasible regions CUT, STAB^2 and COL of the integer problem to simple spectrahedral sets. Now we will use small subgraphs to get closer to original feasible regions again.

For $I \subseteq N$ we denote with X_I the principal submatrix of X corresponding to the rows and columns in I. Furthermore let G_I be the induced subgraph of G on the set of vertices I and let $k_I = |I|$ be the cardinality of I.

We first look at the exact subgraph relaxations for Max-Cut. The *exact subgraph constraint* (ESC) on $I \subseteq N$, introduced in [1] by Adams, Anjos, Rendl and Wiegele, requires that the matrix X_I corresponding to the subgraph G_I lies in the convex hull of the cut matrices of G_I, that is

$$X_I \in \mathrm{CUT}_{|I|}.$$

In this case we say that X is *exact* on I.

Now we want the ESCs to be fulfilled not only for one but for a certain selection of subgraphs. We denote with J the set of subgraphs which we require to be exact and get the following SDP relaxation with ESCs for Max-Cut.

$$\max\{\langle L, X \rangle : \ X \in \mathcal{X}^E, \ X_I \in \mathrm{CUT}_{|I|} \ \forall I \in J\} \tag{4}$$

We proceed analogously for the stable set problem in a graph G. The ESC of a subgraph G_I for the stable set problem requires that $X_I \in \text{STAB}^2(G_I)$ holds and the SDP with ESCs for the stable set problem is

$$\max\{\text{trace}(X) : X \in \mathcal{X}^S, \ X_I \in \text{STAB}^2(G_I) \ \forall I \in J\}. \tag{5}$$

Turning to the coloring problem, we analogously impose additional constraints of the form $X_I \in \text{COL}(G_I)$ to obtain the SDP with ESCs

$$\min\left\{ t : \ \begin{pmatrix} t & \mathbb{1}^T \\ \mathbb{1} & X \end{pmatrix} \in \mathcal{X}^C, \ X_I \in \text{COL}(G_I) \ \forall I \in J \right\}. \tag{6}$$

Note that in the case of the stable set and the coloring problem the polytopes $\text{STAB}^2(G_I)$ and $\text{COL}(G_I)$ depend on the subgraph G_I, whereas in Max-Cut the polytope $\text{CUT}_{|I|}$ only depends on the number of vertices of the subgraph.

From a theoretical point of view, we obtain the k-th level of the exact subgraph hierarchy of [1] if we use $J = \{I \subseteq N : |I| = k\}$ in the relaxations (4), (5) and (6) respectively. We denote the corresponding objective function values with z_{mc}^k, z_{ss}^k and z_c^k. So the k-th level of the hierarchy is obtained by forcing all subgraphs on k vertices to be exact in the basic SDP relaxation.

In the case of the stable set and the Max-Cut problem we have $z_{ss}^n = \alpha(G)$ (see [7]) and $z_{mc}^n = z_{mc}$. For coloring $z_c^n \leqslant \chi(G)$ holds. Let z_{ce}^k be the resulting value if we add the inequalities $t \geqslant \sum_{i=1}^{t_I} [\lambda_I]_i |S_i^I|$ where $|S_i^I|$ is the number of colors used for the SSPM S_i^I and $\lambda_I \in \Delta_{t_I}$ is a variable for the convex combination for each subgraph I to the SDP for z_c^k. Then $z_{ce}^n = \chi(G)$ holds. Since the focus of this paper are computational results we are interested only in the computational results we omit the details and further theoretical investigations.

An important feature of this hierarchy is that the size of the matrix variable remains n or $n + 1$ on all levels of the hierarchy and only more linear variables and constraints (enforcing the ESCs, hence representing convex hull conditions) are added on higher levels. So it is possible to approximate z_{mc}^k, z_{ss}^k and z_c^k by forcing only some subgraphs of order k to be exact. This is our key ingredient to computationally obtain tight bounds on z_{mc}, $\alpha(G)$ and $\chi(G)$.

From a practical point of view solving the relaxations (4), (5) and (6) with standard interior point (IP) solvers like SDPT3 [21] or MOSEK [16] is very time consuming. In Table 1 we list computation times (in seconds) for one specific Max-Cut and one specific stable set instance. We vary the number of ESCs for subgraphs of order 3, 4 and 5, so we solve (4) and (5) for different J. We choose J such that the total number of equality constraints induced by the convex hull formulation of the ESCs b ranges between 6000 and 15000. Since the matrix order n is fixed to $n = 100$, the overall computation time depends essentially on the number of constraints, independent of the specific form of the objective function. Aside from the ESC constraints, we have n additional equations for Max-Cut and $n+m+1$ additional equations for the stable set problem. Here m denotes the number of edges of the graph. We have $m = 722$ in the example graph. Clearly the running times get huge for a large number of ESC. Furthermore MATLAB requires 12 GB of memory for $b = 15000$, showing also memory limitations.

Note that it is argued in [1] that $z_{mc}^4 = z_{mc}^3$, so we omit subgraphs of order $k_I = 4$ for Max-Cut. This is because in the back of our minds our final algorithm to determine the best possible bounds first includes ESCs of size k, starting for example with $k = 3$. As soon as we do not find violated ESCs of size k anymore, we repeat this for size $k + 1$.

4   Partial Lagrangian Dual

To summarize we are interested in solving relaxations (4), (5) and (6) with a potentially large number of ESCs, where using interior point solvers is too time consuming. In this section we will first establish a unified formulation of the relaxations (4), (5) and (6). Then we will build the partial Lagrangian dual of this formulation, where only the ESCs are dualized. This model will be particularly amenable for the bundle method, because it will be straightforward to obtain a subgradient of the model when evaluating it at a certain point.

In order to unify the notation for the three problems observe that the ESCs $X_I \in \mathrm{CUT}_{|I|}$, $X_I \in \mathrm{STAB}^2(G_I)$ and $X_I \in \mathrm{COL}(G_I)$ can be represented as

$$X_I = \sum_{i=1}^{t_I} \lambda_i C_i^I, \quad \lambda \in \Delta_{t_I}, \tag{7}$$

where C_i^I is the i-th cut, stable set or coloring matrix of the subgraph G_I and t_I is their total number.

A formal description of ESC in (7) requires some additional notation. First we introduce the projection $\mathcal{P}_I \colon \mathcal{S}_n \mapsto \mathcal{S}_{k_I}$, mapping X to the submatrix X_I. Second we define a map $\mathcal{A}_I \colon \mathcal{S}_{k_I} \mapsto \mathbb{R}^{t_I}$, such that its adjoint map $\mathcal{A}_I^\top \colon \mathbb{R}^{t_I} \mapsto \mathcal{S}_{k_I}$ is given by $\mathcal{A}_I^\top(\lambda) = \sum_{i=1}^{t_I} \lambda_i C_i^I$ and produces a linear combination of the cut, stable set or coloring matrices. Thus we can rewrite (7) as

$$\mathcal{A}_I^\top(\lambda_I) - \mathcal{P}_I(X) = 0, \quad \lambda_I \in \Delta_{t_I}. \tag{8}$$

The left-hand side of the matrix equality is a symmetric matrix, of which some entries (depending on which problem we consider) are zero for sure, so we do not have to include all $k_I \times k_I$ equality constraints into the SDP. Let b_I be the number of equality constraints we have to include. Note that $b_I = \binom{k_I}{2}$, $b_I = \binom{k_I+1}{2} - m_I$ and $b_I = \binom{k_I}{2} - m_I$ for the Max-Cut, stable set and coloring problem respectively, if m_I denotes the number of edges of G_I. This is because in the case of the stable set problem we also have to include equations for the entries of the main diagonal contrary to Max-Cut and the coloring problem. Then we define a linear map $\mathcal{M}_I \colon \mathbb{R}^{b_I} \mapsto \mathcal{S}_{k_I}$ such that the adjoint operator $\mathcal{M}_I^\top \colon \mathcal{S}_{k_I} \mapsto \mathbb{R}^{b_I}$ extracts the b_I positions, for which we have to include the equality constraints, into a vector. So eventually we can rephrase (8) equivalently as

$$\mathcal{M}_I^\top(\mathcal{A}_I^\top(\lambda_I) - \mathcal{P}_I(X)) = 0, \quad \lambda_I \in \Delta_{t_I},$$

which are b_I+1 equalities and t_I inequalities. In consequence all three relaxations (4), (5) and (6) have the generic form

$$z = \max\{\langle C, \widehat{X} \rangle : \ \widehat{X} \in \mathcal{X}, \ \lambda_I \in \Delta_{t_I}, \ \mathcal{M}_I^\top(\mathcal{A}_I^\top(\lambda_I) - \mathcal{P}_I(X)) = 0 \ \forall I \in J\}, \quad (9)$$

where C, $\mathcal{X}$, $\mathcal{A}_I$, $\mathcal{M}_I$ and b_I have to be defined problem specific. Furthermore $\widehat{X} = X$ in the case of Max-Cut and stable set and $\widehat{X} = \begin{pmatrix} t & \mathbb{1}^T \\ \mathbb{1} & X \end{pmatrix}$ for coloring, but for the sake of understandability we will just use X in the following.

The key idea to get a handle on problem (9) is to consider the partial Lagrangian dual where the ESCs (without the constrains $\lambda_I \in \Delta_{t_I}$) are dualized. We introduce a vector of multipliers y_I of size b_I for each I and collect them in $y = (y_I)_{I \in J}$ and also collect $\lambda = (\lambda_I)_{I \in J}$. The Lagrangian function becomes

$$\mathcal{L}(X, \lambda, y) = \langle C, X \rangle + \sum_{I \in J} \langle y_I, \mathcal{M}_I^\top(\mathcal{A}_I^\top(\lambda_I) - \mathcal{P}_I(X)) \rangle$$

and standard duality arguments (Rockafellar [19, Corollary 37.3.2]) yield

$$z = \min_{y} \max_{\substack{X \in \mathcal{X} \\ \lambda_I \in \Delta_{t_I}}} \mathcal{L}(X, \lambda, y). \quad (10)$$

For a fixed set of multipliers y the inner maximization becomes

$$\max_{\substack{X \in \mathcal{X} \\ \lambda_I \in \Delta_{t_I}}} \left\langle C - \sum_{I \in J} \mathcal{P}_I^\top \mathcal{M}_I(y_I), X \right\rangle + \sum_{I \in J} \langle \mathcal{A}_I \mathcal{M}_I(y_I), \lambda_I \rangle.$$

This maximization is interesting in at least two aspects. First, it is separable in the sense that the first term depends only on X and the second one only on the separate λ_I. Moreover, if we denote the linear map $\mathcal{A}_I \mathcal{M}_I(y_I) \colon \mathbb{R}^{b_I} \mapsto \mathbb{R}^{t_I}$ with $\mathcal{D}_I$, the second term has an explicit solution, namely

$$\max_{\lambda_I \in \Delta_{t_I}} \langle \mathcal{D}_I(y_I), \lambda_I \rangle = \max_{1 \leqslant i \leqslant t_I} [\mathcal{D}_I(y_I)]_i. \quad (11)$$

In order to consider the first term in more detail, we define the following function. Let $b = \sum_{I \in J} b_I$ be the dimension of y. Then $h \colon \mathbb{R}^b \to \mathbb{R}$ is defined as

$$h(y) = \max_{X \in \mathcal{X}} \left\langle C - \sum_{I \in J} \mathcal{P}_I^\top \mathcal{M}_I(y_I), X \right\rangle = \left\langle C - \sum_{I \in J} \mathcal{P}_I^\top \mathcal{M}_I(y_I), X^* \right\rangle, \quad (12)$$

where X^* is a maximizer over the set $\mathcal{X}$ for y fixed. Note that $h(y)$ is convex but non-smooth, but (12) shows that $g_I = -\mathcal{M}_I^T \mathcal{P}_I(X^*)$ is a subgradient of h with respect to y_I. By combining (11) and (12) we can reformulate the partial Lagrangian dual (10) to

$$z = \min_{y} \left\{ h(y) + \sum_{I \in J} \max_{1 \leqslant i \leqslant t_I} [\mathcal{D}_I(y_I)]_i \right\}. \quad (13)$$

The formulation (13) of the original relaxations (4), (5) and (6) fits perfectly into the bundle method setting described by Frangioni and Gorgone in [6], hence we suggest to approach this problem using the bundle method.

5   Solving (13) with the Bundle Method

The bundle method is an iterative procedure for minimizing a convex non-smooth function and firstly maintains the *current center* $\overline{y}$, which represents the current estimate to the optimal solution, throughout the iterations. Secondly it maintains the bundle of the form $\mathcal{B} = \{(y_1, h_1, g_1, X_1), \ldots, (y_r, h_r, g_r, X_r)\}$. Here $y_1, \ldots, y_r$ are the points which we use to set up our subgradient model. Moreover $h_i = h(y_i)$, g_i is a subgradient of h at y_i and X_i is a maximizer of h at y_i as in (12).

At the start we select $y_1 = \overline{y} = 0$ and evaluate h at $\overline{y}$, which yields the bundle $\mathcal{B} = \{(y_1, g_1, h_1, X_1)\}$. A general iteration consists of the two steps determining the new *trial point* and evaluating the *oracle*. For determining a new trial point $\widetilde{y}$ the subgradient information of the bundle $\mathcal{B}$ translates into the subgradient model $h(y) \geqslant h_j + \langle g_j, y - y_j \rangle$ for all $j = 1, \ldots, r$. It is common to introduce $e_j = h(\overline{y}) - h_j - \langle g_j, \overline{y} - y_j \rangle$ for $j = 1, \ldots, r$ and with $\overline{h} = h(\overline{y})$ the subgradient model becomes

$$h(y) \geqslant \max_{1 \leqslant j \leqslant r} \left\{ \overline{h} - e_j + \langle g_j, y - \overline{y} \rangle \right\}. \tag{14}$$

The right-hand side above is convex, piecewise linear and minorizes h. In each iteration of the bundle method we minimize the right-hand side of (14) instead of h, but ensure that we do not move too far from $\overline{y}$ by adding a penalty term of the form $\frac{1}{2}\mu \|y - \overline{y}\|^2$ for a parameter $\mu \in \mathbb{R}_+$ to the objective function. With the auxiliary variables $w \in \mathbb{R}$ and $v_I \in \mathbb{R}$ for all $I \in J$ to model the maximum terms and with $v = (v_I)_{I \in J} \in \mathbb{R}^q$ and $q = |J|$ we end up with

$$\min_{y,w,v} \quad w + \sum_{I \in J} v_I + \frac{1}{2}\mu \|y - \overline{y}\|^2 \tag{15}$$

$$st \quad w \geqslant \overline{h} - e_j + \langle g_j, y - \overline{y} \rangle \qquad \forall j = 1, \ldots, r$$

$$v_I \geqslant [\mathcal{D}_I(y_I)]_i \qquad \forall i = 1, \ldots, t_I \quad \forall I \in J.$$

This is a convex quadratic problem in $1 + q + b$ variables with $r + \sum_{I \in J} t_I$ linear inequality constraints. Its solution $(\widetilde{y}, \widetilde{w}, \widetilde{v})$ includes the new trial point $\widetilde{y}$. Problems of this type can be solved efficiently in various ways, see [7] for further details. In our implementation we view (15) as a rotated second order cone program with one second-order cone constraint and solve it with MOSEK.

The second step in each bundle iteration is to evaluate the dual function h at $\widetilde{y}$. In our case determining $h(\widetilde{y})$ means solving the basic SDP relaxation as introduced in Sect. 2 with a modified objective function. Hence in the case of Max-Cut the oracle can be evaluated very quickly, whereas evaluating the oracle is computationally more expensive for the stable set and the coloring problem.

The bundle iteration finishes by deciding whether $\widetilde{y}$ becomes the new center (serious step, roughly speaking if the increase of the objective function is good) or not (null step). In either case the new point is included in the bundle, some other elements of the bundle are possibly removed, the bundle parameter μ is updated and a new iteration starts.

6   Computational Results and Conclusions

We close with a small sample of computational results and start with comparing our bundle method with interior point methods. In our context we are mostly interested to improve the upper bounds quickly, so we do not run the bundle method described in Sect. 5 until we reach a minimizer, but stop after a fixed number of iterations, say 30. In Table 1 one sees that the running times decrease drastically if we use the bundle method. For $b \approx 15000$ it takes the bundle method only around 8% of the MOSEK running time to get as close as 95% to the optimal value, which is sufficient for our purposes. One sees that our bundle method scales much better for increasing $|J|$.

If we are given a graph and want to get an approximation on z^k_{mc}, z^k_{ss} and z^k_c, then we iteratively perform a fixed number, say 30, iterations of the bundle method and then update the set J. We denote the exact subgraph bounds (ESB) obtained in this way with s^k_{mc}, s^k_{ss} and s^k_c.

For the sake of brevity we will only outline how to determine J heuristically, see [7] for details. Let X^* be the current solution of (4), (5) or (6). We use the fact that the inner product of X^*_I and particular matrices of size k_I is potentially small whenever X^*_I is not in $\mathrm{STAB}^2(G_I)$. Minimizing this inner product over all subgraphs of order k_I would yield a quadratic assignment problem, so we repeatedly use a local search heuristic for fixed particular matrices in order to obtain potential subgraphs. Then we calculate the projection distances from X^*_I to $\mathrm{STAB}^2(G_I)$ for all these subgraphs and include those in J which have the largest distances and hence are violated most.

Finally we present several computational results for obtained ESBs. Note that we refrain from comparing the running times of our bundle method with the running time of inter point methods, because interior point methods would reach their limit very soon. Hence the bounds presented can only be obtained with our methods in reasonable time.

When considering Max-Cut the graphs in Table 2 are from the Biq Mac library [2] with $n = 100$ vertices. The edge density is 10%, 50% and 90%. The first 3 instances have positive weights and the remaining 3 have also some negative weights. The column labeled 3 provides the deviation (in %) of the ESB with $k = 3$ from z_{mc}. Thus if p is the value in the column labeled 3, then $s^3_{mc} = (1 + p/100)z_{mc}$. The columns labeled 5 and 7 are to be understood in a similar way for $k = 5$ and $k = 7$. We note that the improvement of the bound from column 3 to column 7 is quite substantial in all cases. We also point out that the relative gap is much larger if also negative edge weights are present.

In Table 3 we look at graphs from the Beasley collection [2] with $n = 250$. These instances were used by Rendl, Rinaldi and Wiegele [18] in a Branch-and-Bound setting. We only consider the 'hardest' instances from [18] where the Branch-and-Bound tree has more than 200 nodes. The table provides the gap at the root node and also the number of nodes in the Branch-and-Bound tree as reported in [18]. The column 7-gap contains the gap after solving our new relaxation with ESCs up to size $k = 7$. We find it remarkable that the first instance is solved to optimality and the gap in the second instance is reduced by 75% compared to the original gap. This implies that using our ESBs would expectedly reduce the very high number of required Branch-And-Bound nodes tremendously.

We conclude that for Max-Cut our ESB constitute a substantial improvement compared to the previously used strongest bounds based on SDP with triangle inequalities. These correspond to the column 3-gap.

For the calculations for the stable set and the coloring problem all instances are chosen in such a way that $\vartheta(G)$ does not coincide and is not very close to $\alpha(G)$ and $\chi(G)$ respectively.

The instances for the stable set problem are taken partly from the DIMACS challenge [5] with some additional instances from [7] with n ranging from 26 to 200. Table 4 contains the new bounds. Here the starting point is the relaxation $\vartheta(G)$. We carry out 10 cycles of adding ESCs. In each cycle we add at most 200 ESCs, so in the final round we have no more than 2000 ESCs. The column heading indicates the order of the subgraphs. Here the improvement of the bounds is smaller than in the Max-Cut case, but we see that including larger subgraphs leads to much tighter bounds. In Table 5 we show that our approach also reduces the largest found projection distance over all subgraphs G_I of X_I to the corresponding $\mathrm{STAB}^2(G_I)$ in the course of the cycles. This indicates that the violation of the subgraphs decreases over the cycles and less and less subgraphs do not fulfill the ESCs. For example the value 0.000 for the graph spin5 for s_{ss}^2 at the end of the cycles means that we did not find a violated subgraph of order 2 anymore.

Results for a selection of coloring instances from [17] are provided in Tables 6 and 7. As in the stable set case there is only little improvement using small subgraphs ($k = 2$ or 3). The inclusion of larger subgraphs ($k = 6$) shows the potential of the exact subgraph approach.

Summarizing, we offer the following conclusions from these preliminary computational results.

• Our computational approach based on the partial Lagrangian dual is very efficient in handling also a large number of ESCs. The dual function evaluation separates the SDP part from the ESCs and therefore opens the way for large-scale computations. The minimization of the dual function is carried out as a convex quadratic optimization problem without any SDP constraints, and therefore is also suitable for a large number of ESCs.

• On the practical side we consider the small ESCs for Max-Cut a promising new way to tighten bounds for this problem. It will be a promising new project to explore these bounds also in a Branch-and-Bound setting.

• Our computational results for stable set and coloring confirm the theoretical hardness results for these problems. Here the improvement of the relaxations is small for $k \leqslant 3$ but including larger subgraphs yields a noticeable improvement of the bounds. It will be a challenge to extend our approach to larger subgraphs.

A   Tables

Table 1. The running times for one Max-Cut and one stable set instance with different fixed sets of ESCs. The graphs of order $n = 100$ are from the Erdős-Rényi model.

	#ESC of size			b	Interior point		Our bundle		% of MOSEK	
					Time (sec)		Time (sec)			
	3	4	5		MOSEK	SDPT3	Oracle	Overall	Time	Value
MC	2000		0	6000	18.37	49.22	1.01	6.05	32.93	97.20
	2000		300	9000	55.24	134.78	1.18	9.33	16.90	95.02
	4000		0	12000	104.56	289.78	1.71	11.13	10.64	93.66
	3000		600	15000	184.43	525.85	1.56	14.83	8.04	94.54
SS	1050	0	0	5914	23.54	79.25	7.86	10.65	45.22	98.25
	1050	212	63	8719	50.11	174.33	10.61	16.52	32.96	97.89
	2100	0	0	11780	126.40	388.07	7.43	12.27	9.71	93.65
	1575	318	212	14653	241.29	648.83	10.79	20.21	8.38	94.44

Table 2. The deviation of the ESB to z_{mc} for several Max-Cut instances.

Name	3	5	7	z_{mc}
pw01-100.1	0.40	0.00	0.00	2060
pw05-100.1	0.90	0.51	0.39	8045
pw09-100.1	0.58	0.38	0.31	13417
w01-100.1	0.13	0.00	0.00	719
w05-100.1	3.91	1.41	0.85	1606
w09-100.1	8.06	5.66	5.09	2096

Table 3. The gap of the ESB to z_{mc} for two Max-Cut instances.

Name	BBnodes	Root gap	7-gap	z_{mc}
beas-250-6	223	1.02	0.00	41014
beas-250-8	4553	2.19	0.49	35726

Table 4. Tighten $\vartheta(G)$ towards $\alpha(G)$ for several instances for 10 cycles.

Name	n	m	$\vartheta(G)$	s_{ss}^2	s_{ss}^3	s_{ss}^4	s_{ss}^5	s_{ss}^6	$\alpha(G)$
CubicVT26_5	26	39	11.82	11.82	11.00	10.98	10.54	10.46	10
Circulant47_030	47	282	14.30	14.30	13.61	13.21	13.24	13.14	13
G_50_0_5	50	308	13.56	13.46	13.13	12.96	12.82	12.67	12
hamming6_4	64	1312	5.33	4.00	4.00	4.00	4.00	4.00	4
spin5	125	375	55.90	55.90	50.42	50.17	50.00	50.00	50
keller4	171	5100	14.01	13.70	13.54	13.50	13.49	13.49	11
sanr200_0_9	200	2037	49.27	49.04	48.94	48.86	48.78	48.75	42
c_fat200_5	200	11427	60.35	60.34	58.00	58.00	58.00	58.00	58

Table 5. Maximum found projection distance of X_I to $\text{STAB}^2(G_I)$ for the computations of Table 4.

Name	n	Beginning			End		
		s_c^2	s_c^4	s_c^6	s_c^2	s_c^4	s_c^6
CubicVT26_5	26	0.000	0.102	0.193	0.000	0.029	0.013
G_50_0_5	50	0.087	0.093	0.118	0.000	0.013	0.024
spin5	125	0.000	0.084	0.269	0.000	0.046	0.006
sanr200_0_9	200	0.044	0.062	0.107	0.072	0.028	0.020

Table 6. Tighten $\vartheta(G)$ towards $\chi(G)$ for several instances for 10 cycles.

Name	n	m	$\vartheta(G)$	s_c^2	s_c^3	s_c^4	s_c^5	s_c^6	$\chi(G) \leqslant$
myciel4	23	71	2.53	2.53	2.90	2.91	3.28	3.29	5
myciel5	47	236	2.64	2.64	3.05	3.09	3.45	3.45	6
mug88_1	88	146	3.00	3.00	3.00	3.00	3.00	3.00	4
1_FullIns_4	93	593	3.12	3.12	3.25	3.37	3.80	3.80	5
myciel6	95	755	2.73	2.73	3.02	3.09	3.57	3.51	7
myciel7	191	2360	2.82	2.82	3.02	3.08	3.63	3.50	8
2_FullIns_4	212	1621	4.06	4.06	4.32	4.38	4.66	4.68	6
flat300_26_0	300	21633	16.99	17.04	17.12	17.10	17.12	17.12	26

Table 7. Maximum found projection distance of X_I to $\mathrm{COL}(G_I)$ for the computations of Table 6.

Name	n	Beginning			End		
		s_c^2	s_c^4	s_c^6	s_c^2	s_c^4	s_c^6
myciel4	23	0.000	0.365	0.760	0.000	0.000	0.000
1_FullIns_4	93	0.009	0.349	0.629	0.000	0.158	0.203
myciel7	191	0.000	0.356	0.621	0.000	0.207	0.272
flat300_26_0	300	0.127	0.279	0.360	0.143	0.142	0.091

References

1. Adams, E., Anjos, M.F., Rendl, F., Wiegele, A.: A hierarchy of subgraph projection-based semidefinite relaxations for some NP-hard graph optimization problems. INFOR Inf. Syst. Oper. Res. **53**(1), 40–47 (2015)
2. Biq Mac Library. http://biqmac.aau.at/. Accessed 18 Nov 2018
3. Boros, E., Crama, Y., Hammer, P.L.: Upper-bounds for quadratic 0-1 maximization. Oper. Res. Lett. **9**(2), 73–79 (1990)
4. Delorme, C., Poljak, S.: Laplacian eigenvalues and the maximum cut problem. Math. Program. Ser. A **62**(3), 557–574 (1993)
5. DIMACS Implementation Challenges (1992). http://dimacs.rutgers.edu/Challenges/. Accessed 18 Nov 2018
6. Frangioni, A., Gorgone, E.: Bundle methods for sum-functions with "easy" components: applications to multicommodity network design. Math. Program. **145**(1), 133–161 (2014)
7. Gaar, E.: Efficient Implementation of SDP Relaxations for the Stable Set Problem. Ph.D. thesis, Alpen-Adria-Universität Klagenfurt (2018)
8. Goemans, M.X., Williamson, D.P.: Improved approximation algorithms for maximum cut and satisfiability problems using semidefinite programming. J. Assoc. Comput. Mach. **42**(6), 1115–1145 (1995)
9. Grötschel, M., Lovász, L., Schrijver, A.: Geometric Algorithms and Combinatorial Optimization, Algorithms and Combinatorics. Study and Research Texts, vol. 2. Springer, Berlin (1988)
10. Guruswami, V., Khanna, S.: On the hardness of 4-coloring a 3-colorable graph. SIAM J. Discrete Math. **18**(1), 30–40 (2004)
11. Håstad, J.: Clique is hard to approximate within $n^{1-\epsilon}$. Acta Math. **182**(1), 105–142 (1999)
12. Khot, S.: Improved inapproximability results for MaxClique, chromatic number and approximate graph coloring. In: 42nd IEEE Symposium on Foundations of Computer Science, Las Vegas, NV, pp. 600–609. IEEE Computer Society, Los Alamitos (2001)
13. Lasserre, J.B.: An Explicit exact SDP relaxation for nonlinear 0-1 programs. In: Aardal, K., Gerards, B. (eds.) IPCO 2001. LNCS, vol. 2081, pp. 293–303. Springer, Heidelberg (2001). https://doi.org/10.1007/3-540-45535-3_23
14. Lovász, L.: On the Shannon capacity of a graph. IEEE Trans. Inf. Theory **25**(1), 1–7 (1979)

15. Lovász, L., Schrijver, A.: Cones of matrices and set-functions and 0-1 optimization. SIAM J. Optim. **1**(2), 166–190 (1991)
16. MOSEK ApS: The MOSEK optimization toolbox for MATLAB manual. Version 8.0. (2017). http://docs.mosek.com/8.0/toolbox/index.html
17. Nguyen, T.H., Bui, T.: Graph coloring benchmark instances. https://turing.cs.hbg.psu.edu/txn131/graphcoloring.html. Accessed 18 Nov 2018
18. Rendl, F., Rinaldi, G., Wiegele, A.: Solving max-cut to optimality by intersecting semidefinite and polyhedral relaxations. Math. Program. Ser. A **121**(2), 307–335 (2010)
19. Rockafellar, R.T.: Convex Analysis. Princeton Mathematical Series, vol. 28. Princeton University Press, Princeton (1970)
20. Sherali, H.D., Adams, W.P.: A hierarchy of relaxations between the continuous and convex hull representations for zero-one programming problems. SIAM J. Discrete Math. **3**(3), 411–430 (1990)
21. Tütüncü, R.H., Toh, K.C., Todd, M.J.: Solving semidefinite-quadratic-linear programs using SDPT3. Math. Program. Ser. B **95**(2), 189–217 (2003)

Dynamic Flows with Adaptive Route Choice

Lukas Graf[(✉)] and Tobias Harks

Institute of Mathematics, Augsburg University, Augsburg, Germany
{lukas.graf,tobias.harks}@math.uni-augsburg.de

Abstract. We study dynamic network flows and investigate *instantaneous dynamic equilibria (IDE)* requiring that for any positive inflow into an edge, this edge must lie on a currently shortest path towards the respective sink. We measure path length by current waiting times in queues plus physical travel times. As our main results, we show (1) existence of IDE flows for multi-source single sink networks, (2) finite termination of IDE flows for multi-source single sink networks assuming bounded and finitely lasting inflow rates, and, (3) the existence of a complex multi-commodity instance where IDE flows exist, but all of them are caught in cycles and persist forever.

1 Introduction

Dynamic network flows have been studied for decades in the optimization and transportation literature, see the classical book of Ford and Fulkerson [5] or the more recent surveys of Skutella [17] and Peeta [13]. A fundamental model describing the dynamic flow propagation process is the so-called *fluid queue model*, see Vickrey [19]. Here, one is given a digraph $G = (V, E)$, where edges $e \in E$ are associated with a queue with positive service capacity $\nu_e \in \mathbb{Z}_+$ and a physical travel time $\tau_e \in \mathbb{Z}_+$. If the total inflow into an edge $e = vw \in E$ exceeds the queue service capacity ν_e, a queue builds up and agents need to wait in the queue before they are forwarded along the edge. The total travel time along e is thus composed of the waiting time spent in the queue plus the physical travel time τ_e. A schematic illustration of the inflow and outflow mechanics of an edge e is given in Fig. 1. The fluid queue model has been mostly studied from a game-theoretic perspective, where it is assumed that agents act selfishly and travel along shortest routes under prevailing conditions. This behavioral model is known as *dynamic equilibrium* and has been analyzed in the transportation science literature for decades, see [6,12,20]. In the past years, however, several new exciting developments have emerged: Koch and Skutella [10] elegantly characterized dynamic equilibria by their derivatives which gives a template for their computation. Subsequently, Cominetti, Correa and Larré [3] derived alternative characterizations and proved existence and uniqueness in terms of experienced

The research of the authors was funded by the Deutsche Forschungsgemeinschaft (DFG, German Research Foundation) - HA 8041/1-1 and HA 8041/4-1.

A. Lodi and V. Nagarajan (Eds.): IPCO 2019, LNCS 11480, pp. 219–232, 2019.
https://doi.org/10.1007/978-3-030-17953-3_17

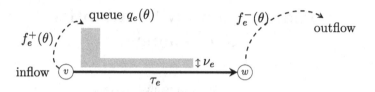

Fig. 1. An edge $e = vw$ with a nonempty queue.

travel times of equilibria even for multi-commodity networks (see also [1,4,16] for further recent work on the fluid queueing model).

The concept 'dynamic equilibrium' assumes *complete knowledge and simultaneous route choice* by all travelers. Complete knowledge requires that a traveler is able to exactly forecast future travel times along the chosen path effectively anticipating the whole evolution of the flow propagation process across the network. While this assumption has been justified by letting travelers learn good routes over several trips, this concept may not accurately reflect the behavioral changes caused by the wide-spread use of navigation devices. As also discussed in Marcotte et al. [11], Hamdouch et al. [9] or Unnikrishnan and Waller [18], drivers may not always learn good routes over several trips but are informed in real-time about the current traffic situations and, if beneficial, reroute instantaneously no matter how good or bad that route was in hindsight (for a more detailed discussion we refer to the full version [8]).

In this paper, we consider an adaptive route choice model, where at every node (intersection), travelers may alter their route depending on the current network conditions, that is, based on current travel times and queuing delays. We assume that, if a traveler arrives at the end of an edge, she may change the current route and opt for a currently shorter one. This type of reasoning does neither rely on private information of travelers nor on the capability of unraveling the future flow propagation process. We term a dynamic flow an *instantaneous dynamic equilibrium (IDE)*, if for every point in time and every edge with positive inflow (of some commodity), this edge lies on a currently shortest path towards the respective sink. In the following we illustrate IDE in comparison to classical dynamic equilibrium with an example.

1.1 An Example

Consider the network in Fig. 2 (left). There are two source nodes s_1 and s_2 with constant inflow rates $u_1(\theta) \equiv 3$ for $\theta \in [0,1)$ and $u_2(\theta) \equiv 4$ for $\theta \in [1,2)$. Commodity 1 (red) has two simple paths connecting s_1 with the sink t. Since both have equal length ($\sum_e \tau_e = 3$), in an IDE, commodity 1 can use both of them. In Fig. 2, the flow takes the direct edge to t with a rate of one, while edge $s_1 v$ is used at a rate of two. This is actually the only split possible in an IDE, since any other split (different in more than just a subset of measure zero of $[0,1)$) would result in a queue forming on one of the two edges, which would make the respective path longer than the other one. At time $\theta = 1$, the inflow at

s_1 stops and a new inflow of commodity 2 (blue) at s_2 starts. This new flow again has two possible paths to t, however here, the direct path ($\sum_e \tau_e = 1$) is shorter than the alternative ($\sum_e \tau_e = 4$). So all flow enters edge $s_2 t$ and starts to form a queue. At time $\theta = 2$, the first flow particles of commodity 1 arrive at s_2 with a rate of 2. Since the flow of commodity 2 has built up a queue of length 3 on edge $s_2 t$ by this time, the estimated travel times $\sum_e (\tau_e + q_e(\theta))$ are the same on both simple s_2-t paths. Thus, the red flow is split evenly between both possible paths. This results in the queue-length on edge $s_2 t$ remaining constant and therefore this split gives us an IDE flow for the interval $[2, 3)$. At time $\theta = 3$, red particles will arrive at s_1 again, thus having traveled a full cycle ($s_1 - v - s_2 - s_1$). This example shows that IDE flows may involve a flow decomposition along cycles.[1] In contrast, the (classical) dynamic equilibrium flow will just send more of the red flow along the direct path (s_1, t) since the future queue growth at edge $s_2 t$ of the alternative path is already anticipated.

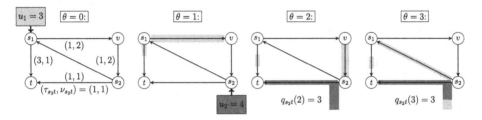

Fig. 2. The evolution of an IDE flow over the time horizon $[0, 3]$. (Color figure online)

1.2   Related Work

In the transportation science literature, the idea of an instantaneous user or dynamic equilibrium has already been proposed since the late 80's, see Ran and Boyce [14, § VII-IX], Boyce, Ran and LeBlanc [2,15], Friesz et al. [7]. These works develop an optimal control-theoretic formulation and characterize instantaneous user equilibria by Pontryagin's optimality conditions. However, the underlying equilibrium concept of Boyce, Ran and LeBlanc [2,15] and Friesz et al. [7] is different from ours. While the verbally written concept of IDE is similar to the one we use here, the mathematical definition of an IDE in [2,7,15] requires that instantaneous travel times are minimal for *used path* towards the sink. A path is used, if every arc of the path has positive flow. As, for instance, the authors in Boyce, Ran and LeBlanc [2, p. 130] admit: "Specifically with our definition of a used route, it is possible that no route is ever 'used' because vehicles stop entering the route before vehicles arrive at the last link on the route. Thus, for some networks every flow can be in equilibrium." Ran and Boyce [14, § VII, pp. 148] present a link-based definition of IDE. They define node labels at nodes $v \in V$ indicating the current shortest travel time from the source node to some

---

[1] Note that cycles may occur even in instances with only a single commodity (e.g. in the same graph with $u_1 = 8$, $\nu_{sv} = 7$ and $\nu_{vw} = 7$).

intermediate node v and require that whenever edge vw has positive flow, edge vw must be contained in a shortest s-w path. This is different from our definition of IDE, because we require that whenever there is positive inflow into an edge vw, it must be contained in a currently shortest v-t path, where t is the sink of the considered inflow. Another important difference to our model is that [2,7,14,15] assume a *finite time horizon* on which the control problems are defined, thus, only describing the flow trajectories over the given time horizon. Our results show that this assumption can be too restrictive: there are multi-commodity instances with finitely lasting bounded inflows that admit IDE flows cycling forever.

1.3 Our Results

We define a notion of instantaneous dynamic equilibrium (IDE) stating that a dynamic flow is an IDE, if at any point in time, for every edge with positive inflow (of some commodity), this edge lies on a currently shortest path towards the respective sink. Our first main result (Theorem 1) shows that IDE exist for multi-source single sink networks with piecewise constant inflow rates. The existence proof relies on a constructive method extending any IDE flow up to time θ to an IDE flow on a strictly larger interval $\theta + \epsilon$ for some $\epsilon > 0$. The key insight for the extension procedure relies on solving a sequence of nonlinear programs, each associated with finding the right outflow split for given node inflows. With the extension property, Zorn's Lemma implies the existence of IDE on the whole $\mathbb{R}_{\geq 0}$. Given that, unlike the classical dynamic equilibrium, IDE flows may involve cycling behavior (see the example in Fig. 2), we turn to the issue of whether it is possible that positive flow volume remains forever in the network (assuming finitely lasting bounded inflows). Our second main result (Theorem 2) shows that for multi-source single sink networks, there exists a finite time $T > 0$ at which the network is cleared, that is, all flow particles have reached their destination within the time horizon $[0, T]$. We then turn to general multi-commodity networks. Here, we show (Theorem 3) that for bounded and finitely lasting inflow rates, termination in finite time is *not* guaranteed anymore. We construct a quite complex instance where IDE flows exist, but all IDE flows are caught in cycles and travel forever.

2 The Flow Model

In the following, we describe a fluid queuing model as used before in Koch and Skutella [10] and Cominetti, Correa and Larré [3]. We are given a digraph[2] $G = (V, E)$ with queue service rates $\nu_e \in \mathbb{Z}_+, e \in E$ and travel times $\tau_e \in \mathbb{Z}_+$ for all $e \in E$. There is a finite set of commodities $I = \{1, \ldots, n\}$, each with a commodity-specific source node $s_i \in V$ and a common sink node $t \in V$.[3] The (infinitesimally small) agents of every commodity $i \in I$ are generated according

---

[2] Or a directed multi-graph. All results from this papers hold there as well.

[3] Without loss of generality we will always assume that all source nodes and the sink t are distinct from each other. Moreover, t is reachable from every other vertex $v \in V$.

to a right-constant inflow rate function $u_i : [r_i, R_i) \to \mathbb{R}_{\geq 0}$, where we say that a function $g : [a, b) \to \mathbb{R}$ is *right-constant* if for every $x \in [a, b)$ there exists an $\varepsilon > 0$ such that g is constant on $[x, x + \varepsilon)$, i.e. for all $y \in [x, x + \varepsilon)$ we have $g(y) = g(x)$. The time points $r_i \geq 0$ and $R_i > r_i$ are the release and ending time of commodity i, respectively. A *flow over time* is a tuple $f = (f^+, f^-)$, where $f^+, f^- : \mathbb{R}_{\geq 0} \times E \to \mathbb{R}_{\geq 0}$ are integrable functions modeling the inflow rate $f_e^+(\theta)$ and outflow rate $f_e^-(\theta)$ of an edge $e \in E$ at time $\theta \geq 0$. The flow conservation constraints are modeled as

$$\sum_{e \in \delta_v^+} f_e^+(\theta) - \sum_{e \in \delta_v^-} f_e^-(\theta) = b_v(\theta), \tag{1}$$

where $\delta_v^+ := \{\, vu \in E \,\}$ and $\delta_v^- := \{\, uv \in E \,\}$ are the sets of outgoing edges from v and incoming edges into v, respectively, and $b_v(\theta)$ is the balance at node v, which needs to be equal to $u_i(\theta)$, if $v = s_i$ and $\theta \in [r_i, R_i)$, non-positive for $v = t$ and any θ and equal to zero in all other cases. The queue length of edge e at time θ is given by

$$q_e(\theta) = F_e^+(\theta) - F_e^-(\theta + \tau_e), \tag{2}$$

where $F_e^+(\theta) := \int_0^\theta f_e^+(z)dz$ and $F_e^-(\theta) := \int_0^\theta f_e^-(z)dz$ denote the cumulative inflow and outflow, respectively. We implicitly assume that $f_e^-(\theta) = 0$ for all $\theta \in [0, \tau_e)$. Together with Constraint (3) this will imply that the queue length is always non-negative. We assume that the queue operates at capacity which can be modeled by

$$f_e^-(\theta + \tau_e) = \begin{cases} \nu(e), & \text{if } q_e(\theta) > 0 \\ \min\{\, f_e^+(\theta), \nu(e) \,\}, & \text{if } q_e(\theta) = 0 \end{cases} \quad \text{for all } e \in E, \theta \in \mathbb{R}_{\geq 0}. \tag{3}$$

It has been shown in Cominetti et al. [3] that this condition is in fact equivalent to the following equation describing the queue length dynamics:

$$q_e'(\theta) = \begin{cases} f_e^+(\theta) - \nu_e, & \text{if } q_e(\theta) > 0 \\ [f_e^+(\theta) - \nu_e]_+, & \text{if } q_e(\theta) = 0. \end{cases} \tag{4}$$

We assume that, whenever an agent arrives at an intermediate node v at time θ, she is given the information about the current queue lengths $q_e(\theta)$ and travel times $\tau_e, e \in E$, and, based on this information, she computes a shortest v-t path and enters the first edge on this path. We define the *instantaneous travel time* of an edge e at time θ as $c_e(\theta) = \tau_e + q_e(\theta)/\nu_e$, where $q_e(\theta)/\nu_e$ is the current waiting time to be spent in the queue of edge e. We can now define node labels $\ell_v(\theta)$ corresponding to current shortest path distances from v to the sink t. For $v \in V$ and $\theta \in \mathbb{R}_{\geq 0}$, define $\ell_t(\theta) = 0$ and $\ell_v(\theta) = \min_{e=vw \in E}\{\ell_w(\theta) + c_e(\theta)\}$ for all $v \neq t$. We say that edge $e = vw$ is *active* at time θ, if $\ell_v(\theta) = \ell_w(\theta) + c_e(\theta)$. We denote by $E_\theta \subseteq E$ the set of active edges. Now we are ready to formally define an instantaneous dynamic equilibrium for the continuous flow version.

Definition 1. *A flow f is an* instantaneous dynamic equilibrium (IDE)*, if it satisfies:*

$$\text{For all } \theta \in \mathbb{R}_{\geq 0}, e \in E: f_e^+(\theta) > 0 \Rightarrow e \in E_\theta. \qquad (5)$$

In words, a flow f is an IDE, if, whenever flow enters an edge $e = vw$ at some point θ, this edge must be contained in a currently shortest path from v to t.

3   Existence of IDE Flows

We now describe an algorithm computing an IDE for multi-source single-sink networks. Let $f = (f^+, f^-)$ denote a flow over time. We denote by $b_v^-(\theta) := \sum_{e \in \delta_v^-} f_e^-(\theta) + \sum_{i \in I: s_i = v} u_i(\theta)$ the current inflow at vertex v at time θ. Moreover, let $\delta_v^-(\theta) := \delta_v^- \cap E_\theta$ denote those outgoing edges of v that are active at time θ.

The main idea of our algorithm works as follows. Starting from time $\theta = 0$ we compute inductively a sequence of intervals $[0, \theta_1), [\theta_1, \theta_2), \dots$ with $0 < \theta_i < \theta_{i+1}$ and corresponding *constant* inflows $(f_e^+(\theta))_e$ for $\theta \in [\theta_i, \theta_{i+1})$ that form together with the corresponding edge outflows $(f_e^-(\theta))_e$ an IDE. Suppose we are given an *IDE flow up to time θ_k*, that is, a tuple (f^+, f^-) of right-constant functions $f_e^+ : [0, \theta_k) \to \mathbb{R}_{\geq 0}$ and $f_e^- : [0, \theta_k + \tau_e) \to \mathbb{R}_{\geq 0}$ satisfying Constraints (1), (3) and (5). Note that this is enough information to compute $F_e^+(\theta_k)$ and $F_e^-(\theta_k + \tau_e)$ and thus also $q_e(\theta_k), c_e(\theta_k)$ and $\ell_v(\theta_k)$ for all $e \in E$ and $v \in V$. We now describe how to extend this flow to the interval $[\theta_k, \theta_k + \varepsilon)$ for some $\varepsilon > 0$. Assume that $b_v^-(\theta)$ is constant for $\theta \in [\theta_k, \theta_k + \varepsilon)$ for some $v \in V$ and $\varepsilon > 0$. Moreover let $\delta_v^-(\theta_k) = \{vw_1, vw_2, \dots, vw_k\}$ for some $k \geq 1$ and define $[k] := \{1, \dots, k\}$. Thus, we have $\ell_v(\theta_k) = c_{vw_i}(\theta_k) + \ell_{w_i}(\theta_k)$ for all $i \in [k]$. Assume that labels of nodes $w_i, i \in [k]$ change linearly after θ_k, that is, there are constants $a_{w_i} \in \mathbb{R}$ for $i \in [k]$ with $\ell_{w_i}(\theta) = \ell_{w_i}(\theta_k) + a_{w_i}(\theta - \theta_k)$ for all $\theta \in [\theta_k, \theta_k + \varepsilon)$. Our goal is to find constant inflows $f_{vw_i}^+(\theta), i \in [k], \theta \in [\theta_k, \theta_k + \varepsilon)$ satisfying the supply $b_v^-(\theta)$ and, for some $\varepsilon' > 0$, fulfilling the following invariant for all $i \in [k]$ and $\theta \in [\theta_k, \theta_k + \varepsilon')$:

$$c_{vw_i}(\theta) + \ell_{w_i}(\theta) \leq c_{vw_j}(\theta) + \ell_{w_j}(\theta) \text{ for all } i, j \in [k] \text{ with } f_{vw_i}^+(\theta) > 0. \qquad (6)$$

If the inflow $f_{vw_i}^+$ is constant, then by Eq. (4) the queue length q_{vw_i} has piecewise constant derivative and, thus, is itself piecewise linear. This implies that the instantaneous travel time c_{vw_i} is piecewise linear as well, with derivative $c'_{vw_i}(\theta) = \frac{q'_{vw_i}(\theta)}{\nu_{vw_i}}$ and, in particular, linear on $[\theta_k, \theta_k + \varepsilon')$ for some $\varepsilon' > 0$. Since the invariant is fulfilled at $\theta = \theta_k$ and the ℓ_{w_i} are assumed to be linear on the interval $[\theta_k, \theta_k + \varepsilon)$, a sufficient condition for constant inflows to satisfy (6) for all $\theta \in [\theta_k, \theta_k + \varepsilon')$ is the following: For all $i \in [k]$ the constant inflows satisfy at time θ_k (and, thus, for all $\theta \in [\theta_k, \theta_k + \varepsilon')$):

$$c'_{vw_i}(\theta_k) + \ell'_{w_i}(\theta_k) \leq c'_{vw_j}(\theta_k) + \ell'_{w_j}(\theta_k) \text{ for all } i, j \in [k] \text{ with } f_{vw_i}^+(\theta_k) > 0. \qquad (7)$$

This condition simply makes sure that whenever an edge vw_i has positive inflow, the remaining distance towards t grows from θ_k onwards at the lowest speed.

We will now define an optimization problem in variables $x_{vw_i}, i \in [k]$ for which an optimal solution exists and satisfies the conditions defined in (7). The proof can be found in Appendix A (Lemma 1).

$$\min_{x_{vw_i} \geq 0, i \in [k]} \sum_{i=1}^{k} \int_{0}^{x_{vw_i}} \frac{g_{vw_i}(z)}{\nu_{vw_i}} + a_{w_i} dz \quad \text{s.t.:} \quad \sum_{i=1}^{k} x_{vw_i} = b_v^-(\theta_k), \quad (\text{OPT-}b_v^-(\theta_k))$$

where $g_{vw_i}(z) := \begin{cases} z - \nu_{vw_i}, & \text{if } q_{vw_i}(\theta_k) > 0 \\ [z - \nu_{vw_i}]_+, & \text{if } q_{vw_i}(\theta_k) = 0. \end{cases}$ Hence, $g_{vw_i}(f_{vw_i}^+(\theta_k))$ is the derivative of q_{vw_i} at θ_k (cf. Eq. (4)).

This way we can extend a given IDE flow up to time $\theta_k + \varepsilon'$ for a single node v by solving Eq. ($\text{OPT-}b_v^-(\theta_k)$) and setting $f_{vw_i}(\theta) := x_{vw_i}$ for some suitable short interval $[\theta_k, \theta_k + \varepsilon')$, provided that the flow is already extended for all nodes with strictly smaller label $\ell_w(\theta_k)$. To do that for all nodes, we simply order them by their current labels at time θ_k and then iteratively solve the above optimization problem for each node, beginning with the one with the smallest label. A more detailed explanation of this procedure is given in Appendix A (Lemma 2).

Theorem 1. *For any multi-source single sink network with right-constant inflow rate functions, there exists an IDE flow f with right-constant functions f_e^+ and $f_e^-, e \in E$.*

The proof can be found in Appendix A. In the full version [8], we give an example that IDE need not be unique.

4   Termination of IDE Flows

In this section, we investigate the question, whether an IDE flow actually vanishes within finite time, that is, if the finitely lasting and bounded inflow reaches the sink within finite time.

Definition 2. *A flow f terminates, if there exists a $\hat{\theta} \geq \theta_0 := \max \{ R_i \mid i \in I \}$ such that by time $\hat{\theta}$ the total volume of flow in the network is zero, i.e.*

$$G(\hat{\theta}) := \sum_{e \in E}(F_e^+(\hat{\theta}) - F_e^-(\hat{\theta})) = \sum_{i \in I} \int_0^{\hat{\theta}} u_i(\theta)d\theta - \sum_{e \in \delta_t^-} F_e^-(\theta) + \sum_{e \in \delta_t^+} F_e^+(\theta) = 0.$$

Theorem 2. *For multi-source single-sink networks, any IDE flow terminates.*

We will only sketch the three main steps of the proof here – for the detailed proof, see [8]. As our first step, we show that in an acyclic network, all flows over time terminate (IDE or not). To show this, we take a topological order on V (with t as the last element) and show that whenever there is a node v with no flow on edges between nodes that come before v (for all times after some θ_1), then, there exists $\theta_2 \geq \theta_1$ such that no flow will arrive at v after θ_2. In the second step, we show that flows with total volume $G(\theta) < 1$ at time $\theta \geq \theta_0$ must terminate. This follows, because for a remaining flow volume less than one, the total length of all queues is less than 1 as well and, thus, an IDE flow can only use edges that lie on a shortest path to t with respect to τ_e. Since these edges form an acyclic subgraph (independent of the time θ) such a flow terminates by step 1. For the third step, we take a generic IDE flow in an arbitrary multi-source single sink network and assume by contradiction that there exist edges e such that for any $\theta \in \mathbb{R}_{\geq 0}$, there exists a time $\theta' \geq \theta$ with $F_e^+(\theta') - F_e^-(\theta') \geq \frac{1}{|E|}$. From these edges we take the closest one to t and show – similarly to the first step – that there exists some time θ'' such that all flow on this edge will travel on a direct path to t (after time θ''). Altogether, this implies that eventually more flow volume arrives at t than the totally generated volume (at the sources), a contradiction. Thus, there exists some time θ^* after which the total amount in the network is less than 1 and, hence, the flow terminates by the second step.

Remark 1. For the entire proof to work, we only need the assumption of boundedness and finite support of inflow rates u_i, thus, the result holds for more general inflow functions.

5   Multi-commodity Networks

We now generalize the model to multi-source multi-sink networks. A multi-commodity flow over time is a tuple $f = ((f_{i,e}^+)_{i \in I, e \in E}, (f_{i,e}^-)_{i \in I, e \in E})$, where $f_{i,e}^+, f_{i,e}^- : \mathbb{R}_{\geq 0} \to \mathbb{R}_{\geq 0}$ are integrable functions for all $i \in I$ and $e \in E$ that satisfy corresponding balance constraints for each $v \in V$. Queue lengths depend on the *aggregate* cumulative inflows and outflows, respectively. For $i \in I, v \in V$ and $\theta \in \mathbb{R}_{\geq 0}$, we define commodity-specific node labels $\ell_v^i(\theta)$ as in the single sink case except that t_i is used as sink node. We say that edge $e = vw$ is *active* for $i \in I$ at time θ, if $\ell_v^i(\theta) = \ell_w^i(\theta) + c_e(\theta)$. Let $E_\theta^i \subseteq E$ be the set of active edges.

Definition 3. *A multi-commodity flow f is an* instantaneous dynamic equilibrium *if for all $i \in I, \theta \in \mathbb{R}_{\geq 0}$ and $e \in E$ it satisfies $f_{i,e}^+(\theta,) > 0 \Rightarrow e \in E_\theta^i$.*

Together with Leon Sering, we are currently working on a proof that IDE always exist for multi-commodity networks, which will be included in the full version of this paper. Regarding termination, however, we can already show that there are instances in which there exists an IDE flow and any IDE flow does not terminate.

Theorem 3. *There is a multi-commodity network with two sinks and all edge travel times and capacities equal to 1, where any IDE flow does not terminate.*

To construct such an instance we make use of several gadgets. The first one, gadget A, will serve as the main building block and is depicted in Fig. 4. It consists of two cycles with one common edge v_1v_2 and one commodity with constant inflow rate of 2 on the interval $[0, 1)$ at node v_1 with a sink node t outside the gadget and reachable from the nodes v_2, v_5 and v_7 via some paths

P_2, P_5 and P_7, respectively. Our goal will be to embed this gadget into a larger instance in such a way, that for any IDE flow, the flow inside gadget A will exhibit the following flow pattern for all $h \in \mathbb{N}$ (see Fig. 3):

1. **On the interval $[5h, 5h + 1)$:** All flow generated at v_1 (for $h = 0$) or arriving at v_1 (for $h > 0$) enters the edge to v_2 at a rate of 2, half of it directly starting to travel along the edge, half of it building up a queue of length 1 at time $5h + 1$.
2. **On the interval $[5h + 1, 5h + 2)$:** The flow arriving at node v_2 enters the edge to v_3 because v_2, v_3, v_4, v_5, P_5 is currently the shortest path to t. The length of the queue of edge $v_1 v_2$ decreases until it reaches 0 at time $5h + 2$.
3. **On the interval $[5h + 2, 5h + 3)$:** The flow arriving at node v_2 enters the edge to v_6 because v_2, v_6, v_7, P_7 is currently the shortest path to t.
4. **On the interval $[5h + 4, 5h + 5)$:** The flows arriving at nodes v_5 and v_7 enter the respective edges towards node v_1 because v_5, v_1, v_2, P_2 and v_7, v_1, v_2, P_2 are currently the shortest paths to get to t.
5. **On the interval $[5h + 5, 5h + 6)$:** There is a total inflow of 2 at node v_1, which enters the edge to v_2. Thus, the pattern repeats.

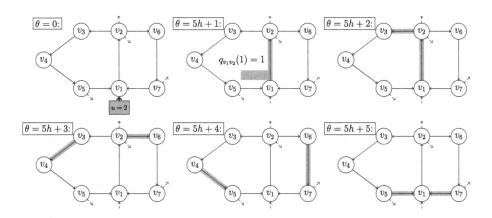

Fig. 3. The desired flow pattern in gadget A at times $\theta = 0, 1, 2, 3, 4, 5, \ldots$.

The effect of this behavior is that other particles outside the gadget, who want to travel through this gadget along the central vertical path, will estimate an additional waiting time as indicated by the diagram displayed inside gadget A in Fig. 4 (next to the vertical red path). Now, in order to actually guarantee the described behavior, we need to embed gadget A into a larger instance in such a way, that for any IDE flow the following assumptions hold:

1. The only paths leaving A are the four dashed paths indicated in Fig. 4.
2. The three (blue) paths P_2, P_5 and P_7 are of the same length L (w.r.t. τ_e).
3. For all $h \in \mathbb{N}$ and all $\theta \in [5h + 1, 5h + 2), (5h + 2, 5h + 5]$ the unique shortest paths are given in the description for the flow pattern above.

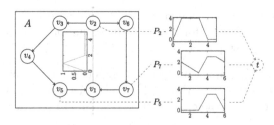

Fig. 4. Gadget A (the dashed paths and nodes are not part of the gadget). The (red) diagram inside the box A indicates the waiting time on edge $v_1 v_2$ (and therefore on the (red) vertical path through the gadget), provided that the flow inside this gadget follows the flow pattern indicated in Fig. 3. The (blue) diagrams on the right indicate the desired waiting times on the paths P_2, P_5 and P_7, respectively (Color figure online).

In order to satisfy the assumptions 1–3., we will now construct three types of gadgets B_2, B_5 and B_7 for the three paths P_2, P_5 and P_7, each of equal length on which any IDE flow induces waiting times as shown by the respective diagrams on the right side in Fig. 4. To build these gadgets we need time shifted versions of gadget A, which we denote by A^{+k}. Such a gadget is constructed the same way as gadget A above, with the only difference that the support of the inflow rate function u_i is shifted to the interval $[k \bmod 5, k \bmod 5 + 1)$. Gadget B_2 now consists of the concatenation of four gadgets of type A^{+0}, four gadgets of type A^{+1} and four gadgets of type A^{+2} in series along their vertical paths through them with three edges between each two gadgets (see Fig. 5 in Appendix A). Similarly, gadget B_5 consists of three copies of A^{+3}-type gadgets, three copies of A^{+4}-type gadgets and additional $6 \cdot 4$ edges to ensure that the vertical path has the same length as the one of gadget B_2. Finally, gadget B_7 consists of three copies of A^{+3}-type gadgets, three copies of A^{+4}-type gadgets, two copies of A^{+5}-type gadgets, one copy of A^{+6}-type gadgets and additional $3 \cdot 4$ edges. We again use the notation B_j^{+k} to refer to a time shifted version of gadget B_j – i.e. with all used gadgets A shifted by additional k time steps. Next, we build a gadget C by just taking one copy of each B_j^{+k} for all $j \in \{2, 5, 7\}$ and $k = 0, 1, 2, 3, 4$ (see Fig. 6 in Appendix A). Finally, taking two copies of this gadget, C and C', and two additional nodes, t and t', where t will be the sink node for all players in C and t' the sink node for all players in C', we can build our entire graph as indicated by Fig. 7 in Appendix A. We connect the top edges of the gadgets B_j^{+k} in gadget C' with the sink t and use those gadgets' respective vertical paths as the P_j^{+k} paths for gadget C and vice versa.

In order to prove the correctness of our construction (i.e. that any IDE flow on this instance does not terminate) we need the following important observation:

Observation 1. *If a flow in some A^{+k}-type gadget (with $k \in \{0, 1, 2, 3, 4\}$) follows the desired flow pattern for all unit time intervals between k and some $\theta \in \mathbb{N}_0, \theta \geq k$, the induced waiting time on edge $v_1 v_2$ of this gadget will follow* *the waiting time function indicated by the diagram in Fig. 4 (shifted by k) for the next unit time interval $[\theta, \theta+1)$, independent of the evolution of the flow in this interval. The same is true for all B_j^{+k}-type gadgets.*

With this observation we can prove Theorem 2 by induction on the number of passed unit time intervals. We assume that a given IDE flow follows the flow pattern described at the beginning of the construction and indicated in Fig. 3 for all unit time intervals up to some $\theta \in \mathbb{N}_0$. Then by Observation 1 we know that at least the waiting time pattern will continue to hold for the next unit time interval. So for each node v within a generic A^{+k}-type gadget, we can identify the shortest v-t path on the next interval and only need to verify that its first edge is indeed the one the flow is supposed to enter. This shows that the flow pattern will hold for all times and, in particular, that the flow never terminates.

Remark 2. It is even possible to modify the network in such a way, that only a single source (and multiple sinks) is necessary.

A   Omitted Proofs and Figures of Sects. 3 and 5

Lemma 1. *There exists an optimal solution $x_{vw_i}, i \in [k]$ to OPT-$b_v^-(\theta_k)$ so that $f_{vw_i}^+(\theta_k) = x_{vw_i}, i \in [k]$ satisfies (7).*

Proof. The objective function is continuous and the feasible region is non-empty and compact, thus, by the theorem of Weierstraß an optimal solution exists. Assigning a multiplier $\lambda \in \mathbb{R}$ to the equality constraint, we obtain $x_{vw_i} > 0 \Rightarrow \frac{g_{vw_i}(x_{vw_i})}{\nu_{vw_i}} + a_{w_i} + \lambda = 0$, $x_{vw_i} = 0 \Rightarrow \frac{g_{vw_i}(x_{vw_i})}{\nu_{vw_i}} + a_{w_i} + \lambda \geq 0$, implying (7). □

Lemma 2. *Let $f = (f^+, f^-)$ be an IDE flow up to time $\theta_k \geq 0$ and suppose there are constant inflow rate functions $b_v^- : [\theta_k, \theta_k + \varepsilon) \to \mathbb{R}_{\geq 0}$ for some $\varepsilon > 0$ and all nodes $v \in V$ (in particular, this means $\varepsilon \leq \min\{\tau_e \mid e \in E\}$). Then there exists some $\varepsilon' > 0$ such that we can extend f to an IDE flow up to time $\theta_k + \varepsilon'$ with all functions f_e^+ constant on the interval $[\theta_k, \theta_k + \varepsilon')$ and all functions f_e^- right-constant on the intervals $[\theta_k + \tau_e, \theta_k + \tau_e + \varepsilon')$.*

Proof. First, we sort the nodes by their labels $\ell_v(\theta_k)$ and will now define the outflows using Lemma 1 for each node, beginning with the one with the smallest label. This first one will always be t (with label $\ell_t(\theta_k) = 0$) for which we can define $f_e^+(\theta) = f_e^-(\theta + \tau_e) = 0$ for all outgoing edges $e \in \delta_t^+$ and all times $\theta \in [\theta_k, \theta_k + \varepsilon)$. Now we take some node v such that for all nodes w with strictly smaller label at time θ_k and all edges $e \in \delta_w^+$ we have already defined f_e^+ on some interval $[\theta_k, \theta_k + \varepsilon')$ and f_e^- on some interval $[\theta_k + \tau_e, \theta_k + \tau_e + \varepsilon')$ in such a way that on the interval $[\theta_k, \theta_k + \varepsilon')$ we have

1. the labels $\ell_w(\theta)$ change linearly,
2. no additional edges are added to the sets $\delta_w^+(\theta)$ of active edges leaving w,

3. the f_e^+ are constant and the f_e^- right-constant for all $e \in \delta_w^+$ and
4. the functions f_e^+ and f_e^- for $e \in \delta_w^+$ satisfy Constraints (1), (3) and (5).

Let $\delta_v^+(\theta_k) := \{vw_1, vw_2, \ldots, vw_k\}$ be the set of active edges at v at time θ_k. Then, at time θ_k, each w_i must have a strictly smaller label than v. Hence, they satisfy Properties 1–4. We can now apply Lemma 1 to determine the flows $f_{vw_i}^+(\theta_k)$. Additionally, we set $f_e^+(\theta_k) = 0$ for all non-active edges leaving v, i.e. all $e \in \delta_v^+ \setminus \delta_v^+(\theta_k)$. Assuming that this flow remains constant on the whole interval $[\theta_k, \theta_k + \varepsilon')$, we can determine the first time $\hat{\theta} \geq \theta_k$, where an additional edge $vw \in \delta_v^+$ or $wv \in \delta_v^-$ becomes newly active. This can only happen after some positive amount of time has passed, i.e., for some $\hat{\theta} > \theta_k$, because: (i) at time θ_k the edge was non-active and therefore $\ell_v(\theta_k) > c_{vw}(\theta_k) + \ell_w(\theta_k)$ or $\ell_w(\theta_k) > c_{wv}(\theta_k) + \ell_v(\theta_k)$, respectively, (ii) all labels change linearly (and thus continuously) and (iii) c_{vw} or c_{wv} is changing piecewise linearly, since the length of its queue does so as well (as both f_{vw}^+ and f_{wv}^- are piecewise constant). If the difference $\hat{\theta} - \theta_k$ is smaller than the current ε', we take it as our new ε', otherwise we keep it as it is. In both cases, we extend f_e^+ to the interval $[\theta_k, \theta_k + \varepsilon')$ for all $e \in \delta_v^+$ by setting $f_e^+(\theta) = f_e^+(\theta_k)$ for all $\theta \in [\theta_k, \theta_k + \varepsilon')$. This guarantees that the label of v changes linearly on this interval, no additional edges become active and the functions f_e^+ are constant. Also f_e^+ satisfies Constraints (1) and (5) by definition. Finally, we define f_e^- by setting $f_e^-(\theta + \tau_e) := \nu_e$, if $q_e(\theta_k) + (\theta - \theta_k)(f_e^+(\theta_k) - \nu_e) > 0$, and $f_e^-(\theta + \tau_e) := f_e^+(\theta)$ else. Then, f_e^- is right-constant and together with f_e^+ satisfies Constraint (3). In summary, using this procedure we can extend f node by node to an IDE flow up to $\theta_k + \varepsilon'$ for some $\varepsilon' > 0$. $\square$

Proof (Proof of Theorem 1). Let $\mathfrak{F}$ be the set of tupels (f, θ), with $\theta \in \mathbb{R}_{\geq 0} \cup \{\infty\}$ and f a IDE flow over time up to time θ with right-constant functions f_e^+ and f_e^-. We define a partial order on $\mathfrak{F}$ by $(f, \theta) \leq (f', \theta') :\Leftrightarrow \theta \leq \theta'$ and $f'|_{[0,\theta)} \equiv f$. Now, $\mathfrak{F}$ is non-empty, since the 0-flow is obviously an IDE flow up to time 0, and for any chain $(f^{(1)}, \theta_1), (f^{(2)}, \theta_2), \ldots$ in $\mathfrak{F}$, we can define an upper bound $(\hat{f}, \hat{\theta})$ to this chain by setting $\hat{\theta} := \sup\{\theta_k\}$ and

$$\hat{f}_e^+ : [0, \hat{\theta}) \to \mathbb{R}_{\geq 0}, \theta \mapsto f_e^{(k),+}(\theta) \text{ with } k \text{ s.t } \theta < \theta_k$$
$$\hat{f}_e^- : [0, \hat{\theta} + \tau_e) \to \mathbb{R}_{\geq 0}, \theta \mapsto f_e^{(k),-}(\theta) \text{ with } k \text{ s.t } \theta < \theta_k + \tau_e.$$

This is well defined and an IDE flow up to $\hat{\theta}$, since for every θ it coincides with some IDE flow $f^{(k)}$ and therefore is an IDE flow up to θ itself. By Zorn's lemma, we get the existence of a maximal element $(f^*, \theta^*) \in \mathfrak{F}$. If we had $\theta^* < \infty$, we could apply the extension property (Lemma 2) to f^*, a contradiction to its maximality. So we must have $\theta^* = \infty$ and, hence, f^* is an IDE flow on $\mathbb{R}_{\geq 0}$. $\square$

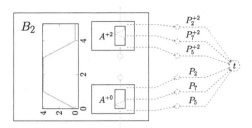

Fig. 5. Gadget B_2 consisting of four copies of each of the types A^{+0}, A^{+1}, A^{+2}. The diagram inside the box of gadget B_2 indicates the waiting time on the vertical path through gadget B_2, provided that within all of the used gadgets A, the flow follows the flow pattern from Fig. 3. The dashed parts are not part of the gadget.

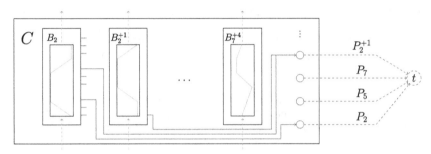

Fig. 6. Gadget C

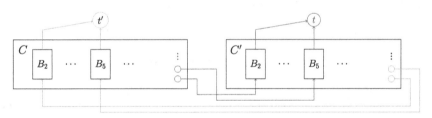

Fig. 7. The graph

References

1. Bhaskar, U., Fleischer, L., Anshelevich, E.: A Stackelberg strategy for routing flow over time. Games Econ. Behav. **92**, 232–247 (2015)
2. Boyce, D.E., Ran, B., LeBlanc, L.J.: Solving an instantaneous dynamic user-optimal route choice model. Transp. Sci. **29**(2), 128–142 (1995)
3. Cominetti, R., Correa, J.R., Larré, O.: Dynamic equilibria in fluid queueing networks. Oper. Res. **63**(1), 21–34 (2015)
4. Cominetti, R., Correa, J.R., Olver, N.: Long term behavior of dynamic equilibria in fluid queuing networks. In: Proceedings of the Integer Programming and Combinatorial Optimization - 19th International Conference, IPCO 2017, Waterloo, ON, Canada, 26–28 June 2017, pp. 161–172 (2017)

5. Ford, L.R., Fulkerson, D.R.: Flows in Networks. Princeton University Press, Princeton (1962)
6. Friesz, T.L., Bernstein, D., Smith, T.E., Tobin, R.L., Wie, B.W.: A variational inequality formulation of the dynamic network user equilibrium problem. Oper. Res. **41**(1), 179–191 (1993)
7. Friesz, T.L., Luque, J., Tobin, R.L., Wie, B.: Dynamic network traffic assignment considered as a continuous time optimal control problem. Oper. Res. **37**(6), 893–901 (1989)
8. Graf, L., Harks, T.: Dynamic flows with adaptive route choice. arXiv (2018), https://arxiv.org/abs/1811.07381
9. Hamdouch, Y., Marcotte, P., Nguyen, S.: A strategic model for dynamic traffic assignment. Networks Spat. Econ. **4**(3), 291–315 (2004)
10. Koch, R., Skutella, M.: Nash equilibria and the price of anarchy for flows over time. Theory Comput. Syst. **49**(1), 71–97 (2011)
11. Marcotte, P., Nguyen, S., Schoeb, A.: A strategic flow model of traffic assignment in static capacitated networks. Oper. Res. **52**(2), 191–212 (2004)
12. Meunier, F., Wagner, N.: Equilibrium results for dynamic congestion games. Transp. Sci. **44**(4), 524–536 (2010). An updated version (2014) is available on Arxiv
13. Peeta, S., Ziliaskopoulos, A.: Foundations of dynamic traffic assignment: the past, the present and the future. Networks Spat. Econ. **1**(3), 233–265 (2001)
14. Ran, B., Boyce, D.: Dynamic Urban Transportation Network Models: Theory and Implications for Intelligent Vehicle-Highway Systems. Lecture Notes in Economics and Mathematical Systems. Springer, Heidelberg (1996). https://doi.org/10.1007/978-3-662-00773-0
15. Ran, B., Boyce, D.E., LeBlanc, L.J.: A new class of instantaneous dynamic user-optimal traffic assignment models. Oper. Res. **41**(1), 192–202 (1993)
16. Sering, L., Vargas-Koch, L.: Nash flows over time with spillback. In: Proceedings of the 30th Annual ACM-SIAM Symposium on Discrete Algorithms. ACM (to appear, 2019)
17. Skutella, M.: An introduction to network flows over time. In: Research Trends in Combinatorial Optimization, Bonn Workshop on Combinatorial Optimization, Bonn, Germany, 3–7 November 2008, pp. 451–482 (2008)
18. Unnikrishnan, A., Waller, S.: User equilibrium with recourse. Networks Spat. Econ. **9**(4), 575–593 (2009)
19. Vickrey, W.S.: Congestion theory and transport investment. Am. Econ. Rev. **59**(2), 251–260 (1969)
20. Zhu, D., Marcotte, P.: On the existence of solutions to the dynamic user equilibrium problem. Transp. Sci. **34**(4), 402–414 (2000)

The Markovian Price of Information

Anupam Gupta[1], Haotian Jiang[2(✉)], Ziv Scully[1], and Sahil Singla[3]

[1] Carnegie Mellon University, Pittsburgh, PA 15213, USA
{anupamg,zscully}@cs.cmu.edu
[2] University of Washington, Seattle, WA 98195, USA
jhtdavid@uw.edu
[3] Princeton University, Princeton, NJ 08544, USA
singla@cs.princeton.edu

Abstract. Suppose there are n Markov chains and we need to pay a per-step *price* to advance them. The "destination" states of the Markov chains contain rewards; however, we can only get rewards for a subset of them that satisfy a combinatorial constraint, e.g., at most k of them, or they are acyclic in an underlying graph. What strategy should we choose to advance the Markov chains if our goal is to maximize the total reward *minus* the total price that we pay?

In this paper we introduce a Markovian price of information model to capture settings such as the above, where the input parameters of a combinatorial optimization problem are given via Markov chains. We design optimal/approximation algorithms that jointly optimize the value of the combinatorial problem and the total paid price. We also study *robustness* of our algorithms to the distribution parameters and how to handle the *commitment* constraint.

Our work brings together two classical lines of investigation: getting optimal strategies for Markovian multi-armed bandits, and getting exact and approximation algorithms for discrete optimization problems using combinatorial as well as linear-programming relaxation ideas.

Keywords: Multi-armed bandits · Gittins index · Probing algorithms

1 Introduction

Suppose we are running an oil company and are deciding where to set up new drilling operations. There are several candidate sites, but the value of drilling each site is a random variable. We must therefore *inspect* sites before drilling. Each inspection gives more information about a site's value, but the inspection process is costly. Based on laws, geography, or availability of equipment, there are constraints on which sets of drilling sites are feasible. We ask:

> What adaptive inspection strategy should we adopt to find a feasible set of sites to drill which maximizes, in expectation, the value of the chosen (drilled) sites minus the total inspection cost of all sites?

© Springer Nature Switzerland AG 2019
A. Lodi and V. Nagarajan (Eds.): IPCO 2019, LNCS 11480, pp. 233–246, 2019.
https://doi.org/10.1007/978-3-030-17953-3_18

Let us consider the optimization challenges in this problem:

(i) Even if we could fully inspect each site for free, choosing the best feasible set of sites is a *combinatorial optimization* problem.

(ii) Each site may have *multiple stages* of inspection. The costs and possible outcomes of later stages may depend on the outcomes of earlier stages. We use a *Markov chain* for each site to model how our knowledge about the value of the site stochastically evolves with each inspection.

(iii) Since a site's Markov chain model may not exactly match reality, we want a *robust* strategy that performs well even under small changes in the model parameters.

(iv) If there is competition among several companies, it may not be possible to do a few stages of inspection at a given site, abandon that site's inspection to inspect other sites, and then later return to further inspect the first site. In this case the problem has additional "take it or leave it" or *commitment* constraints, which prevent interleaving inspection of multiple sites.

While each of the above aspects has been individually studied in the past, no prior work addresses all of them. In particular, aspects (i) and (ii) have not been simultaneously studied before. In this work we advance the state of the art by solving the (i)-(ii)-(iii) and the (i)-(ii)-(iv) problems.

To study aspects (i) and (ii) together, in Sect. 2 we propose the *Markovian Price of Information* (MARKOVIAN PoI) model. The MARKOVIAN PoI model unifies prior models which address (i) or (ii) alone. These prior models include those of Kleinberg et al. [17] and Singla [18], who study the combinatorial optimization aspect (i) in the so-called *price of information* model, in which each site has just a single stage of inspection; and those of Dimitriu et al. [8] and Kleinberg et al. [17, Appendix G], who consider the multiple stage inspection aspect (ii) for the problem of selecting just a single site.

Our main results[1] show how to solve combinatorial optimization problems, including both maximization and minimization problems, in the MARKOVIAN PoI model. We give two methods of transforming classic algorithms, originally designed for the FREE-INFO (inspection is free) setting, into *adaptive* algorithms for the MARKOVIAN PoI setting. These adaptive algorithms respond dynamically to the random outcomes of inspection.

- In Sect. 3.3 we transform "greedy" α-approximation algorithms in the FREE-INFO setting into α-approximation adaptive algorithms in the MARKOVIAN PoI setting (Theorem 1). For example, this yields optimal algorithms for matroid optimization (Corollary 1).

- In Sect. 4 we show how to slightly modify our α-approximations for the MARKOVIAN PoI setting in Theorem 1 to make them robust to small changes in the model parameters (Theorem 2).

---

[1] Due to space constraints, we omit full proofs in this extended abstract. The full version is available at https://arxiv.org/abs/1902.07856.

- In Sect. 5 we use *online contention resolution schemes* (OCRSs) [10] to transform LP based FREE-INFO maximization algorithms into adaptive MARKOVIAN PoI algorithms while respecting the commitment constraints. Specifically, a $1/\alpha$-selectable OCRS yields α-approximation with commitment (Theorem 3).

The general idea behind our first result (Theorem 1) is the following. A FRUGAL combinatorial algorithm (Definition 8) is, roughly speaking, "greedy": it repeatedly selects the feasible item of greatest marginal value. We show how to adapt *any* FRUGAL algorithm to the MARKOVIAN PoI setting:

- Instead of using a fixed value for each item i, we use a *time-varying "proxy" value* that depends on the state of i's Markov chain.
- Instead of immediately selecting the item i of greatest marginal value, we *advance i's Markov chain one step.*

The main difficulty lies in choosing each item's proxy value, for which simple heuristics can be suboptimal. We use a quantity for each state of each item's Markov chain called its *grade*, and an item's proxy value is its *minimum grade so far*. A state's grade is closely related to the Gittins index from the multi-armed bandit literature, which we discuss along with other related work in Sect. 6.

2   The Markovian Price of Information Model

To capture the evolution of our knowledge about an item's value, we use the notion of a Markov system from [8] (who did not consider values at the destinations).

Definition 1 (Markov System). *A Markov system $\mathcal{S} = (V, P, s, T, \boldsymbol{\pi}, \mathbf{r})$ for an element consists of a discrete Markov chain with state space V, a transition matrix $P = \{p_{u,v}\}$ indexed by $V \times V$ (here $p_{u,v}$ is the probability of transitioning from u to v), a starting state s, a set of absorbing destination states $T \subseteq V$, a non-negative probing price $\pi^u \in \mathbb{R}_{\geq 0}$ for every state $u \in V \setminus T$, and a value $r^t \in \mathbb{R}$ for each destination state $t \in T$. We assume that every state $u \in V$ reaches some destination state.*

We have a collection J of *ground elements*, each associated with its own Markov system. An element is *ready* if its Markov system has reached one of its absorbing destination states. For a ready element, if ω is the (random) *trajectory* of its Markov chain then $d(\omega)$ denotes its associated destination state. We now define the MARKOVIAN PoI game, which consists of an objective function on J.

Definition 2 (MARKOVIAN PoI Game). *Given a set of ground elements J, constraints $\mathcal{F} \subseteq 2^J$, an objective function $f : 2^J \times \mathbb{R}^{|J|} \to \mathbb{R}$, and a Markov system $\mathcal{S}_i = (V_i, P_i, s_i, T_i, \boldsymbol{\pi}_i, \mathbf{r}_i)$ for each element $i \in J$, the MARKOVIAN PoI game is the following. At each time step, we either advance a Markov system $\mathcal{S}_i$ from its current state $u \in V_i \setminus T_i$ by incurring price π_i^u, or we end the game by selecting a subset of ready elements $\mathbb{I} \subseteq J$ that are feasible—i.e., $\mathbb{I} \in \mathcal{F}$.*

A common choice for f is the *additive* objective $f(\mathbb{I}, \mathbf{x}) = \sum_{i \in \mathbb{I}} x_i$.

Let $\boldsymbol{\omega}$ denote the *trajectory profile* for the MARKOVIAN PoI game: it consists of the random trajectories ω_i taken by all the Markov chains i at the end of the game. To avoid confusion, we write the selected feasible solution $\mathbb{I}$ as $\mathbb{I}(\boldsymbol{\omega})$. A utility/disutility optimization problem is to give a strategy for a MARKOVIAN PoI game while optimizing both the objective and the total price.

Utility Maximization (UTIL-MAX): A MARKOVIAN PoI game where the constraints $\mathcal{F}$ are *downward-closed* (i.e., *packing*) and the values $\mathbf{r}_i$ are non-negative for every $i \in J$ (i.e., $\forall t \in T_i$, $r_i^t \geq 0$, and can be understood as a reward obtained for selecting i). The goal is to find a strategy ALG maximizing *utility*:

$$U^{\max}(\text{ALG}) \triangleq \mathbb{E}_{\boldsymbol{\omega}}\left[\underbrace{f\left(\mathbb{I}(\boldsymbol{\omega}), \{r_i^{d(\omega_i)}\}_{i \in \mathbb{I}(\boldsymbol{\omega})}\right)}_{\text{value}} - \underbrace{\sum_i \sum_{u \in \omega_i} \pi_i^u}_{\text{total price}} \right]. \quad (1)$$

Since the empty set is always feasible, the optimum utility is non-negative.

We also define a minimization variant of the problem that is useful to capture covering combinatorial problems such as minimum spanning trees and set cover.

Disutility Minimization (DISUTIL-MIN): A MARKOVIAN PoI game where the constraints $\mathcal{F}$ are *upward-closed* (i.e., *covering*) and the values $\mathbf{r}_i$ are non-negative for every $i \in J$ (i.e., $\forall t \in T_i$, $r_i^t \geq 0$, and can be understood as a cost we pay for selecting i). The goal is to find a strategy ALG minimizing *disutility*:

$$U^{\min}(\text{ALG}) \triangleq \mathbb{E}_{\boldsymbol{\omega}}\left[f\left(\mathbb{I}(\boldsymbol{\omega}), \{r_i^{d(\omega_i)}\}_{i \in \mathbb{I}(\boldsymbol{\omega})}\right) + \sum_i \sum_{u \in \omega_i} \pi_i^u \right].$$

We will assume that the function f is non-negative when all $\mathbf{r}_i$ are non-negative. Hence, the disutility of the optimal policy is non-negative.

In the special case where all the Markov chains for a MARKOVIAN PoI game are formed by a *directed acyclic graph* (DAG), we call the corresponding optimization problem DAG-UTIL-MAX or DAG-DISUTIL-MIN.

3 Adaptive Utility Maximization via Frugal Algorithms

FRUGAL algorithms, introduced in Singla [18], capture the intuitive notion of "greedy" algorithms. There are many known FRUGAL algorithms, e.g., optimal algorithms for matroids and $O(1)$-approx algorithms for matchings, vertex cover, and facility location. These FRUGAL algorithms were designed in the traditional *free information* (FREE-INFO) setting, where each ground element has a fixed value. Can we use them in the MARKOVIAN PoI world?

Our main contribution is a technique that adapts *any* FRUGAL algorithm to the MARKOVIAN PoI world, achieving the *same approximation ratio* as the original algorithm. The result applies to *semiadditive* objective functions f, which are those of the form $f(\mathbb{I}, \mathbf{x}) = \sum_{i \in \mathbb{I}} x_i + h(\mathbb{I})$ for some $h : 2^J \to \mathbb{R}$.

Theorem 1. *For a semiadditive objective function* val, *if there exists an α-approximation* FRUGAL *algorithm for a* UTIL-MAX *problem over some packing constraints* $\mathcal{F}$ *in the* FREE-INFO *world, then there exists an α-approximation strategy for the corresponding* UTIL-MAX *problem in the* MARKOVIAN PoI *world.*

We prove an analogous result for DISUTIL-MIN in the full version. The following corollaries immediately follow from known FRUGAL algorithms [18].

Corollary 1. *In the* MARKOVIAN PoI *world, we have:*

- *An optimal algorithm for both* UTIL-MAX *and* DISUTIL-MIN *for matroids.*
- *A 2-approx for* UTIL-MAX *for matchings and a k-approx for a k-system.*
- *A* min$\{\theta, \log n\}$-*approx for* DISUTIL-MIN *for set-cover, where θ is the maximum number of sets in which a ground element is present.*
- *A 1.861-approx for* DISUTIL-MIN *for facility location.*
- *A 3-approx for* DISUTIL-MIN *for prize-collecting Steiner tree.*

Before proving Theorem 1, we define a *grade* for every state in a Markov system in Sect. 3.1, much as in [8]. This grade is a variant of the popular *Gittins index*. In Sect. 3.2, we use the grade to define a *prevailing cost* and an *epoch* for a trajectory. In Sect. 3.3, we use these definitions to prove Theorem 1. We consider UTIL-MAX throughout, but analogous definitions and arguments hold for DISUTIL-MIN.

3.1   Grade of a State

To define the *grade* τ^v of a state $v \in V$ in Markov system $\mathcal{S} = (V, P, s, T, \boldsymbol{\pi}, \mathbf{r})$, we consider the following Markov game called τ-*penalized* $\mathcal{S}$, denoted $\mathcal{S}(\tau)$. Roughly, $\mathcal{S}(\tau)$ is the same as $\mathcal{S}$ but with a *termination penalty*, which is a constant $\tau \in \mathbb{R}$.

Suppose $v \in V$ denotes the current state of $\mathcal{S}$ in the game $\mathcal{S}(\tau)$. In each move, the player has two choices: (a) *Halt* that immediately ends the game, and (b) *Play* that changes the state, price, and value as follows:

- If $v \in V \setminus T$, the player pays price π^v, the current state of $\mathcal{S}$ changes according to the transition matrix P, and the game continues.
- If $v \in T$, then the player receives *penalized value* $r^v - \tau$, where τ is the aforementioned termination penalty, and the game ends.

The player wishes to maximize his *utility*, which is the expected value he obtains minus the expected price he pays. We write $U^v(\tau)$ for the utility attained by optimal play starting from state $v \in V$.

The utility $U^v(\tau)$ is clearly non-increasing in the penalty τ, and one can also show that it is continuous [8, Section 4]. In the case of large penalty $\tau \to +\infty$, it is optimal to halt immediately, achieving $U^v(\tau) = 0$. In the opposite extreme $\tau \to -\infty$, it is optimal to play until completion, achieving $U^v(\tau) \to +\infty$. Thus, as we increase τ from $-\infty$ to $+\infty$, the utility $U^v(\tau)$ becomes 0 at some critical value $\tau = \tau^v$. This critical value τ^v that depends on state v is the *grade*.

Definition 3 (Grade). *The* grade *of a state v in Markov system $\mathcal{S}$ is $\tau^v \triangleq \sup\{\tau \in \mathbb{R} \mid U^v(\tau) > 0\}$. For a* UTIL-MAX *problem, we write the grade of a state v in Markov system $\mathcal{S}_i$ corresponding to element i as τ_i^v.*

The quantity grade of a state is well-defined from the above discussion. We emphasize that it is independent of all other Markov systems. Put another way, the grade of a state is the penalty τ that makes the player *indifferent* between halting and playing. It is known how to compute grade efficiently [8, Section 7].

3.2   Prevailing Cost and Epoch

We now define a *prevailing cost* [8] and an *epoch*. The prevailing cost of Markov system $\mathcal{S}$ is its minimum grade at any point in time.

Definition 4 (Prevailing Cost). *The* prevailing cost *of Markov system $\mathcal{S}_i$ in a trajectory ω_i is $Y^{\max}(\omega_i) = \min_{v \in \omega_i}\{\tau_i^v\}$. For trajectory profile $\boldsymbol{\omega}$, denote $Y^{\max}(\boldsymbol{\omega})$ the list of prevailing costs for each Markov system.*

Put another way, the prevailing cost is the maximum termination penalty for the game $\mathcal{S}(\tau)$ such that for every state along ω the player does not want to halt.

Observe that the prevailing cost of a trajectory can only decrease as it extends further. In particular, it decreases whenever the Markov system reaches a state with grade smaller than each of the previously visited states. We can therefore view the prevailing cost as a non-increasing piecewise constant function of time. This motivates us to define an epoch.

Definition 5 (Epoch). *An* epoch *for a trajectory ω is any maximal continuous segment of ω where the prevailing cost does not change.*

Since the grade can be computed efficiently, we can also compute the prevailing cost and epochs of a trajectory efficiently.

3.3   Adaptive Algorithms for Utility Maximization

In this section, we prove Theorem 1 that adapts a FRUGAL algorithm in FREE-INFO world to a probing strategy in the MARKOVIAN PoI world. This theorem concerns *semiadditive functions*, which are useful to capture non-additive objectives of problems like facility location and prize-collecting Steiner tree.

Definition 6 (Semiadditive Function [18]). *A function $f(\mathbb{I}, \mathbf{X}) : 2^J \times \mathbb{R}^{|J|} \to \mathbb{R}$ is* semiadditive *if there exists a function $h : 2^J \to \mathbb{R}$ s.t. $f(\mathbb{I}, \mathbf{x}) = \sum_{i \in \mathbb{I}} x_i + h(\mathbb{I})$.*

All additive functions are semiadditive with $h(\mathbb{I}) = 0$ for all $\mathbb{I}$. To capture the facility location problem on a graph $G = (J, E)$ with metric (J, d), clients $C \subseteq J$, and facility opening costs $\mathbf{x} : J \to \mathbb{R}_{\geq 0}$, we can define $h(\mathbb{I}) = \sum_{j \in C} \min_{i \in \mathbb{I}} d(j, i)$. Notice h only depends on the identity of facilities $\mathbb{I}$ and not their opening costs.

The proof of Theorem 1 takes two steps. We first give a randomized reduction to upper bound the utility of the optimal strategy in the MARKOVIAN PoI world with the optimum of a *surrogate problem* in the FREE-INFO world. Then, we transform a FRUGAL algorithm into a strategy with utility close to this bound.

Upper Bounding the Optimal Strategy Using a Surrogate. The *main idea* in this section is to show that for UTIL-MAX, no strategy (in particular, optimal) can derive more utility from an element $i \in J$ than its prevailing cost. Here, the prevailing cost of i is for a random trajectory to a destination state in Markov system $\mathcal{S}_i$. Since the optimal strategy can only select a feasible set in $\mathcal{F}$, this idea naturally leads to the following FREE-INFO *surrogate problem*: imagine each element's value is exactly its (random) prevailing cost, the goal is to select a set feasible in $\mathcal{F}$ to maximize the total value. In Lemma 1, we show that the expected optimum value of this surrogate problem is an upper bound on the optimum utility for UTIL-MAX. First, we formally define the surrogate problem.

Definition 7 (Surrogate Problem). *Given a* UTIL-MAX *problem with semi-additive objective* val *and packing constraints $\mathcal{F}$ over universe J, the corresponding* surrogate problem *over J is the following. It consists of constraints $\mathcal{F}$ and (random) objective function $\tilde{f} : 2^J \to \mathbb{R}$ given by $\tilde{f}(\mathbb{I}) = \mathsf{val}(\mathbb{I}, \mathbf{Y}^{\max}(\omega))$, where $\mathbf{Y}^{\max}(\omega)$ denotes the prevailing costs over a random trajectory profile ω consisting of independent random trajectories for each element $i \in J$ to a destination state. The goal is to select $\mathbb{I} \in \mathcal{F}$ to maximize $\tilde{f}(\mathbb{I})$.*

Let $\mathrm{SUR}(\omega) \overset{\Delta}{=} \max_{\mathbb{I} \in \mathcal{F}} \{\mathsf{val}(\mathbb{I}, \mathbf{Y}^{\max}(\omega))\}$ denote the optimum value of the surrogate problem for trajectory profile ω. We now upper bound the optimum utility in the MARKOVIAN PoI world (proved in full version). Our proof borrows ideas from the "prevailing reward argument" in [8].

Lemma 1. *For a* UTIL-MAX *problem with objective* val *and packing constraints $\mathcal{F}$, let* OPT *denote the utility of the optimal strategy. Then,*

$$\mathrm{OPT} \;\leq\; \mathbb{E}_\omega[\mathrm{SUR}(\omega)] \;=\; \mathbb{E}_\omega\big[\max_{\mathbb{I} \in \mathcal{F}}\{\mathsf{val}(\mathbb{I}, \mathbf{Y}^{\max}(\omega))\}\big],$$

where the expectation is over a random trajectory profile ω that has every Markov system reaching a destination state.

Designing an Adaptive Strategy Using a Frugal Algorithm. A FRUGAL algorithm selects elements one-by-one and irrevocably. Besides greedy algorithms, its definition also captures "non-greedy" algorithms such as primal-dual algorithms that do not have the reverse-deletion step [18].

Definition 8 (FRUGAL Packing Algorithm). *For a combinatorial optimization problem on universe J in the FREE-INFO world with packing constraints $\mathcal{F} \subseteq 2^J$ and objective $f : 2^J \to \mathbb{R}$, we say Algorithm $\mathcal{A}$ is FRUGAL if there exists* a marginal-value *function* $g(\mathbf{Y}, i, y) : \mathbb{R}^J \times J \times \mathbb{R} \to \mathbb{R}$ *that is increasing in* y, *and for which the pseudocode is given by Algorithm 1. Note that this algorithm always returns a feasible solution if* $\emptyset \in \mathcal{F}$.

Algorithm 1. FRUGAL Packing Algorithm $\mathcal{A}$

1: Start with $M = \emptyset$ and $v_i = 0$ for each element $i \in J$.
2: For each element $i \notin M$, compute $v_i = g(\mathbf{Y}_M, i, Y_i)$. Let $j =$ arg max$_{i \notin M\ \&\ M \cup i \in \mathcal{F}}\{v_i\}$.
3: If $v_j > 0$ then add j into M and go to Step 2. Otherwise, return M.

The following lemma shows that a FRUGAL algorithm can be converted to a strategy with the same utility in the MARKOVIAN PoI world.

Lemma 2. *Given a* FRUGAL *packing Algorithm* $\mathcal{A}$, *there exists an adaptive strategy* ALG$_\mathcal{A}$ *for the corresponding* UTIL-MAX *problem in* MARKOVIAN PoI *world with utility at least* $\mathbb{E}_\omega[\mathsf{val}(\mathcal{A}(\mathbf{Y}^{\max}(\omega)), \mathbf{Y}^{\max}(\omega))]$, *where* $\mathcal{A}(\mathbf{Y}^{\max}(\omega)$ *is the solution returned by* $\mathcal{A}$ *for objective* $f(\mathbb{I}) = \mathsf{val}(\mathbf{Y}^{\max}(\omega), \mathbb{I})$.

The strategy for Lemma 2 is in Algorithm 2 but the full proof is deferred.

Algorithm 2. ALG$_\mathcal{A}$ for UTIL-MAX in MARKOVIAN PoI

1: Start with $M = \emptyset$ and $v_i = 0$ for all elements i.
2: For each element $i \notin M$, set $g(\mathbf{Y}_M^{\max}, i, \tau_i^{u_i})$ where u_i is the current state of i.
3: Consider the element $j =$ arg max$_{i \notin M\ \&\ M \cup i \in \mathcal{F}}\{v_i\}$.
4: If $v_j > 0$, then if $\mathcal{S}_j$ is not in a destination state then proceed $\mathcal{S}_j$ by one step and go to Step 2. Else, when $v_j > 0$ but $\mathcal{S}_j$ is in a destination state t_j, select j into M and go to Step 2.
5: Else, if every element $i \notin M$ has $v_i \leq 0$ then return set M.

Proof (Proof of Theorem 1). From Lemma 2, the utility of ALG$_\mathcal{A}$ is at least $\mathbb{E}_\omega[\mathsf{val}(\mathcal{A}(\mathbf{Y}^{\max}(\omega)), \mathbf{Y}^{\max}(\omega))]$. Since Algorithm $\mathcal{A}$ is an α-approx algorithm in the FREE-INFO world, it follows

$$\mathbb{E}_\omega[\mathsf{val}(\mathcal{A}(\mathbf{Y}^{\max}(\omega)), \mathbf{Y}^{\max}(\omega))] \geq \frac{1}{\alpha} \cdot \mathbb{E}_\omega\Big[\max_{\mathbb{I}\in\mathcal{F}}\{\mathsf{val}(\mathbb{I}, \mathbf{Y}^{\max}(\omega))\}\Big].$$

Using the upper bound on optimal utility OPT $\leq \mathbb{E}_\omega[\max_{\mathbb{I}\in\mathcal{F}}\{\mathsf{val}(\mathbb{I},\mathbf{Y}^{\max}(\omega))\}]$ from Lemma 1, we have utility of ALG$_\mathcal{A}$ is at least $\frac{1}{\alpha} \cdot$ OPT.

In the full version, a similar approach is used for the DISUTIL-MIN problem with semi-additive function. This shows that for both UTIL-MAX or DISUTIL-MIN problem with semi-additive function, a FRUGAL algorithm can be transformed from FREE-INFO to MARKOVIAN PoI world while retaining its performance.

4 Robustness in Model Parameters

In practical applications, the parameters of Markov systems (i.e., transition probabilities, values, and prices) are not known exactly but are *estimated by* statistical sampling. In this setting, the *true parameters*, which govern how each Markov system evolves, differ from the estimated parameters that the algorithm uses to make decisions. This raises a natural question: how well does an adapted FRUGAL algorithm do when the true and the estimated parameters differ? We would hope to design a *robust* algorithm, meaning small estimation errors cause only small error in the utility objective.

In the important special case where the Markov chain corresponding to each element is formed by a *directed acyclic graph* (DAG), an adaptation of our strategy in Theorem 1 is robust. This DAG assumption turns out to be necessary as similar results do not hold for general Markov chains. In particular, we prove the following generalization of Theorem 1 under the DAG assumption.

Theorem 2 (Informal statement). *If there exists an α-approximation* FRUGAL *algorithm $\mathcal{A}$ ($\alpha \geq 1$) for a packing problem with a semiadditive objective function, then it suffices to estimate the true model parameters of a* DAG-MARKOVIAN PoI *game within an additive error of ϵ/poly, where* poly *is some polynomial in the size of the input, to design a strategy with utility at least $\frac{1}{\alpha} \cdot \mathrm{OPT} - \epsilon$, where* OPT *is the utility of the optimal policy that knows all the* true *model parameters.*

Specifically, our strategy $\widehat{\mathrm{ALG}}_\mathcal{A}$ for Theorem 2 is obtained from the strategy in Theorem 1 by making use of the following idea: each time we advance an element's Markov system, we slightly increase the estimated grade of every state in that Markov system. This ensures that whenever we advance a Markov system, we advance through an entire epoch and remain optimal in the "teasing game".

Our analysis of $\widehat{\mathrm{ALG}}_\mathcal{A}$ works roughly as follows. We first show that under the DAG assumption, close estimates of the model parameters of a Markov system can be used to closely estimate the grade of each state. We can therefore assume that close estimates of all grades are given as input. Next we define the "shifted" prevailing cost corresponding to the "shifted" grades. This allows us to equate the utility of $\widehat{\mathrm{ALG}}_\mathcal{A}$ by the utility of running $\mathcal{A}$ in the "modified" surrogate problem where the input to $\mathcal{A}$ is the "shifted" prevailing costs instead of the *true* prevailing costs. Finally, we prove that the "shifted" prevailing costs are close to the real prevailing costs and thus the "modified" surrogate problem is close to the surrogate problem. This allows us to bound the utility of running $\mathcal{A}$ in the "modified" surrogate problem by the optimal strategy to the surrogate problem. Combining with Lemma 1 finishes the proof of Theorem 2.

An analogous result for DISUTIL-MIN also holds.

5   Handling Commitment Constraints

Consider the MARKOVIAN PoI model defined in Sect. 2 with an additional restriction that whenever we abandon advancing a Markov system, we need to *immediately* and *irrevocably* decide if we are selecting this element into the final solution $\mathbb{I}$. Since we only select ready elements, any element that is not ready when we abandon its Markov system is automatically discarded. We call this

242   A. Gupta et al.

constraint *commitment*. The benchmark for our algorithm is the optimal policy *without* the commitment constraint. For single-stage probing, such commitment constraints have been well studied, especially in the context of stochastic matchings [2,4].

We study UTIL-MAX in the DAG model with the commitment constraint. Our algorithms make use of the *online contention resolution schemes* (OCRSs) proposed in [10]. OCRSs address our problem in the FREE-INFO world[2] (i.e., we can see the realization of the r.v.s for free, but there is the commitment constraint). Constant factor "selectable" OCRSs are known for several constraint families: $\frac{1}{4}$ for matroids, $\frac{1}{2e}$ for matchings, and $\Omega(\frac{1}{k})$ for intersection of k matroids [10]. We show how to adapt them to MARKOVIAN PoI with commitment.

Theorem 3. *For an additive objective, if there exists a $1/\alpha$-selectable OCRS ($\alpha \geq 1$) for a packing constraint $\mathcal{F}$, then there exists an α-approximation algorithm for the corresponding DAG-UTIL-MAX problem with commitment.*

The proof of this result uses a new LP relaxation (inspired from [13]) to bound the optimum utility of a MARKOVIAN PoI game *without* commitment (see Sect. A.1). Although this relaxation is not exact even for Pandora's box (and cannot be used to design optimal strategies in Corollary 1), it turns out to suffice for our approximation guarantees. In Sect. A.2, we use an OCRS to round this LP with only a small loss in the utility, while respecting the commitment constraint.

Remark 1. We do not consider DISUTIL-MIN problem under commitment because it captures prophet inequalities in a minimization setting where no polynomial approximation is possible even for i.i.d. r.v.s [9, Theorem 4].

6 Related Work

Our work is related to work on multi-armed bandits in the scheduling literature. The Gittins index theorem [12] provides a simple optimal strategy for several scheduling problems where the objective is to maximize the long-term exponentially discounted reward. This theorem turned out to be fundamental and [19–21] gave alternate proofs. It can be also used to solve Weitzman's Pandora's box. The reader is referred to the book [11] for further discussions on this topic. Influenced by this literature, [8] studied scheduling of Markovian jobs, which is a minimization variant of the Gittins index theorem without any discounting. Their paper is part of the inspiration for our MARKOVIAN PoI model.

The Lagrangian variant of stochastic probing considered in [13] is similar to our MARKOVIAN PoI model. However, their approach using an LP relaxation to design a probing strategy is fundamentally different from our approach using

---

[2] In fact, OCRSs consider a variant where the adversary chooses the order in which the elements are tried. This handles the present problem where we may choose the order.

a FRUGAL algorithm. E.g., unlike Corollary 1, their approach cannot give *optimal* probing strategies for matroid constraints due to an integrality gap. Also, their approach does not work for DISUTIL-MIN. In Appendix A, we extend their techniques using OCRSs to handle the commitment constraint for UTIL-MAX.

There is also a large body of work in related models where information has a price [1,3,5–7,14–16]. Finally, as discussed in the introduction, the works in [17] and [18] are directly relevant to this paper. The former's primary focus is on *single item* settings and its applications to auction design, and the latter studies price of information in a *single-stage* probing model. Our contributions concern selecting *multiple items* in *multi-stage* probing model, in some sense unifying these two lines of work.

Acknowledgements. A. Gupta and S. Singla were supported in part by NSF awards CCF1536002, CCF-1540541, and CCF-1617790, and the Indo-US Joint Center for Algorithms Under Uncertainty. H. Jiang was supported in part by CCF-1740551, CCF-1749609, and DMS-1839116. Z. Scully was supported by an ARCS Foundation scholarship and the NSF GRFP under Grant Nos. DGE-1745016 and DGE-125222.

A   Details for Handling Commitment Constraints

In this section we handle commitment constraints from Sect. 5 to prove Theorem 3. In Sect. A.1, we give an LP relaxation to upper bound the optimum utility without the commitment constraint. In Sect. A.2, we apply an OCRS to round the LP solution to obtain an adaptive policy, while satisfying the commitment constraint.

A.1   Upper Bounding the Optimum Utility

Define the following variables, where i is an index for the Markov systems.

- y_i^u: probability we reach state u in Markov system $\mathcal{S}_i$ for $u \in V_i \setminus T_i$.
- z_i^u: probability we play $\mathcal{S}_i$ when it is in state u for $u \in V_i \setminus T_i$.
- $x_i = \sum_{u \in T_i} z_i^u$: probability $\mathcal{S}_i$ is selected into the final solution when in a destination state.
- $P_{\mathcal{F}}$ is a convex relaxation containing all feasible solutions for packing $\mathcal{F}$.

We can now formulate the following LP, which is inspired from [13].

$$\max_{\mathbf{z}} \quad \sum_i \left(\sum_{u \in T_i} r_i^u z_i^u - \sum_{u \in V_i \setminus T_i} \pi_i^u z_i^u \right)$$

$$\text{subject to} \quad y_i^{s_i} = 1 \qquad\qquad\qquad \forall i \in J$$

$$y_i^u = \sum_{v \in V_i} (P_i)_{uv} z_i^v \qquad \forall i \in J, \forall u \in V_i \setminus s_i$$

$$x_i = \sum_{u \in T_i} z_i^u \qquad\qquad \forall i \in J$$

$$z_i^u \leq y_i^u \qquad\qquad\qquad \forall i \in J, \forall u \in V_i$$

$$\mathbf{x} \in P_{\mathcal{F}}$$

$$x_i, y_i^u, z_i^u \geq 0 \qquad\qquad \forall i \in J, \forall u \in V_i$$

The first four constraints characterize the dynamics in advancing the Markov systems. The fifth constraint encodes the packing constraint $\mathcal{F}$. We denote the optimal solution of this LP as $(\mathbf{x}, \mathbf{y}, \mathbf{z})$. We can efficiently solve the above LP for packing constraints such as matroids, matchings, and intersection of k matroids.

If we interpret the variables y_i^u, x_i, and z_i^u as the probabilities corresponding to the optimal strategy without commitment, it forms a feasible solution to the LP. This implies the following claim.

Lemma 3. *The optimum utility without commitment is at most the LP value.*

A.2   Rounding the LP Using an OCRS

Before describing our rounding algorithm, we define an OCRS. Intuitively, it is an online algorithm that given a random set ground elements, selects a feasible subset of them. Moreover, if it can guarantee that every i is selected w.p. at least $\frac{1}{\alpha} \cdot x_i$, it is called $\frac{1}{\alpha}$-selectable.

Definition 9 (OCRS [10]). *Given a point $x \in P_{\mathcal{F}}$, let $R(x)$ denote a random set containing each i independently w.p. x_i. The elements i reveal one-by-one whether $i \in R(x)$ and we need to decide irrevocably whether to select an $i \in R(x)$ into the final solution before the next element is revealed. An OCRS is an online algorithm that selects a subset $I \subseteq R(x)$ such that $I \in \mathcal{F}$.*

Definition 10 ($\frac{1}{\alpha}$-Selectability [10]). *Let $\alpha \geq 1$. An OCRS for $\mathcal{F}$ is $\frac{1}{\alpha}$-selectable if for any $x \in P_{\mathcal{F}}$ and all i, we have $\Pr[i \in I \mid i \in R(x)] \geq \frac{1}{\alpha}$.*

Our algorithm ALG uses OCRS as an oracle. It starts by fixing an arbitrary order π of the Markov systems. (Our algorithm works even when an adversary decides the order of the Markov systems.) Then at each step, the algorithm considers the next element i in π and queries the OCRS whether to select element i if it is ready. If OCRS decides to select i, then ALG advances the Markov system such that it plays from each state u with independent probability z_i^u/y_i^u. This guarantees that the destination state is reached with probability x_i. If OCRS is not going to select i, then ALG moves on to the next element in π. A formal description of the algorithm can be found in Algorithm 3.

Algorithm 3. Algorithm ALG for Handling the Commitment Constraint

---

1: Fix an arbitrary order π of the items. Set $M = \emptyset$ and pass $\mathbf{x}$ to OCRS.
2: Consider the next element i in the order of π. Query OCRS whether to add i to M if i is ready.
   (a) If OCRS would add i to M, then keep advancing the Markov system: play from each current state $u \in V_i \setminus T_i$ independently w.p. z_i^u/y_i^u, and otherwise go to Step 2. If a destination state t is reached then add i to M w.p. z_i^t/y_i^t.
   (b) Go to Step 2.

We show below that ALG has a utility of at least $1/\alpha$ times the LP value.

Lemma 4. *The utility of* ALG *is at least* $1/\alpha$ *times the LP optimum.*

Since by Lemma 3 the LP optimum is an upper bound on the utility of any policy without commitment, this proves Theorem 3. We now prove Lemma 4.

Proof (Proof of Lemma 4). Recollect that we call a Markov system ready if it reaches an absorbing destination state. We first notice that once ALG starts to advance a Markov system i, then by Step 2 of Algorithm 3, element i is ready with probability exactly x_i. This agrees with what ALG tells the OCRS. Since the OCRS is $1/\alpha$-selectable, the probability that any Markov system $\mathcal{S}_i$ begins advancing is $1/\alpha$. Here the probability is both over the random choice of the OCRS and the randomness due to the Markov systems. Conditioning on the event that $\mathcal{S}_i$ begins advancing, the probability that it is selected into the final solution on reaching a destination state $t \in T_i$ is exactly z_i^t. Hence, the conditioned utility from Markov system $\mathcal{S}_i$ is exactly

$$\sum_{u \in T_i} r_i^u z_i^u - \sum_{u \in V_i \setminus T_i} \pi_i^u z_i^u. \tag{2}$$

By removing the conditioning and by linearity of expectation, the utility of ALG is at least $\frac{1}{\alpha} \cdot \sum_i \left(\sum_{u \in T_i} r_i^u z_i^u - \sum_{u \notin T_i} \pi_i^u z_i^u \right)$, which proves this lemma.

References

1. Abbas, A.E., Howard, R.A.: Foundations of Decision Analysis. Pearson Higher Ed. London (2015)
2. Bansal, N., Gupta, A., Li, J., Mestre, J., Nagarajan, V., Rudra, A.: When LP is the cure for your matching woes: improved bounds for stochastic matchings. Algorithmica **63**(4), 733–762 (2012)
3. Charikar, M., Fagin, R., Guruswami, V., Kleinberg, J.M., Raghavan, P., Sahai, A.: Query strategies for priced information. J. Comput. Syst. Sci. **64**(4), 785–819 (2002). https://doi.org/10.1006/jcss.2002.1828
4. Chen, N., Immorlica, N., Karlin, A.R., Mahdian, M., Rudra, A.: Approximating matches made in heaven. In: Albers, S., Marchetti-Spaccamela, A., Matias, Y., Nikoletseas, S., Thomas, W. (eds.) ICALP 2009. LNCS, vol. 5555, pp. 266–278. Springer, Heidelberg (2009). https://doi.org/10.1007/978-3-642-02927-1_23
5. Chen, Y., Immorlica, N., Lucier, B., Syrgkanis, V., Ziani, J.: Optimal data acquisition for statistical estimation. arXiv preprint arXiv:1711.01295 (2017)
6. Chen, Y., Hassani, S.H., Karbasi, A., Krause, A.: Sequential information maximization: When is greedy near-optimal? In: Conference on Learning Theory, pp. 338–363 (2015)
7. Chen, Y., Javdani, S., Karbasi, A., Bagnell, J.A., Srinivasa, S.S., Krause, A.: Submodular surrogates for value of information. In: AAAI, pp. 3511–3518 (2015)
8. Dumitriu, I., Tetali, P., Winkler, P.: On playing golf with two balls. SIAM J. Discrete Math. **16**(4), 604–615 (2003)
9. Esfandiari, H., Hajiaghayi, M., Liaghat, V., Monemizadeh, M.: Prophet secretary. SIAM J. Discrete Math. **31**(3), 1685–1701 (2017)

10. Feldman, M., Svensson, O., Zenklusen, R.: Online contention resolution schemes. In: Proceedings of the Twenty-Seventh Annual ACM-SIAM Symposium on Discrete Algorithms, pp. 1014–1033. Society for Industrial and Applied Mathematics (2016)

11. Gittins, J., Glazebrook, K., Weber, R.: Multi-Armed Bandit Allocation Indices. Wiley, Chichester (2011)

12. Gittins, J., Jones, D.: A dynamic allocation index for the sequential design of experiments. In: Gani, J. (ed.) Progress in Statistics, pp. 241–266. North-Holland, Amsterdam (1974)

13. Guha, S., Munagala, K.: Approximation algorithms for budgeted learning problems. In: STOC, pp. 104–113 (2007), full version as: Approximation Algorithms for Bayesian Multi-Armed Bandit Problems. http://arxiv.org/abs/1306.3525

14. Guha, S., Munagala, K., Sarkar, S.: Information acquisition and exploitation in multichannel wireless systems. In: IEEE Transactions on Information Theory. Citeseer (2007)

15. Gupta, A., Kumar, A.: Sorting and selection with structured costs. In: Proceedings of the 42nd IEEE Symposium on Foundations of Computer Science, 2001, pp. 416–425. IEEE (2001)

16. Kannan, S., Khanna, S.: Selection with monotone comparison costs. In: Proceedings of the Fourteenth Annual ACM-SIAM Symposium on Discrete algorithms, pp. 10–17. Society for Industrial and Applied Mathematics (2003)

17. Kleinberg, R., Waggoner, B., Weyl, G.: Descending Price Optimally Coordinates Search. arXiv preprint arXiv:1603.07682 (2016)

18. Singla, S.: The price of information in combinatorial optimization. In: Proceedings of the Twenty-Ninth Annual ACM-SIAM Symposium on Discrete Algorithms. SIAM (2018)

19. Tsitsiklis, J.N.: A short proof of the Gittins index theorem. The Annals of Applied Probability, pp. 194–199 (1994)

20. Weber, R.: On the Gittins index for multiarmed bandits. Ann. Appl. Probab. **2**(4), 1024–1033 (1992)

21. Whittle, P.: Multi-armed bandits and the Gittins index. J. Roy. Stat. Soc. Ser. B (Methodol.), 143–149 (1980)

On Perturbation Spaces of Minimal Valid Functions: Inverse Semigroup Theory and Equivariant Decomposition Theorem

Robert Hildebrand[1] , Matthias Köppe[2(✉)] , and Yuan Zhou[3]

[1] Grado Department of Industrial and Systems Engineering,
Virginia Tech, Blacksburg, USA
rhil@vt.edu
[2] Department of Mathematics, University of California, Davis, USA
mkoeppe@math.ucdavis.edu
[3] Department of Mathematics, University of Kentucky, Lexington, USA
yuan.zhou@uky.edu

Abstract. The non-extreme minimal valid functions for the Gomory–Johnson infinite group problem are those that admit effective perturbations. For a class of piecewise linear functions for the 1-row problem we give a precise description of the space of these perturbations as a direct sum of certain finite- and infinite-dimensional subspaces. The infinite-dimensional subspaces have partial symmetries; to describe them, we develop a theory of inverse semigroups of partial bijections, interacting with the functional equations satisfied by the perturbations. Our paper provides the foundation for grid-free algorithms for testing extremality and for computing liftings of non-extreme functions. The grid-freeness makes the algorithms suitable for piecewise linear functions whose breakpoints are rational numbers with huge denominators.

Keywords: Integer programs · Cutting planes · Group relaxations

1 Introduction

A powerful method to derive cutting planes for unstructured integer linear optimization problems is to study relaxations with more structure and convenient

The authors wish to thank C.Y. Hong, who worked on a first grid-free implementation in 2013, and Q. Louveaux and R. La Haye for valuable discussions during 2013/14. A preliminary version of the development in this paper appeared in Y.Z.'s 2017 Ph.D. thesis [14]. The authors gratefully acknowledge partial support from the National Science Foundation through grants DMS-0914873 (R.H., M.K.) and DMS-1320051 (M.K., Y.Z.) Part of this work was done while R.H. and M.K. were visiting the Simons Institute for the Theory of Computing in Fall 2017. It was partially supported by the DIMACS/Simons Collaboration on Bridging Continuous and Discrete Optimization through NSF grant CCF-1740425.

A. Lodi and V. Nagarajan (Eds.): IPCO 2019, LNCS 11480, pp. 247–260, 2019.
https://doi.org/10.1007/978-3-030-17953-3_19

properties. The pioneering relaxation in this line of research on general-purpose cutting planes is Gomory's *finite group relaxation*. The valid inequalities for this model arise from cut-generating functions, the valid functions of the Gomory–Johnson infinite group problem [6,7]. A notion of domination gives rise to the family of *minimal (valid) functions*; they have the following characterization, which for the purposes of this paper we will take as a definition. Let $G = \mathbb{R}^k$ and fix a parameter $f \in G \setminus \mathbb{Z}^k$; then a minimal valid function π for the infinite group problem $R_f(G, \mathbb{Z}^k)$ is a function $\pi \colon G \to \mathbb{R}_+$ such that $\pi(0) = 0$, $\pi(f) = 1$, and

$$\pi(x + z) = \pi(x) \qquad \text{for } x \in G, z \in \mathbb{Z}^k \qquad \text{(periodicity)} \tag{1a}$$

$$\Delta\pi(x, y) \geq 0 \qquad \text{for } x, y \in G \qquad \text{(subadditivity)}, \tag{1b}$$

$$\Delta\pi(x, f - x) = 0 \qquad \text{for } x \in G \qquad \text{(symmetry condition)}, \tag{1c}$$

where $\Delta\pi(x, y) = \pi(x) + \pi(y) - \pi(x + y)$ is the *subadditivity slack*. Among the minimal functions, the so-called extreme functions stand out; they are an infinite-dimensional analog of facets. Following [9, sect. 6], we define the space

$$\tilde{\Pi}^\pi = \left\{ \tilde{\pi} \colon \mathbb{R}^k \to \mathbb{R} \mid \exists\, \epsilon > 0 \text{ s.t. } \pi^\pm = \pi \pm \epsilon\tilde{\pi} \text{ are minimal valid} \right\} \tag{2}$$

of *effective perturbation functions* for the minimal valid function π. This is a vector space, a subspace of the space of bounded functions. (In this extended abstract, we restrict ourselves to the case of continuous piecewise linear functions π; then $\tilde{\Pi}^\pi$ consists of bounded continuous functions.) The function π is said to be *extreme* if the space $\tilde{\Pi}^\pi$ is trivial.

Our paper considers the single-row case ($k = 1$). For this case, Basu et al. [2] gave the first algorithm to decide extremality of a piecewise linear function with rational breakpoints in some "grid" (group) $G = \frac{1}{q}\mathbb{Z}$. To obtain this result, Basu et al. [2] use two closely linked techniques: (a) Group actions describe the required symmetries (equivariance) of perturbations. (b) Polyhedral complexes model the structure of the subadditivity slack function $\Delta\pi \colon \mathbb{R}^2 \to \mathbb{R}$.

For a set B of breakpoints, assumed to be $\mathbb{Z}$-periodic, define $\Delta\mathcal{P}_B$ as the $\mathbb{Z}^2$-periodic polyhedral complex resulting by partitioning the plane $\mathbb{R}^2$ by the arrangement of lines $x \in B$, $y \in B$ and $x + y \in B$. This complex defines convex polygons on which $\Delta\pi$ is affine [2,9]. When $B = \frac{1}{q}\mathbb{Z}$, then $p_1 \colon (x, y) \mapsto x$, $p_2 \colon (x, y) \mapsto y$, and $p_3 \colon (x, y) \mapsto x + y$ project all vertices of $\Delta\mathcal{P}_B$ back to the set B; we have *stabilization of breakpoints* due to unimodularity. Going to higher dimension, piecewise linear functions defined on a standard triangulation of $\mathbb{R}^2$ studied in the IPCO 2013 paper [1] and [3,5] also stabilize. However, the non-existence of triangulations with stabilization for $\mathbb{R}^k$, $k \geq 3$ [8] has blocked the path for further generalizations of the approach of [2].

In the present paper, we develop a foundational theory for grid-free algorithms for $k = 1$, paving a new way for generalizations. Under the assumption of a *finitely presented moves closure* (Assumption 4.2), we are able to provide in Theorem 4.9 a "dynamic" stabilization result. It depends on more detailed data of the function than the group G generated by B. In fact, we replace the use of group actions in [2] with a new approach using inverse semigroup actions, which are a better model for the *partial symmetries* that perturbation functions satisfy. We develop a rich theory of these inverse semigroups in Sect. 3. Already for $k = 1$, the grid-free approach of the present paper extends the practical reach of algorithms, enabling computations that are less sensitive to the size of denominators in the input, as well as computations with irrational input.

Moreover, while the earlier result by Basu et al. [2] only guarantees to construct a piecewise linear effective perturbation (when the space is nontrivial), we provide a precise description of the space as a direct sum of certain finite- and infinite-dimensional subspaces (Theorems 4.15 and 4.16). Each of the latter is isomorphic to the space of Lipschitz functions supported on a compact interval that vanish on the boundary, extended equivariantly according to the action of the inverse semigroup. (This also improves upon a coarser decomposition result limited to the grid case [4, Theorem 3.14].) The precise description of the perturbation space of a non-extreme minimal function π enables new lifting algorithms for cutting planes. Given a minimal valid function, we can strengthen it by following improving effective perturbations. By our theorem, the problem of finding such a direction decomposes into subproblems; one finite-dimensional, the others independent variational problems over Lipschitz functions.

2   Functional Equations, Move Ensembles, Equivariance

We begin with the following standard observation.

Lemma 2.1. *Let π be a minimal valid function. If additivity $(\Delta\pi(x,y) = 0)$ holds for some (x,y), then the subadditivity (1b) of minimal valid functions implies that also $\Delta\tilde{\pi}(x,y) = 0$ holds for every effective perturbation $\tilde{\pi} \in \tilde{\Pi}^\pi$.*

Because π is assumed to be piecewise linear, the *set of additivities* $E(\pi) = \{(x,y) \mid \Delta\pi(x,y) = 0\}$ can be structured ("combinatorialized") using the polyhedral complex $\Delta\mathcal{P}_B$ introduced in Sect. 1; see also [2,9]. Setting aside the 0-dimensional additivities (vertices of $\Delta\mathcal{P}_B$, which come back into play in Sect. 4), we represent the remaining additivities by one-parameter families of two types. In the first type, we have

$$\Delta\tilde{\pi}(x,t) = 0 \quad \Leftrightarrow \quad \tilde{\pi}(x) = \tilde{\pi}(x+t) - \tilde{\pi}(t), \quad \text{for } x \in D, \qquad (3)$$

where D is an open interval and $t \in \mathbb{R}$ is a point. When $\tilde{\pi}(t) = 0$ we say that $\tilde{\pi}$ is *invariant* under the action of the *restricted translation move* $\tau_t|_D : D \to D + t$, $x \mapsto x + t$. A second type of one-parameter families of additivities is

$$\Delta\tilde{\pi}(x, r-x) = 0 \quad \Leftrightarrow \quad \tilde{\pi}(x) = -\tilde{\pi}(r-x) + \tilde{\pi}(r), \quad \text{for } x \in D. \qquad (4)$$

Here a negative sign enters in the relation of values at x and $r - x$. We define the *restricted reflection move* $\rho_r|_D : D \to r - D$, $x \mapsto r - x$. By assigning a *character* $\chi(\tau_t) = +1$ and $\chi(\rho_r) = -1$ to the translations and reflections, we can unify these equations as

$$\tilde{\pi}(x) = \chi(\gamma)\,\tilde{\pi}(\gamma(x)) + c^{\tilde{\pi}}_{\gamma|_D} \quad \text{for } x \in D, \qquad (5)$$

where γ is either a translation or a reflection and $c^{\tilde{\pi}}_{\gamma|_D}$ is some constant. We then say that $\tilde{\pi}$ is *equivariant* under the action of $\gamma|_D$ when $c^{\tilde{\pi}}_{\gamma|_D} = 0$. Otherwise, we say that $\tilde{\pi}$ is *affinely equivariant* under the action of $\gamma|_D$.

Definition 2.2 (Initial move ensemble). *We collect all moves $\tau_t|_D$ and $\rho_r|_D$ that arise in this way from Lemma 2.1 in a set Ω^0, which we call the* initial move ensemble. *(For convenience, we also add the inverses of these moves, as well as the empty moves (translations and reflections with domain $\emptyset$) to Ω^0.)*

Definition 2.3. *Whenever we use the notation $\gamma|_D$ in our paper, γ will be a translation or reflection, and the domain $D \subseteq \mathbb{R}$ will be an open interval (or $\emptyset$). These restricted moves form an inverse semigroup [13], which we denote by $\Gamma^{\subseteq}(\mathbb{R})$, with respect to the operations composition $\circ$ and inverse $\gamma|_D{}^{-1} = \gamma^{-1}|_{\gamma(D)}$. Subsets of $\Gamma^{\subseteq}(\mathbb{R})$ are called* (move) ensembles.

Note that the inverse of a move is *not* an inverse in a group-theoretic sense: The compositions $\gamma|_D \circ (\gamma|_D)^{-1} = \tau_0|_{\gamma(D)}$ and $(\gamma|_D)^{-1} \circ \gamma|_D = \tau_0|_D$ are only *partial identities* (restrictions of the identity τ_0 to intervals) and therefore not neutral elements but merely idempotents. This illustrates that inverse semigroup actions, and the symmetries they describe, are more general than group actions.

Definition 2.4. *Let Ω be a move ensemble and let $\tilde{\pi}: \mathbb{R} \to \mathbb{R}$ be a function. We say that $\tilde{\pi}$ is* affinely Ω-equivariant *(in short, $\tilde{\pi}$ respects Ω) provided that for every $\gamma|_D \in \Omega$ there exists a constant $c^{\tilde{\pi}}_{\gamma|_D}$ such that (5) holds.*

3   Inverse Semigroup Actions and Closures

Our strategy for computing the space of effective perturbations is to consider the properties of the following ensemble containing Ω^0, for $\tilde{\pi} \in \tilde{\Pi}^\pi$.

Definition 3.1. *For a function $\tilde{\pi}: \mathbb{R} \to \mathbb{R}$, we denote the ensemble of moves under whose action $\tilde{\pi}$ is affinely equivariant as*

$$\Gamma^{\mathrm{resp}}(\tilde{\pi}) = \left\{ \gamma|_D \in \Gamma^{\subseteq}(\mathbb{R}) \mid \exists c^{\tilde{\pi}}_{\gamma|_D} \in \mathbb{R} \text{ s.t. (5) holds} \right\}.$$

The ensemble $\Gamma^{\mathrm{resp}}(\tilde{\pi})$ satisfies various closure properties. We can enlarge the known initial moves ensemble Ω^0 by forming its closure under the same properties. Below we discuss these properties, starting with the algebraic and order-theoretic axioms. That $\Gamma^{\mathrm{resp}}(\tilde{\pi})$ satisfies them follows directly from (5).

Definition 3.2. *A move ensemble Γ is a* move semigroup *(i.e., an inverse subsemigroup of $\Gamma^{\subseteq}(\mathbb{R})$) if it satisfies the following axioms (closure properties):*

$$\gamma'|_{D'} \circ \gamma|_D \in \Gamma \quad \text{for all} \quad \gamma|_D, \gamma'|_{D'} \in \Gamma, \qquad \text{(composition)}$$

$$(\gamma|_D)^{-1} \in \Gamma \quad \text{for all} \quad \gamma|_D \in \Gamma. \qquad \text{(inv)}$$

For a move ensemble Ω, the move semigroup isemi(Ω) *generated by Ω is the smallest move semigroup containing Ω.*

Definition 3.3. *The* joined ensemble join(Ω) *of an ensemble Ω is the smallest set $\Omega^\vee \supseteq \Omega$ that is* join-closed, *i.e., satisfies*

$$\gamma|_D \in \Omega^\vee \quad \text{if} \quad \exists \{\gamma|_{D^i}\}_{i \in \mathfrak{I}} \subseteq \Omega^\vee \text{ such that } D \subseteq \bigcup_{i \in \mathfrak{I}} D^i. \qquad \text{(join)}$$

In particular, join-closed ensembles $\Omega^\vee$ are closed under restrictions: For $\gamma|_D \in \Omega^\vee$ and an open interval (or empty set) $D' \subseteq D$, we have $\gamma|_{D'} \in \Omega^\vee$. The same is true for the *corestrictions* $_{I'}|(\gamma|_D) = \gamma|_{D \cap \gamma^{-1}(I')}$ to intervals I'. A join-closed ensemble $\Omega^\vee$ has a set $\mathrm{Max}(\Omega^\vee)$ of maximal elements (in the restriction partial order), such that every element is a restriction of a maximal element. The initial ensemble Ω^0 of π already satisfies (join) and (inv), but it is not a semigroup. Next we introduce a topological version of axiom (join); it is satisfied by Ω^0 and $\Gamma^{\mathrm{resp}}(\tilde{\pi})$ because π and $\tilde{\pi} \in \tilde{\Pi}^\pi$ are continuous functions.

Definition 3.4. *The* extended move ensemble extend(Ω) *of a move ensemble Ω is the smallest set $\overline{\Omega}^\vee \supseteq \Omega$ that is* extension-closed, *i.e., satisfies the following:*

$$\gamma|_D \in \overline{\Omega}^\vee \quad \text{if} \quad \exists \{\gamma|_{D^i}\}_{i \in \mathfrak{I}} \subseteq \overline{\Omega}^\vee \text{ such that } D \subseteq \mathrm{cl}(\bigcup_{i \in \mathfrak{I}} D^i). \qquad \text{(extend)}$$

The following closure property also holds because of continuity.

Definition 3.5. *The* limits closure lim(Ω) *of a moves ensemble Ω is the smallest moves ensemble $\bar{\Omega} \supseteq \Omega$ that is* limits-closed, *i.e., satisfies the following axiom, for open intervals $D \subseteq \mathbb{R}$.*

$$\text{If } \gamma^i \to \gamma \text{ (as affine functions) and } \gamma^i|_D \in \bar{\Omega} \text{ for all } i, \text{ then } \gamma|_D \in \bar{\Omega}. \qquad \text{(lim)}$$

The final closure property comes in from the theory of functional equations. For an open set $O \subseteq \mathbb{R}^2$, define the (join-closed) move ensemble moves(O) as the set of $\gamma|_D$ whose graph $\mathrm{Gr}(\gamma|_D)$ is contained in O. Using the fact that $\tilde{\pi}$ is a bounded function, it follows from [4, Theorem 4.3] that $\Gamma^{\mathrm{resp}}(\tilde{\pi})$ satisfies:

Definition 3.6. *A move ensemble $\Omega^\boxtimes$ is a* kaleidoscopic join-closed ensemble *if it satisfies* (join) *and the following axiom, for open intervals $D, I \subseteq \mathbb{R}$:*

$$\{\tau_t|_{D'} \in \mathrm{moves}(D \times I)\} \subseteq \Omega^\boxtimes \quad \text{iff} \quad \{\rho_r|_{D'} \in \mathrm{moves}(D \times I)\} \subseteq \Omega^\boxtimes. \quad \text{(kaleido)}$$

(In fact, if $\tilde{\pi}$ respects moves$(D \times I) \subseteq \Omega^\boxtimes$, then by [4, Theorem 4.3], it follows that $\tilde{\pi}$ is affine on D and I, with the same slope. We say that $D \cup I$ is a *connected covered component* of $\Omega^\boxtimes$.) We now arrive at the main definition.

Definition 3.7. *A* closed move semigroup *is a* limits-closed extension-closed kaleidoscopic join-closed move semigroup, *i.e., a move ensemble that satisfies all the following axioms:* (composition), (inv), (join), (extend), (lim), *and* (kaleido).

Definition 3.8. *We define the* closed move semigroup clsemi(Ω) *generated by a move ensemble Ω (or just* moves closure *of Ω) to be the smallest (by set inclusion) closed move semigroup containing Ω.*

Theorem 3.9. (Moves closure). *Suppose θ is bounded and continuous. If θ is affinely Ω-equivariant, then θ is affinely* clsemi(Ω)*-equivariant.*

Corollary 3.10. *Let π be a continuous piecewise linear minimal valid function. Then π is affinely* clsemi(Ω^0)*-equivariant. Every effective perturbation function $\tilde{\pi} \in \tilde{\Pi}^\pi$ is also affinely* clsemi(Ω^0)*-equivariant.*

4   Decomposition Theorem

Assumption 4.1. $\pi \colon \mathbb{R} \to \mathbb{R}$ *is a minimal valid function that is continuous piecewise linear with breakpoints B, so $B \cap [0,1]$ is finite. We assume that there is no redundant breakpoint (B is minimal).*

4.1   Assumption: Finitely Presented Moves Closure

Let Ω^0 be the initial additive move ensemble of π (Sect. 2).

Assumption 4.2. *The moves closure* $\mathrm{clsemi}(\Omega^0)$ *has a finite presentation by moves and components, i.e., there exists a finite set Ω and a finite set $\mathcal{C} = \{C_1, C_2, \ldots, C_k\}$ of connected covered components C_i, each of which is a finite union of open intervals, such that*

$$\mathrm{clsemi}(\Omega^0) = \mathrm{jmoves}(\Omega, \mathcal{C}) := \mathrm{join}(\Omega \cup \textstyle\bigcup_{i=1}^{k} \mathrm{moves}(C_i \times C_i)).$$

Without loss of generality, the components C_i can be taken as maximal and pairwise disjoint. (Fig. 1 (right) shows an example.)

Assumption 4.2 holds in the following case.

Theorem 4.3 (Finite presentation of the moves closure, rational case). *Let π be a piecewise linear function whose breakpoints are rational, i.e., $B \subseteq G = \frac{1}{q}\mathbb{Z}$ for some $q \in \mathbb{N}$. Then* $\mathrm{clsemi}(\Omega^0)$ *has a finite presentation $(\Omega, \mathcal{C})$, where (i) the endpoints of all domains and the values t and r of moves $\tau_t, \rho_r|_D \in \Omega$ lie in $G \cap [0,1]$, (ii) the endpoints of all maximal intervals of all $C_i \in \mathcal{C}$ lie in $G \cap [0,1]$.*

The proof of Theorem 4.3 can be made fully algorithmic (and grid-free). We can compute $\mathrm{clsemi}(\Omega^0)$ using a completion algorithm that manipulates finite presentations, maintaining properties (i) and (ii), using only the algebraic and order-theoretic axioms and (extend). There are only finitely many finite presentations satisfying (i) and (ii); this implies the finiteness of the algorithm.

The following result, applied to the polygons in $E(\pi)$ on which $\Delta\pi = 0$ (the additive two-dimensional faces), provides the initialization.

Proposition 4.4 (Patching lemma; akin to [5, Lemma 2.7]). *Let $O \subseteq \mathbb{R}^2$ be a connected open set. Let $D = \{x \mid (x,y) \in O\}$ and $I = \{y \mid (x,y) \in O\}$. Then* $\mathrm{join}(\mathrm{isemi}(\mathrm{moves}(O))) = \mathrm{moves}((D \cup I) \times (D \cup I)).$

However, finitely presented closures $\mathrm{clsemi}(\Omega^0)$ as required in Assumption 4.2 arise in a more general setting than the rational case of Theorem 4.3, through the interplay of the entire list of closure properties from Sect. 3. Our key theorem using the analytic properties is the following: Rectangles appear in the closure whenever there is a convergent sequence of moves.

Theorem 4.5 (Limits imply components). *Let $\bar{\Gamma}^{\boxtimes}$ be a limits-closed kaleidoscopic join-closed move semigroup. Assume that $\gamma|_D$ is the limit move of a sequence $\{\gamma^i|_D\}_{i \in \mathbb{N}}$ of moves in $\bar{\Gamma}^{\boxtimes}$ with $\gamma^i \neq \gamma$ for every i. Let $I = \gamma(D)$. Then* $\mathrm{moves}((D \cup I) \times (D \cup I)) \subseteq \bar{\Gamma}^{\boxtimes}.$

This is related to an observation in [2] regarding the function bhk_irrational ⟍⟍. Our Theorem 4.5 is much more general and does not require the specific arithmetic context of [2]. (It is also key to the proof of Theorem 4.8 below.)

Empirically, for all families of piecewise linear minimal valid functions in the literature (see [10] for an electronic compendium), even if the breakpoints are irrational, the closure $\mathrm{clsemi}(\Omega^0)$ has a finite presentation.

Assumption 4.6. *We are given a finite presentation $(\Omega, \mathcal{C})$ of $\mathrm{clsemi}(\Omega^0)$ in reduced form, i.e., each move $\gamma|_D \in \Omega$ is maximal in $\mathrm{clsemi}(\Omega^0)$, and the graph $\mathrm{Gr}(\gamma|_D)$ is not covered by the union of open rectangles $C_i \times C_i$, $i = 1, \dots, k$.*

The finite presentation shown in Fig. 1 (right) has this property.

4.2 Properties of the Finitely Presented Moves Closure

Let $\mathcal{C} := C_1 \cup C_2 \cup \dots \cup C_k$ denote the open set of points in $(0, 1)$ that are covered. We will refer to the open set $U := (0, 1) \setminus \mathrm{cl}(\mathcal{C})$ as the set of points in $(0, 1)$ that are *uncovered*. Then we have the partition $[0, 1] = \mathcal{C} \cup X \cup U$, where

$$X := \{0\} \cup \partial \mathcal{C} \cup \{1\} = \{0\} \cup \partial U \cup \{1\}. \qquad (6)$$

(In Fig. 1, $X = \{0, \frac{1}{24}, \frac{1}{8}, \frac{1}{6}, \frac{1}{4}, \dots, 1\}$.) Recall from Sect. 2 that we set aside the 0-dimensional additivities (vertices of $\Delta \mathcal{P}_B$). Now they come back. Let

$$V := \big\{\, p_i(x, y) \mid (x, y) \in \mathrm{vert}(\Delta \mathcal{P}_B), \ \Delta \pi(x, y) = 0, \ i = 1, 2, 3 \,\big\} \cap [0, 1]. \qquad (7)$$

We define the *orbit* of $V \cap U$ under Ω, a finite set by Assumptions 4.1/4.2/4.6,

$$Y := \Omega(V \cap U) = \big\{\, \gamma|_D(x) \mid x \in V \cap U, \ x \in D \text{ and } \gamma|_D \in \Omega \,\big\}. \qquad (8)$$

(In Fig. 1, $V \cap U = \{\frac{1}{3}, \frac{5}{12}, \frac{1}{2}, \frac{7}{12}\}$. This set is already closed under the action of Ω, as $\rho_{11/12}(\frac{1}{3}) = \frac{7}{12}$ and $\rho_{11/12}(\frac{5}{12}) = \frac{1}{2}$. Thus $Y = V \cap U$ in the example.)

Remark 4.7. It follows from the closure properties that the orbit stays within U. More generally, consider the ensemble $\Omega|_U$, consisting of the restrictions of the moves of Ω to U. Then the images of all restricted moves stay in U as well. In other words, (a) the restriction $\Omega|_U$ equals the double restriction $_U|\Omega|_U$, (b) $\Omega|_U$ is a finite move ensemble.

By Assumptions 4.1/4.2/4.6, all elements of Ω are maximal in $\mathrm{clsemi}(\Omega^0)$. It follows that all elements of $\Omega|_U$ are maximal in $\mathrm{clsemi}(\Omega^0)|_U$.

After these preliminaries, we are able to state the main theorem.

Theorem 4.8 (Structure and generation theorem for finitely presented moves closures). *Under Assumptions 4.1/4.2/4.6,*

(a) $\mathrm{clsemi}(\Omega^0) = \mathrm{extend}(\mathrm{clsemi}(\Omega^0|_U) \cup \mathrm{clsemi}(\Omega^0|_{\mathcal{C}}))$.
(b) $\Omega|_U = \mathrm{Max}(\mathrm{extend}(\mathrm{isemi}(\Omega^0|_U)))$.
(c) $a, b, \gamma(a), \gamma(b) \in X \cup Y$ *for any* $\gamma|_{(a,b)} \in \Omega|_U$.

We emphasize that the theorem does *not* depend on an algorithm to compute the moves closure. The proof appears in the appendix.

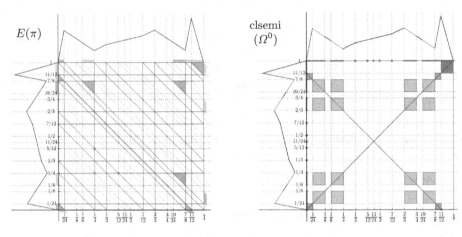

Fig. 1. The function π = equiv7_example_xyz_2() provided by the software [12] (*graph at the left and top borders of both diagrams*). *Left*, its complex $\Delta\mathcal{P}$ (*solid gray lines*) with the additive faces forming $E(\pi)$ *colored in green*. *Right*, the moves closure clsemi(Ω^0) of π, as computed by the command igp.equiv7_mode = True; igp.extremality_test(igp.equiv7_example_xyz_2(), True, show_all_perturbations=True). It has a finite presentation by $\Omega = \{\tau_0|_{(0,1)}, \rho_{11/12}|_{(0,11/12)}\}$ (*blue and red line segments of slopes ± 1*) and a set $\mathcal{C} = \{C_1, C_2, C_3\}$ of (maximal) connected covered components $C_1 = (\frac{11}{12}, 1)$ (the *lavender square* shows $C_1 \times C_1$), $C_2 = (0, \frac{1}{24}) \cup (\frac{7}{8}, \frac{11}{12})$ (*coral*), and $C_3 = (\frac{1}{24}, \frac{1}{8}) \cup (\frac{1}{6}, \frac{1}{4}) \cup (\frac{2}{3}, \frac{3}{4}) \cup (\frac{19}{24}, \frac{7}{8})$ (*lime*). The set $C \cup B' = C \cup X \cup Y \cup Z$ of covered points and refined breakpoints is marked in *magenta on the left and top borders*. (Color figure online)

4.3   Refined Breakpoints B', Finite-Dimensional Perturbations

In addition to the finite sets X and Y, we define

$$Z := \{\, x \mid x \in U, \ x = \rho|_D(x) \text{ for some reflection move } \rho|_D \in \Omega \,\}, \qquad (9)$$

the set of *uncovered character conflicts*. (In Fig. 1, $Z = \{\frac{11}{24}\}$.) Under Assumptions 4.1/4.2/4.6, the sets X, Y, Z are finite and closed under the action of clsemi(Ω^0). We then define a *refined set of breakpoints*, $B' := (X \cup Y \cup Z) + \mathbb{Z}$. Because B was chosen minimal, $B \cap C = \emptyset$ and thus $B \subseteq B'$. Hence, the polyhedral complex $\mathcal{T} := \mathcal{P}_{B'}$ is a refinement of $\mathcal{P}_B$, so π is piecewise linear over $\mathcal{T}$.

Theorem 4.9 (Breakpoint stabilization theorem). *Let (x, y) be an additive vertex of $\Delta\mathcal{T}$ with $x, y \in [0, 1]$. Let $z = x + y$. Then, $x, y, z \in B' \cup (C + \mathbb{Z})$.*

Definition 4.10. *Define the* space of finite-dimensional perturbations *as*

$$\tilde{\Pi}^\pi_{\mathcal{T}} := \{\, \tilde{\pi} \in \tilde{\Pi}^\pi \mid \tilde{\pi} \text{ is continuous piecewise linear over } \mathcal{T} \,\}. \qquad (10)$$

Lemma 4.11. *We have $\tilde{\pi}_{\mathcal{T}} \in \tilde{\Pi}^\pi_{\mathcal{T}}$ if and only if $\tilde{\pi}_{\mathcal{T}}$ is continuous piecewise linear over $\mathcal{T}$ and satisfies the following conditions.*

(i) $\tilde{\pi}_T(0) = 0$ *and* $\tilde{\pi}_T(f) = 0$*;*
(ii) $\tilde{\pi}_T$ *is* $\mathbb{Z}$*-periodic;*
(iii) *For any vertex* (x, y) *of* ΔT, $\Delta\pi(x, y) = 0$ *implies* $\Delta\tilde{\pi}_T(x, y) = 0$.

4.4   Connected Uncovered Components U_i, Equivariant Perturbations

Definition 4.12. *Define the* space of equivariant perturbations

$$\tilde{\Pi}^\pi_{\text{zero}(T)} := \{\, \tilde{\pi} \in \tilde{\Pi}^\pi \mid \tilde{\pi}(t) = 0,\ \forall t \in B' \,\}.$$

Definition 4.13. *Define the set* $U' := U \setminus B'$ *of refined uncovered points. Consider the ensemble* $\Omega|_{U'}$ *of maximal moves restricted to* U'. *Partition* U' *into the (maximal) connected uncovered components* $\{U_1, \ldots, U_l\}$, *as follows. Each component* U_i *is a maximal subset of* U' *that is the disjoint union of the uncovered intervals* $I_1, \ldots, I_p \subseteq U'$ *such that any two* I_j *and* I_k *($1 \le j, k \le p$) are connected by a maximal move* $\gamma|_{I_k} \in \Omega|_{U'}$ *with domain* I_k *and image* $I_j = \gamma(I_k)$. *Pick* $D_i \in \{I_1, \ldots, I_p\}$ *arbitrarily as the* fundamental domain, *and write* $I_j = \gamma_{i,j}(D_i)$ *where* $\gamma_{i,j}|_{D_i} \in \Omega|_{U'}$ *for* $j = 1, \ldots, p$.

This is well-defined; the ensemble $\Omega|_{U'}$ only has moves $\gamma|_D$ whose domain D and image $\gamma(D)$ are both contained in the same U_i, for $i = 1, 2, \ldots, l$. Each component $U_i \subseteq U'$ can be written as $U_i = \bigcup \gamma_{i,j}(D_i)$.

Theorem 4.14 (Characterization of the equivariant perturbations supported on an uncovered component). *Under Assumptions 4.1/4.2/4.6 and notation of Definition 4.13, let* $i \in \{1, \ldots, l\}$ *and let* $\tilde{\pi}_i \colon \mathbb{R} \to \mathbb{R}$ *be a* $\mathbb{Z}$*-periodic function such that* $\tilde{\pi}_i(x) = 0$ *for* $x \notin U_i$. *Then* $\tilde{\pi}_i \in \tilde{\Pi}^\pi_{\text{zero}(T)}$ *if and only if*

(i) $\tilde{\pi}_i$ *is Lipschitz continuous on* $\text{cl}(D_i)$;
(ii) $\tilde{\pi}_i(x) = 0$ *for* $x \in \partial D_i$;
(iii) $\tilde{\pi}_i(x) = \chi(\gamma_{i,j})\tilde{\pi}_i(\gamma_{i,j}(x))$ *for* $x \in D_i$, $j = 1, \ldots, p$.

For $i = 1, \ldots, l$, denote the space of functions $\tilde{\pi}_i$ as in the theorem by $\tilde{\Pi}^\pi_{U_i}$. It is independent of the choice of fundamental domain.

Theorem 4.15 (Direct sum decomposition of equivariant perturbations by uncovered components). *We have the direct sum decomposition* $\tilde{\Pi}^\pi_{\text{zero}(T)} = \tilde{\Pi}^\pi_{U_1} \oplus \cdots \oplus \tilde{\Pi}^\pi_{U_l}$, *i.e., if* $\tilde{\pi} \in \tilde{\Pi}^\pi_{\text{zero}(T)}$, *then it has a unique decomposition* $\tilde{\pi} = \tilde{\pi}_1 + \tilde{\pi}_2 + \cdots + \tilde{\pi}_l$ *such that* $\tilde{\pi}_i \in \tilde{\Pi}^\pi_{U_i}$ *for* $i = 1, \ldots, l$.

Each $\tilde{\pi}_i$ ($i = 1, 2, \ldots, l$) satisfies the conditions in Theorem 4.14, and thus is supported on the connected uncovered component U_i and is obtained by choosing an arbitrary Lipschitz continuous template on the fundamental domain D_i, then by extending equivariantly to the other intervals through the moves in $\Omega|_{U'}$.

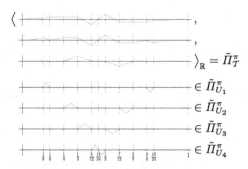

Fig. 2. Direct sum decomposition of the space $\tilde{\Pi}^\pi$ of effective perturbations for the function π from Fig. 1. *Top,* basis of the space $\tilde{\Pi}_{\mathcal{T}}$ of finite-dimensional perturbations. *Bottom,* representatives of the equivariant perturbation spaces $\tilde{\Pi}^\pi_{U_i}$ for the 4 connected uncovered components U_i.

4.5   Decomposition Theorem for Effective Perturbations

The following main theorem generalizes [4, Lemma 3.14] to our setting, which does not require assuming $B' = \frac{1}{q}\mathbb{Z}$. Figure 2 illustrates the decomposition.

Theorem 4.16 (Perturbation decomposition theorem). *Under Assumptions 4.1/4.2/4.6, we have the direct sum decomposition $\tilde{\Pi}^\pi = \tilde{\Pi}^\pi_{\mathcal{T}} \oplus \tilde{\Pi}^\pi_{\mathrm{zero}(\mathcal{T})}$, i.e., for an effective perturbation $\tilde{\pi} \in \tilde{\Pi}^\pi$, there exists a unique pair $\tilde{\pi}_{\mathcal{T}} \in \tilde{\Pi}^\pi_{\mathcal{T}}$ and $\tilde{\pi}_{\mathrm{zero}(\mathcal{T})} \in \tilde{\Pi}^\pi_{\mathrm{zero}(\mathcal{T})}$ such that $\tilde{\pi} = \tilde{\pi}_{\mathcal{T}} + \tilde{\pi}_{\mathrm{zero}(\mathcal{T})}$.*

Proof. Let $\tilde{\pi} \in \tilde{\Pi}^\pi$ be an effective perturbation. Let $\tilde{\pi}_{\mathcal{T}}$ be the unique continuous piecewise linear function over $\mathcal{T}$ such that $\tilde{\pi}_{\mathcal{T}}(t) = \tilde{\pi}(t)$ for every $t \in B'$. Define $\tilde{\pi}_{\mathrm{zero}(\mathcal{T})} = \tilde{\pi} - \tilde{\pi}_{\mathcal{T}}$. Note that $\tilde{\pi}_{\mathcal{T}}$ is the unique continuous piecewise linear function over $\mathcal{T}$ such that $\tilde{\pi}_{\mathrm{zero}(\mathcal{T})}(t) = 0$ for every $t \in B'$. It is left to show that $\tilde{\pi}_{\mathcal{T}}, \tilde{\pi}_{\mathrm{zero}(\mathcal{T})} \in \tilde{\Pi}^\pi$. We first show that $\tilde{\pi}_{\mathcal{T}} \in \tilde{\Pi}^\pi$, by applying Lemma 4.11. It suffices to show that $\tilde{\pi}_{\mathcal{T}}$ satisfies condition (iii) of Lemma 4.11. Let (x, y) be an additive vertex of $\Delta\mathcal{T}$. By Lemma 2.1, $\Delta\pi(x, y) = 0$ implies that $\Delta\tilde{\pi}(x, y) = 0$. Since (x, y) is an additive vertex of $\Delta\mathcal{T}$, Theorem 4.9 implies that $x, y, z \in B' \cup C$, where $z = x + y$. We have $\tilde{\pi}_{\mathcal{T}}(t) = \tilde{\pi}(t)$ for $t = x, y$ or z, and hence $\Delta\tilde{\pi}_{\mathcal{T}}(x, y) = \Delta\tilde{\pi}(x, y) = 0$. Therefore, $\tilde{\pi}_{\mathcal{T}} \in \tilde{\Pi}^\pi$. The vector space $\tilde{\Pi}^\pi$ contains both $\tilde{\pi}$ and $\tilde{\pi}_{\mathcal{T}}$, so $\tilde{\pi}_{\mathrm{zero}(\mathcal{T})} = \tilde{\pi} - \tilde{\pi}_{\mathcal{T}} \in \tilde{\Pi}^\pi$. $\square$

Finally, we show that, up to possible finite-dimensional affine linear relations between slopes on covered intervals, the moves closure is exactly the inverse semigroup of all moves that are respected by π and its effective perturbations.

Theorem 4.17. *Under Assumptions 4.1/4.2/4.6, we have that*

$$\mathrm{clsemi}(\Omega^0)|_U = \Gamma^{\mathrm{resp}}(\pi + \tilde{\Pi}^\pi)|_U = \Gamma^{\mathrm{resp}}(\tilde{\Pi}^\pi)|_U, \tag{11}$$

where $\Gamma^{\mathrm{resp}}(\tilde{\Pi}^\pi) = \bigcap_{\tilde{\pi} \in \tilde{\Pi}^\pi} \Gamma^{\mathrm{resp}}(\tilde{\pi})$.

Remark 4.18 (Discontinuous case). We have proved all results in a more general setting that allows discontinuous piecewise linear functions that are one-sided continuous at 0. Axiom (extend) is modified to take the set of continuity of π into account. We use additional lemmas regarding the existence of limits. The space of equivariant perturbations stays the same space of Lipschitz functions; all discontinuities go into the finite-dimensional perturbations, which then consists of piecewise linear functions with possible jumps at the refined breakpoints. The results do *not* extend to the two-sided discontinuous case; see [11].

A   Some Omitted Proofs and Results

We summarize some results on the interactions of the closure axioms. While the inverse semigroup generated by a join-closed ensemble is *not* automatically join-closed, the following holds: Applying (join) or the stronger (extend) to an inverse semigroup preserves the inverse semigroup properties,

$$\mathrm{join}(\mathrm{isemi}(\varOmega)) = \mathrm{isemi}(\mathrm{join}(\mathrm{isemi}(\varOmega))), \tag{12a}$$

$$\mathrm{extend}(\mathrm{isemi}(\varOmega)) = \mathrm{isemi}(\mathrm{extend}(\mathrm{isemi}(\varOmega))). \tag{12b}$$

Also applying limit and then join to a semigroup retains the semigroup properties , i.e., $\mathrm{join}(\mathrm{lim}(\mathrm{join}(\mathrm{isemi}(\varOmega))))$ is a semigroup.

For some of the individual closure properties, *generation theorems* are available. Let $\varOmega^{\mathrm{inv}}$ be the union of the move ensemble $\varOmega$ and the inverses of all of its moves; this is clearly the smallest ensemble containing $\varOmega$ that satisfies (inv). Also $\mathrm{isemi}(\varOmega)$, defined as the smallest inverse semigroup containing $\varOmega$, is, of course, the set of all finite compositions $\gamma^k|_{D_k} \circ \cdots \circ \gamma^1|_{D_1}$ of moves $\gamma^i|_{D_i} \in \varOmega^{\mathrm{inv}}$. The joined ensemble $\mathrm{join}(\varOmega)$ and the extended ensemble $\mathrm{extend}(\varOmega)$ consist of the following moves:

$$\mathrm{join}(\varOmega) = \left\{\, \gamma|_D \mid D \subseteq C_\gamma \,\right\}, \tag{13a}$$

$$\mathrm{extend}(\varOmega) = \left\{\, \gamma|_D \mid D \subseteq \mathrm{cl}(C_\gamma) \,\right\}, \quad \text{where } C_\gamma := \bigcup\{\, I \mid \gamma|_I \in \varOmega \,\}. \tag{13b}$$

In particular, joined ensembles and extended ensembles have maximal elements in the restriction partial order, which is not true for arbitrary ensembles.

For closures with respect to more complex combinations of our axioms, no generation theorem is available. In particular, we have no generation theorem for the moves closure. It is unknown if $\mathrm{clsemi}(\varOmega)$ can be obtained by a finite number of applications of closures with respect to the individual axioms similar to (12). Instead we reason about the moves closure in an indirect way, as follows: Let $\mathbb{L}$ be the family of closed move semigroups containing $\varOmega$. Then $\mathrm{clsemi}(\varOmega) = \bigcap \mathbb{L} = \bigcap_{\varOmega' \in \mathbb{L}} \varOmega'$; this holds because of the structure of our axioms.

In addition to the various closure properties, we use the following lemma in the proof of the structure and generation theorem for finitely presented moves closures (Theorem 4.8). Recall that the initial moves ensemble $\varOmega^0$ is join-closed.

Lemma A.1 (Filtration of isemi($\Omega^0|_U$) **by word length; maximal moves).** *Under Assumptions 4.1/4.2/4.6, for* $k \in \mathbb{N}$, *let*

$$\Omega^0|_U{}^k = \left\{ \gamma^k|_{D^k} \circ \cdots \circ \gamma^1|_{D^1} \mid \gamma^i|_{D^i} \in \Omega^0|_U \text{ for } 1 \le i \le k \right\}.$$

Then $\Omega^0|_U{}^1 \subseteq \Omega^0|_U{}^2 \subseteq \ldots$ *and* isemi($\Omega^0|_U$) = $\bigcup_{k \in \mathbb{N}} \Omega^0|_U{}^k$. *For each* $k \in \mathbb{N}$, *the ensemble* $\Omega^0|_U{}^k$ *is equal to the set of restrictions of the ensemble* Max($\Omega^0|_U{}^k$) *of its maximal elements in the restriction partial order, which is a finite set. For* $\gamma|_{(a,b)} \in$ Max($\Omega^0|_U{}^k$), *we have* $a, b, \gamma(a), \gamma(b) \in X \cup Y$.

Proof of Theorem 4.8 (Structure and generation theorem)

Part (a). Let Ω' denote the right hand side of the equation in part (a). Clearly, $\Omega^0 \subseteq \Omega' \subseteq$ clsemi(Ω^0). We now show that Ω' is a closed move semigroup. By Remark 4.7-(a), we have that clsemi($\Omega^0|_U$) $\subseteq$ restrict($\Omega|_U$) $\subseteq$ moves($U \times U$), where restrict($\Omega|_U$) is the set of restrictions of moves $\gamma|_D \in \Omega$ to domains that are open subintervals of $D \cap U$ (or $\emptyset$); and clsemi($\Omega^0|_C$) = $\bigcup_{i=1}^k$ moves($C_i \times C_i$) $\subseteq$ moves($C \times C$), where the open sets U and C are disjoint. Thus, we have that clsemi($\Omega^0|_U$) $\cup$ clsemi($\Omega^0|_C$) is a move semigroup, under Assumption 4.2. It follows from (12b) that Ω' is a move semigroup that satisfies (extend). Note that for any open intervals D and I such that moves($D \times I$) $\subseteq$ clsemi(Ω^0), we have moves($D \times I$) $\subseteq$ clsemi($\Omega^0|_C$). Therefore, Ω' also satisfies (kaleido) and (lim) by Theorem 4.5. We conclude that Ω' is a closed move semigroup. Hence, part (a) holds.

Part (b). By restricting the moves ensembles on both sides of the equation in part (a) to domain U, we obtain that

$$\text{restrict}(\Omega|_U) = \text{clsemi}(\Omega^0)|_U = \text{clsemi}(\Omega^0|_U) \tag{14}$$

Next, we show that

$$\text{clsemi}(\Omega^0|_U) = \text{extend}(\text{isemi}(\Omega^0|_U)). \tag{15}$$

It follows from (12b) that extend(isemi($\Omega^0|_U$)) is a move semigroup that satisfies (extend) (and also (join)). Since

$$\text{extend}(\text{isemi}(\Omega^0|_U)) \subseteq \text{clsemi}(\Omega^0|_U) = \text{restrict}(\Omega|_U), \tag{16}$$

where the equality follows from (14), and $\Omega|_U$ is a finite move ensemble by Remark 4.7-(b), we obtain that the move semigroup extend(isemi($\Omega^0|_U$)) also satisfies (kaleido) and (lim). Therefore, extend(isemi($\Omega^0|_U$)) is a closed move semigroup which contains $\Omega^0|_U$. Since clsemi($\Omega^0|_U$) is the smallest closed move semigroup containing $\Omega^0|_U$, we have clsemi($\Omega^0|_U$) $\subseteq$ extend(isemi($\Omega^0|_U$)). Together with (16), we conclude that (15) holds. Since Ω has only maximal moves, (14) and (15) imply the equation in part (b).

Part (c). Let $\gamma|_{(a,b)} \in \Omega|_U$. By symmetry, it suffices to show that $a, b \in X \cup Y$. Consider $x = a$ or $x = b$. Part (b) implies that

$$\Omega|_U = \mathrm{Max}(\mathrm{extend}(\mathrm{join}(\mathrm{isemi}(\Omega^0|_U)))).$$

Together with (13b), we know that x is the limit of a sequence $\{x^j\}_{j \in \mathbb{N}}$, where x^j is an endpoint of the domain D^j of a move $\gamma|_{D^j} \in \mathrm{Max}(\mathrm{join}(\mathrm{isemi}(\Omega^0|_U)))$. By Lemma A.1 and (13a), for any $j \in \mathbb{N}$, we have that D^j is a maximal subinterval of $\bigcup \{ D \mid \gamma|_D \in \bigcup_{k \in \mathbb{N}} \mathrm{Max}(\Omega^0|_U{}^k) \}$. Thus for every $j \in \mathbb{N}$, there exists a sequence $\{x_k^j\}_{k \in \mathbb{N}}$ such that each x_k^j is an endpoint of the domain of a move $\gamma|_{D_k^j} \in \mathrm{Max}(\Omega^0|_U{}^k)$, and $x_k^j \to x^j$ as $k \to \infty$. We obtain that $x_k^k \to x$ as $k \to \infty$, where each $x_k^k \in X \cup Y$ by Lemma A.1. Since $X \cup Y$ is a finite discrete set under Assumption 4.2, we obtain that $x \in X \cup Y$. $\qquad\square$

References

1. Basu, A., Hildebrand, R., Köppe, M.: Equivariant perturbation in Gomory and Johnson's Infinite Group Problem: II. The unimodular two-dimensional case. In: Goemans, M., Correa, J. (eds.) IPCO 2013. LNCS, vol. 7801, pp. 62–73. Springer, Heidelberg (2013). https://doi.org/10.1007/978-3-642-36694-9_6

2. Basu, A., Hildebrand, R., Köppe, R.: Equivariant perturbation in Gomory and Johnson's infinite group problem. I. The one-dimensional case. Math. Oper. Res. **40**(1), 105–129 (2014). https://doi.org/10.1287/moor.2014.0660

3. Basu, A., Hildebrand, R., Köppe, M.: Equivariant perturbation in Gomory and Johnson's infinite group problem. IV. The general unimodular two-dimensional case, Manuscript (2016)

4. Basu, A., Hildebrand, R., Köppe, M.: Light on the infinite group relaxation I: foundations and taxonomy. 4OR **14**(1), 1–40 (2016). https://doi.org/10.1007/s10288-015-0292-9

5. Basu, A., Hildebrand, R., Köppe, M.: Equivariant perturbation in Gomory and Johnson's infinite group problem–III: foundations for the k-dimensional case with applications to k = 2. Math. Program. **163**(1), 301–358 (2017). https://doi.org/10.1007/s10107-016-1064-9

6. Gomory, R.E., Johnson, E.L.: Some continuous functions related to corner polyhedra I. Math. Program. **3**, 23–85 (1972). https://doi.org/10.1007/BF01584976

7. Gomory, R.E., Johnson, E.L.: Some continuous functions related to corner polyhedra. Math. Program. **3**, 359–389 (1972). https://doi.org/10.1007/BF01585008

8. Hildebrand, R.: On polyhedral subdivisions closed under group operations, Manuscript (2013)

9. Hong, C.Y., Köppe, M., Zhou, Y.: Equivariant perturbation in Gomory and Johnson's infinite group problem (V). Software for the continuous and discontinuous 1-row case. Optim. Methods Softw. **33**(3), 475–498 (2018). https://doi.org/10.1080/10556788.2017.1366486

10. Köppe, M., Zhou, Y.: An electronic compendium of extreme functions for the Gomory-Johnson infinite group problem. Oper. Res. Lett. **43**(4), 438–444 (2015). https://doi.org/10.1016/j.orl.2015.06.004

11. Köppe, M., Zhou, Y.: Equivariant perturbation in Gomory and Johnson's infinite group problem. VI. The curious case of two-sided discontinuous minimal valid functions. Discrete Optim. **30**, 51–72 (2018). https://doi.org/10.1016/j.disopt.2018.05.003

12. Köppe, M., Zhou, Y., Hong, C.Y., Wang, J.: Cutgeneratingfunctionology: Sage code for computation and experimentation with cut-generating functions, in particular the Gomory-Johnson infinite group problem (2019). https://github.com/mkoeppe/cutgeneratingfunctionology, version 1.3

13. Lawson, M.V.: Inverse semigroups: The theory of partial symmetries. World Scientific (1998)

14. Zhou, Y.: Infinite-dimensional relaxations of mixed-integer optimization problems, Ph.D. thesis, University of California, Davis, Graduate Group in Applied Mathematics, May 2017. https://search.proquest.com/docview/1950269648

On Compact Representations of Voronoi Cells of Lattices

Christoph Hunkenschröder$^{(\boxtimes)}$ ⓘ, Gina Reuland, and Matthias Schymura$^{(\boxtimes)}$ ⓘ

École Polytechnique Fédérale de Lausanne, 1015 Lausanne, Switzerland
{christoph.hunkenschroder,matthias.schymura}@epfl.ch,
ginareuland@gmail.com

Abstract. In a seminal work, Micciancio & Voulgaris (2010) described a deterministic single-exponential time algorithm for the Closest Vector Problem (CVP) on lattices. It is based on the computation of the Voronoi cell of the given lattice and thus may need exponential space as well. We address the major open question whether there exists such an algorithm that requires only polynomial space.

To this end, we define a lattice basis to be c-compact if every facet normal of the Voronoi cell is a linear combination of the basis vectors using coefficients that are bounded by c in absolute value. Given such a basis, we get a polynomial space algorithm for CVP whose running time naturally depends on c. Thus, our main focus is the behavior of the smallest possible value of c, with the following results: There always exist c-compact bases, where c is bounded by n^2 for an n-dimensional lattice; there are lattices not admitting a c-compact basis with c growing sublinearly with the dimension; and every lattice with a zonotopal Voronoi cell has a 1-compact basis.

Keywords: Closest Vector Problem · Lattices · Voronoi cells

1 Introduction

An n-dimensional lattice is the integral linear span of n linearly independent vectors, $\Lambda = \{Bz : z \in \mathbb{Z}^n\}$, $B \in \mathbb{R}^{d \times n}$. If not stated otherwise, we always assume $d = n$, that is, the lattice has full rank.

Two widely investigated and important problems in the Algorithmic Geometry of Numbers, Cryptography, and Integer Programming are the Shortest Vector Problem and the Closest Vector Problem. Given a lattice Λ, the Shortest Vector Problem (SVP) asks for a shortest non-zero vector in Λ. For a target vector $t \in \mathbb{R}^n$, the Closest Vector Problem (CVP) asks for a lattice vector $z^\star$ minimizing the Euclidean length $\|t - z\|$ among all $z \in \Lambda$. We will only recall some milestones of the algorithmic development, for a more detailed overview we refer to the work of Hanrot, Pujol & Stehlé [15], as well as to the more recent Gaussian Sampling Algorithms, the most recent one by Aggarwal & Stephens-Davidowitz [1].

In the 1980's, Kannan presented two algorithms solving SVP and CVP in bit-complexity $n^{\mathcal{O}(n)}$ and polynomial space [17]. Although the constants involved in

© Springer Nature Switzerland AG 2019
A. Lodi and V. Nagarajan (Eds.): IPCO 2019, LNCS 11480, pp. 261–274, 2019.
https://doi.org/10.1007/978-3-030-17953-3_20

the running time had been improved, it took roughly fifteen years until a significantly better algorithm was discovered. In 2001, Ajtai, Kumar & Sivakumar [2] gave a randomized algorithm for the Shortest Vector Problem, only taking $2^{\mathcal{O}(n)}$ time. However, in addition to the randomness, they also had to accept exponential space dependency for their improved running time. Though their algorithm is not applicable to the Closest Vector Problem in its full generality, they show in a follow-up work that for any fixed ε, it can be used to approximate CVP up to a factor of $(1 + \varepsilon)$ with running time depending on $1/\varepsilon$ [3]. These authors moreover posed the question whether randomness or exponential space is necessary for a running time better than $n^{\mathcal{O}(n)}$. It took again around a decade until this question was partially answered by Micciancio & Voulgaris [23], who obtained a deterministic $2^{\mathcal{O}(n)}$ algorithm for both problems. Their algorithm is based on computing the Voronoi cell $\mathcal{V}_\Lambda$ of the lattice, the region of all points at least as close to the origin as to any other lattice point. But as the Voronoi cell is a polytope with up to $2(2^n - 1)$ facets, the Micciancio-Voulgaris algorithm needs exponential space for storing the Voronoi cell in the worst (and generic) case. Since storing the Voronoi cell in a different, "more compact," way than by facet-description would lead to a decreased space requirement, they raise the question whether such a representation exists in general.

Our main objective is to propose such a compact representation of the Voronoi cell and to investigate its merits towards a single-exponential time and polynomial space algorithm for the CVP. As being closer to the origin than to a certain lattice vector v expresses in the inequality $2\,x^{\mathsf{T}}v \leq \|v\|^2$, the facets of $\mathcal{V}_\Lambda$ can be stored as a set $\mathcal{F}_\Lambda \subseteq \Lambda$ of lattice vectors, which are called the *Voronoi relevant vectors*, or facet vectors. We say that a basis B of a lattice Λ is *c-compact*, if each Voronoi relevant vector of Λ can be represented in B with coefficients bounded by c in absolute value. Hence, by iterating over $(2c + 1)^n$ vectors, we include the set $\mathcal{F}_\Lambda$. With $c(\Lambda)$, we denote the smallest c such that there exists a c-compact basis of Λ. As a consequence of the ideas in [23] and this notion of compactness we get (Corollary 2): Given a c-compact basis of a lattice $\Lambda \subseteq \mathbb{R}^n$, we can solve the Closest Vector Problem in time $(2c + 1)^{\mathcal{O}(n)}\,\mathrm{poly}(n)$ and polynomial space.

Thus, the crucial question is: How small can we expect $c(\Lambda)$ to be for an arbitrary lattice? If $c(\Lambda)$ is constant, then the above yields asymptotically the same running time as the initial Micciancio-Voulgaris algorithm, but uses only polynomial space. Of course, this only holds under the assumption that we know a c-compact basis of Λ. This observation has consequences for the variant of CVP with preprocessing, which we discuss in Sect. 4.

As an example of a large family of lattices, we prove in Sect. 2.3, that zonotopal lattices are as compact as possible: If the Voronoi cell of Λ is a zonotope, then $c(\Lambda) = 1$, and a 1-compact basis can even be found among the Voronoi relevant vectors. Moreover, every lattice of rank at most four has a 1-compact basis (see Corollary 1). However, starting with dimension five there are examples of lattices with $c(\Lambda) > 1$, and thus we want to understand how large this compactness constant can be in the worst case. Motivated by applications in crystallography, the desire for good upper bounds on $c(\Lambda)$ was already formulated in [10, 11], and results of Seysen [26] imply that $c(\Lambda) \in n^{\mathcal{O}(\log n)}$. We improve this to $c(\Lambda) \leq n^2$ and, on the negative side, we identify a family of lattices without a $o(n)$-compact basis (Sects. 2.1 and 2.2).

In Sect. 3, we relax the notion of a c-compact basis as follows. Denote by $\bar{c}(\Lambda)$ the smallest constant $\bar{c}$ such that there is *any* square matrix W with $\mathcal{F}_\Lambda \subseteq \{Wz : z \in \mathbb{Z}^n, \|z\|_\infty \leq \bar{c}\}$. Hence, in general, the matrix W generates a superlattice of Λ. This relaxation is motivated by the fact that, given a basis, membership to a lattice can be checked in polynomial time. Thus if $\bar{c}(\Lambda)$ is much smaller than $c(\Lambda)$, this additional check is faster than iterating over a larger set. Regarding the relaxed compactness constant we prove that for every lattice Λ, we have $\bar{c}(\Lambda) \in \mathcal{O}(n \log n)$, and that there are lattices $\Lambda \subseteq \mathbb{R}^n$ with $c(\Lambda) / \bar{c}(\Lambda) \in \Omega(n)$.

In summary, our contribution can be described as follows: If we are given a $c(\Lambda)$-compact basis of a lattice, then we can modify the algorithm of Micciancio & Voulgaris to obtain a polynomial space algorithm for CVP. In whole generality, the time complexity of this algorithm cannot be better than $n^{\mathcal{O}(n)}$, as in Kannan's work. However, we provide evidence that there are large and interesting classes of lattices, for which this improves to single-exponential time. We think that it is worth to study the proposed compactness concept further. In particular, it would be interesting to understand the size of the compactness constant for a generic lattice, and to conceive an efficient algorithm to find a c-compact basis.

An extended version of this work is available on the arXiv preprint server [16].

2   The Notion of a c-compact Basis

Given a lattice $\Lambda \subseteq \mathbb{R}^n$, its *Voronoi cell* is defined by

$$\mathcal{V}_\Lambda = \{x \in \mathbb{R}^n : \|x\| \leq \|x - z\| \text{ for all } z \in \Lambda\},$$

where $\|\cdot\|$ denotes the Euclidean norm. It consists of all points that are at least as close to the origin than to any other lattice point of Λ. The Voronoi cell turns out to be a centrally symmetric polytope having outer description $\mathcal{V}_\Lambda = \{x \in \mathbb{R}^n : 2 x^\mathsf{T} z \leq \|z\|^2 \text{ for all } z \in \Lambda\}$. A vector $v \in \Lambda$ is called *weakly Voronoi relevant* if the corresponding inequality $2 x^\mathsf{T} v \leq \|v\|^2$ defines a supporting hyperplane of $\mathcal{V}_\Lambda$, and it is called *(strictly) Voronoi relevant* if it is moreover facet-defining. Let $\mathcal{F}_\Lambda$ and $\mathcal{C}_\Lambda$ be the set of strictly and weakly Voronoi relevant vectors of Λ, respectively. The central definition of this work is the following.

Definition 1. *A basis B of a lattice Λ is called* c-compact, *if*

$$\mathcal{F}_\Lambda \subseteq \{Bz : z \in \mathbb{Z}^n, \|z\|_\infty \leq c\}.$$

Moreover, the compactness constant *of Λ is defined as*

$$c(\Lambda) = \min\{c \geq 0 : \Lambda \text{ possesses a } c\text{-compact basis}\}.$$

As discussed in the introduction, the notion of a c-compact basis provides a compact representation of the Voronoi cell $\mathcal{V}_\Lambda$, the complexity of which depends on the value of the constant c. Before we set out to study the compactness constant in detail, we offer various equivalent definitions that serve as auxiliary tools and that also provide a better understanding of the underlying concept.

To this end, let $\Lambda^* = \{y \in \mathbb{R}^n : y^\mathsf{T} z \in \mathbb{Z} \text{ for all } z \in \Lambda\}$ be the *dual lattice* of Λ, and let $K^* = \{x \in \mathbb{R}^n : x^\mathsf{T} y \leq 1 \text{ for all } y \in K\}$ be the *polar body* of a compact convex set $K \subseteq \mathbb{R}^n$ containing the origin in its interior. The basic properties we need are the following: If B is a basis of Λ, then $B^{-\mathsf{T}}$ is a basis of Λ^*, usually called the *dual basis* of B. For a matrix $A \in \mathrm{GL}_n(\mathbb{R})$ and a compact convex set K as above, we have $(AK)^* = A^{-\mathsf{T}} K^*$. We refer to Gruber's textbook [14] for details and more information on these concepts.

Lemma 1. *Let $B = \{b_1, \ldots, b_n\}$ be a basis of a lattice $\Lambda \subseteq \mathbb{R}^n$. The following are equivalent:*

(i) B is c-compact,
(ii) $c \cdot \mathrm{conv}(\mathcal{F}_\Lambda)^*$ contains the dual basis $B^{-\mathsf{T}}$ of Λ^*,
(iii) writing $B^{-\mathsf{T}} = \{b_1^\star, \ldots, b_n^\star\}$, we have $\mathcal{F}_\Lambda \subseteq \{x \in \Lambda : |x^\mathsf{T} b_i^\star| \leq c, \forall 1 \leq i \leq n\}$,
(iv) $\mathcal{F}_\Lambda \subseteq c P_B$, where $P_B = \sum_{i=1}^n [-b_i, b_i]$.

Proof. (i) $\iff$ (ii): By definition, B is c-compact if and only if $\mathcal{F}_\Lambda \subseteq \{Bz : z \in \mathbb{Z}^n, \|z\|_\infty \leq c\}$. This means that $Q = \mathrm{conv}(\mathcal{F}_\Lambda) \subseteq B[-c, c]^n$. Taking polars, we see that this is equivalent to $B^{-\mathsf{T}} \frac{1}{c} C_n^* \subseteq Q^*$, where $C_n^* = \mathrm{conv}\{\pm e_1, \ldots, \pm e_n\}$ is the standard crosspolytope. Since the columns of $B^{-\mathsf{T}}$ constitute a basis of the dual lattice Λ^*, the proof is finished.

(i) $\iff$ (iii): $B = \{b_1, \ldots, b_n\}$ is c-compact if and only if the representation $v = \sum_{i=1}^n \alpha_i b_i$ of any Voronoi relevant vector $v \in \mathcal{F}_\Lambda$ satisfies $|\alpha_i| \leq c$, for all $1 \leq i \leq n$. By the definition of the dual basis, we have $\alpha_i = v^\mathsf{T} b_i^\star$, which proves the claim.

(i) $\iff$ (iv): By definition, $\mathcal{F}_\Lambda \subseteq c P_B$ if and only if for every $v \in \mathcal{F}_\Lambda$, there are coefficients $\alpha_1, \ldots, \alpha_n \in \mathbb{R}$ such that $v = \sum_{i=1}^n \alpha_i b_i$ and $|\alpha_i| \leq c$. These coefficients are unique, and since B is a basis of Λ, they are integral, that is $\alpha_i \in \mathbb{Z}$. Thus, the inclusion we started with is equivalent to saying that B is c-compact. $\qquad\square$

Part (iv) of the above lemma shows that the compactness constant $c(\Lambda)$ is the minimum c such that $\mathcal{F}_\Lambda \subseteq c P_B$, for some basis B of Λ. In this definition, the concept has been introduced already by Engel, Michel & Senechal [11] together with the variant $\chi(\Lambda)$, where one replaces $\mathcal{F}_\Lambda$ by the larger set $\mathcal{C}_\Lambda$ of weakly Voronoi relevant vectors. Motivated by applications in crystallography, a reoccurring question posed in [10,11] is to give good upper bounds on these lattice invariants $c(\Lambda)$ and $\chi(\Lambda)$.

Results of Seysen [26] on simultaneous lattice reduction of the primal and dual lattice imply that $c(\Lambda) \leq \chi(\Lambda) \in n^{\mathcal{O}(\log n)}$. This is however the only bound that we are aware of.

2.1 A Polynomial Upper Bound

In the sequel, we occassionally need Minkowski's *successive minima* of a convex body K and a lattice Λ in $\mathbb{R}^n$. For $1 \leq i \leq n$, the ith successive minimum is defined as

$$\lambda_i(K, \Lambda) = \min\left\{\lambda \geq 0 : \lambda K \text{ contains } i \text{ linearly independent points of } \Lambda\right\}.$$

Minkowski's development of his Geometry of Numbers was centered around the study of these important lattice parameters (we refer to Gruber's handbook [14] for background). With this notion, Lemma 1(ii) provides a lower bound on the compactness constant of a given lattice. Indeed, we have $c(\Lambda) \geq \lambda_n(Q^\star, \Lambda^\star)$, where $Q = \mathrm{conv}(\mathcal{F}_\Lambda)$.

Our first result aims for an explicit upper bound on $c(\Lambda)$ only depending on the dimension of the lattice. To this end, we first need an auxiliary result.

Lemma 2. *Let Λ be a lattice with Voronoi cell $\mathcal{V}_\Lambda$. Then, $\lambda_1(\mathcal{V}_\Lambda^\star, \Lambda^\star) \leq \frac{2n}{\pi}$, that is, there is a dual lattice vector $y^\star \in \Lambda^\star$ such that $\mathcal{V}_\Lambda \subseteq \left\{x \in \mathbb{R}^n : |x^\mathsf{T} y^\star| \leq \frac{2n}{\pi}\right\}$.*

Proof. Since $\lambda_i(\mathcal{V}_\Lambda, \Lambda) = 2$, for all $1 \leq i \leq n$, this follows from the transference bound $\lambda_1(\mathcal{V}_\Lambda, \Lambda)\lambda_1(\mathcal{V}_\Lambda^\star, \Lambda^\star) \leq \frac{4n}{\pi}$ (cf. [18, Lem. (1.2)], [19, Cor. 1.6]). $\square$

Theorem 1. *For every lattice $\Lambda \subseteq \mathbb{R}^n$, there exists an n^2-compact basis.*

Proof. We prove by induction on the dimension that there exists a basis $D = \{y_1, \ldots, y_n\}$ of $\Lambda^\star$ such that $\mathcal{V}_\Lambda \subseteq \left\{x \in \mathbb{R}^n : |x^\mathsf{T} y_i| \leq \frac{1}{2}n^2, \ 1 \leq i \leq n\right\}$.

Since every Voronoi relevant vector lies in the boundary of $2\mathcal{V}_\Lambda$, its inner product with each y_i is then bounded by n^2. Hence, the basis of Λ that is dual to D is an n^2-compact basis by Lemma 1(iii).

If $n = 1$, the claimed containment is trivially true, hence let $n \geq 2$. Let y_1 be a shortest vector of $\Lambda^\star$ with respect to the norm $\|\cdot\|_{\mathcal{V}_\Lambda^\star}$. By Lemma 2, we have $\mathcal{V}_\Lambda \subseteq \left\{x \in \mathbb{R}^n : |x^\mathsf{T} y_1| \leq \frac{2n}{\pi}\right\}$. Let $\Lambda' = \Lambda \cap \{x \in \mathbb{R}^n : x^\mathsf{T} y_1 = 0\}$, and observe that the orthogonal projection $\pi : \mathbb{R}^n \to \{x \in \mathbb{R}^n : x^\mathsf{T} y_1 = 0\}$ fulfills $\pi(\Lambda^\star) = (\Lambda')^\star$, where we dualize with respect to the linear span of Λ' (cf. [20, Ch. 1]). By induction hypothesis, there is a basis $D' = \{y_2', \ldots, y_n'\}$ of $(\Lambda')^\star$, such that $\mathcal{V}_{\Lambda'} \subseteq \left\{x \in \mathbb{R}^n : x^\mathsf{T} y_1 = 0 \text{ and } |x^\mathsf{T} y_i'| \leq \frac{1}{2}(n-1)^2, \ 2 \leq i \leq n\right\}$. As $\Lambda' \subseteq \Lambda$, we have $\mathcal{V}_\Lambda \subseteq \mathcal{V}_{\Lambda'} + \mathrm{lin}\{y_1\}$. Moreover, as $(\Lambda')^\star$ is the projection of $\Lambda^\star$ along y_1, there exist $\alpha_i \in [-1/2, 1/2)$ such that $y_i = y_i' + \alpha_i y_1 \in \Lambda^\star$ for $2 \leq i \leq n$, and $D = \{y_1, \ldots, y_n\}$ is a basis of $\Lambda^\star$. Hence,

$$\mathcal{V}_\Lambda \subseteq \left\{x \in \mathbb{R}^n : |x^\mathsf{T} y_1| \leq \frac{2n}{\pi}, \ |x^\mathsf{T} y_i'| \leq \frac{1}{2}(n-1)^2, \ 2 \leq i \leq n\right\}$$
$$\subseteq \left\{x \in \mathbb{R}^n : |x^\mathsf{T} y_1| \leq \frac{2n}{\pi}, \ |x^\mathsf{T} y_i| \leq \frac{1}{2}(n-1)^2 + \frac{n}{\pi}, \ 2 \leq i \leq n\right\}$$
$$\subseteq \left\{x \in \mathbb{R}^n : |x^\mathsf{T} y_i| \leq \frac{1}{2}n^2, \ 1 \leq i \leq n\right\},$$

finishing the proof. $\square$

Remark 1. As also the weakly Voronoi relevant vectors $\mathcal{C}_\Lambda$ lie in the boundary of $2\mathcal{V}_\Lambda$, the basis from the previous proof also shows $\chi(\Lambda) \leq n^2$, for every lattice Λ.

2.2 Lattices Without Sublinearly-Compact Bases

In this part, we identify an explicit family of lattices whose compactness constant grows at least linearly with the dimension. While the pure existence of such a family also follows from Proposition 4(iii) below, the class of lattices discussed in this section also allows to discriminate between the compactness constant and a relaxed variant, which will be introduced in the next section.

For any $a \in \mathbb{N}$ and $n \in \mathbb{N}$, we define the lattice

$$\Lambda_n(a) = \{z \in \mathbb{Z}^n : z_1 \equiv \cdots \equiv z_n \mod a\}. \tag{1}$$

As the characterization of the facet vectors, as well as the proof of the following theorem is rather technical, we refer to Appendix for the details.

Theorem 2. *Let $n \in \mathbb{N}_{\geq 4}$, $a = \lceil n/2 \rceil$. Then, the lattice $\Lambda_n = \Lambda_n(a)$ has compactness constant $c(\Lambda_n) \geq \lceil \frac{n}{4} \rceil$.*

2.3 Compact Bases and Zonotopal Lattices

For the sake of brevity, we call a 1-compact basis of a lattice just a *compact basis*. A class of lattices that allow for a compact representation of their Voronoi cells are the lattices of *Voronoi's first kind*. They correspond to those lattices Λ that comprise the first reduction domain in Voronoi's reduction theory (see [28,29]). These lattices have been characterized in [7] by possessing an *obtuse superbasis*, which is a set of vectors $\{b_0, \ldots, b_n\} \subseteq \Lambda$ that generates Λ, and that fulfills the superbasis condition $b_0 + \ldots + b_n = 0$ and the obtuseness condition $b_i^\mathsf{T} b_j \leq 0$, for all $i \neq j$. Given an obtuse superbasis, for each Voronoi relevant vector $v \in \Lambda$ there is a strict non-empty subset $S \subseteq \{0, 1, \ldots, n\}$ such that $v = \sum_{i \in S} b_i$.

Proposition 1. *(i) Every lattice of Voronoi's first kind has a compact basis.*
(ii) Every lattice of rank at most three has a compact basis.
(iii) For $n \geq 4$, the checkerboard lattice $D_n = \{x \in \mathbb{Z}^n : \mathbf{1}^\mathsf{T} x \in 2\mathbb{Z}\}$ *is not of Voronoi's first kind, but has a compact basis.*
(iv) There exists a lattice $\Lambda \subseteq \mathbb{R}^5$ with $c(\Lambda) \geq 2$.

Proof. (i): Every obtuse superbasis contains in fact a compact basis. Indeed, using the representation of a Voronoi relevant vector above and writing $b_0 = -\sum_{i=1}^{n} b_i$, we get $v = \sum_{i \in S} b_i = -\sum_{i \notin S} b_i$. One of the terms does not use b_0.
(ii): Every lattice of dimension at most three is of Voronoi's first kind (cf. [7]).
(iii): Bost and Künnemann [6, Prop. B.2.6] showed that for $n \geq 4$, the lattice D_n is not of Voronoi's first kind. The set $B = \{b_1, \ldots, b_n\}$ with $b_1 = e_1 + e_n$, and $b_i = e_i - e_{i-1}$ for $2 \leq i \leq n$, is a basis of D_n. The vectors $2e_i \pm 2e_j$ are contained in $2 D_n$, for all i, j. Hence, if $\pm v$ are the unique shortest vectors in $v + 2\Lambda$, they are of the form $\{\pm(e_i \pm e_j) : 1 \leq i < j \leq n\}$. A routine calculation shows that all these vectors are a $\{-1, 0, 1\}$-combination of the basis B.
(iv): This follows immediately from Theorem 2 with the lattice $\Lambda_5(3)$.   □

We now explore to which extent these initial observations on lattices with compact bases can be generalized.

A *zonotope* Z in $\mathbb{R}^n$ is a Minkowski sum of finitely many line segments, that is, $Z = \sum_{i=1}^r [a_i, b_i]$, for some $a_i, b_i \in \mathbb{R}^n$. The vectors $b_1 - a_1, \ldots, b_r - a_r$ are usually called the *generators* of Z. We call a lattice *zonotopal* if its Voronoi cell is a zonotope. A generic zonotopal lattice has typically high combinatorial complexity. An explicit example is the root lattice A_n^*; its zonotopal Voronoi cell is generated by $\binom{n+1}{2}$ vectors and it has exactly the maximum possible $2(2^n - 1)$ facets (cf. [8, Ch. 4 & Ch. 21]).

It turns out that every lattice of Voronoi's first kind is zonotopal, but starting from dimension four, the class of zonotopal lattices is much richer (cf. Vallentin's thesis [28, Ch. 2] and [13]). In the following, we prove that every zonotopal lattice possesses a compact basis, thus extending Proposition 1(i) significantly.

Theorem 3. *Every zonotopal lattice has a compact basis. It can be found among its Voronoi relevant vectors.*

Proof. Let Λ be a zonotopal lattice in $\mathbb{R}^n$, and let $Z = \mathcal{V}_\Lambda$ be its Voronoi cell. The general idea of our proof is the following: Using Erdahl's [12] structural results on zonotopes that tile space by translation, we can find a dicing which induces the same tiling of $\mathbb{R}^n$ as the Delaunay tiling of Λ. By the duality of the Delaunay and the Voronoi tiling this provides us with additional structure that is used to identify a compact basis among the Voronoi relevant vectors. For details we refer to the Appendix. □

Our next result is in a similar spirit. It shows that if we are able to add a zonotope to a Voronoi cell and obtain a Voronoi cell again, then the compactness constant can only decrease. For its statement, we write $Z(U) = \sum_{i=1}^r [-u_i, u_i]$ for the possibly lower-dimensional zonotope spanned by the set of vectors $U = \{u_1, \ldots, u_r\}$. Recall, that $\chi(\Lambda)$ denotes the compactness constant for representing the set of weakly Voronoi relevant vectors of Λ.

Proposition 2. *Let $\Lambda \subseteq \mathbb{R}^n$ be a lattice such that its Voronoi cell admits a decomposition $\mathcal{V}_\Lambda = \mathcal{V}_\Gamma + Z(U)$, for some full-dimensional lattice Γ and vectors $U \subseteq \mathbb{R}^n$. Then, we have $\chi(\Lambda) \leq \chi(\Gamma)$.*

Proof. It suffices to prove the claim for the case $r = 1$. Indeed, if $Z(U)$ is generated by more than one generator, we just repeat the process successively. Hence, in the following we assume that $\mathcal{V}_\Lambda = \mathcal{V}_\Gamma + [-u, u]$, for some non-zero vector $u \in \mathbb{R}^n$. Dutour Sikirić et al. [9, Lem. 1 & Lem. 3] give a characterization of the weakly Voronoi relevant vectors of Λ in terms of those of Γ: First of all, there is a dual lattice vector $e_u \in \Gamma^*$ such that $\Lambda = A_u \Gamma$, where $A_u x = x + 2(e_u^\mathsf{T} x) u$, for $x \in \mathbb{R}^n$. Then, $z = A_u w \in \Lambda$ is weakly Voronoi relevant if and only if w is weakly Voronoi relevant for Γ, and $e_u^\mathsf{T} w \in \{0, \pm 1\}$.

Now, let $B = \{b_1, \ldots, b_n\}$ be a basis of Γ such that for every weakly Voronoi relevant vector $w \in \mathcal{C}_\Gamma$, we have $w = \sum_{i=1}^n \gamma_i b_i$, for some coefficients $|\gamma_i| \leq \chi(\Gamma)$. Thus, if $z = A_u w$ is weakly Voronoi relevant for Λ, then $z = \sum_{i=1}^n \gamma_i (A_u b_i)$, and $A_u B$ is a basis of Λ. As a consequence, $\chi(\Lambda) \leq \chi(\Gamma)$. □

As a corollary we settle the question on the largest possible compactness constant of a four-dimensional lattice. For the proof we refer to Appendix.

Corollary 1. *Every lattice of rank at most four has a compact basis.*

3   Relaxing the Basis Condition

The compact representation problem for the set of Voronoi relevant vectors does not need B to be a basis of the lattice Λ. In fact, it suffices that we find linearly independent vectors $W = \{w_1, \ldots, w_n\}$ that allow to decompose each Voronoi relevant vector as an integer linear combination with small coefficients, as the membership to a lattice can easily be decided by solving a system of linear equations. If the constant reduces drastically by this relaxation, the additional check is still faster.

Definition 2. *Let $\Lambda \subseteq \mathbb{R}^n$ be a lattice. A set of linearly independent vectors $W = \{w_1, \ldots, w_n\} \subseteq \mathbb{R}^n$ is called c-compact for Λ, if*

$$\mathcal{F}_\Lambda \subseteq \{w_1 z_1 + \ldots + w_n z_n : z \in \mathbb{Z}^n, \|z\|_\infty \leq c\}.$$

We define the relaxed compactness constant *of Λ as*

$$\bar{c}(\Lambda) = \min\{c \geq 0 : \text{there is a } c\text{-compact set } W \text{ for } \Lambda\}.$$

If every Voronoi relevant vector is an integral combination of W, then so is *every* lattice vector. That is, a c-compact set W for Λ gives rise to a superlattice $\Gamma = W\mathbb{Z}^n \supseteq \Lambda$. The compactness constants $\bar{c}(\Lambda)$ and $c(\Lambda)$ are related as follows.

Proposition 3. *For every lattice Λ in $\mathbb{R}^n$ and $Q = \operatorname{conv}(\mathcal{F}_\Lambda)$, we have*

$$\bar{c}(\Lambda) = \lambda_n(Q^\star, \Lambda^\star) \qquad and \qquad \bar{c}(\Lambda) \leq c(\Lambda) \leq n\,\bar{c}(\Lambda).$$

Proof. The identity $\bar{c}(\Lambda) = \lambda_n(Q^\star, \Lambda^\star)$ follows by arguments analogous to those establishing the equivalence of (i) and (ii) in Lemma 1. The inequality $\bar{c}(\Lambda) \leq c(\Lambda)$ is a direct consequence of the definition of these parameters.

By definition of the n-th successive minimum, there are linearly independent vectors $v_1, \ldots, v_n \in (\lambda_n(Q^\star, \Lambda^\star) \cdot Q^\star) \cap \Lambda^\star$. By induction on the dimension one can show that the parallelepiped $P = \sum_{i=1}^n [0, v_i]$ contains a basis of $\Lambda^\star$. Since P is contained in $n\,\lambda_n(Q^\star, \Lambda^\star) \cdot Q^\star$, the inequality $c(\Lambda) \leq n\,\bar{c}(\Lambda)$ follows. $\square$

While the relaxation to representing $\mathcal{F}_\Lambda$ by a set W rather than by lattice bases may reduce the respective compactness constant by $\mathcal{O}(n)$, there is still a class of lattices that show that in the worst case the relaxed compactness constant can be linear in the dimension as well. In combination with Theorem 2, the second part of the following result moreover shows that the factor n in Proposition 3 is tight up to a constant.

Proposition 4. *(i) For every lattice $\Lambda \subseteq \mathbb{R}^n$, we have $\bar{c}(\Lambda) \in \mathcal{O}(n \log n)$.*
(ii) For $a = \lceil \frac{n}{2} \rceil$, let $\Lambda_n = \Lambda_n(a)$ be the lattice defined in (1). For every $n \in \mathbb{N}$, we have $\bar{c}(\Lambda_n) \leq 3$, whereas $c(\Lambda_n) \geq \lceil \frac{n}{4} \rceil$, for $n \geq 4$.
(iii) There are self-dual lattices $\Lambda \subseteq \mathbb{R}^n$ with relaxed compactness constant $\bar{c}(\Lambda) \in \Omega(n)$.

Proof. (i) The polytope $Q = \operatorname{conv}(\mathcal{F}_\Lambda)$ is centrally symmetric, all its vertices are points of Λ, and $\operatorname{int}(Q) \cap \Lambda = \{0\}$. Therefore, we have $\lambda_1(Q, \Lambda) = 1$. Proposition 3 and the transference theorem of Banaszczyk [4] thus imply that there is an absolute constant $\gamma > 0$ such that

$$\bar{c}(\Lambda) = \lambda_n(Q^\star, \Lambda^\star) = \lambda_1(Q, \Lambda) \cdot \lambda_n(Q^\star, \Lambda^\star) \leq \gamma\, n \log n. \qquad (2)$$

(ii) In view of Proposition 3, we have to find n linearly independent points of $\Lambda_n^\star$ in $3\,Q^\star$. To this end, we define $y_i := \frac{1}{a}(e_i - e_n)$, for $1 \leq i \leq n - 1$. Furthermore, let $y_n = \frac{1}{a}\mathbf{1}$, if n is even, and $y_n = (\{\frac{1}{a}\}^{n-1}, \frac{2}{a})$, if n is odd. We claim that the vectors $y_1, \ldots, y_n$ do the job. They are clearly linearly independent, and since $\Lambda_n(a)^\star = \{z \in \frac{1}{a}\mathbb{Z}^n : \mathbf{1}^\mathsf{T} z \in \mathbb{Z}\}$ they belong to Λ_n. The characterization of Voronoi relevant vectors of Λ_n in Lemma 3 allows to verify $|y_i^\mathsf{T} v| \leq 3$, for all $1 \leq i \leq n$ and $v \in \mathcal{F}_{\Lambda_n}$.

(iii) Let Λ be a self-dual lattice and let $\mathcal{V}_\Lambda$ be its Voronoi cell. Each Voronoi relevant vector $v \in \mathcal{F}_\Lambda$ provides a facet of $\mathcal{V}_\Lambda$ via the inequality $v^\mathsf{T} x \leq \frac{1}{2}\|v\|^2$, as well as a facet of $Q^\star$ via the inequality $v^\mathsf{T} x \leq 1$ (this defines indeed a facet, as v is a vertex of Q – the polar of $Q^\star$). As $\|v\| \geq \lambda_1(B_n, \Lambda)$, for every $c < \lambda_1(B_n, \Lambda)^2$, we have that $c \cdot Q^\star$ is contained in the interior of twice the Voronoi cell of $\Lambda^\star = \Lambda$, and hence contains no non-trivial dual lattice point. Therefore, $\bar{c}(\Lambda) \geq \lambda_1(B_n, \Lambda)^2$.

Conway & Thompson (see [24, Ch. 2, §9]) proved that there are self-dual lattices Λ in $\mathbb{R}^n$ with minimal norm $\lambda_1(B_n, \Lambda) \geq \left\lfloor \frac{1}{\sqrt{\pi}} \left(\frac{5}{3} \Gamma \left(\frac{n}{2} + 1 \right) \right)^{\frac{1}{n}} \right\rfloor$. Stirling's approximation then gives that $\bar{c}(\Lambda) \in \Omega(n)$. $\qquad \square$

Based on the common belief that the best possible upper bound in (2) is linear in n, we conjecture that $\bar{c}(\Lambda) \in \mathcal{O}(n)$, and even $c(\Lambda) \in \mathcal{O}(n)$, for every lattice $\Lambda \subseteq \mathbb{R}^n$.

4   Algorithmic Point of View

When it comes to computing a $c(\Lambda)$-compact basis, not much is known. Lemma 1 suggests to take the polar of $\operatorname{conv}(\mathcal{F}_\Lambda)$, and then to look for a dual basis in a suitable dilate thereof. However, in order to do this, we need a description of the Voronoi relevant vectors in the first place. Therefore, we rather discuss how to incorporate an already known c-compact basis into the algorithm of Micciancio and Voulgaris [23].

Their algorithm consists of two main parts. In a preprocessing step, it computes the Voronoi cell $\mathcal{V}_\Lambda$, which can be done in time $2^{\mathcal{O}(n)}$ in a recursive manner. Given a c-compact basis B this part is immediate as B grants a superset of $\mathcal{F}_\Lambda$

by definition. Once the Voronoi cell $\mathcal{V}_\Lambda$ is computed, a vector $p \in \Lambda$ is closest to a target vector t if and only if $t - p \in \mathcal{V}_\Lambda$. In the second part, they iteratively identify a Voronoi relevant vector $v \in \mathcal{F}_\Lambda$ whose induced facet inequality $2x^\mathsf{T}v \leq \|v\|^2$ is violated by t. Replacing t by the shorter vector $t - v$ and keeping track of the successively found vectors v, yields a lattice vector $p \in \Lambda$ such that $t - p \in \mathcal{V}_\Lambda$ after finitely many steps. This technique previously known as the *iterative slicer* [27], was refined in [23] to estimate the number of necessary steps by $2^n \operatorname{poly}(n)$. More sophisticated arguments, as presented in [5] allow to further decrease the number of iterations.

Corollary 2. *Assume that we are given a c-compact basis B of a lattice $\Lambda \subseteq \mathbb{R}^n$. For any target point $t \in \mathbb{R}^n$, a closest lattice vector to t can be found in time $\mathcal{O}((2c + 1)^n \, 2^n \operatorname{poly}(n))$ and space polynomial in the input size.*

Proof. Theorem 4.2 and Remark 4.4 in [23] state that a closest vector can be found in time $\mathcal{O}(|V| \cdot 2^n \operatorname{poly}(n))$, where V is a superset of the Voronoi relevant vectors $\mathcal{F}_\Lambda$. We set $V = \{Bz : z \in \mathbb{Z}^n, \|z\|_\infty \leq c\} \supseteq \mathcal{F}_\Lambda$.

The reduction to polynomial space follows from [23, Rem. 4.3]: Their algorithm may need exponential space because they store $\mathcal{F}_\Lambda$. As a subset of V it is however described just by the polynomial-size data (B, c). $\square$

The Micciancio-Voulgaris algorithm naturally can be presented as an algorithm for the Closest Vector Problem with Preprocessing (CVPP). In this variant of CVP, we may precompute the lattice for an arbitrary amount of time and store some additional information. Only then the target vector is revealed to us, and the additional information can be used to find a closest vector faster. In practice, we might have to solve CVP on the same lattice with several target vectors, hence we might benefit from spending more time for preprocessing.

Considered in this setting, our results compress the information after the preprocessing step into polynomial space. However, it is unclear how to compute a c-compact basis *without* computing the Voronoi cell first.

Problem 1. Can we compute a basis B of Λ attaining $c(\Lambda)$ in single-exponential time and polynomial space?

McKilliam et al. [21] show that for lattices of Voronoi's first kind, CVP can be solved in polynomial time, provided an obtuse superbasis is known. One may wonder whether our representation also allows for solving CVPP faster. However, Micciancio [22] showed that if CVPP can be solved in polynomial time for arbitrary lattices, then $\mathrm{NP} \subseteq \mathrm{P/poly}$ and the polynomial hierarchy collapses.

Acknowledgments. We thank Daniel Dadush and Frank Vallentin for helpful remarks and suggestions. In particular, Daniel Dadush pointed us to the arguments in Theorem 1 that improved our earlier estimate of order $\mathcal{O}(n^2 \log n)$.

This work was supported by the Swiss National Science Foundation (SNSF) within the project *Convexity, geometry of numbers, and the complexity of integer programming (Nr. 163071)*. The paper grew out of the master thesis of the second author [25].

Appendix

Lemma 3. *Let $n \in \mathbb{N}_{\geq 4}$, $a = \lceil n/2 \rceil$, and $\Lambda_n = \Lambda_n(a)$. A vector $v \in \Lambda_n$ is Voronoi relevant if and only if $v = \pm \mathbf{1}$, or there exists $\emptyset \neq S \subsetneq \{1, \dots, n\}$ s.t.*

$$v_i = a - \ell \ (i \in S), \quad v_j = -\ell \ (j \notin S), \quad and \quad \ell \in \{\lfloor \tfrac{a|S|}{n} \rfloor, \lceil \tfrac{a|S|}{n} \rceil\}. \tag{3}$$

Proof (Sketch). Voronoi characterized a strictly Voronoi relevant vector v in a lattice Λ by the property that $\pm v$ are the only shortest vectors in the co-set $v + 2\Lambda$ (cf. [8, p. 477]). We use this crucially to show that Voronoi relevant vectors different from $\pm \mathbf{1}$ are characterized by (3).

The vectors $\pm \mathbf{1}$ are Voronoi relevant as they are shortest vectors of the lattice; if two linearly independent shortest vectors v_1, v_2 were in the same co-set $v_1 + 2\Lambda_n$, then $(v_1 + v_2)/2$ would be a strictly shorter vector. To analyze any shortest vector u of some co-set $v + 2\Lambda_n$, $v \in \Lambda_n$, we make the following two observations. First, as $2ae_i \in 2\Lambda_n$, we have $u \in [-a, a]^n$. Due to the definition of Λ_n, either $u \in \{0, \pm a\}^n$, or $u \in [-a + 1, a - 1]^n$. In the first case, if we have at least two non-zero entries, we can flip the sign of one entry and obtain a vector of the same length in the same co-set, but linearly independent. Hence, that co-set does not have any Voronoi relevant vectors. In the other case, again due to $v_i \equiv v_j \bmod a$ for any lattice vector, $u \in \{a - \ell, -\ell\}^n$ for some $1 \leq \ell < n$. Considering the norm of u as a function in ℓ and bearing in mind that $\mathbf{1} \in 2\Lambda_n$, we see that $\|u\|^2$ is minimized precisely for the choices of ℓ given in (3). Due to this line of thought, in order to show that each vector u of shape (3) is indeed Voronoi relevant, it suffices to show that any vector in $\{-a, 0, a\}^n$ is either longer than u, or in another residue class. $\qquad\square$

Proof (Theorem 2). For brevity, we write $c = c(\Lambda_n)$, $Q = \mathrm{conv}(\mathcal{F}_{\Lambda_n})$. As $\mathbf{1} \in \Lambda_n$, there exists a $w \in \Lambda_n^\star$ with $\mathbf{1}^\mathsf{T} w = 1$, implying that each basis of $\Lambda_n^\star$ contains a vector y such that $\mathbf{1}^\mathsf{T} y$ is an odd integer. In particular, by Lemma 1, we know that $cQ^\star$ contains such a y. As $Q^\star$ is centrally symmetric, assume $\mathbf{1}^\mathsf{T} y \geq 1$. Further, since $\Lambda_n^\star$ is invariant under permutation of the coordinates, we may assume that $y_1 \geq y_2 \geq \cdots \geq y_n$. Let us outline our arguments: We split $\mathbf{1}^\mathsf{T} y$ into two parts, by setting $A := \sum_{i=1}^k y_i$, and $B := \sum_{i>k}^n y_i$, where $k = \lceil n/2 \rceil$. We show that $A \geq B + 1$, and construct a Voronoi relevant vector $v \in \Lambda_n$ by choosing $S = \{1, \dots, k\}$ and $\ell = \lfloor ak/n \rfloor$. Hence, $(a - \ell), \ell \approx n/4$ and we obtain $v^\mathsf{T} y \gtrsim \tfrac{n}{4} A - \tfrac{n}{4} B \geq n/4$ by distinguishing the four cases $n \bmod 4$.

For showing $A \geq B + 1$, consider y_k. As $y \in \Lambda_n^\star$, there is an integer z such that we can write $y_k = \tfrac{z}{a}$. Note that we have $A \geq k y_k = z$ and $B \leq (n-k)\tfrac{z}{a} \leq z$. Let $\alpha, \gamma \geq 0$ such that $A = z + \alpha$ and $B = z - \gamma$. As $A + B = 2z + \alpha - \gamma$ has to be an odd integer, we have $|\alpha - \gamma| \geq 1$, implying $\alpha \geq 1$ or $\gamma \geq 1$. Therefore, in fact we have $A \geq \max\{B + 1, 1\}$. $\qquad\square$

We now give the details of the proof of Theorem 3. A *dicing* $\mathfrak{D}$ in $\mathbb{R}^n$ is an arrangement consisting of families of infinitely many equally-spaced hyperplanes with the following properties: (i) there are n families with linearly independent normal vectors, and (ii) every vertex of the arrangement is contained in a hyper-plane of each family. The vertex set of a dicing forms a lattice $\Lambda(\mathfrak{D})$. Erdahl [12, Thm. 3.1] represents a dicing $\mathfrak{D}$ as a set $D = \{\pm d_1, \ldots, \pm d_r\}$ of hyperplane normals and a set $E = \{\pm e_1, \ldots, \pm e_s\} \subseteq \Lambda(\mathfrak{D})$ of edge vectors of the arrangement $\mathfrak{D} = \mathfrak{D}(D, E)$ satisfying: (E1) Each pair of edges $\pm e_j \in E$ is contained in a line $d_{i_1}^\perp \cap \ldots \cap d_{i_{n-1}}^\perp$, for some linearly independent $d_{i_1}, \ldots, d_{i_{n-1}} \in D$, and conversely each such line contains a pair of edges; (E2) For each $1 \leq i \leq r$ and $1 \leq j \leq s$, we have $d_i^\mathsf{T} e_j \in \{0, \pm 1\}$.

Proof (Theorem 3). We start by reviewing the *Delaunay tiling* of the lattice Λ. A sphere $B_c(R) = \{x \in \mathbb{R}^n : \|x - c\|^2 \leq R^2\}$ is called an *empty sphere* of Λ (with center $c \in \mathbb{R}^n$ and radius $R \geq 0$), if every point in $B_c(R) \cap \Lambda$ lies on the boundary of $B_c(R)$. A *Delaunay polytope* of Λ is defined as the convex hull of $B_c(R) \cap \Lambda$, and the family of all Delaunay polytopes induces a tiling $\mathcal{D}_\Lambda$ of $\mathbb{R}^n$ which is the Delaunay tiling of Λ. This tiling is in fact dual to the Voronoi tiling.

Erdahl [12, Thm. 2] shows that the Voronoi cell of a lattice is a zonotope if and only if its Delaunay tiling is a dicing. More precisely, the tiling $\mathcal{D}_\Lambda$ induced by the Delaunay polytopes of Λ is equal to the tiling induced by the hyperplane arrangement of a dicing $\mathfrak{D} = \mathfrak{D}(D, E)$ with normals $D = \{\pm d_1, \ldots, \pm d_r\}$ and edge vectors $E = \{\pm e_1, \ldots, \pm e_s\}$. By the duality of the Delaunay and the Voronoi tiling, an edge of $\mathcal{D}_\Lambda$ containing the origin corresponds to a facet normal of the Voronoi cell $\mathcal{V}_\Lambda$. Therefore, the edge vectors E are precisely the Voronoi relevant vectors of Λ.

Now, choosing n linearly independent normal vectors, say $d_1, \ldots, d_n \in D$, the properties (E1) and (E2) imply the existence of edge vectors, say $e_1, \ldots, e_n \in E$, such that $d_i^\mathsf{T} e_j = \delta_{ij}$, with δ_{ij} being the Kronecker delta. Moreover, the set $B = \{e_1, \ldots, e_n\}$ is a basis of $\{x \in \mathbb{R}^n : d_i^\mathsf{T} x \in \mathbb{Z}, 1 \leq i \leq n\}$, which by property E2) equals the whole lattice Λ. Hence, $\{d_1, \ldots, d_n\}$ is the dual basis of B and every Voronoi relevant vector $v \in \mathcal{F}_\Lambda = E$ fulfills $d_i^\mathsf{T} v \in \{0, \pm 1\}$. In view of Lemma 1 (iii), this means that B is a compact basis of Λ consisting of Voronoi relevant vectors, as desired. $\qquad\square$

Proof (Corollary 1). By Proposition 1(ii), every lattice of rank ≤ 3 has a compact basis. Thus, let $\Lambda \subseteq \mathbb{R}^4$ be of full rank. If Λ is zonotopal, then by Theorem 3 $c(\Lambda) = 1$. Voronoi's reduction theory shows that if Λ is not zonotopal, then its Voronoi cell $\mathcal{V}_\Lambda$ has the 24-cell as a Minkowski summand (cf. [28, Ch. 3]). Up to isometries and scalings, the only lattice whose Voronoi cell is combinatorially equivalent to the 24-cell, is the root lattice D_4. Thus, we have a decomposition $\mathcal{V}_\Lambda = \mathcal{V}_\Gamma + Z(U)$, for some generators $U = \{u_1, \ldots, u_r\} \subseteq \mathbb{R}^4$ and a lattice Γ that is isometric to D_4. Hence, by Proposition 2, we get $c(\Lambda) \leq \chi(\Lambda) \leq \chi(\Gamma) = \chi(D_4)$. Engel et al. [11] computed that $\chi(D_4) = 1$, which finishes our proof. $\qquad\square$

References

1. Aggarwal, D., Stephens-Davidowitz, N.: Just take the average! an embarrassingly simple 2^n-time algorithm for SVP (and CVP). In: OASIcs-OpenAccess Series in Informatics, vol. 61. Schloss Dagstuhl-Leibniz-Zentrum fuer Informatik (2018). https://doi.org/10.4230/OASIcs.SOSA.2018.12
2. Ajtai, M., Kumar, R., Sivakumar, D.: A sieve algorithm for the shortest lattice vector problem. In: Proceedings of the Thirty-Third Annual ACM Symposium on Theory of Computing, pp. 601–610. ACM (2001). https://doi.org/10.1145/380752.380857
3. Ajtai, M., Kumar, R., Sivakumar, D.: Sampling short lattice vectors and the closest lattice vector problem. In: Proceedings 17th IEEE Annual Conference on Computational Complexity, pp. 53–57. IEEE (2002). https://doi.org/10.1109/CCC.2002.1004339
4. Banaszczyk, W.: Inequalities for convex bodies and polar reciprocal lattices in R^n. II. Application of K-convexity. Discrete Comput. Geom. **16**(3), 305–311 (1996). https://doi.org/10.1007/BF02711514
5. Bonifas, N., Dadush, D.: Short paths on the Voronoi graph and closest vector problem with preprocessing. In: Proceedings of the Twenty-Sixth Annual ACM-SIAM Symposium on Discrete Algorithms, pp. 295–314. SIAM, Philadelphia (2015). https://doi.org/10.1137/1.9781611973730.22
6. Bost, J.B., Künnemann, K.: Hermitian vector bundles and extension groups on arithmetic schemes I. Geometry of numbers. Adv. Math. **223**(3), 987–1106 (2010). https://doi.org/10.1016/j.aim.2009.09.005
7. Conway, J.H., Sloane, N.J.A.: Low-dimensional lattices. VI. Voronoï reduction of three-dimensional lattices. Proc. Roy. Soc. Lond. Ser. A **436**(1896), 55–68 (1992). https://doi.org/10.1098/rspa.1992.0004
8. Conway, J.H., Sloane, N.J.A.: Sphere packings, lattices and groups, Grundlehren der Mathematischen Wissenschaften. Fundamental Principles of Mathematical Sciences, 3rd edn., vol. 290. Springer, New York (1999). https://doi.org/10.1007/978-1-4757-6568-7
9. Dutour Sikirić, M., Grishukhin, V., Magazinov, A.: On the sum of a parallelotope and a zonotope. Eur. J. Combin. **42**, 49–73 (2014). https://doi.org/10.1016/j.ejc.2014.05.005
10. Engel, P.: Mathematical problems in modern crystallography. Comput. Math. Appl. **16**(5–8), 425–436 (1988). https://doi.org/10.1016/0898-1221(88)90232-5
11. Engel, P., Michel, L., Senechal, M.: New geometric invariants for Euclidean lattices. In: Réseaux euclidiens, designs sphériques et formes modulaires, Monogr. Enseign. Math., vol. 37, pp. 268–272. Enseignement Math., Geneva (2001)
12. Erdahl, R.M.: Zonotopes, dicings, and Voronoi's conjecture on parallelohedra. Eur. J. Combin. **20**(6), 527–549 (1999). https://doi.org/10.1006/eujc.1999.0294
13. Erdahl, R.M., Ryshkov, S.S.: On lattice dicing. Eur. J. Combin. **15**(5), 459–481 (1994). https://doi.org/10.1006/eujc.1994.1049
14. Gruber, P.M.: Convex and discrete geometry, Grundlehren der Mathematischen Wissenschaften. Fundamental Principles of Mathematical Sciences, vol. 336. Springer, Berlin (2007). https://doi.org/10.1007/978-3-540-71133-9
15. Hanrot, G., Pujol, X., Stehlé, D.: Algorithms for the shortest and closest lattice vector problems. In: Chee, Y.M., et al. (eds.) IWCC 2011. LNCS, vol. 6639, pp. 159–190. Springer, Heidelberg (2011). https://doi.org/10.1007/978-3-642-20901-7_10

16. Hunkenschröder, C., Reuland, G., Schymura, M.: On compact representations of Voronoi cells of lattices. https://arxiv.org/abs/1811.08532 (2018)
17. Kannan, R.: Minkowski's convex body theorem and integer programming. Math. Oper. Res. **12**(3), 415–440 (1987). https://doi.org/10.1287/moor.12.3.415
18. Kannan, R., Lovász, L.: Covering minima and lattice-point-free convex bodies. Ann. Math. (2) **128**(3), 577–602 (1988). https://doi.org/10.2307/1971436
19. Kuperberg, G.: From the mahler conjecture to gauss linking integrals. Geom. Funct. Anal. **18**(3), 870–892 (2008). https://doi.org/10.1007/s00039-008-0669-4
20. Martinet, J.: Perfect lattices in Euclidean spaces, Grundlehren der Mathematischen Wissenschaften. Fundamental Principles of Mathematical Sciences, vol. 327. Springer, Heidelberg (2003). https://doi.org/10.1007/978-3-662-05167-2
21. McKilliam, R.G., Grant, A., Clarkson, I.V.L.: Finding a closest point in a lattice of Voronoi's first kind. SIAM J. Discrete Math. **28**(3), 1405–1422 (2014). https://doi.org/10.1137/140952806
22. Micciancio, D.: The hardness of the closest vector problem with preprocessing. IEEE Trans. Inform. Theory **47**(3), 1212–1215 (2001). https://doi.org/10.1109/18.915688
23. Micciancio, D., Voulgaris, P.: A deterministic single exponential time algorithm for most lattice problems based on Voronoi cell computations. SIAM J. Comput. **42**(3), 1364–1391 (2013). https://doi.org/10.1137/100811970
24. Milnor, J., Husemoller, D.: Symmetric Bilinear Forms. Springer, Heidelberg (1973). ergebnisse der Mathematik und ihrer Grenzgebiete, Band 73
25. Reuland, G.: A Compact Representation of the Voronoi Cell. École Polytechnique Fédérale de Lausanne, January 2018. Master thesis
26. Seysen, M.: A measure for the non-orthogonality of a lattice basis. Combin. Probab. Comput. **8**(3), 281–291 (1999). https://doi.org/10.1017/S0963548399003764
27. Sommer, N., Feder, M., Shalvi, O.: Finding the closest lattice point by iterative slicing. SIAM J. Discrete Math. **23**(2), 715–731 (2009). https://doi.org/10.1137/060676362
28. Vallentin, F.: Sphere coverings, lattices, and tilings (in low dimensions). Ph.D. thesis, Technical University Munich, Germany (2003). http://nbn-resolving.de/urn/resolver.pl?urn:nbn:de:bvb:91-diss2003112600173, 128 p
29. Voronoi, G.: Nouvelles applications des paramètres continus à la théorie des formes quadratiques. Deuxième mémoire. Recherches sur les parallélloèdres primitifs. J. Reine Angew. Math. **134**, 198–287 (1908). https://doi.org/10.1515/crll.1908.134.198

An Efficient Characterization of
Submodular Spanning Tree Games

Zhuan Khye Koh[1]([✉]) and Laura Sanità[2]

[1] Department of Mathematics, London School of Economics,
London WC2A 2AE, UK
z.koh3@lse.ac.uk
[2] Department of Combinatorics and Optimization, University of Waterloo,
Waterloo, ON N2L 3G1, Canada
lsanita@uwaterloo.ca

Abstract. *Cooperative games* are an important class of problems in
game theory, where the goal is to distribute a value among a set of players who are allowed to cooperate by forming coalitions. An outcome of
the game is given by an allocation vector that assigns a value share to
each player. A crucial aspect of such games is *submodularity* (or *convexity*). Indeed, convex instances of cooperative games exhibit several nice
properties, e.g. regarding the existence and computation of allocations
realizing some of the most important solution concepts proposed in the
literature. For this reason, a relevant question is whether one can give a
polynomial time characterization of submodular instances, for prominent
cooperative games that are in general non-convex.

In this paper, we focus on a fundamental and widely studied cooperative game, namely *the spanning tree game*. An efficient recognition of
submodular instances of this game was not known so far, and explicitly
mentioned as an open question in the literature. We here settle this open
problem by giving a polynomial time characterization of submodular
spanning tree games.

Keywords: Spanning trees · Cooperative games ·
Submodular functions

1 Introduction

Cooperative games are among the most studied classes of problems in game
theory, with plenty of applications in economics, mathematics, and computer
science. In such games, the goal is to distribute cost (or revenue) among a set
of participants, usually called *players*, who are allowed to cooperate. Formally,
we are given a set of players N, and a characteristic function $\nu : 2^N \to \mathbb{R}$, with

This work was supported by the NSERC Discovery Grant Program and an Early
Researcher Award by the Province of Ontario.
Z. K. Koh—This work was done while the author was at the University of Waterloo.

A. Lodi and V. Nagarajan (Eds.): IPCO 2019, LNCS 11480, pp. 275–287, 2019.
https://doi.org/10.1007/978-3-030-17953-3_21

$\nu(\emptyset) = 0$. Here, $\nu(S)$ represents the cost paid (revenue received) by the subset of players S if they choose to form a coalition. An outcome of the game is given by an *allocation* $y \in \mathbb{R}^N$ such that $\sum_{v \in N} y_v = \nu(N)$, which assigns a cost (revenue) share to each player. Of course, there are a number of criteria for evaluating how "good" an allocation is, such as *stability*, *fairness*, and so on.

Probably the most popular solution concept for cooperative games is the *core*. It is the set of stable outcomes where no subset of players has an incentive to form a coalition to deviate. In a cooperative cost game, this translates naturally to the following constraint: $\sum_{v \in S} y_v \leq \nu(S)$, for all $S \subseteq N$. Intuitively, if this constraint is violated for some set S, the total cost currently paid by the players in S is more than the total cost $\nu(S)$ they would have to pay if they form a coalition – this incentivizes these players to deviate from the current allocation. Besides the core, there are several other crucial solution concepts which have been defined in the literature, e.g. the *Shapley value*, the *nucleolus*, the *kernel*, the *bargaining set*, and the *von Neumann-Morgenstern solution set* (we refer to [2] for details). Many fundamental questions involving such solution concepts have been investigated in the past few decades: Which cooperative game instances admit an allocation realizing a particular solution concept? Can we efficiently compute it? Can we test whether a given allocation belong to such sets?

Submodularity (or *convexity*) is a crucial property which yields interesting answers to some of the questions above. An instance of a cooperative cost game is called submodular if the characteristic function ν is submodular, meaning that

$$\forall A, B \subseteq N, \; \nu(A) + \nu(B) \geq \nu(A \cup B) + \nu(A \cap B). \qquad (*)$$

Submodular games exhibit a large number of desirable properties. In particular, (i) a core solution always exists and can be computed in polynomial time [14]; (ii) testing whether an allocation belongs to the core is equivalent to separating over the extended polymatroid of ν, which can be performed efficiently [8]; (iii) computing the nucleolus can be done efficiently [10]; (iv) there is a nice "snow-balling" effect that arises when the game is played cooperatively, meaning that joining a coalition becomes more attractive as the coalition grows, and so the value of the so-called grand coalition $\nu(N)$ is always reached [14]. We refer to [11,14] for other interesting properties of submodular games involving other crucial solution concepts. Given these observations, it is not surprising that some researchers have investigated whether it is possible to give an efficient characterization of submodular instances, for prominent cooperative games that are in general non-convex. Such characterizations are known, for example, for the minimum coloring game and the minimum vertex cover game [13], as well as for some communication games [12].

This paper focuses on one of the most fundamental cooperative games, namely the *spanning tree game*. This game was introduced more than 40 years ago [1,3], and since then it has been widely studied in the literature. To get an intuition about the problem, consider the following setting. A set of clients N would like to be connected to a central source r which can provide a service to them. The clients wish to build a network connecting them to the source r, at minimum cost. An obvious way to solve this problem is to compute a minimum spanning tree connecting $N \cup \{r\}$. But how should the clients fairly split the cost of the tree among them? Formally, an instance of the spanning tree game is described by an edge-weighted complete graph $G = (V, E)$ where $V = N \cup \{r\}$. The set of players is given by N, and the characteristic function $\nu(S)$ is equal to the cost of a minimum spanning tree in the subgraph induced by $S \cup \{r\}$.

Despite being one of the most studied cooperative games, the existence of an efficient characterization of submodularity for the spanning tree game has remained elusive so far. Granot and Huberman [7] proved that spanning tree games are *permutationally convex* (which is a generalization of submodularity). Their result implies that a core solution always exists for such games, despite being non-convex in general (this was first proven by the same authors in [6]). However, other nice properties of submodular games do not generalize: for general spanning tree games, testing core membership is coNP-hard [4], and computing the nucleolus is NP-hard [5]. Trudeau [15] gave a sufficient condition for an instance of the game to be submodular. An important step forward was made by Kobayashi and Okamoto [9], who gave a characterization of submodularity for instances of the spanning tree game where the edge weights are restricted to take only two values. For general weights, they stated some necessary (but not always sufficient) as well as some sufficient (but not always necessary) conditions for an instance to be submodular. Whether a polynomial time characterization of submodularity exists for spanning tree games is left as an open question. In fact, they stated twice in their paper:

"We feel that recognizing a submodular minimum-cost spanning tree game is coNP-complete, but we are still far from proving such a result."

Our Results and Techniques. In this paper, we finally settle this open question: we give a polynomial time characterization of submodular spanning tree games.

Our characterization uses combinatorial techniques and it is based on two main ingredients. The first one, described in Sect. 3, is a generalization of Kobayashi and Okamoto's result [9]. When the edges can have only two distinct weights, they proved that the only obstruction to submodularity comes from the presence of certain cycles in the graph induced by the cheaper edges. When dealing with more weight values, say $w_1 < w_2 < \cdots < w_k$, things become necessarily more complicated. We can still prove that an obstruction to submodularity is given by certain cycles, which we call *violated*, but (a) our definition of violated cycles is more involved than the one in [9], and (b) we have to look for such cycles not just in one induced graph, but in each graph induced by the edges of weight at most w_i, for all $i < k$.

Furthermore, the presence of violated cycles is not anymore the only obstruction to submodularity. Roughly speaking, violated cycles capture how the edges of a certain weight should relate to the cheaper ones, but we still need a condition that takes into account the "magnitude" of distinct weight values, when $k > 2$. This leads to the second main ingredient of our characterization, described in Sect. 4. We show that, under the assumption of not having violated cycles, we can identify polynomially many subsets of vertices which could yield the highest possible violation to the submodularity inequality (∗). We can then efficiently test the submodularity of our instance by checking whether the inequality (∗) is satisfied on this family of subsets of vertices. Combining these two ingredients yields a polynomial time characterization of submodularity for spanning tree games, as described in Sect. 5.

2   Preliminaries and Notation

For a subset $S \subseteq V$, let $\mathsf{mst}(S)$ denote the weight of a minimum spanning tree in $G[S]$, where $G[S]$ is the subgraph of G induced by S. Given a subgraph H of G, let $w(H)$ denote the sum of edge weights in H, i.e. $\sum_{e \in E(H)} w(e)$. For an edge set F, we will also use $w(F)$ to indicate the sum of edge weights in F. For a vertex $u \in V$, $N_H(u)$ is the neighborhood of u in H, while $\delta_H(u)$ is the set of edges incident to u in H. Given a pair of vertices $u, v \in N$, let $\mathcal{S}_{uv}$ denote the family of vertex subsets which contain r but not u or v, i.e. $\mathcal{S}_{uv} := \{S \subseteq V : r \in S \text{ and } u, v \notin S\}$. Define the function $f_{uv} : \mathcal{S}_{uv} \to \mathbb{R}$ as

$$f_{uv}(S) := \mathsf{mst}(S \cup u) + \mathsf{mst}(S \cup v) - \mathsf{mst}(S) - \mathsf{mst}(S \cup \{u, v\}).$$

It is easy to see that the spanning tree game on G is submodular if and only if $f_{uv}(S) \geq 0$ for all $u, v \in N$ and $S \in \mathcal{S}_{uv}$. Let $w_1 < w_2 < \cdots < w_k$ be the edge-weights of G. For each $i \in \{1, 2, \ldots, k\}$, define the graph $G_i := (V, E_i)$ where $E_i := \{e \in E : w(e) \leq w_i\}$. Note that $G_k = G$. For a vertex $u \in V$, denote $N_i(u)$ as the neighborhood of u in G_i. For an edge $uv \in E$, define the neighborhood of uv in G_i as

$$N_i(uv) := N_i(u) \cap N_i(v).$$

It represents the set of vertices whose edges to u and v have weight at most w_i. Notice that $u, v \notin N_i(uv)$. We will also need the following graph theory terminology. A *hole* is an induced cycle with at least four vertices. A *diamond* is the complete graph K_4 minus one edge. We will refer to the vertices of degree 2 in a diamond as *tips*. Lastly, the following property of minimum spanning trees will be useful to us (we omit its straightforward proof).

Lemma 1. *Let T be a minimum spanning tree of G. For every subset $S \subseteq V$, there exists a minimum spanning tree of $G[S]$ which contains $E(T[S])$.*

3   Violated Cycles

In this section, we will prove that a submodular spanning tree game does not contain violated cycles, which will be defined later. First, we need to introduce the concept of *well-covered* cycles.

Definition 1. Given a cycle C and a chord $f = uv$, let P_1 and P_2 denote the two u-v paths in C. The cycles $P_1 + f$ and $P_2 + f$ are called the *subcycles of C formed by f*. We say that f *covers* C if $w(f) \geq w(e)$ for all $e \in E(P_1)$ or for all $e \in E(P_2)$. If C is covered by all of its chords, then it is *well-covered*.

Next, we define the following two simple structures. We then proceed to show that a submodular spanning tree game does not contain either of them.

Definition 2. A hole is *bad* if at least one of its vertices is not adjacent to r. An induced diamond is *bad* if its hamiltonian cycle is well-covered but at least one of its tips is not adjacent to r.

The proof of the next lemma can be found in the appendix.

Lemma 2. *If the spanning tree game on G is submodular, then there are no (a) bad holes or (b) bad induced diamonds in G_i for all $i < k$.*

We are now ready to define the main object of study in this section:

Definition 3. A *violated* cycle is a well-covered cycle which contains at least a pair of non-adjacent vertices and at least a vertex not adjacent to r.

Observe that bad holes and hamiltonian cycles of bad induced diamonds are examples of violated cycles (we consider a hole to be well-covered). The next lemma extends the scope of Lemma 2 to include violated cycles. When $k = 2$, this coincides with the condition given by Kobayashi and Okamoto [9] because every cycle in G_1 is well-covered.

Lemma 3. *If the spanning tree game on G is submodular, then there are no violated cycles in G_i for all $i < k$.*

Proof. We will prove the contrapositive. Let j be the smallest integer such that G_j contains a violated cycle. By our choice of j, there are no violated cycles in G_i for all $i < j$. Let C be a smallest violated cycle in G_j. Then, $\max_{e \in E(C)} w(e) = w_j$. We first prove the following claim:

Claim. For any chord f, the subcycles of C formed by f are well-covered.

Proof. Let C_1 and C_2 denote the subcycles of C formed by f. For the purpose of contradiction, suppose C_2 is not well-covered. Let $g = uv$ be the cheapest chord in C_2 such that $w(g) < w(f)$ and $w(g) < w(h)$ for some edge $h \in E(C_2)$, where f and h lie in different subcycles of C_2 formed by g (see Fig. 1 for an example). This chord exists because C is well-covered but C_2 is not. Consider the subcycles C_3 and C_4 of C formed by g, where f is a chord of the former. Observe that C_3 is well-covered because $w(g) < w(h)$, while C_4 is well-covered due to our choice of g. Moreover, we have $w(g) \geq w(e)$ for all $e \in E(C_3)$ as C is well-covered. Let $w(g) = w_\ell$ for some $\ell < j$. Then, C_3 is still present in G_ℓ but not f because $w(g) < w(f)$. Thus, the vertices of C_3 are adjacent to r in G_ℓ because there are no violated cycles in G_ℓ. In particular, we have $ru, rv \in E_\ell$. Next, since C is a violated cycle in G_j, there exists a vertex $s \in V(C_4) \setminus V(C_3)$ such that $rs \notin E_j$. This implies that the vertices of C_4 are pairwise adjacent in G_j, as otherwise it is a smaller violated cycle than C. In particular, we have $su, sv \in E_j$. Now, consider the 4-cycle D defined by $E(D) := \{ru, rv, su, sv\}$. It is well-covered because $w(g) = w_\ell$ and $ru, rv \in E_\ell$. As $rs \notin E_j$, it is a violated cycle in G_j. However, it is smaller than C because C_3 has at least 4 vertices. We have arrived at a contradiction.   □

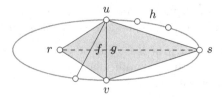

Fig. 1. The ellipse represents the violated cycle C in the previous claim. The shaded region highlights the smaller violated cycle D. The dashed edge indicates $rs \notin E_j$.

Our goal is to show the existence of a bad hole or a bad induced diamond in G_j. Then, we can invoke Lemma 2 to conclude that the game is not submodular. We may assume there is at least one chord in C, otherwise it is a bad hole. First, consider the case when $r \in V(C)$. Let $s \in V(C)$ where $rs \notin E_j$. For any chord f in C, observe that r and s lie in different subcycles of C formed by f. This is because the subcycles are well-covered by the previous claim, so the one which contains both r and s will contradict the minimality of C. Now, let g be a chord of C. Let C_r and C_s denote the subcycles of C formed by g where $r \in V(C_r)$ and $s \in V(C_s)$. The vertices of C_r are adjacent to r due to the minimality of C. Thus, C_r is a triangle. Otherwise, there exists a chord in C_r incident to r, and it forms a subcycle of C which contains both r and s. By an analogous argument, C_s is also a triangle. Therefore, C is a bad induced diamond in G_j.

Next, consider the case when $r \notin V(C)$. From this point forward, we may assume that every smallest violated cycle in G_j does not contain r. Otherwise, we are back in the first case again. With this additional assumption, non-adjacency within C implies non-adjacency with r, as shown by the following claim.

Claim. For any pair of vertices $u, v \in V(C)$ such that $uv \notin E_j$, we have $ru \notin E_j$ or $rv \notin E_j$.

Proof. For the purpose of contradiction, suppose $ru, rv \in E_j$. Let $s \in V(C)$ such that $rs \notin E_j$. Let P_{su} and P_{sv} denote the edge-disjoint s-u and s-v paths in C respectively. Let u' and v' be the closest vertex to s on P_{su} and P_{sv} respectively such that $ru', rv' \in E_j$ (see Fig. 2 for an example). Without loss of generality, let $w(ru') \geq w(rv')$. Denote $P_{su'}$ and $P_{sv'}$ as the s-u' and s-v' subpaths of P_{su} and P_{sv} respectively. Now, consider the cycle $D := P_{su'} + P_{sv'} + ru' + rv'$. Observe that it contains r and is no bigger than C. Furthermore, it does not contain a chord incident to r by our choice of u' and v'. To arrive at a contradiction, it is left to show that D is well-covered, as this would imply D is violated. Suppose for a contradiction, that D is not well-covered. Then, there exists a chord g in D such that $w(g) < w(ru')$ and $w(g) < w(h)$ for some $h \in E(D)$, where ru' and h lie in different subcycles of D formed by g. This chord exists because C is well-covered but D is not. Let C_1 and C_2 denote the subcycles of C formed by g, where $h \in E(C_2)$. Note that C_1 is well-covered because $w(g) < w(h)$. Moreover, we also have $w(g) \geq w(e)$ for all $e \in E(C_1)$ because C is well-covered. Let $w(g) = w_\ell$ for some $\ell < j$. Then, C_1 is still present in G_ℓ but not ru'. Since

C_1 also contains u, v and $uv \notin E_\ell$, it is a violated cycle in G_ℓ. However, this is a contradiction because there are no violated cycles in G_ℓ. □

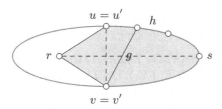

Fig. 2. The ellipse represents the violated cycle C in the previous claim. The shaded region highlights the violated cycle D. The dashed edges indicate non-adjacency in G_j. In this example, $u = u'$ and $v = v'$.

The remaining proof proceeds in a similar fashion to the first case. Let $u, v \in V(C)$ such that $uv \notin E_j$. By the claim above, we know that $ru \notin E_j$ or $rv \notin E_j$. For any chord f in C, observe that u and v lie in different subcycles of C formed by f. This is because the subcycles are well-covered, so the one which contains both u and v will contradict the minimality of C. Now, let g be a chord of C. Let C_u and C_v denote the subcycles of C formed by g where $u \in V(C_u)$ and $v \in V(C_v)$. The vertices of C_u are pairwise adjacent due to the minimality of C. Thus, C_u is a triangle. Otherwise, there exists a chord in C_u incident to u, and it forms a subcycle of C which contains both u and v. By an analogous argument, C_v is also a triangle. Therefore, C is a bad induced diamond in G_j. □

Observe that we have proven something stronger. Namely, if G_j contains a violated cycle, then there exists an $i \leq j$ such that G_i contains a bad hole or a bad induced diamond. Thus, if there are no bad holes or bad induced diamonds in G_i for all $i < k$, then there are no violated cycles in these subgraphs too. As a result, verifying the condition in Lemma 3 reduces to searching for bad holes and bad induced diamonds in G_i, which can be done efficiently. To look for bad holes, one could check if there exists a hole through a given vertex v for all $v \in N$ where $rv \notin E_i$. To look for bad induced diamonds, a naive implementation would involve examining all vertex subsets of size 4, which still runs in polynomial time.

4 Candidate Edges and Expensive Neighborhood

In the previous section, we have shown that violated cycles are an obstruction to submodularity. Moreover, their existence can be tested in polynomial time. In light of this fact, we now focus on graphs which do not contain violated cycles. For the sake of brevity, we will use $(\star)$ to denote the following property:

There are no violated cycles in G_i for all $i < k$.

The goal of this section is to study the behaviour of f_{uv} assuming ($\star$) holds. As a first step, the following lemma sheds light on how a minimum spanning tree changes under vertex removal.

Lemma 4. *Assume ($\star$) holds. Let T be a minimum spanning tree of $G[S]$ where $r \in S \subseteq V$. For any $s \neq r$, there exists a minimum spanning tree of $G[S \setminus s]$ which contains $E(T \setminus s)$ and additionally, only uses edges from $G[N_T(s) \cup r]$.*

Proof. Pick a vertex $s \in S \setminus r$. By Lemma 1, there exists a minimum spanning tree of $G[S \setminus s]$ which contains $E(T \setminus s)$. Let T' be such a tree which uses the most edges from $G[N_T(s) \cup r]$. We will show that T' is our desired tree. For the purpose of contradiction, suppose T' has an edge uv where $uv \notin E(T)$ and $u \notin N_T(s) \cup r$. Note that u and v lie in different components of $T \setminus s$. Let P_{su} and P_{sv} denote the unique s-u and s-v paths in T respectively. Then, $C := P_{su} \cup P_{sv} \cup uv$ is a well-covered cycle in G_i where $w(uv) = w_i$. Let u' and v' be the vertices adjacent to s in P_{su} and P_{sv} respectively. By our choice of T', $w(u'v') > w(uv)$. Since uv is the most expensive edge in C, the vertices of C are not pairwise adjacent in G_i. So they are adjacent to r in G_i. However, adding ru' or rv' to T' creates a fundamental cycle which uses the edge uv. Swapping it with uv creates another minimum spanning tree of $G[S \setminus s]$ which contains $E(T \setminus s)$ and uses more edges from $G[N_T(s) \cup r]$. We have arrived at a contradiction. $\square$

Given a pair of vertices $u, v \in N$ where $w(uv) = w_i$, the following definition distinguishes the neighbours of u, v in G from the neighbours of u, v in G_i.

Definition 4. For an edge $uv \in E$, if $w(uv) = w_i$, the *expensive neighborhood* of uv is defined as

$$\hat{N}(uv) := N_k(uv) \setminus N_i(uv).$$

In other words, the expensive neighborhood of an edge uv is the set of vertices $s \notin \{u, v\}$ such that $\max\{w(su), w(sv)\} > w(uv)$. It turns out that the function f_{uv} always returns zero when evaluated on a set which does not lie entirely in the expensive neighborhood of uv.

Lemma 5. *Assume ($\star$) holds. Let $u, v \in N$ and $S \in \mathcal{S}_{uv}$. If $S \not\subseteq \hat{N}(uv)$, then $f_{uv}(S) = 0$.*

Proof. Let T be a minimum spanning tree of $G[S \cup \{u, v\}]$. First, we show that we can assume $uv \notin E(T)$. Since $S \not\subseteq \hat{N}(uv)$, there exists a vertex $s \in S$ such that $\max\{w(su), w(sv)\} \leq w(uv)$. If $uv \in E(T)$, then by rooting T at s, u is either a child or a parent of v. Adding su to T in the former and sv in the latter creates a fundamental cycle which contains uv. Thus, we can replace uv with this new edge to obtain the desired tree. Now, by Lemma 4, there exists a minimum spanning tree T' of $G[S \cup v]$ which contains $E(T \setminus u)$ and additionally, only uses edges from $G[N_T(u) \cup r]$. Since $v \notin N_T(u) \cup r$, the neighborhood of v is identical in both trees, i.e. $N_T(v) = N_{T'}(v)$.

Consider the forest $T \setminus v$. Let $p \in N_T(v)$ such that p and r lie in the same component of $T \setminus v$ (see Fig. 3 for an example). Note that $p = r$ if $r \in N_T(v)$. We claim that p and r also lie in the same component of the forest $T' \setminus v$. We may assume that $p \neq r$, as otherwise the claim is trivially true. Moreover, we may assume that u lies on the unique p-r path in T. Otherwise, we are done because the same path is present in $T' \setminus v$. Let C_r denote the component of $T \setminus v$ which contains p, r and u. By Lemma 4, the endpoints of every edge in $E(T') \setminus E(T \setminus u)$ lie in C_r. This proves the claim.

Using Lemma 4, we can construct a minimum spanning tree of $G[S \cup u]$ by deleting v from T and adding a set of edges F from $G[N_T(v) \cup r]$. Note that $pr \notin F$ as p and r lie in the same component of $T \setminus v$. Since p and r also lie in the same component of $T' \setminus v$ and $N_T(v) = N_{T'}(v)$, deleting v from T' and adding F creates a minimum spanning tree of $G[S]$. Thus, we get

$$
\begin{aligned}
f_{uv}(S) &= \mathsf{mst}(S \cup u) + \mathsf{mst}(S \cup v) - \mathsf{mst}(S) - \mathsf{mst}(S \cup \{u, v\}) \\
&= \Big(\mathsf{mst}(S \cup u) - w(T) \Big) - \Big(\mathsf{mst}(S) - w(T') \Big) \\
&= \Big(w(F) - w(\delta_T(v)) \Big) - \Big(w(F) - w(\delta_{T'}(v)) \Big) = 0
\end{aligned}
$$

as desired. □

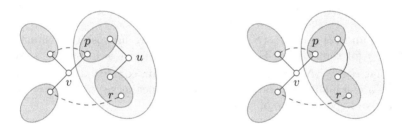

Fig. 3. The left image depicts an example of the minimum spanning tree T in $G[S \cup \{u, v\}]$. The right image depicts an example of the minimum spanning tree T' in $G[S \cup v]$. The solid edges belong to the trees while dashed edges belong to the edge set F.

We can now focus solely on vertex sets which lie entirely in the expensive neighborhood of uv. Observe that if $r \notin \hat{N}(uv)$, then $S \not\subseteq \hat{N}(uv)$ for all $S \in \mathcal{S}_{uv}$. Thus, we do not have to check these edges as $f_{uv}(S) = 0$ for all $S \in \mathcal{S}_{uv}$ by the previous lemma. This motivates the following definition:

Definition 5. An edge $uv \in E$ is called a *candidate edge* if $r \in \hat{N}(uv)$.

With a mild assumption, we can show that the function f_{uv} is inclusion-wise nonincreasing in the expensive neighborhood of uv.

Lemma 6. *Assume* ($\star$) *holds and* $f_{xy}(\hat{N}(xy)) \geq 0$ *for every candidate edge* xy. *Let* uv *be a candidate edge and* $S \in \mathcal{S}_{uv}$ *such that* $S \subseteq \hat{N}(uv)$. *For any* $s \neq r$, $f_{uv}(S) \leq f_{uv}(S \setminus s)$.

Proof. Pick a vertex $s \in S \setminus r$. Without loss of generality, assume $w(su) \geq w(sv)$. Then, $w(su) > w(uv)$ because $s \in \hat{N}(uv)$. However, these two inequalities also imply that $v \notin \hat{N}(su)$. It follows that the set $(S \setminus s) \cup v$ is not contained in the expensive neighborhood of su. By Lemma 5,

$$0 = f_{su}((S \setminus s) \cup v)$$
$$= \mathsf{mst}(S \cup v) + \mathsf{mst}((S \setminus s) \cup \{u, v\}) - \mathsf{mst}((S \setminus s) \cup v) - \mathsf{mst}(S \cup \{u, v\}).$$

Rearranging yields

$$\mathsf{mst}(S \cup v) - \mathsf{mst}(S \cup \{u, v\}) = \mathsf{mst}((S \setminus s) \cup v) - \mathsf{mst}((S \setminus s) \cup \{u, v\}). \quad (1)$$

Since uv is a candidate edge, let $w(uv) = w_i$ for some $i < k$. We will proceed by induction on i. For the base case $i = k-1$, we have $w_{k-1} = w(uv) < w(su) = w_k$. Since $\hat{N}(su) = \emptyset$, the set $S \setminus s$ is not contained in the expensive neighborhood of su because $r \in S \setminus s$. By Lemma 5,

$$0 = f_{su}(S \setminus s) = \mathsf{mst}(S) + \mathsf{mst}((S \setminus s) \cup u) - \mathsf{mst}(S \setminus s) - \mathsf{mst}(S \cup u).$$

Rearranging yields

$$\mathsf{mst}(S \cup u) - \mathsf{mst}(S) = \mathsf{mst}((S \setminus s) \cup u) - \mathsf{mst}(S \setminus s). \quad (2)$$

Adding (1) and (2) gives $f_{uv}(S) = f_{uv}(S \setminus s)$. Now, suppose the lemma is true for all $i \geq j$ for some $j < k$. For the inductive step, let $w(uv) = w_{j-1}$. We may assume that $S \setminus s \subseteq \hat{N}(su)$, as otherwise we obtain equality again. This implies that su is a candidate edge because $r \in S \setminus s$. Since $w(su) > w(uv) = w_{j-1}$, by the inductive hypothesis we obtain

$$0 \leq f_{su}(\hat{N}(su))$$
$$\leq f_{su}(S \setminus s) = \mathsf{mst}(S) + \mathsf{mst}((S \setminus s) \cup u) - \mathsf{mst}(S \setminus s) - \mathsf{mst}(S \cup u)$$

where the first inequality is due to our assumption. Then, by rearranging and adding it to (1), we obtain $f_{uv}(S) \leq f_{uv}(S \setminus s)$ as desired.   □

5   Characterization of Submodularity

We are finally ready to give an efficient characterization of submodular spanning tree games.

Theorem 1. *The spanning tree game on G is submodular if and only if:*

(i) There are no violated cycles in G_i for all $i < k$.
(ii) For every candidate edge uv, $f_{uv}(\hat{N}(uv)) \geq 0$.

Furthermore, these conditions can be verified in polynomial time.

Proof. For necessity, assume the game is submodular. Then, Condition (i) is satisfied by Lemma 3 while Condition (ii) is satisfied trivially. For sufficiency, assume Conditions (i) and (ii) hold. Let $u, v \in N$ and $S \in \mathcal{S}_{uv}$. If $S \nsubseteq \hat{N}(uv)$, then $f_{uv}(S) = 0$ by Lemma 5. On the other hand, if $S \subseteq \hat{N}(uv)$, then uv is a candidate edge. By Lemma 6,

$$f_{uv}(S) \geq f_{uv}(\hat{N}(uv)) \geq 0.$$

Therefore, the game is submodular.

Finally, Condition (ii) can clearly be verified in polynomial time, and Condition (i) can be verified in polynomial time as discussed at the end of Sect. 3. □

Appendix

Proof of Lemma 2(a)

We will prove the contrapositive. Let C be a bad hole in G_i for some $i < k$. Consider the following cases:

Case 1: C contains r. Let u, v be the vertices adjacent to r in C. Let P be the path obtained by deleting r, u, v from C. Let u', v' be the endpoints of P where $uu', vv' \in E(C)$. Note that $u' = v'$ if P is a singleton. Let $S = V(P) \cup r$. Then,

$$\mathsf{mst}(S) \geq w(P) + w_{i+1}$$
$$\mathsf{mst}(S \cup u) = w(P) + w(ru) + w(uu')$$
$$\mathsf{mst}(S \cup v) = w(P) + w(rv) + w(vv')$$
$$\mathsf{mst}(S \cup \{u, v\}) \geq w(P) + w(ru) + w(uu') + w(rv) + w(vv') - w_i$$

which yields: $\mathsf{mst}(S \cup u) + \mathsf{mst}(S \cup v) - \mathsf{mst}(S) - \mathsf{mst}(S \cup \{u, v\}) \leq w_i - w_{i+1} < 0$.

Case 2: C does not contain r. Let $s = \arg\min_{x \in V(C)} w(rx)$. Let u, v be the vertices adjacent to s in C. Let P be the path obtained by deleting s, u, v from C. Let u', v' be the endpoints of P where $uu', vv' \in E(C)$. Let $S = V(P) \cup \{r, s\}$. Observe that if r is adjacent to two non-adjacent vertices of C, then we are done because there is a bad hole containing r. We are left with the following subcases:

Subcase 2.1: r is adjacent to at most one vertex of C. We have

$$\mathsf{mst}(S) \geq w(P) + w(rs) + w_{i+1}$$
$$\mathsf{mst}(S \cup u) = w(P) + w(rs) + w(su) + w(uu')$$
$$\mathsf{mst}(S \cup v) = w(P) + w(rs) + w(sv) + w(vv')$$
$$\mathsf{mst}(S \cup \{u, v\}) \geq w(P) + w(rs) + w(su) + w(uu') + w(sv) + w(vv') - w_i$$

which yields: $\mathsf{mst}(S \cup u) + \mathsf{mst}(S \cup v) - \mathsf{mst}(S) - \mathsf{mst}(S \cup \{u, v\}) \leq w_i - w_{i+1} < 0$.

Subcase 2.2: r is adjacent to two vertices of C. Suppose $rs, ru \in E_i$. Then,

$$\mathsf{mst}(S) \geq w(P) + w(rs) + w_{i+1}$$
$$\mathsf{mst}(S \cup u) = w(P) + w(rs) + \min\{w(ru), w(su)\} + w(uu')$$
$$\mathsf{mst}(S \cup v) = w(P) + w(rs) + w(sv) + w(vv')$$
$$\mathsf{mst}(S \cup \{u,v\}) \geq w(P) + w(rs) + \min\{w(ru), w(su)\} + w(uu') + w(sv) + w(vv') - w_i$$

which yields: $\mathsf{mst}(S \cup u) + \mathsf{mst}(S \cup v) - \mathsf{mst}(S) - \mathsf{mst}(S \cup \{u,v\}) \leq w_i - w_{i+1} < 0$.

□

Proof of Lemma 2(b)

We will prove the contrapositive. Let D be a bad induced diamond in G_i for some $i < k$. Consider the following cases:

Case 1: D contains r. Note that r is a tip of D. Let s be the other tip and u, v be the non-tip vertices of D. Let $S = \{r, s\}$. Then,

$$\mathsf{mst}(S) \geq w_{i+1}$$
$$\mathsf{mst}(S \cup u) = w(ru) + w(su)$$
$$\mathsf{mst}(S \cup v) = w(rv) + w(sv)$$
$$\mathsf{mst}(S \cup \{u,v\}) \geq w(ru) + w(su) + w(rv) + w(sv) - w_i$$

which yields: $\mathsf{mst}(S \cup u) + \mathsf{mst}(S \cup v) - \mathsf{mst}(S) - \mathsf{mst}(S \cup \{u,v\}) \leq w_i - w_{i+1} < 0$.

Case 2: D does not contain r. Let s, t be the tips of D where $w(rs) \leq w(rt)$. Note that $rt \notin E_i$. Let u, v be the non-tip vertices of D where $w(ru) \leq w(rv)$. Let $S = \{r, s, t\}$. Consider the following subcases:

Subcase 2.1: r is adjacent to at most one vertex of D. Note that $rv \notin E_i$. So,

$$\mathsf{mst}(S) \geq w(rs) + w_{i+1}$$
$$\mathsf{mst}(S \cup u) = \min\{w(rs), w(ru)\} + w(su) + w(tu)$$
$$\mathsf{mst}(S \cup v) = \min\{w(rs), w(rv)\} + w(sv) + w(tv)$$
$$\mathsf{mst}(S \cup \{u,v\}) \geq \min\{w(rs), w(ru)\} + w(su) + w(tu) + w(sv) + w(tv) - w_i$$

which yields: $\mathsf{mst}(S \cup u) + \mathsf{mst}(S \cup v) - \mathsf{mst}(S) - \mathsf{mst}(S \cup \{u,v\}) \leq w_i - w_{i+1} < 0$.

Subcase 2.2: r is adjacent to two vertices of D. Observe that if $ru, rv \in E_i$, then we are done because there is a bad induced diamond containing r. So, let $rv \notin E_i$. This implies $rs, ru \in E_i$. We may also assume $w(su) < \max\{w(rs), w(ru)\}$. Otherwise, $\{rs, ru, su, sv, uv\}$ is a bad induced diamond containing r. Then,

$$\mathsf{mst}(S) \geq w(rs) + w_{i+1}$$
$$\mathsf{mst}(S \cup u) = \min\{w(rs), w(ru)\} + w(su) + w(tu)$$
$$\mathsf{mst}(S \cup v) = w(rs) + w(sv) + w(tv)$$
$$\mathsf{mst}(S \cup \{u,v\}) \geq \min\{w(rs), w(ru)\} + w(su) + w(tu) + w(sv) + w(tv) - w_i$$

which yields: $\mathsf{mst}(S \cup u) + \mathsf{mst}(S \cup v) - \mathsf{mst}(S) - \mathsf{mst}(S \cup \{u,v\}) \le w_i - w_{i+1} < 0$.

Subcase 2.3: r is adjacent to three vertices of D. Let $w(rv) = w_j$ for some $j \le i$. Consider the induced diamond $\{ru, rv, tu, tv, uv\}$. If it is well-covered, then we are done because it is bad and contains r. So let $\max \{w(su), w(sv)\} \le w(uv) < w(rv)$. We may also assume $w(su) < \max \{w(rs), w(ru)\}$. Otherwise, $\{rs, ru, su, sv, uv\}$ is a bad induced diamond in G_{j-1} which contains r. Then,

$$\mathsf{mst}(S) \ge w(rs) + w_{i+1}$$
$$\mathsf{mst}(S \cup u) = \min \{w(rs), w(ru)\} + w(su) + w(tu)$$
$$\mathsf{mst}(S \cup v) = \min \{w(rs), w(rv)\} + w(sv) + w(tv)$$
$$\mathsf{mst}(S \cup \{u,v\}) \ge \min \{w(rs), w(ru)\} + w(su) + w(tu) + w(sv) + w(tv) - w_i$$

which yields: $\mathsf{mst}(S \cup u) + \mathsf{mst}(S \cup v) - \mathsf{mst}(S) - \mathsf{mst}(S \cup \{u,v\}) \le w_i - w_{i+1} < 0$.

$\square$

References

1. Bird, C.G.: On cost allocation for a spanning tree: a game theoretic approach. Networks **6**(4), 335–350 (1976)
2. Chalkiadakis, G., Elkind, E., Wooldridge, M.: Computational Aspects of Cooperative Game Theory. Synthesis Lectures on Artificial Intelligence and Machine Learning. Morgan & Claypool Publishers, San Rafael (2011)
3. Claus, A., Kleitman, D.J.: Cost allocation for a spanning tree. Networks **3**(4), 289–304 (1973)
4. Faigle, U., Kern, W., Fekete, S.P., Hochstättler, W.: On the complexity of testing membership in the core of min-cost spanning tree games. Int. J. Game Theory **26**(3), 361–366 (1997)
5. Faigle, U., Kern, W., Kuipers, J.: Note computing the nucleolus of min-cost spanning tree games is NP-hard. Int. J. Game Theory **27**(3), 443–450 (1998)
6. Granot, D., Huberman, G.: Minimum cost spanning tree games. Math. Program. **21**(1), 1–18 (1981)
7. Granot, D., Huberman, G.: The relationship between convex games and minimum cost spanning tree games: a case for permutationally convex games. SIAM J. Algebraic Discrete Methods **3**(3), 288–292 (1982)
8. Grötschel, M., Lovász, L., Schrijver, A.: Geometric Algorithms and Combinatorial Optimization, Algorithms and Combinatorics, vol. 2. Springer, Heidelberg (1993). https://doi.org/10.1007/978-3-642-78240-4
9. Kobayashi, M., Okamoto, Y.: Submodularity of minimum-cost spanning tree games. Networks **63**(3), 231–238 (2014)
10. Kuipers, J.: A polynomial time algorithm for computing the nucleolus of convex games. Report M 96–12, Maastricht University (1996)
11. Maschler, M., Peleg, B., Shapley, L.S.: The kernel and bargaining set for convex games. Int. J. Game Theory **1**(1), 73–93 (1971)
12. van den Nouweland, A., Borm, P.: On the convexity of communication games. Int. J. Game Theory **19**(4), 421–430 (1991)
13. Okamoto, Y.: Submodularity of some classes of the combinatorial optimization games. Math. Methods Oper. Res. **58**(1), 131–139 (2003)
14. Shapley, L.S.: Cores of convex games. Int. J. Game Theory **1**(1), 11–26 (1971)
15. Trudeau, C.: A new stable and more responsive cost sharing solution for minimum cost spanning tree problems. Games Econ. Behav. **75**(1), 402–412 (2012)

The Asymmetric Traveling Salesman Path LP Has Constant Integrality Ratio

Anna Köhne, Vera Traub$^{(\boxtimes)}$, and Jens Vygen

Research Institute for Discrete Mathematics, University of Bonn, Bonn, Germany
{koehne,traub,vygen}@or.uni-bonn.de

Abstract. We show that the classical LP relaxation of the asymmetric traveling salesman path problem (ATSPP) has constant integrality ratio. If ρ_{ATSP} and ρ_{ATSPP} denote the integrality ratios for the asymmetric TSP and its path version, then $\rho_{\text{ATSPP}} \leq 4\rho_{\text{ATSP}} - 3$.

We prove an even better bound for node-weighted instances: if the integrality ratio for ATSP on node-weighted instances is ρ_{ATSP}^{NW}, then the integrality ratio for ATSPP on node-weighted instances is at most $2\rho_{\text{ATSP}}^{NW} - 1$. Moreover, we show that for ATSP node-weighted instances and unweighted digraph instances are almost equivalent. From this we deduce a lower bound of 2 on the integrality ratio of unweighted digraph instances.

1 Introduction

In the asymmetric traveling salesman path problem (ATSPP), we are given a directed graph $G = (V, E)$, two vertices $s, t \in V$, and weights $c : E \rightarrow \mathbb{R}_{\geq 0} \cup \{\infty\}$. We look for a sequence $s = v_0, v_1, \ldots, v_k = t$ that contains every vertex at least once (an s-t-tour); the goal is to minimize $\sum_{i=1}^{k} c(v_{i-1}, v_i)$. Equivalently, we can assume that G is complete and the triangle inequality $c(u, v) + c(v, w) \geq c(u, w)$ holds for all $u, v, w \in V$, and require the sequence to contain every vertex exactly once.

The special case $s = t$ is known as the asymmetric traveling salesman problem (ATSP). In a recent breakthrough, Svensson, Tarnawski, and Végh [11] found the first constant-factor approximation algorithm for ATSP, and they also proved that its standard LP relaxation has constant integrality ratio.

Feige and Singh [4] showed that any α-approximation algorithm for ATSP implies a $(2\alpha + \varepsilon)$-approximation algorithm for ATSPP (for any $\varepsilon > 0$). Hence ATSPP also has a constant-factor approximation algorithm. In this paper we prove a similar relation for the integrality ratios. This answers an open question by Friggstad, Gupta, and Singh [5].

Given that the upper bound on the integrality ratio by Svensson, Tarnawski, and Végh [11] is a large constant that will probably be improved in the future, such a blackbox result seems particularly desirable. Any improved upper bound on the integrality ratio for ATSP then immediately implies a better bound for the path version.

© Springer Nature Switzerland AG 2019
A. Lodi and V. Nagarajan (Eds.): IPCO 2019, LNCS 11480, pp. 288–298, 2019.
https://doi.org/10.1007/978-3-030-17953-3_22

1.1   The Linear Programming Relaxation

The classical linear programming relaxation for ATSPP (for $s \neq t$) is

$$\min c(x)$$
$$
\begin{aligned}
\text{s.t.} \quad & x(\delta^-(s)) - x(\delta^+(s)) = -1 \\
& x(\delta^-(t)) - x(\delta^+(t)) = 1 \\
& x(\delta^-(v)) - x(\delta^+(v)) = 0 && \text{for } v \in V \setminus \{s,t\} \\
& x(\delta(U)) \geq 2 && \text{for } \emptyset \neq U \subseteq V \setminus \{s,t\} \\
& x_e \geq 0 && \text{for } e \in E
\end{aligned}
$$
(ATSPP LP)

Here (and henceforth) we write $c(x) := \sum_{e \in E} c(e) x_e$, $x(F) := \sum_{e \in F} x_e$, $\delta^+(U) := \{(u,v) \in E : u \in U, v \in V \setminus U\}$, $\delta^-(U) := \delta^+(V \setminus U)$, $\delta(U) := \delta^-(U) \cup \delta^+(U)$, $\delta^+(v) := \delta^+(\{v\})$, and $\delta^-(v) := \delta^-(\{v\})$. For an instance $\mathcal{I}$ we denote by $\text{LP}_\mathcal{I}$ the value of an optimum solution to (ATSPP LP) and by $\text{OPT}_\mathcal{I}$ the value of an optimum integral solution. If the instance is clear from the context, we will sometimes simply write LP and OPT. Note that the integral solutions of (ATSPP LP) are precisely the incidence vectors of multi-digraphs (V, F) that are connected and become Eulerian by adding one edge (t, s). Hence they correspond to walks from s to t that visit all vertices, in other words: s-t-tours.

The integrality ratio of (ATSPP LP), denoted by ρ_{ATSPP}, is the maximal ratio of an optimum integral solution and an optimum fractional solution; more precisely $\sup_\mathcal{I} \frac{\text{OPT}_\mathcal{I}}{\text{LP}_\mathcal{I}}$, where the supremum goes over all instances $\mathcal{I} = (G, c, s, t)$ with $s \neq t$ for which the denominator is nonzero and finite. Nagarajan and Ravi [9] proved that $\rho_{\text{ATSPP}} = O(\sqrt{n})$, where $n = |V|$. This bound was improved to $O(\log n)$ by Friggstad, Salavatipour, and Svitkina [6] and to $O(\log n / \log \log n)$ by Friggstad, Gupta, and Singh [5]. In this paper we prove that the integrality ratio of (ATSPP LP) is in fact constant.

Let ρ_{ATSP} denote the integrality ratio of the classical linear programming relaxation for ATSP:

$$\min c(x)$$
$$
\begin{aligned}
\text{s.t.} \quad & x(\delta^-(v)) - x(\delta^+(v)) = 0 && \text{for } v \in V \\
& x(\delta(U)) \geq 2 && \text{for } \emptyset \neq U \subsetneq V \\
& x_e \geq 0 && \text{for } e \in E
\end{aligned}
$$
(ATSP LP)

Svensson, Tarnawski, and Végh [11] proved that ρ_{ATSP} is a constant. By an infinite sequence of instances, Charikar, Goemanns, and Karloff [2] showed that $\rho_{\text{ATSP}} \geq 2$. It is obvious that $\rho_{\text{ATSPP}} \geq \rho_{\text{ATSP}}$: split an arbitrary vertex of an ATSP instance into two copies, one (called s) inheriting the outgoing edges, and one (called t) inheriting the entering edges; add an edge (t, s) of cost zero and with $x_{(t,s)} := x(\delta^+(s)) - 1$. Figure 1 displays a simpler family of examples, due to Friggstad, Gupta, and Singh [5], showing that $\rho_{\text{ATSPP}} \geq 2$.

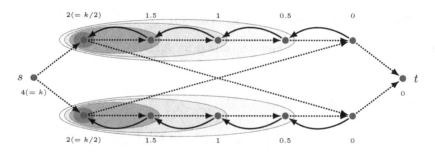

Fig. 1. Example with integrality ratio approaching 2 as the number of vertices increases. Setting $x_e := \frac{1}{2}$ for all shown edges defines a feasible solution of (ATSPP LP). If the $2k$ curved edges have cost 1 and the dotted edges have cost 0, we have LP $= c(x) = k$, but any s-t-tour costs at least $2k - 1$. (In the figure, $k = 4$.) Setting $y_U = \frac{1}{2}$ for the vertex sets indicated by the ellipses and a_v as shown in blue defines an optimum solution of (ATSPP DUAL). (Color figure online)

1.2 Our Results and Techniques

Our main result says that $\rho_{\text{ATSPP}} \leq 4\rho_{\text{ATSP}} - 3$. Together with [11], this implies a constant integrality ratio for (ATSPP LP).

Similarly as Feige and Singh [4], we transform our ATSPP instance to an ATSP instance by adding a feedback path from t to s and work with an integral solution to this ATSP instance. This may use the feedback path several times and hence consist of several s-t-walks in the original instance. We now merge these to a single s-t-walk that contains all vertices. In contrast to Feige and Singh [4], the merging procedure cannot use an optimum s-t-tour, but only an LP solution. Our merging procedure is similar to one step of the approximation algorithm for ATSP by Svensson, Tarnawski, and Végh [11], but our analysis is more involved. The main difficulty is that the reduction of ATSP to so-called "laminarly-weighted" instances used by Svensson, Tarnawski, and Végh [11] does not work for the path version.

In Sect. 3, we describe our merging procedure and obtain a first bound on the cost of our single s-t-walk that contains all vertices. However, this bound still depends on the difference of two dual LP variables corresponding to the vertices s and t. In Sect. 4 we give a tight upper bound on this value, which will imply our main result $\rho_{\text{ATSPP}} \leq 4\rho_{\text{ATSP}} - 3$.

The main lemma that we use to prove this bound essentially says that adding an edge (t, s) of cost equal to the LP value does not change the value of an optimum LP solution. Note that using the new edge (t, s) with value one or more is obviously pointless, but it is not obvious that this edge will not be used at all.

For node-weighted instances we obtain a better result: if the integrality ratio for ATSP on node-weighted instances is $\rho_{\text{ATSP}}^{\text{NW}}$, then the integrality ratio for ATSPP on node-weighted instances is at most $2\rho_{\text{ATSP}}^{\text{NW}} - 1$. Svensson [10] showed that $\rho_{\text{ATSP}}^{\text{NW}} \leq 13$.

Boyd and Elliot-Magwood [1] describe a family of node-weighted instances that shows $\rho_{\text{ATSP}}^{\text{NW}} \geq 2$. In Sect. 5 we observe that for ATSP node-weighted instances behave in the same way as unweighted instances. Hence for ATSP there is a family of unweighted digraphs whose integrality ratio tends to 2. Therefore such a family exists also for ATSPP.

In this version some proofs are omitted or only sketched. For full proofs we refer to [8].

2   Preliminaries

Given an instance (G, c, s, t) and an optimum solution x^* to (ATSPP LP), we may assume that $G = (V, E)$ is the support graph of x^*; so $x_e^* > 0$ for all $e \in E$. (This is because omitting edges e with $x_e^* = 0$ does not change the optimum LP value and can only increase the cost of an optimum integral solution.) We consider the dual LP of (ATSPP LP):

$$
\begin{aligned}
\max \quad & a_t - a_s + \sum_{\emptyset \neq U \subseteq V \setminus \{s,t\}} 2y_U \\
\text{s.t.} \quad & a_w - a_v + \sum_{U:e \in \delta(U)} y_U \leq c(e) \quad \text{for } e = (v, w) \in E \qquad \text{(ATSPP DUAL)} \\
& y_U \geq 0 \qquad \text{for } \emptyset \neq U \subseteq V \setminus \{s, t\}.
\end{aligned}
$$

The *support* of y is the set of nonempty subsets U of $V \setminus \{s, t\}$ for which $y_U > 0$. We denote it by $\text{supp}(y)$. We say that a dual solution (a, y) has *laminar support* if for any two nonempty sets $A, B \in \text{supp}(y)$ we have $A \cap B = \emptyset$, $A \subseteq B$, or $B \subseteq A$. See Fig. 1 for an example. We recall some well-known properties of primal and dual LP solutions (cf. [8,11]):

Proposition 1. *Let (a, y) be an optimum solution to (ATSPP DUAL). Then there is a vector y' such that (a, y') is an optimum solution to (ATSPP DUAL) and has laminar support.*

Proposition 2. *Let (G, c, s, t) be an instance of ATSPP, where G is the support graph of an optimum solution x^* to (ATSPP LP). Let (a, y) be an optimum solution of (ATSPP DUAL). Let $U \in \{V\} \cup \text{supp}(y)$. Then the strongly connected components of $G[U]$ can be numbered $U_1, \ldots, U_l$ such that $\delta^-(U) = \delta^-(U_1)$, $\delta^+(U) = \delta^+(U_l)$, and $\delta^+(U_i) = \delta^-(U_{i+1}) \neq \emptyset$ for $i = 1, \ldots, l-1$. If $U = V$, then $s \in U_1$ and $t \in U_l$.*

Proposition 3. *Let (G, c, s, t) be an instance of ATSPP, where G is the support graph of an optimum solution to (ATSPP LP). Let (a, y) be an optimum solution to (ATSPP DUAL) with laminar support. Let $\bar{U} \in \{V\} \cup \text{supp}(y)$ and $v, w \in \bar{U}$. If w is reachable from v in the induced subgraph $G[\bar{U}]$, then there is a v-w-path in $G[\bar{U}]$ that enters and leaves every set $U \in \text{supp}(y)$ at most once.*

3 Bounding the Integrality Ratio

We first transform an instance and a solution to (ATSPP LP) to an instance and a solution to (ATSP LP) and work with an integral solution of this ATSP instance. The following lemma is essentially due to Feige and Singh [4]. For completeness, we prove it here again for our setting.

Lemma 1. *Let $d \geq 0$ be a constant. Then $\rho_{\text{ATSPP}} \leq (d+1)\rho_{\text{ATSP}} - d$ if the following condition holds for every instance $\mathcal{I} = (G, c, s, t)$ of ATSPP where G is the support graph of an optimum solution to (ATSPP LP): If there are s-t-walks $P_1, \ldots, P_k$ ($k > 0$) of total cost L in G, there is a single s-t-walk P in G with cost $c(P) \leq L + d(k-1) \cdot \text{LP}$ which contains all vertices of $P_1, \ldots, P_k$.*

Proof. Let $\mathcal{I} = (G, c, s, t)$ be an instance of ATSPP and x^* be an optimum solution to (ATSPP LP); so $\text{LP} = c(x^*)$. We may assume that G is the support graph of x^*. Consider the instance $\mathcal{I}' = (G', c')$ of ATSP that arises from $\mathcal{I}$ as follows. We add a new vertex v to G and two edges (t, v) and (v, s) with weights $c'(t, v) = d \cdot \text{LP}$ and $c'(v, s) = 0$. Then there is a feasible solution of (ATSP LP) for $\mathcal{I}'$ with cost $(d+1) \cdot \text{LP}$ (extend x^* by setting $x^*_{(t,v)} = x^*_{(v,s)} = 1$). Hence there is a solution to ATSP for $\mathcal{I}'$ with cost at most $(d+1)\rho_{\text{ATSP}} \cdot \text{LP}$. Let R be such a solution. Then R has to use (t, v) and (v, s) at least once, since it has to visit v. By deleting all copies of (t, v) and (v, s) from R, we get $k > 0$ s-t-walks in G with total cost at most $(d+1)\rho_{\text{ATSP}} \cdot \text{LP} - dk \cdot \text{LP}$ such that every vertex of G is visited by at least one of them. Our assumption now guarantees the existence of a single s-t-walk P with cost $c(P) \leq (d+1)\rho_{\text{ATSP}} \cdot \text{LP} - dk \cdot \text{LP} + d(k-1) \cdot \text{LP} = ((d+1)\rho_{\text{ATSP}} - d) \cdot \text{LP}$ in G, which contains every vertex of G. This walk is a solution of ATSPP for $\mathcal{I}$ and thus we have $\rho_{\text{ATSPP}} \leq (d+1)\rho_{\text{ATSP}} - d$ as proposed.   □

We now describe a merging procedure similar to one step ("inducing on a tight set") of the approximation algorithm for ATSP by Svensson, Tarnawski, and Végh [11].

Lemma 2. *Let (G, c, s, t) be an instance of ATSPP, where G is the support graph of an optimum solution to (ATSPP LP). Let (a, y) be an optimum solution to (ATSPP DUAL) with laminar support.*

Let $k > 0$ and $P_1, \ldots, P_k$ be s-t-walks in G with total cost L. Then there is a single s-t-walk P in G which contains every vertex of $P_1, \ldots, P_k$ and has cost at most $L + (k-1)(\text{LP} + 2(a_s - a_t))$.

Proof. Let $V_1, \ldots, V_l$ be the vertex sets of the strongly connected components of G in their topological order, which is unique by Proposition 2. Let P_i^j be the section of P_i that visits vertices in V_j (for $i = 1, \ldots, k$ and $j = 1, \ldots, l$). By Proposition 2 applied to $U = V$, none of these sections of P_i is empty. (Such a section might consist of a single vertex and no edges, but it has to contain at least one vertex.)

We consider paths R_i^j in G for $j = 1, \ldots, l$ that we will use to connect the walks $P_1^j, \ldots, P_k^j$ to a single walk visiting all vertices in V_j. See Fig. 3. If j is odd, let R_i^j (for $i = 1, \ldots k - 1$) be a path from the last vertex of P_i^j to the first vertex of P_{i+1}^j. If j is even, let R_i^j (for $i = 2, \ldots, k$) be a path from the last vertex of P_i^j to the first vertex of P_{i-1}^j. (Such paths exists because $G[V_j]$ is strongly connected.) By Proposition 3 we can choose the paths R_i^j such that they do not enter or leave any element of supp(y) more than once.

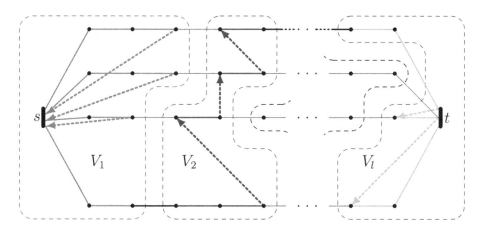

Fig. 2. Construction of P. The s-t-walks $P_1, \ldots, P_k$ are shown with solid lines. (Here, P_1 is the topmost walk and P_k is shown in the bottom.) The vertex sets $V_1, \ldots, V_l$ of the strongly connected components are indicated by the dashed lines. The red, blue, and green solid paths show the walks P_i^j, i.e. the sections of the walks P_i within the strongly connected components of G. The dotted arrows indicate the paths R_i^j. (Color figure online)

We now construct our s-t-walk P that will visit every vertex of $P_1, \ldots, P_k$. We start by setting $P = s$ and then add for $j = 1, \ldots, l$ all the vertices in V_j to P as follows. If j is odd, we append P_i^j and R_i^j for $i = 1$ to $i = k - 1$ and at last P_k^j. If j is even, we append P_i^j and R_i^j for $i = k$ to $i = 2$ and at last P_1^j. Note that when moving from one connected component V_i to the next component V_{i+1}, we use an edge from either P_1 (if i is even) or P_k (if i is odd). Then P is, indeed, an s-t walk in G and contains every vertex of $P_1, \ldots, P_k$. We now bound the cost of the walk P. For every edge $e = (v, w)$ of P we have by complementary slackness

$$c(e) = a_w - a_v + \sum_{U:e\in\delta(U)} y_U.$$

For an s-t-walk R in G we have

$$c(R) = \sum_{(v,w)\in E(R)} \left(a_w - a_v + \sum_{U:(v,w)\in\delta(U)} y_U \right) = a_t - a_s + c^y(R), \qquad (1)$$

where the cost function c^y is defined as $c^y(e) := \sum_{U:e\in\delta(U)} y_U$. Hence, to bound the cost of the s-t-walk P, we can bound $c^y(P)$ and then subtract a_s and add a_t.

P is constructed from pieces of $P_1, \dots, P_k$ and the paths R_i^j. Each of the paths R_i^j can only contain vertices of V_j. Two paths R_i^j and $R_{i'}^{j'}$, such that $j \neq j'$, can never both enter or both leave the same element of supp(y): otherwise they would contain vertices of the same strongly connected component of G by Proposition 2. Thus every element of supp(y) is entered at most $k-1$ times and left at most $k-1$ times on all the paths R_i^j used in the construction of P, and the total c^y cost of these paths is at most $(k-1)\sum_U 2y_U = (k-1)(\text{LP}+a_s-a_t)$. The c^y cost of the edges of $P_1, \dots, P_k$ is

$$\sum_{i=1}^k c^y(P_i) = \sum_{i=1}^k (c(P_i) - a_t + a_s) = L + k \cdot a_s - k \cdot a_t.$$

Consequently, we have

$$\begin{aligned} c(P) &= a_t - a_s + c^y(P) \\ &\leq a_t - a_s + L + k \cdot a_s - k \cdot a_t + (k-1)(\text{LP}+a_s-a_t) \\ &= L + (k-1)(\text{LP} + 2(a_s - a_t)) \end{aligned}$$

as claimed.   □

Svensson, Tarnawski, and Végh [11] reduced ATSP to so-called laminarly-weighted instances. In a laminarly-weighted instance we have $a = 0$ (and (a,y) has laminar support). For such instances Lemmas 1 and 2 would immediately imply our main result (even with better constants). However, the reduction to laminarly-weighted instances for ATSP does not yield an analogous statement for the path version. Instead, we will prove that $a_s - a_t \leq \text{LP}$ for some optimum dual LP solution (Sect. 4).

Let us first consider a simpler special case.

Definition 1. *An instance (G, c, s, t) of ATSPP or an instance (G, c) of ATSP is called* node-weighted *if there are nonnegative node weights $(c_v)_{v\in V}$ such that $c(v, w) = c_v + c_w$ for every edge (v, w).*

Note that node-weighted instances are not necessarily symmetric because it might happen that an edge (v, w) exists, but (w, v) does not exist. For node-weighted instances an argument similar to the proof of Lemma 2 can be used to prove the following bound.

Theorem 1. *Let ρ_{ATSP}^{NW} be the integrality ratio for ATSP on node-weighted instances and ρ_{ATSPP}^{NW} be the integrality ratio for ATSPP on node-weighted instances. Then*

$$\rho_{ATSPP}^{NW} \leq 2\rho_{ATSP}^{NW} - 1.$$

4   Bounding the Difference of a_s and a_t

Now we bound the difference of the dual variables a_s and a_t by LP. Using Lemmas 1 and 2, this will imply our main result $\rho_{\text{ATSPP}} \leq 4\rho_{\text{ATSP}} - 3$.

First, we give an equivalent characterization of the minimum value of $a_s - a_t$ in any optimum dual solution. This will not be needed to prove our main result, but might help to get some intuition.

Lemma 3. *Let $\mathcal{I} = (G, c, s, t)$ be an instance of ATSPP and let $\Delta \geq 0$. Now consider the instance $\mathcal{I}' = (G + e', c, s, t)$, where we add an edge $e' = (t, s)$ with $c(e') := \Delta$. Then $\text{LP}_{\mathcal{I}} \geq \text{LP}_{\mathcal{I}'}$. Moreover, $\text{LP}_{\mathcal{I}} = \text{LP}_{\mathcal{I}'}$ if and only if there exists an optimum solution (a, y) of (ATSPP DUAL) for the instance $\mathcal{I}$ with $a_s - a_t \leq \Delta$.*

We will now work with an optimum dual solution (a, y) with $a_s - a_t$ minimum. Note that this minimum is attained because for every feasible dual solution (a, y) we have $a_s - a_t \geq -\text{LP}$. By Proposition 1, we can assume in addition that (a, y) has laminar support.

Lemma 4. *Let (G, c, s, t) be an instance of ATSPP, where G is the support graph of an optimum solution to (ATSPP LP). Let (a, y) be an optimum solution of (ATSPP DUAL) such that $a_s - a_t$ is minimum. Let $\bar{U} \subseteq V \setminus \{s, t\}$ such that every s-t-path in G enters (and leaves) $\bar{U}$ at least once. Then $y_{\bar{U}} = 0$.*

Proof. (sketch) Suppose $y_{\bar{U}} > 0$ and let $\varepsilon := y_{\bar{U}}$. Let R be the set of vertices reachable from s in $G - \bar{U}$. We define a dual solution $(\bar{a}, \bar{y})$ as follows:

$$\bar{y}(U) := \begin{cases} y_U - \varepsilon & \text{if } U = \bar{U} \\ y_U & \text{else} \end{cases} \qquad \bar{a}_v := \begin{cases} a_v - 2\varepsilon & \text{if } v \in R \\ a_v - \varepsilon & \text{if } v \in \bar{U} \\ a_v & \text{else.} \end{cases}$$

We can show that $(\bar{a}, \bar{y})$ is also an optimum solution to (ATSPP DUAL). Since $\bar{a}_s - \bar{a}_t < a_s - a_t$, this yields a contradiction.   □

Lemma 5. *Let (G, c, s, t) be an instance of ATSPP, where G is the support graph of an optimum solution to (ATSPP LP). Let (a, y) be an optimum solution to (ATSPP DUAL) that has laminar support and minimum $a_s - a_t$.*

Then G contains two s-t-paths P_1 and P_2 such that for every set $U \in \text{supp}(y)$ we have $|E(P_1) \cap \delta(U)| + |E(P_2) \cap \delta(U)| \leq 2$.

Proof. (sketch) By Lemma 4, for every set $U \in \text{supp}(y)$ there is an s-t-path in G that visits no vertex in U. We contract all maximal sets $U \in \text{supp}(y)$. Using a variant of Menger's theorem, we can find two s-t-paths in G such that each vertex arising from the contraction of a set $U \in \text{supp}(y)$ is visited by at most one of the two paths.

Now we revert the contraction of the sets $U \in \text{supp}(y)$. We complete the edge sets of the two s-t-paths we found before (which are not necessarily connected anymore after undoing the contraction), to paths P_1 and P_2 with the desired properties. To see that this is possible, let v be the end vertex of an edge entering a contracted set $U \in \text{supp}(y)$ and let w be the start vertex of an edge leaving U. Then by Proposition 2, the vertex w is reachable from v in $G[U]$ and by Proposition 3, we can choose a v-w-path in $G[U]$ that enters and leaves every set $U' \in \text{supp}(y)$ with $U' \subsetneq U$ at most once. □

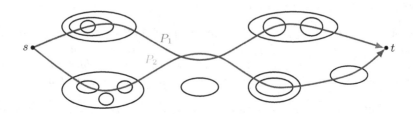

Fig. 3. The paths P_1 and P_2 as in Lemma 5. In black the vertex sets $U \in \text{supp}(y)$ are shown. The paths P_1 and P_2 are not necessarily disjoint but they never both cross the same set U with $y_U > 0$.

We finally show our main lemma.

Lemma 6. *Let $\mathcal{I} = (G, c, s, t)$ be an instance of ATSPP, where G is the support graph of an optimum solution to (ATSPP LP). Then there is an optimum solution (a, y) of (ATSPP DUAL) with laminar support and $a_s - a_t \leq \text{LP}$.*

Proof. Let (a, y) be an optimum solution to (ATSPP DUAL) that has laminar support and minimum $a_s - a_t$. Note that such an optimum dual solution exists by Proposition 1. We again define the c^y cost of an edge e to be $c^y(e) = \sum_{U: e \in \delta(U)} y_U$. By Lemma 5, G contains two s-t-paths P_1 and P_2 such that $c^y(P_1) + c^y(P_2) \leq \sum_{\emptyset \neq U \subseteq V \setminus \{s,t\}} 2 \cdot y_U$. Then, using (1),

$$
\begin{aligned}
0 &\leq c(P_1) + c(P_2) \\
&= c^y(P_1) - (a_s - a_t) + c^y(P_2) - (a_s - a_t) \\
&\leq \sum_{\emptyset \neq U \subseteq V \setminus \{s,t\}} 2 \cdot y_U - 2(a_s - a_t),
\end{aligned}
$$

implying

$$
a_s - a_t \leq \sum_{\emptyset \neq U \subseteq V \setminus \{s,t\}} 2 \cdot y_U - (a_s - a_t) = \text{LP}.
$$

□

We remark (although we will not need it) that Lemma 6 also holds for general instances. To adapt the proof, work with the subgraph G' of G that contains all edges of G for which the dual constraint is tight. Now G' plays the role of G in the proof, and by choosing ε small enough in the proof of Lemma 4 we maintain dual feasibility also for the edges that are not in G'.

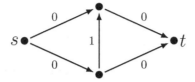

Fig. 4. Example with no optimum dual solutions with $a_s - a_t <$ LP: The numbers next to the arcs denote their cost. For this instance we have LP $= 1$. However adding an edge (t, s) with cost $\gamma < 1$ would result in an instance with LP $= \gamma$. By Lemma 3 there cannot be an optimum dual solution where $a_s - a_t < 1 =$ LP.

By Lemma 3, this also shows that adding an edge (t, s) of cost equal to the LP value does not change the value of an optimum LP solution.

The instance in Fig. 4 shows that the bound $a_s - a_t \le$ LP is tight. Note that the bound is also tight for the instance in Fig. 1 in which $x_e^* > 0$ for all edges e, and in which the integrality ratio is arbitrarily close to the best known lower bound of 2.

We will now prove our main result.

Theorem 2. *Let ρ_{ATSP} be the integrality ratio of (ATSP LP). Then the integrality ratio ρ_{ATSPP} of (ATSPP LP) is at most $4\rho_{ATSP} - 3$.*

Proof. Let (G, c, s, t) be an instance of ATSPP, where G is the support graph of an optimum solution to (ATSPP LP). By Lemma 6, there is an optimum dual solution (a, y) with laminar support and $a_s - a_t \le$ LP. Using Lemma 2, this implies that the condition of Lemma 1 is fulfilled for $d = 3$. This shows $\rho_{ATSPP} \le 4\rho_{ATSP} - 3$. ☐

5 Node-Weighted and Unweighted Instances

In the full version of the paper we observe that, for ATSP, node-weighted instances are not much more general than unweighted instances.

Theorem 3. *The integrality ratio of (ATSP LP) is the same for unweighted and for node-weighted instances. For any constants $\alpha \ge 1$ and $\varepsilon > 0$, there is a polynomial-time $(\alpha + \varepsilon)$-approximation algorithm for node-weighted instances if there is a polynomial-time α-approximation algorithm for unweighted instances.*

In particular, our construction implies that the node-weighted instances from Boyd and Elliot-Magwood [1] can be transformed to unweighted instances whose integrality ratio tends to 2. It seems that previously only unweighted instances with integrality ratio at most $\frac{3}{2}$ were known (e.g. [7]).

By splitting an arbitrary vertex into two copies s and t, both inheriting all incident edges, this also yields a family of unweighted digraph instances of ATSPP whose integrality ratio tends to two. We summarize:

Corollary 1. *The integrality ratio for unweighted digraph instances is at least two, both for (ATSP LP) and (ATSPP LP).*

References

1. Boyd, S., Ellitot-Magwood, P.: Computing the integrality gap of the asymmetric traveling salesman problem. Electron. Notes Discrete Math. **19**, 241–247 (2005)
2. Charikar, M., Goemans, M.X., Karloff, H.: On the integrality ratio for the asymmetric traveling salesman problem. Math. Oper. Res. **31**, 245–252 (2006)
3. Edmonds, J.: Optimum branchings. J. Res. Nat. Bur. Stand. B **71**, 233–240 (1967)
4. Feige, U., Singh, M.: Improved approximation ratios for traveling salesperson tours and paths in directed graphs. In: Charikar, M., Jansen, K., Reingold, O., Rolim, J.D.P. (eds.) APPROX/RANDOM -2007. LNCS, vol. 4627, pp. 104–118. Springer, Heidelberg (2007). https://doi.org/10.1007/978-3-540-74208-1_8
5. Friggstad, Z., Gupta, A., Singh, M.: An improved integrality gap for asymmetric TSP paths. Math. Oper. Res. **41**, 745–757 (2016)
6. Friggstad, Z., Salavatipour, M.R., Svitkina, Z.: Asymmetric traveling salesman path and directed latency problems. SIAM J. Comput. **42**, 1596–1619 (2013)
7. Gottschalk, C.: Approximation algorithms for the traveling salesman problem in graphs and digraphs. Master's Thesis, Research Institute for Discrete Mathematics, University of Bonn (2013)
8. Köhne, A., Traub, V., Vygen, J.: The asymmetric traveling salesman path LP has constant integrality ratio. arXiv:1808.06542 (2018)
9. Nagarajan, V., Ravi, R.: The directed minimum latency problem. In: Goel, A., Jansen, K., Rolim, J.D.P., Rubinfeld, R. (eds.) APPROX/RANDOM -2008. LNCS, vol. 5171, pp. 193–206. Springer, Heidelberg (2008). https://doi.org/10.1007/978-3-540-85363-3_16
10. Svensson, O.: Approximating ATSP by relaxing connectivity. In: Proceedings of the 56th Annual IEEE Symposium on Foundations of Computer Science (FOCS 2015), pp. 1–19 (2015)
11. Svensson, O., Tarnawski, J., Végh, L.: A constant-factor approximation algorithm for the asymmetric traveling salesman problem. In: Proceedings of the 50th Annual ACM Symposium on Theory of Computing (STOC 2018), pp. 204–213 (2018)

Approximate Multi-matroid Intersection via Iterative Refinement

André Linhares[1], Neil Olver[2,3], Chaitanya Swamy[1], and Rico Zenklusen[4(✉)]

[1] Department of Combinatorics and Optimization,
University of Waterloo, Waterloo, Canada
{alinhare,cswamy}@uwaterloo.ca
[2] Department of Econometrics and Operations Research,
Vrije Universiteit Amsterdam, Amsterdam, The Netherlands
n.olver@vu.nl
[3] CWI, Amsterdam, The Netherlands
[4] Department of Mathematics, ETH Zurich, Zurich, Switzerland
ricoz@math.ethz.ch

Abstract. We introduce a new iterative rounding technique to round a point in a matroid polytope subject to further matroid constraints. This technique returns an independent set in one matroid with limited violations of the other ones. On top of the classical steps of iterative relaxation approaches, we iteratively refine/split involved matroid constraints to obtain a more restrictive constraint system, that is amenable to iterative relaxation techniques. Hence, throughout the iterations, we both tighten constraints and later relax them by dropping constraints under certain conditions. Due to the refinement step, we can deal with considerably more general constraint classes than existing iterative relaxation/rounding methods, which typically round on one matroid polytope with additional simple cardinality constraints that do not overlap too much.

We show how our rounding method, combined with an application of a matroid intersection algorithm, yields the first 2-approximation for finding a maximum-weight common independent set in 3 matroids. Moreover, our 2-approximation is LP-based, and settles the integrality gap for the natural relaxation of the problem. Prior to our work, no upper bound better than 3 was known for the integrality gap, which followed from the greedy algorithm. We also discuss various other applications of our techniques, including an extension that allows us to handle a mixture of matroid and knapsack constraints.

1 Introduction

Matroids are among the most fundamental and well-studied structures in combinatorial optimization. Recall that a *matroid M* is a pair $M = (N, \mathcal{I})$, where N

A. Linhares and C. Swamy—Research supported by NSERC grant 327620-09 and an NSERC DAS Award.

N. Olver—Supported by NWO VIDI grant 016.Vidi.189.087.

R. Zenklusen—Supported by Swiss National Science Foundation grant 200021_165866.

A. Lodi and V. Nagarajan (Eds.): IPCO 2019, LNCS 11480, pp. 299–312, 2019.
https://doi.org/10.1007/978-3-030-17953-3_23

is a finite ground set and $\mathcal{I} \subseteq 2^N$ is a family of sets, called *independent sets*, such that (i) $\emptyset \in \mathcal{I}$, (ii) if $A \in \mathcal{I}$ and $B \subseteq A$, then $B \in \mathcal{I}$, and (iii) if $A, B \in \mathcal{I}$ with $|A| > |B|$, then there is an element $e \in A \setminus B$ such that $B \cup \{e\} \in \mathcal{I}$. We make the standard assumption that a matroid is specified via an *independence oracle*, which, given $S \subseteq N$ as input, returns if $S \in \mathcal{I}$. Matroids capture many interesting problems, and matroid-optimization algorithms provide a powerful tool in the design and analysis of efficient algorithms. A key matroid optimization problem is *matroid intersection*, wherein we seek a maximum-weight set that is independent in *two* matroids, for which various efficient algorithms are known, and we also have a celebrated min-max theorem and a polyhedral understanding of the problem. The versatility of matroid intersection comes from the fact that the intersection of matroids allows for describing a very broad family of constraints.

Unfortunately, as soon as the intersection of 3 or more matroids is considered, already the unweighted version of determining a maximum cardinality common independent set becomes APX-hard. Due to its fundamental nature, and many natural special cases, the problem of optimizing over 3 or more matroids has received considerable attention. In particular, there is extensive prior work ranging from the study of maximum cardinality problems [15], the maximization of submodular functions over the intersection of multiple matroids (see [4, 8, 11, 16, 17] and the references therein), to various interesting special cases like k-dimensional matching (see [3, 6, 7, 12, 13] and the references therein; many of these results apply also to the k-set packing problem which generalizes k-dimensional matching).

Nevertheless, there are still basic open questions regarding the approximability of the optimization over 3 or more matroids. Perhaps the most basic problem of this type is the *weighted 3-matroid intersection* problem, defined as follows.

Weighted 3-Matroid Intersection. Given matroids $M_i = (N, \mathcal{I}_i)$, for $i = 1, 2, 3$, on a common ground set N, and a weight vector $w \in \mathbb{R}^N$, solve

$$\max \left\{ w(I) : I \in \mathcal{I}_1 \cap \mathcal{I}_2 \cap \mathcal{I}_3 \right\},$$

where we use the shorthand $w(S) := \sum_{e \in S} w(e)$ for any set $S \subseteq N$.

The *unweighted 3-matroid intersection* problem, which is also sometimes called the *cardinality* version of 3-matroid intersection, is the special case where $w(e) = 1$ for all $e \in N$, so $w(S) = |S|$ for $S \subseteq N$.

The 3-matroid intersection problem has a natural and canonical LP-relaxation:

$$\max \left\{ w^T x : x \in P_{\mathcal{I}_1} \cap P_{\mathcal{I}_2} \cap P_{\mathcal{I}_3} \right\}, \tag{LP$_{\text{3-mat}}$}$$

where, for a matroid $M = (N, \mathcal{I})$, we denote by $P_{\mathcal{I}} \subseteq [0, 1]^N$ the matroid polytope of M, which is the convex hull of all characteristic vectors of sets in $\mathcal{I}$. It has a well known inequality description given by

$$P_{\mathcal{I}} = \left\{ x \in \mathbb{R}_{\geq 0}^N : x(S) \leq r(S) \ \forall S \subseteq N \right\},$$

where $r : 2^N \longrightarrow \mathbb{Z}_{\geq 0}$ is the *rank function* of M, which, for $S \subseteq N$, is defined by $r(S) := \max\{|I| : I \in \mathcal{I}, I \subseteq S\}$. The rank function is submodular, and $r(S)$ can be computed for any $S \subseteq N$ using an independence oracle. It will therefore often be convenient to assume that a matroid M is specified via its *rank oracle* that, given $S \subseteq N$ as input, returns $r(S)$. In particular, one can efficiently optimize any linear function over $P_\mathcal{I}$ given a rank oracle (or equivalently an independence oracle). The above LP-relaxation extends naturally to the *k-matroid intersection* problem, which is the extension of 3-matroid intersection to k matroids.

Whereas (LP$_{3\text{-mat}}$), and its extension (LP$_{k\text{-mat}}$) to k-matroid intersection, are well-known LP-relaxations, there remain various gaps in our understanding of these relaxations. It is widely known that the greedy algorithm is a k-approximation for k-matroid intersection. Moreover, this approximation is relative to the optimal value of (LP$_{k\text{-mat}}$), which leads to the current best upper bound of k on the integrality gap of (LP$_{k\text{-mat}}$), for all $k \geq 3$. However, the best lower bound on the integrality gap of (LP$_{k\text{-mat}}$) is $k-1$, whenever $k-1$ is a prime power; this is known to be achievable in instances where the involved matroids are partition matroids, and for unweighted instances [3,9,15,18].

Significant progress on approximating k-matroid intersection was achieved by Lee, Sviridenko, and Vondrák [17], who presented, for any fixed $\epsilon > 0$, a local search procedure with running time exponential in ϵ that leads to a $(k - 1 + \epsilon)$-approximation (i.e., the weight of the set returned is at least (optimum)$/(k-1+\epsilon)$). Unfortunately, apart from its high running time dependence on ϵ, this approach does not shed any insights on (LP$_{k\text{-mat}}$), as the above guarantee is not relative to $OPT_{\text{LP}_{k\text{-mat}}}$. Further progress on understanding the quality of the LP-relaxations has only been achieved in special cases. In particular, for *unweighted k-matroid intersection*, Lau, Ravi and Singh [15] give an LP-based $(k - 1)$-approximation through iterative rounding. Their proof is based on identifying an element with "large" fractional value, picking it, and altering the fractional solution so that it remains feasible; the last step crucially uses the fact that the instance is unweighted to control the loss in the LP objective value. For the intersection of k unitary partition matroids, a problem also known as *k-dimensional matching*, Chan and Lau [3] were able to obtain a $(k - 1)$-approximation based on (LP$_{k\text{-mat}}$), and Parekh and Pritchard [18] later obtained the same approximation factor for the intersection of k (not necessarily unitary) partition matroids.

Although it is generally believed that a $(k - 1)$-approximation for k-matroid intersection should exist, and that the integrality gap of (LP$_{k\text{-mat}}$) is equal to the known lower bound of $k-1$, this has remained open even for 3-matroid intersection (prior to our work). Recall that in this case, the best known upper and lower bounds on the integrality gap of (LP$_{3\text{-mat}}$) are 3 (via the classical greedy algorithm) and 2 respectively. Moreover, the only method to beat the trivial 3-approximation of the greedy algorithm is the non-LP based and computationally quite expensive $(2 + \epsilon)$-approximation in [17]. One main reason for the limited progress is the lack of techniques for rounding points in the intersection of multiple matroid polytopes with sufficiently strong properties. In particular, one technical difficulty that is encountered is that the tight constraints (even at an extreme point) may have large overlap, and we do not know of ways for dealing with this.

Our Results. We introduce a new iterative rounding approach to handle the above difficulties, that allows for dealing with a very general class of optimization problems involving matroids. Before delving into the details of this technique, we highlight its main implication in the context of 3-matroid intersection.

Theorem 1. *There is an LP-relative 2-approximation for weighted 3-matroid intersection. That is, for any instance, we can efficiently find a common independent set R with $w(R) \geq OPT_{\text{LP3-mat}}/2$; thus, the integrality gap of (LP$_{3\text{-mat}}$) is at most 2.*

This is the *first* 2-approximation for 3-matroid intersection (with general weights). Moreover, our result *settles* the integrality gap of (LP$_{3\text{-mat}}$) due to the known matching integrality gap lower bound of 2.

The chief new technical ingredient that leads to Theorem 1, and results for other applications discussed in Sect. 3, is an approximation result based on a novel *iterative refinement* technique (see Sect. 2) for problems of the following type. Let $N = N_0$ be a finite ground set, and $M_i = (N_i, \mathcal{I}_i)$ for $i = 0, \ldots, k$ be $k+1$ matroids with rank functions $\{r_i\}$, where $N_i \subseteq N$, and $w \in \mathbb{R}^N$ be a weight vector (note that *negative* weights are allowed). We consider the problem

$$\max \left\{ w(I) : \ I \in \mathcal{B}_0, \quad I \cap N_i \in \mathcal{I}_i \ \forall i \in [k] \right\}, \tag{1}$$

where $\mathcal{B}_0$ is the set of all bases of M_0 and $[k] := \{1, \ldots, k\}$. The reason we consider matroids M_i for $i \in [k]$ defined on ground sets N_i that are subsets of N, is because, as we show below, we obtain guarantees depending on how strongly the sets N_i overlap; intuitively, problem (1) becomes easier as the overlap between $N_1, \ldots, N_k$ decreases, and our guarantee improves correspondingly.

We cannot hope to solve (1) optimally, as this would enable one to solve the NP-hard k-matroid intersection problem. Our goal will be to find a basis of M_0 of large weight that is "approximately independent" in the matroids $M_1, \ldots, M_k$.

How to quantify "approximate independence"? Perhaps the two notions that first come to mind are additive and multiplicative violation of the rank constraints. Whereas additive violations are common in the study of degree-bounded MST problems, which can be cast as special cases of (1), it turns out that such a guarantee is impossible to obtain (in polytime) for (1). More precisely, we show in Appendix A (via a replication idea) that, even for $k = 2$, if we could find in polytime a basis B of M_0 satisfying $|B| \leq r_i(B) + \alpha$ for $i = 1, 2$ for $\alpha = O(|N|^{1-\epsilon})$ for any $\epsilon > 0$, then we could efficiently find a basis of M_0 that is independent in M_1, M_2; the latter problem is easily seen to be NP-hard via a reduction from Hamiltonian path. We therefore consider multiplicative violation of the rank constraints. We say that $S \subseteq N$ is α-*approximately independent*, or simply α-*independent*, for a matroid $M = (N, \mathcal{I})$, if $|T| \leq \alpha \cdot r(T) \ \forall T \subseteq S$ (equivalently, $\chi^S \in \alpha P_\mathcal{I}$, where χ^S is the characteristic vector of S). This is much stronger than simply requiring that $|S| \leq \alpha \cdot r(S)$, and it is easy to give examples where this weaker notion admits sets that one would consider to be quite far from being independent. An appealing feature of the stronger definition is that, using the min-max result for matroid-intersection (or via matroid partition; see, e.g., [5]), it follows easily that if $\alpha \in \mathbb{Z}_{\geq 1}$, then S is α-independent iff S can be partitioned into at most α independent sets of M. We now state the guarantee we obtain for (1) precisely. We consider the following canonical LP relaxation of (1):

$$\max \left\{ w^T x : \ x \in \mathbb{R}_{\geq 0}^N, \ x|_{\mathcal{B}_0} \in P_{\mathcal{B}_0}, \ x|_{N_i} \in P_{\mathcal{I}_i} \ \forall i \in [k] \right\}, \qquad (\text{LP}_{\text{mat}})$$

where for a set $S \subseteq N$, we use $x|_S \in \mathbb{R}^S$ to denote the restriction of x to S, and $P_{\mathcal{B}_0} := P_{\mathcal{I}_0} \cap \{x \in \mathbb{R}^N : x(N) = r_0(N)\}$ is the matroid base polytope of M_0. For ease of notation, we will sometimes write $x \in P_{\mathcal{I}_i}$ and $R \in \mathcal{I}_i$ instead of $x|_{N_i} \in P_{\mathcal{I}_i}$ and $R \cap N_i \in \mathcal{I}_i$, respectively. Our main result for (1), based on a new iterative rounding algorithm for (LP_{mat}) described in Sect. 2, is the following.

Theorem 2. *Let* $q_1, \ldots, q_k \in \mathbb{Z}_{\geq 1}$ *such that*

$$\sum_{i \in [k]: e \in N_i} q_i^{-1} \leq 1 \qquad \forall e \in N. \qquad (2)$$

If (LP_{mat}) *is feasible, then one can efficiently compute* $R \subseteq N$ *such that (i)* $R \in \mathcal{B}_0$; *(ii)* $w(R) \geq OPT_{\text{LP}_{\text{mat}}}$; *and (iii)* R *is* q_i-*independent in* $M_i \ \forall i \in [k]$.

Note that, in particular, taking $q_i = \max_{e \in N} \left| \{j \in [k] : e \in N_j\} \right|$ for all $i \in [k]$ satisfies (2). Thus, we violate the constraints imposed by the other matroids $M_1, \ldots, M_k$ by a multiplicative factor depending on how strongly the N_is overlap.

While we have stated Theorem 2 in terms of *bases* of M_0, the following natural variant is easily deduced from it (we defer the proof to Appendix B).

Corollary 3. *Theorem 2 also holds when* R *is required only to be an independent set in* M_0 *(as opposed to a basis), and we replace* $P_{\mathcal{B}_0}$ *in* (LP_{mat}) *by* $P_{\mathcal{I}_0}$.

A variety of problem settings can be handled via Theorem 2 and Corollary 3 in a unified way. We first show how to obtain a crisp, simple proof of Theorem 1.

Proof of Theorem 1. Given matroids $M_i = (N, \mathcal{I}_i)$ for $i = 0, 1, 2$, and a weight vector $w \in \mathbb{R}^N$, we first solve $(\text{LP}_{3\text{-mat}})$ to obtain an optimal solution x^*. Now we utilize Corollary 3 with the same three matroids, and $q_1 = q_2 = 2$. Clearly, these q-values satisfy (2), and x^* is a feasible solution to (LP_{mat}), when we replace $P_{\mathcal{B}_0}$ by $P_{\mathcal{I}_0}$. Thus, we obtain a set $A \in \mathcal{I}_0$ with $w(A) \geq w^T x^*$ and $\chi^A \in 2P_{\mathcal{I}_1} \cap 2P_{\mathcal{I}_2}$.

It is well known that $P_{\mathcal{I}_1} \cap P_{\mathcal{I}_2}$ is a polytope with integral extreme points (see, e.g., [5]). So since $\chi^A/2 \in P_{\mathcal{I}_1} \cap P_{\mathcal{I}_2}$, by using an algorithm for (weighted) matroid intersection applied to matroids M_1 and M_2 restricted to A, we can find a set $R \subseteq A$ such that $R \in \mathcal{I}_1 \cap \mathcal{I}_2$ and $w(R) \geq w^T \chi^A/2 \geq w^T x^*/2$. Finally, since $R \subseteq A$ and $A \in \mathcal{I}_0$, we also have that $R \in \mathcal{I}_0$. $\qquad \square$

Beyond 3-matroid intersection, Theorem 2 is applicable to various constrained (e.g., degree-bounded) spanning tree problems; we expand on this below. In Sect. 3, we discuss an application in this direction, wherein we seek a min-cost spanning tree satisfying matroid-independence constraints on the edge-sets of a given disjoint collection of node sets. Using Theorem 2, we obtain a spanning tree with a multiplicative factor-2 violation of the matroid constraints.

In Sect. 3, we also present a noteworthy extension of Theorem 2 with t *knapsack constraints* in addition to k matroid constraints, where we obtain bounded multiplicative violations of all involved constraints. The only other such result we are aware of that applies to a *mixture* of matroid and knapsack constraints is by Gupta et al. [10]; their result in our setting yields an $O(kt)$-approximation with no constraint violation, which is incomparable to our result.

Related Work and Connections. By choosing M_0 to be a graphic matroid, problem (1) generalizes many known constrained spanning tree problems, including degree-bounded spanning trees, and generalizations thereof considered by Bansal et al. [2], Király et al. [14], and Zenklusen [21]. Theorem 2 thus yields a unified way to deal with various spanning tree problems considered in the literature, where the soft/degree constraints are violated by at most a constant factor. However, as noted earlier, whereas the above works obtain stronger, additive violation results, such guarantees are not possible for our general problem (1) (see Appendix A). This hardness (of obtaining small additive violations) carries over to the spanning tree application that we consider in Sect. 3 (which generalizes the matroidal degree-bounded spanning tree problem considered in [21]).

To showcase how Theorem 2 can be used for such problems, consider the minimum degree-bounded spanning tree problem, where given is a graph $G = (V, E)$ with edge weights $w : E \to \mathbb{R}$ and degree bounds $B_v \in \mathbb{Z}_{\geq 1}$ for $v \in V$. The nominal problem asks to find a spanning tree $T \subseteq E$ with $|T \cap \delta(v)| \leq B_v$ for $v \in V$ minimizing $w(T)$, where $\delta(v)$ denotes the set of edges incident with v. Here one can apply Theorem 2 with M_0 being the graphic matroid of G, and for each $v \in V$ we define a uniform matroid M_v with ground set $\delta(v)$ and rank B_v. Theorem 2 with $q_v = 2$ $\forall v \in V$ and negated edge weights leads to a spanning tree T with $|T \cap \delta(v)| \leq 2B_v$ $\forall v \in V$ and weight no more than the optimal LP-weight. Whereas this is a simple showcase example, Theorem 2 can be used in a similar way for considerably more general constraints than just degree constraints.

Finally, we highlight a main difference of our approach compared to prior techniques. Prior techniques for related problems, as used, for example, by Singh and Lau [20], Király et al. [14], and Bansal et al. [2], successively drop constraints of a relaxation. Also, interesting variations have been suggested that do not just drop constraints but may relax constraints by replacing a constraint by a weaker family (see work by Bansal etal. [1]). In contrast, our method does not just relax constraints, but also strengthens the constraint family in some iterations, so as to simplify it and enable one to drop constraints later on.

2   Our Rounding Technique

Our rounding technique heavily relies on a simple yet very useful "splitting" procedure for matroids, which we call *matroid refinement*.

Matroid Refinement. Let $M = (N, \mathcal{I})$ be a matroid with rank function $r : 2^N \rightarrow \mathbb{Z}_{\geq 0}$, and let $S \subsetneq N$, $S \neq \emptyset$. *Refining* M with respect to S yields the two matroids $M_1 = M|_S$ obtained by restricting M to S, and $M_2 = M/S$ obtained by contracting S in M. Formally, the independent sets of the two matroids $M_1 = (S, \mathcal{I}_1), M_2 = (N \setminus S, \mathcal{I}_2)$ are given by $\mathcal{I}_1 = \{I \subseteq S : I \in \mathcal{I}\}$, and $\mathcal{I}_2 = \{I \subseteq N \setminus S : I \cup I_S \in \mathcal{I}\}$, where $I_S \in \mathcal{I}$ is a maximum cardinality independent subset of S. It is well-known that the definition of $\mathcal{I}_2$ does not depend on which set I_S is chosen. The rank functions $r_1 : 2^S \rightarrow \mathbb{Z}_{\geq 0}$ and $r_2 : 2^{N \setminus S} \rightarrow \mathbb{Z}_{\geq 0}$ of M_1 and M_2, respectively, are given by

$$r_1(A) = r(A) \ \forall A \subseteq S, \text{ and } r_2(B) = r(B \cup S) - r(S) \ \forall B \subseteq N \setminus S. \quad (3)$$

We refer the reader to [19, Volume B] for more information on matroid restrictions and contractions. The following lemma describes two basic yet important relations between a matroid $M = (N, \mathcal{I})$ and its refinements $M_1 = M|_S$ and $M_2 = M/S$. These relations easily follow from well-known properties of matroids; we omit the proofs here, but include them in the full version.

Lemma 4. *(i)   If $x \in \mathbb{R}^N$ satisfies $x|_S \in P_{\mathcal{I}_1}$ and $x|_{N \setminus S} \in P_{\mathcal{I}_2}$, then $x \in P_{\mathcal{I}}$.
(ii)   Let $x \in P_{\mathcal{I}}$ be such that $x(S) = r(S)$. Then $x|_S \in P_{\mathcal{I}_1}$ and $x|_{N \setminus S} \in P_{\mathcal{I}_2}$.*

Intuitively, matroid refinement serves to partly decouple the matroid independence constraints for M, thereby allowing one to work with somewhat "simpler" matroids subsequently, and we leverage this carefully in our algorithm.

An Algorithm Based on Iterative Refinement and Relaxation. Algorithm 1 describes our method to prove Theorem 2. Recall that the input is an instance of problem (1), which consists of $k + 1$ matroids $M_i = (N_i, \mathcal{I}_i)$ for $i = 0, \ldots, k$, where each N_i is a subset of a finite ground set $N = N_0$, and a weight vector $w \in \mathbb{R}^N$. We are also given integers $q_i \geq 1$ for $i \in [k]$ satisfying (2).

Algorithm 1 starts by solving the natural LP-relaxation in step 2 to obtain an optimal extreme point x^*. As is common in iterative rounding algorithms, we delete all elements of value 0 and fix all elements of value 1 through contractions in step 3. Apart from these standard operations, we *refine* the matroids in step 5, as long as there is a matroid $M' = (N', \mathcal{I}')$ in our collection $\mathcal{M}$ with a nontrivial x^*-*tight* set $S \subseteq N'$, i.e., $x^*(S) = r'(S)$ and $S \notin \{\emptyset, N'\}$. Notice that after step 5, the q-values for the matroids in the new collection $\mathcal{M}$ continue to satisfy (2). Step 6 is our relaxation step, where we drop a matroid $M' = (N', \mathcal{I}')$ if $|N'| - x^*(N') < q_{M'}$. This is the step that results in a violation of the matroid constraints, but, as we show, the above condition ensures that even if we select all elements of N' in the solution, the violation is still within the prescribed bounds.

Algorithm 1. Iterative refinement/relaxation algorithm for Theorem 2

1. Initialize $\mathcal{M} \leftarrow \{M_1, \ldots, M_k\}$, $q_{M_i} \leftarrow q_i$ for all $i \in [k]$.
2. Compute an optimal basic solution x^* to (LP$_{\text{mat}}$) for the matroids $\{M_0\} \cup \mathcal{M}$.
3. Delete all $e \in N$ with $x^*(e) = 0$ and contract all $e \in N$ with $x^*(e) = 1$ from all relevant matroids, updating also the ground set N.
4. If $N = \emptyset$: **return** the set of all elements contracted so far.
5. **While** there is a matroid $M' = (N', \mathcal{I}') \in \mathcal{M}$ with associated rank function r', s.t. $\exists \emptyset \neq S \subsetneq N'$ with $x^*(S) = r'(S)$:
   (Refinement.) Set $M_1' = M'|_S$, $M_2' = M'/S$, and $q_{M_1'} = q_{M_2'} = q_{M'}$.
   Update $\mathcal{M} \leftarrow (\mathcal{M} \setminus \{M'\}) \cup \{M_1', M_2'\}$.
6. Find a matroid $M' = (N', \mathcal{I}') \in \mathcal{M}$ with associated rank function r', such that $x^*(N') = r'(N')$ and $|N'| - x^*(N') < q_{M'}$; remove M' from $\mathcal{M}$. Go to step 2.

Moreover, we will show in the proof of Lemma 6 that, whenever Algorithm 1 is at step 6, there is a matroid $M' = (N', \mathcal{I}') \in \mathcal{M}$ that can be dropped, i.e., $x^*(N') = r'(N')$ and $|N'| - x^*(N') < q_{M'}$. We remark that, in step 6, one could also drop *all* matroids $M' \in \mathcal{M}$ fulfilling this condition, instead of just a single one, without impacting the correctness of the algorithm.

One can find an x^*-tight set $\emptyset \neq S \subsetneq N'$ (if one exists) in step 5 by minimizing the submodular function $r'(A) - x^*(A)$ over the sets $\emptyset \neq A \subsetneq N'$. Depending on the matroids involved, faster specialized approaches can be employed.

It is perhaps illuminating to consider the combined effect of all the refinement steps and step 6 corresponding to a given basic optimal solution x^*. Using standard uncrossing techniques, one can show that for each matroid $M' = (N', \mathcal{I}') \in \mathcal{M}$, there is a nested family of sets $\emptyset \subsetneq S_1 \subsetneq \ldots \subsetneq S_p \subseteq N'$ whose rank constraints span the x^*-tight constraints of M'. Let $p' := p$ if $p = 0$ or $S_p \neq N'$, and let $p' := p - 1$ otherwise; so any S_i with $i \in [p']$ can be used to refine M'. The combined effect of steps 5 for M' can be seen as replacing M' by the matroids $(M'|_{S_\ell})/S_{\ell-1}$ for $\ell = 1, \ldots, p' + 1$, where $S_0 := \emptyset$ and $S_{p'+1} := N'$. Step 6 chooses some $M' \in \mathcal{M}$ and a "ring" $S_\ell \setminus S_{\ell-1}$ of its nested family satisfying $|S_\ell \setminus S_{\ell-1}| - x^*(S_\ell \setminus S_{\ell-1}) < q_{M'}$, and drops the matroid created for this ring.

Analysis. Lemma 5 shows that if Algorithm 1 terminates, then it returns a set with the desired properties. In Lemma 6, we show that the algorithm terminates in a polynomial number of iterations. In particular, we show that in step 6, there will always be a matroid in our collection that we can drop.

Lemma 5. *Suppose that Algorithm 1 returns a set $R \subseteq N$. Then, R satisfies the properties stated in Theorem 2.*

Proof. Note that $R \in \mathcal{B}_0$, as M_0 is only modified via deletions or contractions. Moreover, $w(R) \geq OPT$, where OPT is the optimal value of (LP$_{\text{mat}}$) for the input instance. Indeed, if x^* is the current optimal solution, and we update our instance (via deletions, contractions, refinements, or dropping matroids), then

x^* restricted to the new ground set remains feasible for (LP$_{\text{mat}}$) for the new instance. This is immediate for deletions and contractions, and if we drop a matroid; it holds for refinements due to Lemma 4 (ii). So if the optimal value of (LP$_{\text{mat}}$) decreases, this is only because we contract elements with $x^*(e) = 1$, which we include in R. It follows that $w(R) \geq OPT$.

It remains to show that R satisfies property (iii) of Theorem 2, i.e., R is q_i-independent in M_i for all $i \in [k]$. To this end, consider the state of the algorithm at a point during its execution right before step 2. Hence, the instance may already have been modified through prior refinements, contractions, deletions, and relaxations. We claim that the invariant below holds throughout the algorithm:

> If a subset R' of the current ground set satisfies the properties of Theorem 2 with respect to the current instance, then the set R consisting of R' and all elements contracted so far fulfills the properties of Theorem 2 with respect to the original instance.

To show the claim, it suffices to show that the invariant is preserved whenever we change the instance in the algorithm. Note that $R = R'$ unless the change involves contracting an element. First, one can observe that if the instance changes by deleting an element of value 0 or contracting an element of value 1, then the invariant is preserved. Next, consider step 5, where we refine $M' = (N', \mathcal{I}') \in \mathcal{M}$ to obtain $M'|_S = (S, \mathcal{I}'_1)$ and $M'/S = (N' \setminus S, \mathcal{I}'_2)$ whose q-values are set to $q_{M'}$. We are given that $\chi^{R'}|_S \in q_{M'} P_{\mathcal{I}'_1}$ and $\chi^{R'}|_{N' \setminus S} \in q_{M'} P_{\mathcal{I}'_2}$, and we have $R = R'$. So by Lemma 4 (i), we have $\chi^R / q_{M'} \in P_{\mathcal{I}'}$, or equivalently $\chi^R \in q_{M'} P_{\mathcal{I}'}$.

Finally, consider the case where a matroid $M' = (N', \mathcal{I}') \in \mathcal{M}$ gets dropped in step 6. We have $R = R'$, and we need to show that $\chi^R|_{N'} \in q_{M'} P_{\mathcal{I}'}$. Let x^* be the optimal solution used in the algorithm when M' was dropped. We have $|N'| - x^*(N') < q_{M'}$, and since $x^*(N') = r'(N')$, and both $|N'|$ and $q_{M'}$ are integral, this implies $|N'| - x^*(N') \leq q_{M'} - 1$. So N' can be partitioned into a basis of M', which has size $r'(N') = x^*(N') \geq |N'| - (q_{M'} - 1)$, and at most $q_{M'} - 1$ other singleton sets. Each singleton $\{e\}$ is independent in M', since $0 < x^*(e) \leq r'(\{e\})$ as $x^*|_{N'} \in P_{\mathcal{I}'}$. Therefore, N' can be partitioned into at most $q_{M'}$ independent sets of M'. Intersecting these sets with R shows that $R \cap N'$ can be partitioned into at most $q_{M'}$ independent sets of M'. □

We now prove that the algorithm terminates. Note that refinements guarantee that whenever the algorithm is at step 6, then for any $M' = (N', \mathcal{I}') \in \mathcal{M}$, only the constraint of $P_{\mathcal{I}'}$ corresponding to N' may be x^*-tight. This allows us to leverage ideas similar to those in [2,14] to show that step 6 is well defined.

Lemma 6. *Algorithm 1 terminates in at most $(2k + 1)|N|$ iterations.*

Proof. We show that whenever the algorithm is at step 6, then at least one matroid in our collection can be dropped. This implies the above bound on the number of iterations as follows. There can be at most $|N|$ deletions or contractions. Each matroid $M_i = (N_i, \mathcal{I}_i)$ in our input spawns at most $|N_i|$ refinements, as each refinement of a matroid creates two matroids with disjoint (nonempty) ground sets. This also means that step 6 can be executed at most $k|N|$ times.

We focus on showing that step 6 is well defined. Consider the current collection of matroids $\mathcal{M}$. (Recall that $\mathcal{M}$ does not contain the current version of M_0.) Let x^* be the current basic solution, which is not integral; otherwise every element would have been deleted or contracted in step 3 and we would have terminated in step 4. Since we deleted all elements e with $x^*(e) = 0$, the current ground set N satisfies $N = \mathrm{supp}(x^*) := \{e \in N : x^*(e) > 0\}$.

Consider a full-rank subsystem of $(\mathrm{LP}_{\mathrm{mat}})$, $Ax = b$, consisting of linearly independent, x^*-tight constraints. By standard uncrossing arguments, we may assume that the constraints of $Ax = b$ coming from a single matroid correspond to a nested family of sets. The system $Ax = b$ must contain some constraint corresponding to a matroid $M' \in \mathcal{M}$. Otherwise, we would have a full-rank system consisting of constraints coming from only one matroid, namely M_0, which would yield a unique integral solution; but x^* is not integral. Furthermore, for a matroid $M' = (N', \mathcal{I}') \in \mathcal{M}$, the only constraint of $P_{\mathcal{I}'}$ that can be x^*-tight corresponds to N', as otherwise, M' would have been refined in step 5. So a matroid $M' \in \mathcal{M}$ gives rise to at most one row of A, which we denote by $A_{M'}$ if it exists. Let $\emptyset \subsetneq S_1 \subsetneq \ldots \subsetneq S_p \subseteq N_0 = N$ denote the nested family of sets that give rise to the constraints of M_0 in our full-rank system.

Consider the following token-counting argument. Each $e \in N$ gives $x^*(e)$ tokens to the row of A corresponding to the smallest set S_ℓ containing e (if one exists). It also supplies $\left(1 - x^*(e)\right)/q_{M'}$ tokens to every row $A_{M'}$ corresponding to a matroid $M' \in \mathcal{M}$ whose ground set contains e. Since the q-values satisfy (2), every $e \in N$ supplies at most one unit of token in total to the rows of A. Every row of A corresponding to a set S_ℓ receives $x^*(S_\ell) - x^*(S_{\ell-1})$ tokens, where $S_0 := \emptyset$. This is positive and integer, and thus at least 1. We claim that there is some $e \in N$ that supplies strictly less than one token unit. Given this, it must be that there is a row $A_{M'}$ corresponding to a matroid $M' = (N', \mathcal{I}') \in \mathcal{M}$ that receives less than 1 token unit; thus $|N'| - x^*(N') < q_{M'}$ as desired.

Finally, we prove the claim. If every element supplies *exactly* one token unit, then it must be that: (i) $S_p = N$, (ii) inequality (2) is tight for all $e \in N$, and (iii) for every $e \in N$, every matroid $M' = (N', \mathcal{I}') \in \mathcal{M}$ with $e \in N'$ gives rise to a row $A_{M'}$. But then $\sum_{M' \in \mathcal{M}} \frac{1}{q_{M'}} \cdot A_{M'} = \chi^N$, which is the row of A corresponding to the constraint of M_0 for the set S_p. This contradicts that A has full rank. □

3   Further Applications and Extensions

Generalized Matroidal Degree-Bounded Spanning Tree (GMDST). In this problem, we are given an undirected graph $G = (V, E)$ with edge costs $c \in \mathbb{R}^E$, disjoint node-sets $S_1, \ldots, S_k$, and matroids $M_i = (\delta(S_i), \mathcal{I}_i)$ for all $i \in [k]$, where $\delta(S_i)$ is the set of edges of G that cross S_i. We want to find a spanning tree T of minimum cost such that $T \cap \delta(S_i) \in \mathcal{I}_i$ for all $i \in [k]$. This generalizes the matroidal degree-bounded MST problem considered by [21], wherein each node $\{v\}$ is an S_i set. Clearly, each edge belongs to at most 2 ground sets of the matroids $\{M_i\}_{i \in [k]}$. Thus, by taking M_0 to be the graphic matroid and setting $w = -c$, Theorem 2 leads to a tree T of cost at most the optimum such that $T \cap \delta(S_i)$ is 2-independent in M_i for all $i \in [k]$.

We remark that, whereas [21] obtains an $O(1)$-*additive* violation of the matroid constraints for matroidal degree-bounded MST problem, such a polytime additive guarantee is not possible for GMDST unless $P = NP$. This follows from the same replication idea used in Appendix A to rule out small additive violations for (1).

Extension to Knapsack Constraints. We can consider a generalization of (1), where, in addition to the matroids $M_0, \ldots, M_k$ (over subsets of N) and weight vector $w \in \mathbb{R}^N$, we have t knapsack constraints, indexed by $i = k + 1, \ldots, k + t$. The i-th knapsack constraint is specified by a ground set $N_i \subseteq N$, cost vector $c^i \in \mathbb{R}_{\geq 0}^{N_i}$, and budget $U_i \geq 0$. The goal is to find a maximum-weight set R such that $R \in \mathcal{B}_0 \cap \mathcal{I}_1 \cap \ldots \cap \mathcal{I}_k$, and satisfying $c^i(R \cap N_i) \leq U_i$ for all $i = k + 1, \ldots, k + t$.

We consider the natural LP-relaxation ($\mathrm{LP}_{\mathrm{matkn}}$) for this problem, and extend Theorem 2 to obtain the following result; we sketch the proof in Appendix B.

Theorem 7. *Let* $q_1, \ldots, q_{k+t} \in \mathbb{Z}_{\geq 1}$ *be such that* $\sum_{i \in [k+t]: e \in N_i} \frac{1}{q_i} \leq 1$ *for all* $e \in N$. *If* ($\mathrm{LP}_{\mathrm{matkn}}$) *is feasible, then one can efficiently compute* $R \subseteq N$ *such that (i)* $R \in \mathcal{B}_0$; *(ii)* $w(R) \geq OPT_{\mathrm{LP}_{\mathrm{matkn}}}$; *(iii)* R *is* q_i-*independent in* M_i *for all* $i \in [k]$; *and (iv)* $c^i(R \cap N_i) \leq U_i + q_i \cdot \left(\max_{e \in N_i} c_e^i \right)$ *for all* $i \in \{k+1, \ldots, k+t\}$.

A   Impossibility of Achieving Small Additive Violations

We show that Theorem 2 for problem (1) cannot be strengthened to yield a basis of M_0 that has small additive violation for the matroid constraints of $M_1, \ldots, M_k$, even when $k = 2$.

We first define additive violation precisely. Given a matroid $M = (N, \mathcal{I})$ with rank function r, we say that a set $R \subseteq N$ is μ-*additively independent* in M if $|R| - r(R) \leq \mu$; equivalently, we can remove at most μ elements from R to obtain an independent set in M. Unlike results for degree-bounded spanning trees, or matroidal degree-bounded MST [21], we show that small additive violation is not possible in polytime (assuming $P \neq NP$) even for the special case of (1) where $k = 2$, so we seek a basis of M_0 that is independent in M_1, M_2.

Theorem 8. *Let* $f(n) = O(n^{1-\varepsilon})$, *where* $\varepsilon > 0$ *is a constant. Suppose we have a polytime algorithm* $\mathcal{A}$ *for (1) that returns a basis* B *of* M_0 *satisfying* $|B| \leq r_i(B) + f(|N|)$ *for* $i = 1, 2$ *(where* r_i *is the rank function of* M_i). *Then we can find in polytime a basis of* M_0 *that is independent in* M_1, M_2.

The problem of finding a basis of M_0 that is independent in M_1, M_2 is NP-hard, as shown by an easy reduction from the directed Hamiltonian path problem. Thus, Theorem 8 shows that it is NP-hard to obtain an additive violation for problem (1) that is substantially better than linear violation.

Proof of Theorem 8. Choose t large enough so that $t > 2f(t|N|)$. Since $f(n) = O(n^{1-\varepsilon})$, this is achieved by some $t = \text{poly}(|N|)$. For each $i \in \{0, 1, 2\}$, let M_i' be the direct sum of t copies of M_i. Let N' be the ground set of these matroids, which consists of t disjoint copies of N, which we label $N_1, \ldots, N_t$.

Clearly, the instance (M_0', M_1', M_2') is feasible iff the original instance is feasible. Suppose that running $\mathcal{A}$ on the replicated instance yields a basis R' of M_0' that has the stated additive violation for the matroids M_1', M_2'. Hence, there are two sets $Q_1, Q_2 \subseteq R'$ with $|Q_1|, |Q_2| \le f(t|N|)$, such that $R' \setminus Q_i$ is independent in M_i' for $i = 1, 2$. Hence, $R' \setminus (Q_1 \cup Q_2)$ is independent in both M_1' and M_2'. Because $|Q_1 \cup Q_2| \le 2f(t|N|) < t$, we have by the pigeonhole principle that there is one $j \in [t]$ such that $(Q_1 \cup Q_2) \cap N_j = \emptyset$. This implies that $R = R' \cap N_j = (R' \setminus (Q_1 \cup Q_2)) \cap N_j$, when interpreted on the ground set N, is independent in both M_1 and M_2. Moreover, the elements of R, when interpreted on the ground set N, are a basis in M_0 because R' is a basis in M_0'. Hence, R is the desired basis without any violations.   □

B   Omitted Proofs

Proof of Corollary 3. Extend N by adding a set F of $r(N_0)$ additional elements with 0 weight, where r is the rank function of M_0. We modify M_0 to a matroid $\widehat{M_0}$ on the ground set $N_0 \cup F$, given by the rank function $\widehat{r}(S) := \min\{r(S \cap N_0) + |S \cap F|, r(N_0)\}$. That is, $\widehat{M_0}$ is the union of M_0 with a free matroid on F, but then truncated to have rank $r(N_0)$. Let $P_{\widehat{\mathcal{B}_0}}$ be the matroid base polytope of $\widehat{M_0}$. It is now easy to see that if $x \in \mathbb{R}^{N_0 \cup F}$ lies in $P_{\widehat{\mathcal{B}_0}}$, then $x|_{N_0} \in P_{\mathcal{I}_0}$. Moreover, we can extend $x \in \mathbb{R}^{N_0}$ with $x \in P_{\mathcal{I}_0}$ to $x' \in \mathbb{R}^{N_0 \cup F}$ so that $x' \in P_{\widehat{\mathcal{B}_0}}$ and $x'|_{N_0} = x$. The corollary thus follows by applying Theorem 2 to $\widehat{M_0}, M_1, \ldots, M_k$.   □

Proof sketch of Theorem 7. We first state the LP-relaxation (LP$_{\text{matkn}}$).

$$\max \left\{ w^T x : \ x \in \mathbb{R}^N_{\ge 0}, \quad x \in P_{\mathcal{B}_0}, \quad x|_{N_i} \in P_{\mathcal{I}_i} \ \forall i \in [k], \right.$$
$$\left. (c^i)^T x|_{N_i} \le U_i \ \forall i = k+1, \ldots, k+t \right\}. \qquad \text{(LP}_{\text{matkn}})$$

The algorithm leading to Theorem 7 is quite similar to Algorithm 1, and so is its analysis, and we highlight the main changes.

In the algorithm, whenever we contract an element e, for each knapsack constraint with $e \in N_i$, we now update $U_i \leftarrow U_i - c_e^i$ and drop e from N_i. After performing all possible deletions, contractions, and refinements, we now either drop a matroid $M' \in \mathcal{M}'$ in step 6 as before, *or*, we drop a knapsack constraint for some $i \in \{k+1, \ldots, k+t\}$ if $|N_i| - x^*(N_i) \le q_i$.

To prove termination, we need only argue that we can always drop a matroid constraint, or a knapsack constraint in step 6 (modified as above). This follows from the same token-counting argument as in the proof of Lemma 6. Recall that if $Ax = b$ is a full-rank subsystem of $(\mathrm{LP_{matkn}})$ consisting of linearly independent x^*-tight constraints, then we may assume that the rows of A corresponding to the M_0-constraints form a nested family $\mathcal{C}$. We define a token-assignment scheme, where each $e \in N$ supplies $x^*(e)$ tokens to the row of A corresponding to the smallest set in $\mathcal{C}$ containing e (if one exists), and $(1 - x^*(e))/q_{M'}$ to each row $A_{M'}$ coming from a matroid $M' \in \mathcal{M}$ in our collection whose ground set contains e. *Additionally*, every $e \in N$ now also supplies $(1 - x^*(e))/q_i$ tokens to each row of A originating from a knapsack constraint whose ground set contains e. Under this scheme, as before, given the constraint on our q-values, it follows that every $e \in N$ supplies at most 1 token unit. Also, as before, each row of A corresponding to an M_0 constraint receives at least 1 token unit. So either there is some row $A_{M'}$ coming from a matroid in $\mathcal{M}$ that receives strictly less than 1 token-unit, or there must be some row of A corresponding to a knapsack constraint that receives at most 1 token-unit; the latter case corresponds to a knapsack constraint i with $|N_i| - x^*(N_i) \le q_i$.

The proof of parts (i)–(iii) is exactly as before. To prove part (iv), consider the i-th knapsack constraint. Note that the only place where we possibly introduce a violation in the knapsack constraint is when we drop the constraint. If x^* is the optimal solution just before we drop the constraint, then we know that $(c^i)^T x^*|_{N_i} \le U_i$. (Note that N_i and U_i refer to the updated ground set and budget.) It follows that if S denotes the set of elements included from this residual ground set N_i, then the additive violation in the knapsack constraint is

$$c^i(S) - U_i \le c^i(N_i) - U_i \le \Big(\max_{e \in N_i} c^i_e\Big)\big(|N_i| - x^*(N_i)\big) \le q_i \cdot \Big(\max_{e \in N_i} c^i_e\Big). \qquad \square$$

References

1. Bansal, N., Khandekar, R., Könemann, J., Nagarajan, V., Peis, B.: On generalizations of network design problems with degree bounds. In: Proceedings of Integer Programming and Combinatorial Optimization (IPCO), pp. 110–123 (2010)
2. Bansal, N., Khandekar, R., Nagarajan, V.: Additive guarantees for degree-bounded directed network design. SIAM J. Comput. **39**(4), 1413–1431 (2009)
3. Chan, Y.H., Lau, L.C.: On linear and semidefinite programming relaxations for hypergraphic matching. Math. Program. Ser. A **135**, 123–148 (2012)
4. Chekuri, C., Vondrák, J., Zenklusen, R.: Submodular function maximization via the multilinear relaxation and contention resolution schemes. SIAM J. Comput. **43**(6), 1831–1879 (2014)
5. Cook, W.J., Cunningham, W.H., Pulleyblank, W.R., Schrijver, A.: Combinatorial Optimization. Wiley, New York (1998)
6. Cygan, M.: Improved approximation for 3-dimensional matching via bounded pathwidth local search. In: Proceedings of 54th IEEE Symposium on Foundations of Computer Science (FOCS), pp. 509–518 (2013)

7. Cygan, M., Grandoni, F., Mastrolilli, M.: How to sell hyperedges: the hypermatching assignment problem. In: Proceedings of the 24th Annual ACM-SIAM Symposium on Discrete Algorithms (SODA), pp. 342–351 (2013)
8. Fisher, M.L., Nemhauser, G.L., Wolsey, L.A.: An analysis of approximations for maximizing submodular set functions - II. Math. Program. Study **8**, 73–87 (1978)
9. Füredi, Z.: Maximum degree and fractional matchings in uniform hypergraphs. Combinatorica **1**(2), 155–162 (1981)
10. Gupta, A., Nagarajan, V., Ravi, R.: Robust and maxmin optimization under matroid and knapsack uncertainty sets. ACM Trans. Algorithms **12**(1), 121 (2015)
11. Gupta, A., Roth, A., Schoenebeck, G., Talwar, K.: Constrained non-monotone submodular maximization: offline and secretary algorithms. In: Saberi, A. (ed.) WINE 2010. LNCS, vol. 6484, pp. 246–257. Springer, Heidelberg (2010). https://doi.org/10.1007/978-3-642-17572-5_20
12. Halldórsson, M.M.: Approximating discrete collections via local improvements. In: Proceedings of the 6th Annual ACM-SIAM Symposium on Discrete Algorithms (SODA), pp. 160–169 (1995)
13. Hurkens, C.A.J., Schrijver, A.: On the size of systems of sets every t of which have an SDR, with an application to the worst-case ratio of heuristics for packing problems. SIAM J. Discrete Math. **2**(1), 68–72 (1989)
14. Király, T., Lau, L.C., Singh, M.: Degree bounded matroids and submodular flows. In: Proceedings of Integer Programming and Combinatorial Optimization (IPCO), pp. 259–272 (2008)
15. Lau, L.C., Ravi, R., Singh, M.: Iterative Methods in Combinatorial Optimization, 1st edn. Cambridge University Press, New York (2011)
16. Lee, J., Mirrokni, V., Nagarajan, V., Sviridenko, M.: Maximizing nonmonotone submodular functions under matroid or knapsack constraints. SIAM J. Discrete Math. **23**(4), 2053–2078 (2010)
17. Lee, J., Sviridenko, M., Vondrák, J.: Submodular maximization over multiple matroids via generalized exchange properties. Math. Oper. Res. **35**(4), 795–806 (2010)
18. Parekh, O., Pritchard, D.: Generalized hypergraph matching via iterated packing and local ratio. In: Bampis, E., Svensson, O. (eds.) WAOA 2014. LNCS, vol. 8952, pp. 207–223. Springer, Cham (2015). https://doi.org/10.1007/978-3-319-18263-6_18
19. Schrijver, A.: Combinatorial Optimization, Polyhedra and Efficiency. Springer, Heidelberg (2003)
20. Singh, M., Lau, L.C.: Approximating minimum bounded degree spanning trees to within one of optimal. In: Proceedings of the 39th Annual ACM Symposium on Theory of Computing (STOC), pp. 661–670 (2007)
21. Zenklusen, R.: Matroidal degree-bounded minimum spanning trees. In: Proceedings of the 23rd Annual ACM-SIAM Symposium on Discrete Algorithms (SODA), pp. 1512–1521 (2012)

An Exact Algorithm for Robust Influence Maximization

Giacomo Nannicini[1(✉)], Giorgio Sartor[2], Emiliano Traversi[3],
and Roberto Wolfler-Calvo[3,4]

[1] IBM T. J. Watson Research Center, Yorktown Heights, NY, USA
nannicini@us.ibm.com
[2] SINTEF Digital, Oslo, Norway
giorgio.sartor@sintef.no
[3] LIPN, Université Paris 13, Villetaneuse, France
{traversi,wolfler}@lipn.fr
[4] Department of Mathematics and Computer Science,
University of Cagliari, Cagliari, Italy

Abstract. We propose a Branch-and-Cut algorithm for the robust influence maximization problem. The influence maximization problem aims to identify, in a social network, a set of given cardinality comprising actors that are able to influence the maximum number of other actors. We assume that the social network is given in the form of a graph with node thresholds to indicate the resistance of an actor to influence, and arc weights to represent the strength of the influence between two actors. In the robust version of the problem that we study, the node thresholds are affected by uncertainty and we optimize over a worst-case scenario within a given robustness budget. Numerical experiments show that we are able to solve to optimality instances of size comparable to other exact approaches in the literature for the non-robust problem, but in addition to this we can also tackle the robust version with similar performance.

Keywords: Influence maximization · Integer programming · Robust optimization

1 Introduction

Social networks are an integral part of social analysis, because they play an important role in the spread of, e.g., information, innovation, or purchase decisions. A social network is defined as a graph with actors (or groups of actors) corresponding to nodes, and arcs corresponding to interactions between actors. Interactions may represent different concepts such as friendship, mentor-apprentice, one- or two-way communication, and so on. Recent years have witnessed growing interest in the definition and study of mathematical models to represent the propagation of *influence* – broadly defined – in a social network, as well as in the identification of the actors that can play an important role in facilitating such propagation. This paper concerns the identification of such actors.

© Springer Nature Switzerland AG 2019
A. Lodi and V. Nagarajan (Eds.): IPCO 2019, LNCS 11480, pp. 313–326, 2019.
https://doi.org/10.1007/978-3-030-17953-3_24

The influence maximization problem is defined on a graph with an associated *diffusion process* that models the spread of influence on the graph. A node is defined as *active* if it is affected by the diffusion process. A subset of the nodes are selected as *seeds*, and their role is to initialize the diffusion process. Influence propagates in a breadth-first manner starting from the seeds. Several rules can be used to model the activation of a node. A commonly used model associates an *activation threshold* to each node, and a nonnegative weight to each arc representing the strength of the interaction; this paper uses such a model. The condition under which a node is activated by its neighbors is often described by an *activation function*, several types of which are discussed in the literature. The Influence Maximization Problem (IMP) is defined as the problem of identifying a subset of nodes of a given cardinality that maximizes the number of nodes activated at the end of the influence propagation process.

Literature Review. The idea of identifying the set of nodes that maximizes influence on a network dates back to [5,15]. Several variants of the IMP have been presented in the literature; we refer to the recent surveys [2,12,13] for an extensive analysis of these variants.

The majority of the literature models the diffusion of influence on the graph using a *threshold* model or a *cascade* model, see e.g., [11]. In the threshold model, a node becomes active if and only if a function of the weights on arcs incoming from activated neighbors is larger than the node threshold. In the cascade model, a node is activated if at least one of its neighbors is active. We further distinguish between *deterministic*, *stochastic* and *robust* models. In deterministic models, the graph parameters (i.e., weights and thresholds) are given and immutable. In the stochastic model, some of them are random variables and we optimize the expected number of active nodes. In the robust model, some parameters are uncertain and we optimize the worst case over a given uncertainty set. This distinction is crucial, because the stochastic version of the problem leads to a monotone submodular maximization problem under reasonable assumptions [14]. Hence, it admits an efficient $(1 - \frac{1}{e})$-approximation using a greedy algorithm, see e.g., [11,14]. The deterministic and robust version are not known to admit such an approximation in general and they tend to be much harder to solve in practice (but see [4] for an approximation algorithm for a robust version of IMP under some conditions, that still requires submodularity).

The papers [11,12] study the greedy approach for the threshold and cascade model in the stochastic setting. [17] proposes an exact cutting plane approach for the same class of models, using strong optimality cuts exploiting submodularity. The papers [4,10] present a greedy algorithm for a robust version of the cascade models, optimizing a measure of regret regarding the set of chosen seeds.

Among the variants of IMP, we mention the Target Set Selection Problem (TSSP) [1]) and the (Generalized) Least Cost Influence Problem (GLCIP) in [8,9]. The TSSP looks for the minimum-cost set of seed nodes that guarantees activation of a given percentage of the total number of nodes. The TSSP and the IMP are in some sense two different formulations for the same problem [1]: in TSSP, the total number of activated nodes is a constraint and the number of seed nodes is the objective function, while for IMP it is the other way around. GLCIP is a generalization of TSSP that allows *incentives* to decrease node activation thresholds paying a cost [8]. Both [1] and [8] use integer programming formulations with an exponential size.

Contributions of this Paper. We present an exact algorithm for the deterministic and robust IMP assuming a linear threshold model; we conjecture that many results could be generalized to other activation functions. The algorithm that we propose is based on a mixed-integer linear program (MILP) and Branch-and-Cut. The model of uncertainty for the robust IMP is akin to the Γ-robustness of [3]: we assume that the node activation thresholds are allowed to vary within a certain range, but the total amount of deviation from the nominal problem data is limited by a robustness parameter; our goal is to choose seeds so as to optimize the total influence on the graph, assuming the worst-case realization of the problem data (i.e., node thresholds) allowed by the given robustness parameter. To the best of our knowledge, this is the first time that an exact algorithm for a robust version of IMP is proposed in the literature. Furthermore, even the non-robust version of the MILP used in this paper is novel. Our algorithm for the robust IMP is not simply the application of the ideas of [3] to the non-robust model: indeed, for reasons that will become apparent after discussing the mathematical model for IMP in more detail, it is not clear how to apply the procedure of [3] to our model. We therefore propose a full Branch-and-Cut algorithm that uses a cut separation procedure to "robustify" a MILP for IMP.

The MILP that we propose for IMP originates from a bilevel formulation of the problem, where the inner problem (a linear problem with an exponential number of constraints and a provably integer optimum) is dualized, leading to a quadratic problem with binary and linear variables. This formulation is linearized with the use of indicator constraints. The final model contains an exponential number of variables, that could be generated with a column generation procedure. The number of variables depends on the density of the graph and the arc weights. To make the problem robust, we use an exponential number of cuts that can be separated solving a sub-MILP.

We test the proposed Branch-and-Cut algorithm on a set of instances comprising social network graphs taken from [8]. We show that our integer programming formulation for the non-robust model is competitive with the exact algorithm of [8] for the related GLCIP problem. Furthermore, we are able to solve the robust IMP to optimality for instances of similar size.

2   Problem Formulation

To formulate the problem of maximizing the influence on a graph $G = (V, E)$ with arc weights w_{ij}, we start by considering the problem of computing the amount of influence spread once the activation seeds are given. Assume w.l.o.g. that $V = \{1, \ldots, n\}$, and denote by $\delta^-(j)$ and $\delta^+(j)$ the instar and outstar of node j. In the rest of this paper, we will use $y \in \{0, 1\}^n$ as the incidence vector of the seeds. Given seeds y and a vector of node thresholds t, we define the set of active nodes as the set returned by Algorithm 1, and the corresponding influence as its cardinality.

1: $A_0 \leftarrow \{j \in V : y_j = 1\}$
2: **for** $k = 1, \ldots, n$ **do**
3:     $A_k \leftarrow A_{k-1}$
4:     **for** every $j \in V : A_{k-1} \cap \delta^-(j) \neq \emptyset$ **do**
5:         **if** $\sum_{i' \in \delta^-(j):i' \in A_{k-1}} w_{i'j} \geq t_j$ **then** $A_k \leftarrow A_k \cup \{j\}$
    return A_n

Algorithm 1: Function INFLUENCESPREAD(y, t).

The activation function used in Algorithm 1 is known as the *linear threshold model*. If the node activation thresholds t_j are given as input, the model is deterministic. This paper studies a robust counterpart of linear threshold model, in which the activation threshold t_j of each node j can deviate by some fraction Δ from its nominal value, and the total amount of threshold variations is upper bounded by a given number B. Because we want to optimize over a worst-case scenario, the activation thresholds can only increase. Given a vector of seeds $\bar{y} \in \{0,1\}^n$, the total amount of influence that spreads on the graph under this robust setting is the optimum of the following problem:

$$
\left.
\begin{aligned}
\mathrm{RI}_{x,\theta}(\bar{y}) := \quad & \min_{x,\theta} \sum_{j \in V} x_j \\
\forall j \in V \quad \sum_{i \in \delta^-(j)} w_{ij}x_i - \theta_j + \epsilon - & \sum_{i \in \delta^-(j)} w_{ij}x_j \leq t_j \\
\forall j \in V \quad & x_j \geq \bar{y}_j \\
& \sum_{j \in V} \theta_j \leq B \\
\forall j \in V \quad & 0 \leq \theta_j \leq \Delta t_j \\
\forall j \in V \quad & x_j \in \{0,1\}.
\end{aligned}
\right\}
$$

In the above formulation, ϵ is a small enough value ensuring that if $x_j = 0$ then $\sum_{i \in \delta^-(j)} w_{ij}x_i < t_j$. In our experiments we use $\epsilon = 10^{-d}/2$, where d is the number of digits of precision of the problem data after the decimal dot. The constraints $\sum_{j \in V} \theta_j \leq B, 0 \leq \theta_j \leq \Delta t_j \ \forall j \in V$ define a polyhedral uncertainty set in a manner similar to [3], although we consider r.h.s. uncertainty rather than on the constraint matrix.

Proposition 1. *The optimum of* $\mathrm{RI}_{x,\theta}(\bar{y})$ *is the total influence spread on* G *from seeds* $\bar{y}$ *under our robustness model, and the variables* x_j *indicate which nodes are active at the end of the influence propagation process.*

All proofs are given in the Appendix. The above formulation does not suffer from self-activating loops (which require additional caution, see e.g., [8]), and it has n binary variables only. Because of Proposition 1, the robust IMP with q activation seeds can be solved as the following bilevel optimization problem:

$$\left.\begin{array}{c} \max_y \ \mathrm{RI}_{x,\theta}(y) \\ \sum_{j \in V} y_j = q \\ \forall j \in V \qquad y_j \in \{0,1\}. \end{array}\right\} \qquad \text{(R-IMP)}$$

Notice that if we fix $\theta = 0$, solving $\mathrm{RI}_{x,\theta=0}(\bar{y})$ yields $x_j = 1$ exactly for the nodes returned by $\textsc{InfluenceSpread}(\bar{y},t)$. It is easy to show by counterexample that $\mathrm{RI}_{x,\theta}(y)$, taken as a set function of the incidence vector y, is not submodular even for fixed θ. The robust approach of [3] is difficult to apply here because we do not even have a single-level formulation for the problem. To obtain a single-level formulation, one possibility would be to consider a time-expanded graph with n^2 nodes to represent the n iterations of Algorithm 1. We instead keep a model with n binary variables, and rely on lazy constraints for robustness.

3 Activation Set Formulation for the Non-robust Problem

Our first step toward solving (R-IMP) is a formulation for the non-robust counterpart of the problem. In this section we therefore assume that $\theta = 0$ for simplicity. The node activation threshold at node j is then t_j. For all $j \in V$, let $\mathcal{C}_j : \{S \subseteq \delta^-(j) : \sum_{i \in S} w_{ij} \geq t_j, S \text{ is minimal}\}$. It is obvious that a node j is active if and only if there exists $S \in \mathcal{C}_j$ such that all nodes in S are active. This is called a *minimal activation set*. The concept of activation set was first introduced in the recent paper [8]. We developed the idea independently and we use our simpler definition, but it is easy to verify that the definition above corresponds to the minimal influencing set of [8] in the context of the linear threshold model and no incentives. We can reformulate $\mathrm{RI}_{x,\theta=0}(\bar{y})$ using minimal activation sets.

$$\left.\begin{array}{c} \mathrm{AS}_x(\bar{y}) := \min \sum_{j \in V} x_j \\ \forall j \in V, \forall S \in \mathcal{C}_j \ \sum_{i \in S} x_i - x_j \leq |S| - 1 \\ \forall j \in V \qquad\qquad x_j \geq \bar{y}_j. \end{array}\right\} \qquad (1)$$

Proposition 2. *If $\bar{y}$ is a 0-1 vector, the optimal solution x^* to $\mathrm{AS}_x(\bar{y})$ is integer and $x_j = 1$ if and only if $j \in A_n$ as returned by $\textsc{InfluenceSpread}(\bar{y},t)$.*

We can use $\mathrm{AS}_x(\bar{y})$ to obtain a single-level linear formulation for the restriction of (OPT) in which $\theta = 0$. More specifically, we consider the following problem:

$$\left.\begin{array}{c} \max_y \ \mathrm{AS}_x(y) \\ \sum_{j \in V} y_j = q \\ \forall j \in V \qquad y_j \in \{0,1\}. \end{array}\right\} \qquad \text{(IMP-}\theta 0\text{)}$$

We first take the dual of the inner problem $\mathrm{AS}_x(\bar{y})$ for a fixed $\bar{y}$. The dual is:

$$\left.\begin{array}{c} \max_{\pi,\mu} \ \sum_{j \in V} \sum_{S \in \mathcal{C}_j} (|S| - 1)\pi_S + \sum_{j \in V} \mu_j \bar{y}_j \\ \forall j \in V \ \sum_{k \in \delta^+(j)} \sum_{S \in \mathcal{C}_k : j \in S} \pi_S - \sum_{S \in \mathcal{C}_j} \pi_S + \mu_j \leq 1 \\ \forall j \in V, \forall S \in \mathcal{C}_j \qquad\qquad \pi_S \leq 0 \\ \forall j \in V \qquad\qquad\qquad \mu_j \geq 0. \end{array}\right\}$$

The solution of this problem has value equal to that of its primal problem and therefore, by Proposition 2, to INFLUENCESPREAD$(\bar{y}, t)$ whenever $\bar{y} \in \{0,1\}^n$. It follows that a valid formulation for (IMP-θ0) is the following:

$$
\left.
\begin{array}{ll}
\max_{\pi,\mu,y} & \sum_{j \in V} \sum_{S \in \mathcal{C}_j} (|S| - 1)\pi_S + \sum_{j \in V} \mu_j y_j \\
\forall j \in V \; \sum_{k \in \delta^+(j)} \sum_{S \in \mathcal{C}_k : j \in S} \pi_S - \sum_{S \in \mathcal{C}_j} \pi_S + \mu_j \leq 1 & \\
& \sum_{j \in V} y_j = q \\
\forall j \in V, \forall S \in \mathcal{C}_j & \pi_S \leq 0 \\
\forall j \in V & \mu_j \geq 0 \\
\forall j \in V & y_j \in \{0,1\}.
\end{array}
\right\}
$$

This is a quadratic problem, but we can easily reformulate it as a linear problem with indicator constraints. To achieve this result, we simply notice that whenever $y_j = 0$, μ_j has no contribution in the objective function; since μ_j appears in a single constraint and increasing μ_j reduces the feasible region for the remaining variables, there exists an optimal solution in which $y_j = 0$ implies $\mu_j = 0$. Thus, we obtain the following formulation for (IMP-θ0):

$$
\left.
\begin{array}{ll}
\max_{\pi,\mu,y} & \sum_{j \in V} \sum_{S \in \mathcal{C}_j} (|S| - 1)\pi_S + \sum_{j \in V} \mu_j \\
\forall j \in V \; \sum_{k \in \delta^+(j)} \sum_{S \in \mathcal{C}_k : j \in S} \pi_S - \sum_{S \in \mathcal{C}_j} \pi_S + \mu_j \leq 1 & \\
\forall j \in V & y_j = 0 \Rightarrow \mu_j = 0 \\
& \sum_{j \in V} y_j = q \\
\forall j \in V, \forall S \in \mathcal{C}_j & \pi_S \leq 0 \\
\forall j \in V & \mu_j \geq 0 \\
\forall j \in V & y_j \in \{0,1\}.
\end{array}
\right\}
$$
$$\text{(DUAL-}\theta 0)$$

The advantage with respect to (R-IMP) is of course that we now have a single-level problem, rather than bilevel. To achieve this result we had to fix $\theta = 0$. This restriction will be lifted in the next section.

We prove an additional result that we could not successfully exploit from an empirical point of view, but may be interesting for future research. The proof is based on total dual integrality.

Proposition 3. *For every $\bar{y} \in \{0,1\}^n$, the polyhedron corresponding to the LP obtained by fixing $y = \bar{y}$ in (DUAL-θ0) is an integral polyhedron.*

4 Branch-and-Cut for Robust Influence Maximization

For the robust version of the problem it is no longer sufficient to consider only the case $\theta = 0$. In fact, we would like to optimize over θ as in the original formulation (R-IMP), but it is not obvious how to use the dualization trick described in the previous section when θ is not fixed. This is because the activation sets may change with θ, since θ directly affects the node activation thresholds. To overcome this difficulty, we propose to work with a modification of (DUAL-θ0) that includes dual π_S variables for all the activation sets that may be minimal for any of the possible values of θ, ensuring via cuts that the objective function value for a given indicator vector of seed nodes $\bar{y}$ corresponds to $\mathrm{RI}_{x,\theta}(\bar{y})$, i.e., the minimum possible influence spread when optimizing over θ.

To implement this idea we must define an appropriate collection of activation sets. Define $\mathcal{C}_j^e : \{S \subseteq \delta^-(j) : \exists \theta_j \in [0, \Delta t_j]$ such that $\sum_{i \in S} w_{ij} \geq t_j + \theta_j, S$ is minimal$\}$. We call this an *extended collection of activation sets*. An algorithm to compute $\mathcal{C}_j^e$ is given in Algorithm 2; for a given node j, the first call to the recursive function in Algorithm 2 should be $\mathrm{RECGENEXT}(j, S \leftarrow \emptyset, \mathrm{first\_to\_add} \leftarrow 1, \mathcal{C}_j^e \leftarrow \emptyset)$. This will produce $\mathcal{C}_j^e$.

1: **if** first_to_add $> |\delta^-(j)|$ **then return false**
2: **if** $\sum_{i \in S} w_{ij} \geq t$ **then**
3:     $\mathcal{C}_j^e \leftarrow \mathcal{C}_j^e \cup S$
4:     **if** $\sum_{i \in S} w_{ij} \geq t_j + \Delta t_j$ **then return true** /* *If S is an activation set for any possible threshold of node j, we do not need to add more items to the set* */
5: any_generated $\leftarrow$ **false**
6: **for** $i \leftarrow$ first_to_add **to** $|\delta^-(j)|$ **do**
7:     next_generated $\leftarrow \mathrm{RECGENEXT}(j, S \cup \{i\}, i+1, \mathcal{C}_j^e)$ /* *Check if adding the next largest item would make S a valid activation set* */
8:     **if** next_generated = **false then break else** any_generated $\leftarrow$ **true**
    return any_generated

Algorithm 2: Function $\mathrm{RECGENEXT}(j, S, \mathrm{first\_to\_add}, \mathcal{C}_j^e)$. We assume w.l.o.g. (up to relabeling) that $\delta^-(j) := \{1, \ldots, |\delta^-(j)|\}$, and $\delta^-(j)$ is sorted by decreasing value of w_{ij}. This guarantees that the generated sets are minimal.

We then define a family of 2^n inequalities that, together with a modification of (DUAL-θ0), define a problem equivalent to (R-IMP). For every $\bar{y} \in \{0,1\}^n$, let $\theta^{\bar{y}}$ be the optimum value of θ for problem $\mathrm{RI}_{x,\theta}(\bar{y})$. Let $\mathcal{C}_j^{\bar{y}} := \{S \subseteq \delta^-(j) : \sum_{i \in S} w_{ij} \geq t_j + \theta_j^{\bar{y}}, S$ is minimal$\} \subseteq \mathcal{C}_j^e$ be the set of activation sets that are minimal for node thresholds $t_j + \theta_j^{\bar{y}}$. Consider the following problem:

$$
\begin{aligned}
\max_{\pi,\mu,y,z} \quad & z \\
\forall \bar{y} \in \{0,1\}^n \quad & \sum_{j \in V} \sum_{S \in \mathcal{C}_j^{\bar{y}}} (|S| - 1)\pi_S + \\
& \sum_{j \in V} \sum_{S \in \mathcal{C}_j^e \setminus \mathcal{C}_j^{\bar{y}}} |S|\pi_S + \sum_{j \in V} \mu_j - z \geq 0 \\
\forall j \in V \quad & \sum_{k \in \delta^+(j)} \sum_{S \in \mathcal{C}_k^e : j \in S} \pi_S - \sum_{S \in \mathcal{C}_j^e} \pi_S + \mu_j \leq 1 \\
\forall j \in V \quad & y_j = 0 \Rightarrow \mu_j = 0 \\
& \sum_{j \in V} y_j = q \\
\forall j \in V, \forall S \in \mathcal{C}_j^e \quad & \pi_S \leq 0 \\
\forall j \in V \quad & \mu_j \geq 0 \\
\forall j \in V \quad & y_j \in \{0,1\}.
\end{aligned}
\right\}
$$

$$(\text{R-IMP-LAZY})$$

Proposition 4. *The optimum of* (R-IMP-LAZY) *is equal to the optimum of* (R-IMP), *and the value of the y variables at the respective optima coincide.*

In (R-IMP-LAZY), the first set of inequalities ensures that in the dual only some activation set constraints are binding, see the proof of Proposition 4. The algorithm for (R-IMP) that we propose is the following. First, generate the extended collection $\mathcal{C}_j^e$ of activation covers, using Algorithm 2. It is easy to see that each activation set $S \in \mathcal{C}_j^e$ is valid in the range $[t_j, \min\{(1 + \Delta)t_j, \sum_{i \in S} w_{ij}\}]$ and it is minimal if the threshold is strictly greater than $\sum_{i \in S} w_{ij} - \min_{i \in S} w_{ij}$. Then, implement problem (R-IMP-LAZY) with a Branch-and-Cut solver, defining the first set of constraints as lazy constraints, i.e., constraints which are generated when an integer solution is found, rather than at problem definition. The callback for lazy constraints keeps a list L of the points $\bar{y}$ for which the corresponding constraints have already been generated, and proceeds as follows:

- Given a candidate solution $\tilde{y}$, check if $\tilde{y} \in L$; if so, return without generating additional cuts.
- Solve $\mathrm{RI}_{x,\theta}(\tilde{y})$, and let $\tilde{\theta}$ be the optimal value of θ.
- For every $j \in V$, construct $\mathcal{C}_j^{\tilde{y}}$ by comparing $\tilde{\theta}_j$ with the ranges for validity and minimality associated with each of the activation sets.
- Add the constraint $\sum_{j \in V} \sum_{S \in \mathcal{C}_j^{\tilde{y}}}(|S| - 1)\pi_S + \sum_{j \in V} \sum_{S \in \mathcal{C}_j^e \setminus \mathcal{C}_j^{\tilde{y}}}(|S|)\pi_S + \sum_{j \in V} \mu_j - z \geq 0$ to the problem, and add $\tilde{y}$ to L.

The solver terminates with an optimal solution for (R-IMP-LAZY), which is then equal to (R-IMP) by Proposition 4. The hope is that termination is achieved adding a comparatively small number of lazy constraints, rather than all the 2^n constraints involving $\bar{y}$ in (R-IMP-LAZY). We remark that we need to solve a MILP to find the optimal $\tilde{y}$ in each call of the lazy constraint generator. Nevertheless, given two vectors $\bar{y}, \bar{y}'$ we expect that $\theta^{\bar{y}}$ and $\theta^{\bar{y}'}$ have many components that are equal to each other, due to the budget constraint on θ. Hence, the cut generated for some $\bar{y}$ is likely to help for other $\bar{y}'$ as well.

We remark that $\mathcal{C}_j^e$ may have exponentially large cardinality. However, its elements and the related variables can be generated via column generation. In this paper, the full sets $\mathcal{C}_j^e$ are generated explicitly.

5   Computational Results

We test our approach on a large collection of graphs taken from the literature. The Branch-and-Cut algorithm is written in Python using IBM ILOG CPLEX 12.8.0 as MILP solver. We activate the numerical emphasis setting because of some difficulties faced during our numerical evaluation (on some graphs, model (DUAL-θ0) can have solutions with very large values). Related to this, we remark that reformulating the indicator constraints with big-M constraints is likely to fail: we tried a similar approach, but the big-M values required for validity are too large in practice. CPLEX may delete lazy constraints at times, hence our implementation adds previously generated constraints if a new incumbent violates them. All experiments are executed on a single core of an Intel Xeon E5-4620 at 2.2 GHz, 4 GB RAM, and a time limit of 1 h.

We consider the class of directed graphs SW [8,16], with a number of nodes chosen from $\{50, 75, 100\}$ and an average node degree chosen from $\{4, 8, 12\}$. We use the same SW instances as in [8]. This includes 10 instances for each choice of number of nodes and average node degrees. We additionally test a directed version of the Erdos-Rényi random graph [7]: the conclusions of the study are similar and are not reported for space reasons.

Table 1. Computing time, optimality gap, and number of instances solved to optimality for type SW graphs, averaged across different average node degrees.

		B	0	1			10			100		
n	s	Δ	0	0.1	0.5	1	0.1	0.5	1	0.1	0.5	1
50	0.02	time (s)	0.1	0.9	0.5	0.6	1.0	0.7	0.7	0.5	0.7	1.0
		gap (%)	0.0	0.0	0.0	0.0	0.0	0.0	0.0	0.0	0.0	0.0
		opt (%)	100.0	100.0	100.0	100.0	100.0	100.0	100.0	100.0	100.0	100.0
	0.05	time (s)	9.7	24.3	10.4	8.3	25.0	19.0	20.5	21.9	18.8	17.3
		gap (%)	0.0	0.0	0.0	0.0	0.0	0.0	0.0	0.0	0.0	0.0
		opt (%)	100.0	100.0	100.0	100.0	100.0	100.0	100.0	100.0	100.0	100.0
	0.1	time (s)	233.1	766.6	417.9	314.8	1197.6	1466.1	969.0	1102.3	1738.6	1664.9
		gap (%)	0.0	1.7	0.0	0.0	2.5	1.4	0.0	3.5	3.6	15.6
		opt (%)	100.0	95.0	100.0	100.0	95.0	95.0	100.0	95.0	95.0	85.0
75	0.02	time (s)	43.2	258.0	8.5	3.8	285.7	9.2	5.4	297.5	8.7	5.3
		gap (%)	0.0	0.0	0.0	0.0	0.0	0.0	0.0	0.0	0.0	0.0
		opt (%)	100.0	96.7	96.7	96.7	96.7	96.7	96.7	96.7	96.7	96.7
	0.05	time (s)	1522.5	1997.4	1292.5	500.4	2411.8	2210.1	878.5	2350.7	2275.8	1113.7
		gap (%)	433.3	486.2	0.0	0.0	539.5	63.8	0.0	558.0	57.2	0.8
		opt (%)	66.7	66.7	96.7	96.7	43.3	66.7	96.7	43.3	66.7	93.3
	0.1	time (s)	1592.8	2989.2	3458.8	3515.6	3600.0	3600.0	3600.2	3569.6	3600.0	3600.0
		gap (%)	102.5	227.5	283.5	262.1	355.4	469.3	437.0	339.5	543.2	576.1
		opt (%)	66.7	23.3	6.7	6.7	0.0	0.0	0.0	3.3	0.0	0.0
100	0.02	time (s)	134.1	818.6	16.9	7.4	817.2	19.2	10.4	835.7	19.3	10.0
		gap (%)	0.0	170.0	0.0	0.0	271.1	0.0	0.0	186.1	0.0	0.0
		opt (%)	100.0	93.3	100.0	100.0	93.3	100.0	100.0	93.3	100.0	100.0
	0.05	time (s)	2563.0	3600.0	3578.1	3600.0	3600.0	3600.0	3600.0	3600.0	3600.0	3600.0
		gap (%)	598.7	809.6	589.6	534.1	930.9	976.4	815.7	893.8	1020.5	987.4
		opt (%)	36.7	0.0	3.3	0.0	0.0	0.0	0.0	0.0	0.0	0.0
	0.1	time (s)	1386.9	3428.8	3600.0	3600.0	3600.0	3600.0	3600.0	3512.0	3600.0	3600.0
		gap (%)	109.9	275.4	336.0	409.1	352.8	562.9	595.1	349.6	616.7	703.5
		opt (%)	66.7	6.7	0.0	0.0	0.0	0.0	0.0	3.3	0.0	0.0

We solve the robust and non-robust IMP on each graph using different sets of parameters. The number of seeds is chosen as a fraction $s \in \{0.02, 0.05, 0.1\}$

of n, rounded to the nearest integer. The robustness budget B is chosen from $\{0,1,10,100\}$, where 0 corresponds to the non-robust case (i.e., the deterministic linear threshold model); the maximum threshold deviation Δ, only applicable to the robust case, is chosen $\in \{0.1,0.5,1\}$. We have a total of 2400 SW instances.

Table 1 shows the average computation time, the average gap and the percentage of instances solved to optimality for each combination of parameters. In our tests CPLEX generates very few cuts – mostly implied bounds, but almost no other cut. This may be related to the numerical difficulties mentioned above. The complexity of the instance quickly increases with the number of nodes and with the number of seeds; this is not surprising and agrees with similar findings presented in [8]. The robust version seems harder than the non-robust counterpart, but at least for smaller instances the computation time is similar. In some cases, the robust problem actually solves faster than the non-robust one. For the instances with $n = 100$, $s \geq 0.05$, the robust problem is rarely solved to optimality within one hour. In our tests, problems with high average degree are more difficult to solve, probably because the cardinality of the collections of extended covers $\mathcal{C}_j^e$ increases rapidly, leading to thousands of variables.

Table 2. Values of the best incumbent found for type SW graphs, averaged across different average node degrees.

		B	0	1			10			100		
n	s	Δ	0	0.1	0.5	1	0.1	0.5	1	0.1	0.5	1
50	0.02		5.5	3.5	3.5	3.3	3.1	2.1	1.9	3.1	2.1	1.9
	0.05		23.0	12.8	12.0	10.5	10.2	6.5	5.3	10.2	6.5	5.1
	0.1		40.3	27.5	22.6	18.1	21.6	10.8	9.4	21.6	10.2	8.2
75	0.02		12.1	6.4	6.3	5.9	5.7	3.8	3.3	5.7	3.8	3.2
	0.05		28.9	17.5	13.3	11.9	12.1	7.6	6.8	12.1	7.3	6.2
	0.1		57.7	40.0	30.8	27.5	24.0	14.7	13.8	26.0	13.1	11.1
100	0.02		10.9	5.8	5.8	5.5	5.4	3.6	3.2	5.4	3.6	3.2
	0.05		41.1	21.2	18.9	15.4	15.3	9.0	8.5	15.7	8.8	7.3
	0.1		77.2	43.0	32.9	26.9	30.3	16.5	15.3	32.1	15.3	13.1

We briefly compare to the results of [8], which use the same type of graphs SW although they solve the related GLCIP problem, rather than IMP. For the non-robust problem, we are able to solve to optimality problems of similar size (we solve all instances with $n = 50$ and 2/3 of the instances with $n = 75, 100$, [8] solves roughly 3/4 of the instances with $n = 50$ and 1/3 of the instances with $n = 75, 100$).

Table 2 shows that the price of robustness is quite high: protecting against a small amount of uncertainty ($B = 1$, $\Delta = 0.1$) decreases the total number of activated nodes by 41% already. To further investigate the change in the optimal solutions, we measure the Hamming distance between the set of seeds computed by the non-robust model and the robust model. We find that the difference is significant, e.g., for $n = 50$ on average 83% of the seeds change.

Conclusions and Future Research. We present an exact Branch-and-Cut algorithm for the robust and deterministic influence maximization problem. To solve the deterministic version of the problem we introduce a new formulation originating from a bilevel problem, exploiting the dual of some (integral) subproblems. The formulation is extended to the robust case by adding an exponential number of cuts that are separated in a Branch-and-Cut framework. We numerically show that the proposed algorithm is capable of finding optimal robust solutions on graphs from the literature.

This paper discusses uncertainty on the node activation threshold; it is possible that our approach can be adapted to handle uncertainty on the edge weights as well. This is not pursued in this paper. Furthermore, our numerical evaluation suggests that finding classes of valid cuts may be highly beneficial. Because this may be difficult with the current formulation, finding alternative formulations that still allow "robustification" cuts is an interesting research direction for the future. With the current formulation, column generation for the π_S variables appears to be necessary to scale up to larger instances.

Proofs

Proof. **Proposition 1.** It suffices to show that for a given choice of the adversarial threshold modification θ_j, problem $\mathrm{RI}_{x,\theta}(\bar{y})$ computes the total influence spread as if we had applied $\textsc{InfluenceSpread}(\bar{y}, t + \theta)$. Notice that $\mathrm{RI}_{x,\theta}(\bar{y})$ is a minimization problem and each x_j is lower bounded by two quantities only: $\bar{y}_j$, and $\frac{\sum_{i \in \delta^-(j)} w_{ij} x_i - t_j - \theta_j + \epsilon}{\sum_{i \in \delta^-(j)} w_{ij}}$. The latter quantity is > 0 if and only if $\sum_{i \in \delta^-(j)} w_{ij} x_i > t_j + \theta_j$, implying that $x_j = 1$ if and only if its activation rule is triggered by its neighbors. If we apply $\textsc{InfluenceSpread}(\bar{y}, t + \theta)$, it is easy to see by induction over the main loop that for each $x_j \in A_k$ there is an implied lower bound $x_j \geq 1$, and for all nodes $\notin A_n$ there is no such implied lower bound. It follows that in the optimal solution $x_j = 1$ if and only if $j \in A_n$.

Proof. **Proposition 2.** Every x_j is lower bounded by $\bar{y}_j$ and by $\sum_{i \in S} x_i - |S| + 1$ for some subset of nodes S adjacent to node j.

We first show by induction for $k = 1, \ldots, n$ that for every node $j \in A_k$ in $\textsc{InfluenceSpread}(\bar{y}, t)$, we have an implied lower bound $x_j \geq 1$ in $\mathrm{AS}_x(\bar{y})$.

For $k = 1$ the claim is obvious because of the constraints $x_j \geq \bar{y}_j$. To go from $k - 1$ to k, notice that if node j is added to A_k at step k of $\textsc{InfluenceSpread}(\bar{y}, t)$, it must be that $\sum_{i \in \delta^-(j): i \in A_{k-1}} w_{ij} \geq t_j$. By the induction hypothesis for all $i \in A_{k-1}$ we have $x_i \geq 1$, hence $\sum_{i \in \delta^-(j)} w_{ij} x_i \geq t_j$. By definition of minimal activation set, there must exist some $S \in \mathcal{C}_j$, say $\bar{S}$, such that $\bar{S} \subseteq A_{k-1}$. Then the corresponding constraint $\sum_{i \in \bar{S}} x_i - x_j \leq |\bar{S}| - 1$ in the formulation $\mathrm{AS}_x(\bar{y})$ reads $|\bar{S}| - x_j \leq |\bar{S}| - 1$, implying $x_j \geq 1$.

Finally, for every $j \notin A_n$, all the constraints $\sum_{i \in S} x_i - x_j \leq |S| - 1$ are slack because there does not exist $S \in \mathcal{C}_j, S \subseteq A_k$ for some k. Hence, the implied lower bound for x_j is 0. Since we are minimizing $\sum_{j \in V} x_j$, at the optimum $x_j = 1$ if and only if $j \in A_n$.

Proof. **Proposition** 3. We show that for a given 0–1 vector $\bar{y}$, the remaining system in (DUAL-$\theta 0$) is total dual integral. This implies that it defines an integral polyhedron [6].

The discussion in this section shows that the dual of (DUAL-$\theta 0$) for fixed $y = \bar{y}$ is the problem $\mathrm{AS}_x(\bar{y})$ defined in (1). To show total dual integrality of the desired system, we need to show that for any integer value of the r.h.s. of the first set of constraints in $\mathrm{AS}_x(\bar{y})$, either $\mathrm{AS}_x(\bar{y})$ is infeasible, or it has an optimal solution that is integer.

Let b be a given vector of integer r.h.s. values for the first set of constraints, which are indexed by $j \in V, S \in \mathcal{C}_j$. First, notice that if $b_{j,S} < 0$ for any j, S, the problem is infeasible; hence, we only need to consider the case $b \geq 0$. We show how to construct an integer optimal solution.

Define $x^0 := \bar{y}$. Apply the following algorithm: for $k = 1, \ldots, n$, (i) set $x_j^k \leftarrow 0 \,\forall j$; (ii) for $j \in V, S \in \mathcal{C}_j$, set $x_j^k \leftarrow \max\{x_j^k, \sum_{i \in S} x_i^{k-1} - b_{j,S}\}$. It is clear that this defines an integral vector x^n. We now show that this solution is optimal. Let x^* be an optimal solution for the problem with r.h.s. b. We first show by induction that $x^k \leq x^*$. For $k = 0$ this is obvious as $x \geq \bar{y} = x^0$ is among the constraints. Assume $x^{k-1} \leq x^*$ and suppose $x_h^k > x_h^*$ for some h. Since x_h^k is initially 0, it must be that for some S, x_h^k is set to $\sum_{i \in S} x_i^{k-1} - b_{h,S} > x_h^*$ for the first time. But $\sum_{i \in S} x_i^{k-1} - b_{h,S} \leq \sum_{i \in S} x_i^* - b_{h,S} \leq x_h^*$, because x^* satisfies the constraints; this is a contradiction. It follows that $x^k \leq x^*$ for all $k = 1, \ldots, n$. It is easy to check that x^n is feasible by construction, and therefore it must be optimal.

Proof. **Proposition** 4. Let π^*, y^*, μ^*, z^* be the optimal solution of (R-IMP-LAZY). Consider the following LP, obtained by fixing $y = y^*$, keeping only one of the constraints involving z for a given value $\bar{y}$ in (R-IMP-LAZY), and eliminating the z variable (which is unnecessary if it appears in one constraint only):

$$\begin{aligned}
\max_{\pi,\mu,y} \quad & \sum_{j \in V} \sum_{S \in \mathcal{C}_j^{\bar{y}}} (|S|-1)\pi_S + \\
& \sum_{j \in V} \sum_{S \in \mathcal{C}_j^e \setminus \mathcal{C}_j^{\bar{y}}} |S|\pi_S + \sum_{j \in V} \mu_j \\
\forall j \in V \quad & \sum_{k \in \delta^+(j)} \sum_{S \in \mathcal{C}_k^e : j \in S} \pi_S - \sum_{S \in \mathcal{C}_j^e} \pi_S + \mu_j \leq 1 \\
\forall j \in V, y_j^* = 0 \quad & \mu_j = 0 \\
\forall j \in V, \forall S \in \mathcal{C}_j^e \quad & \pi_S \leq 0 \\
\forall j \in V \quad & \mu_j \geq 0.
\end{aligned} \qquad (2)$$

This problem is feasible as the all-zero solution is feasible. Using the reverse of the transformations discussed in Sect. 3, we can show that the dual of the above problem is equivalent to the following LP:

$$
\left.
\begin{array}{rc}
\min & \sum_{j \in V} x_j \\
\forall j \in V, \forall S \in \mathcal{C}_j^{\bar{y}} & \sum_{i \in S} x_i - x_j \leq |S| - 1 \\
\forall j \in V, \forall S \in \mathcal{C}_j^e \setminus \mathcal{C}_j^{\bar{y}} & \sum_{i \in S} x_i - x_j \leq |S| \\
\forall j \in V & x_j \geq y_j^* \\
\forall j \in V & x_j \leq 1 \\
\forall j \in V & x_j \geq 0.
\end{array}
\right\}
\tag{3}
$$

The constraints with r.h.s. value $|S|$ are redundant and can be dropped. As a result, with the same argument used for Proposition 2, the optimum value of (3) and therefore (2) is equal to $\mathrm{RI}_{x,\theta=\theta^{\bar{y}}}(y^*)$, i.e., the influence spread with seeds determined by y^* and node thresholds equal to $t + \theta^{\bar{y}}$.

Now notice that the objective function of (R-IMP-LAZY) corresponding to π^*, y^*, μ^*, z^* is equal to the minimum objective function of all the problems (2), for all $\bar{y} \in \{0,1\}^n$. In other words, (R-IMP-LAZY) yields the following value:

$$
\min_{\bar{y} \in \{0,1\}^n} \mathrm{RI}_{x,\theta=\theta^{\bar{y}}}(y^*).
$$

By definition of $\theta^{\bar{y}}$, this minimum is attained for $\bar{y} = y^*$. Hence, the optimum of (R-IMP-LAZY) has value $\mathrm{RI}_{x,\theta=\theta^{y^*}}(y^*)$, i.e., the influence spread with seeds y^*, when the node thresholds are chosen to be the worst possible within the allowed set of node thresholds. Since we are maximizing, this is equivalent to solving (R-IMP).

References

1. Ackerman, E., Ben-Zwi, O., Wolfovitz, G.: Combinatorial model and bounds for target set selection. Theoret. Comput. Sci. **411**(44–46), 4017–4022 (2010)
2. Banerjee, S., Jenamani, M., Pratihar, D.K.: A survey on influence maximization in a social network. arXiv preprint arXiv:1808.05502 (2018)
3. Bertsimas, D., Sim, M.: The price of robustness. Oper. Res. **52**(1), 35–53 (2004)
4. Chen, W., Lin, T., Tan, Z., Zhao, M., Zhou, X.: Robust influence maximization. In: Proceedings of the 22nd ACM SIGKDD International Conference on Knowledge Discovery and Data Mining, pp. 795–804. ACM (2016)
5. Domingos, P., Richardson, M.: Mining the network value of customers. In: Proceedings of the Seventh ACM SIGKDD International Conference on Knowledge Discovery and Data Mining, KDD 2001, pp. 57–66. ACM, New York (2001)
6. Edmonds, J., Giles, R.: A min-max relation for submodular functions on graphs. In: Annals of Discrete Mathematics, vol. 1, pp. 185–204. Elsevier (1977)
7. Erdos, P., Rényi, A.: On the evolution of random graphs. Publ. Math. Inst. Hung. Acad. Sci **5**(1), 17–60 (1960)
8. Fischetti, M., Kahr, M., Leitner, M., Monaci, M., Ruthmair, M.: Least cost influence propagation in (social) networks. Math. Program. **170**(1), 293–325 (2018)

9. Gunnec, D.: Integrating social network effects in product design and diffusion. Ph.D. thesis (2012)
10. He, X., Kempe, D.: Stability and robustness in influence maximization. ACM Trans. Knowl. Discovery Data (TKDD) **12**(6), 66 (2018)
11. Kempe, D., Kleinberg, J., Tardos, E.: Maximizing the spread of influence through a social network. In: Proceedings of the Ninth ACM SIGKDD International Conference on Knowledge Discovery and Data Mining, KDD 2003, pp. 137–146. ACM, New York (2003)
12. Kempe, D., Kleinberg, J., Tardos, E.: Maximizing the spread of influence through a social network. Theory Comput. **11**(4), 105–147 (2015)
13. Li, Y., Fan, J., Wang, Y., Tan, K.L.: Influence maximization on social graphs: a survey. IEEE Trans. Knowl. Data Eng. **30**, 1852–1872 (2018)
14. Mossel, E., Roch, S.: On the submodularity of influence in social networks. In: Proceedings of the Thirty-Ninth Annual ACM Symposium on Theory of Computing, pp. 128–134. ACM (2007)
15. Richardson, M., Domingos, P.: Mining knowledge-sharing sites for viral marketing. In: Proceedings of the Eighth ACM SIGKDD International Conference on Knowledge Discovery and Data Mining, pp. 61–70. ACM (2002)
16. Watts, D.J., Strogatz, S.H.: Collective dynamics of small-world networks. Nature **393**(6684), 440 (1998)
17. Wu, H.H., Küçükyavuz, S.: A two-stage stochastic programming approach for influence maximization in social networks. Comput. Optim. Appl. **69**(3), 563–595 (2018)

A New Contraction Technique
with Applications
to Congruency-Constrained Cuts

Martin Nägele[✉] and Rico Zenklusen

Department of Mathematics, ETH Zurich, Zurich, Switzerland
`martin.naegele@ifor.math.ethz.ch`, `ricoz@math.ethz.ch`

Abstract. Minimum cut problems are among the most classical problems in Combinatorial Optimization and are used in a wide set of applications. Some of the best-known efficiently solvable variants include global minimum cuts, minimum s-t cuts, and minimum odd cuts in undirected graphs. We study a problem class that can be seen to generalize the above variants, namely finding congruency-constrained minimum cuts, i.e., we consider cuts whose number of vertices is congruent to r modulo m, for some integers r and m. Apart from being a natural generalization of odd cuts, congruency-constrained minimum cuts exhibit an interesting link to a long-standing open problem in Integer Programming, namely whether integer programs described by an integer constraint matrix with bounded subdeterminants can be solved efficiently.

We develop a new contraction technique inspired by Karger's celebrated contraction algorithm for minimum cuts, that, together with further insights, leads to a polynomial time randomized approximation scheme for congruency-constrained minimum cuts for any constant modulus m. Instead of contracting edges of the original graph, we use splitting-off techniques to create an auxiliary graph on a smaller vertex set, which is used for performing random edge contractions. This way, a well-structured distribution of candidate pairs of vertices to be contracted is obtained, where the involved pairs are generally not connected by an edge. As a byproduct, our technique reveals new structural insights into near-minimum odd cuts, and, more generally, near-minimum congruency-constrained cuts.

1 Introduction

Cuts in undirected graphs are a basic structure in Combinatorial Optimization with a multitude of applications. The global minimum cut problem, the minimum s-t cut problem, and the minimum odd cut problem are among the best known efficiently solvable minimum cut variants, and have been the cradle of many exciting algorithmic techniques. We study a generalization of these problems

R. Zenklusen—Supported by Swiss National Science Foundation grant 200021_165866.

© Springer Nature Switzerland AG 2019
A. Lodi and V. Nagarajan (Eds.): IPCO 2019, LNCS 11480, pp. 327–340, 2019.
https://doi.org/10.1007/978-3-030-17953-3_25

that we call *congruency-constrained minimum cut* (CCMC), where a congruency constraint on the vertices in the cut is imposed, as described in the box below.[1]

Congruency-Constrained Minimum Cut (CCMC): Let $G = (V, E)$ be an undirected graph with edge weights $w \colon E \to \mathbb{R}_{\geqslant 0}$ and let $\gamma \colon V \to \mathbb{Z}_{\geqslant 0}$. Let $m \in \mathbb{Z}_{>0}$ and $r \in \mathbb{Z}_{\geqslant 0}$. The task is to find a minimizer of

$$\min \left\{ w(\delta(C)) \,\middle|\, \emptyset \subsetneq C \subsetneq V, \ \sum_{v \in C} \gamma(v) \equiv r \pmod{m} \right\}. \qquad \text{(CCMC)}$$

We call m the *modulus* of the problem, and we will typically consider m to be constant. Moreover, allowing general γ-values—instead of setting $\gamma(v) = 1$ for all $v \in V$, i.e., requiring that $|C| \equiv r \pmod{m}$—is merely for convenience. Indeed, the case of general γ-values can be reduced to the unit case by replacing each vertex v by a clique of $(\gamma(v) \bmod m)$-many vertices with large edge values.[2]

Apart from generalizing well-known cut problems, we are interested in the study of (CCMC) due to a link to an intriguing open question in Integer Programming, namely whether integer linear programs (ILPs) defined by an integer constraint matrix with bounded subdeterminants can be solved efficiently. Recently it was shown in [1] that ILPs of the form $\min\{c^T x \mid Ax \leqslant b, \ x \in \mathbb{Z}^n\}$ can be solved efficiently if $A \in \mathbb{Z}^{m \times n}$ is *bimodular*, i.e., A has full column-rank and the determinant of any $n \times n$ submatrix of A is in $\{-2, -1, 0, 1, 2\}$. This result implies that if A is totally bimodular, i.e., all subdeterminants of A are in $\{-2, -1, 0, 1, 2\}$, then the corresponding ILP can be solved in polynomial time even without the requirement of A having full column rank (see [1] for details). This extends the well-known fact that ILPs with a totally unimodular constraint matrix can be solved efficiently; here, the absolute value of subdeterminants is bounded by 1. Only very limited techniques are known for attacking the question whether ILPs remain efficiently solvable in the Δ-modular case for some constant $\Delta > 2$, i.e., $\text{rank}(A) = n$ and any $n \times n$ subdeterminant of A is in $\{-\Delta, -\Delta + 1, \ldots, \Delta\}$. Interestingly, to approach the bimodular case, classical combinatorial optimization problems with congruency constraints play a crucial role, and the problem can be reduced to certain types of congruency-constrained cut and flow problems (see [1]). In particular, (CCMC) with modulus m can

---

[1] The minimum odd cut problem is captured by (CCMC) by choosing $m = 2$, $r = 1$, and $\gamma(v) = 1$ for all $v \in V$. Global minimum cuts correspond to $m = 1$, r arbitrary, and $\gamma(v) = 0$ for all $v \in V$, and s-t cuts can be modeled as minimum $\{s, t\}$-odd cuts, i.e., $m = 2$, $r = 1$, $\gamma(s) = \gamma(t) = 1$, and $\gamma(v) = 0$ for all $v \in V \setminus \{s, t\}$.

[2] Here, $(\gamma(v) \bmod m)$ denotes the smallest positive integer congruent to $\gamma(v)$ modulo m. Reducing modulo m is crucial to obtain a blow-up bounded by m, which, as mentioned, we will typically assume to be constant.

be reduced to m-modular ILPs.[3] Hence, if one believes that Δ-modular ILPs can be solved efficiently for $\Delta = O(1)$, then (CCMC) should admit an efficient algorithm. Conversely, despite the fact that, for $\Delta \geq 3$, further gaps have to be overcome to reduce Δ-modular ILPs to congruency-constrained cut and flow problems, the results in [1] give hope that congruency-constrained cuts may be a useful building block for attacking Δ-modular ILPs, besides merely being a special case thereof.

Not much is known in terms of techniques to deal with congruency constraints in Combinatorial Optimization beyond parity constraints ($m = 2$). These constraints introduce an algebraic component to the underlying problem, which is a main additional hurdle to overcome. Some progress has been achieved for moduli m that are constant prime powers, where it was shown in [17] that submodular function minimization under congruency constraints with such moduli can be solved efficiently. As the cut function is submodular, this implies that (CCMC) can be solved efficiently for m being a constant prime power. However, the techniques in [17] do not extend to general constant moduli m, due to intrinsic additional complications appearing when m has two different prime divisors.

The goal of this paper is to show that contraction techniques, in the spirit of Karger's algorithm for global minimum cuts [11,12], can be employed to approach (CCMC). A naive way of using Karger for (CCMC) faces several hurdles, which we exemplify through the (CCMC) instance in Fig. 1, parameterized by an even number n and a weight $M \geq 1$.[4] It consists of n paths of length 2 between two vertices u and w. An optimal cut is highlighted in gray. Karger's algorithm returns any global minimum cut in a graph $G = (V, E)$ with probability $\Omega(|V|^{-2})$, implying that there are at most $O(|V|^2)$ global minimum cuts. However, for $M = 1$, the (CCMC) problem in Fig. 1 has expo-

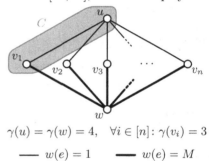

$\gamma(u) = \gamma(w) = 4, \quad \forall i \in [n]: \gamma(v_i) = 3$

——— $w(e) = 1$ ——— $w(e) = M$

Fig. 1. A (CCMC) instance with $m = 6$ and $r = 1$. Its optimal value is $n + M - 1$, achieved by the highlighted cut C.

nentially many optimal solutions. Hence, such a result cannot hold here. Moreover, if one of the n many u-w paths gets contracted, then the problem turns

---

[3] If M is the incidence matrix of the digraph $H = (V, A)$ obtained by bidirecting G,

$$\min \left\{ \sum_{a \in A} w_a y_a \;\middle|\; \begin{array}{l} Mx - y \leq 0,\; \sum_{v \in V} \gamma(v)x_v + zm = r,\; x_s = 1,\; x_t = 0, \\ x \in \{0,1\}^V,\; y \in \{0,1\}^A,\; z \in \mathbb{Z}. \end{array} \right\}$$

solves the minimum s-t cut problem in G with congruency constraint $\gamma(C) \equiv r \pmod{m}$, where the cut corresponds to the set $C = \{v \in V \mid x_v = 1\}$. Moreover, the constraint matrix of the above ILP can be seen to be m-modular. Analogously to how global min cut problems can be reduced to min s-t cut problems, every (CCMC) problem can be reduced to solving linearly many problems of the above type.

[4] Even n ensures that $S = \{w, v_1, v_2, \ldots, v_n\}$ is infeasible, i.e., $\gamma(S) \not\equiv 1 \pmod{6}$.

infeasible. It is not clear how to fix this. Even if we forbid contractions that make the instance infeasible, it is likely that in many of the u-w paths, one would contract the edge of weight 1. It is not hard to verify that Karger-type contractions would with high probability lead to a cut that is about twice as large as the minimum cut if M is chosen large (and this factor of 2 can be boosted further).

To overcome these and further hurdles, substantial changes seem necessary, and we introduce new techniques to employ contraction algorithms in our context. A key difference of our method to Karger's algorithm, and other contraction algorithms in a similar spirit (see a recent result of Chandrasekaran, Xu and Yu [4] for a nice adaptation of Karger's algorithm to the hypergraph k-cut problem), is that we do not contract edges of the graph. Instead, we define a distribution over pairs of vertices to contract that may not be connected by an edge. Moreover, we only look for contractions among certain vertices, namely those $v \in V$ fulfilling $\gamma(v) \not\equiv 0 \pmod{m}$. We show that splitting-off techniques from Graph Theory can be leveraged to design an efficient procedure to sample from a distribution of vertex pairs to contract with strong properties.

1.1   Our Results

Our main result for (CCMC) via our new contraction technique is the following.

Theorem 1. (CCMC) *with constant modulus m admits a PRAS.*

Recall that a *PRAS* (*polynomial time randomized approximation scheme*) is an efficient procedure that, for any fixed $\varepsilon > 0$, returns with high probability, by which we mean with probability at least $1 - 1/|V|$, a $(1+\varepsilon)$-approximate solution.

Moreover, for a constant composite modulus m that is the product of only two primes, we obtain an exact procedure.

Theorem 2. (CCMC) *with a constant modulus that is the product of two primes admits an efficient randomized algorithm that w.h.p. returns an optimal solution.*

This is in stark contrast to prior procedures, in particular for congruency-constrained submodular function minimization [17], which employ techniques that face hard barriers for moduli beyond prime powers.

Finally, in a similar spirit to Karger's algorithm for global minimum cuts, our contraction algorithm allows us to derive structural results on near-minimum congruency-constrained cuts. Whereas Karger's analysis shows that there are only polynomially many cuts of value at most a constant factor higher than the minimum cut, we cannot hope for results of this type: The example in Fig. 1 shows that (CCMC) problems can have exponentially many optimal solutions. For prime moduli, we show that near-minimum (CCMC) cuts are near-minimum cuts (without congruency constraint) in one of only a polynomial number of minimum s-t cut or global minimum cut instances. These instances are defined on *contractions of G*, i.e., graphs obtained from $G = (V, E)$ by successively contracting nonempty node sets $S \subseteq V$. When *contracting* a set S, all vertices of S are replaced by a single vertex v_S with $\gamma(v_S) := \sum_{v \in S} \gamma(v)$, all edges with both endpoints in S are deleted, and each edge connecting a vertex in S to a vertex $u \in V \setminus S$ is replaced by an edge between u and v_S of the same weight. By construction, a cut C in a contraction of G naturally corresponds to a cut $\overline{C}$ in G of the same weight with $\gamma(C) = \gamma(\overline{C})$, and thus, we can identify these cuts.

Theorem 3. *Consider a* (CCMC) *problem on* $G = (V, E)$ *with constant prime modulus* m *and optimal value* OPT, *and let* $\rho \geq 1$ *be a constant. Then there is an efficient randomized method returning* $\mathrm{poly}(|V|)$ *many minimum cut instances, each of which is either a minimum s-t cut or global minimum cut instance, defined on contractions of* G, *such that the following holds with high probability. A cut* $C \subseteq V$, $C \neq \emptyset$, *is a solution to* (CCMC) *of value at most* $\rho \cdot$ OPT *if and only if* C *is a feasible solution of value at most* $\rho \cdot$ OPT *in one of the minimum cut instances (without congruency constraint).*

Due to space constraints, the focus of this extended abstract lies on Theorem 1. Proofs of Theorems 2 and 3 are deferred to a long version of the paper.

1.2 Further Discussion on Related Results

Work on minimum cut problems with constraints of congruency type date back to the early '80s, when Padberg and Rao [19] presented a method to efficiently find a minimum cut among all cuts with an odd number of vertices. Barahona and Conforti [2] later showed that efficient minimization is also possible over all cuts with an even number of vertices. Later works by Grötschel, Lovász, and Schrijver [10], and by Goemans and Ramakrishnan [8] generalized these results, by showing that even any submodular function can be minimized over so-called triple families and, more generally, parity families. Submodular functions generalize cut functions, and triple as well as parity families capture congruency constraints with modulus 2. More generally, these approaches even allow for minimizing over all cuts $C \subseteq V$ of cardinality *not* congruent to r modulo m, for any integers r and m, which turns out to be a much simpler constraint than requiring a cardinality congruent to r modulo m. Indeed, (CCMC) for unbounded m quickly leads to NP-hard problems, as one could model an arbitrary cardinality constraint through a congruency constraint. In particular, if $G = (V, E)$ is a graph with an even number of vertices, then seeking a minimum cut C with $|C| \equiv 0 \pmod{|V|/2}$ captures the well-known minimum bisection problem. Khot [13] showed that, unless NP has randomized sub-exponential time algorithms, the minimum bisection problem does not admit a polynomial time approximation scheme. Hence, it seems unlikely that a PRAS can be obtained for (CCMC) without a bound on the modulus.

We briefly mention further works linked to matrices with bounded subdeterminants. This includes the problem of finding a maximum weight independent set in a graph with constant odd-cycle packing number, for which a PTAS was obtained by Bock, Faenza, Moldenhauer, Vargas, and Jacinto [3]. This problem readily reduces to ILPs with bounded subdeterminants, due to an observation of Grossman, Kulkarni, and Schochteman [9]. Another recent result by Eisenbrand

and Vempala [6] is a randomized simplex-type algorithm for linear programming that is strongly polynomial whenever all subdeterminants of the constraint matrix defining the LP are bounded by a polynomial in the dimension of the problem. Furthermore, there has been interesting recent progress on the problem of approximating the largest subdeterminant of a matrix (see Di Summa, Eisenbrand, Faenza, and Moldenhauer [5], and Nikolov [18]).

2   An Overview of Our Approach

As mentioned, the core of our approach is a contraction procedure inspired by Karger's global minimum cut algorithm, where we sample vertex pairs to be contracted from a certain distribution. In fact, the analysis of Karger's random contraction algorithm exploits that, whenever a random edge is contracted in a graph $G = (V, E)$, this contraction is *bad* with probability at most $k/|V|$ for some constant $k \in \mathbb{Z}_{>0}$. More precisely, in the analysis, an arbitrary minimum cut C is fixed, and a contraction is *bad* if it contracts two vertices on different sides of C. The probability of bad contractions being at most $k/|V|$ implies that by contracting until only k vertices remain, and then enumerating all cuts among those vertices, each minimum cut is found with probability at least $1/\binom{|V|}{k}$.

For (CCMC), an important observation is that it suffices to decide which vertices in $V_{\neq 0} := \{v \in V \mid \gamma(v) \not\equiv 0 \pmod{m}\}$ are part of a solution. Indeed, for any cut C, the value of $\gamma(C)$ is determined by the intersection $C \cap V_{\neq 0}$. Moreover, for any $U \subseteq V_{\neq 0}$, the value

$$\nu(U) := \min \left\{ w(\delta(C)) \mid \emptyset \subsetneq C \subsetneq V, \ C \cap V_{\neq 0} = U \right\}$$

and a minimizer C_U can be obtained efficiently by a minimum cut computation in a contraction of G.[5] As $C_U \cap V_{\neq 0} = U$, we have $\gamma(C_U) \equiv \gamma(U) \pmod{m}$.

Due to the above, instead of performing contractions over the full graph, as done in Karger's algorithm, we only contract pairs in $V_{\neq 0}$, with the goal to reduce $V_{\neq 0}$ to a constant-size set. If we achieve this, it suffices to enumerate over all $U \subseteq V_{\neq 0}$ with $\gamma(U) \equiv r \pmod{m}$, minimize $\nu(U)$, and return a corresponding cut C_U. The theorem below is a key technical result of this paper, and shows that a suitable distribution over vertex pairs in $V_{\neq 0}$ to contract exists whenever the sum $\sum_{v \in V_{\neq 0}} \nu(\{v\})$ is large enough.

Theorem 4. *Let* $\mathcal{I} = (G, w, \gamma, m, r)$ *be a* (CCMC) *instance on* $G = (V, E)$. *Let* $\alpha > 0$ *and* $c > 0$ *with* $\sum_{v \in V_{\neq 0}} \nu(\{v\}) > \frac{2\alpha}{c} \cdot |V_{\neq 0}|$. *Then, there is a distribution* $\mathcal{D}$ *over pairs in* $V_{\neq 0}$ *such that* $\Pr_{\{u,v\} \sim \mathcal{D}} \left[|\{u, v\} \cap C| = 1 \right] \leqslant c/|V_{\neq 0}|$ *for any feasible solution* C *of* $\mathcal{I}$ *with* $w(\delta(C)) \leqslant \alpha$. *Moreover, there is an efficient procedure to sample from* $\mathcal{D}$.

---

[5] If $U \notin \{\emptyset, V_{\neq 0}\}$, then C_U can be computed by contracting U and $V_{\neq 0} \setminus U$ in G, and by determining a minimum cut in the contracted graph that separates the two vertices corresponding to the contracted sets. If $U \in \{\emptyset, V_{\neq 0}\}$, then $\nu(U)$ is obtained by contracting $V_{\neq 0}$ and finding a global minimum cut in the contracted graph.

To prove Theorem 4, we use weighted splitting-off techniques on G to construct a weighted auxiliary graph H on the vertex set $V_{\neq 0}$. We show that by choosing edges of H with probabilities proportional to the edge weights, a distribution with the properties highlighted in Theorem 4 is obtained. Details of the proof are discussed in Sect. 3 and Appendix A.2.

Theorem 4 with $\alpha = \mathrm{OPT}$ (or α slightly larger than OPT) implies that, whenever $\sum_{v \in V_{\neq 0}} \nu(\{v\})$ is large compared to OPT, a contraction step has good success probability, similar to Karger's analysis. Otherwise, instead of performing a contraction, we approximately reduce the problem to another (CCMC) instance with smaller modulus. More precisely, if $\sum_{v \in V_{\neq 0}} \nu(\{v\})$ is sufficiently small, there are many vertices $v \in V_{\neq 0}$ where the smallest cut $C_{\{v\}} \subseteq V$ separating v from $V_{\neq 0} \setminus \{v\}$ has weight no more than $\beta = \kappa \cdot \mathrm{OPT}$ for a small constant κ. Such cuts are useful to modify a cut with wrong residue class. Indeed, consider a cut C with small weight $w(\delta_G(C))$, but $\gamma(C) \not\equiv r \pmod{m}$. Then, $\overline{C} := C \triangle C_{\{v\}}$ satisfies $\gamma(\overline{C}) \equiv \gamma(C) \pm \gamma(v)$ (where the sign depends on whether $v \in C$), while the weight $w(\delta(\overline{C}))$ increased by at most β compared to $w(\delta(C))$; we recall that β is small with respect to OPT. Our plan is that if we have enough small cuts $C_{\{v\}}$, we can simplify the congruency constraint to one with smaller modulus, because the small cuts of type $C_{\{v\}}$ allow for moving solutions into the right residue class. This idea leads to the following notion of a *reduction family*.

Definition 1 (Reduction family). *Let* $\mathcal{I} = (G, w, \gamma, m, r)$ *be a (CCMC) instance on the graph* $G = (V, E)$. *For* $\beta \in \mathbb{R}_{>0}$ *and* $q \in [m-1]$, *a family* $\mathcal{R}(\beta, q) \subseteq 2^V$ *is a reduction family for* $\mathcal{I}$ *if*

(i) $\mathcal{R}(\beta, q) = \{R_1, R_2, \ldots, R_{2m_q - 1}\}$ with $m_q := \frac{m}{\gcd(m,q)}$,[6]
(ii) for each $i \in [2m_q - 1]$, there is one vertex $u_i \in R_i$ with $\gamma(u_i) \equiv q \pmod{m}$, and $\gamma(u) \equiv 0 \pmod{m}$ for all other $u \in R_i \setminus \{u_i\}$,
(iii) the vertices $u_1, \ldots u_{2m_q - 1}$ are distinct, and
(iv) $w(\delta(R_i)) \leqslant \beta$ for all $i \in [2m_q - 1]$.

A reduction family $\mathcal{R}(\beta, q)$ allows for correcting the residue class $\gamma(C)$ of a solution C by any multiple of q modulo m, with losses in terms of cut weight controlled by the parameter β. Given a reduction family $\mathcal{R}(\beta, q)$, it is thus sufficient to find a solution C' satisfying $\gamma(C') \equiv r \pmod{m'}$ for $m' = \gcd(m, q)$. This is formalized in the following lemma, with a proof given in Appendix A.1.

Lemma 1 (Reduction lemma). *Let $\mathcal{R}(\beta, q)$ be a reduction family for a (CCMC) instance (G, w, γ, m, r), and let $m' = \gcd(m, q)$. Given a cut $C' \subsetneq V$, $C' \neq \emptyset$, with $\gamma(C') \equiv r \pmod{m'}$, one can efficiently obtain a cut $C \subsetneq V$, $C \neq \emptyset$, such that (i) $w(\delta(C)) \leqslant w(\delta(C')) + \left(\frac{m}{m'} - 1\right)\beta$, and (ii) $\gamma(C) \equiv r \pmod{m}$.*

The above reduction lemma applied with a reduction family $\mathcal{R}(\beta, q)$ allows for reducing the modulus from m to a divisor m' of m, which is strictly smaller than m, as $0 < q < m$. We call such a reduction to a smaller modulus through a

---

[6] $\gcd(m, q)$ denotes the greatest common divisor of m and q.

reduction family a *reduction step*. The next theorem shows that reduction families exist (and can be found efficiently) whenever Theorem 4 fails to guarantee a distribution with the desired properties for Karger-type contraction steps, i.e., whenever $\sum_{v \in V_{\neq 0}} \nu(\{v\})$ is small. In this case, there are many vertices $v \in V_{\neq 0}$ for which $\nu(\{v\})$ is small, i.e., the cut $C_{\{v\}}$ has small value. A subset of these cuts can then be used as a reduction family. This idea is concretized in Theorem 5 below, a proof of which we defer to a full version of this extended abstract.

Theorem 5. *Let $\mathcal{I}$ be a* (CCMC) *instance with modulus m and let $B > 0$. Assume that $|V_{\neq 0}| \geqslant 4m^2$ and $\sum_{v \in V_{\neq 0}} \nu(\{v\}) \leqslant B \cdot |V_{\neq 0}|$. Then, for some $q \in [m-1]$, one can efficiently obtain a reduction family $\mathcal{R}(2B, q)$ for $\mathcal{I}$.*

A reduction step reduces the modulus m to a divisor strictly smaller than m, hence we can perform at most $\log_2(m)$ many reduction steps, and might end up solving a problem with modulus 1, i.e., an unconstrained minimum cut problem.

Altogether, the ingredients discussed above lead to Algorithm 1. This algorithm requires a guess α for the value of the optimal solution, which we can assume to know up to a factor of $(1 + \varepsilon)$ by trying all polynomially many values

$$\alpha \in \{0\} \cup \left\{ (1+\varepsilon)^i \cdot w_{\min} \,\middle|\, 0 \leqslant i \leqslant \lceil \log_{1+\varepsilon}(w_{\text{tot}}/w_{\min}) \rceil \right\}, \tag{1}$$

where $w_{\min} := \min\{w(e) \mid e \in E, \ w(e) \neq 0\}$ and $w_{\text{tot}} := w(E)$.

Algorithm 1. Contraction-Reduction algorithm for (CCMC).

Input: (CCMC) instance $\mathcal{I} = (G, w, \gamma, m, r)$ on $G = (V, E)$, error parameter
    $\rho > 0$, optimal value guess $\alpha > 0$.

while $|V_{\neq 0}| > \max\left\{4m^2, \lceil \frac{4m}{\rho} \rceil\right\}$ **and** $\sum_{v \in V_{\neq 0}} \nu(\{v\}) > \frac{\rho\alpha}{2m} \cdot |V_{\neq 0}|$ **do**
    1. Sample a pair $\{u, v\}$ from the distribution $\mathcal{D}$ guaranteed by Theorem 4.
    2. Modify G by contracting the set $\{u, v\}$.

if $|V_{\neq 0}| \leqslant \max\left\{4m^2, \lceil \frac{4m}{\rho} \rceil\right\}$ **then**
    1. For every $S \subseteq V_{\neq 0}$ with $\gamma(S) \equiv r \pmod{m}$, let
        $C_S \in \arg\min\{w(\delta(C)) \mid \emptyset \subsetneq C \subsetneq V, \ C \cap V_{\neq 0} = S\}$.
    2. Among all cuts C_S obtained in step 1, let C be one of smallest value
        $w(\delta(C))$.
  return *Cut that corresponds to C in input graph before contractions.*
else
    1. Use Theorem 5 to get reduction family $\mathcal{R}(\beta, q)$ for $\beta = \frac{\rho\alpha}{m}$ and some
        $q \in [m-1]$.
    2. Let $m' = \gcd(m, q)$. Apply Algorithm 1 recursively to $\mathcal{I}' = (G, w, \gamma, m', r)$
        with error parameter ρ and optimal value guess α to obtain a solution
        C' of $\mathcal{I}'$.
    3. Apply Lemma 1 to get a solution C of $\mathcal{I}$ from C' and $\mathcal{R}(\beta, q)$.
  return *Cut that corresponds to C in input graph before contractions.*

While $|V_{\neq 0}|$ is large, Algorithm 1 contracts two vertices of $V_{\neq 0}$ whenever the conditions of Theorem 4 are met with $c = \frac{4m}{\rho}$. Note that every contraction step reduces the number of vertices in $V_{\neq 0}$ by one or two, depending on whether $\gamma(u) + \gamma(v) \not\equiv 0 \pmod{m}$ or not. The if-block in Algorithm 1 performs the enumeration step described earlier once there are at most $\max\left\{4m^2, \left\lceil\frac{4m}{\rho}\right\rceil\right\}$ vertices left in $V_{\neq 0}$. If neither of the above is possible, then Theorem 5 and Lemma 1 allow for a reduction step, which is executed in the else-block, where we recursively invoke Algorithm 1 on an instance with strictly smaller modulus. Combining the above insights, we can prove the following guarantee for Algorithm 1.

Theorem 6. *Consider a* (CCMC) *instance* (G, w, γ, m, r) *with optimal solution value* OPT. *Let* $\alpha \geqslant$ OPT *and* $\rho > 0$. *Algorithm 1 is an efficient procedure that, by using* α *as an optimal value guess and* ρ *as error parameter, returns a solution with value at most* OPT $+ \rho\alpha\log_2 m$ *with probability at least* $1/\binom{|V|}{\lceil 4m/\rho\rceil}$.

Proof. The only randomized step of Algorithm 1 occurs in the while-loop, where pairs $\{u, v\}$ for contraction are sampled. For the analysis, we fix an optimal solution C_0 of $\mathcal{I}$, and first assume that no contraction is bad w.r.t. C_0, i.e., that no contraction step contracts two vertices on different sides of C_0 throughout Algorithm 1. Under this assumption, we prove by induction on m that Algorithm 1 returns a cut C satisfying $w(\delta(C)) \leqslant$ OPT $+ \rho\alpha\log_2 m$.

If $m = 1$, then $V_{\neq 0} = \emptyset$, hence the algorithm directly executes the if-block, where an unconstrained minimum cut problem is solved, giving an exact solution. This reflects that for $m = 1$, (CCMC) is an unconstrained minimum cut problem.

Now let $m > 1$. If no bad contraction is performed, C_0 remains feasible once the while-loop terminated, and α remains an upper bound on the optimal solution value in the new contracted graph. If $|V_{\neq 0}| \leqslant \max\left\{4m^2, \left\lceil\frac{4m}{\rho}\right\rceil\right\}$, then, in the if-block, all remaining options are enumerated, and an optimal solution is found. Else, we have $|V_{\neq 0}| \geqslant 4m^2$ and $\sum_{v \in V_{\neq 0}} \nu(\{v\}) \leqslant \frac{\rho\alpha}{2m} \cdot |V_{\neq 0}|$, hence by Theorem 5 with $B = \frac{\rho\alpha}{2m}$, a reduction family $\mathcal{R}(\frac{\rho\alpha}{m}, q)$ can be found efficiently. We have $q \in [m-1]$ by Theorem 5, so $m' = \gcd(m, q) < m$. Thus, by the inductive assumption, the recursive application of Algorithm 1 in step 2 of the else-block returns a solution $C' \subsetneq V$, $C' \neq \emptyset$, of $\mathcal{I}'$ with

$$\gamma(C') \equiv r \pmod{m'} \quad \text{and} \quad w(\delta(C')) \leqslant \text{OPT} + \rho\alpha\log_2(m'). \qquad (2)$$

Note that in the inequality, we used OPT$(\mathcal{I}') \leqslant$ OPT, which follows from the fact that C_0 remains feasible for $\mathcal{I}'$. By (2) and Lemma 1, the solution C of $\mathcal{I}$ constructed in step 3 is a cut, satisfies $\gamma(C) \equiv r \pmod{m}$, and

$$w(\delta(C)) \leqslant w(\delta(C')) + \left(\frac{m}{m'} - 1\right)\frac{\rho\alpha}{m} \leqslant \text{OPT} + \rho\alpha(\log_2 m' + 1) \leqslant \text{OPT} + \rho\alpha\log_2 m,$$

where the last inequality follows from $m' \leqslant m/2$, as m' is a divisor of m and strictly smaller than m. This concludes the induction. Thus, if no bad contraction steps are performed, a solution of value at most OPT $+ \rho\alpha\log_2 m$ is returned.

We now show that with probability $1/\binom{|V|}{\lceil 4m/\rho\rceil}$, no contraction step is bad w.r.t. C_0 throughout all recursive calls of Algorithm 1. Contraction steps are performed if $\sum_{v \in V_{\neq 0}} \nu(\{v\}) > \frac{\rho\alpha}{2m} \cdot |V_{\neq 0}|$, hence by Theorem 4 with $c = \frac{4m}{\rho}$,

$$\Pr[\text{a random contraction is bad w.r.t. } C_0] \leqslant \frac{4m}{\rho \cdot |V_{\neq 0}|} \leqslant \frac{k}{|V_{\neq 0}|}, \qquad (3)$$

where $k := \lceil \frac{4m}{\rho} \rceil$. The same bound can be derived in all recursive calls of Algorithm 1 w.r.t. the corresponding modulus m' used in that call. As $m' < m$ for any recursive call with modulus m', the upper bound $k/|V_{\neq 0}|$ from (3) holds at any stage. Let $s_1, s_2, \ldots, s_\ell \in \mathbb{Z}$, in this order, be the sizes of $V_{\neq 0}$ when contraction steps are performed. Note that every contraction step reduces $|V_{\neq 0}|$ by at least 1, reduction steps do not increase $|V_{\neq 0}|$, and $|V_{\neq 0}| > k$ whenever contraction steps are performed. This implies $|V| \geqslant s_1 > s_2 > \ldots > s_\ell > k$. Consequently,

$$\Pr[\text{no contraction is bad w.r.t. } C_0] \geqslant \prod_{i=1}^{\ell}\left(1 - \frac{k}{s_i}\right) \geqslant \prod_{i=k+1}^{|V|}\left(1 - \frac{k}{i}\right) = \frac{1}{\binom{|V|}{k}},$$

as desired. Finally, all operations in Algorithm 1 can be performed in polynomial time (using running time guarantees from Theorems 4, 5, and Lemma 1). As observed earlier, solving a (CCMC) instance with modulus m requires at most $\log_2(m)$ many recursive calls to Algorithm 1, hence Algorithm 1 is efficient. □

Guessing the optimal solution value up to a factor $(1+\varepsilon)$ and repeating Algorithm 1 polynomially often independently implies our main result, Theorem 1.

Proof of Theorem 1. For all polynomially many values of α given in (1), we run Algorithm 1 with $\rho = \frac{\varepsilon}{(1+\varepsilon)\log_2(m)}$ for $\binom{|V|}{k} \log |V|$ many times independently, where $k = \lceil 4m/\delta \rceil$, and we return the best solution found over all iterations. By Theorem 6, for $\alpha \in [\text{OPT}, (1+\varepsilon)\text{OPT})$, a single iteration returns a $(1+\varepsilon)$-approximate solution with probability at least $1/\binom{|V|}{k}$. Hence, among all iterations with this α, a $(1+\varepsilon)$-approximate solution is found with probability at least

$$1 - \left(1 - 1/\binom{|V|}{k}\right)^{\binom{|V|}{k}\cdot\log|V|} \geqslant 1 - \exp(-\log|V|) = 1 - 1/|V|. \qquad \square$$

3   Good Contraction Distributions Through Splitting-Off

To obtain a good distribution for Karger-type contractions (Theorem 4), we construct a weighted auxiliary graph $H = (V_{\neq 0}, F)$, and then select a pair of vertices $f \in F$ for contraction in G with probabilities proportional to the edge weights in H. The construction of H is based on splitting-off techniques, which, loosely speaking, allow for modifying a given graph such that certain connectivity properties are preserved. Our interest lies in preserving the values $\nu(\{v\}) = \mu_{G,w}(\{v\}, V_{\neq 0}\setminus\{v\})$ for all $v \in V_{\neq 0}$, where we use the notation $\mu_{G,w}(A, B) := \min\{w(\delta(C)) \mid A \subseteq C \subseteq V \setminus B\}$. Generalizing a splitting-off theorem of Lovász [14] to a weighted setting in combination with algorithmic ideas of Frank [7], we obtain the following theorem (see Appendix A.2 for more details).

Theorem 7. *Let $G = (V, E)$ be a graph with edge weights $w\colon E \to \mathbb{R}_{\geqslant 0}$, and let $Q \subseteq V$. There is a strongly polynomial time algorithm to obtain a graph $H = (Q, F)$ and edge weights $w_H\colon F \to \mathbb{R}_{\geqslant 0}$ such that*

(i) $w_H(\delta_H(q)) = \mu_{G,w}(\{q\}, Q \setminus \{q\})$ *for all $q \in Q$, and*
(ii) $w_H(\delta_H(C \cap Q)) \leqslant w(\delta_G(C))$ *for all $C \subseteq V$.*

We now show how Theorem 7 is used to prove Theorem 4.

Proof of Theorem 4. Apply Theorem 7 to (G, w) with $Q = V_{\neq 0}$ to obtain the graph $H = (V_{\neq 0}, F)$ with weights w_H. The distribution $\mathcal{D}$ over vertex pairs $\{u, v\}$ we use is given by choosing $\{u, v\} \in F$ with probability proportional to $w_H(\{u, v\})$. This is clearly an efficient sampling procedure. By Theorem 7(i),

$$2 \cdot w_H(F) = \sum_{v \in V_{\neq 0}} w_H(\delta_H(v)) = \sum_{v \in V_{\neq 0}} \mu_{G,w}(\{v\}, Q \setminus \{v\}) = \sum_{v \in V_{\neq 0}} \nu(\{v\}).$$

If C is a solution of $\mathcal{I}$ with $w(\delta(C)) \leqslant \alpha$, then by choice of $\mathcal{D}$ and the above,

$$\Pr_{\{u,v\} \sim \mathcal{D}}\left[|\{u, v\} \cap C| = 1\right] = \frac{w_H(\delta_H(C \cap V_{\neq 0}))}{w_H(F)} \leqslant \frac{2 \cdot w(\delta_G(C))}{\sum\limits_{v \in V_{\neq 0}} \nu(\{v\})} \leqslant \frac{c}{|V_{\neq 0}|}, \quad (4)$$

as desired, where the inequalities are due to Theorem 7(ii), $w(\delta_G(C)) \leqslant \alpha$, and the assumption $\sum_{v \in V_{\neq 0}} \nu(\{v\}) > \frac{2\alpha}{c} \cdot |V_{\neq 0}|$ in Theorem 4. $\qquad \square$

A   Missing Proofs

A.1   Proof of Lemma 1

Let $\mathcal{R}(\beta, q) = \{R_1, R_2, \ldots, R_{2m_q - 1}\}$ with distinct $u_i \in R_i$ for all $i \in [2m_q - 1]$ as given in item (ii) of Definition 1. We distinguish two cases: Either, there are m_q many vertices among the u_i with $u_i \in C'$, or there are m_q many with $u_i \notin C'$.

In the first case, assume w.l.o.g. that $u_1, \ldots, u_{m_q} \in C'$, and let $U_k := \bigcup_{i=1}^k R_i$ for $k \in \{0, \ldots, m_q - 1\}$. We show that for some k, the set $C_k := C' \triangle U_k$ has the desired properties. First observe that all C_k are cuts, as $C_0 = C'$ is a cut, and $u_1 \notin C_k \ni u_{m_q}$ for $k \in [m_q - 1]$. Moreover, $k \leqslant m_q - 1$ implies

$$w(\delta(C_k)) \leqslant w(\delta(C')) + \sum_{i=1}^k w(\delta(R_i)) \leqslant w(\delta(C')) + (m_q - 1)\beta. \quad (5)$$

Using $m_q = \frac{m}{\gcd(m,q)} = \frac{m}{m'}$, we see that (5) is precisely point (i) of Lemma 1 for C_k. To conclude, we show that there exists k such that C_k satisfies $\gamma(C_k) \equiv r$ (mod m), i.e., point (ii). Using that $\gamma(u) \equiv 0$ (mod m) for all $u \in R_i \setminus \{u_i\}$, and $u_i \in C'$ for all $i \in [m_q]$, we obtain $\gamma(C_k) \equiv \gamma(C') - \sum_{i=1}^k \gamma(u_i) \equiv \gamma(C') - kq$ (mod m). It thus suffices to find $k \in \{0, \ldots, m_q - 1\}$ with $\gamma(C') - kq \equiv r$ (mod m), or equivalently,

$$kq \equiv \gamma(C') - r \quad (\text{mod } m). \quad (6)$$

By assumption, $\gamma(C') - r \equiv 0 \pmod{m'}$, so $r' := \frac{\gamma(C')-r}{m'} \in \mathbb{Z}$, and $q' := \frac{q}{m'} \in \mathbb{Z}$ because $m' = \gcd(m, q)$. Dividing (6) by m', we obtain the equivalent equation $kq' \equiv r' \pmod{m_q}$, which has a solution $k \in \{0, \dots, m_q - 1\}$ as $\gcd(q', m_q) = 1$.

The second case, i.e., $u_1, \dots, u_{m_q} \notin C'$, is similar: C_k always is a cut because $C_0 = C'$ is a cut, and $u_1 \in C_k \not\ni u_{m_q}$ for $k \geq 1$. Equation (5) remains true and implies point (i). For point (ii), we use $\gamma(C_k) \equiv \gamma(C') + \sum_{i=1}^{k} \gamma(u_i)$, and the above analysis results in $kq' \equiv -r' \pmod{m_q}$, admitting a solution $k \in \{0, \dots, m_q - 1\}$.

Finally, given $\mathcal{R}(\beta, q)$ and C', checking which of the two cases applies can be done in polynomial time, as well as solving the respective congruence equation for k. Thus, a cut C with the desired properties can be obtained efficiently. $\square$

A.2   Sketch of proof of Theorem 7

As indicated earlier, Theorem 7 is a consequence of splitting-off techniques from Graph Theory, a fundamental tool dating back to the '70s [14–16]. Typically, a graph is modified by repeatedly *splitting off* two edges from a vertex v, i.e., replacing two non-parallel edges $\{v, x\}$ and $\{v, y\}$ by a new edge $\{x, y\}$, or deleting two parallel edges incident to v. Denoting $\mu_G(A, B) := \min\{|\delta_G(C)| \mid A \subseteq C \subseteq V \setminus B\}$ for a graph $G = (V, E)$ and $A, B \subseteq V$, Lovász proved the following.

Theorem 8 (Lovász [14]). *Let $G = (V, E)$ be Eulerian, let $Q \subseteq V$, and let $v \in V \setminus Q$. For every edge $\{v, x\} \in E$, there exists another edge $\{v, y\} \in E$ such that the graph G' arising from G by splitting off $\{v, x\}$ and $\{v, y\}$ from v satisfies*

$$\mu_G(\{q\}, Q \setminus \{q\}) = \mu_{G'}(\{q\}, Q \setminus \{q\}) \qquad \forall q \in Q.$$

Iterative applications of Theorem 8 for fixed $Q \subseteq V$ and $v \in V \setminus Q$ result in a new graph on the vertex set $V \setminus v$ only, without changing the value of minimum cuts separating a single vertex q from $Q \setminus \{q\}$, for all $q \in Q$. We aim for a generalization of this statement to a weighted setting, where the graph $G = (V, E)$ has edge weights $w \colon E \to \mathbb{R}_{\geqslant 0}$, a splitting operation consists of decreasing the weight on two edges $\{v, x\}$ and $\{v, y\}$ by some $\varepsilon > 0$ while increasing the weight on the edge $\{x, y\}$ by ε,

---

Algorithm 2. Fractionally splitting off a single vertex.

Input: Graph $G = (V, E)$ with edge weights $w \colon E \to \mathbb{R}_{\geqslant 0}$, $Q \subseteq V$, $v \in V \setminus Q$.

foreach $x, y \in N_G(v) := \{z \in V \setminus \{v\} \mid \{v, z\} \in E\}$, $x \neq y$ **do**
  foreach $q \in Q$ **do**
    Calculate the min cut sizes
    $$c_1^q = \mu_{G,w}(\{q\}, Q \setminus \{q\}), \quad c_2^q = \mu_{G,w}(\{q, v\}, (Q \setminus \{q\}) \cup \{x, y\}),$$
    $$\text{and} \quad c_3^q = \mu_{G,w}(\{q, x, y\}, (Q \setminus \{q\}) \cup \{v\}).$$
  Split off ε from $e_1 = \{v, x\}$ and $e_2 = \{v, y\}$, where
  $$\varepsilon = \min_{q \in Q} \min\left\{ (c_2^q - c_1^q)/2, (c_3^q - c_1^q)/2, w(e_1), w(e_2) \right\}.$$

return *Modified graph G with vertex v deleted and modified weights w.*

and we want the weighted cut values $\mu_{G,w}(\{q\}, Q \setminus \{q\})$ to be invariant. We claim that this is achieved by Algorithm 2. We highlight that efficient weighted versions of other splitting-off results (than Theorem 8) have already been considered by Frank [7], and our method is heavily inspired by Frank's approach.

In the inner for loop in Algorithm 2, if $q \in \{x, y\}$, then $c_2^q = \mu_{G,w}(\{q, v\}, (Q \setminus \{q\}) \cup \{x, y\})$ is the value of an infeasible cut problem (because both arguments of $\mu_{G,w}$ contain q), which we interpret as ∞.

In each iteration of the outer for loop in Algorithm 2, we split off $\varepsilon \geqslant 0$ from $\{v, x\}$ and $\{v, y\}$, with ε chosen maximal so that all weights remain non-negative and the connectivities of interest are preserved. This choice of ε implies that once the outer for loop terminated, there is no pair of edges incident to v from which a positive weight can be split off. Uniformly scaling all weights of this remaining graph to even integral weights (which we interpret as edge multiplicities) and employing Theorem 8, we can prove that there can only be a single edge with positive weight incident to v in the remaining graph, which we can thus safely delete without affecting connectivities within $V \setminus \{v\}$.

The following lemma summarizes the guarantees that we thereby obtain for Algorithm 2. A formal proof is deferred to a long version of this paper.

Lemma 2. *Let $G = (V, E)$ be a graph with edge weights $w \colon E \to \mathbb{R}_{\geqslant 0}$, let $Q \subsetneq V$ and $v \in V \setminus Q$. On this input, Algorithm 2 returns, in running time dominated by $\mathcal{O}(|V|^3)$ many minimum cut computations in (G, w), a graph $H = (V \setminus \{v\}, F)$ with edge weights $w_H \colon F \to \mathbb{R}$ such that*

(i) $\mu_{H,w_H}(\{q\}, Q \setminus \{q\}) = \mu_{G,w}(\{q\}, Q \setminus \{q\})$ for all $q \in Q$, and
(ii) $w_H(\delta_H(C \setminus \{v\})) \leqslant w(\delta_G(C))$ for all $C \subseteq V$.

Applying Lemma 2 iteratively for all $v \in V \setminus Q$ immediately yields Theorem 7.

References

1. Artmann, S., Weismantel, R., Zenklusen, R.: A strongly polynomial algorithm for bimodular integer linear programming. In: Proceedings of the 49th Annual ACM Symposium on Theory of Computing (STOC), pp. 1206–1219 (2017)
2. Barahona, F., Conforti, M.: A construction for binary matroids. Discrete Math. **66**(3), 213–218 (1987)
3. Bock, A., Faenza, Y., Moldenhauer, C., Ruiz-Vargas, A.J.: Solving the stable set problem in terms of the odd cycle packing number. In: Proceedings of the 34th IARCS Annual Conference on Foundations of Software Technology and Theoretical Computer Science (FSTTCS), pp. 187–198 (2014)
4. Chandrasekaran, K., Xu, C., Yu, X.: Hypergraph k-cut in randomized polynomial time. In: Proceedings of the 29th Annual ACM-SIAM Symposium on Discrete Algorithms (SODA), pp. 1426–1438 (2018)
5. Di Summa, M., Eisenbrand, F., Faenza, Y., Moldenhauer, C.: On largest volume simplices and sub-determinants. In: Proceedings of the 26th Annual ACM-SIAM Symposium on Discrete Algorithms (SODA), pp. 315–323 (2015)
6. Eisenbrand, F., Vempala, S.: Geometric random edge. Math. Program. **164**(1), 325–339 (2017)

7. Frank, A.: Augmenting graphs to meet edge-connectivity requirements. SIAM J. Discrete Math. **5**(1), 25–53 (1992)

8. Goemans, M.X., Ramakrishnan, V.S.: Minimizing submodular functions over families of sets. Combinatorica **15**(4), 499–513 (1995)

9. Grossman, J.W., Kulkarni, D.M., Schochetman, I.E.: On the minors of an incidence matrix and its smith normal form. Linear Algebra Appl. **218**, 213–224 (1995)

10. Grötschel, M., Lovász, L., Schrijver, A.: Corrigendum to our paper "The ellipsoid method and its consequences in combinatorial optimization". Combinatorica **4**(4), 291–295 (1984)

11. Karger, D.R.: Global min-cuts in $\mathcal{RNC}$, and other ramifications of a simple min-cut algorithm. In: Proceedings of the 4th Annual ACM-SIAM Symposium on Discrete Algorithms (SODA), pp. 21–30 (1993)

12. Karger, D.R., Stein, C.: A new approach to the minimum cut problem. J. ACM **43**(4), 601–640 (1996)

13. Khot, S.: Ruling out PTAS for graph min-bisection, dense k-subgraph, and bipartite clique. SIAM J. Comput. **36**(4), 1025–1071 (2006)

14. Lovász, L.: On some connectivity properties of Eulerian graphs. Acta Mathematica Academiae Scientiarum Hungarica **28**(1), 129–138 (1976)

15. Lovász, L.: Combinatorial Problems and Exercises. North-Holland, Amsterdam (1979)

16. Mader, W.: A reduction method for edge-connectivity in graphs. Ann. Discrete Math. **3**, 145–164 (1978)

17. Nägele, M., Sudakov, B., Zenklusen, R.: Submodular minimization under congruency constraints. In: Proceedings of the 29th Annual ACM-SIAM Symposium on Discrete Algorithms (SODA), pp. 849–866 (2018)

18. Nikolov, A.: Randomized rounding for the largest simplex problem. In: Proceedings of the 47th Annual ACM Symposium on Theory of Computing (STOC), pp. 861–870 (2015)

19. Padberg, M.W., Rao, M.R.: Odd minimum cut-sets and b-matchings. Math. Oper. Res. **7**(1), 67–80 (1982)

Sparsity of Integer Solutions
in the Average Case

Timm Oertel[1], Joseph Paat[2(✉)], and Robert Weismantel[2]

[1] School of Mathematics, Cardiff University, Cardiff, UK
[2] Institute for Operations Research, ETH Zürich, Zürich, Switzerland
joseph.paat@ifor.math.ethz.ch

Abstract. We examine how sparse feasible solutions of integer programs are, on average. Average case here means that we fix the constraint matrix and vary the right-hand side vectors. For a problem in standard form with m equations, there exist LP feasible solutions with at most m many nonzero entries. We show that under relatively mild assumptions, integer programs in standard form have feasible solutions with $O(m)$ many nonzero entries, on average. Our proof uses ideas from the theory of groups, lattices, and Ehrhart polynomials. From our main theorem we obtain the best known upper bounds on the integer Carathéodory number provided that the determinants in the data are small.

1 Introduction

Let $m, n \in \mathbb{Z}_{\geq 1}$ and $A \in \mathbb{Z}^{m \times n}$. We always assume that A has full row rank. We also view A as a set of its column vectors. So, $W \subseteq A$ implies that W is a subset of the columns of A.

We aim to find a sparse integer vector in the set

$$P(A, b) := \{x \in \mathbb{Z}_{\geq 0}^n : Ax = b\},$$

where $b \in \mathbb{Z}^m$. That is, we aim at finding a solution z such that $|\operatorname{supp}(z)|$ is as small as possible, where $\operatorname{supp}(x) := \{i \in \{1, \ldots, m\} : x_i \neq 0\}$ for $x \in \mathbb{R}^n$. To this end, we define the *support function of* (A, b) to be

$$\sigma(A, b) := \min\{|\operatorname{supp}(z)| : z \in P(A, b)\}.$$

If $P(A, b) = \emptyset$, then $\sigma(A, b) := \infty$. We define the *support function of* A to be

$$\sigma(A) := \max\{\sigma(A, b) : b \in \mathbb{Z}^m \text{ and } \sigma(A, b) < \infty\}.$$

The question of determining $\sigma(A)$ generalizes problems that have been open for decades. A notable special case is the so-called integer Carathéodory number, i.e. the minimum number of Hilbert basis elements in a rational pointed polyhedral cone required to represent an integer point in the cone. We say that A has the *Hilbert basis property* if its columns correspond to a Hilbert basis of $\operatorname{cone}(A)$.

© Springer Nature Switzerland AG 2019
A. Lodi and V. Nagarajan (Eds.): IPCO 2019, LNCS 11480, pp. 341–353, 2019.
https://doi.org/10.1007/978-3-030-17953-3_26

For A with the Hilbert basis property, Cook et al. [8] showed that $\sigma(A) \leq 2m-1$ and Sebő showed that $\sigma(A) \leq 2m-2$ [12]. Bruns et al. [7] provide an example of A with the Hilbert basis property with $\frac{7}{6}m \leq \sigma(A)$. However, for matrices with the Hilbert basis property, the true value of $\sigma(A)$ is unknown.

For general choices of A, Eisenbrand and Shmonin [10] showed that $\sigma(A) \leq 2m \log_2(4m\|A\|_\infty)$, where $\|\cdot\|_\infty$ is the max norm. Aliev et al. [1] and Aliev et al. [2] improved the previous result and showed that

$$\sigma(A) \leq m + \log_2(g^{-1}\sqrt{\det(AA^\mathsf{T})}) \leq 2m \log_2(2\sqrt{m}\|A\|_\infty), \tag{1}$$

where $g = \gcd\{|\det(B)| : B$ is an $m \times m$ submatrix of $A\}$. It turns out that the previous upper bound is close to the true value of $\sigma(A)$. In fact, for every $\epsilon > 0$, Aliev et al. [1] provide an example of A for which $m \log_2(\|A\|_\infty)^{1/(1+\epsilon)} \leq \sigma(A)$.

In this paper, we consider $\sigma(A, b)$ for *most choices* of b. We formalize this 'average case' using the *asymptotic support function of A* defined by

$$\sigma^{\mathrm{asy}}(A) := \min\left\{k \in \mathbb{Z} : \lim_{t\to\infty} \frac{|\{b \in \{-t,...,t\}^m : \sigma(A,b) \leq k\}|}{|\{b \in \{-t,...,t\}^m : P(A,b) \neq \emptyset\}|} = 1\right\}.$$

Note that $\sigma^{\mathrm{asy}}(A) \leq \sigma(A) \leq |A|$.

The value $\sigma^{\mathrm{asy}}(A)$ can be thought of as the smallest k such that almost all feasible integer programs with constraint matrix A have solutions with support of cardinality at most k. The function $\sigma^{\mathrm{asy}}(\cdot)$ was introduced by Bruns and Gubeladze in [6], where it was shown that $\sigma^{\mathrm{asy}}(A) \leq 2m - 3$ for matrices with the Hilbert basis property. In general, an average case analysis of the support question has not been provided in the literature. Average case behavior of integer programs has been studied in specialized settings, see, e.g., [9] for packing problems in $0, 1$ variables and [3] for problems with only one constraint. However, to the best of our knowledge, there are no other studies available that are concerned with the average case behavior of integer programs, in general.

Our analysis reveals that the *sizes* of the $m \times m$ minors of A affect sparsity. It turns out that *the number of factors in the prime decomposition* of the minors also affects sparsity. Moreover, for matrices with *large* minors but *few* factors, there exist solutions whose support depends on the number of factors rather than the size of the minors. Recall that a prime is a natural number greater than or equal to 2 that is divisible only by itself and 1. We now formalize these parameters related to the minors of a matrix.

Let $W \in \mathbb{Z}^{m \times d}$ be of full row rank, where $d \in \mathbb{Z}_{\geq 1}$. Denote the set of absolute values of the $m \times m$ minors by

$$\Delta(W) := \{|\det(W')| : W' \text{ is an invertible } m \times m \text{ submatrix of } W\},$$

and denote the set of 'number of prime factors' in each minor by

$$\Phi(W) := \left\{t \in \mathbb{Z}_{\geq 1} : \begin{array}{l} W' \text{ an invertible } m \times m \text{ submatrix of } W, \text{ and} \\ |\det(W')| = \prod_{i=1}^t \alpha_i \text{ with } \alpha_1, \ldots, \alpha_t \text{ prime} \end{array}\right\}. \tag{2}$$

If $\Phi(W)$ consists of only one element (e.g., when $W \in \mathbb{Z}^{m \times m}$), then we denote the element by $\phi(W)$. If $W \in \mathbb{Z}^{m \times m}$ and $|\det(W)| = 1$, then $\phi(W) = 0$. We denote the maximum and minimum of these sets by

$$\delta^{\max}(W) := \max(\Delta(W)), \qquad \delta^{\min}(W) := \min(\Delta(W)),$$
$$\phi^{\max}(W) := \max(\Phi(W)), \quad \text{and} \quad \phi^{\min}(W) := \min(\Phi(W)).$$

Our first main result bounds σ^{asy} using these parameters.

Theorem 1. *Let $A \in \mathbb{Z}^{m \times n}$ and $W \subseteq A$ such that $\mathrm{cone}(A) = \mathrm{cone}(W)$. Then*

(i) $\sigma^{asy}(A) \leq m + \phi^{\max}(W) \leq m + \log_2(\delta^{\max}(W))$,

(ii) $\sigma^{asy}(A) \leq 2m + \phi^{\min}(W) \leq 2m + \log_2(\delta^{\min}(W))$.

Theorem 1 guarantees that the average support $\sigma^{\mathrm{asy}}(A)$ is linear in m in two special cases: (a) the minimum minor of A is on the order of 2^m or (b) there is a prime minor. We emphasize that *(ii)* uses the *minimum* values $\phi^{\min}$ and $\delta^{\min}$, which can be bounded by sampling any $m \times m$ invertible submatrix of A. Thus, $\sigma^{\mathrm{asy}}(A)$ can be bounded by finding a single $m \times m$ invertible submatrix of A.

Note that the bound in (1) includes the term g. Our proof of Theorem 1 can be adjusted to prove $\sigma^{\mathrm{asy}}(A) \leq m + \log_2(g^{-1}\delta^{\max}(W))$ and $\sigma^{\mathrm{asy}}(A) \leq 2m + \log_2(g^{-1}\delta^{\min}(W))$. We omit this analysis here to simplify the exposition. However, it should be mentioned that

$$\delta^{\max}(A) \leq (\textstyle\sum_{\delta \in \Delta(A)} \delta^2)^{1/2} = \sqrt{\det(AA^\mathsf{T})},$$

where the equation follows from the so-called Cauchy-Binet formula. Therefore, if A has two nonzero $m \times m$ minors, then Theorem 1 *(i)* improves (1), on average.

A corollary of Theorem 1 is that if A has the Hilbert basis property, then the extreme rays of $\mathrm{cone}(A)$ provide enough information to bound $\sigma^{\mathrm{asy}}(A)$.

Corollary 1. *Let $V \subseteq \mathbb{Z}^m$ and $H \subseteq \mathbb{Z}^{m \times t}$. Assume that H has the Hilbert basis property and $\mathrm{cone}(H) = \mathrm{cone}(V)$. Then*

$$\sigma^{asy}(H) \leq m + \phi^{\max}(V) \leq m + \log_2(\delta^{\max}(V)).$$

If $\delta^{\max}(V) < 2^{m-3}$, then the bound in Corollary 1 improves the bound in [6].

By modifying a construction in [1], we obtain two interesting examples of $\sigma^{\mathrm{asy}}(A)$. The first example shows that Theorem 1 *(i)* gives a tight bound. The second example shows that Theorem 1 *(ii)* gives a tight bound and that $\sigma^{\mathrm{asy}}(A)$ can be significantly smaller than $\sigma(A)$.

Theorem 2. *For every $m \in \mathbb{Z}_{\geq 1}$ and $d \in \mathbb{Z}_{\geq 1}$, there is a matrix $A \in \mathbb{Z}^{m \times n}$ such that $\phi^{\max}(A) = d$ and $\sigma^{asy}(A) = m + d$.*

For every $m \in \mathbb{Z}_{\geq 1}$ and $d \in \mathbb{Z}_{\geq m+3}$, there is a matrix $B \in \mathbb{Z}^{(m+1) \times n}$ such that $\phi^{\min}(B) = 0$ and $\sigma^{asy}(B) = 2m + 2 < m + d = \sigma(B)$.

344   T. Oertel et al.

The proof of Theorem 1 is based on a combination of group theory, lattice theory, and Ehrhart theory. On a high level, the combination of group and lattice theory bears similarities to papers of Gomory [11] and Aliev et al. [2]. Gomory investigated the value function of an IP and proved its periodicity when the right-hand side vector is sufficiently large. Aliev et al. showed periodicity for the function $\sigma(A, b)$ provided again that b is sufficiently large. Our refined analysis allows us to quantify the number of right-hand sides for which the support function is small. This new contribution requires not only group and lattice theory, but also Ehrhart theory.

Sections 2 and 3 provide background on groups and subcones. In Sect. 4 we use the average support for each subcone to prove Theorem 1. We prove Theorem 2 in Appendix A.

2   The Group Structure of a Parallelepiped

Let $W \in \mathbb{Z}^{m \times m}$ be an invertible matrix, which we also view as a set of m linearly independent column vectors. Let $\Pi(W)$ denote the integer vectors in the fundamental parallelepiped generated by W:

$$\Pi(W) := \{z \in \mathbb{Z}^m : z = W\lambda \text{ for } \lambda \in [0, 1)^m\}.$$

For each $b \in \mathbb{Z}^m$, there is a unique $g \in \Pi(W)$ such that $b = g + Wz$, where $z \in \mathbb{Z}^m$ [5, Lemma 2.1, page 286]. Thus, we can define a *residue function* $\rho_W : \mathbb{Z}^m \to \Pi(W)$ by

$$\rho_W(b) = \rho_W(g + Wz) \mapsto g. \tag{3}$$

The image of $\mathbb{Z}^m$ under ρ_W (i.e., $\Pi(W)$) creates a group $G_W(\mathbb{Z}^m)$ using the operation $+_{G_W} : \Pi(W) \times \Pi(W) \to \Pi(W)$ defined by

$$g +_{G_W} h \mapsto \rho_W(g + h).$$

The identity of $G_W(\mathbb{Z}^m)$ is the zero vector in $\mathbb{Z}^m$, and

$$|G_W(\mathbb{Z}^m)| = |\det(W)|, \tag{4}$$

see, e.g., [5, Corollary 2.6, page 286]. Equation (4) implies $G_W(\mathbb{Z}^m)$ is finite.

The choice of notation for $G_W(\mathbb{Z}^m)$ is to emphasize that it is the group generated by the residues of all integer linear combinations of vectors in $\mathbb{Z}^m$. We can also consider the group generated by any subset of vectors in $\mathbb{Z}^m$. Given $B \subseteq \mathbb{Z}^m$, we denote the subgroup of $G_W(\mathbb{Z}^m)$ generated by B by

$$G_W(B) := \{\rho_W(Bz) : z \in \mathbb{Z}^{|B|}\}. \tag{5}$$

If $B = \emptyset$, then $G_W(B) := \{0\}$. The set $G_W(B)$ is a subgroup of $G_W(\mathbb{Z}^m)$ because $\{Bz : z \in \mathbb{Z}^{|B|}\}$ is a sublattice of $\mathbb{Z}^m$.

We collect some basic properties about the group $G_W(B)$.

Lemma 1. *Let* $W \in \mathbb{Z}^{m \times m}$ *be an invertible matrix. For every* $B \subseteq \mathbb{Z}^m$, $G_W(B) = \{\rho_W(Bz) : z \in \mathbb{Z}_{\geq 0}^{|B|}\}$.

Proof. For each $z \in \mathbb{Z}^{|B|}$, we can write Bz as

$$Bz = \sum_{b \in B: z_b \geq 0} z_b b + \sum_{b \in B: z_b < 0} z_b b.$$

Thus, it suffices to show $\rho_W(-b) \in \{\rho_W(By) : y \in \mathbb{Z}^{|B|}_{\geq 0}\} =: C$ for each $b \in B$. If $\rho_W(b) = 0$, then $\rho_W(-b) = \rho_W(b) = 0 \in C$. If $\rho_W(b) \neq 0$, then because $G_W(B)$ is finite there exists $\tau \in \mathbb{Z}_{\geq 2}$ with $\rho_W(\tau b) = 0$. Note that $\rho_W((\tau-1)b)+\rho_W(b) = 0 = \rho_W(b) + \rho_W(-b)$, so $\rho_W(-b) = \rho_W((\tau-1)b) \in C$. □

Lemma 2. *Let $W \in \mathbb{Z}^{m \times m}$ be an invertible matrix and $B \subseteq \mathbb{Z}^m$. If $t \in \mathbb{Z}_{\geq 0}$ with $t \geq \phi(W)$, then there exist $w^1, \ldots, w^t \in B$ (possibly with repetitions) such that $G_W(\{w^1, \ldots, w^t\}) = G_W(B)$.*

Proof. Set $s := \phi(W)$. First, we show that for each $r \in \{0, \ldots, s\}$ there exist $w^1, \ldots, w^r \in B$ (possibly with repetitions) such that

$$\begin{aligned} \text{either} \quad & G_W(\{w^1, \ldots, w^r\}) = G_W(B) \\ \text{or} \quad & G_W(\emptyset) \subsetneq G_W(\{w^1\}) \subsetneq \ldots \subsetneq G_W(\{w^1, \ldots, w^r\}). \end{aligned} \tag{6}$$

We prove (6) by induction on r. The result is vacuously true for $r = 0$, so assume that (6) holds for $r \in \mathbb{Z}_{\geq 0}$ and consider $r + 1$. Define

$$G^r := G_W(\{w^1, \ldots, w^r\}). \tag{7}$$

By the induction hypothesis, there exist $w^1, \ldots, w^r \in B$ such that (6) holds. If $G^r = G_W(B)$, then $w^{r+1} := w^r$ proves (6) for $r + 1$. If $G^r \subsetneq G_W(B)$, then $G^0 \subsetneq \ldots \subsetneq G^r$ by (6) and induction. Recall $\rho_W(\cdot)$ from (3). If $\rho_W(b) \in G^r$ for every $b \in B$, then $G_W(B) \subseteq G^r$ and $|G_W(B)| \leq |G^r| < |G_W(B)|$, which is a contradiction. Thus, there exists $w^{r+1} \in B$ such that $\rho_W(w^{r+1}) \notin G^r$. The sequence $G^0, \ldots, G^r, G^{r+1} := G_W(\{w^1, \ldots, w^{r+1}\})$ satisfies (6), which proves (6).

Let $G^1, \ldots, G^s$ be chosen to satisfy (6). If $G^s = G_W(B)$, then set $w^{s+1} = \ldots = w^t := w^s$ to conclude $G_W(\{w^1, \ldots, w^t\}) = G_W(B)$. It is left to consider the case when $G^s \subsetneq G_W(B)$. We claim that this leads to a contradiction.

By (2) and (4), $|G_W(\mathbb{Z}^m)| = \prod_{i=1}^s \alpha_i$ for primes $\alpha_1, \ldots, \alpha_s$. By (7), $G^1, \ldots, G^s$ are subgroups of $G_W(\mathbb{Z}^m)$, so $|G^1|, \ldots, |G^s|$ divide $|G_W(\mathbb{Z}^m)|$ (see, e.g., [4, Chapter 2]). Also, $G^s \subsetneq G_W(B)$ and (6) imply that $G^1 \subsetneq \ldots \subsetneq G^s$. Hence, $1 < |G^1| < \ldots < |G^s|$ and $|G^i|$ divides $|G^{i+1}|$ for each $i \in \{1, \ldots, s-1\}$. This implies that $|G^s|$ has at least s many prime factors. However, $|G^s| < |G_W(B)| \leq |G_W(\mathbb{Z}^m)|$, and $|G_W(\mathbb{Z}^m)|$ only has s many prime factors. Thus, $|G^i| = |G_W(\mathbb{Z}^m)|$ for some $i \in \{1, \ldots, s\}$, which contradicts $G^i = G_W(\mathbb{Z}^m) \supseteq G_W(B)$. □

3   Lattice Points in Cones

A set $\Lambda \subseteq \mathbb{Z}^m$ is a lattice if $0 \in \Lambda$, $x + y \in \Lambda$ for $x, y \in \Lambda$, and if $x \in \Lambda$ then $-x \in \Lambda$ (see, e.g., [5, Chapter VII]). So, Λ is a subgroup of $\mathbb{Z}^m$. We assume that a lattice contains m linearly independent vectors. For $B \subseteq \mathbb{R}^m$ and $x \in \mathbb{R}^m$, set $B + x := \{b + x : b \in B\}$.

We use following lemma to find suitable translated subcones in which $\sigma(A, \cdot)$ is bounded. The proof of Lemma 3 is in Appendix B.

Lemma 3. *Let $v^1, \ldots, v^m \in \mathbb{Z}^m$ be linearly independent vectors and set $K := \mathrm{cone}(\{v^1, \ldots, v^m\})$. For $t \in \mathbb{Z}_{\geq 0}$ and $x^1, \ldots, x^t \in \mathbb{Z}^m$, there is a $z = \sum_{i=1}^m k_i v^i \in K \cap \mathbb{Z}^m$, where $k_1, \ldots, k_m \in \mathbb{Z}_{\geq 0}$, such that $K + z \subseteq K \cap \bigcap_{i=1}^t (K + x^i)$.*

Let $W \subseteq \mathbb{Z}^m$. For each $x \in \mathrm{cone}(W)$, Carathéodory's Theorem implies that there is a linearly independent set $W^i \subseteq W$ such that $x \in \mathrm{cone}(W^i)$. Thus,

$$\mathrm{cone}(W) = \bigcup_{i=1}^s \mathrm{cone}(W^i), \tag{8}$$

where $s \in \mathbb{Z}_{\geq 1}$ and $W^1, \ldots, W^s \subseteq W$ are the linearly independent subsets of W. The following lemma states that for a given lattice Λ, 'most' of the points in $\mathrm{cone}(W) \cap \Lambda$ are found in translations of the subcones $\mathrm{cone}(W^1), \ldots, \mathrm{cone}(W^s)$.

Lemma 4. *Let $W \subseteq \mathbb{Z}^m$ be such that $\mathrm{cone}(W)$ is m-dimensional. Let $s \in \mathbb{Z}_{\geq 1}$ and $W^1, \ldots, W^s \subseteq W$ be as in (8). Let $\Lambda \subseteq \mathbb{Z}^m$ be a lattice and assume that $W^1, \ldots, W^s \subseteq \Lambda$. For each $i \in \{1, \ldots, s\}$, choose any $k_w \in \mathbb{Z}_{\geq 0}$ for each $w \in W^i$, and define $z^i := \sum_{w \in W^i} k_w w$. Then*

$$\lim_{t \to \infty} \frac{|\{-t, \ldots, t\}^m \cap \bigcup_{i=1}^s (\Lambda \cap (\mathrm{cone}(W^i) + z^i))|}{|\{-t, \ldots, t\}^m \cap \bigcup_{i=1}^s (\Lambda \cap \mathrm{cone}(W^i))|} = 1. \tag{9}$$

Proof. For $i \in \{1, \ldots, s\}$ set $K^i := \mathrm{cone}(W^i)$. The fraction in (9) equals

$$1 - \frac{|\{-t, \ldots, t\}^m \cap \bigcap_{i=1}^s (\Lambda \cap [K^i \setminus (K^i + z^i)])|}{|\{-t, \ldots, t\}^m \cap \bigcup_{i=1}^s (\Lambda \cap K^i)|}.$$

Thus, in order to prove (9), it is enough to prove

$$\lim_{t \to \infty} \frac{|\{-t, \ldots, t\}^m \cap \bigcap_{i=1}^s (\Lambda \cap [K^i \setminus (K^i + z^i)])|}{|\{-t, \ldots, t\}^m \cap \bigcup_{i=1}^s (\Lambda \cap K^i)|} = 0. \tag{10}$$

By assumption, $\mathrm{cone}(W)$ is m-dimensional. Thus, we may assume that the sets $W^1, \ldots, W^s$ each have m linearly independent vectors.

Let $i \in \{1, \ldots, s\}$ and $L^i \subseteq \Lambda \cap K^i$ be the Λ points that are coordinate-wise at most one more than z^i in the coordinate system defined by W^i:

$$L^i := \left\{ \sum_{w \in W^i} \beta_w w : \beta_w \in \mathbb{R} \text{ and } 0 \leq \beta_w \leq k_w + 1 \; \forall \; w \in W^i \right\} \cap \Lambda.$$

The set L^i is finite.

The numerator of (10) considers $\Lambda \cap [K^i \setminus (K^i + z^i)]$, so take $y \in \Lambda \cap [K^i \setminus (K^i + z^i)]$. We claim that

$$y \in r + \left\{ \sum_{w \in I} \lambda_w w : \lambda_w \in \mathbb{R}_{\geq 0} \; \forall \; w \in I \right\}, \tag{11}$$

where $r \in L^i$ and $I \subseteq W^i$ with $|I| \leq m - 1$. Write y as $y = \sum_{w \in W^i} \gamma_w w$, where $\gamma_w \in \mathbb{R}_{\geq 0}$ for each $w \in W^i$ and $\gamma_{\bar{w}} < k_{\bar{w}}$ for some $\bar{w} \in W^i$. We have $y - \tau w \in \Lambda$ for each $w \in W^i \setminus \{\bar{w}\}$ and $\tau \in \mathbb{Z}$ because $W^i \subseteq \Lambda$ and $y \in \Lambda$. In particular, $y - \lfloor \gamma_w \rfloor w \in \Lambda \cap K^i$ and $y - \sum_{w \in V} \lfloor \gamma_w \rfloor w \in L^i$, where $V := \{w \in W^i : \gamma_w > k_w + 1\}$. This proves (11). Note that we use the fact that L^i is defined by $\beta_w \leq k_w + 1$ rather than $\beta_w \leq k_w$: if L^i was defined by $\beta_w \leq k_w$, then in the extreme case $0 = k_w$ and $\gamma_w \in (0, 1)$, the vector $y - \lfloor \gamma_w \rfloor w = y$ is not in L^i.

We use the fact that $|I| < m$ to show $\Lambda \cap [K^i \setminus (K^i + z^i)]$ is contained in finite union of lower dimensional spaces. Although we showed $|I| \leq m - 1$, we can assume $|I| = m - 1$ by extending it arbitrarily to have $m - 1$ columns and setting $\lambda_w = 0$ for these new columns. Hence,

$$\bigcap_{i=1}^{s} \Lambda \cap [K^i \setminus (K^i + z^i)]$$

$$\subseteq \bigcap_{i=1}^{s} \bigcup_{r \in L^i} \bigcup_{\substack{I \subseteq W^i \\ |I| = m - 1}} r + \left\{ \sum_{w \in I} \lambda_w w : \lambda_w \in \mathbb{R}_{\geq 0} \; \forall \; w \in I \right\}. \tag{12}$$

For each $i \in \{1, \dots, s\}$ and $I \subseteq W^i$ with $|I| = m - 1$, define the polytope

$$P^{(i,I)} := \left\{ \sum_{w \in I} \lambda_w w : \lambda_w \in [0, 1] \; \forall \; w \in I \right\}.$$

By assumption, $w \in \Lambda$ for each $w \in I$, so the vertices of $P^{(i,I)}$ are in Λ. Ehrhart theory then implies that there is a polynomial $\pi^{(i,I)}(t)$ of degree $m - 1$ such that

$$\pi^{(i,I)}(t) = |tP^{(i,I)} \cap \Lambda| = \left| \left\{ \sum_{w \in I} \lambda_w w : \lambda_w \in [0, t] \; \forall \; w \in I \right\} \cap \Lambda \right|$$

for each $t \in \mathbb{Z}_{\geq 1}$. The leading coefficient of $\pi^{(i,I)}$ is the $(m - 1)$ dimensional volume of $P^{(i,I)}$, which is positive, see [5, Chapter VIII]. Similarly, for the polytope

$$P^i := \left\{ \sum_{w \in W^i} \lambda_w w : \lambda_w \in [0, 1] \; \forall \; w \in W^i \right\}$$

there exists a polynomial $\pi^i(t)$ of degree m with positive leading coefficient such that for each $t \in \mathbb{Z}_{\geq 1}$

$$\pi^i(t) = |tP^i \cap \Lambda| = \left| \left\{ \sum_{w \in W^i} \lambda_w w : \lambda_w \in [0, t] \; \forall \; w \in W^i \right\} \cap \Lambda \right|.$$

Define

$$d := \max \left\{ \left\| r + \sum_{w \in I} w \right\|_\infty : i \in \{1, \dots, s\}, \; r \in L^i, \; I \subseteq W^i \text{ with } |I| \leq m - 1 \right\}.$$

We show that the values in (10) go to zero as $t \to \infty$ by bounding the fraction

$$\frac{|\{-td, \dots, td\}^m \cap \bigcap_{i=1}^{s} (\Lambda \cap [K^i \setminus (K^i + z^i)])|}{|\{-td, \dots, td\}^m \cap \bigcup_{i=1}^{s} (\Lambda \cap K^i)|}$$

for each $t \in \mathbb{Z}_{\geq 0}$. By the definition of d, $tP^i \subseteq \{-td, \dots, td\}^m \cap K^i$ for every $i \in \{1, \dots, s\}$. So for each $i \in \{1, \dots, s\}$, say $i = 1$, it follows that

$$\pi^1(t) = |tP^1 \cap \Lambda| \leq |\{-td, \dots, td\}^m \cap \Lambda \cap K^1| \leq \left| \{-td, \dots, td\}^m \cap \bigcup_{i=1}^{s} (\Lambda \cap K^i) \right|.$$

Hence,

$$\frac{1}{|\{-td,...,td\}^m \cap \bigcup_{i=1}^{s}(\Lambda \cap K^i)|} \leq \frac{1}{\pi^1(t)}. \tag{13}$$

If $i \in \{1,\ldots,s\}$ and $y \in \{-td,\ldots,td\}^m \cap \Lambda \cap [K^i \setminus (K^i + z^i)]$, then, by (12), $y = r + \sum_{w \in I} \lambda_w w$ for $r \in L^i$, $I \subseteq W^i$ with $|I| = m-1$, and $\lambda_w \in \mathbb{R}_{\geq 0}$ for each $w \in I$. This implies that

$$\|\textstyle\sum_{w \in I} \lambda_w w\|_\infty = \|y - r\|_\infty \leq \|y\|_\infty + \|r\|_\infty \leq td + d = (t+1)d.$$

Hence,

$$\{-td,\ldots,td\}^m \cap \Lambda \cap [K^i \setminus (K^i + z^i)] \subseteq \bigcup_{r \in L^i} \bigcup_{\substack{I \subseteq W^i \\ |I| = m-1}} r + (t+1)dP^{(i,I)}.$$

If $r \in L^i$, then by the definition of L^i, $r \in \Lambda$. This implies that the number of Λ points in $r + (t+1)dP^{(i,I)}$ is equal to $\pi^{(i,I)}((t+1)d)$. So, for each $i \in \{1,\ldots,s\}$,

$$|\{-td,\ldots,td\}^m \cap \Lambda \cap [K^i \setminus (K^i + z^i)]| \leq \sum_{r \in L^i} \sum_{\substack{I \subseteq W^i \\ |I| = m-1}} \pi^{(i,I)}((t+1)d). \tag{14}$$

The polynomial on the right-hand side of (14), call it $\psi(t+1)$, is of degree $m-1$ and has a positive leading coefficient. Also, by (13) and (14),

$$\frac{|\{-td,...,td\}^m \cap \bigcap_{i=1}^{s}(\Lambda \cap [K^i \setminus (K^i + z^i)])|}{|\{-td,...,td\}^m \cap \bigcup_{i=1}^{s}(\Lambda \cap K^i)|} \leq \frac{\psi(t+1)}{\pi^1(t)}.$$

Recall that π^1 is of degree m, ψ is of degree $m-1$, and ψ and π^1 have positive leading coefficients. Moreover, the limit as $t \to \infty$ is the same as $td \to \infty$. Hence,

$$\lim_{t \to \infty} \frac{|\{-t,...,t\}^m \cap \bigcap_{i=1}^{s}(\Lambda \cap [K^i \setminus (K^i + z^i)])|}{|\{-t,...,t\}^m \cap \bigcup_{i=1}^{s}(\Lambda \cap K^i)|}$$

$$= \lim_{t \to \infty} \frac{|\{-td,...,td\}^m \cap \bigcap_{i=1}^{s}(\Lambda \cap [K^i \setminus (K^i + z^i)])|}{|\{-td,...,td\}^m \cap \bigcup_{i=1}^{s}(\Lambda \cap K^i)|} = \lim_{t \to \infty} \frac{\psi(t+1)}{\pi^1(t)} = 0. \qquad \square$$

4   Proof of Theorem 1

The assumption $\mathrm{cone}(A) = \mathrm{cone}(W)$ indicates that we can write $\mathrm{cone}(A)$ as

$$\mathrm{cone}(A) = \bigcup_{i=1}^{s} \mathrm{cone}(W^i), \tag{15}$$

where $s \in \mathbb{Z}_{\geq 1}$ and $W^1,\ldots,W^s \subseteq W$ are linearly independent sets; see (8). Also, A has full row rank, so we assume that $W^1,\ldots,W^s$ each contain m linearly independent vectors. For $i \in \{1,\ldots,s\}$, let $K^i := \mathrm{cone}(W^i)$.

First, we prove $\sigma^{\mathrm{asy}}(A) \leq m + \phi^{\max}(A)$. In order to do this, we find a lattice Λ and points $z^1 \in K^1, \ldots, z^s \in K^s$ such that

$$\sigma(A, b) \leq m + \phi^{\max}(W) \quad \forall\, b \in (\Lambda \cap (K^1 + z^1)) \cup \ldots \cup (\Lambda \cap (K^s + z^s))$$

and Λ contains every $b \in \mathbb{Z}^m$ such that $P(A, b) \neq \emptyset$. With these values, we will be able to apply Lemma 4 to prove the desired result.

Fix $i \in \{1, \ldots, s\}$ and set $\phi^i := \phi(W^i)$. Let $G_{W^i}(\mathbb{Z}^m)$ be the group defined in Sect. 2. In view of Lemma 2, there exist $w^1, \ldots, w^t \in A$ with $t \leq \phi^i$ and

$$G_{W^i}(\{w^1, \ldots, w^t\}) = G_{W^i}(A).$$

We emphasize that the choice of $w^1, \ldots, w^t$ depends on W^i. Define the lattice

$$\Lambda^i := \left\{ \sum_{h \in G_{W^i}(A)} k_h h + \sum_{w \in W^i} p_w w : k_h \in \mathbb{Z} \,\forall\, h \in G_{W^i}(A),\ p_w \in \mathbb{Z} \,\forall\, w \in W^i \right\}.$$

In Lemma 6, we show that Λ^i does not depend on i. Lemma 1 implies that $\Lambda^i \supseteq \{g \in G_{W^i}(\mathbb{Z}^m) : \exists\, b \in \mathbb{Z}^m \text{ such that } \rho_{W^i}(b) = g \text{ and } P(A, b) \neq \emptyset\}$. Thus,

$$\text{if } b \notin \Lambda^i \text{ (equivalently, if } \rho_{W^i}(b) \notin G_{W^i}(A)), \text{ then } P(A, b) = \emptyset. \qquad (16)$$

Lemma 5. *There exists $z^i \in \Lambda^i \cap K^i$ that satisfies the following: for every $b \in (K^i + z^i) \cap \mathbb{Z}^m$, either $b \notin \Lambda^i$ (so $P(A, b) = \emptyset$) by (16)) or $\sigma(A, b) \leq m + \phi^i$. The vector z^i satisfies $z^i = \sum_{w \in W^i} k_w w$, where $k_w \in \mathbb{Z}_{\geq 0}$ for each $w \in W^i$.*

Proof of Lemma. For each $g \in G_{W^i}(A) = G_{W^i}(\{w^1, \ldots, w^t\})$, there exists $x^g \in \mathbb{Z}^m$ such that

$$x^g - g = \sum_{w \in W^i} \tau_w w \quad \text{and} \quad x^g = \sum_{w \in W^i} q_w w + \sum_{\ell=1}^{t} p_\ell w^\ell, \qquad (17)$$

where $\tau_w \in \mathbb{Z}$ and $q_w \in \mathbb{Z}_{\geq 0}$ for each $w \in W^i$ and $p_1, \ldots, p_t \in \mathbb{Z}_{\geq 0}$. By Lemma 3, there exists $z^i \in \Lambda^i \cap K^i$ such that $\rho_{W^i}(z^i) = 0$ and $K^i + z^i \subseteq K^i \cap \bigcap_{g \in G_{W^i}(A)} (K^i + x^g)$. Let $b \in (K^i + z^i) \cap \mathbb{Z}^m$ such that $P(A, b) \neq \emptyset$. By (16), there is a $g \in G_{W^i}(A)$ such that $\rho_{W^i}(b) = g$. So, by (17),

$$\begin{aligned} b = g + \sum_{w \in W^i} \bar{\tau}_w w &= x^g + \sum_{w \in W^i} (\bar{\tau}_w - \tau_w) w \\ &= \sum_{\ell=1}^{t} p_\ell w^\ell + \sum_{w \in W^i} (q_w + \bar{\tau}_w - \tau_w) w, \end{aligned} \qquad (18)$$

where $\bar{\tau}_w \in \mathbb{Z}$ for each $w \in W^i$. Note $\bar{\tau}_w - \tau_w \in \mathbb{Z}_{\geq 0}$ for each $w \in W^i$ because $b \in K^i + z^i \subseteq K^i + x^g$. Thus, $P(A, b) \neq \emptyset$ and $\sigma(A, b) \leq |W^i| + t \leq m + \phi^i$. $\square$

Lemma 6. *For every pair $i, j \in \{1, \ldots, s\}$, the lattices Λ^i and Λ^j are equal.*

Proof of Lemma. It is enough to show that $\Lambda^1 \subseteq \Lambda^2$. Let $x \in \Lambda^1$. By Lemmas 3 and 5, there is a point $y \in (K^1 + z^1) \cap \Lambda^1$ such that $\rho_{W^1}(y) = \rho_{W^1}(x)$. Also, by Lemma 5, $P(A, y) \neq \emptyset$. Hence, by (16), $y \in \Lambda^2$. Similarly, $w \in \Lambda^2$ for each $w \in W^1$. These inclusions along with $\rho_{W^1}(y) = \rho_{W^1}(x)$ imply $x \in \Lambda^2$. $\square$

Set $\Lambda := \Lambda^1 = \ldots = \Lambda^s$. Lemma 5 implies that

$$\bigcup_{i=1}^{s}(\Lambda \cap (K^i + z^i)) \subseteq \{b \in \mathbb{Z}^m : \sigma(A, b) \le m + \phi^{\max}\}.$$

By (15) and (16), it follows that

$$\{b \in \mathbb{Z}^m : P(A, b) \ne \emptyset\} \subseteq \operatorname{cone}(A) \cap \Lambda = \bigcup_{i=1}^{s} \Lambda \cap K^i$$

Hence, for each $t \in \mathbb{Z}_{\ge 1}$, it follows that

$$\begin{aligned}
&\frac{|\{b \in \{-t, \ldots, t\}^m : \sigma(A, b) \le m + \phi^{\max}\}|}{|\{b \in \{-t, \ldots, t\}^m : P(A, b) \ne \emptyset\}|} \\
&\ge \frac{|\{b \in \{-t, \ldots, t\}^m \cap (\bigcup_{i=1}^{s} \Lambda \cap (K^i + z^i))\}|}{|\{b \in \{-t, \ldots, t\}^m \cap (\bigcup_{i=1}^{s} \Lambda \cap K^i)\}|}.
\end{aligned} \tag{19}$$

By Lemma 4, it follows that $\sigma^{\mathrm{asy}}(A) \le m + \phi^{\max}(W)$. Also, the inequality $\phi^i \le \log_2(|\det(W^i)|)$ for each $i \in \{1, \ldots, s\}$ implies $\phi^{\max}(W) \le \log_2(\delta^{\max}(W))$ and $\sigma^{\mathrm{asy}}(A) \le m + \phi^{\max}(W) \le m + \log_2(\delta^{\max}(W))$.

Consider the inequality $\sigma^{\mathrm{asy}}(A) \le 2m + \phi^{\min}(W)$. Without loss of generality, $\phi^1 \le \ldots \le \phi^s$. Let $z^1 \in K^1 \cap \Lambda$ be given from Lemma 5. Let $i \in \{2, \ldots, s\}$. Using Lemma 3 and the fact that $K^1 + z^1$ is m-dimensional, the representative set $\{x^g : g \in G_{W^i}(A)\}$ from (17) can be chosen in $K^1 + z^1$. Let $b \in K^i + z^i$. By (18), there exists a $g \in G_{W^i}(A)$ such that

$$b = x^g + \sum_{w \in W^i}(\bar{\tau}_w - \tau_w)w,$$

where $\bar{\tau}_w - \tau_w \in \mathbb{Z}_{\ge 0}$ for each $w \in W^i$. The point x^g is in $K^1 + z^1$, so $P(A, x^g) \ne \emptyset$ and there are $w^1, \ldots, w^{m+\phi_1} \in A$ such that $x^g = \sum_{i=1}^{m+\phi_1} q_\ell w^\ell$, where $q_1, \ldots, q_{m+\phi_1} \in \mathbb{Z}_{\ge 0}$. So,

$$b = \sum_{\ell=1}^{m+\phi_1} q_\ell w^\ell + \sum_{w \in W^i}(\bar{\tau}_w - k_w)w.$$

Thus, $P(A, b) \ne \emptyset$ and $\sigma(A, b) \le 2m + \phi^1 = 2m + \phi^{\min}(W)$. Hence, $\sigma^{\mathrm{asy}}(A) \le 2m + \phi^{\min}(W)$.

Finally, assume $\log_2(\delta^{\min}(W)) = \log_2(W^2)$. Observe that $\phi(W^2) \le \log_2(W^2)$, so $\sigma^{\mathrm{asy}}(A) \le 2m + \phi^{\min}(W) \le 2m + \log_2(\delta^{\min}(W))$. □

A Proof of Theorem 2

We construct both matrices A and B using a submatrix $\tilde{A}$, which we construct first. Let $d \in \mathbb{Z}_{\ge 1}$ and $p_1 < \ldots < p_d$ be prime. For $i \in \{1, \ldots, d\}$, define $q_i := \prod_{j=1, j \ne i}^{d} p_j$ and $\delta := \prod_{j=1}^{d} p_j$. Define the matrix $\tilde{A} := [q_1, \ldots q_d, -\delta]$. The matrix $\tilde{A}$ has $d+1$ columns, so $\sigma^{\mathrm{asy}}(\tilde{A}) \le 1 + d$. The matrix $\tilde{A}$ is similar to the example in [1, Theorem 2] and the theory of so-called *primorials*. We claim

if $b \in \mathbb{Z}_{<0}$ and $b \equiv 1 \bmod \delta$, then $P(\tilde{A}, b) \ne \emptyset$ and $\sigma(\tilde{A}, b) = 1 + d$. (20)

Note that $\gcd(q_1,\ldots,q_d) = 1$. The *Frobenius number* of $\{q_1,\ldots,q_d\}$ is the largest integer that cannot be written as a positive integer linear combination of $q_1,\ldots,$ and q_d. Hence, if we choose $\bar{b} \in \mathbb{Z}_{\geq 1}$ to be the Frobenius number of $\{q_1,\ldots,q_d\}$, then $b \geq \bar{b}+1$ implies $P(\tilde{A},b) \neq \emptyset$. If $b \equiv 1 \bmod \delta$, then b is not divisible by p_i for any $i \in \{1,\ldots,d\}$. Thus, if $b \geq \bar{b}+1$ and $b \equiv 1 \bmod \delta$, then $\sigma(\tilde{A},b) = d$. Finally, observe that if $b < 0$, then $b + k\delta > \bar{b}$ for large enough $k \in \mathbb{Z}_{\geq 1}$. The only negative column of $\tilde{A}$ is $-\delta$, so $\sigma(\tilde{A},b) = 1+d$. This proves (20).

Now we define the matrix A. Let $m \in \mathbb{Z}_{\geq 1}$ and define

$$A := \begin{bmatrix} I^{m-1} & 0^{(m-1)\times(d+1)} \\ 0^{1\times(m-1)} & \tilde{A} \end{bmatrix} \in \mathbb{Z}^{m\times(m+d)},$$

where $I^k \in \mathbb{Z}^{k\times k}$ is the identity matrix and $0^{k\times s} \in \mathbb{Z}^{k\times s}$ is the all zero matrix for $k,s \in \mathbb{Z}_{\geq 1}$. Note that $\phi^{\max}(A) = d$. If $b \in \mathbb{Z}_{\geq 0}^{m-1} \times \mathbb{Z}_{<0}$ is such that the last component is equivalent to $1 \bmod \delta$, then $\sigma(A,b) = m+d$ by the arguments above. Now, the set of $b \in \mathbb{Z}^m$ such that $P(A,b) \neq \emptyset$ is contained in $\mathbb{Z}_{\geq 0}^{m-1} \times \mathbb{Z}$. So, for every $t \in \mathbb{Z}_{\geq 1}$, the set of feasible solutions in $\{-t\delta,\ldots,t\delta\}^m$ contains $t(t\delta-1)^{m-1}$ points b such that $\sigma(A,b) = m+d$. Moreover, if $t \in \mathbb{Z}_{\geq \bar{b}}$, then $P(A,b) \neq \emptyset$ for every $b \in \{0,\ldots,t\delta\}^{m-1} \times \{-t\delta,\ldots,t\delta\}$. Therefore,

$$\begin{aligned} &\lim_{t\to\infty} \frac{|\{b \in \{-t,...,t\} : \sigma(A,b) \leq (m-1)+d\}|}{|\{b \in \{-t,...,t\} : P(A,b) \neq \emptyset\}|} \\ =\ &\lim_{t\to\infty} \frac{|\{b \in \{-t\delta,...,t\delta\} : \sigma(A,b) \leq (m-1)+d\}|}{|\{b \in \{-t\delta,...,t\delta\} : P(A,b) \neq \emptyset\}|} \\ \leq\ &\lim_{t\to\infty} \frac{(2t\delta+1)(t\delta+1)^{m-1} - t(t\delta+1)^{m-1}}{(2t\delta+1)(t\delta+1)^{m-1}} < 1. \end{aligned}$$

Using this and the fact that A has $m+d$ columns, we have $\sigma^{\mathrm{asy}}(A) = m+d$.

Now we define the matrix B. Let $A \in \mathbb{Z}^{m\times(m+d)}$ be as above. Let $e^{1\times(m+1)} \in \mathbb{Z}^{1\times(m+1)}$ be the all ones matrix and $U \in \mathbb{Z}^{m\times(m+1)}$. Assume

$$\left| \det\left(\begin{bmatrix} U \\ e^{1\times(m+1)} \end{bmatrix} \right) \right| = 1$$

and set

$$B := \begin{bmatrix} U & A \\ e^{1\times(m+1)} & 0^{1\times(m+d)} \end{bmatrix} \in \mathbb{Z}^{(m+1)\times(2m+1+d)}.$$

Note that $\phi^{\min}(B) = 0$, so Theorem 1 *(ii)* implies that $\sigma^{\mathrm{asy}}(B) \leq 2m+2$. Let $b \in \mathbb{Z}^m \times \{0\}$ be such that $P(B,b) \neq \emptyset$. If $z \in P(B,b)$, then the first $m+1$ components of z are zero. So, similarly to above, there are $b \in \mathbb{Z}^{m+1}$ such that $\sigma(B,b) = m+d$. Hence, $\sigma^{asy}(B) \leq 2m+2 < m+d = \sigma(B)$. □

B   Proof of Lemma 3

Assume that $t = 2$. Let $x := x^1$ and $y := x^2$. First, we show that $K \cap (K+x) \cap (K+y) \neq \emptyset$. Since $v^1,\ldots,v^m$ are linearly independent, K is a full-dimensional simplicial cone. Hence, there exist linearly independent vectors $a^1, \ldots, a^m \in \mathbb{R}^m$ such that $K = \{w \in \mathbb{R}^m : (a^i)^\mathsf{T} w \leq 0 \; \forall \, i \in \{1, \ldots, m\}\}$ and linearly independent vectors $r^1, \ldots, r^m \in K$ such that $(a^i)^\mathsf{T} r^i < 0$ for each $i \in \{1, \ldots, m\}$.

There is a set $J \subseteq \{1, \cdots, m\}$ such that $(a^j)^\mathsf{T}(x - y) > 0$ for each $j \in J$ and $(a^j)^\mathsf{T}(x - y) \leq 0$ for each $j \in \{1, \ldots, m\} \setminus J$. For $j \in \{1, \ldots, m\}$, set

$$
\lambda_j := \begin{cases} \max\left\{0, -\dfrac{(a^j)^\mathsf{T} x}{(a^j)^\mathsf{T} r^j}\right\}, & \text{if } j \in \{1, \ldots, m\} \setminus J \\[3mm] \max\left\{-\dfrac{(a^j)^\mathsf{T}(x-y)}{(a^j)^\mathsf{T} r^j}, -\dfrac{(a^j)^\mathsf{T} x}{(a^j)^\mathsf{T} r^j}\right\}, & \text{if } j \in J. \end{cases}
$$

Note that $\lambda_1, \ldots, \lambda_m \in \mathbb{R}_{\geq 0}$, so $x + \sum_{j=1}^m \lambda_j r^j \in K + x$. For each $i \in \{1, \ldots, m\}$, it follows that

$$
(a^i)^\mathsf{T}\left(x + \sum_{j=1}^m \lambda_j r^j - y\right) \leq (a^i)^\mathsf{T}(x - y) + \lambda_i (a^i)^\mathsf{T} r^i \leq 0.
$$

So, $x + \sum_{j=1}^m \lambda_j r^j - y \in K$ and $x + \sum_{j=1}^m \lambda_j r^j \in K + y$. Finally, for each $i \in \{1, \ldots, m\}$, it follows that

$$
(a^i)^\mathsf{T}\left(x + \sum_{j=1}^m \lambda_j r^j\right) \leq (a^i)^\mathsf{T} x + \lambda_i (a^i)^\mathsf{T} r^i \leq 0.
$$

Hence, $x + \sum_{j=1}^m \lambda_j r^j \in K$ and $K \cap (K + x) \cap (K + y) \neq \emptyset$.

Let $w \in K \cap (K + x) \cap (K + y)$. Then $K + w \subseteq K \cap (K + x) \cap (K + y)$. Because K is full-dimensional, there exists a point $z \in (K + w) \cap \mathbb{Z}^m$ such that $z = \sum_{i=1}^m k_i v^i$ for $k_i \in \mathbb{Z}_{\geq 0}$. Note that $z \in K + w \subseteq K$ and

$$
K + z \subseteq K + w \subseteq K + (K \cap (K + x) \cap (K + y)) \subseteq K \cap (K + x) \cap (K + y).
$$

For $t \geq 3$, the result follows by induction. □

References

1. Aliev, I., De Loera, J., Eisenbrand, F., Oertel, T., Weismantel, R.: The support of integer optimal solutions. SIAM J. Optim. **28**, 2152–2157 (2018)
2. Aliev, I., De Loera, J., Oertel, T., O'Neil, C.: Sparse solutions of linear diophantine equations. SIAM J. Appl. Algebra Geom. **1**, 239–253 (2017)
3. Aliev, I., Henk, M., Oertel, T.: Integrality gaps of integer knapsack problems. In: Eisenbrand, F., Koenemann, J. (eds.) IPCO 2017. LNCS, vol. 10328, pp. 25–38. Springer, Cham (2017). https://doi.org/10.1007/978-3-319-59250-3_3
4. Artin, M.: Algebra. Prentice Hall, Englewood Cliffs (1991)
5. Barvinok, A.: A Course in Convexity, vol. 54. American Mathematical Society, Providence (2002)
6. Bruns, W., Gubeladze, J.: Normality and covering properties of affine semigroups. J. für die reine und angewandte Mathematik **510**, 151–178 (2004)

7. Bruns, W., Gubeladze, J., Henk, M., Martin, A., Weismantel, R.: A counterexample to an integer analogue of Carathéodory's theorem. J. für die reine und angewandte Mathematik **510**, 179–185 (1999)
8. Cook, W., Fonlupt, J., Schrijver, A.: An integer analogue of Carathéodory's theorem. J. Comb. Theory Ser. B **40**(1), 63–70 (1986)
9. Dyer, M., Frieze, A.: Probabilistic analysis of the multidimensional knapsack problem. Math. Oper. Res. **14**, 162–176 (1989)
10. Eisenbrand, F., Shmonin, G.: Carathéodory bounds for integer cones. Oper. Res. Lett. **34**, 564–568 (2006)
11. Gomory, R.: On the relation between integer and noninteger solutions to linear programs. Proc. Natl. Acad. Sci. **53**, 260–265 (1965)
12. Sebő, A.: Hilbert bases, Carathéodory's theorem and combinatorial optimization. In: Proceedings of the 1st Integer Programming and Combinatorial Optimization Conference, pp. 431–455 (1990)

A Generic Exact Solver for Vehicle Routing and Related Problems

Artur Pessoa[3], Ruslan Sadykov[1,2($\boxtimes$)], Eduardo Uchoa[3],
and François Vanderbeck[1,2]

[1] Inria Bordeaux — Sud-Ouest, Talence, France
ruslan.sadykov@inria.fr
[2] University of Bordeaux, Talence, France
[3] Universidade Federal Fluminense, Niterói, Brazil

Abstract. Major advances were recently obtained in the exact solution of Vehicle Routing Problems (VRPs). Sophisticated Branch-Cut-and-Price (BCP) algorithms for some of the most classical VRP variants now solve many instances with up to a few hundreds of customers. However, adapting and reimplementing those successful algorithms for other variants can be a very demanding task. This work proposes a BCP solver for a generic model that encompasses a wide class of VRPs. It incorporates the key elements found in the best recent VRP algorithms: ng-path relaxation, rank-1 cuts with limited memory, and route enumeration; all generalized through the new concept of "packing set". This concept is also used to derive a new branch rule based on accumulated resource consumption and to generalize the Ryan and Foster branch rule. Extensive experiments on several variants show that the generic solver has an excellent overall performance, in many problems being better than the best existing specific algorithms. Even some non-VRPs, like bin packing, vector packing and generalized assignment, can be modeled and effectively solved.

Keywords: Integer programming · Column generation · Routing

1 Introduction

Since its introduction by Dantzig and Ramser [25], the Vehicle Routing Problem (VRP) has been one of the most widely studied in combinatorial optimization. Google Scholar indicates that 728 works containing both words "vehicle" and "routing" in the title were published only in 2017. VRP relevance stems from its direct use in the real systems that distribute goods and provide services, vital to the modern economies. Reflecting the large variety of conditions in those systems, the VRP literature is spread into dozens, perhaps hundreds, of variants. For example, there are variants that consider capacities, time windows, heterogeneous fleets, pickups and deliveries, optional customer visits, arc routing, etc.

In recent years, big advances in the exact solution of VRPs had been accomplished. A milestone was certainly the Branch-Cut-and-Price (BCP) algorithm

© Springer Nature Switzerland AG 2019
A. Lodi and V. Nagarajan (Eds.): IPCO 2019, LNCS 11480, pp. 354–369, 2019.
https://doi.org/10.1007/978-3-030-17953-3_27

of [45,47], that could solve Capacitated VRP (CVRP) instances with up to 360 customers, a large improvement upon the previous record of 150 customers. That algorithm exploits many elements introduced by several authors, combining and enhancing them. In particular, the new concept of *limited memory cut* proved to be pivotal. Improvements of the same magnitude were later obtained for a number of classical variants like VRP with Time Windows (VRPTW) [46], Heterogeneous Fleet VRP (HFVRP) and Multi Depot VRP (MDVRP) [50], and Capacitated Arc Routing (CARP) [49]. For all those variants, instances with about 200 customers are now likely to be solved, perhaps in hours or even days. However, there is something even more interesting: typical instances with about 100 customers, that a few years ago would take hours, are solved in less than 1 min. This means that many more real world instances can now be tackled by exact algorithms in reasonable times.

Unhappily, designing and coding each one of those complex and sophisticated BCPs has been a highly demanding task, measured on several work-months of a skilled team. In effect, this prevents the practical use of those algorithms in real world problems, that actually, seldom correspond exactly to one of the most classical variants. This work presents a framework that can handle most VRP variants found in the literature and can be used to model and solve many other new variants. In order to obtain state-of-the-art BCP performance, some key elements found in the best specific VRP algorithms had to be generalized. The new concept of *packing set* was instrumental for that.

The quest for general exact VRP algorithms can be traced back to Balinski and Quandt [8], where a set partitioning formulation valid for many variants was proposed. That formulation had only turned practical in the 1980's and 1990's, when the Branch-and-Price (BP) method was developed. At that time, it was recognized that the pricing subproblems could often be modeled as Resource Constrained Shortest Path (RCSP) problems and solved by labeling algorithms, leading to quite generic methods (for example, [27]). However, those BP algorithms only worked well on problems with "tightly constrained" routes, like VRPTW with narrow time windows. Many variants, including CVRP, were much better handled by Branch-and-Cut (BC) algorithms using problem-specific cuts (for example, [43]). In the late 2010's, after works like [4,5,14,28,33,38,56], it became clear that the combination of cut and column generation performs better than pure BP or pure BC on almost all problems. Until today, BCP remains the dominant VRP approach. A first attempt of a generic BCP was presented in [6], where 7 variants, all of them particular cases of the HFVRP, could be solved. Recently, [58] proposed a BCP for several particular cases of the HFVRP with time windows. The framework now proposed is far more generical than that.

2   The Basic Model

2.1   Graphs for RCSP Generation

Define directed graphs $G^k = (V^k, A^k)$, $k \in K$. Let $V = \cup_{k \in K} V^k$ and $A = \cup_{k \in K} A^k$. The graphs are not necessarily simple and may even have loops. Vertices and arcs in all graphs are distinct and carry the information about which graph they belong: a vertex $v \in V$ belongs to $G^{k(v)}$ and an arc $a \in A$ belongs to $G^{k(a)}$. Each graph has special source and sink vertices: v_{source}^k and v_{sink}^k. Define a set R of resources, divided into *main resources* R^M and *secondary resources* R^N. For each r in R and $a \in A$, $q_{a,r} \in \mathbb{R}$ is the consumption of resource r in arc a. If $r \in R^N$, consumptions are unrestricted in sign. However, for $r \in R^M$, consumptions should be non-negative. Moreover, for any $k \in K$ there should not exist a cycle in G^k where the consumption of all main resources are zero. Finally, there are finite accumulated resource consumption intervals $[l_{a,r}, u_{a,r}]$, $a \in A$. Since in most applications these intervals are more naturally defined on vertices, we may define intervals $[l_{v,r}, u_{v,r}]$, $v \in V$, meaning that $[l_{a,r}, u_{a,r}] = [l_{v,r}, u_{v,r}]$ for every arc $a \in \delta^-(v)$ (i.e., entering v).

A resource constrained path $p = (v_{\text{source}}^k = v_0, a_1, v_1, \ldots, a_{n-1}, v_{n-1}, a_n, v_n = v_{\text{sink}}^k)$ over a graph G^k, having $n \geq 1$ arcs, is feasible if: for every $r \in R$, the accumulated resource consumption $S_{j,r}$ at visit j, $0 \leq j \leq n$, where $S_{0,r} = 0$ and $S_{j,r} = \max\{l_{a_j,r}, S_{j-1,r} + q_{a_j,r}\}$, does not exceed $u_{a_j,r}$. Note that some feasible paths may not be elementary, some vertices or arcs being visited more than once. For each $k \in K$, let P^k denote the set of all feasible resource constrained paths in G^k. We will assume that each set P^k is finite, either because G^k is acyclic or because at least one main resource is defined on it. Define $P = \cup_{k \in K} P^k$. Again, a general path $p \in P$ carries the information of its graph, $G^{k(p)}$.

2.2   Formulation

The problem should be formulated as follows. There are variables x_j, $1 \leq j \leq n_1$, and variables y_s, $1 \leq s \leq n_2$. The first $\bar{n}_1$ x variables and the first $\bar{n}_2$ y variables are defined to be integer. Equations (1a) and (1b) define a general objective function and m general constraints over those variables, respectively. Constraints (1b) may even contain exponentially large families of cuts, provided that suitable procedures are given for their separation. However, by simplicity, we continue the presentation as if all the m constraints are explicitly defined. For each variable x_j, $1 \leq j \leq n_1$, $M(x_j) \subseteq A$ defines its *mapping* into a non-empty subset of the arcs. We remark that mappings do not need to be disjoint, the same arc can mapped to more than one variable x_j. Define $M^{-1}(a)$ as $\{j | 1 \leq j \leq n_1; a \in M(x_j)\}$. As not all arcs need to belong to some mapping, some M^{-1} sets may be empty. For each path $p \in P$, let λ_p be a non-negative integer variable; coefficient h_a^p indicates how many times a appears in p. The relation between variables x and λ is given by (1c). For each $k \in K$, L^k and U^k are given lower and upper bounds on number of paths in a solution.

$$\text{Min} \qquad \sum_{j=1}^{n_1} c_j x_j + \sum_{s=1}^{n_2} f_s y_s \qquad\qquad\qquad (1a)$$

$$\text{S.t.} \qquad \sum_{j=1}^{n_1} \alpha_{ij} x_j + \sum_{s=1}^{n_2} \beta_{is} y_s \geq d_i, \qquad i = 1, \ldots, m, \qquad (1b)$$

$$x_j = \sum_{k \in K} \sum_{p \in P^k} \left(\sum_{a \in M(x_j)} h_a^p \right) \lambda_p, \qquad j = 1 \ldots, n_1, \qquad (1c)$$

$$L^k \leq \sum_{p \in P^k} \lambda_p \leq U^k, \qquad k \in K, \qquad (1d)$$

$$\lambda_p \in \mathbb{Z}_+, \qquad p \in P, \qquad (1e)$$

$$x_j \in \mathbb{N}, \ y_s \in \mathbb{N} \qquad j = 1, \ldots, \bar{n}_1, \ s = 1, \ldots, \bar{n}_2. \quad (1f)$$

Eliminating the x variables and relaxing the integrality constraints, the following LP is obtained:

$$\text{Min} \qquad \sum_{k \in K} \sum_{p \in P^k} \left(\sum_{j=1}^{n_1} c_j \sum_{a \in M(j)} h_a^p \right) \lambda_p + \sum_{s=1}^{n_2} f_s y_s \qquad (2a)$$

$$\text{S.t.} \sum_{k \in K} \sum_{p \in P^k} \left(\sum_{j=1}^{n_1} \alpha_{ij} \sum_{a \in M(x_j)} h_a^p \right) \lambda_p + \sum_{s=1}^{n_2} \beta_{is} y_s \geq d_i, \quad i = 1, \ldots, m, \ (2b)$$

$$L^k \leq \sum_{p \in P^k} \lambda_p \leq U^k, \qquad k \in K, \qquad (2c)$$

$$\lambda_p \geq 0, \qquad p \in P. \qquad (2d)$$

Master LP (2a)–(2d) is solved by column generation. Let π_i, $1 \leq i \leq m$, denote the dual variables of Constraints (2b), ν_+^k and ν_-^k, $k \in K$, are the dual variables of Constraints (2c). The reduced cost of an arc $a \in A$ is defined as:

$$\bar{c}_a = \sum_{j \in M^{-1}(a)} c_j - \sum_{i=1}^{m} \sum_{j \in M^{-1}(a)} \alpha_{ij} \pi_i.$$

The reduced cost of a path $p = (v_0, a_1, v_1, \ldots, a_{n-1}, v_{n-1}, a_n, v_n) \in P^k$ is:

$$\bar{c}(p) = \sum_{j=1}^{n} \bar{c}_{a_j} - \nu_+^k - \nu_-^k.$$

So, the pricing subproblems correspond to finding, for each $k \in K$, a path $p \in P^k$ with minimum reduced cost.

3 Generalizing State-of-the-Art Elements: Packing Sets

Formulation (1a)–(1f) can be used to model most VRP variants (and also many other non-VRPs). It can be solved by a standard BP algorithm (or a standard robust BCP algorithm [52], if (1b) contains separated constraints), where the

RCSP subproblems are handled by a labeling dynamic programming algorithm. However, its performance on the more classic VRP variants would be very poor when compared to the best existing specific algorithms. One of the main contributions of this work is a generalization of the key additional elements found in those state-of-the-art algorithms, leading to the construction of a powerful and still quite generic BCP algorithm.

In order to do that, we introduce a new concept. Let $\mathcal{B} \subset 2^A$ be a collection of mutually disjoint subsets of A such that the constraints:

$$\sum_{a \in B} \sum_{p \in P} h_a^p \lambda_p \leq 1, \quad B \in \mathcal{B}, \tag{3}$$

are satisfied by at least one optimal solution (x^*, y^*, λ^*) of Formulation (1a)–(1f). In those conditions, we say that $\mathcal{B}$ defines a collection of *packing sets*. Note that a packing set can contain arcs from different graphs and not all arcs in A need to belong to some packing set. The definition of a proper $\mathcal{B}$ is part of the modeling. It does not follow automatically from the analysis of (1a)–(1f).

3.1 *ng*-Paths

One of the weaknesses of linear relaxation (2a)–(2d) when modeling classical VRPs is the existence of non-elementary paths in P. In those cases, one would like to eliminate those paths. However, this would make the pricing subproblems much harder, intractable in many cases. A good compromise between formulation strength and pricing difficulty can be obtained by the so-called *ng*-paths, introduced in Baldacci et al. [7].

In our more general context, ideally, we would like to keep only routes that do not use more than one arc in the same packing set. Instead, we settle for generalized *ng*-paths defined as follows. For each arc $a \in A$, let $NG(a) \subseteq \mathcal{B}$ denote the *ng*-set of a. An *ng*-path may use two arcs belonging to the same packing set B, but only if the subpath between those two arcs passes by an arc a such that $B \notin NG(a)$. The *ng*-sets may be determined a priori or dynamically, like in [17] and [54].

3.2 Limited Memory Rank-1 Cuts

The Rank-1 Cuts (R1Cs) [48] are a generalization of the Subset Row Cuts proposed by Jepsen et al. [38]. Here, they are further generalized as follows. Consider a collection of packing sets $\mathcal{B}$ and non-negative multipliers ρ_B for each $B \in \mathcal{B}$. A Chvátal-Gomory rounding of Constraints (3) yields:

$$\sum_{p \in P} \left\lfloor \sum_{B \in \mathcal{B}} \rho_B \sum_{a \in B} h_a^p \right\rfloor \lambda_p \leq \left\lfloor \sum_{B \in \mathcal{B}} \rho_B \right\rfloor. \tag{4}$$

R1Cs are potentially strong, but each added cut makes the pricing subproblems significantly harder. The limited memory technique [45] is essential for mitigating that negative impact. In our context, a R1C characterized by its multipliers ρ is associated to a memory set $A(\rho) \subseteq A$. Variables λ_p corresponding to paths p passing by arcs $a \notin A(\rho)$ may have their coefficients decreased in (4). However, if the memory sets are adjusted in such a way that variables λ_p with positive values in the current linear relaxation have their best possible coefficients, the resulting limited memory R1C (lm-R1C) is as effective as the original R1C.

3.3 Path Enumeration

The path enumeration technique was proposed by Baldacci et al. [5], and later improved by Contardo and Martinelli [23]. It consists in trying to enumerate into a table all paths in a certain set P^k that can possibly be part of an improving solution. After a successful enumeration, the corresponding pricing subproblem k can be solved by inspection, saving time. If the enumeration has already succeeded for all $k \in K$ and the total number of paths in the tables is not too large (say, less than 10,000) the overall problem may be even finished by a MIP solver.

In our context, we try to enumerate paths p without more than one arc in the same packing set, and with $\bar{c}(p) < UB - LB$, where UB is the best known integer solution cost, and LB the value of the current linear relaxation. Moreover, if two paths p and p' lead to variables λ_p and λ'_p with identical coefficients in (2b)–(2c), we drop the one with a larger cost.

3.4 Branching

Branching constraints over x and y variables do not change the structure of the pricing subproblems. In many models they suffice for correctness. However, there are models where Constraints (1e) need to be explicitly enforced. Branching over individual λ variables is permitted, but should be avoided due to a big negative impact in the pricing and also due to highly unbalanced branch trees [64].

The model has the option of branching using a generalization of the Ryan and Foster rule [57]. Choose distinct sets B and B' in $\mathcal{B}$. Let $P(B, B') \subseteq P$ be the subset of the paths that contain arcs in both B and B'. The branch is over the value of $\sum_{p \in P(B,B')} \lambda_p$. The pricing still becomes harder, but branch trees are more balanced.

We included in the model a new way of branching that does not increase the pricing difficulty. For chosen $B \in \mathcal{B}$, $r \in R^M$ and for a certain threshold value t^*: in the left child make $u_{a,r} = t^*$, for all $a \in B$; in the right child make $l_{a,r} = t^*$. In other words, the branch is over the accumulated consumption of resource r on arcs in B. In principle, this branching it is not complete: some fractional λ solutions can not be eliminated by it. However, it may work very well in practice.

4   Model Examples

We selected 4 problems to exemplify how problems are modeled in our solver. First, a simple didactic model; then a case where branching over the λ variables is necessary; the third model illustrates the use of secondary resources; the fourth model relies on a non-trivial transformation of the original problem.

4.1   Generalized Assignment Problem (GAP)

Data: Set T of tasks; set K of machines; capacity Q^k, $k \in K$; assignment cost c_t^k and machine load w_t^k, $t \in T$, $k \in K$.
Goal: Find an assignment of tasks to machines such that the total load in each machine does not exceed its capacity, with minimum total cost.
Model: Graph $G^k = (V^k, A^k)$ for each $k \in K$, $V^k = \{v_t^k : t = 0, \ldots, |T|\}$, $A^k = \{a_{t+}^k = (v_{t-1}^k, v_t^k), a_{t-}^k = (v_{t-1}^k, v_t^k) : t = 1, \ldots, |T|\}$, $v_{\text{source}}^k = v_0^k$, $v_{\text{sink}}^k = v_{|T|}^k$ (see Fig. 1). $R = R^M = \{1\}$; $q_{a_{t+}^k, 1} = w_t^k, q_{a_{t-}^k, 1} = 0, t \in T$; $[l_{v_t^k, 1}, u_{v_t^k, 1}] = [0, Q^k]$, $t \in T \cup \{0\}$. Binary variables x_t^k, $t \in T$, $k \in K$. The formulation is:

$$\text{Min} \sum_{t \in T} \sum_{k \in K} c_t^k x_t^k \tag{5a}$$

$$\text{S.t.} \sum_{k \in K} x_t^k = 1, \quad t \in T; \tag{5b}$$

$L^k = 0$, $U^k = 1$, $k \in K$; $M(x_t^k) = \{a_{t+}^k\}, t \in T, k \in K$ (by abuse of notation we use the variable itself instead of its linear index as the argument of mapping M). $\mathcal{B} = \cup_{t \in T}\{\{a_{t+}^k : k \in K\}\}$. Branching is over the x variables.

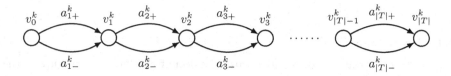

Fig. 1. GAP model graph, RCSPs correspond to binary knapsack solutions.

4.2   Vector Packing (VPP)/Bin Packing (BPP)

Data: Set T of items; set D of dimensions; bin capacities Q^d, $d \in D$; item weight w_t^d, $t \in T$, $d \in D$. (Bin packing is the case where $|D| = 1$).
Goal: Find a packing using the minimum number of bins, such that, for each dimension, the total weight of the items in a bin does not exceed its capacity.
Model: A single graph $G = (V, A)$ (we omit the index k in such cases), $V = \{v_t : t = 0, \ldots, |T|\}$, $A = \{a_{t+} = (v_{t-1}, v_t), a_{t-} = (v_{t-1}, v_t) : t = 1, \ldots, |T|\}$, $v_{\text{source}} = v_0$, $v_{\text{sink}} = v_{|T|}$. $R = R^M = D$; $q_{a_{t+}, d} = w_t^d, q_{a_{t-}, d} = 0, t \in T, d \in D$; $[l_{v_t, d}, u_{v_t, d}] = [0, Q^d]$, $t \in T \cup \{0\}, d \in D$. Binary variables x_t, $t \in T \cup \{0\}$. The formulation is:

$$\text{Min} \quad x_0 \tag{6a}$$

$$\text{S.t.} \ x_t = 1, \quad t \in T; \tag{6b}$$

$L = 0$, $U = \infty$; $M(x_0) = \{a_{1+}, a_{1-}\}$, $M(x_t) = \{a_{t+}\}, t \in T$. $\mathcal{B} = \cup_{t \in T}\{\{a_{t+}\}\}$. Branching on λ variables; first over accumulated resource consumption and, if still needed, by Ryan and Foster rule.

4.3 Pickup and Delivery VRPTW (PDPTW)

Data: Directed graph $G' = (V', A')$, where $V' = \{0\} \cup P' \cup D'$, $P' = \{1, \ldots, n\}$ is the set of pickup vertices and $D' = \{n+1, \ldots, 2n\}$ the set of corresponding deliveries (a pickup at i correspond to a delivery at $i+n$); vehicle capacities Q; traveling cost c_a and time (traveling time plus service time) t_a, $a \in A'$; positive demands d_v, $v \in P'$ ($d_v = -d_{v-n}$, $v \in D'$); and time windows $[l'_v, u'_v]$, $v \in V'$.

Goal: Find a set of routes such that each pickup or delivery vertex is visited exactly once, any visit to a pickup vertex implies that the corresponding delivery vertex is visited later by the same route, the accumulated demand of visited nodes never exceed the capacity along a route, and the accumulated sum of traversal and waiting times until reaching each node falls within its time window (waiting times are added to meet lower bounds), minimizing the total sum of traversal costs.

Model: A single graph $G = (V, A)$, $V = V' \cup \{2n+1\}$, $A = (A' \setminus \{(v, 0) : v \in D'\}) \cup \{(v, 2n+1) : v \in D'\}$ (assume $c_{(v,2n+1)} = c_{(v,0)}$ and $t_{(v,2n+1)} = t_{(v,0)}$), $v_{\text{source}} = v_0$, $v_{\text{sink}} = v_{2n+1}$. $R^M = \{n+2\}$; $R^N = \{1, \ldots, n+1\}$; $q_{(v,v'),v'} = 1$, if $v' \in P'$, $q_{(v,v'),v'-n} = -1$, if $v' \in D'$, and $q_{(v,v'),n+1} = d_{v'}$, $(v, v') \in A$; $q_{a,n+2} = t_a$, $a \in A$; all other resource consumptions are zero; $u_{v,r} = 1$, $r = 1, \ldots, n$, $u_{v,n+1} = u_{2n+1,n+1} = Q$ and $(l_{v,n+2}, u_{v,n+2}) = (l'_v, u'_v)$, $v \in P' \cup D'$; all other resource bounds are zero. Binary variables x_a, $a \in A$. The formulation is:

$$\text{Min} \quad \sum_{a \in A} c_a x_a \tag{7a}$$

$$\text{S.t.} \sum_{a \in \delta^-(v)} x_a = 1, \quad v \in P'; \tag{7b}$$

$L = 0$, $U = \infty$; $M(x_a) = \{a\}$, $a \in A$. $\mathcal{B} = \cup_{v \in V}\{\delta^-(v)\}$. Branching on x variables.

4.4 Capacitated Arc Routing (CARP)

Data: Undirected graph $G' = (V', E)$, $V' = \{0, \ldots, n\}$, 0 is the depot vertex; positive cost c_e and non-negative demand d_e, $e \in E$, set of required edges $S = \{e \in E \mid d_e > 0\}$; vehicle capacity Q.

Goal: Find a minimum cost set of routes, closed walks starting and ending at the depot, that serve the demands in all required edges. Edges in a route can be traversed either serving or deadheading (not servicing). The sum of the demands of the served edges in a route can not exceed capacity.

Model: For $i, j \in V'$, let $D(i, j) \subseteq E$ be the set of edges in a chosen cheapest path from i to j, with cost $C(i, j) = \sum_{e \in D(i,j)} c_e$. Define a dummy required edge $r_0 = (0, 0')$ and $S_0 = S \cup \{r_0\}$. For each $r = (w_1, w_2) \in S_0$, define $o(r, w_1) = w_2$ and $o(r, w_2) = w_1$.

The model has a single graph $G = (V, A)$, $V = \{v_r^w : r \in S_0, w \in r\}$, $A = \{(v_{r_1}^{w_1}, v_{r_2}^{z_1}), (v_{r_1}^{w_1}, v_{r_2}^{z_2}), (v_{r_1}^{w_2}, v_{r_2}^{z_1}), (v_{r_1}^{w_2}, v_{r_2}^{z_2}) : r_1 = (w_1, w_2), r_2 = (z_1, z_2) \in S_0\}$, $v_{\text{source}} = v_{r_0}^0$, $v_{\text{sink}} = v_{r_0}^{0'}$; $R = R^M = \{1\}$; for $a = (v_{r_1}^w, v_{r_2}^z) \in A$, $q_{a,1} = d_{r_2}$; $l_{v,1} = 0, u_{v,1} = Q, v \in V$. Binary variables x_a, $a \in A$. For $a = (v_{r_1}^w, v_{r_2}^z) \in A$,

$c_a = C(w, o(r_2, z)) + c_{r_2}$. The formulation is:

$$\text{Min} \quad \sum_{a \in A} c_a x_a \tag{8a}$$

$$\text{S.t.} \sum_{a \in \delta^-(\{v_r^{w_1}, v_r^{w_2}\})} x_a = 1, \quad r = (w_1, w_2) \in S, \tag{8b}$$

plus Rounded Capacity Cuts [43] and Lifted Odd-Cutsets [9,11]; $L = 0$, $U = \infty$; $M(x_a) = \{a\}$, $a \in A$. $\mathcal{B} = \cup_{r=(w_1,w_2) \in S}\{\delta^-(\{v_r^{w_1}, v_r^{w_2}\})\}$. Branching on aggregation of x variables (see Appendix A).

5   Computational Experiments

The generic BCP solver optimization algorithms were coded in C++ over the BaPCod package [63]. IBM CPLEX Optimizer version 12.8.0 was used as the LP solver in column generation and as the solver for the enumerated MIPs. The experiments were run on a 2 Deca-core Ivy-Bridge Haswell Intel Xeon E5-2680 v3 server running at 2.50 GHz. The 128 GB of available RAM was shared between 8 copies of the algorithm running in parallel on the server. Each instance is solved by one copy of the algorithm using a single thread. The models are defined using either a C++ interface or a Julia–JuMP [29] based interface.

A description of the main algorithms used in the BCP solver can not be presented here by lack of space. However, they are generalizations and enhancements of already published algorithms. In some cases the original algorithms had to be significantly revised, to avoid that the generalizations introduce excessive performance overheads. Pricing problems are solved by a bucket graph based labeling algorithm [58], including a bucket arc elimination procedure based on reduced costs. Automatic dual price smoothing [51] is employed to stabilize the column generation convergence. Path enumeration is performed using an extension of the algorithm from [5,47]. Multi-phase strong branching [47,55] is used to reduce the search tree size. Restricted master and diving heuristics [59] are built-in to improve the primal solution during the search.

In Table 1, we show computational results for 13 problems. The first column is the problem acronym, second column refers to data sets, the third indicates the number of instances. Next is the time limit per instance. The last three columns show the results obtained by our generic solver, as well as by two other algorithms, those with the best (to our knowledge) published results for the data set. For each algorithm, we give the number of instances solved within the time limit, the average time in brackets (geometric mean time if the time limit is 10 hours or more), and its reference. For instances not solved, the time limit is considered as the solution time. Best results are marked in bold. Note that the generic solver uses a single parameterization per problem, not per data set. Additional information about experiments is available in Appendix A.

The results presented in Table 1 show that the generic BCP significantly outperforms the state-of-the-art for VRPTW, TOP, CTOP, CPTP, VRPSL, and VPP. A noticeably better performance is achieved for CVRP and HFVRP. For

Table 1. Generic solver vs best specific solvers on 13 problems.

Problem	Data set	#	T.L	Gen. BCP	Best Publ.	2nd Best
CVRP	E-M [20,21]	12	10 h	12 (61 s)	**12 (49 s)** [47]	10 (432 s) [23]
	X [62]	58	60 h	**36 (147 m)**	34 (209 m) [62]	–
VRPTW	Solomon Hardest [61]	14	1 h	**14 (5 m)**	13 (17 m) [46]	9 (39 m) [7]
	Homberger 200 [34]	60	30 h	**56 (21 m)**	50 (70 m) [46]	7 (-) [39]
HFVRP	BaldacciMingozzi [6]	40	1 h	**40 (144 s)**	39 (287 s) [50]	34 (855 s) [6]
MDVRP	Cordeau [24]	11	1 h	**11 (6 m)**	11 (7 m) [50]	9 (25 m) [23]
PDPTW	RopkeCordeau [56]	40	1 h	**40 (5 m)**	33 (17 m) [35]	32 (14 m) [4]
	LiLim [41]	30	1 h	3 (56 m)	**23 (20 m)** [4]	18 (27 m) [35]
TOP	Chao class 4 [19]	60	1 h	**55 (8 m)**	39 (15 m) [13]	30 (-) [31]
CTOP	Archetti [2]	14	1 h	**13 (7 m)**	6 (35 m) [1]	7 (34 m) [2]
CPTP	Archetti open [2]	28	1 h	**24 (9 m)**	0 (1 h) [16]	0 (1 h) [1]
VRPSL	Bulhoes [16]	180	2 h	**159 (16 m)**	49 (90 m) [16]	–
GAP	OR-Lib, type D [10]	6	2 h	5 (40 m)	**5 (30 m)** [53]	5 (46 m) [3]
	Nauss [44]	30	1 h	**25 (23 m)**	1 (58 m) [36]	0 (1 h) [44]
VPP	Classes 1,4,5,9 [18]	40	1 h	**38 (8 m)**	13 (50 m) [37]	10 (53 m) [15]
BPP	Falkenauer T [32]	80	10 m	80 (16 s)	**80 (1 s)** [15]	80 (24 s) [12]
	Hard28 [60]	28	10 m	28 (17 s)	**28 (7 s)** [12]	26 (14 s) [15]
	AI [26]	250	1 h	**160 (25 m)**	116 (35 m) [12]	100 (40 m) [15]
	ANI [26]	250	1 h	103 (35 m)	**164 (35 m)** [22]	51 (48 m) [12]
CARP	Eglese [30]	24	30 h	**22 (36 m)**	22 (43 m) [49]	10 (237 m) [9]

MDVRP, GAP, BPP and CARP, the generic BCP is comparable to the best performing algorithms in the literature. Results are mixed for PDPTW. Worse performance for LiLim instances can be explained by the fact that the generic BCP does not incorporate some labeling algorithm acceleration techniques specific to PDPTW. For the RopkeCordeau instances however, generic state-of-the-art BCP elements mitigate the effect of lacking ad-hoc enhancements.

6   Conclusions

We proposed a new generic way of modeling VRPs and related problems, so that they can be solved by an algorithm that already includes many state-of-the-art elements. It combines old modeling concepts (like the use of RCSPs for defining the valid routes) with a new one, the packing sets. The experiments show that the generic solver has a performance either comparable or better than the specific algorithms for all VRP variants tested. The cases where the performance was much better can be explained by the fact that previous authors often did not use some advanced BCP elements due to the complexity of their implementation. However, if generic BCP solvers become publicly and/or commercially available, we believe that their use may become as standard as that of MIP solvers nowadays.

We plan to release the presented generic solver for academic use after additional testing and documentation. It will include the optimization algorithms in a pre-compiled library and a Julia–JuMP user interface. Modeling a typical VRP variant, like those in our tests, requires around 100 lines of Julia code. This means that a user can already have a good working algorithm in a day. More work on computational experiments for parameter tuning may be needed for an improved performance. Then separation routines for problem specific cuts can be added for top performance.

A   Additional Information on the Experiments

We provide additional information about the experiments reported in Table 1, including relevant modeling decisions, datasets and remarks.

Note that the performance of exact algorithms is sensitive to the initial primal bound value given by the user before execution. We tried to be as fair as possible in this regard. Unless stated otherwise for the problems below we use the same bounds (usually took from the heuristic literature) as in previous works.

CVRP (Capacitated Vehicle Routing Problem)*:* The model is defined over undirected edge variables and separates Rounded Capacity Cuts (RCCs) [40], using the procedure in CVRPSEP [42]. A packing set is defined for each customer and contains all incoming arcs to the corresponding node in the graph. Branching is done over edge variables. The considered E-M instances are the 12 hardest ones, those considered in [47]. The considered X instances are those with less than 400 customers.

VRPTW (Vehicle Routing Problem with Time Windows)*:* The same model as CVRP except that only time is defined as a graph resource, capacity is enforced by RCCs. The considered Solomon instances (all with 100 customers) are the hardest ones according to [46].

HFVRP (Heterogeneous Fleet Vehicle Routing Problem)*:* The model is defined over undirected edge variables. Each graph G^k (with capacity resource) corresponds to a vehicle type. Branching is on the number of paths in P^k, assignment of packing sets to graphs, and on edge variables. Instances with 50, 75, and 100 customers are considered.

MDVRP (Multi-Depot Vehicle Routing problem)*:* The model is defined over undirected edge variables. Each graph G^k corresponds to a depot. Branching is the same as for HFVRP. Only instances with one capacity resource are considered (without time constraints).

PDPTW: The model is precisely defined in Sect. 4.3.

TOP/CTOP (Team Orienteering Problem)*:* The model contains binary variables y that are not mapped to any RCSP graph, so they appear directly in Formulation 2. Those variables indicate which customers will be visited. The problem is to maximize the total profit of visited customers. For TOP, one (time) resource is defined. In CTOP, an additional capacity resource is considered. Branching is on y and edge variables. No initial upper bound is defined. Instances of class 4, the most difficult one according to [13], are considered for TOP. Only basic instances from [2] are considered for CTOP, as well as open ones.

CPTP (Capacitated Profitable Tour Problem): Similar to CTOP, except that there is no time resource. The objective is the difference between the total profit and the transportation cost. Only open instances from [2] are considered.

VRPSL (VRP with Service Level constraints): Generalization of CVRP in which a service weight is defined for each customer. For each predefined group of customers, total service weight of visited customers should not be below a threshold. The model contains edge and y variables. For each group, a knapsack constraint over y variables is defined. Branching is both on y and edge variables.

GAP: The model is precisely defined in Sect. 4.1. Instances of the most difficult type D are considered. For OR-Library instances, we took best known solution values as initial upper bounds. We used Nauss instances with $|T| = 90, 100$ and $|K| = 25, 30$. Initial bounds for them were calculated by us using problem specific strong diving heuristic from [58]. Its time is included in the reported time.

BPP/VPP: The model is precisely defined in Sect. 4.2. The branching over the accumulated resource consumption showed to be effective so that Ryan and Foster branch rule was never needed. For VPP, we took only largest instances (200 items) with 2 resources of classes 1, 4, 5, and 9, the most difficult ones according to [37]. No initial bound is given for VPP. For BPP, we used the initial primal bound equal to the rounded up column generation dual bound plus one. Such solutions are easily obtainable by very simple heuristics.

CARP: The model is defined in Sect. 4.4. The branching is done on aggregation of x variables: (1) corresponding to node degrees in the original graph; (2) corresponding to whether required two edges are served immediately one after another by the same route or not. The Eglese dataset is used in all recent works on CARP.

B   Open CVRP Instances Solved

According to CVRPLIB (http://vrp.atd-lab.inf.puc-rio.br) there were 52 open CVRP instances in the X set [62]. We started long runs of the generic solver on the most promising ones, using a specially calibrated parameterization. We could solve 5 instances to optimality for the first time, as indicated in Table 2. Improved best known solutions are underlined.

Table 2. Detailed results on the open instances solved.

Instance	Prev. BKS	Root LB	Nodes	Total time	OPT
X-n284-k15	20226	20168	940	11.0 days	**20215**
X-n322-k28	29834	29731	1197	5.6 days	**29834**
X-n393-k38	38260	38194	1331	5.8 days	**38260**
X-n469-k138	221909	221585	8964	15.2 days	**221824**
X-n548-k50	86701	86650	337	2.0 days	**86700**

References

1. Archetti, C., Bianchessi, N., Speranza, M.: Optimal solutions for routing problems with profits. Discrete Appl. Math. **161**(4–5), 547–557 (2013)
2. Archetti, C., Feillet, D., Hertz, A., Speranza, M.G.: The capacitated team orienteering and profitable tour problems. J. Oper. Res. Soc. **60**(6), 831–842 (2009)
3. Avella, P., Boccia, M., Vasilyev, I.: A computational study of exact knapsack separation for the generalized assignment problem. Comput. Optim. Appl. **45**(3), 543–555 (2010)
4. Baldacci, R., Bartolini, E., Mingozzi, A.: An exact algorithm for the pickup and delivery problem with time windows. Oper. Res. **59**(2), 414–426 (2011)
5. Baldacci, R., Christofides, N., Mingozzi, A.: An exact algorithm for the vehicle routing problem based on the set partitioning formulation with additional cuts. Math. Program. **115**, 351–385 (2008)
6. Baldacci, R., Mingozzi, A.: A unified exact method for solving different classes of vehicle routing problems. Math. Program. **120**(2), 347–380 (2009)
7. Baldacci, R., Mingozzi, A., Roberti, R.: New route relaxation and pricing strategies for the vehicle routing problem. Oper. Res. **59**(5), 1269–1283 (2011)
8. Balinski, M., Quandt, R.: On an integer program for a delivery problem. Oper. Res. **12**(2), 300–304 (1964)
9. Bartolini, E., Cordeau, J.F., Laporte, G.: Improved lower bounds and exact algorithm for the capacitated arc routing problem. Math. Program. **137**(1), 409–452 (2013)
10. Beasley, J.E.: OR-library: distributing test problems by electronic mail. J. Oper. Res. Soc. **41**(11), 1069–1072 (1990)
11. Belenguer, J., Benavent, E.: The capacitated arc routing problem: valid inequalities and facets. Comput. Optim. Appl. **10**(2), 165–187 (1998)
12. Belov, G., Scheithauer, G.: A branch-and-cut-and-price algorithm for one-dimensional stock cutting and two-dimensional two-stage cutting. Eur. J. Oper. Res. **171**(1), 85–106 (2006)
13. Bianchessi, N., Mansini, R., Speranza, M.G.: A branch-and-cut algorithm for the team orienteering problem. Int. Trans. Oper. Res. **25**(2), 627–635 (2018)
14. Bode, C., Irnich, S.: Cut-first branch-and-price-second for the capacitated arc-routing problem. Oper. Res. **60**(5), 1167–1182 (2012)
15. Brandão, F., Pedroso, J.P.: Bin packing and related problems: general arc-flow formulation with graph compression. Comput. Oper. Res. **69**, 56–67 (2016)
16. Bulhoes, T., Hà, M.H., Martinelli, R., Vidal, T.: The vehicle routing problem with service level constraints. Eur. J. Oper. Res. **265**(2), 544–558 (2018)

17. Bulhoes, T., Sadykov, R., Uchoa, E.: A branch-and-price algorithm for the minimum latency problem. Comput. Oper. Res. **93**, 66–78 (2018)

18. Caprara, A., Toth, P.: Lower bounds and algorithms for the 2-dimensional vector packing problem. Discrete Appl. Math. **111**(3), 231–262 (2001)

19. Chao, I.M., Golden, B.L., Wasil, E.A.: The team orienteering problem. Eur. J. Oper. Res. **88**(3), 464–474 (1996)

20. Christofides, N., Eilon, S.: An algorithm for the vehicle-dispatching problem. Oper. Res. Q. **20**, 309–318 (1969)

21. Christofides, N., Mingozzi, A., Toth, P.: The vehicle routing problem. In: Christofides, N., Mingozzi, A., Toth, P. (eds.) Combinatorial Optimization, pp. 315–338. Wiley, Chichester (1979)

22. Clautiaux, F., Hanafi, S., Macedo, R., Émilie Voge, M., Alves, C.: Iterative aggregation and disaggregation algorithm for pseudo-polynomial network flow models with side constraints. Eur. J. Oper. Res. **258**(2), 467–477 (2017)

23. Contardo, C., Martinelli, R.: A new exact algorithm for the multi-depot vehicle routing problem under capacity and route length constraints. Discrete Optim. **12**, 129–146 (2014)

24. Cordeau, J.F., Gendreau, M., Laporte, G.: A tabu search heuristic for periodic and multi-depot vehicle routing problems. Networks **30**(2), 105–119 (1997)

25. Dantzig, G., Ramser, J.: The truck dispatching problem. Manage. Sci. **6**(1), 80–91 (1959)

26. Delorme, M., Iori, M., Martello, S.: Bin packing and cutting stock problems: mathematical models and exact algorithms. Eur. J. Oper. Res. **255**(1), 1–20 (2016)

27. Desaulniers, G., Desrosiers, J., loachim, I., Solomon, M.M., Soumis, F., Villeneuve, D.: A unified framework for deterministic time constrained vehicle routing and crew scheduling problems. In: Crainic, T.G., Laporte, G. (eds.) Fleet Management and Logistics. CRT, pp. 57–93. Springer, Boston (1998). https://doi.org/10.1007/978-1-4615-5755-5_3

28. Desaulniers, G., Lessard, F., Hadjar, A.: Tabu search, partial elementarity, and generalized k-path inequalities for the vehicle routing problem with time windows. Transp. Sci. **42**(3), 387–404 (2008)

29. Dunning, I., Huchette, J., Lubin, M.: JuMP: a modeling language for mathematical optimization. SIAM Rev. **59**(2), 295–320 (2017)

30. Eglese, R.W., Li, L.Y.O.: Efficient routeing for winter gritting. J. Oper. Res. Soc. **43**(11), 1031–1034 (1992)

31. El-Hajj, R., Dang, D.C., Moukrim, A.: Solving the team orienteering problem with cutting planes. Comput. Oper. Res. **74**, 21–30 (2016)

32. Falkenauer, E.: A hybrid grouping genetic algorithm for bin packing. J. Heurist. **2**, 5–30 (1996)

33. Fukasawa, R., et al.: Robust branch-and-cut-and-price for the capacitated vehicle routing problem. Math. Program. **106**(3), 491–511 (2006)

34. Gehring, H., Homberger, J.: Parallelization of a two-phase metaheuristic for routing problems with time windows. J. Heurist. **8**(3), 251–276 (2002)

35. Gschwind, T., Irnich, S., Rothenbächer, A.K., Tilk, C.: Bidirectional labeling in column-generation algorithms for pickup-and-delivery problems. Eur. J. Oper. Res. **266**(2), 521–530 (2018)

36. Gurobi Optimization, Inc.: Gurobi optimizer reference manual, version 7.5 (2017). http://www.gurobi.com

37. Heßler, K., Gschwind, T., Irnich, S.: Stabilized branch-and-price algorithms for vector packing problems. Eur. J. Oper. Res. **271**(2), 401–419 (2018)

38. Jepsen, M., Petersen, B., Spoorendonk, S., Pisinger, D.: Subset-row inequalities applied to the vehicle-routing problem with time windows. Oper. Res. **56**(2), 497–511 (2008)

39. Kallehauge, B., Larsen, J., Madsen, O.: Lagrangian duality applied to the vehicle routing problem with time windows. Comput. Oper. Res. **33**(5), 1464–1487 (2006)

40. Laporte, G., Nobert, Y.: A branch and bound algorithm for the capacitated vehicle routing problem. Oper.-Res.-Spektrum **5**(2), 77–85 (1983)

41. Li, H., Lim, A.: A metaheuristic for the pickup and delivery problem with time windows. Int. J. Artif. Intell. Tools **12**(02), 173–186 (2003)

42. Lysgaard, J.: CVRPSEP: a package of separation routines for the capacitated vehicle routing problem. Aarhus School of Business, Department of Management Science and Logistics (2003)

43. Lysgaard, J., Letchford, A.N., Eglese, R.W.: A new branch-and-cut algorithm for the capacitated vehicle routing problem. Math. Program. **100**(2), 423–445 (2004)

44. Nauss, R.M.: Solving the generalized assignment problem: an optimizing and heuristic approach. INFORMS J. Comput. **15**(3), 249–266 (2003)

45. Pecin, D., Pessoa, A., Poggi, M., Uchoa, E.: Improved branch-cut-and-price for capacitated vehicle routing. In: Lee, J., Vygen, J. (eds.) IPCO 2014. LNCS, vol. 8494, pp. 393–403. Springer, Cham (2014). https://doi.org/10.1007/978-3-319-07557-0_33

46. Pecin, D., Contardo, C., Desaulniers, G., Uchoa, E.: New enhancements for the exact solution of the vehicle routing problem with time windows. INFORMS J. Comput. **29**(3), 489–502 (2017)

47. Pecin, D., Pessoa, A., Poggi, M., Uchoa, E.: Improved branch-cut-and-price for capacitated vehicle routing. Math. Program. Comput. **9**(1), 61–100 (2017)

48. Pecin, D., Pessoa, A., Poggi, M., Uchoa, E., Santos, H.: Limited memory rank-1 cuts for vehicle routing problems. Oper. Res. Lett. **45**(3), 206–209 (2017)

49. Pecin, D., Uchoa, E.: Comparative analysis of capacitated arc routing formulations for designing a new branch-cut-and-price algorithm. Transp. Sci. (2019, to appear)

50. Pessoa, A., Sadykov, R., Uchoa, E.: Enhanced branch-cut-and-price algorithm for heterogeneous fleet vehicle routing problems. Eur. J. Oper. Res. **270**, 530–543 (2018)

51. Pessoa, A., Sadykov, R., Uchoa, E., Vanderbeck, F.: Automation and combination of linear-programming based stabilization techniques in column generation. INFORMS J. Comput. **30**(2), 339–360 (2018)

52. Poggi de Aragão, M., Uchoa, E.: Integer program reformulation for robust branch-and-cut-and-price. In: Wolsey, L. (ed.) Annals of Mathematical Programming in Rio, Búzios, Brazil, pp. 56–61 (2003)

53. Posta, M., Ferland, J.A., Michelon, P.: An exact method with variable fixing for solving the generalized assignment problem. Comput. Optim. Appl. **52**, 629–644 (2012)

54. Roberti, R., Mingozzi, A.: Dynamic ng-path relaxation for the delivery man problem. Transp. Sci. **48**(3), 413–424 (2014)

55. Røpke, S.: Branching decisions in branch-and-cut-and-price algorithms for vehicle routing problems. In: Presentation in Column Generation 2012 (2012)

56. Ropke, S., Cordeau, J.F.: Branch and cut and price for the pickup and delivery problem with time windows. Transp. Sci. **43**(3), 267–286 (2009)

57. Ryan, D.M., Foster, B.A.: An integer programming approach to scheduling. In: Wren, A. (ed.) Computer Scheduling of Public Transport: Urban Passenger Vehicle and Crew Scheduling, pp. 269–280. North-Holland, Amsterdam (1981)

58. Sadykov, R., Uchoa, E., Pessoa, A.: A bucket graph based labeling algorithm with application to vehicle routing. Technical report L-2017-7, Cadernos do LOGIS-UFF, Niterói, Brazil, October 2017

59. Sadykov, R., Vanderbeck, F., Pessoa, A., Tahiri, I., Uchoa, E.: Primal heuristics for branch-and-price: the assets of diving methods. INFORMS J. Comput. (2018)

60. Schoenfield, J.E.: Fast, exact solution of open bin packing problems without linear programming. Technical report, US Army Space and Missile Defense Command (2002)

61. Solomon, M.M.: Algorithms for the vehicle routing and scheduling problems with time window constraints. Oper. Res. **35**(2), 254–265 (1987)

62. Uchoa, E., Pecin, D., Pessoa, A., Poggi, M., Subramanian, A., Vidal, T.: New benchmark instances for the capacitated vehicle routing problem. Eur. J. Oper. Res. **257**(3), 845–858 (2017)

63. Vanderbeck, F., Sadykov, R., Tahiri, I.: BaPCod – a generic Branch-And-Price Code (2018). https://realopt.bordeaux.inria.fr/?page_id=2

64. Vanderbeck, F., Wolsey, L.A.: Reformulation and decomposition of integer programs. In: Jünger, M., et al. (eds.) 50 Years of Integer Programming 1958–2008, pp. 431–502. Springer, Heidelberg (2010). https://doi.org/10.1007/978-3-540-68279-0_13

Earliest Arrival Transshipments in Networks with Multiple Sinks

Miriam Schlöter[(✉)]

Institute for Operations Research, ETH Zürich, Zürich, Switzerland
miriam.schloeter@ifor.math.ethz.ch

Abstract. We study a classical flow over time problem that captures the essence of evacuation planning: given a network with capacities and transit times on the arcs and sources/sinks with supplies/demands, an *earliest arrival transshipment (EAT)* sends the supplies from the sources to the sinks such that the amount of flow which has reached the sinks is maximized for every point in time simultaneously. In networks with only a single sink earliest transshipments do exist for every choice of supplies and demands. This is why so far a lot of effort has been put into the development of efficient algorithms for computing EATs in this class of networks, whereas not much is known about EATs in networks with *multiple sinks*, aside from the fact that they don't exist in general.

We make huge progress regarding EATs in networks with multiple sinks by formulating the first exact algorithm that decides whether a given *tight* EAT problem has solution *and* that computes the EAT in case of existence. Our algorithm only works on the originally given network without requiring any form of expansion and thus just requires polynomial space. Complementing this algorithm we show that in multiple sink networks it is, already for tight instances, $\mathcal{NP}$-hard to decide whether an EAT does exist for a specific choice of supplies and demands.

Keywords: Flows over time · Submodular function minimization · Earliest arrival flows · Earliest arrival transshipments · Evacuation

1 Introduction

In 2017 the United States and several Caribbean islands were hit by two severe hurricanes. On both occasions it was necessary to evacuate millions of people from the dangers of these tropical storms. However, in many places the evacuation was disorganized and inefficient which resulted in a large number of unnecessary injuries and deaths. Due to climate change the number and intensity of such natural disasters will probably increase in the future, which makes it even more essential to develop well-planned evacuation strategies. Flows over time, which were introduced by Ford and Fulkerson in the 1950s [13,14], have been used successfully to model evacuation scenarios in the past [3,7,19]. One main

This work was supported by the DFG SPP 1736 "Algorithms for Big Data".

A. Lodi and V. Nagarajan (Eds.): IPCO 2019, LNCS 11480, pp. 370–384, 2019.
https://doi.org/10.1007/978-3-030-17953-3_28

aim when disaster strikes is to bring endangered people to safe areas as quickly as possible but often this is not enough to ensure optimal evacuation. For example, meteorologists are still not able to precisely predict when a hurricane will hit the coast. In such situations an evacuation strategy should thus also have the property that it saves as many people as possible no matter when the actual tragedy occurs. This property is captured by *earliest arrival transshipments (EATs)*. The results we present in this paper are deep insights into the structure and efficient computation of EATs in networks with multiple sinks.

Introduction to Flow Over Time and Related Work. Throughout this paper $\mathbb{R}_+$ and $\mathbb{Z}_+$ denote the sets of non-negative reals and integers, respectively. We consider a *dynamic network* $\mathcal{N} = (D = (V, A), u, \tau, S^+, S^-)$ that consists of a directed graph $D = (V, A)$, a *capacity* function $u \colon A \to \mathbb{Z}_+$, a *transit time* function $\tau \colon A \to \mathbb{Z}_+$ and disjoint sets of *sources* $S^+ \subseteq V$ and *sinks* $S^- \subseteq V$. The union $S^+ \cup S^-$ is called the set of *terminals*. We assume that the sources have no ingoing and the sinks no outgoing arcs. In this setting an arc's capacity bounds the rate at which flow may enter it, while the transit time of an arc specifies the time flow needs to travel from its tail to its head. Moreover, we denote by $b \colon S^+ \cup S^- \to \mathbb{Z}$ a *supply/demand-function* with *supplies* $b(s) > 0$ for all $s \in S^+$ and *demands* $b(t) < 0$ for all $t \in S^-$ such that $b(S^+ \cup S^-) := \sum_{x \in S^+ \cup S^-} b(x) = 0$.

A *flow over time* f is a Lebesque integrable function[1] $f \colon A \times \mathbb{R}_+ \to \mathbb{R}_+$ that specifies for each arc a and each point in time $\theta \geq 0$ the flow rate at which flow enters the arc – and leaves it at time $\theta + \tau_a$. Additionally, a flow over time f is required to respect the capacity of every arc at every point time and to fulfill flow conservation, i.e. at every point in time and for every note $v \in V \setminus (S^+ \cup S^-)$ the flow entering and leaving the node v must cancel out. Given $T \geq 0$ we say that f has *time horizon* T if no flow remains in the network after time T.

Ford and Fulkerson [13,14] also introduced the concept of the *time-expanded network* $\mathcal{N}^T$ corresponding to a dynamic network $\mathcal{N}$ and an integral time horizon T that gets rid of the transit time. It consists of one copy of the nodes of $\mathcal{N}$ for each point in time $\theta \in \{1, \dots, T\}$, called *time layers*. For each arc a in $\mathcal{N}$ the time-expanded network contains copies of the same capacity connecting any two layers at distance τ_a. Many flow over time problems can be reduced to classical *static* flow problems in the time-expanded network. However, the size of the time-expanded network is exponential (pseudo-polynomial) in the input-size. When considering a flow over time problem it is often the first goal to find a *polynomial space* algorithm that does not rely on time-expansion and only works on the original network. We achieve this aim for tight EAT problems.

Given a dynamic network $\mathcal{N}$ and a supply/demand function $b \colon S^+ \cup S^- \to \mathbb{Z}$, the aim of the *earliest arrival transshipment problem* $(\mathcal{N}, b)_{\text{EAT}}$ is to find a flow over time f that satisfies the supplies/demands and has the property that the amount of flow that has arrived at the sinks up to time θ is maximized for each $\theta \geq 0$. Such a flow f is called *earliest arrival transshipment (EAT)*. Gale [15]

---

[1] There exist two different models for flows over time – a discrete and a continuous model. We consider the continuous model but the presented results also hold in the discrete case (see also [12]).

studied the EAT problem in networks with a single source and a single sink, also denoted as *earliest arrival flow problem*, and showed existence of earliest arrival flows (EAFs) (see also [31]). The first polynomial space algorithms for computing EAFs, which rely on the successive shortest path algorithm, are given in [29,40]. In [6] it is shown that the earliest arrival flow problem is $\mathcal{NP}$-hard. In networks with several sinks EATs do *not* exist for every choice of supplies/demands [9]. On the other hand, in [32] it is observed that in networks with multiple sources *but only a single sink* EATs do always exist. This is the reason why so far a lot of research has been focused on the efficient computation of EATs in dynamic networks with only a single sink [1,9,11,17,33,34,39]. In particular, there is a polynomial space algorithm to compute EATs in such networks [34]. Regarding EATs in multiple sink networks not many results have been achieved. Schmidt and Skutella [35] characterize networks with multiple sources and sinks with *only zero transit times* in which every EAT problem has a solution and there are efficient algorithms that compute *approximations* of EATs in multiple sink networks [16]. See [2,9,18,20,23,24,28,34] for the latest results about a related flow over time problem, the *quickest transshipment problem*.

Our Contribution. The only result that is known about EATs in multiple sink networks is that they in general do not exist. In particular, the problem of checking whether there exists an EAT solving a given problem and of computing such an EAT efficiently in case of existence is still open. We make huge progress in this context by formulating the first polynomial space algorithm for these problems for the special case of *tight* EAT problems in dynamic networks with multiple sources *and* multiple sinks. An EAT problem $(\mathcal{N}, b)_{\text{EAT}}$ is *tight* if the maximum amount of flow that can be sent from S^+ to S^- until time T (disregarding supplies/demands) is equal to $b(S^+)$. Here, T is the minimal time needed to fulfill all supplies/demands. Complementing this algorithm we derive that it is, already for tight instances, $\mathcal{NP}$-*hard to decide whether a given EAT problem has a solution*. We remark that all our results carry over to general (non-tight) EAT problems in dynamic networks with multiple sinks *and a single source*. The ideas and techniques for tight EAT problems can also be applied to this more general case. However, doing so in detail is beyond the scope of this paper.

As our first result we derive the earliest arrival pattern corresponding to tight EAT problems (Sect. 3), that is, the function that describes the time-dependent maximum flow value. Knowing the pattern is essential for the construction of our algorithm (Sect. 4). Overall, we show that we can check whether a tight EAT problem has a solution by just minimizing a suitably defined submodular function that we can evaluate in polynomial space. To obtain the EAT in case of existence we exploit that many algorithms for submodular function minimization not only compute the minimum of the given function but also a dual optimality certificate which is a vector inside the submodular function's base polytope. This vector is given as a convex combination of vertices of the base polytope that in our case we show to correspond to so-called *lex-max earliest arrival flows* for which we present a polynomial space algorithm to compute them. Thus, we essentially achieve a solution to a tight EAT problem as a convex combination of lex-max EAFs. Although the output size of our algorithm is necessarily exponential in the input size, it runs in polynomial space producing the output sequentially. *Due to the space limitation some proofs of non-trivial results and some technical details will be skipped in this extended abstract. We will always state when we do so.*

2   Preliminaries and Notation

For a finite set U, $\mathbb{R}^U$ is defined to be the $|U|$-dimensional real vector space with components indexed by the elements in U. Furthermore, we define $x(X) := \sum_{u \in X} x(u)$ for $X \subseteq U$, and by 2^U we denote the power set of the set U.

Submodular Function Minimization. Given a finite ground set U, a set function $g \colon 2^U \to \mathbb{R}$ is *submodular* if for all $X \subseteq Y \subseteq U$ and $x \in U \setminus Y$ it satisfies $g(X \cup \{x\}) - g(X) \geq g(Y \cup \{x\}) - g(Y)$, and *submodular function minimization (SFM)* is the problem of computing the minimum value as well as a minimizer $X^* \subseteq U$ of a submodular function. In many algorithms for SFM the *base polytope* $\mathcal{B}(g)$ of a submodular function g plays a central role,

$$\mathcal{B}(g) := \{x \in \mathbb{R}^U \mid x(X) \leq g(X) \text{ for all } X \subseteq U \text{ and } x(U) = g(U)\}.$$

It is possible to optimize over $\mathcal{B}(g)$ in strongly polynomial time [8,37]:

Theorem 1 ([8,37]). *Given a submodular function $g \colon 2^U \to \mathbb{R}$ and $w \in \mathbb{R}^U$, a vector $x^* \in \mathcal{B}(g)$ with $w^T x^* = \max\{w^T x \mid x \in \mathcal{B}(g)\}$ can be found via a greedy algorithm: find a total order $\prec$ on U such that $u \prec u'$ if $w(u) \geq w(u')$ for $u, u' \in U$ and define $x^* \in \mathbb{R}^U$ by $x^*(u) := g(\{u' \in U \mid u' \prec u\} \cup \{u\}) - g(\{u' \in U \mid u' \prec u\})$.*

Theorem 1 implies that each total order $\prec$ on U induces a corresponding vertex $v^\prec$ of $\mathcal{B}(g)$ and vice versa. The following theorem is also central for SFM. For $x \in \mathbb{R}^U$ define the vector $x^- \in \mathbb{R}^U$ by $x^-(u) := \min\{x(u), 0\}$ for each $u \in U$.

Theorem 2 (Edmonds [8]). *For a submodular function $g \colon 2^U \to \mathbb{R}$ with the property that $g(\emptyset) = 0$, it holds that $\min_{X \subseteq U} g(X) = \max_{x \in B(g)} x^-(U)$.*

Many combinatorial algorithms for SFM rely on Theorem 2 and a lot of them, e.g., [5,10,21,22,30,36], make use of the same key idea, the so-called *framework of Cunningham* [4,5]: while minimizing a submodular function g, these algorithms also compute the optimal point $x^* \in \mathcal{B}(g)$ maximizing $x^-(U)$ (see Theorem 2) as a convex combination of vertices of $\mathcal{B}(g)$. The currently fastest known strongly polynomial time algorithm for SFM using this framework is Orlin's algorithm [30], which we denote by $\mathrm{SFM_{Orlin}}$ (the fastest algorithm for SFM is due to Lee et al. [27]). It returns a minimizer $X^* \subseteq U$ of g, the optimal point $x^* \in \mathcal{B}(g)$ given as a minimal convex combination of vertices of $\mathcal{B}(g)$, and the minimal value $v_{\min}$ of g. Note that a submodular function can only be minimized in (strongly) polynomial running time if we can evaluate g in (strongly) polynomial time.

Flows Over Time – Additional Notation. The *value* of a flow over time f in a dynamic network $\mathcal{N}$ at time $\theta \geq 0$, denoted by $|f|_\theta$, is the overall amount of flow that has reached the sinks of $\mathcal{N}$ until time θ. For each $v \in V$ we denote the *net amount of flow* that has *left* the node v up to time θ in a flow over time f by $\mathrm{net}_f(v, \theta)$. If $v \in S^-$, then we have $\mathrm{net}_f(v, \theta) \leq 0$ for all $\theta \geq 0$ and $-\mathrm{net}_f(v, \theta)$ is the amount of flow that has arrived at the sink v until time θ.

A dynamic network $\mathcal{N}$ can also be regarded as a *static* network. In this case we usually consider the transit times τ as costs on the arcs. If x is a static flow in $\mathcal{N}$, we denote by $\mathcal{N}_x = (D_x, u_x, \tau, S^+, S^-)$ the corresponding residual network with $\tau(\overleftarrow{a}) := -\tau(a)$ and $u_x(\overleftarrow{a}) := u_x(a) - x(a)$. Here $\overleftarrow{a} = (v, u)$ is the backwards arc of $a = (u, v) \in A$ and D_x is the residual graph of D.

The flows over time returned by the main algorithm presented in this paper are of a special structure: they are *generalized temporally repeated* flows over time. In a dynamic network $\mathcal{N}$ we denote by $\overleftrightarrow{\mathcal{P}}$ the set of all paths from S^+ to S^- that might use all arcs in forward and backward direction. Let $x \colon A \to \mathbb{R}_+$ be a static flow in $\mathcal{N}$. A *generalized temporally repeated flow* f_x corresponding to x is obtained out of a *generalized* path decomposition $(x_P)_{P \in \overleftrightarrow{\mathcal{P}}}$ of x by sending flow at rate x_P into each flow carrying path $P \in \overleftrightarrow{\mathcal{P}}$ starting from time 0 on. The flow over time f_x is not necessarily feasible as $f_x(a, \theta)$ can be negative.

When considering the time-expanded network $\mathcal{N}^T$ corresponding to a dynamic network $\mathcal{N}$ and an integral time horizon T the copy of a node $v \in V$ in layer θ is called v^θ. The sources/sinks of $\mathcal{N}^T$ are given by $S^+ \cup S^-$ which are added to $\mathcal{N}^T$ by connecting the nodes in $S^+ \cup S^-$ to its respective copies in each time layer by arcs with infinite capacity. There is a nice correspondence between static flows in $\mathcal{N}^T$ and flows over time with time horizon T in $\mathcal{N}$. In fact, each feasible static flow x^T in $\mathcal{N}^T$ yields a flow over time f in $\mathcal{N}$ with time horizon T such that

$$\mathrm{net}_{x^T}(\{v^1, v^2, \ldots, v^\theta\}) = \mathrm{net}_f(v, \theta) \ \forall \ v \in S^+ \cup S^-, \ \theta \in \{1, 2, \ldots, T\}. \quad (1)$$

See [25] for a proof of this fact. Note that the time-expanded network can similarly be constructed for arbitrary rational time horizons $T = p/q$, $p, q \in \mathbb{Z}_{>0}$ by creating a time layer for each $\theta \in \{1/q, \ldots, p/q\}$. This can be done such that (1) still holds. For a thorough introduction to flows over time see [38].

Earliest Arrival Transshipments. Let $\mathcal{N} = (D = (V, A), u, \tau, S^+, S^-)$ be a dynamic network and $b \colon S^+ \cup S^- \to \mathbb{Z}$ a supply/demand function. Consider the corresponding EAT problem $(\mathcal{N}, b)_{\mathrm{EAT}}$. We say that $T \geq 0$ is the *minimal feasible time horizon* for $(\mathcal{N}, b)_{\mathrm{EAT}}$ if T is the minimal time needed to be able to fulfill all supplies/demands. The *earliest arrival pattern* $p^* \colon \mathbb{R}_+ \to \mathbb{R}_+$ of $(\mathcal{N}, b)_{\mathrm{EAT}}$ is defined by setting $p^*(\theta)$ to be the maximal amount of flow that can be sent from S^+ to S^- before time $\theta \geq 0$ while respecting the supplies/demands.

For $\theta \geq 0$ the function $o^\theta \colon 2^{S^+ \cup S^-} \to \mathbb{R}_+$ is obtained by setting $o^\theta(X)$ to be the maximal amount of flow that can be sent from X to $S^- \setminus X$ before time θ (disregarding supplies/demands) for all $X \subseteq S^+ \cup S^-$. The value $o^\theta(X)$ can be computed via one minimum-cost flow computation using an algorithm of Ford and Fulkerson [13]. We call an EAT problem $(\mathcal{N}, b)_{\mathrm{EAT}}$ with minimal feasible time horizon T *tight* if $o^T(S^+) = b(S^+)$. Hoppe and Tardos [20] observe that o^θ is a submodular function for every $\theta \geq 0$.

In dynamic networks with a single source s and a single sink t an EAT with time horizon T can be derived using the successive shortest path algorithm (SSPA) from s to t. Hereby, the transit times are regarded as costs [29,40]. The generalized temporally repeated flow corresponding to the paths and their flow values occurring in the SSPA-computation is an earliest arrival flow. The pattern of such an earliest arrival flow is $\theta \mapsto o^\theta(S^+)$.

3 The Earliest Arrival Pattern

The construction of our algorithm that checks whether a *tight* EAT problem $(\mathcal{N}, b)_{\mathrm{EAT}}$ has a solution and computes the solution in case of existence heavily relies on the structure of the *earliest arrival pattern* p^* corresponding to such problems. Assume that T is the minimal feasible time horizon for a given tight EAT problem $(\mathcal{N}, b)_{\mathrm{EAT}}$, i.e., $o^T(S^+) = b(S^+)$. In this tight case respecting the supplies and demands does not reduce the maximal amount of flow that can be sent from the sources to the sinks until time T. It turns out that this is true for the whole earliest arrival pattern.

Theorem 3. *Let $(\mathcal{N}, b)_{\mathrm{EAT}}$ be a tight EAT problem with minimal feasible time horizon T. We have $p^*(\theta) = o^\theta(S^+)$ for all $\theta \leq T$.*

A similar statement for EATs in dynamic networks with *multiple sources and a single sink* was shown in [1]. The proof for Theorem 3 is essentially a generalization of the one given in [1]. This is why we skip the proof of Theorem 3 here.

4 Computing Tight Earliest Arrival Transshipments

In this section we present the main result of this paper: a polynomial space algorithm that checks whether a given *tight* EAT problem $(\mathcal{N}, b)_{\mathrm{EAT}}$ has a solution and that computes this solution in case of existence. Throughout most of this section we consider tight EAT problems in dynamic networks with *a single source*. Only at the end of this section all our results will be generalized to tight problems in networks with multiple sources *and* sinks. Let $\mathcal{N} = (D, u, \tau, \{s\}, S^-)$ denote a dynamic network with multiple sinks S^- and a *single source* s, and assume that $(\mathcal{N}, b)_{\mathrm{EAT}}$ is *tight*, i.e., $o^T(\{s\}) = -b(S^-)$ where T is the minimal feasible time horizon of $(\mathcal{N}, b)_{\mathrm{EAT}}$. Note that T can be determined in polynomial time by one parametric submodular function minimization of $o^\theta - b$ according to a feasibility criterion of Klinz [26]. Since $\mathcal{N}$ is a single sink network, we can assume that b is just defined on S^-. The supply of s is implicitly given by $-b(S^-)$.

Before we state the structural main result that our algorithm is based on, we need two additional definitions. First, EAT problems are heavily connected to a suitably defined submodular function. This function is defined similarly to the function o^θ, but it additionally incorporates the earliest arrival pattern of a tight EAT problem, $\theta \mapsto o^\theta(\{s\})$. Given a dynamic network $\mathcal{N}$ with a single source s and a time horizon $\theta \geq 0$ we define the set function $\gamma^\theta : 2^{S^-} \to \mathbb{R}$ as follows:

$$X \mapsto \begin{array}{l} \text{maximum amount of flow that can arrive at the sinks in } X \text{ until time } \theta \text{ in a} \\ \text{flow over time } f \text{ in } \mathcal{N} \text{ with time horizon } \theta \text{ and } |f|_{\theta'} = o^{\theta'}(\{s\}) \text{ for all } \theta' < \theta. \end{array}$$

A flow over time f with time horizon θ and pattern $\theta' \mapsto o^{\theta'}(\{s\})$ for all $\theta' \in [0, \theta)$ that also fulfills $-\operatorname{net}_f(X, \theta') = \gamma^{\theta'}(X)$ for some $\theta' < \theta$ and $X \subseteq S^-$ is said to *satisfy* $\gamma^{\theta'}(X)$. Note that the function γ^θ is completely independent of the demands on the sinks. It turns out that γ^θ is in fact submodular.

Lemma 1. *Let $\mathcal{N}$ be a dynamic network with only a single source s. The corresponding set function γ^θ is submodular for every $\theta \geq 0$.*

The proof of Lemma 1 is deferred to Appendix A. The second ingredient for our structural main result is a special class of flows over time that is similar to lexicographically maximal flows over time [20,29] but again also incorporates the earliest arrival pattern of tight EAT problems.

Definition 1. *Let $\mathcal{N}$ be a dynamic network with a single source s, and $\prec$ a total order on the set of sinks S^-. We call a flow over time f with time horizon T a lexicographically maximal earliest arrival flow (lex-max EAF) with respect to $\prec$ if f fulfills the following conditions: (i) The flow over time f has pattern $\theta \mapsto o^\theta(\{s\})$ for $\theta \in [0, T)$. (ii) The amount of flow sent into the sinks is maximized lexicographically in decreasing order $\prec$ while respecting the pattern.*

Since flows over time with pattern $\theta \mapsto o^\theta(\{s\})$ do exist, e.g., earliest arrival flows from s to S^+, lex-max earliest arrival flows do also exist for every choice of parameters. With the definition of γ^θ and of lex-max earliest arrival flows we are now ready to state the structural main result of our paper.

Theorem 4 (Structure of EATs). *A tight EAT problem $(\mathcal{N}, b)_{\text{EAT}}$ in a dynamic network $\mathcal{N}$ with a single source and minimal feasible time horizon T has a solution if and only if $-b \in \mathcal{B}(\gamma^T)$. If $(\mathcal{N}, b)_{\text{EAT}}$ has a solution, it can be achieved as a convex combination of at most $|S^-|$ lex-max EAFs with time horizon T.*

To prove this theorem, several steps are necessary. As a first observation note that if $-b \notin \mathcal{B}(\gamma^T)$, then $(\mathcal{N}, b)_{\text{EAT}}$ does not have a solution: by definition of $\mathcal{B}(\gamma^T)$, $-b \notin \mathcal{B}(\gamma^T)$ implies that $-b(X) > \gamma^T(X)$ for some $X \subseteq S^-$ which contradicts the existence of an EAT solving $(\mathcal{N}, b)_{\text{EAT}}$ by the definition of γ^T and the fact that the pattern of a tight EAT problem $(\mathcal{N}, b)_{\text{EAT}}$ is given by $\theta \mapsto o^\theta(\{s\})$ (see Theorem 3). In order to prove Theorem 4, we still need to show the converse statement. For this purpose we deduce that the vertices of $\mathcal{B}(\gamma^T)$ are given by the excess vectors of lex-max earliest arrival flows with time horizon T.

This also immediately implies that in case of existence, i.e. if $-b \in \mathcal{B}(\gamma^T)$, an EAT can be obtained as a convex combination of lex-max earliest arrival flows.

Thus, it is our next goal to derive the excess vector of a lex-max earliest arrival flow. We start with a simple observation: let $\mathcal{N}$ be a dynamic network with only a single source, $\prec$ a total order on S^- and $T \geq 0$. If a flow over time f with time horizon T fulfills $-\operatorname{net}_f(\{t' \in S^- \mid t \preceq t'\}, T) = \gamma^T(\{t' \in S^- \mid t \preceq t'\})$ for all $t \in S^-$, then the flow over time f is a lex-max EAF with respect to $\prec$ and T by the definition of γ^T. The existence of flows over time fulfilling this property can be derived from a connection between flows over time f in $\mathcal{N}$ satisfying $\gamma^T(X)$ for some $X \subseteq S^-$ and certain static lex-max flows in the time-expanded network. This is done in Appendix A. Overall, we can state the following lemma about the excess vector of a lex-max earliest arrival flow.

Lemma 2. *Let $\mathcal{N}$ be a dynamic network with a single source s, $T \geq 0$ a time horizon and $\prec$ a total order on S^-. A lex-max earliest arrival flow f with respect to $\prec$ and time horizon T fulfills $-\operatorname{net}_f(\{t' \in S^- \mid t \preceq t'\}) = \gamma^T(\{t' \in S^- \mid t \preceq t'\})$.*

With Lemma 2 it is easy to derive a correspondence between the vertices of $\mathcal{B}(\gamma^T)$ and lex-max earliest arrival flows with time horizon T. Recall that by Theorem 1 the vertices of $\mathcal{B}(\gamma^T)$ correspond to total orders $\prec$ on S^-.

Corollary 1. *Let $\mathcal{N}$ be a dynamic network with a single source, $T \geq 0$ a time horizon and $\prec$ a total order on S^-. If $f_\prec$ is the lex-max EAF with time horizon T in $\mathcal{N}$ and $v^\prec$ the vertex of $\mathcal{B}(\gamma^T)$ corresponding to $\prec$, it holds that $\operatorname{net}_{f_\prec}(s, T) = -v^\prec(t)$ for all $t \in S^-$.*

Corollary 1 is an immediate consequence of Theorem 1 and Lemma 2. Using Corollary 1 we are now also able to prove Theorem 4.

Proof of Theorem 4. We already argued that $-b \in \mathcal{B}(\gamma^T)$ is a necessary condition for the existence of a solution of $(\mathcal{N}, b)_{\mathrm{EAT}}$. We will now prove the converse direction. For this purpose, choose some $x \in \mathcal{B}(\gamma^T)$. Thus, we can obtain x as convex combination of vertices of $\mathcal{B}(\gamma^T)$. Then, by Corollary 2, a corresponding convex combination of lex-max earliest arrival flows gives us a solution to the problem $(\mathcal{N}, -x)_{\mathrm{EAT}}$. That the number of elements in the convex combination is at most $|S^-|$ follows with Carathéodory's theorem.   □

When given a tight earliest arrival transshipment problem $(\mathcal{N}, b)_{\mathrm{EAT}}$ with time horizon $T \geq 0$ in a dynamic network $\mathcal{N}$ with a single source, we can by Theorem 4 check the existence of a solution by testing whether $-b \in \mathcal{B}(\gamma^T)$. This can be done by minimizing the the submodular function $\gamma^T + b$. If its minimum is at most zero, then $-b$ is in the base polytope, if the minimum is strictly smaller than zero, it is not. In order to be able to minimize this submodular function using only polynomial space, we need to be able to *evaluate* it in polynomial space. Additionally, in order to be able to compute a solution to $(\mathcal{N}, b)_{\mathrm{EAT}}$ as a convex combination of lex-max EAFs, we also need a polynomial space algorithm to compute such flows over time. With Algorithm 1 we present an algorithm that solves both problems: To compute a lex-max EAF with respect to some given order $\prec$ on S^- and some time horizon $T \geq 0$, the first objective is to compute a flow over time f with pattern $\theta \mapsto o^\theta(\{s\})$ for $\theta \in [0, T)$. It was shown by Wilkinson [40] that the successive shortest path algorithm (SSPA) can be used to compute a flow over time with this pattern. This is why the base of Algorithm 1 is the SSPA. Since we want the flow arriving at the sinks to respect the order $\prec$, our implementation of the SSPA has the additional feature that in each iteration a shortest path is chosen with respect to the order $\prec$. By $d(\mathcal{N}, u, v)$ we denote the length of a shortest path between $u, v \in V$ in $\mathcal{N}$.

Algorithm 1. Computation of lex-max earliest arrival flows

Input : A dynamic network $\mathcal{N} = (D = (V, A), u, \tau, \{s\}, S^-)$, a time horizon $T \geq 0$ and a total order $\prec$ on S^-
Output: A lex-max earliest arrival flow f in $\mathcal{N}$ with respect to T and $\prec$
$x_P \leftarrow 0$ for all $P \in \overleftrightarrow{\mathcal{P}}$
$x \leftarrow$ static flow from s to S^- with generalized path decomposition $(x_P)_{P \in \overleftrightarrow{\mathcal{P}}}$
while $d(\mathcal{N}_x, s, S^-) < T$ **do**
  $l \leftarrow d(\mathcal{N}_x, s, S^-)$
  for $i = k, k-1, \ldots, 1$ **do**
    while $d(\mathcal{N}_x, s, t_i) = l$ **do**
      $P \leftarrow$ shortest s-t_i path in $\mathcal{N}_x$ and $\gamma \leftarrow \min\{u(a) \mid a \in P\}$
      augment x along P by γ

$f \leftarrow$ gen. temporally repeated flow with time horizon T given by $(x_P)_{P \in \overleftrightarrow{\mathcal{P}}}$.

Theorem 5 (Correctness of Algorithm 1). *Let $\mathcal{N}$ be a dynamic network with a single source, $T \geq 0$ a time horizon, and $\prec$ a total order on the sinks S^-. The flow over time f returned by Algorithm 1 with respect to these parameters fulfills $-\operatorname{net}_f(\{t' \in S^- \mid t \preceq t'\}, T) = \gamma^T(\{t' \in S^- \mid t \preceq t'\})$ for all $t \in S^-$ and is thus a lex-max EAF with respect to T and $\prec$.*

It is clear that Algorithm 1 computes a flow over time with the correct pattern (see [29,40]). That the returned flow fulfills the other property of a lex-max EAF follows directly from the fact that the shortest paths are chosen with respect to the given order. We skip the proof of this theorem here. Algorithm 1 can also be used to evaluate the submodular function γ^T at an arbitrary $X \subseteq S^-$ by choosing a total order $\prec$ on S^- with $t \prec t'$ for all $t \in S^- \setminus X$ and $t' \in X$.

Corollary 2. *Using Algorithm 1, γ^T can be evaluated in polynomial space.*

By incorporating our polynomial space algorithm for evaluating γ^T into a strongly polynomial time SFM algorithm, we can hence test whether $(\mathcal{N}, b)_{\mathrm{EAT}}$ has a solution using only polynomial space. It turns out that a suitable convex combination of lex-max EAFs solving $(\mathcal{N}, b)_{\mathrm{EAT}}$ is essentially also computed while minimizing the submodular function $\gamma^T + b$.

Corollary 3. *By minimizing $\gamma^T + b$ with the help of Algorithm 1, we can check in polynomial space whether a tight EAT problem $(\mathcal{N}, b)_{\text{EAT}}$ with time horizon T in a network with a single source has a solution. If an SFM algorithm using Cunningham's framework is used, also a convex combination of lex-max EAFs solving $(\mathcal{N}, b)_{\text{EAT}}$ is computed during the minimization, in case of existence.*

Proof. Only the last statement remains to be proved. When minimizing the submodular function $\gamma^T + b$, e.g., with the algorithm of Orlin, then, besides the minimal value of the submodular function, also a convex combination of vertices of $\mathcal{B}(\gamma^T + b)$ giving the vector $x^* = \text{argmax}\{x^-(S^-) \mid x \in \mathcal{B}(\gamma^T)\}$ is computed (see Theorem 2). In particular, we get $d \leq |S^-|$ total orders $\prec_1, \ldots, \prec_d$ corresponding to the suitable vertices of $\mathcal{B}(\gamma^T + b)$ and convex coefficients $\lambda_1, \ldots, \lambda_d \geq 0$. Denote by $f_1, \ldots, f_d$ the lex-max EAFs in $\mathcal{N}$ with time horizon T with respect to the total orders $\prec_1, \ldots, \prec_d$, respectively. Corollary 1, the observation that x^* is the zero vector, and the fact that $\mathcal{B}(\gamma^T) = \mathcal{B}(\gamma^T + b) - b$ imply that f fulfills $\text{net}_f(t, T) = b(t)$ for all $t \in S^-$. Since by construction f has pattern $\theta \mapsto o^{\theta}(\{s\})$ for $\theta \in [0, T]$, the flow over time f is an EAT solving $(\mathcal{N}, b)_{\text{EAT}}$. The above arguing shows that while doing SFM of $\gamma^T + b$ also a convex combination of lex-max EAFs solving the problem $(\mathcal{N}, b)_{\text{EAT}}$ is determined, in case of existence of a solution. The specific lex-max EAFs have to be computed by Algorithm 1.   □

Summarizing, we thus obtain a *polynomial space* method that checks whether $(\mathcal{N}, b)_{\text{EAT}}$ has a solution and computes it in case of existence.

Tight EATs in General Networks. We conclude by sketching how our results can be generalized to tight EATs in general dynamic networks. Assume that $(\mathcal{N}, b)_{\text{EAT}}$ is a tight EAT problem with minimal feasible time horizon T in a dynamic network $\mathcal{N}$ with multiple sources *and* multiple sinks. Denote by $\mathcal{N}_s$ the dynamic network in which a super source s is attached to all sources in S^+ by arcs with zero transit time and infinite capacity. The source s is the new single source of $\mathcal{N}_s$. By b_s we denote the restriction of b to the sinks in S^-. Similarly, we can define $\mathcal{N}_t$ and b_t by attaching a super sink t to the sinks in S^-. Thus, we obtain two new tight EAT problems $(\mathcal{N}_s, b_s)_{\text{EAT}}$ and $(\mathcal{N}_t, b_t)_{\text{EAT}}$ with the same minimal feasible time horizon T as before.

Theorem 6. *A tight EAT problem $(\mathcal{N}, b)_{\text{EAT}}$ in a dynamic network with multiple sources and multiple sinks has a solution if and only if $(\mathcal{N}_s, b_s)_{\text{EAT}}$ has a solution.*

Proof. If $(\mathcal{N}, b)_{\text{EAT}}$ has a solution, then a flow over time f solving this problem can simply be transformed into a flow over time solving $(\mathcal{N}_s, b_s)_{\text{EAT}}$. For the other direction let f_s and f_t be flows over time solving $(\mathcal{N}_s, b_s)_{\text{EAT}}$ and $(\mathcal{N}_t, b_t)_{\text{EAT}}$, respectively. Denote by x_s and x_t the corresponding (maximum) flows in the time-expanded network (see (1)). By a result of Minieka [29] both static flows can be "glued" together to obtain a maximum flow x in the time-expanded network which hat the departure pattern of x_t and the arrival pattern of x_s. The flow over time f induced by x by construction fulfills all supplies and demands and has pattern $\theta \mapsto o^\theta(S^+)$ as required (see Theorem 3). $\qquad\qquad\square$

By Theorem 6 we can use Theorem 4 to check if a general tight EAT problem $(\mathcal{N}, b)_{\mathrm{EAT}}$ has a solution. To compute a solution to $(\mathcal{N}, b)_{\mathrm{EAT}}$ in polynomial space, we combine the results from this paper with [34]. In [34] it is shown that $(\mathcal{N}_t, b_t)_{\mathrm{EAT}}$ can be solved by a convex combination of lex-max flows over time. The main idea for the polynomial space computation of a flow solving $(\mathcal{N}, b)_{\mathrm{EAT}}$ is to combine a convex combination of lex-max flows over time solving $(\mathcal{N}_t, b_t)_{\mathrm{EAT}}$ and a convex combination of lex-max EAFs solving $(\mathcal{N}_s, b_s)_{\mathrm{EAT}}$ by incorporating Algorithm 1 into the algorithm in [20] for computing lex-max flows over time. The details of this construction are beyond the scope of this paper.

Finally, we state that by reducing from Partition using techniques from [6] we can show the following hardness result:

Theorem 7. *Let $(\mathcal{N}, b)_{\mathrm{EAT}}$ be a tight EAT Problem with two sinks. It is $\mathcal{NP}$-hard to decide whether there exists an EAT solving $(\mathcal{N}, b)_{\mathrm{EAT}}$.*

A   Appendix

Preliminaries for the Proof of Lemma 1. To prove Lemma 1 we need the notion of static lex-max flows. Let $\mathcal{N}$ be a dynamic network and $\prec$ a total order on $S^+ \cup S^-$. A *static lexicographically maximum (lex-max) flow* x in $\mathcal{N}$ with respect to $\prec$ is a static flow which maximizes the net amount of flow sent out of $S^+ \cup S^-$ in increasing order $\prec$. For the sinks this means that the flow into them is maximized in decreasing order $\prec$. If we denote by $\max_{\mathcal{N}}(S, X)$ the value of a static maximum flow from $S \subseteq S^+$ to $X \subseteq S^-$ in network $\mathcal{N}$, then it is due to Minieka [29] that

$$\begin{aligned} \mathrm{net}_x(\{s' \in S^+ \cup S^- \mid s' \preceq s\}) \\ = \max_{\mathcal{N}}(\{s' \in S^+ \mid s' \preceq s\}, \{s' \in S^- \mid s \preceq s'\}) \; \forall \; s \in S^+ \cup S^-. \end{aligned} \tag{2}$$

The proof of Lemma 1 strongly relies on Lemma 3, a connection between flows over time in $\mathcal{N}$ satisfying $\gamma^\theta(X)$ for some rational $\theta = p, q$ with $p, q \in \mathbb{Z}_{>0}$ and some $X \subseteq S^-$, and certain static lex-max flows in the time-expanded network. We consider the time-expanded network $\mathcal{N}^\theta$ corresponding to our given dynamic network $\mathcal{N}$ with only a single source s and denote by s^* the single super source of $\mathcal{N}^\theta$. Additionally, in each layer θ' we attach a super sink $t_X^{\theta'}$ to nodes $t^{\theta'}$ with $t \in X$ by an arc $(t^{\theta'}, t_X^{\theta'})$ with infinite capacity. Similarly, we attach a super sink $t_{S^- \setminus X}^{\theta'}$ and moreover we add an overall super sink $t_*^{\theta'}$ to each layer by infinite capacity arcs $(t_{S^- \setminus X}^{\theta'}, t_*^{\theta'})$ and $(t_X^{\theta'}, t_*^{\theta'})$. We consider $\{t_X^{1/q}, t_{S^- \setminus X}^{1/q}, \ldots, t_X^{p/q}, t_{S^- \setminus X}^{p/q}\}$ or $\{t_*^{1/q}, \ldots, t_*^{p/q}\}$ as sets of sinks of $\mathcal{N}^T$.

Lemma 3. *A flow over time f in a dynamic network $\mathcal{N}$ with only a single source s with rational time horizon $T = p/q$, $p, q \in \mathbb{Z}_{>0}$, that satisfies $\gamma^T(X)$ for some $X \subseteq S^-$ fulfills the following properties: (1) We have $\mathrm{net}_f(X, \theta) = -\gamma^\theta(X)$*

for all $\theta \leq T$. (2) The flow over time f induces a static lex-max flow in $\mathcal{N}^T$ with respect to the total order $s^* \prec t^{p/q}_{S^- \backslash X} \prec t^{p/q}_X \prec t^{(p-1)/q}_{S^- \backslash X} \prec t^{(p-1)/q}_X \prec \ldots \prec t^{1/q}_{S^- \backslash X} \prec t^{1/q}_X$.

The proof of Lemma 3 mostly relies on (1) and (2) and we skip it here.

Proof of Lemma 1. Lemma 3, (1) and (2), imply for an integral $T \geq 0$,

$$\gamma^T(X) \stackrel{(2)}{=} \sum_{i=1}^{T} \max_{\mathcal{N}^T}(s^*, \{t^1_*, \ldots, t^i_*, t^i_X\}) - \max_{\mathcal{N}^T}(s^*, \{t^1_*, \ldots, t^i_*\}).$$

Note that $\max_{\mathcal{N}^T}(s^*, \{t^1_*, \ldots, t^i_*\})$ is independent of X and thus in order to show submodularity, it suffices to show that $g^\theta : 2^{S^-} \to \mathbb{R}$ with

$$g^\theta(X) := \max_{\mathcal{N}^T}(s^*, \{t^1_*, \ldots, t^\theta_*, t^\theta_X\}) \text{ for } X \subseteq S^-,$$

is submodular on S^- for all $\theta \in \{1, \ldots, T\}$. To show this fact, fix some integral θ, $A \subseteq B \subseteq S^-$, and $v \in S^- \setminus B$. We redefine the time-expanded network $\mathcal{N}^\theta$. In time layer θ we now attach a super sink t^θ_v to v^θ, a super sink t^θ_A to the copies of sinks in A in layer θ, a super sink $t^\theta_{B \backslash A}$ to the copies of the sinks in $B \setminus A$, and a super sink $t^\theta_{S^- \backslash (B \cup \{v\})}$ to the copies of the remaining sinks in layer θ. Then, $g^\theta(A \cup \{v\}) - g^\theta(A)$ is exactly the amount of flow that reaches t^θ_v in a lex-max flow in $\mathcal{N}^\theta$ with respect to the order $\prec$ given by

$$s^* \prec t^\theta_{S^- \backslash (B \cup \{v\})} \prec t^\theta_{B \backslash A} \prec t^\theta_v \prec t^\theta_A \prec t^{\theta-1}_* \prec \ldots \prec t^1_*.$$

Similarly, $g^\theta(B \cup \{v\}) - g^\theta(B)$ is exactly the amount of flow that reaches t^θ_v in a static lex-max flow in $\mathcal{N}^\theta$ with respect to

$$s^* \prec t^\theta_{S^- \backslash (B \cup \{v\})} \prec t^\theta_v \prec t^\theta_{B \backslash A} \prec t^\theta_A \prec t^{\theta-1}_* \prec \ldots \prec t^1_*.$$

Since the sink t^θ_v has a higher priority with respect to $\prec$ in the first order, we obtain $g^\theta(B \cup \{v\}) - g^\theta(B) \leq g^\theta(A \cup \{v\}) - g^\theta(A)$, and thus submodularity. The submodularity for arbitrary rational time horizons follows by using a finer discretization of time and for irrational time horizons by continuity. □

Preliminaries for the Proof of Lemma 2. It turns out that Lemma 2 is a direct consequence of the following lemma which is the reverse direction of Lemma 3.

Lemma 4. *Let $\mathcal{N}$ be a dynamic network and $T = p/q$ with $p, q \in \mathbb{Z}_{>0}$ a rational time horizon. A static lex-max flow x in $\mathcal{N}^T$ with respect to the total order*

$$s^* \prec t^{p/q}_{S^- \backslash X} \prec t^{p/q}_X \prec t^{(p-1)/q}_{S^- \backslash X} \prec t^{(p-1)/q}_X \prec \ldots \prec t^{1/q}_{S^- \backslash X} \prec t^{1/q}_X,$$

induces a flow over time with time horizon T satisfying $\gamma^\theta(X)$ for each $\theta \in [0, T)$.

Proof. For the simplification of notation we only consider integral time horizons in our proof. Let f be the flow over time induced by the static lex-max flow x. By Lemma 3, (1) and (2) we obtain for our flow over time f,

$$- \operatorname{net}_f(X, \theta)$$

$$\overset{(1),(2)}{=} \sum_{i=1}^{\theta} (\max_{\mathcal{N}^T}(s^*, \{t_*^1, \ldots, t_*^{i-1}, t_X^i\}) - \max_{\mathcal{N}^T}(s^*, \{t_*^1, \ldots, t_*^{i-1}\}))$$

$$\overset{\text{Lem. 3}}{=} \gamma^\theta(X) \text{ for all } \theta \in \{0, 1, \ldots, T\}.$$

By definition of f and (1) it also holds that $|f|_\theta = o^\theta(\{s\})$ for all $\theta \in \{0, 1, \ldots, T\}$. The function $\theta \mapsto o^\theta(\{s\})$ is piecewise linear [29,40] and since all transit times are integral, breakpoints of this function only occur at integral points in time. However, by the construction of f from x (see [25]) the function $\theta \mapsto |f|_\theta$ is also a piecewise linear function with breakpoints only occurring at integral points in time. Thus, we have $|f|_\theta = o^\theta(\{s\})$ for all $\theta \leq T$ and hence by our arguing above the flow over time f satisfies $\gamma^\theta(X)$ for each integral $\theta \in \{1, 2, \ldots, T\}$. By the first statement of Lemma 3, $\gamma^\theta(X)$ is satisfied by f for all $\theta \in [0, T)$. □

We can now prove Lemma 2.

Proof of Lemma 2. Let $\prec$ be given by $t_1 \prec t_2 \prec \ldots \prec t_k$ for $S^- = \{t_1, t_2, \ldots, t_k\}$. Lemma 4 implies that a static lex-max flow in $\mathcal{N}^T$ with respect to a total order $\prec'$ given by $s^* \prec' t_1^{p/q} \prec' \ldots \prec' t_k^{p/q} \prec' t_1^{(p-1)/q} \prec' \ldots \prec' t_k^{(p-1)/q} \prec' \ldots \prec' t_1^{1/q} \prec' \ldots \prec' t_k^{1/q}$ induces a flow over time f with the required properties. □

References

1. Baumann, N., Skutella, M.: Earliest arrival flows with multiple sources. Math. Oper. Res. **34**, 499–512 (2009). https://doi.org/10.1287/moor.1090.0382
2. Burkard, R.E., Dlaska, K., Klinz, B.: The quickest flow problem. Zeitschrift für Oper. Res. **37**(1), 31–58 (1993). https://doi.org/10.1007/BF01415527
3. Chalmet, L.G., Francis, R.L., Saunders, P.B.: Network models for building evacuation. Fire Technol. **18**(1), 90–113 (1982). https://doi.org/10.1007/BF02993491
4. Cunningham, W.H.: Testing membership in matroid polyhedra. J. Comb. Theory Ser. B **36**(2), 161–188 (1984). https://doi.org/10.1016/0095-8956/(84)90023-6
5. Cunningham, W.H.: On submodular function minimization. Combinatorica **5**(3), 185–192 (1985). https://doi.org/10.1007/BF02579361
6. Disser, Y., Skutella, M.: The simplex algorithm is NP-mighty. In: Indyk, P. (ed.) Proceedings of the Twenty-Sixth Annual ACM-SIAM Symposium on Discrete Algorithms, pp. 858–872. Society for Industrial and Applied Mathematics SIAM (2015). https://doi.org/10.1137/1.9781611973730.59
7. Dressler, D., et al.: On the use of network flow techniques for assigning evacuees to exits. Procedia Eng. **3**, 205–215 (2010). https://doi.org/10.1016/j.proeng.2010.07.019

8. Edmonds, J.: Submodular functions, matroids and certain polyhedra. In: Proceedings of the Calgary International Conference on Combinatorial Structures and Their Applications, pp. 69–87 (1970)
9. Fleischer, L.: Faster algorithms for the quickest transshipment problem. SIAM J. Optim. **12**(1), 18–35 (2001). https://doi.org/10.1137/S1052623497327295
10. Fleischer, L., Iwata, S.: Improved algorithms for submodular function minimization and submodular flow. In: Proceedings of the Thirty-Second Annual ACM Symposium on Theory of Computing, pp. 107–116. Association for Computing Machinery ACM (2000). http://doi.acm.org/10.1145/335305.335318
11. Fleischer, L., Skutella, M.: Quickest flows over time. SIAM J. Comput. **36**(6), 1600–1630 (2007). https://doi.org/10.1137/S0097539703427215
12. Fleischer, L., Tardos, É.: Efficient continuous-time dynamic network flow algorithms. Oper. Res. Lett. **23**(3–5), 71–80 (1998). https://doi.org/10.1016/S0167-6377(98)00037-6
13. Ford, L.R., Fulkerson, D.R.: Constructing maximal dynamic flows from static flows. Oper. Res. **6**(3), 419–433 (1958). https://doi.org/10.1287/opre.6.3.419
14. Ford, L.R., Fulkerson, D.R.: Flows in Networks. Princeton University Press, Princeton (1962). https://press.princeton.edu/titles/9233.html
15. Gale, D.: Transient flows in networks. Mich. Math. J. **6**(1), 59–63 (1959). https://doi.org/10.1307/mmj/1028998140
16. Groß, M., Kappmeier, J.-P.W., Schmidt, D.R., Schmidt, M.: Approximating earliest arrival flows in arbitrary networks. In: Epstein, L., Ferragina, P. (eds.) ESA 2012. LNCS, vol. 7501, pp. 551–562. Springer, Heidelberg (2012). https://doi.org/10.1007/978-3-642-33090-2_48
17. Hajek, B., Ogier, R.G.: Optimal dynamic routing in communication networks with continuous traffic. Networks **14**(3), 457–487 (1984). https://doi.org/10.1002/net.3230140308
18. Hall, A., Hippler, S., Skutella, M.: Multicommodity flows over time: efficient algorithms and complexity. In: Baeten, J.C.M., Lenstra, J.K., Parrow, J., Woeginger, G.J. (eds.) ICALP 2003. LNCS, vol. 2719, pp. 397–409. Springer, Heidelberg (2003). https://doi.org/10.1007/3-540-45061-0_33
19. Hamacher, H.W., Tjandra, S.A.: Mathematical modelling of evacuation problems: a state of the art. In: Schreckenberg, M., Sharma, S.D. (eds.) Pedestrian and Evacuation Dynamics, pp. 227–266. Springer, Heidelberg (2002)
20. Hoppe, B., Tardos, É.: The quickest transshipment problem. Math. Oper. Res. **25**(1), 36–62 (2000). https://doi.org/10.1287/moor.25.1.36.15211
21. Iwata, S.: A faster scaling algorithm for minimizing submodular functions. In: Cook, W.J., Schulz, A.S. (eds.) IPCO 2002. LNCS, vol. 2337, pp. 1–8. Springer, Heidelberg (2002). https://doi.org/10.1007/3-540-47867-1_1
22. Iwata, S., Fleischer, L., Fujishige, S.: A combinatorial strongly polynomial algorithm for minimizing submodular functions. J. ACM (JACM) **48**(4), 761–777 (2001). http://doi.acm.org/10.1145/502090.502096
23. Kamiyama, N., Katoh, N., Takizawa, A.: An efficient algorithm for evacuation problems in dynamic network flows with uniform arc capacity. In: Cheng, S.-W., Poon, C.K. (eds.) AAIM 2006. LNCS, vol. 4041, pp. 231–242. Springer, Heidelberg (2006). https://doi.org/10.1007/11775096_22
24. Kamiyama, N., Katoh, N., Takizawa, A.: An efficient algorithm for the evacuation problem in a certain class of networks with uniform path-lengths. Discrete Appl. Math. **157**, 3665–3677 (2009). https://doi.org/10.1016/j.dam.2009.04.007
25. Kappmeier, J.W.: Generalizations of flows over time with applications in evacuations optimization. Ph.D. thesis, TU Berlin (2014)

26. Klinz, B.: Cited as personal comunication (1994) in [20]
27. Lee, Y.T., Sidford, A., Wong, S.C.W.: A faster cutting plane method and its implications for combinatorial and convex optimization. In: 56th Annual Symposium on Foundations of Computer Science. pp. 1049–1065. IEEE (2015). https://doi.org/10.1109/focs.2015.68
28. Mamada, S., Uno, T., Makino, K., Fujishige, S.: An $O(n \log^2 n)$ algorithm for a sink location problem in dynamic tree networks. Discrete Appl. Math. **154**, 2387–2401 (2006). http://www.sciencedirect.com/science/article/pii/S0166218X06001880
29. Minieka, E.: Maximal, lexicographic, and dynamic network flows. Oper. Res. **21**(2), 517–527 (1973). https://doi.org/10.1287/opre.21.2.517
30. Orlin, J.B.: A faster strongly polynomial time algorithm for submodular function minimization. Math. Program. **118**(2), 237–251 (2009). https://doi.org/10.1007/s10107-007-0189-2
31. Philpott, A.: Continuous-time flows in networks. Math. Oper. Res. **15**(4), 640–661 (1990). https://doi.org/10.1287/moor.15.4.640
32. Richardson, D., Tardos, É.: Cited as personal comunication (2002) in [11]
33. Ruzika, S., Sperber, H., Steiner, M.: Earliest arrival flows on series-parallel graphs. Networks **57**(2), 169–173 (2011). https://doi.org/10.1002/net.20398
34. Schlöter, M., Skutella, M.: Fast and memory-efficient algorithms for evacuation problems. In: Klein, P.N. (ed.) Proceedings of the Twenty-Eighth Annual ACM-SIAM Symposium on Discrete Algorithms, SODA 2017, pp. 821–840. Society for Industrial and Applied Mathematics SIAM (2017). https://doi.org/10.1137/1.9781611974782.52
35. Schmidt, M., Skutella, M.: Earliest arrival flows in networks with multiple sinks. Discrete Appl. Math. **164**, 320–327 (2014). https://doi.org/10.1016/j.dam.2011.09.023
36. Schrijver, A.: A combinatorial algorithm minimizing submodular functions in strongly polynomial time. J. Comb. Theory Ser. B **80**(2), 346–355 (2000). https://doi.org/10.1006/jctb.2000.1989
37. Shapley, L.S.: Cores of convex games. Int. J. Game Theory **1**(1), 11–26 (1971). https://doi.org/10.1007/BF01753431
38. Skutella, M.: An introduction to network flows over time. In: Cook, W., Lovász, L., Vygen, J. (eds.) Research Trends in Combinatorial Optimization, pp. 451–482. Springer, Heidelberg (2009). https://doi.org/10.1007/978-3-540-76796-1_21
39. Tjandra, S.A.: Dynamic network optimization with application to the evacuation problem. Ph.D. thesis, TU Kaiserslautern (2003). https://kluedo.ub.uni-kl.de/frontdoor/index/index/year/2003/docId/1407
40. Wilkinson, W.L.: An algorithm for universal maximal dynamic flows in a network. Oper. Res. **19**(7), 1602–1612 (1971). https://doi.org/10.1287/opre.19.7.1602

Intersection Cuts for Factorable MINLP

Felipe Serrano[(⊠)] [iD]

Optimization Department, Zuse Institute Berlin,
Takustr. 7, 14195 Berlin, Germany
serrano@zib.de

Abstract. Given a factorable function f, we propose a procedure that constructs a concave underestimator of f that is tight at a given point. These underestimators can be used to generate intersection cuts. A peculiarity of these underestimators is that they do not rely on a bounded domain. We propose a strengthening procedure for the intersection cuts that exploits the bounds of the domain. Finally, we propose an extension of monoidal strengthening to take advantage of the integrality of the non-basic variables.

Keywords: Mixed-integer nonlinear programming · Intersection cuts · Monoidal strengthening

1 Introduction

In this work we propose a procedure for generating intersection cuts for mixed integer nonlinear programs (MINLP). We consider MINLP of the following form

$$
\begin{aligned}
\max \ & c^{\mathsf{T}} x \\
\text{s.t. } & g_j(x) \le 0, j \in J \\
& Ax = b \\
& x_i \in \mathbb{Z}, i \in I \\
& x \ge 0,
\end{aligned}
\tag{1}
$$

where $J = \{1, \ldots, l\}$ denotes the indices of the nonlinear constraints, $g_j \colon \mathbb{R}^n \to \mathbb{R}$ are assumed to be continuous and factorable (see Definition 1), $A \in \mathbb{R}^{m \times n}$, $c \in \mathbb{R}^n$, $b \in \mathbb{R}^m$, and $I \subseteq \{1, \ldots, n\}$ are the indices of the integer variables. We denote the set of feasible solutions by S and a generic relaxation of S by R, that is, $S \subseteq R$. When R is a translated simplicial cone and C contains its

This work has been supported by the Research Campus MODAL *Mathematical Optimization and Data Analysis Laboratories* funded by the Federal Ministry of Education and Research (BMBF Grant 05M14ZAM). The author thank the Schloss Dagstuhl – Leibniz Center for Informatics for hosting the Seminar 18081 "Designing and Implementing Algorithms for Mixed-Integer Nonlinear Optimization" for providing the environment to develop the ideas in this paper.

© Springer Nature Switzerland AG 2019
A. Lodi and V. Nagarajan (Eds.): IPCO 2019, LNCS 11480, pp. 385–398, 2019.
https://doi.org/10.1007/978-3-030-17953-3_29

apex and no point of S in its interior, valid inequalities for $\text{conv}(R\backslash C)$ are called intersection cuts [3]. See the excellent survey [18] for recent developments and details on intersection cuts for mixed integer linear programs (MILP).

Many applications can be modeled as MINLP [13]. The current state of the art for solving MINLP to global optimality is via linear programming (LP), convex nonlinear programming and (MILP) relaxations of S, together with spatial branch and bound [10, 27, 28, 30, 39, 42]. Roughly speaking, the LP-based spatial branch and bound algorithm works as follows. The initial polyhedral relaxation is solved and yields $\bar{x}$. If the solution $\bar{x}$ is feasible for (1), we obtain an optimal solution. If not, we try to separate the solution from the feasible region. This is usually done by considering each violated constraint separately. Let $g(x) \leq 0$ be a violated constraint of (1). If $g(\bar{x}) > 0$ and g is convex, then $g(\bar{x}) + v^\mathsf{T}(x - \bar{x}) \leq 0$, where $v \in \partial g(\bar{x})$ and $\partial g(\bar{x})$ is the subdifferential of g at $\bar{x}$, is a valid cut. If g_j is non-convex, then a convex underestimator g_{vex}, that is, a convex function such that $g_{vex}(x) \leq g(x)$ over the feasible region, is constructed and if $g_{vex}(\bar{x}) > 0$ the previous cut is constructed for g_{vex}. If the point cannot be separated, then we branch, that is, we select a variable x_k in a violated constraint and split the problem into two problems, one with $x_k \leq \bar{x}_k$ and the other one with $x_k \geq \bar{x}_k$.

Applying the previous procedure to the MILP case, that is (1) with $J = \emptyset$, reveals a problem with this approach. In this case, the polyhedral relaxation is just the linear programming (LP) relaxation. Assuming that $\bar{x}$ is not feasible for the MILP, then there is an $i \in I$ such that $x_i \notin \mathbb{Z}$. Let us treat the constraint $x_i \in \mathbb{Z}$ as a nonlinear non-convex constraint represented by some function as $g(x_i) \leq 0$. Then, $g(\bar{x}_i) > 0$. However, a convex underestimator $\bar{g}$ of g must satisfy that $g_{vex}(z) \leq 0$ for every $z \in \mathbb{R}$, since $g_{vex}(z) \leq g(z) \leq 0$ for every $z \in \mathbb{Z}$ and $g_{vex}(z)$ is convex. Since separation is not possible, we need to branch.

However, for the current state-of-the-art algorithms for MILP, cutting planes are a fundamental component [1]. A classical technique for building cutting planes in MILP is based on exploiting information from the simplex tableau [18]. When solving the LP relaxation, we obtain $x_B = \bar{x}_B + Rx_N$, where B and N are the indices of the basic and non-basic variables, respectively. Since $\bar{x}$ is infeasible for the MILP, there must be some $k \in B \cap I$ such that $\bar{x}_k \notin \mathbb{Z}$. Now, even though $\bar{x}$ cannot be separated from the violated constraint $x_k \in \mathbb{Z}$, the equivalent constraint, $\bar{x}_k + \sum_{j \in N} r_{kj} x_j \in \mathbb{Z}$ can be used to separate $\bar{x}$.

In the MINLP case, this framework generates equivalent non-linear constraints with some appealing properties. The change of variables $x_k = \bar{x}_k + \sum_{j \in N} r_{kj} x_j$ for the basic variables present in a violated nonlinear constraint $g(x) \leq 0$, produces the non-linear constraint $h(x_N) \leq 0$ for which $h(0) > 0$ and $x_N \geq 0$. Assuming that the convex envelope of h exists in $x_N \geq 0$, then we can always construct a valid inequality. Indeed, by [38, Corollary 3], the convex envelope of h is tight at 0. Since an ϵ-subgradient[1] always exists for any $\epsilon > 0$ and $x \in \text{dom } h$ [14], an $\frac{h(0)}{2}$-subgradient, for instance, at 0 will separate it.

---

[1] An ϵ-subgradient of a convex function f at $y \in \text{dom } f$ is v such that $f(x) \geq f(y) - \epsilon + v^\mathsf{T}(x - y)$ for all $x \in \text{dom } f$.

Even when there is no convex underestimator for h, a valid cutting plane does exist. Continuity of h implies that $X = \{x_N \geq 0 : h(x_N) \leq 0\}$ is closed and [17, Lemma 2.1] ensures that $0 \notin \overline{conv}X$, thus, a valid inequality exists. We introduce a technique to construct such a valid inequality. The idea is to build a *concave underestimator* of h, h_{ave}, such that $h_{ave}(0) = h(0) > 0$. Then, $C = \{x_N : h_{ave}(x_N) \geq 0\}$ is an *S-free set*, that is, a convex set that does not contain any feasible point in its interior, and as such can be used to build an intersection cut (IC) [3,24,41].

First Contribution. In Sect. 3, we present a procedure to build concave underestimators for factorable functions that are tight at a given point. The procedure is similar to McCormick's method for constructing convex underestimators, and generalizes Proposition 3.2 and improves Proposition 3.3 of [26]. These underestimators can be used to build intersection cuts. We note that IC from a concave underestimator can generate cuts that cannot be generated by using the convex envelope. This should not be surprising, given that intersection cuts work at the feasible region level, while convex underestimators depend on the graph of the function. A simple example is $\{x \in [0,2] : -x^2 + 1 \leq 0\}$. When separating 0, the intersection cut gives $x \geq 1$, while the convex envelope over $[0,2]$ yields $x \geq 1/2$.

There are many differences between concave underestimators and convex ones. Maybe the most interesting one is that concave underestimators do not need bounded domains to exist. As an extreme example, $-x^2$ is a concave underestimator of itself, but a convex underestimator only exists if the domain of x is bounded. Even though this might be regarded as an advantage, it is also a problem. If concave underestimators are independent of the domain, then we cannot improve them when the domain shrinks.

Second Contribution. In Sect. 4, we propose a strengthening procedure that uses the bounds of the variables to enlarge the S-free set. Our procedure improves on the one used by Tuy [41].

Other techniques for strengthening IC have been proposed, such as, exploiting the integrality of the non-basic variables [6,19,20], improving the relaxation R [7,32,33] and computing the convex hull of $R\backslash C$ [8,17,23,36,37].

Third Contribution. By interpreting IC as disjunctive cuts [4], we extend monoidal strengthening to our setting [6] in Sect. 5. Although its applicability seems to be limited, we think it is of independent interest, especially for MILP.

2   Related Work

There have been many efforts on generalizing cutting planes from MILP to MINLP, we refer the reader to [31] and the references therein. In [31], the authors study how to compute $conv(R\backslash C)$ where R is not polyhedral, but C is a k-branch split. In practice, such sets C usually come from the integrality of the variables. Works that build sets C which do not come from integrality considerations include [9,11,21,22,34,35]. We refer to [12] and the references therein for more details. We would like to point out that the disjunctions built in [9,34,35] can be interpreted as piecewise linear concave underestimators. However, our approach is not suitable for disjunctive cuts built through cut generating LPs [5], since we generate infinite disjunctions, see Sect. 5, so we rely on the classical concept of intersection cuts where R is a translated simplicial cone.

Khamisov [26] studies functions $f : \mathbb{R}^n \to \mathbb{R}$, representable as $f(x) = \max_{y \in R} \varphi(x, y)$ where φ is continuous and concave on x. These functions allow for a concave underestimator at *every* point. He shows that this class of functions is very general, in particular, the class of functions representable as difference of convex functions is a strict subset of this class. He then proposes a procedure to build concave underestimators of composition of functions which is a special case of Theorem 1 below. He also suggests how to build an underestimator for the product of two functions over a compact domain. We simplify the construction for the product and no longer need a compact domain.

Although not directly related to this work, other papers that use underestimators other than convex are [15,16,25].

3   Concave Underestimators

In his seminal paper [29], McCormick proposed a method to build convex underestimators of *factorable* functions.

Definition 1. *Given a set of univariate functions $\mathcal{L}$, e.g., $\mathcal{L} = \{\cos, \cdot^n, \exp, \log, ...\}$, the set of* factorable functions $\mathcal{F}$ *is the smallest set that contains $\mathcal{L}$, the constant functions, and is closed under addition, product and composition.*

As an example, $e^{-(\cos(x^2) + xy/4)^2}$ is a factorable function for $\mathcal{L} = \{\cos, \exp\}$.

Given the inductive definition of factorable functions, to show a property about them one just needs to show that said property holds for all the functions in $\mathcal{L}$, constant functions, and that it is preserved by the product, addition and composition. For instance, McCormick [29] proves, constructively, that every factorable function admits a convex underestimator and a concave overestimator, by showing how to construct estimators for the sum, product and composition of two functions for which estimators are known.

An estimator for the sum of two functions is the sum of the estimators. For the product, McCormick uses the well-known McCormick inequalities. Less known is the way McCormick handles the composition $f(g(x))$. Let f_{vex} be a convex underestimator of f and $z_{\min} = \arg\min f_{vex}(z)$. Let g_{vex} be a convex underestimator of g and g^{ave} a concave overestimator. McCormick shows[2] that $f_{vex}(\text{mid}\{g_{vex}(x), g^{ave}(x), z_{\min}\})$ is a convex underestimator of $f(g(x))$, where $\text{mid}\{x, y, z\}$ is the median between x, y and z. It is well known that the optimum of a convex function over a closed interval is given by such a formula, thus

$$f_{vex}(\text{mid}\{g_{vex}(x), g^{ave}(x), z_{\min}\}) = \min\{f_{vex}(z) : z \in [g_{vex}(x), g^{ave}(x)]\},$$

see also [40].

---

[2] He actually leaves it as an exercise for the reader.

Definition 2. *Let* $\mathcal{X} \subseteq \mathbb{R}^n$ *be convex, and* $f : \mathcal{X} \to \mathbb{R}$ *be a function. We say that* $f_{ave} : \mathcal{X} \to \mathbb{R}$ *is a concave underestimator of* f *at* $\bar{x} \in \mathcal{X}$ *if* f_{ave} *is concave,* $f_{ave}(x) \le f(x)$ *for every* $x \in \mathcal{X}$ *and* $f_{ave}(\bar{x}) = f(\bar{x})$. *Similarly we define a convex overestimator of* f *at* $\bar{x} \in \mathcal{X}$.

Remark 1. For simplicity, we will consider only the case where $\mathcal{X} = \mathbb{R}^n$. This restriction leaves out some common functions like log. One possibility to include these function is to let the range of the function to be $\mathbb{R} \cup \{\pm\infty\}$. Then, $\log(x) = -\infty$ for $x \in \mathbb{R}_-$. Note that other functions like $\sqrt{x}$ can be handled by replacing them by a concave underestimator defined on all $\mathbb{R}$.

We now show that every factorable function admits a concave underestimator at a given point. Since the case for the addition is easy, we just need to specify how to build concave underestimators and convex overestimators for

– the product of two functions for which estimators are known,
– the composition $f(g(x))$ where estimators of f and g are known and f is univariate.

Theorem 1. *Let* $f : \mathbb{R} \to \mathbb{R}$ *and* $g : \mathbb{R}^n \to \mathbb{R}$. *Let* g_{ave}, f_{ave} *be, respectively, a concave underestimator of* g *at* $\bar{x}$ *and of* f *at* $g(\bar{x})$. *Further, let* g^{vex} *be a convex overestimator of* g *at* $\bar{x}$. *Then,* $h : \mathbb{R}^n \to \mathbb{R}$ *given by*

$$h(x) := \min\{f_{ave}(g_{ave}(x)), f_{ave}(g^{vex}(x))\},$$

is a concave underestimator of $f \circ g$ *at* $\bar{x}$.

Remark 2. The generalization of Theorem 1 to the case where f is multivariate in the spirit of [40] is straightforward.

The computation of a concave underestimator and convex overestimator of the product of two functions reduces to the computation of estimators for the square of a function through the polarization identity

$$4f(x)g(x) = (f(x) + g(x))^2 - (f(x) - g(x))^2.$$

Let $h : \mathbb{R}^n \to \mathbb{R}$ for which we know estimators $h_{vex} \le h \le h^{ave}$ at $\bar{x}$. From Theorem 1, a convex overestimator of h^2 at $\bar{x}$ is given by $\max\{h_{vex}^2, h^{ave2}\}$. On the other hand, a concave underestimator of h^2 at $\bar{x}$ can be constructed from the underestimator $h^2(x) \ge h^2(\bar{x}) + 2h(\bar{x})(h(x) - h(\bar{x}))$. From here we obtain

$$\begin{cases} 2h(\bar{x})h^{vex}(x) - h^2(\bar{x}), & \text{if } h(\bar{x}) \le 0 \\ 2h(\bar{x})h_{ave}(x) - h^2(\bar{x}), & \text{if } h(\bar{x}) > 0. \end{cases} \tag{2}$$

Example 1. Let us compute a concave underestimator of $f(x) = e^{-(\cos(x^2)+x/4)^2}$ at 0. Estimators of x^2 are given by $0 \le x^2 \le x^2$. For $\cos(x)$, estimators are $\cos(x) - x^2/2 \le \cos(x) \le 1$. Then, a concave underestimator of $\cos(x^2)$ is, according to Theorem 1, $\min\{\cos(0) - 0^2/2, \cos(x^2) - x^4/2\} = \cos(x^2) - x^4/2$. A convex overestimator is 1. Hence, $\cos(x^2) - x^4/2 + x/4 \le \cos(x^2) + x/4 \le 1 + x/4$.

Given that $-x^2$ is concave, a concave underestimator of $-(\cos(x^2)+x/4)^2$ is $\min\{-(\cos(x^2)-x^4/2+x/4)^2, -(1+x/4)^2\}$. To compute a convex overestimator of $-(\cos(x^2)+x/4)^2$, we compute a concave underestimator of $(\cos(x^2)+x/4)^2$. Since, $\cos(x^2)+x/4$ at 0 is 1, (2) yields $2(\cos(x^2)-x^4/2+x/4)-1$.

Finally, a concave underestimator of e^x at $x=-1$ is just its linearization, $e^{-1}+e^{-1}(x+1)$ and so $e^{-1}+e^{-1}(1+\min\{-(\cos(x^2)-x^4/2+x/4)^2, -(1+x/4)^2\})$ is a concave underestimator of $f(x)$. The intermediate estimators as well as the final concave underestimator are illustrated in Fig. 1.

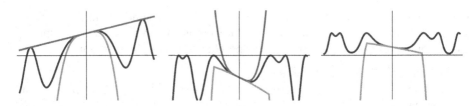

Fig. 1. Concave underestimator (orange) and convex overestimator (green) of $\cos(x^2)+x/4$ (left), $-(\cos(x^2)+x/4)^2$ (middle) and $f(x)$ (right) at $x=0$. (Color figure online)

For ease of exposition, in the rest of the paper we assume that the concave underestimator is differentiable. All results can be extended to the case where the functions are only sub- or super-differentiable.

4   Enlarging the S-free Sets by Using Bound Information

In Sect. 3, we showed how to build concave underestimators which give us S-free sets. Note that the construction does not make use of the bounds of the domain. We can exploit the bounds of the domain by the observation that the concave underestimator only needs to underestimate within the feasible region. However, to preserve the convexity of the S-free set, we must ensure that the underestimator is still concave.

Let $h(x) \leq 0$ be a constraint of (1), assume $x \in [l, u]$ and let h_{ave} be a concave underestimator of h. Throughout this section, $S = \{x \in [l, u] : h(x) \leq 0\}$. In order to construct a concave function $\hat{h}$ such that $\{x : \hat{h}(x) \geq 0\}$ contains $\{x : h_{ave}(x) \geq 0\}$, consider the following function

$$\hat{h}(x) = \min\{h_{ave}(z) + \nabla h_{ave}(z)^\top(x-z) : z \in [l, u], \ h_{ave}(z) \geq 0\}. \quad (3)$$

A similar function was already considered by Tuy [41]. The only difference is that Tuy's strengthening does not use the restriction $h_{ave}(z) \geq 0$, see Fig. 2.

Proposition 1. *Let h_{ave} be a concave underestimator of h at $\bar{x} \in [l, u]$, such that $h(\bar{x}) > 0$. Define $\hat{h}$ as in (3). Then, the set $C = \{x : \hat{h}(x) \geq 0\}$ is a convex S-free set and $C \supseteq \{x : h_{ave}(x) \geq 0\}$.*

In general, evaluating $\hat{h}$ is a difficult problem and there is no closed form formula. However, when h_{ave} is quadratic, the problem in the right hand side of (3) is convex and a cut could be strengthen in polynomial time.

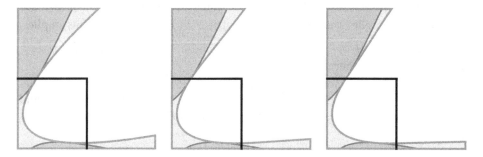

Fig. 2. Feasible region $\{x, y \in [0, 2] : h(x, y) \le 0\}$, where $h = x^2 - 2y^2 + 4xy - 3x + 2y + 1$, in blue together with $h_{ave}(x, y) \le 0$ at $\bar{x} = (1, 1)$ (left), Tuy's strengthening (middle) and $\hat{h} \le 0$ (right) in orange. Region shown is $[0, 4]^2$, $[0, 2]^2$ is bounded by black lines. The difference between the S-free sets can be seen on the top of the picture. (Color figure online)

5   "Monoidal" Strengthening

We show how to strengthen cuts from reverse convex constraints when exactly *one* non-basic variable is integer. Our technique is based on monoidal strengthening applied to disjunctive cuts, see Lemma 1 and the discussion following it. If more than one variable is integer, we can generate one cut per integer variable, relaxing the integrality of all but one variable at a time. However, under some conditions (see Remark 6), we can exploit the integrality of several variables at the same time. Our exposition of the monoidal strengthening technique is slightly different from [6] and is inspired by [43, Section 4.2.3].

Throughout this section, we assume that we already have a concave underestimator, and that we have performed the change of variables described in the introduction. Therefore, we consider the constraint $\{x \in [0, u] : h(x) \le 0\}$ where $h : \mathbb{R}^n \to \mathbb{R}$ is concave and $h(0) > 0$. Let $Y = \{y \in [0, u] : h(y) = 0\}$. The convex S-free set $C = \{x \in [0, u] : h(x) \ge 0\}$ can be written as

$$C = \bigcap_{y \in Y} \{x \in [0, u] : \nabla h(y)^\mathsf{T} x \ge \nabla h(y)^\mathsf{T} y\}.$$

The concavity of h implies that $h(0) \le h(y) - \nabla h(y)^\mathsf{T} y$ for all y in the domain of h. In particular, if $y \in Y$, then $\nabla h(y)^\mathsf{T} y \le -h(0) < 0$. Since all feasible points satisfy $h(x) \le 0$, they must satisfy the infinite disjunction

$$\bigvee_{y \in Y} \frac{\nabla h(y)^\mathsf{T}}{\nabla h(y)^\mathsf{T} y} x \ge 1. \tag{4}$$

The maximum principle [4] implies that with

$$\alpha_j = \max_{y \in Y} \frac{\partial_j h(y)}{\nabla h(y)^\mathsf{T} y}, \tag{5}$$

the cut $\sum_j \alpha_j x_j \geq 1$ is valid. We remark that the maximum exists, since the concavity of h implies that for $y \in Y$, $h(e_j) \leq \partial_j h(y) - \nabla h(y)^\mathsf{T} y$. This implies, together with $\nabla h(y)^\mathsf{T} y \leq -h(0) < 0$, that $\frac{\partial_j h(y)}{\nabla h(y)^\mathsf{T} y} \leq 1 + \frac{h(e_j)}{\nabla h(y)^\mathsf{T} y}$. If $h(e_j) \geq 0$, then $\frac{\partial_j h(y)}{\nabla h(y)^\mathsf{T} y} \leq 1$. Otherwise, $\frac{\partial_j h(y)}{\nabla h(y)^\mathsf{T} y} \leq 1 - \frac{h(e_j)}{h(0)}$.

The application of monoidal strengthening [6, Theorem 3] to a valid disjunction $\bigvee_i \alpha^i x \geq 1$ requires the existence of bounds β_i such that $\alpha^i x \geq \beta_i$ is valid for every feasible point. Let $\beta(y)$ be such a bound for (4). An example of $\beta(y)$ is

$$\beta(y) = \min_{x \in [0,u]} \frac{\nabla h(y)^\mathsf{T} x}{\nabla h(y)^\mathsf{T} y}.$$

Remark 3. If $\beta(y) \geq 1$, then $\nabla h(y)^\mathsf{T} x / \nabla h(y)^\mathsf{T} y \geq 1$ is redundant and can be removed from (4). Therefore, we can assume without loss of generality that $\beta(y) < 1$.

The strengthening derives from the fact that a new disjunction can be obtained from (4) and, with it, a new disjunctive cut. The disjunction on the following Lemma is trivially satisfied, but provides the basis for building non-trivial new disjunctions.

Lemma 1. *Every $x \geq 0$ that satisfies (4), also satisfies*

$$\bigvee_{y \in Y} \frac{\nabla h(y)^\mathsf{T} x}{\nabla h(y)^\mathsf{T} y} + z(y)(1 - \beta(y)) \geq 1, \tag{6}$$

where $z \colon Y \to \mathbb{Z}$ is such that $z \equiv 0$ or there is a $y_0 \in Y$ for which $z(y_0) > 0$.

Remark 4. Even if some disjunctive terms have no lower bound, that is, $\beta(y) = -\infty$ for $y \in Y' \subseteq Y$, Lemma 1 still holds if, additionally, $z(y) = 0$ for all $y \in Y'$. This means that we are not using that disjunction for the strengthening. In particular, if for some variable x_j, α_j is defined by some $y \in Y'$, then this cut coefficient cannot be improved.

Assume now that $x_k \in \mathbb{Z}$ for every $k \in K \subseteq \{1, \ldots, n\}$. One way of constructing a new disjunction is to find a set of functions M such that for any choice of $m^k \in M$ and any feasible assignment of x_k, $z(y) := \sum_{k \in K} x_k m^k(y)$ satisfies the conditions of Lemma 1, that is, z is in

$$Z = \{z \colon Y \to \mathbb{Z} : z \equiv 0 \vee \exists y \in Y, z(y) > 0\}.$$

Once such a family of functions has been identified, the cut $\sum_j \gamma_j x_j \geq 1$ with $\gamma_j = \alpha_j$ if $j \notin K$, and

$$\gamma_k = \inf_{m \in M} \max_{y \in Y} \frac{\partial_k h(y)}{\nabla h(y)^\mathsf{T} y} + m(y)(1 - \beta(y)) \quad \text{for } k \in K, \tag{7}$$

is valid and at least as strong as (5). Any $M \subseteq Z$ such that $(M, +)$ is a monoid, that is, $0 \in M$ and M is closed under addition can be used in (7).

Remark 5. This is exactly what is happening in [6, Theorem 3]. Indeed, in the finite case, that is, when Y is finite, Balas and Jeroslow considered $M = \{m \in \mathbb{Z}^Y : \sum_{y \in Y} m_y \geq 0\}$. Clearly, $(M, +)$ is a monoid and $M \subseteq Z$. Therefore, Lemma 1 implies that $\bigvee_{y \in Y} \alpha^y x + \sum_k m_y^k x_k (1 - \beta_y) \geq 1$ is valid for any choice of $m^k \in M$, which in turn implies [6, Theorem 3].
For an application that uses a different monoid see [2].

The question that remains is how to choose M. For example, the monoid $M = \{m : Y \to \mathbb{Z} : m$ has finite support and $\sum_{y \in Y} m(y) \geq 0\}$ is an obvious candidate for M. However, the problem is how to optimize over such an M, see (7).

We circumvent this problem by considering only one integer variable at a time. Fix $k \in K$. In this setting we can use Z as M, which is *not* a monoid. Indeed, if $z \in Z$, then $x_k z \in Z$ for any $x_k \in \mathbb{Z}_+$. The advantage of using Z is that the solution of (7) is easy to characterize.

With $M = Z$, the cut coefficients (7) of all variables are the same as (5) except for x_k. The cut coefficient of x_k is given by

$$\inf_{z \in Z} \max_{y \in Y} \frac{\partial_k h(y)}{\nabla h(y)^\mathsf{T} y} + z(y)(1 - \beta(y)).$$

To compute this coefficient, observe that one would like to have $z(y) < 0$ for points y such that the objective function of (5) is large. However, z must be positive for at least one point. Therefore,

$$\min_{y \in Y} \frac{\partial_k h(y)}{\nabla h(y)^\mathsf{T} y} + (1 - \beta(y))$$

is the best coefficient we can hope for if $z \not\equiv 0$. This coefficient can be achieved by

$$z(y) = \begin{cases} 1, & \text{if } y \in \arg\min_{y \in Y} \frac{\partial_k h(y)}{\nabla h(y)^\mathsf{T} y} + (1 - \beta(y)), \\ -L, & \text{otherwise} \end{cases} \tag{8}$$

where $L > 0$ is sufficiently large.

Summarizing, we can obtain the following cut:

$$\alpha_j = \begin{cases} \max_{y \in Y} \frac{\partial_j h(y)}{\nabla h(y)^\mathsf{T} y} & \text{if } j \neq k \\ \min\{\max_{y \in Y} \frac{\partial_j h(y)}{\nabla h(y)^\mathsf{T} y}, \min_{y \in Y} \frac{\partial_j h(y)}{\nabla h(y)^\mathsf{T} y} + (1 - \beta(y))\} & \text{if } j = k \end{cases} \tag{9}$$

Remark 6. Let $z^k \in Z$ be given by (8) for each $k \in K$. Assume there is a subset $K_0 \subseteq K$ and a monoid $M \subseteq Z$ such that $z^k \in M$ for every $k \in K_0$. Then, the strengthening can be applied to all x_k for $k \in K_0$.

Alternatively, if there is a constraint enforcing that at most one of the x_k can be non-zero for $k \in K_0$, e.g., $\sum_{k \in K} x_k \leq 1$, then the strengthening can be applied to all x_k for $k \in K_0$.

In the finite case, our application of monoidal strengthening would be dominated by the original technique of [6] by using an appropriate monoid. However, in the presence of extra constraint, such as the one described above, our technique can dominate vanilla monoidal strengthening.

Example 2. Consider the constraint $\{x \in \{0,1,2\} \times [0,5] : h(x) \leq 0\}$, where $h(x_1,x_2) = -10x_1^2 - 1/2x_2^2 + 2x_1x_2 + 4$, see Fig. 3. The IC is given by $\sqrt{5/2}x_1 + 1/(2\sqrt{2})x_2 \geq 1$. Note that $(1/\sqrt{10}, \sqrt{10}) \in Y$ and yields the term $1/\sqrt{10}x_2 \geq 1$ in (4). Since $x_2 \geq 0$, $\beta(1/\sqrt{10}, \sqrt{10}) = 0$. Hence, (9) yields $\alpha_1 \leq \min\{\sqrt{5/2}, 1\} = 1$ and the strengthened inequality is $x_1 + 1/(2\sqrt{2})x_2 \geq 1$.

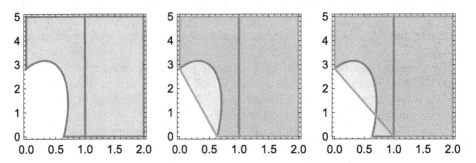

Fig. 3. The feasible region $\{x \in \{0,1,2\} \times [0,5] : h(x) \leq 0\}$ from Example 2 (left), the IC (middle), and the strengthened cut (right).

6   Conclusions

We have introduced a procedure to generate concave underestimators of factorable functions, which can be used to generate intersection cuts, together with two strengthening procedures.

It remains to be seen the practical performance of these intersection cuts. We expect that its generation is cheaper than the generation of disjunctive cuts, given that there is no need to solve an LP. As for the strengthening procedures, they might be too expensive to be of practical use. An alternative is to construct a polyhedral inner approximation of the S-free set and use monoidal strengthening in the finite setting. However, in this case, the strengthening proposed in Sect. 4 has no effect. Nonetheless, as far as the author knows, this has been the first application of monoidal strengthening that is able to exploit further problem structure such as demonstrated in Remark 6 and it might be interesting to investigate further.

Identification of when the proposed concave underestimators generate maximal S-free sets is the subject of current investigation.

Acknowledgments. The author would like to thank Stefan Vigerske, Franziska Schlösser, Sven Wiese, Ambros Gleixner, Dan Steffy and Juan Pablo Vielma for helpful discussions, and Leon Eifler, Daniel Rehfeldt for comments that improved the manuscript. He would also like to thank three anonymous reviewers for valuable comments.

A   Proofs

A.1   Proof of Theorem 1

Proof. Clearly, $h(\bar{x}) = f(g(\bar{x}))$.

To establish $h(x) \leq f(g(x))$, notice that

$$h(x) = \min\{f_{ave}(z) : g_{ave}(x) \leq z \leq g^{vex}(x)\}. \qquad (10)$$

Since $z = g(x)$ is a feasible solution and f_{ave} is an underestimator of f, we obtain that $h(x) \leq f(g(x))$.

Now, let us prove that h is concave. To this end, we again use the representation (10). To simplify notation, we write g_1, g_2 for g_{ave}, g^{vex}, respectively. We prove concavity by definition, that is,

$$h(\lambda x_1 + (1 - \lambda)x_2) \geq \lambda h(x_1) + (1 - \lambda)h(x_2), \text{ for } \lambda \in [0, 1].$$

Let

$$I = [g_1(\lambda x_1 + (1 - \lambda)x_2), g_2(\lambda x_1 + (1 - \lambda)x_2)]$$
$$J = [\lambda g_1(x_1) + (1 - \lambda)g_1(x_2), \lambda g_2(x_1) + (1 - \lambda)g_2(x_2)].$$

By the concavity of g_1 and convexity of g_2 we have $I \subseteq J$. Therefore,

$$h(\lambda x_1 + (1 - \lambda)x_2) = \min\{f_{ave}(z) : z \in I\} \geq \min\{f_{ave}(z) : z \in J\}.$$

Since f_{ave} is concave, the minimum is achieved at the boundary,

$$\min\{f_{ave}(z) : z \in J\} = \min_{i \in \{1,2\}} f_{ave}(\lambda g_i(x_1) + (1 - \lambda)g_i(x_2)).$$

Furthermore, $f_{ave}(\lambda g_i(x_1) + (1 - \lambda)g_i(x_2)) \geq \lambda f_{ave}(g_i(x_1)) + (1 - \lambda)f_{ave}(g_i(x_2))$ which implies that

$$h(\lambda x_1 + (1 - \lambda)x_2) \geq \min_{i \in \{1,2\}} \lambda f_{ave}(g_i(x_1)) + (1 - \lambda)f_{ave}(g_i(x_2))$$
$$\geq \min_{i \in \{1,2\}} \lambda f_{ave}(g_i(x_1)) + \min_{i \in \{1,2\}} (1 - \lambda)f_{ave}(g_i(x_2))$$
$$= \lambda h(x_1) + (1 - \lambda)h(x_2),$$

as we wanted to show.

A.2   Proof of Proposition 1

Proof. The function $\hat{h}$ is concave since it is the minimum of linear functions. This establishes the convexity of C.

To show that $C \supseteq \{x : h_{ave}(x) \geq 0\}$, notice that $h_{ave}(x) = \min_z h_{ave}(z) + \nabla h_{ave}(z)^{\mathsf{T}}(x - z)$. The inclusion follows from observing that the objective function in the definition of $\hat{h}(x)$ is the same as above, but over a smaller domain.

To show that it is S-free, we will show that for every $x \in [l, u]$ such that $h(x) \leq 0$, $\hat{h}(x) \leq 0$.

Let $x_0 \in [l, u]$ such that $h(x_0) \leq 0$. Since h_{ave} is a concave underestimator at $\bar{x}$, $h_{ave}(\bar{x}) > 0$ and $h_{ave}(x_0) \leq 0$. If $h_{ave}(x_0) = 0$, then, by definition, $\hat{h}(x_0) \leq h_{ave}(x_0) = 0$ and we are done. We assume, therefore, that $h_{ave}(x_0) < 0$.

Consider $g(\lambda) = h_{ave}(\bar{x}+\lambda(x_0-\bar{x}))$ and let $\lambda_1 \in (0,1)$ be such that $g(\lambda_1) = 0$. The existence of λ_1 is justified by the continuity of g, $g(0) > 0$ and $g(1) < 0$. Equivalently, $x_1 = \bar{x} + \lambda_1(x_0 - \bar{x})$ is the intersection point between the segment joining x_0 with $\bar{x}$ and $\{x : h_{ave}(x) = 0\}$. The linearization of g at λ_1 evaluated at $\lambda = 1$ is negative, because g is concave, and equals $h_{ave}(x_1) + \nabla h_{ave}(x_1)^T(x_0 - x_1)$. Finally, given that $x_1 \in [l, u]$ and $h_{ave}(x_1) = 0$, x_1 is feasible for (3) and we conclude that $\hat{h}(x_0) < 0$.

A.3 Proof of Lemma 1

Proof. If $z \equiv 0$, then (6) reduces to (4).

Otherwise, let $y_0 \in Y$ such that $z(y_0) > 0$, that is, $z(y_0) \geq 1$. By Remark 3, for every $y \in Y$, it holds $1 - \beta(y) > 0$, and so

$$z(y_0)(1 - \beta(y_0)) \geq 1 - \beta(y_0).$$

Therefore, $\beta(y_0) \geq 1 - z(y_0)(1 - \beta(y_0))$. Since every $x \geq 0$ satisfying (4) satisfies $\frac{\nabla h(y_0)^T x}{\nabla h(y_0)^T y_0} \geq \beta(y_0)$, we conclude that $\frac{\nabla h(y_0)^T x}{\nabla h(y_0)^T y_0} + z(y_0)(1 - \beta(y_0)) \geq 1$ holds.

References

1. Achterberg, T., Wunderling, R.: Mixed integer programming: analyzing 12 years of progress. In: Jünger, M., Reinelt, G. (eds.) Facets of Combinatorial Optimization, pp. 449–481. Springer, Heidelberg (2013). https://doi.org/10.1007/978-3-642-38189-8_18
2. Balas, E., Qualizza, A.: Monoidal cut strengthening revisited. Discrete Optim. **9**(1), 40–49 (2012). https://doi.org/10.1016/j.disopt.2011.11.002
3. Balas, E.: Intersection cuts–a new type of cutting planes for integer programming. Oper. Res. **19**(1), 19–39 (1971). https://doi.org/10.1287/opre.19.1.19
4. Balas, E.: Disjunctive programming. In: Discrete Optimization II, Proceedings of the Advanced Research Institute on Discrete Optimization and Systems Applications of the Systems Science Panel of NATO and of the Discrete Optimization Symposium co-sponsored by IBM Canada and SIAM Banff, Aha. and Vancouver, pp. 3–51. Elsevier BV (1979). https://doi.org/10.1016/s0167-5060(08)70342-x
5. Balas, E., Ceria, S., Cornuéjols, G.: A lift-and-project cutting plane algorithm for mixed 0–1 programs. Math. Program. **58**(1–3), 295–324 (1993). https://doi.org/10.1007/bf01581273
6. Balas, E., Jeroslow, R.G.: Strengthening cuts for mixed integer programs. Eur. J. Oper. Res. **4**(4), 224–234 (1980). https://doi.org/10.1016/0377-2217(80)90106-x

7. Balas, E., Margot, F.: Generalized intersection cuts and a new cut generating paradigm. Math. Program. **137**(1–2), 19–35 (2011). https://doi.org/10.1007/s10107-011-0483-x
8. Basu, A., Cornuéjols, G., Zambelli, G.: Convex sets and minimal sublinear functions. J. Convex Anal. **18**(2), 427–432 (2011)
9. Belotti, P.: Disjunctive cuts for nonconvex MINLP. In: Lee, J., Leyffer, S. (eds.) Mixed Integer Nonlinear Programming, vol. 154, pp. 117–144. Springer, New York (2011). https://doi.org/10.1007/978-1-4614-1927-3_5
10. Belotti, P., Lee, J., Liberti, L., Margot, F., Wchter, A.: Branching and bounds tightening techniques for non-convex MINLP. Optim. Methods Softw. **24**(4–5), 597–634 (2009). https://doi.org/10.1080/10556780903087124
11. Bienstock, D., Chen, C., Muñoz, G.: Outer-product-free sets for polynomial optimization and oracle-based cuts. http://arxiv.org/abs/1610.04604
12. Bonami, P., Linderoth, J., Lodi, A.: Disjunctive cuts for mixed integer nonlinear programming problems. Prog. Comb. Optim. **18**, 521–541 (2011)
13. Boukouvala, F., Misener, R., Floudas, C.A.: Global optimization advances in mixed-integer nonlinear programming, MINLP, and constrained derivative-free optimization, CDFO. Eur. J. Oper. Res. **252**(3), 701–727 (2016). https://doi.org/10.1016/j.ejor.2015.12.018
14. Brondsted, A., Rockafellar, R.T.: On the subdifferentiability of convex functions. Proc. Am. Math. Soc. **16**(4), 605 (1965). https://doi.org/10.2307/2033889
15. Buchheim, C., D'Ambrosio, C.: Monomial-wise optimal separable underestimators for mixed-integer polynomial optimization. J. Glob. Optim. **67**(4), 759–786 (2016). https://doi.org/10.1007/s10898-016-0443-3
16. Buchheim, C., Traversi, E.: Separable non-convex underestimators for binary quadratic programming. In: Bonifaci, V., Demetrescu, C., Marchetti-Spaccamela, A. (eds.) SEA 2013. LNCS, vol. 7933, pp. 236–247. Springer, Heidelberg (2013). https://doi.org/10.1007/978-3-642-38527-8_22
17. Conforti, M., Cornuéjols, G., Daniilidis, A., Lemaréchal, C., Malick, J.: Cut-generating functions and S-free sets. Math. Oper. Res. **40**(2), 276–391 (2015). https://doi.org/10.1287/moor.2014.0670
18. Conforti, M., Cornuéjols, G., Zambelli, G.: Corner polyhedron and intersection cuts. Surv. Oper. Res. Manag. Sci. **16**(2), 105–120 (2011). https://doi.org/10.1016/j.sorms.2011.03.001
19. Conforti, M., Cornuéjols, G., Zambelli, G.: A geometric perspective on lifting. Oper. Res. **59**(3), 569–577 (2011). https://doi.org/10.1287/opre.1110.0916
20. Dey, S.S., Wolsey, L.A.: Two row mixed-integer cuts via lifting. Math. Program. **124**(1–2), 143–174 (2010). https://doi.org/10.1007/s10107-010-0362-x
21. Fischetti, M., Ljubić, I., Monaci, M., Sinnl, M.: Intersection cuts for bilevel optimization. In: Louveaux, Q., Skutella, M. (eds.) IPCO 2016. LNCS, vol. 9682, pp. 77–88. Springer, Cham (2016). https://doi.org/10.1007/978-3-319-33461-5_7
22. Fischetti, M., Ljubić, I., Monaci, M., Sinnl, M.: A new general-purpose algorithm for mixed-integer bilevel linear programs. Oper. Res. **65**(6), 1615–1637 (2017). https://doi.org/10.1287/opre.2017.1650
23. Glover, F.: Polyhedral convexity cuts and negative edge extensions. Zeitschrift fr Oper. Res. **18**(5), 181–186 (1974). https://doi.org/10.1007/bf02026599
24. Glover, F.: Convexity cuts and cut search. Oper. Res. **21**(1), 123–134 (1973). https://doi.org/10.1287/opre.21.1.123
25. Hasan, M.M.F.: An edge-concave underestimator for the global optimization oftwice-differentiable nonconvex problems. J. Glob. Optim. **71**(4), 735–752 (2018). https://doi.org/10.1007/s10898-018-0646-x

26. Khamisov, O.: On optimization properties of functions, with a concave minorant. J. Glob. Optim. **14**(1), 79–101 (1999). https://doi.org/10.1023/a:1008321729949
27. Kılınç, M.R., Sahinidis, N.V.: Exploiting integrality in the global optimization of mixed-integer nonlinear programming problems with BARON. Optim. Methods Softw. **33**(3), 540–562 (2017). https://doi.org/10.1080/10556788.2017.1350178
28. Lin, Y., Schrage, L.: The global solver in the LINDO API. Optim. Methods Softw. **24**(4–5), 657–668 (2009). https://doi.org/10.1080/10556780902753221
29. McCormick, G.P.: Computability of global solutions to factorable nonconvex programs: part i – convex underestimating problems. Math. Program. **10**(1), 147–175 (1976). https://doi.org/10.1007/bf01580665
30. Misener, R., Floudas, C.A.: ANTIGONE: algorithms for coNTinuous/integerglobal optimization of nonlinear equations. J. Glob. Optim. **59**(2–3), 503–526 (2014). https://doi.org/10.1007/s10898-014-0166-2
31. Modaresi, S., Kılınç, M.R., Vielma, J.P.: Intersection cuts fornonlinear integer programming: convexification techniques for structured sets. Math. Program. **155**(1–2), 575–611 (2015). https://doi.org/10.1007/s10107-015-0866-5
32. Porembski, M.: How to extend the concept of convexity cuts to derive deeper cutting planes. J. Glob. Optim. **15**(4), 371–404 (1999). https://doi.org/10.1023/a:1008315229750
33. Porembski, M.: Finitely convergent cutting planes for concave minimization. J. Glob. Optim. **20**(2), 109–132 (2001). https://doi.org/10.1023/a:1011240309783
34. Saxena, A., Bonami, P., Lee, J.: Convex relaxations of non-convex mixed integer quadratically constrained programs: extended formulations. Math. Program. **124**(1–2), 383–411 (2010). https://doi.org/10.1007/s10107-010-0371-9
35. Saxena, A., Bonami, P., Lee, J.: Convex relaxations of non-convex mixed integer quadratically constrained programs: projected formulations. Math. Program. **130**(2), 359–413 (2010). https://doi.org/10.1007/s10107-010-0340-3
36. Sen, S., Sherali, H.D.: Facet inequalities from simple disjunctions in cutting plane theory. Math. Program. **34**(1), 72–83 (1986). https://doi.org/10.1007/bf01582164
37. Sen, S., Sherali, H.D.: Nondifferentiable reverse convex programs and facetial convexity cuts via a disjunctive characterization. Math. Program. **37**(2), 169–183 (1987). https://doi.org/10.1007/bf02591693
38. Tawarmalani, M., Sahinidis, N.V.: Convex extensions and envelopes of lower semicontinuous functions. Math. Program. **93**(2), 247–263 (2002). https://doi.org/10.1007/s10107-002-0308-z
39. Tawarmalani, M., Sahinidis, N.V.: A polyhedral branch-and-cut approach to global optimization. Math. Program. **103**(2), 225–249 (2005). https://doi.org/10.1007/s10107-005-0581-8
40. Tsoukalas, A., Mitsos, A.: Multivariate McCormick relaxations. J. Glob. Optim. **59**(2–3), 633–662 (2014). https://doi.org/10.1007/s10898-014-0176-0
41. Tuy, H.: Concave programming with linear constraints. In: Doklady Akademii Nauk, vol. 159, pp. 32–35. Russian Academy of Sciences (1964)
42. Vigerske, S., Gleixner, A.: SCIP: global optimization of mixed-integer nonlinear programs in a branch-and-cut framework. Optim. Methods Softw. **33**(3), 563–593 (2017). https://doi.org/10.1080/10556788.2017.1335312
43. Wiese, S.: On the interplay of mixed integer linear, mixed integer nonlinear and constraint programming (2016). https://doi.org/10.6092/unibo/amsdottorato/7612

Linear Programming Using Limited-Precision Oracles

Ambros Gleixner[1] and Daniel E. Steffy[2]

[1] Konrad-Zuse-Zentrum für Informationstechnik Berlin,
Takustr. 7, 14195 Berlin, Germany
gleixner@zib.de
[2] Mathematics and Statistics, Oakland University, Rochester, MI, USA
steffy@oakland.edu

Abstract. Linear programming is a foundational tool for many aspects of integer and combinatorial optimization. This work studies the complexity of solving linear programs exactly over the rational numbers through use of an oracle capable of returning limited-precision LP solutions. Under mild assumptions, it is shown that a polynomial number of calls to such an oracle and a polynomial number of bit operations, is sufficient to compute an exact solution to an LP. Previous work has often considered oracles that provide solutions of an arbitrary specified precision. While this leads to polynomial-time algorithms, the level of precision required is often unrealistic for practical computation. In contrast, our work provides a foundation for understanding and analyzing the behavior of the methods that are currently most effective in practice for solving LPs exactly.

Keywords: Linear programming · Oracle complexity · Diophantine approximation · Exact solutions · Symbolic computation · Rational arithmetic · Extended-precision arithmetic · Iterative refinement

1 Introduction

This paper studies algorithms for solving linear programs (LPs) exactly over the rational numbers. The focus lies on methods that employ a limited-precision LP oracle—an oracle that is capable of providing approximate primal-dual solutions. We assume that the reader is familiar with fundamental results and notation related to linear optimization, such as those presented in [15,22]. We define the *encoding length* or *size* of an integer $n \in \mathbb{Z}$ as $\langle n \rangle := 1 + \lceil \log(|n| + 1) \rceil$, using base-two logarithms throughout the paper. For a rational number p/q, $\langle p/q \rangle := \langle p \rangle + \langle q \rangle$. Encoding lengths of vectors and matrices are defined as the sums of the encoding lengths of their entries. To clearly distinguish between the

The first author was supported by the Research Campus MODAL funded by the German Ministry of Education and Research under grant number 05M14ZAM.

A. Lodi and V. Nagarajan (Eds.): IPCO 2019, LNCS 11480, pp. 399–412, 2019.
https://doi.org/10.1007/978-3-030-17953-3_30

size and the value of numbers, we will often explicitly use the term *value* when referring to the numeric value taken by numbers.

In order to compute exact rational solutions to linear programs, many algorithms first compute sufficiently accurate approximate solutions and then exploit techniques to convert these solutions to exact rational solutions such as the following, related to the *Diophantine approximation problem*.

Theorem 1 ([22], **Cor. 6.3b**). *For* $\alpha \in \mathbb{Q}$, $M > 0$ *there exists at most one rational number* p/q *such that* $|p/q - \alpha| < 1/(2Mq)$ *and* $1 \leqslant q \leqslant M$. *There exists a polynomial-time algorithm to test whether this number exists and, if so, to compute this number.*

The underlying algorithm is essentially the extended Euclidean algorithm. Its running time is polynomial in the size of α and M, and is very fast in practice. Hence, if an approximation α of an unknown rational number p/q can be computed with error at most $1/(2M^2)$, where M bounds the unknown denominator q, then the exact value of p/q can be recovered efficiently; we refer to this process as *rational reconstruction*. For LPs it is well known that a priori bounds on the denominators of the entries of any basic primal-dual solution can be computed from Cramer's rule and Hadamard's inequality [22]. Therefore, upon computing a sufficiently accurate approximation of an optimal basic solution, Theorem 1 can be applied componentwise to recover the exact solution.

An alternative, computationally more expensive technique for reconstructing rational vectors that works under milder assumptions, is based on polynomial-time *lattice reduction* algorithms as pioneered by [20]. It is often referred to as *simultaneous Diophantine approximation* and has notably been used by [15] to develop polynomial-time algorithms to solve LPs exactly over the rational numbers. Their algorithm relies on the ellipsoid method [17] to compute highly accurate approximate solutions which are used to reconstruct exact rational solutions. Unfortunately, the levels of precision prescribed when computing the approximate solution would be prohibitively expensive to use in practice.

Example 1. Consider the following small and unremarkable LP.

$$
\begin{aligned}
\max \quad & 2x_1 + 3x_2 + 2x_3 + x_4 + 2x_5 - x_6 \\
\text{s.t.} \quad & x_1 + x_2 + 2x_3 + 3x_4 + x_5 && \leqslant 3 \\
& x_1 - x_2 + x_4 + 3x_5 - 2x_6 && \leqslant 2 \\
& x_1 + 2x_2 + x_3 + 3x_4 + x_6 && \leqslant 4 \\
& x_1, x_2, x_3, x_4, x_5, x_6 \geqslant 0
\end{aligned}
$$

According to [15, Theorem 6.3.2], an exact rational solution to an LP can be found by first calling a weak optimization oracle to find an approximate solution followed by simultaneous Diophantine approximation. The tolerance stated for this purpose is $\varepsilon = \frac{2^{-18n\langle c\rangle - 24n^4\varphi}}{\|c\|_\infty}$, where c is the objective vector and φ is the facet complexity (an upper bound on the encoding length of inequalities sufficient to describe the polyhedron, see [15]). For the above problem, $\varepsilon \approx 10^{-169,059}$.

For problems of practical interest, ε will be even smaller. While ε is suitable for establishing polynomial-time algorithms, it may be far beyond what is feasible for real computations in practice. Other theoretically motivated methods based on interior point methods [6,16,21] suffer from similar limitations.

By contrast, the largest encoding length of any vertex of the above example is merely 27 and the largest denominator across all vertices is 8. Thus, a solution with componentwise difference from a vertex under $1/128$ would be sufficient to apply Theorem 1 to recover the vertex. Also in general, most LPs of practical interest are highly sparse and may have other special characteristics that result in their solutions having encoding length dramatically smaller than the value of derivable worst-case bounds on these values. In this light, one main contribution of this paper is an *output-sensitive* algorithm with running time depending not only on input length, but also on the size of the output, see Sect. 4.

Most algorithms used in practice today for computing exact rational solutions to LPs are based on the simplex method. Directly applying the simplex method in rational arithmetic is often prohibitively expensive due to the cost of arithmetic on rational numbers that can grow very large during the intermediate computations. Methods have been developed to replace these rational computations by integer arithmetic [4,10–12]. However, most successful in practice today is the combined use of floating-point and exact computation [3,8,13,18,19]. For example, the QSopt_ex solver [3] applies the simplex method in floating-point arithmetic and then recomputes the final primal-dual solution from the returned basis using exact arithmetic. If the primal-dual solution is not optimal or feasible, further simplex pivots are executed at increasingly higher levels of precision until an optimal solution is found. This *incremental precision boosting* is often very effective, but can become slow in cases where many extended-precision simplex pivots are needed. A further contribution of this paper is a significant improvement of this approach for which double-precision floating-point pivots suffice, see Sect. 3. The foundation is the recent *iterative refinement* method for linear programming, that has proven effective for computing high-accuracy solutions to LPs [14]. Its properties and necessary adjustments are described in Sect. 2. Sect. 5 analyzes the computational benefits of the new methods in practice.

2   Iterative Refinement with Limited-Precision Oracles

Our starting point is the iterative refinement method proposed in [14], which uses calls to a limited-precision LP solver in order to generate a sequence of increasingly accurate solutions. In the following we give a precise definition of a limited-precision LP oracle which is necessary to evaluate the behavior of the algorithms defined in this paper. It is also helpful to introduce the set $\mathbb{F}(p) := \{n/2^p \in \mathbb{Q} : n \in \mathbb{Z}, |n| \leqslant 2^{2p}\}$ for some fixed $p \in \mathbb{N}$; this can be viewed as a superset of floating-point numbers, that is easier to handle in the subsequent proofs. Double-precision floating-point numbers, for example, are contained in $\mathbb{F}(1074)$.

Definition 1. *We call an oracle a* limited-precision LP oracle *if there exist constants* $p \in \mathbb{N}$, $0 < \eta < 1$, *and* $\sigma > 0$ *such that for any LP*

$$\min\{c^T x : Ax = b, x \geqslant \ell\} \qquad (P)$$

with $A \in \mathbb{Q}^{m \times n}$, $b \in \mathbb{Q}^m$, *and* $c, \ell \in \mathbb{Q}^n$, *the oracle either reports a "failure" or returns an approximate primal–dual solution* $\bar{x} \in \mathbb{F}(p)^n$, $\bar{y} \in \mathbb{F}(p)^m$ *that satisfies*

$$\|A\bar{x} - b\|_\infty \leqslant \eta, \qquad (1a)$$
$$\bar{x} \geqslant \ell - \eta \mathbb{1}, \qquad (1b)$$
$$c - A^T \bar{y} \geqslant -\eta \mathbb{1}, \qquad (1c)$$
$$|(\bar{x} - \ell)^T (c - A^T \bar{y})| \leqslant \sigma, \qquad (1d)$$

when it is given the LP $\min\{\bar{c}^T x : \bar{A}x = \bar{b}, x \geqslant \bar{\ell}\}$, *where* $\bar{A} \in \mathbb{Q}^{m \times n}$, $\bar{c}, \bar{\ell} \in \mathbb{Q}^n$, *and* $\bar{b} \in \mathbb{Q}^m$ *are* A, c, ℓ, *and* b *with all numbers rounded to* $\mathbb{F}(p)$. *We call the oracle a* limited-precision LP-basis oracle *if it additionally returns a basis* $\mathcal{B} \subseteq \{1, \ldots, n\}$ *satisfying*

$$|\bar{x}_i - \ell_i| \leqslant \eta \text{ for all } i \notin \mathcal{B}, \qquad (2a)$$
$$|c_i - \bar{y}^T A_{.i}| \leqslant \eta \text{ for all } i \in \mathcal{B}. \qquad (2b)$$

Relating this definition with the behavior of real-world limited-precision LP solvers, we note that although real-world solvers are not guaranteed to find a solution with residual errors bounded by a fixed constant, these errors could nonetheless be computed and checked, correctly identifying the case of "failure".

Algorithm 1 states the basic iterative refinement procedure from [14]; presentation is simplified by consolidating tolerances and scaling factors. The basic convergence result, restated here as Lemma 1, carries over from [14].

Lemma 1. *Given an LP of form (P) and a limited-precision LP oracle with constants* η *and* σ, *let* $(x_k, y_k, \Delta_k)_{k=1,2,\ldots}$ *be the sequence of primal–dual solutions and scaling factors produced by Algorithm 1 with incremental scaling limit* $\alpha \geqslant 2$. *Let* $\varepsilon := \max\{\eta, 1/\alpha\}$. *Then for all iterations* k, $\Delta_{k+1} \geqslant \Delta_k/\varepsilon$, *and*

$$\|Ax_k - b\|_\infty \leqslant \varepsilon^k, \qquad (3a)$$
$$x_k - \ell \geqslant -\varepsilon^k \mathbb{1}, \qquad (3b)$$
$$c - A^T y_k \geqslant -\varepsilon^k \mathbb{1}, \qquad (3c)$$
$$|(x_k - \ell)^T (c - A^T y_k)| \leqslant \sigma \varepsilon^{2(k-1)}. \qquad (3d)$$

Hence, for any $\tau > 0$, *Algorithm 1 terminates in finite time after at most* $\lceil \max\{\log(\tau)/\log(\varepsilon), \log(\tau\varepsilon/\sigma)/\log(\varepsilon^2)\} \rceil$ *oracle calls.*

This lemma shows that the number of calls to the LP oracle before reaching a positive termination tolerance τ is linear in the encoding length of τ. In order to prove polynomial running time it is also necessary to argue that the encoding length of numbers encountered at intermediate steps of the algorithm do not grow too fast. This gives the following new result. See the appendix for a proof.

Algorithm 1: Iterative Refinement for a Primal and Dual Feasible LP

---

    input   : rational LP data A, b, ℓ, c, termination tolerance $\tau \geqslant 0$
    parameters : incremental scaling limit $\alpha \in \mathbb{N}, \alpha \geqslant 2$
    output   : primal–dual solution $x^* \in \mathbb{Q}^n, y^* \in \mathbb{Q}^m$ within tolerance τ

1 **begin**
2    $\Delta_1 \leftarrow 1$ `/* initial solve */`
3   get $(\bar{A}, \bar{b}, \bar{\ell}, \bar{c}) \approx (A, b, \ell, c)$ in working precision of the oracle
4   call oracle for $\min\{\bar{c}^T x : \bar{A}x = \bar{b}, x \geqslant \bar{\ell}\}$, abort if failure
5    $(x_1, y_1) \leftarrow$ approximate primal–dual solution returned

6   **for** $k \leftarrow 1, 2, \ldots$ **do**   `/* refinement loop */`
7       $\hat{b} \leftarrow b - Ax_k, \hat{\ell} \leftarrow \ell - x_k, \hat{c} \leftarrow c - A^T y_k$ `/* compute residual error */`
8       $\delta_k \leftarrow \max\left\{\max_j |\hat{b}_j|, \max_i \hat{\ell}_i, \max_i -\hat{c}_i, |\sum_i -\hat{\ell}_i \hat{c}_i|\right\}$
9   **if** $\delta_k \leqslant \tau$ **then** return $x^* \leftarrow x_k, y^* \leftarrow y_k$
10      $\delta_k \leftarrow \max\{\delta_k, 1/(\alpha \Delta_k)\}$ `/* scale problem */`
11      $\Delta_{k+1} \leftarrow 2^{\lceil \log(1/\delta_k) \rceil}$ `/* round scaling factor to power of two */`
12   get $(\bar{b}, \bar{\ell}, \bar{c}) \approx \Delta_{k+1}(\hat{b}, \hat{\ell}, \hat{c})$ in working precision of the oracle
                                                    `/* solve for corrector solution */`
13   call oracle for $\min\{\bar{c}^T x : \bar{A}x = \bar{b}, x \geqslant \bar{\ell}\}$, abort if failure
14      $(\hat{x}, \hat{y}) \leftarrow$ approximate primal–dual solution returned
15      $(x_{k+1}, y_{k+1}) \leftarrow (x_k, y_k) + \frac{1}{\Delta_{k+1}}(\hat{x}, \hat{y})$ `/* perform correction */`

Theorem 2. *Algorithm 1 with a limited-precision LP oracle according to Definition 1 runs in oracle-polynomial time, i.e., it requires a polynomial number of oracle calls and a polynomial number of bit operations in the size of the input A, b, ℓ, c, τ.*

3   Oracle Algorithms with Basis Verification

Iterative refinement as stated in Algorithm 1 only terminates in finite time for positive termination tolerance $\tau > 0$. The first extension, presented in this section, assumes a *limited-precision LP-basis oracle* as formalized in Definition 1 and computes exact basic solutions in oracle-polynomial time. Using such an oracle, Algorithm 1 produces a sequence $(x_k, y_k, \mathcal{B}_k)_{k=1,2,\ldots}$, where the bases $\mathcal{B}_k$ correspond to the transformed problems solved at line 13. From [14] we know that these bases also correspond to LP bases for the original problem. We may then ask whether or not this sequence is guaranteed to arrive at an optimal basis. The following lemma helps to answer this question in the affirmative.

Lemma 2. *Given an LP (P) with rational data $A \in \mathbb{Q}^{m \times n}$, $b \in \mathbb{Q}^m$, and $\ell, c \in \mathbb{Q}^n$, the following hold for any basic primal–dual solution x, y: (1) Either x is (exactly) primal feasible or its maximum primal violation has at least the value $1/2^{4\langle A,b \rangle + 5\langle \ell \rangle + 2n^2 + 4n}$; and (2) either y is (exactly) dual feasible or its maximum dual violation has at least the value $1/2^{4\langle A,c \rangle + 2n^2 + 4n}$.*

A detailed proof is omitted, but the basic idea is summarized as follows. Suppose x, y is a basic primal–dual solution with respect to some basis $\mathcal{B}$. It can be shown that the size of the entries in x and y are bounded by a polynomial in $\langle A, b, \ell, c \rangle$ and that all nonzero violations can be expressed as differences of rational numbers with bounded denominator, resulting in the above bounds. The following theorem states the main convergence result.

Theorem 3. *Suppose we are given an LP (P), a fixed $0 < \varepsilon < 1$, and a sequence of primal–dual solutions x_k, y_k with associated bases $\mathcal{B}_k$ such that (3a–3c) and*

$$|(x_k)_i - \ell_i| \leqslant \varepsilon^k \text{ for all } i \notin \mathcal{B}_k, \tag{4a}$$

$$|c_i - y_k^T A_{\cdot i}| \leqslant \varepsilon^k \text{ for all } i \in \mathcal{B}_k \tag{4b}$$

hold for $k = 1, 2, \ldots$. Then there exists a threshold $K = K(A, m, n, b, \ell, c, \varepsilon)$ such that the bases $\mathcal{B}_k$ are optimal for all $k \geqslant K$. The function satisfies the asymptotic bound $K(A, m, n, b, \ell, c, \varepsilon) \in O((m^2 \langle A \rangle + \langle b, \ell, c \rangle + n^2)/\log(1/\varepsilon))$.

Again, a detailed proof is omitted, but the main idea is summarized as follows. The proof uses (4a) and (4b) to show analogs of (3a–3c) hold with right-hand side $2^{4m^2 \langle A \rangle + 2} \varepsilon^k$ for the solutions $\tilde{x}_k, \tilde{y}_k$ associated with bases $\mathcal{B}_k$. Then for $k \geqslant K$, the primal and dual violations of $\tilde{x}_k, \tilde{y}_k$ drop below the minimum thresholds stated in Lemma 2. From then on $\mathcal{B}_k$ must be optimal.

Conditions (3a–3c) require that the primal and dual violations of x_k, y_k converge to zero precisely as is guaranteed by Lemma 1 for the sequence of numeric solutions produced by iterative refinement. The validity of the additional conditions (4a) and (4b) that the numeric solutions become "more and more basic" at the same rate as the primal and dual violations decrease is established in the proof of the main result at the end of this section.

The bound on the number of refinements in Theorem 3 may seem surprisingly large when considering that the best-known iteration complexity for interior point methods is $O(\sqrt{n+m}\langle A, b, \ell, c \rangle)$ [21] and that in each refinement round an LP is solved. One reason for this difference is that iterative refinement converges linearly, while interior point algorithms are a form of Newton's method, which allows for superlinear convergence. Additionally, the low-precision LPs solved by Algorithm 1 may be less expensive in practice than performing interior point iterations in very high-precision arithmetic. Moreover, as observed experimentally in [14], an optimal basis is typically reached after very few refinements. Hence, we do not want to rely on bounds computed a priori, but rather check the optimality of the basis early.

This is achieved by extending Algorithm 1 as follows. Suppose the LP oracle returns a basis $\hat{\mathcal{B}}$ in line 13. The termination criterion on line 9 is replaced by first solving the linear system of equations associated with $\hat{\mathcal{B}}$ for a basic solution $\tilde{x}, \tilde{y}$ in exact arithmetic and then checking its optimality. If this check is successful then the solution is returned, otherwise the solution is discarded and Algorithm 1 continues with the next refinement round. This leads to the main result of this section. The proof is given in the appendix.

Theorem 4. *Suppose we are given a rational, primal and dual feasible LP (P) and a limited-precision LP-basis oracle according to Definition 1. Then Algorithm 1 interleaved with a basis verification, as described above, terminates with an optimal solution to (P) in oracle-polynomial running time.*

4   Rational Reconstruction Algorithms

The algorithm developed in the previous section relies solely on the optimality of the basis information. Except for computing the residual vectors it does not make use of the more and more accurate numerical solutions produced. In this section, we discuss a conceptually different technique that exploits the approximate solutions as starting points in order to reconstruct an exact optimal solution. First we need to show that the sequence of approximate solutions converges.

Lemma 3. *Suppose we are given a rational, primal and dual feasible LP (P) and a limited-precision LP-basis oracle with precision p, and define $C := 2^p$. Let $(x_k, y_k, \Delta_k)_{k=1,2,\dots}$ be the sequence of primal–dual solutions and scaling factors produced by Algorithm 1. Then (x_k, y_k) converges to a rational, basic, and optimal solution $(\tilde{x}, \tilde{y})$ of (P) and $\|(\tilde{x}, \tilde{y}) - (x_k, y_k)\|_\infty \leqslant C \sum_{i=k+1}^\infty \Delta_i^{-1}$.*

A proof of this lemma can be found in the appendix. Note that the statement holds for any upper bound C on the absolute values in the corrector solutions $\hat{x}, \hat{y}$ returned by the oracle in line 14 of Algorithm 1. In practice, this may be much smaller than the largest floating-point representable value in $\mathbb{F}(p)$, 2^p.

Now suppose we know *a priori* a bound M on the denominators in the limit, then we can compute $\tilde{x}, \tilde{y}$ from an approximate solution satisfying $\|(x_k, y_k) - (\tilde{x}, \tilde{y})\|_\infty < 1/(2M^2)$ by applying Theorem 1 componentwise. If the size of M is small, i.e., polynomial in the input size, then iterative refinement produces a sufficiently accurate solution after a polynomial number of refinements. However, the worst-case bounds produced by Hadamard's inequality of this type have been demonstrated to be weak in practice [1, 23]. Computing an approximate solution with error below $1/(2M^2)$ before applying Theorem 1 can thus be unnecessarily expensive. This motivates the following extension of Algorithm 1 to an output-sensitive algorithm that attempts to reconstruct exact solution vectors during early rounds of refinement.

First, we fix a parameter β between 1 and $1/\varepsilon$, $\varepsilon := max\{\eta, 1/\alpha\}$. During Algorithm 1, we attempt reconstruction after line 10. We compute a speculative bound on the denominator as $M_k := \sqrt{\Delta_{k+1}/(2\beta^k)}$. Then the value $1/(2M_k^2)$ equals $\beta^k/\Delta_{k+1} \approx \beta^k \delta_k$ and tries to estimate the error in the solution. If reconstruction attempts fail, the term β^k keeps growing exponentially such that we eventually obtain a true bound on the error. Initially, however, β^k is small in order to account for the many cases where the residual δ_k is a good proxy for the error. Primal feasibility, dual feasibility, and complementary slackness of these heuristically reconstructed solutions must be checked exactly using rational arithmetic. If this check fails, we discard the solution and continue with the next refinement round. Otherwise, it is returned as optimal. The following theorem shows that the algorithm computes an exactly optimal solution to a primal and dual feasible LP under the conditions guaranteed by Lemma 3.

Theorem 5. *Suppose we are given an LP (P), fixed constants $C \geqslant 1$, $0 < \varepsilon < 1$, $1 < \beta < 1/\varepsilon$, and a rational limit point $\tilde{x}, \tilde{y}$ with the denominator of each component at most $\tilde{q}$. Furthermore, suppose a sequence of primal–dual solutions x_k, y_k and scaling factors $\Delta_k \geqslant 1$ satisfies $\|(\tilde{x}, \tilde{y}) - (x_k, y_k)\|_\infty \leqslant C \sum_{i=k+1}^\infty \Delta_i^{-1}$ with $\Delta_1 = 1$ and $\Delta_k/\Delta_{k+1} \leqslant \varepsilon$. Let $M_k := \sqrt{\Delta_{k+1}/(2\beta^k)}$.*
  Then there exists $K \in O(\max\{\langle \tilde{q} \rangle, \langle C \rangle\})$ such that

$$\|(\tilde{x}, \tilde{y}) - (x_k, y_k)\|_\infty < 1/(2M_k^2), \ 1 \leqslant \tilde{q} \leqslant M_k, \tag{5}$$

holds for all $k \geqslant K$, i.e., $\tilde{x}, \tilde{y}$ can be reconstructed from x_k, y_k componentwise in polynomial time using Theorem 1.

Again, a proof is provided in the appendix. The running time is output-sensitive as it depends on the encoding length of the solution. The value of C is a constant bound on the absolute values in the corrector solutions. Although C is independent of the input size and does not affect asymptotic running time, we include it explicitly in order to exhibit the practical dependency on the corrector solutions returned by the oracle.

We now consider the cost associated with reconstructing the solution vectors. The cost of applying the standard extended Euclidean algorithm to perform rational reconstruction on input with encoding length d is $O(d^2)$, asymptotically faster variants exist but are not used here for the sake of simplicity [24]. The following lemma, proven in the appendix, shows that if Algorithm 1 is modified as in the preceding discussion to apply rational reconstruction componentwise for each approximate solution (x_k, y_k) encountered, the cost of this added computation will be polynomial in the number of iterations.

Lemma 4. *The running time of applying rational reconstruction componentwise to x_k and y_k within the k-th refinement round is $O((n + m)k^2)$. Moreover, if it is applied at a geometric frequency, namely at rounds $k = \lceil f \rceil, \lceil f^2 \rceil, \ldots$ for some $f > 1$, and Algorithm 1 terminates at round K then the cumulative time spent on rational reconstruction throughout the algorithm is $O((n + m)K^2)$.*

Now, assuming the conditions laid out in Theorem 5 hold, then the number of iterations that Algorithm 1 interleaved with rational reconstruction performs before computing an exact rational solution is polynomially bounded in the encoding length of the input. Together with this bound on the number of iterations, Lemma 4 gives a polynomial bound on the time spent on rational reconstruction. The arguments from Theorem 2 still apply and limit the growth of the numbers and cost of the other intermediate computations, giving the following.

Theorem 6. *Suppose we are given a primal and dual feasible LP (P) and a limited-precision LP-basis oracle according to Definition 1 with constants*

p, η, and σ. Fix a scaling limit $\alpha \geqslant 2$ and let $\varepsilon := \max\{\eta, 1/\alpha\}$. Then Algorithm 1 interleaved with rational reconstruction using denominator bound $M_k := \sqrt{\Delta_{k+1}/(2\beta^k)}$, $\beta < 1/\varepsilon$, at round k, terminates with an optimal solution in oracle-polynomial running time.

Note that the basis does not need to be known explicitly. Accordingly, the algorithm may even return an optimal solution x^*, y^* that is different from the limit point $\tilde{x}, \tilde{y}$ if it is discovered by rational reconstruction at an early iterate x_k, y_k. In this case, x^*, y^* is not guaranteed to be a basic solution unless one explicitly discards solutions that are not basic in the optimality checks.

5   Computational Experiments

Using the simplex-based LP solver SoPlex [14,25], we analyzed the computational performance of iterative refinement both with basis verification (SoPlex$_{fac}$) and rational reconstruction (SoPlex$_{rec}$). In addition, we compared both methods against the state-of-the-art solver QSopt_ex, version 2.5.10 [2,3].

For basis verification, the exact solution of the primal and dual basis systems relies on a rational LU factorization and two triangular solves for the standard, column-wise basis matrix. For efficiency, basis verification is only called after two refinement steps have not updated the basis information. The rational reconstruction routine is an adaption of code used in [23]. The error correction factor β was set to 1.1. The rational reconstruction frequency f is set to 1.2, i.e., after a failed attempt at reconstructing an optimal solution, reconstruction is paused until 20% more refinement steps have been performed. We employ the DLCM method [5,7] for accelerating the reconstruction of the solution vectors.

As test bed we use an extensive collection of 1,202 primal and dual feasible LPs from several public sources detailed in the electronic supplement of [14]. The experiments were conducted on a cluster of 64-bit Intel Xeon X5672 CPUs with 48 GB main memory, simultaneously running at most one job per node and using a time limit of two hours per instance for each SoPlex and QSopt_ex run.

Overall Results. None of the solvers dominates the others: for each of QSopt_ex, SoPlex$_{fac}$, and SoPlex$_{rec}$ there exist instances that can be solved only by this one solver. Overall, however, the iterative refinement-based methods are able to solve more instances than QSopt_ex, and SoPlex$_{fac}$ exhibits significantly shorter runtimes than the other two methods. Of the 1,202 instances, 1,158 are solved by all three. QSopt_ex solves 1,163 instances, SoPlex$_{rec}$ solves 1,189 instances, and SoPlex$_{fac}$ solves the largest number of instances: 1,191.

On 8 of the 39 instances not solved by QSopt_ex this is due to a memory limit during or after precision boosts and highlights that solving extended-precision LPs may not only be time-consuming, but also require excessive memory. By contrast, the iterative refinement-based methods work with a more memory-efficient double-precision floating-point rounding of the LP and never reach the memory limit. However, SoPlex$_{rec}$ and SoPlex$_{fac}$ could not solve seven instances because the floating-point simplex implementation failed during iterative refinement.

Finally, for the 492 instances that could be solved by all three algorithms, but were sufficiently nontrivial such that one of the solvers took at least two seconds, Table 1 compares average runtimes and number of simplex iterations. The lines starting with 64-bit, 128-bit, and 192-bit filter for the instances corresponding to the final precision level used by QSopt_ex. Overall, SoPlex$_{fac}$ is 1.85 times faster than QSopt_ex and even 2.85 times faster than SoPlex$_{rec}$. On LPs where QSopt_ex found the optimal solution after the double-precision solve (line 64-bit), SoPlex$_{fac}$ is 30% faster although it uses about 40% more simplex iterations than QSopt_ex. When QSopt_ex has to boost the working precision of the floating-point solver (lines 128-bit and 192-bit), the results become even more pronounced, with SoPlex$_{fac}$ being over three times faster than QSopt_ex.

Table 1. Aggregate comparison on instances that could be solved by all and where one solver took at least 2 s. Columns #iter and t report shifted geometric means of simplex iterations and solving times, using a shift of 2 s and 100 simplex iterations, respectively. Column Δt gives the ratios of times with those of QSopt_ex.

QSopt_ex prec	#inst	QSopt_ex		SoPlex$_{fac}$			SoPlex$_{rec}$		
		#iter	t	#iter	t	Δt	#iter	t	Δt
any	492	8025.7	15.6	9740.6	8.5	0.54	9740.6	24.2	1.55
64-bit	324	8368.3	16.1	11683.7	11.3	0.70	11683.7	14.8	0.92
128-bit	163	7217.1	13.9	6757.2	4.3	0.31	6757.2	58.5	4.21
192-bit	5	16950.9	72.5	10763.4	20.5	0.28	10763.4	134.6	1.86

Rational Reconstruction vs. Basis Verification. Table 2 compares SoPlex$_{rec}$ and SoPlex$_{fac}$ in more detail for all 1,186 instances solved by both methods. Here, the lines starting with $[t, 7200]$ filter for subsets of increasingly hard instances for which at least one method took $t = 1, 10$ or 100 s.

Table 2. Comparison of SoPlex$_{fac}$ and SoPlex$_{rec}$ on instances solved by both. Columns #ref, #fac, #rec contain arithm. means of refinements, basis verifications, and reconstruction attempts. Columns t, t_{fac}, t_{rec} report shifted geom. mean times for the total solving process, the basis verifications, and rational reconstruction routines, with a shift of 2 s. Column Δt reports the ratio between the mean solve times.

Test set	#inst	SoPlex$_{fac}$				SoPlex$_{rec}$				
		#ref	#fac	t_{fac}	t	#ref	#rec	t_{rec}	t	Δt
all	1186	2.1	0.95	0.21	2.8	68.3	6.74	1.26	5.2	1.82
$[1, 7200]$	591	2.3	0.98	0.43	8.8	135.1	11.04	3.28	21.8	2.47
$[10, 7200]$	311	2.4	0.99	0.83	24.1	241.6	15.47	9.15	101.9	4.22
$[100, 7200]$	161	2.7	0.98	1.40	42.8	384.9	19.70	22.95	340.3	7.95

Compared to SoPlex$_{\text{rec}}$, the number of refinements for SoPlex$_{\text{fac}}$ is very small because the final, optimal basis is almost always reached by the second round. For 1,123 LPs, SoPlex$_{\text{fac}}$ performs exactly one rational factorization; for 7 instances two factorizations. Notably, there are 61 instances where no factorization is necessary because the approximate solution is exactly optimal. As a result, the average number of factorizations (column "#fac") is slightly below one.

On average, the time for rational factorization and triangular solves (column "t_{fac}") is small compared to the total solving time. In combination with the small number of refinements, this explains why SoPlex$_{\text{fac}}$ is significantly faster. However, for 21 instances, t_{fac} consumes more than 90% of the runtime. On 3 of these instances, SoPlex$_{\text{fac}}$ times out, while they can be solved by SoPlex$_{\text{rec}}$.

Column "t_{rec}" shows that calling rational reconstruction at a geometric frequency succeeds in keeping reconstruction time low also as the number of refinements increases. However, 5 instances solved by SoPlex$_{\text{fac}}$, but not by SoPlex$_{\text{rec}}$, show large denominators in the optimal solution and point to a bottleneck of SoPlex$_{\text{rec}}$. The number of refinements that could be performed within the time limit did not suffice to reach a sufficiently accurate approximate solution.

Appendix

This appendix collects proofs for the main results used in the paper.

Proof of Theorem 2

Proof. Line 10 ensures that at iteration k, $\Delta_k \leq 2^{\lceil \log \alpha \rceil (k-1)}$ and thus $\langle \Delta_k \rangle \leq \lceil \log \alpha \rceil k$. From line 15, the entries in the refined solution vectors x_k, y_k have the form $\sum_{j=1}^{k} \Delta_j^{-1} \frac{n_j}{2^p}$ with $n_j \in \mathbb{Z}$, $|n_j| \leq 2^{2p}$, for $j = 1, \ldots, k$. With $D_j := \log(\Delta_j)$ and $a := \lceil \log(\alpha) \rceil$ this can be rewritten as

$$\sum_{j=1}^{k} 2^{-D_j} \frac{n_j}{2^p} = \Big(\sum_{j=1}^{k} n_j 2^{a(k-1)-D_j} \Big) / 2^{p+a(k-1)}. \qquad (6)$$

The latter is a fraction with integer numerator and denominator. The numerator is bounded by $2^{2p+ak} - 1$. Hence, $\langle x_k \rangle + \langle y_k \rangle \leq (n+m)(2ak+3p+2)$. The numbers stored in $\hat{b}, \hat{l}, \hat{c}$ and δ_k satisfy the same asymptotic bound. By Lemma 1, the maximum number of iterations is $O(\log(1/\tau)) = O(\langle \tau \rangle)$. All in all, the size of the numbers encountered during the algorithm is $O(\langle A, b, \ell, c \rangle + (n+m)\langle \tau \rangle)$.

Finally, the total number of elementary operations is polynomially bounded as follows. The initial rounding of the constraint matrix costs $O(nnz)$ elementary operations, where nnz is the number of nonzero entries in A. For each of the $O(\langle \tau \rangle)$ refinements, computing residual errors, maximum violation, and checking termination involves $O(nnz+n+m)$ operations; the computation of the scaling factors takes constant effort, and the update of the transformed problem and the correction of the primal–dual solution vectors costs $O(n+m)$ operations. $\square$

Proof of Theorem 4

Proof. Let $\varepsilon := \max\{\eta, 1/\alpha\}$, where η is the feasibility tolerance of the LP oracle and α is the scaling limit used in Algorithm 1. Conditions (3a–3c) of Theorem 3 follow directly from Lemma 1. We prove conditions (4a) and (4b) by induction. For $k = 1$, they follow from (2a) and (2b). Suppose they hold for $k \geqslant 1$. Let $\hat{x}, \hat{y}, \hat{\mathcal{B}}$ be the last approximate solution returned by the oracle. Then for all $i \notin \mathcal{B}_{k+1}$

$$
|(x_{k+1})_i - \ell_i| = \left|(x_k)_i + \frac{\hat{x}_i}{\Delta_{k+1}} - \ell_i\right| = |\hat{x}_i + \Delta_{k+1}\underbrace{((x_k)_i - \ell_i)}_{=-\hat{\ell}_i}|/\underbrace{\Delta_{k+1}}_{\geqslant \varepsilon^{-k} \text{ by Lemma 1}}
$$

$$
\leqslant |\hat{x}_i - \Delta_{k+1}\hat{\ell}_i|/\varepsilon^k \overset{(2a)}{\leqslant} \eta\varepsilon^k \leqslant \varepsilon^{k+1},
$$

proving (4a). The induction step for (4b) is analogous. Thus, the sequence of basic solutions $x_k, y_k, \mathcal{B}_k$ satisfies the conditions of Theorem 3 and $\mathcal{B}_k$ is optimal after a polynomial number of refinements. According to Theorem 2, this runs in oracle-polynomial time. The linear systems used to compute the basic solutions exactly over the rational numbers can be solved in polynomial time [9].   □

Proof of Lemma 3

Proof. This result inherently relies on the boundedness of the corrector solutions returned by the oracle. Since their entries are in $\mathbb{F}(p)$, $\|(\hat{x}_k, \hat{y}_k)\|_\infty \leqslant 2^p$. Then $(x_k, y_k) = \sum_{i=1}^k \frac{1}{\Delta_i}(\hat{x}_i, \hat{y}_i)$ constitutes a Cauchy sequence: for any $k, k' \geqslant K$, $\|(x_k, y_k) - (x_{k'}, y_{k'})\|_\infty \leqslant 2^p \sum_{i=K+1}^\infty \varepsilon^i = 2^p \varepsilon^{K+1}/(1-\varepsilon)$, where ε is the rate of convergence from Lemma 1. Thus, a unique limit point $(\tilde{x}, \tilde{y})$ exists. Using proof techniques as for Theorem 3 one can show that $(\tilde{x}, \tilde{y})$ is basic, hence rational.   □

Proof of Theorem 5

Proof. Note that $\Delta_k/\Delta_{k+1} \leqslant \varepsilon$ for all k implies $\Delta_i \geqslant \Delta_j \varepsilon^{j-i}$ for all $j \leqslant i$. Then $M_k = \sqrt{\Delta_{k+1}/(2\beta^k)} \geqslant \tilde{q}$ holds if $\sqrt{1/(2\beta^k \varepsilon^k)} \geqslant \tilde{q}$. This holds for all

$$
k \geqslant K_1 := (2\log \tilde{q} + 1)/\log(1/(\beta\varepsilon)) \in O(\langle \tilde{q}\rangle). \tag{7}
$$

Furthermore, $C \sum_{i=k+1}^\infty \Delta_i^{-1} \leqslant C \sum_{i=k+1}^\infty \varepsilon^{i-k-1}\Delta_{k+1}^{-1} = C/((1-\varepsilon)\Delta_{k+1})$, which is less than $1/(2M_k^2) = \beta^k/\Delta_{k+1}$ for all

$$
k > K_2 := (\log C - \log(1-\varepsilon))/\log \beta \in O(\langle C\rangle). \tag{8}
$$

Hence (5) holds for all $k > K := \max\{K_1, K_2\}$.   □

Proof of Lemma 4

Proof. From the proof of Theorem 2 we know that at the k-th iteration of the algorithm, the encoding length of the components of x_k, y_k are each bounded by $(2\alpha k + 3p + 2)$, which is $O(k)$ as α, p are constants. Together with the fact that rational reconstruction can be performed in quadratic time, as discussed above, the first result is established.

To show the second claim we assume that $f > 1$ and let K be the final index at which rational reconstruction is attempted. Then if we consider the sequence of indices at which rational reconstruction was applied, that sequence is term-wise bounded above by the following sequence (given in decreasing order): $S = (K, \lfloor K/f \rfloor, \lfloor K/f^2 \rfloor, \ldots, \lfloor K/f^a \rfloor)$ where $a = \lceil \log_f K \rceil$. Thus, the cost to perform rational reconstruction at iterations indexed by the sequence S gives an upper bound on the total cost. Again, using the quadratic bound on the cost for rational reconstruction, we arrive at the following cumulative bound, involving a geometric series:

$$O\left((n+m)\sum_{i=0}^{a}\lfloor K/f^i \rfloor^2\right) = O\left((n+m)K^2\sum_{i=0}^{a}f^{-2i}\right) = O((n+m)K^2).$$

This establishes the result. $\square$

References

1. Abbott, J., Mulders, T.: How tight is Hadamard's bound? Exp. Math. **10**(3), 331–336 (2001). http://projecteuclid.org/euclid.em/1069786341
2. Applegate, D.L., Cook, W., Dash, S., Espinoza, D.G.: Qsopt_ex. http://www.dii.uchile.cl/~daespino/ESolver_doc/
3. Applegate, D.L., Cook, W., Dash, S., Espinoza, D.G.: Exact solutions to linear programming problems. Oper. Res. Lett. **35**(6), 693–699 (2007). https://doi.org/10.1016/j.orl.2006.12.010
4. Azulay, D.-O., Pique, J.-F.Ç.: Optimized Q-pivot for exact linear solvers. In: Maher, M., Puget, J.-F. (eds.) CP 1998. LNCS, vol. 1520, pp. 55–71. Springer, Heidelberg (1998). https://doi.org/10.1007/3-540-49481-2_6
5. Chen, Z., Storjohann, A.: A BLAS based C library for exact linear algebra on integer matrices. In: Proceedings of the 2005 International Symposium on Symbolic and Algebraic Computation, ISSAC 2005, pp. 92–99 (2005). https://doi.org/10.1145/1073884.1073899
6. Cheung, D., Cucker, F.: Solving linear programs with finite precision: II. Algorithms. J. Complex. **22**(3), 305–335 (2006). https://doi.org/10.1016/j.jco.2005.10.001
7. Cook, W., Steffy, D.E.: Solving very sparse rational systems of equations. ACM Trans. Math. Softw. **37**(4), 39:1–39:21 (2011). https://doi.org/10.1145/1916461.1916463
8. Dhiflaoui, M., et al.: Certifying and repairing solutions to large LPs: how good are LP-solvers? In: Proceedings of the 14th Annual ACM-SIAM Symposium on Discrete Algorithms, SODA 2003, pp. 255–256. SIAM (2003)
9. Edmonds, J.: Systems of distinct representatives and linear algebra. J. Res. Natl. Bur. Stan. **71B**(4), 241–245 (1967)
10. Edmonds, J., Maurras, J.F.: Note sur les Q-matrices d'Edmonds. RAIRO. Recherche Opérationnelle **31**(2), 203–209 (1997). http://www.numdam.org/item?id=RO_1997__31_2_203_0
11. Escobedo, A.R., Moreno-Centeno, E.: Roundoff-error-free algorithms for solving linear systems via Cholesky and LU factorizations. INFORMS J. Comput. **27**(4), 677–689 (2015). https://doi.org/10.1287/ijoc.2015.0653

12. Escobedo, A.R., Moreno-Centeno, E.: Roundoff-error-free basis updates of LU factorizations for the efficient validation of optimality certificates. SIAM J. Matrix Anal. Appl. **38**(3), 829–853 (2017). https://doi.org/10.1137/16M1089630
13. Gärtner, B.: Exact arithmetic at low cost - a case study in linear programming. Comput. Geom. **13**(2), 121–139 (1999). https://doi.org/10.1016/S0925-7721(99)00012-7
14. Gleixner, A.M., Steffy, D.E., Wolter, K.: Iterative refinement for linear programming. INFORMS J. Comput. **28**(3), 449–464 (2016). https://doi.org/10.1287/ijoc.2016.0692
15. Grötschel, M., Lovász, L., Schrijver, A.: Geometric Algorithms and Combinatorial Optimization. Algorithms and Combinatorics, vol. 2. Springer, Heidelberg (1988). https://doi.org/10.1007/978-3-642-78240-4
16. Karmarkar, N.: A new polynomial-time algorithm for linear programming. Combinatorica **4**(4), 373–395 (1984). https://doi.org/10.1007/BF02579150
17. Khachiyan, L.G.: Polynomial algorithms in linear programming (in Russian). Zhurnal Vychislitel'noi Matematiki i Matematicheskoi Fiziki **20**(1), 51–68 (1980). https://doi.org/10.1016/0041-5553(80)90061-0. English translation: USSR Computational Mathematics and Mathematical Physics, 20(1):53-72, 1980
18. Koch, T.: The final NETLIB-LP results. Oper. Res. Lett. **32**(2), 138–142 (2004). https://doi.org/10.1016/S0167-6377(03)00094-4
19. Kwappik, C.: Exact linear programming. Master's thesis, Universität des Saarlandes, May 1998
20. Lenstra, A.K., Lenstra, H.W., Lovász, L.: Factoring polynomials with rational coefficients. Mathematische Annalen **261**(4), 515–534 (1982). https://doi.org/10.1007/BF01457454
21. Renegar, J.: A polynomial-time algorithm based on Newton's method, for linear programming. Math. Program. **40**(1–3), 59–93 (1988). https://doi.org/10.1007/BF01580724
22. Schrijver, A.: Theory of Linear and Integer Programming. Wiley, New York (1986)
23. Steffy, D.E.: Exact solutions to linear systems of equations using output sensitive lifting. ACM Commun. Comput. Algebra **44**(3/4), 160–182 (2011). https://doi.org/10.1145/1940475.1940513
24. Wang, X., Pan, V.Y.: Acceleration of Euclidean algorithm and rational number reconstruction. SIAM J. Comput. **2**(32), 548–556 (2003)
25. Zuse Institute Berlin: SoPlex. Sequential object-oriented simPlex. http://soplex.zib.de/

Computing the Nucleolus of Weighted Cooperative Matching Games in Polynomial Time

Jochen Könemann[1], Kanstantsin Pashkovich[2], and Justin Toth[1(✉)]

[1] University of Waterloo, Waterloo, ON N2L 3G1, Canada
{jochen,wjtoth}@uwaterloo.ca
[2] University of Ottawa, Ottawa, ON K1N 6N5, Canada
kpashkov@uottawa.ca

Abstract. We provide an efficient algorithm for computing the nucleolus for an instance of a weighted cooperative matching game. This resolves a long-standing open question posed in [Faigle, Kern, Fekete, Hochstättler, Mathematical Programming, 1998].

Keywords: Combinatorial optimization · Algorithmic game theory · Matchings

1 Introduction

Imagine a network of players that form partnerships to generate value. For example, maybe a tennis league pairing players to play exhibition matches [3], or people making trades in an exchange network [39]. These are examples of what are called *matching games*. In a (weighted) matching game, we are given a graph $G = (V, E)$, weights $w : E \to \mathbb{R}_{\geq 0}$, the player set is the set V of nodes of G, and $w(uv)$ denotes the value earned when u and v collaborate. Each coalition $S \subseteq V$ is assigned a *value* $\nu(S)$ so that $\nu(S)$ is equal to the value of a maximum weight matching on the induced subgraph $G[S]$. The special case of matching games where $w = \mathbb{1}$ is the all-ones vector, and G is bipartite is called an *assignment game* and was introduced in a classical paper by Shapley and Shubik [39], and was later generalized to general graphs by Deng, Ibaraki, and Nagamochi [11].

We are interested in what a fair redistribution of the total value $\nu(V)$ to the players in the network looks like. The field of *cooperative game theory* gives us the language to make this question formal. A vector $x \in \mathbb{R}^V$ is called an *allocation* if

This work was done in part while the second author was visiting the Simons Institute for the Theory of Computing. Supported by DIMACS/Simons Collaboration on Bridging Continuous and Discrete Optimization through NSF grant #CCF-1740425.
We acknowledge the support of the Natural Sciences and Engineering Research Council of Canada (NSERC). Cette recherche a été financée par le Conseil de recherches en sciences naturelles et en génie du Canada (CRSNG).

© Springer Nature Switzerland AG 2019
A. Lodi and V. Nagarajan (Eds.): IPCO 2019, LNCS 11480, pp. 413–426, 2019.
https://doi.org/10.1007/978-3-030-17953-3_31

$x(V) = \nu(V)$ (where we use $x(V)$ as a short-hand for $\sum_{i \in V} x(i)$ as usual). Given such an allocation, we let $x(S) - \nu(S)$ be the *excess* of coalition $S \subseteq V$. This quantity can be thought of as a measure of the satisfaction of coalition S. A fair allocation should maximize the bottleneck excess, i.e. maximize the minimum excess, and this can be accomplished by an LP:

$$\max \quad \varepsilon \qquad\qquad\qquad\qquad\qquad\qquad\qquad (P)$$
$$\text{s.t.} \quad x(S) \geq \nu(S) + \varepsilon \qquad \text{for all} \quad S \subseteq V$$
$$x(V) = \nu(V)$$
$$x \geq 0.$$

Let ε^* be the optimum value of (P), and define $P(\varepsilon^*)$ to be the set of allocations x such that (x, ε^*) is feasible for (P). The set $P(\varepsilon^*)$ is known as the *leastcore* [33] of the given cooperative game, and the special case when $\varepsilon^* = 0$, $P(0)$ is the well-known *core* [21] of (V, ν). Intuitively, allocations in the core describe payoffs in which no coalition of players could profitably deviate from the *grand coalition* V.

Why stop at maximizing the bottleneck excess? Consider an allocation which, subject to maximizing the smallest excess, maximizes the second smallest excess, and subject to that maximizes the third smallest excess, and so on. This process of successively optimizing the excess of the worst-off coalitions yields our primary object of interest, the *nucleolus*. For an allocation $x \in \mathbb{R}^V$, let $\theta(x) \in \mathbb{R}^{2^V-2}$ be the vector obtained by sorting the list of excess values $x(S) - \nu(S)$ for any $\varnothing \neq S \subset V$ in non-decreasing order[1]. The *nucleolus*, denoted $\eta(V, \nu)$ and defined by Schmeidler [38], is the unique allocation that lexicographically maximizes $\theta(x)$:

$$\eta(V, \nu) := \arg \operatorname{lex} \max\{\theta(x) : x \in P(\varepsilon^*)\}.$$

We refer the reader to Appendix B for an example instance of the weighted matching game with its nucleolus. We now have sufficient terminology to state our main result:

Theorem 1. *Given a graph $G = (V, E)$ and weights $w : E \to \mathbb{R}$, the nucleolus $\eta(V, \nu)$ of the corresponding weighted matching game can be computed in polynomial time.*

Despite its intricate definition the concept of the nucleolus is surprisingly ancient. Its history can be traced back to a discussion on bankruptcy division in the Babylonian Talmud [1]. Modern research interest in the nucleolus stems not only from its geometric beauty [33], or several practical applications (e.g., see [5,32]), but from the strange way problems of computing the nucleolus fall in the complexity landscape, seeming to straddle the NP vs P boundary.

---

[1] It is common within the literature, for instance in [26], to exclude the coalitions for $S = \varnothing$ and $S = V$ in the definition of the nucleolus. On the other hand, one could also consider the definition of the nucleolus with all possible coalitions, including $S = \varnothing$ and $S = V$. We note that the two definitions of the nucleolus are equivalent in all instances of matching games except for the trivial instance of a graph consisting of two nodes joined by a single edge.

Beyond being one of the most fundamental problems in combinatorial optimization, starting with the founding work of Kuhn on the Hungarian method for the assignment problem [29], matching problems have historically teetered on the cusp of hardness. For example, prior to Edmonds' celebrated Blossom Algorithm [12,13] it was not clear whether Maximum Matching belonged in P. For another example, until Rothvoß' landmark result [37] it was thought that the matching polytope could potentially have sub-exponential extension complexity. In cooperative game theory, matchings live up to their historical pedigree of representing a challenging problem class. The long standing open problem in this area was whether the nucleolus of a weighted matching game instance can be computed in polynomial time. The concept of the nucleolus has been known since 1969 [38], and the question was posed as an important open problem in multiple papers. In 1998, Faigle, Kern, Fekete, and Hochstättler [15] mention the problem in their work on the nucleon, a multiplicative-error analog to the nucleolus which they show is polynomial time computable. Kern and Paulusma state the question of computing the nucleolus for general matching games as an important open problem in 2003 [26]. In 2008, Deng and Fang [9] conjectured this problem to be NP-hard, and in 2017 Biró, Kern, Paulusma, and Wojuteczky [4] reaffirmed this problem as an interesting open question. Theorem 1 settles the question, providing a polynomial-time algorithm to compute the nucleolus of a general instance of a weighted cooperative matching game.

Prior to our work, the nucleolus was known to be polynomial-time computable only in structured instances of the matching game. Solymosi and Raghavan [40] showed how to compute the nucleolus in an (unweighted) assignment game instance in polynomial time. Kern and Paulusma [26] later provided an efficient algorithm to compute the nucleolus in general unweighted matching game instances. Paulusma [35] extended the work in [26] and gave an efficient algorithm to compute the nucleolus in matching games where edge weights are induced by node potentials. Farczadi [20] finally extended Paulusma's framework further using the concept of *extendible allocations*. We note also that it is easy to compute the nucleolus in weighted instances of the matching game with non-empty core. For such instances, the leastcore has a simple compact description that does not include constraints for coalitions of size greater than 2. Thus it is relatively straightforward to adapt the iterative algorithm of Maschler [33] to a polynomial-time algorithm for computing the nucleolus (e.g., see [20, Chapter 2.3] for the details, Sect. 1.2 for an overview).

In a manner analogous to how we have defined matching games, a wide variety *combinatorial optimization games* can be defined [11]. In such games, the value of a coalition S of players is succinctly given as the optimal solution to an underlying combinatorial optimization problem. It is natural to conjecture that the complexity of computing the nucleolus in an instance of such a game would fall in lock-step with the complexity of the underlying problem. Surprisingly this is not the case. For instance, computing the nucleolus is known to be NP-hard for network flow games [10], weighted threshold games [14], and spanning tree games [16,19]. On the other hand, polynomial time algorithms are known for finding the nucleolus of special cases of flow games, certain classes of matching games, fractional matching games, and convex games [2,6,10,17,20,22,23,26,30,34–36,40].

The nucleolus is known to lie in the prekernel [38], a solution concept representing allocations which, speaking intuitively, reflect a balance of power between players. The prekernel of a cooperative game is known to be non-convex and even disconnected in general [28,41]. Despite this, Faigle, Kern and Kuipers [17] showed how to compute a point in the intersection of prekernel and leastcore in polynomial time under the reasonable assumption that the game has a polynomial time oracle to compute the minimum excess coalition for a given allocation. Later the same authors [18] refine their result to computing a point in the intersection of the core and lexicographic kernel, a set which is also known to contain the nucleolus. Bateni et al. [2] pose as an open question the existence of an efficiently computable, balanced and *unique* way of sharing profit in a network bargaining setting. The nucleolus is always unique [38], and balanced in the sense of lying in the leastcore intersect prekernel. Theorem 1 therefore resolves the latter open question left in [2].

1.1 Leastcore and Core of Matching Games

It is straightforward to see that (P) can be rewritten equivalently as

$$\max \ \varepsilon \qquad\qquad\qquad (P_1)$$
$$\text{s.t. } x(M) \geq w(M) + \varepsilon \qquad \text{for all} \quad M \in \mathcal{M}$$
$$x(V) = \nu(G)$$
$$x \geq 0,$$

where $\mathcal{M}$ is the set of all matchings M on G, and $x(M)$ is a shorthand for $x(V(M))$.

The separation problem for the linear program (P_1) can be reduced to finding a maximum weight matching in the graph G with edge weights $w(uv) - x(uv)$, $uv \in E$ (where we use $x(uv)$ as a shorthand for $x(u) + x(v)$). Since the maximum weight matching can be found in polynomial time [12], we know that the linear program (P_1) can be solved in polynomial time as well [25].

We use ε_1 to denote the optimal value of (P_1) and $P_1(\varepsilon_1)$ for the set of allocations x such that (x, ε_1) is feasible for the leastcore linear program (P_1). In general, for a value ε and a linear program Q on variables in $\mathbb{R}^V \times \mathbb{R}$ we denote by $Q(\varepsilon)$ the set $\{x \in \mathbb{R}^V : (x, \varepsilon) \text{ is feasible for } Q\}$.

Note that $\varepsilon_1 \leq 0$. Indeed, $\varepsilon \leq 0$ in any feasible solution (x, ε) to (P_1) as otherwise $x(M)$ would need to exceed $w(M)$ for all matchings M. In particular this would also hold for a maximum weight matching on G, implying that $x(V) > \nu(G)$. If $\varepsilon_1 = 0$ then the core of the cooperative matching game is non-empty. One can see that $\varepsilon_1 = 0$ if and only if the value of a maximum weight matching on G with weights w equals the value of a maximum weight fractional matching. This follows since $x \in P_1(\varepsilon_1)$ is a fractional weighted node cover of value $\nu(G)$ when $\varepsilon_1 = 0$. When $\varepsilon_1 < 0$, we say that the cooperative matching game has an empty core. The matching game instance given in Appendix B has an empty core.

In this paper, we assume that the cooperative matching game (G, w) has an empty core, as computing the nucleolus is otherwise well-known to be doable in polynomial time [35].

1.2   Maschler's Scheme

As discussed, our approach to proving Theorem 1 relies on Maschler's scheme. The scheme requires us to solve a linear number of LPs: $\{(P_j)\}_{j \geq 1}$ that we now define. (P_1) is the leastcore LP that we have already seen in Sect. 1.1. LPs (P_j) for $j \geq 2$ are defined inductively. Crucial in their definition is the notion of *fixed coalitions* that we introduce first. For a polyhedron $Q \subseteq \mathbb{R}^V$ we denote by $\text{Fix}(Q)$ the collection of sets $S \subseteq V$ such that $x(S)$ is constant over the polyhedron Q, i.e.

$$\text{Fix}(Q) := \{ S \subseteq V : \quad x(S) = x'(S) \quad \text{for all} \quad x, x' \in Q \}.$$

With this we are now ready to state LP (P_j) for $j \geq 2$:

$$\max \quad \varepsilon \qquad\qquad\qquad\qquad\qquad\qquad\qquad\qquad\qquad (P_j)$$
$$\text{s.t.} \quad x(S) - \nu(S) \geq \varepsilon \qquad \text{for all} \quad S \subset V, S \notin \text{Fix}(P_{j-1}(\varepsilon_{j-1}))$$
$$x \in P_{j-1}(\varepsilon_{j-1}),$$

where ε_j be the optimal value of the linear program (P_j). Let j^* be the minimum number j such that $P_j(\varepsilon_j)$ contains a single point. This point is the nucleolus of the game [8]. It is well-known [33] that $P_{j-1}(\varepsilon_{j-1}) \subset P_j(\epsilon_j)$ and $\varepsilon_{j-1} < \varepsilon_j$ for all j. Since the dimension of the polytope describing feasible solutions of (P_j) decreases in every round until the dimension becomes zero, we have $j^* \leq |V|$ [33], [35, Pages 20–24].

Therefore, in order to find the nucleolus of the cooperative matching game efficiently it suffices to solve each linear program (P_j), $j = 1, \ldots, j^*$ in polynomial time. We accomplish this by providing polynomial-size formulations for (P_j) for all $j \geq 1$.

In Sect. 2 we introduce the concept of *universal matchings* which are fundamental to our approach, and give a compact formulation for the first linear program in Maschler's Scheme, the leastcore. We also present our main technical lemma, Lemma 5, which provides a crucial symmetry condition on the values allocations can take over the vertices of blossoms in the graph decomposition we use to describe the compact formulation. In Sect. 3 we describe the successive linear programs in Maschler's Scheme and provide a compact formulation for each one in a matching game.

2   Leastcore Formulation

In this section we provide a polynomial-size description of (P_1). It will be useful to define a notation for *excess*: for any $x \in P_1(\varepsilon_1)$ and $M \in \mathcal{M}$ let $\text{excess}(x, M) := x(M) - w(M)$.

2.1   Universal Matchings, Universal Allocations

For each $x \in P_1(\varepsilon_1)$ we say that a matching $M \in \mathcal{M}$ is an x-tight matching whenever $\mathrm{excess}(x, M) = \varepsilon_1$. We denote by $\mathcal{M}^x$ the set of x-tight matchings.

A *universal matching* $M \in \mathcal{M}$ is a matching which is x-tight for all $x \in P_1(\varepsilon_1)$. We denote the set of universal matchings on G by $\mathcal{M}_{uni}$. A *universal allocation* $x^* \in P_1(\varepsilon_1)$ is a leastcore point whose x^*-tight matchings are precisely the set of universal matchings, i.e. $\mathcal{M}^{x^*} = \mathcal{M}_{uni}$.

Lemma 1. *There exists a universal allocation* $x^* \in P_1(\varepsilon_1)$.

Proof. Indeed, it is straightforward to show that every x^* in the relative interior (see [42, Lemma 2.9(ii)]) of $P_1(\varepsilon_1)$ is a universal allocation. If the relative interior is empty then $P_1(\varepsilon_1)$ is a singleton, which trivially contains a universal allocation. In the arxiv version [27] we provide a combinatorial proof.   ∎

Lemma 2. *A universal allocation* $x^* \in P_1(\varepsilon_1)$ *can be computed in polynomial time.*

Proof. A point x^* in the relative interior of $P_1(\varepsilon_1)$ can be found in polynomial time using the ellipsoid method (Theorem 6.5.5 [24], [7]). Since any allocation x^* from the relative interior of $P_1(\varepsilon_1)$ is a universal allocation, this implies the statement of the lemma.   ∎

Given a non-universal allocation x and a universal allocation x^*, we observe that $\mathcal{M}^{x^*} \subset \mathcal{M}^x$ and so $\theta(x^*)$ is strictly lexicographically greater than $\theta(x)$. Thus the nucleolus is a universal allocation. We emphasize that $\mathcal{M}^{x^*} = \mathcal{M}_{uni}$ is invariant under the (not necessarily unique) choice of universal allocations x^*. Henceforth we fix a universal allocation $x^* \in P_1(\varepsilon_1)$.

2.2   Description for Convex Hull of Universal Matchings.

By the definition of universal allocation x^*, a matching M is universal if and only if it is x^*-tight. Thus, M is a universal matching if and only if its characteristic vector lies in the optimal face of the matching polytope corresponding to (the maximization of) the linear objective function assigning weight $-\mathrm{excess}(x^*, uv) = w(uv) - x^*(uv)$ to each edge $uv \in E$. Let $\mathcal{O}$ be the set of node sets $S \subseteq V$ such that $|S| \geq 3$, $|S|$ is odd. Edmonds [12] gave a linear description of the matching polytope of G as the set of $y \in \mathbb{R}^E$ satisfying:

$$
\begin{aligned}
y(\delta(v)) &\leq 1 && \text{for all} \quad v \in V \\
y(E(S)) &\leq (|S| - 1)/2 && \text{for all} \quad S \in \mathcal{O} \\
y &\geq 0 \, .
\end{aligned}
$$

Thus, a matching $M \in \mathcal{M}$ is universal if and only if it satisfies the constraints

$$
\begin{aligned}
M \cap \delta(v) &= 1 && \text{for all} \quad v \in W \\
M \cap E(S) &= (|S| - 1)/2 && \text{for all} \quad S \in \mathcal{L} \\
M \cap \{e\} &= 0 && \text{for all} \quad e \in F \, ,
\end{aligned}
\tag{1}
$$

where W is some subset of V, $\mathcal{L}$ is a subset of $\mathcal{O}$, and F is a subset of E. Using an uncrossing argument, as in [31, Pages 141–150], we may assume that the collection of sets $\mathcal{L}$ is a laminar family of node sets; i.e., for any two distinct sets $S, T \in \mathcal{L}$, either $S \cap T = \varnothing$ or $S \subseteq T$ or $T \subseteq S$.

Lemma 3. *For every node* $v \in V$ *there exists* $M \in \mathcal{M}_{uni}$ *such that* v *is exposed by* M. *Hence,* $W = \varnothing$.

Proof. Assume for a contradiction that there exists a node $v \in V$ such that $v \in W$.

First, note that there always exists a non-universal matching $M \in \mathcal{M} \backslash \mathcal{M}_{uni}$ since otherwise the empty matching would be universal, and thus

$$0 = x^*(\varnothing) = w(\varnothing) + \varepsilon_1,$$

implying that the core of the given matching game instance is non-empty.

Suppose first that there exists a node $u \in V$ exposed by some matching $M' \in \mathcal{M}_{uni}$ such that $x_u^* > 0$. We define

$$\delta_0 := \min\{\operatorname{excess}(x^*, M) - \varepsilon_1 : \quad M \in \mathcal{M} \backslash \mathcal{M}_{uni}\}.$$

Recall that $\mathcal{M}_{uni}$ is the set of maximum weight matchings on G with respect to the node weights $w(uv) - x^*(uv)$, $uv \in E$, i.e. $\mathcal{M}_{uni}$ is the set of x^*-tight matchings. Moreover, recall that $\operatorname{excess}(x^*, M) = \varepsilon_1$ for $M \in \mathcal{M}_{uni}$. Thus, we have $\delta_0 > 0$.

We define $\delta := \min\{\delta_0, x_u^*\} > 0$ and a new allocation x' as follows:

$$x_r' := \begin{cases} x_r^* + \delta, & \text{if } r = v \\ x_r^* - \delta, & \text{if } r = u \\ x_r^*, & \text{otherwise.} \end{cases}$$

Since all universal matchings contain v, the excess with respect to x' of any universal matching is no smaller than their excess with respect to x^*. Therefore, by our choice of δ, (x', ε_1) is a feasible, and hence optimal, solution for (P_1). But M' is not x'-tight, since M' covers v and exposes u. This contradicts that M' is a universal matching.

Now consider the other case: for all $u \in V$ if u is exposed by a universal matching then $x_u^* = 0$. Then, for every universal matching $M \in \mathcal{M}_{uni}$ we have

$$\varepsilon_1 = \operatorname{excess}(x^*, M) = x^*(V) - w(M) = \nu(G) - w(M).$$

Since $\nu(G)$ is the maximum weight of a matching on G with respect to the weights w, we get that $\varepsilon_1 \geq 0$. Thus x^* is in the core, contradicting our assumption that the core is empty. ∎

2.3   Description of Leastcore

We denote inclusion-wise maximal sets in the family $\mathcal{L}$ as $S_1^*, S_2^*, \ldots, S_k^*$. We define the edge set E^+ to be the set of edges in G such that at most one of its nodes is in S_i^* for every $i \in [k] := \{1, \ldots, k\}$, i.e.

$$E^+ := E \setminus \left(\bigcup_{i=1}^{k} E(S_i^*) \right).$$

Lemma 4. *For every choice of $v_i \in S_i^*$, $i \in [k]$, there exists a universal matching $M \in \mathcal{M}_{uni}$ such that the node set covered by M is as follows*

$$\bigcup_{i=1}^{k} S_i^* \setminus \{v_i\}.$$

Proof. By Lemma 3, we know that for every $i \in [k]$ there exists a universal matching $M_{v_i} \in \mathcal{M}_{uni}$ such that v_i is exposed by M_{v_i}. Now, for every $i \in [k]$, let us define

$$M_i := E(S_i^*) \cap M_{v_i}.$$

Since M_i satisfies all laminar family constraints in $\mathcal{L}$ for subsets of S_i^* we have that

$$\bigcup_{i=1}^{k} M_i$$

is a matching satisfying all the constraints (1), and hence is a universal matching covering the desired nodes.   ∎

For each $i \in [k]$ fix a unique *representative* node $v_i^* \in S_i^*$. By Lemma 4, there exists a universal matching M^* covering precisely $\bigcup_{i \in [k]} S_i^* \setminus \{v_i^*\}$. For any $x \in P_1(\varepsilon_1)$ and $S \subseteq V$ we use $\mathrm{diff}(x, S)$ to denote

$$\mathrm{diff}(x, S) := x(S) - x^*(S).$$

For single nodes we use the shorthand $\mathrm{diff}(x, v) = \mathrm{diff}(x, \{v\})$. We now prove the following crucial structure result on allocations in the leastcore.

Lemma 5. *For every leastcore allocation x, i.e. for every $x \in P_1(\varepsilon_1)$, we have that*

(i) for all $i \in [k]$, for all $u \in S_i^*$: $\mathrm{diff}(x, u) = \mathrm{diff}(x, v_i^*)$,
(ii) for all $e \in E^+$: $\mathrm{excess}(x, e) \geq 0$.

Proof. Consider $u \in S_i^*$, and note that we may use Lemma 4 to choose a universal matching M_u covering precisely

$$S_i^* \setminus \{u\} \cup \bigcup_{j \neq i} S_j^* \setminus \{v_j^*\}.$$

Hence we have $V(M_u) \cup \{u\} = V(M^*) \cup \{v_i^*\}$, and since M^* and M_u are universal, $x(M^*) = x^*(M^*)$ and $x(M_u) = x^*(M_u)$. Using these observations we see that

$$
\begin{aligned}
sym(x, u) &= x(u) + x(M_u) - (x^*(u) - x^*(M_u)) \\
&= x(v_i^*) + x(M^*) - (x^*(v_i^*) - x^*(M^*)) = \mathrm{diff}(x, v_i^*).
\end{aligned}
$$

showing (i).

Now we prove (ii). Consider $e \in E^+$ where $e = \{u, v\}$. Since $e \notin E(S_i^*)$ for all $i \in [k]$, we can choose a universal matching M exposing u and v by Lemma 4. Thus $M \cup \{e\}$ is also a matching. Notice M is x-tight, and so we have

$$
\mathrm{excess}(x, e) = \underbrace{\mathrm{excess}(x, M \cup \{e\})}_{\geq \varepsilon_1} - \underbrace{\mathrm{excess}(x, M)}_{=\varepsilon_1} \geq 0
$$

as desired.   ∎

Lemma 6. *Let* $x \in P_1(\varepsilon_1)$ *and let* $M \in \mathcal{M}$ *be a matching such that* $M \subseteq \bigcup_{i \in [k]} E(S_i^*)$. *Then there exists* $M' \subseteq M^*$ *such that* $\mathrm{excess}(x, M') \leq \mathrm{excess}(x, M)$ *and for all* $i \in [k]$, $|M' \cap E(S_i^*)| = |M \cap E(S_i^*)|$.

Proof. See the arxiv version [27].   ∎

Recall that x^* is a fixed universal allocation in $P_1(\varepsilon_1)$. Let $E^* \subseteq E$ denote the union of universal matchings, i.e. $E^* = \bigcup_{M \in \mathcal{M}_{uni}} M$. We now define linear program $(\overline{P}_1)$.

$$
\max \quad \varepsilon \tag{$\overline{P}_1$}
$$

$$
\begin{aligned}
\text{s.t.} \quad & \mathrm{diff}(x, u) = \mathrm{diff}(x, v_i^*) && \text{for all} \quad u \in S_i^*, i \in [k] && (2) \\
& \mathrm{excess}(x, e) \leq 0 && \text{for all} \quad e \in E^* \\
& \mathrm{excess}(x, e) \geq 0 && \text{for all} \quad e \in E^+ \\
& \mathrm{excess}(x, M^*) = \varepsilon \\
& x(V) = \nu(G) \\
& x \geq 0.
\end{aligned}
$$

Let $\overline{\varepsilon}_1$ be the optimal value of the linear program $(\overline{P}_1)$. We now show that $\overline{P}_1(\overline{\varepsilon}_1)$ is indeed a compact description of the leastcore $P_1(\varepsilon_1)$.

Theorem 2. *We have* $\varepsilon_1 = \overline{\varepsilon}_1$ *and* $P_1(\varepsilon_1) = \overline{P}_1(\overline{\varepsilon}_1)$.

Proof. See Appendix A.

3   Computing the Nucleolus

The last section presented a polynomial-size formulation for the leastcore LP (P_1). In this section we complete our polynomial-time implementation of Maschler's scheme by showing that (P_j) has the following compact reformulation:

$$\max \quad \varepsilon \qquad\qquad\qquad (\overline{P}_j)$$

$$\begin{aligned}
\text{s.t.} \quad \text{excess}(x,e) &\geq \varepsilon - \varepsilon_1 && \text{for all} \quad e \in E^+,\ e \notin \text{Fix}(\overline{P}_{j-1}(\overline{\varepsilon}_{j-1})) \\
x(v) &\geq \varepsilon - \varepsilon_1 && \text{for all} \quad v \in V,\ v \notin \text{Fix}(\overline{P}_{j-1}(\overline{\varepsilon}_{j-1})) \\
\text{excess}(x,e) &\leq \varepsilon_1 - \varepsilon && \text{for all} \quad e \in E^*,\ e \notin \text{Fix}(\overline{P}_{j-1}(\overline{\varepsilon}_{j-1})) \\
x &\in \overline{P}_{j-1}(\overline{\varepsilon}_{j-1}),
\end{aligned}$$

where $\overline{\varepsilon}_j$ is the optimal value of the linear program $(\overline{P}_j)$.

Theorem 3. *For all $j = 1, \ldots, j^*$, we have $\varepsilon_j = \overline{\varepsilon}_j$ and $P_j(\varepsilon_j) = \overline{P}_j(\overline{\varepsilon}_j)$.*

Proof. We refer the reading to the arxiv version [27] for the proof.

With Theorem 3 we can replace each linear program (P_j) with $(\overline{P}_j)$ in Maschler's Scheme. Since the universal allocation x^*, the node sets S_i^*, $i \in [k]$, the edge set E^+, and the edge set E^* can all be computed in polynomial time, we have shown that the nucleolus of any cooperative matching game with empty core can be computed in polynomial time. Therefore we have shown Theorem 1.

Open Questions. Matching Games generalize naturally to b-matching games, where instead the underlying optimization problem is to find an edge subset M with $|M \cap \delta(v)| \leq b_v$ for each node v. Biro, Kern, Paulusma, and Wojuteczky [4] showed that the core-separation problem when $b_v > 2$ for some vertex v, is coNP-Hard. Despite this, the complexity of computing the nucleolus of these games is open.

Our algorithm relies heavily on the ellipsoid method. When the core is non-empty, there is a combinatorial algorithm for finding the nucleolus [3]. It would be interesting to develop a combinatorial algorithm for nucleolus computation of matching games in general.

Acknowledgements. The authors thank Umang Bhaskar, Daniel Dadush, and Linda Farczadi for stimulating and insightful discussions related to this paper.

Appendix

A   Proof of Theorem 2

Proof. First, we show that $P_1(\varepsilon_1) \subseteq \overline{P}_1(\varepsilon_1)$. Consider $x \in P_1(\varepsilon_1)$. By Lemma 5(i) we have

$$\text{diff}(x,u) = \text{diff}(x, v_i^*) \qquad \text{for all} \quad u \in S_i^*,\ i \in [k].$$

Lemma 5(ii) shows that $\mathrm{excess}(x, e) \geq 0$ for all $e \in E^+$, and $\mathrm{excess}(x, M^*) = \varepsilon_1$ holds by the universality of M^*. It remains to show that

$$\mathrm{excess}(x, e) \leq 0 \quad \text{for all} \quad e \in E^*.$$

Suppose for contradiction there exists $e \in E^*$ such that $\mathrm{excess}(x, e) > 0$. By the definition of E^*, there exists a universal matching M' containing e. Since M' is universal, $\mathrm{excess}(x, M') = \varepsilon_1$. But by our choice of e,

$$\mathrm{excess}(x, M' \setminus \{e\}) < \mathrm{excess}(x, M') = \varepsilon_1$$

contradicting that x is in $P_1(\varepsilon_1)$. Thus we showed that (x, ε_1) is feasible for $(\overline{P}_1)$, i.e. we showed that $P_1(\varepsilon_1) \subseteq \overline{P}_1(\varepsilon_1)$.

To complete the proof we show that $\overline{P}_1(\overline{\varepsilon}_1) \subseteq P_1(\overline{\varepsilon}_1)$. Let x be an allocation in $\overline{P}_1(\overline{\varepsilon}_1)$. Due to the description of the linear program (P_1), it is enough to show that for every matching $M \in \mathcal{M}$ we have

$$\mathrm{excess}(x, M) \geq \overline{\varepsilon}_1 .$$

Since $\mathrm{excess}(x, e) \geq 0$ for all $e \in E^+$, it suffices to consider only the matchings M, which are unions of matchings on the graphs $G[S_i^*]$, $i \in [k]$. Let $t_i := |M \cap E(S_i^*)|$. By Lemma 6 applied to x^* there exists $M' \subseteq M^*$ such that $\mathrm{excess}(x^*, M) \geq \mathrm{excess}(x^*, M')$ and $|M' \cap E(S_i^*)| = t_i$, for all $i \in [k]$. Then due to constraints (2) in $(\overline{P}_1)$ we have

$$\mathrm{excess}(x, M) = \underbrace{\sum_{i=1}^{k} 2t_i \, \mathrm{diff}(x, v_i^*)}_{=\mathrm{diff}(x, M')} + \underbrace{\mathrm{excess}(x^*, M)}_{\geq \mathrm{excess}(x^*, M')}$$

$$\geq \mathrm{excess}(x, M') \geq \mathrm{excess}(x, M^*) = \overline{\varepsilon}_1 ,$$

where the last inequality follows since $M' \subseteq M^*$ and $\mathrm{excess}(x, e) \leq 0$ for all $e \in E^*$.

Thus, we showed that $P_1(\varepsilon_1) \subseteq \overline{P}_1(\varepsilon_1)$ and $\overline{P}_1(\overline{\varepsilon}_1) \subseteq P_1(\overline{\varepsilon}_1)$. Recall, that ε_1 and $\overline{\varepsilon}_1$ are the optimal values of the linear programs (P_1) and $(\overline{P}_1)$ respectively. Thus, we have $\varepsilon_1 = \overline{\varepsilon}_1$ and $P_1(\varepsilon_1) = \overline{P}_1(\overline{\varepsilon}_1)$. ∎

B   Example of a Matching Game With Empty Core

Consider the example in Fig. 1. This graph $G = (V, E)$ is a 5-cycle with two adjacent edges 15 and 12 of weight 2, and the remaining three edges of weight 1. Since the maximum weight matching value is $\nu(G) = 3$, but the maximum weight fractional matching value is $\frac{7}{2}$, the core of this game is empty. The allocation x^* defined by $x^*(1) = \frac{7}{5}$ and $x^*(2) = x^*(3) = x^*(4) = x^*(5) = \frac{2}{5}$ lies in the leastcore. Each edge has the same excess, $-\frac{1}{5}$, and any coalition of four vertices yields a minimum excess coalition with excess $-\frac{2}{5}$. Hence the leastcore value of this game is $\varepsilon_1 = -\frac{2}{5}$.

In fact, we can see that x^* is the nucleolus of this game. To certify this we can use the result of Schmeidler [38] that the nucleolus lies in the intersection of the leastcore and the prekernel. For this example, the prekernel condition that for all $i \neq j \in V$,

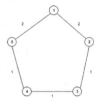

$$\max\{x(S \cup \{i\}) - \nu(S \cup \{i\}) : S \subseteq V \backslash \{j\}\}$$
$$= \max\{x(S \cup \{j\}) - \nu(S \cup \{j\}) : S \subseteq V \backslash \{i\}\}$$

Fig. 1. Matching game with empty core

reduces to the condition that the excess values of non-adjacent edges are equal. Since G is an odd cycle, this implies that all edges has equal excess, i.e.

$$\mathrm{excess}(x, 12) = \mathrm{excess}(x, 23) = \mathrm{excess}(x, 34)$$
$$= \mathrm{excess}(x, 45) = \mathrm{excess}(x, 15).$$

Combining the four equations above with the leastcore condition that $x(V) = \nu(G)$ we obtain a system of equations with the unique solution x^*. Hence the intersection of the leastcore and prekernel is precisely $\{x^*\}$, and so by Schmeidler, x^* is the nucleolus.

References

1. Aumann, R.J., Maschler, M.: Game theoretic analysis of a bankruptcy problem from the talmud. J. Econ. Theory **36**(2), 195–213 (1985)
2. Bateni, M.H., Hajiaghayi, M.T., Immorlica, N., Mahini, H.: The cooperative game theory foundations of network bargaining games. In: Abramsky, S., Gavoille, C., Kirchner, C., Meyer auf der Heide, F., Spirakis, P.G. (eds.) ICALP 2010. LNCS, vol. 6198, pp. 67–78. Springer, Heidelberg (2010). https://doi.org/10.1007/978-3-642-14165-2_7
3. Biró, P., Kern, W., Paulusma, D.: Computing solutions for matching games. Int. J. Game Theory **41**, 75–90 (2012)
4. Biró, P., Kern, W., Paulusma, D., Wojuteczky, P.: The stable fixtures problem with payments. Games Econ. Behav. **11**(9), 245–268 (2017)
5. Brânzei, R., Solymosi, T., Tijs, S.: Strongly essential coalitions and the nucleolus of peer group games. Int. J. Game Theory **33**(3), 447–460 (2005)
6. Chen, N., Lu, P., Zhang, H.: Computing the nucleolus of matching, cover and clique games. In: AAAI (2012)
7. Dadush, D.: Personal communication (2017)
8. Davis, M., Maschler, M.: The kernel of a cooperative game. Naval Res. Logist. Q. **12**(3), 223–259 (1965)
9. Deng, X., Fang, Q.: Algorithmic cooperative game theory. In: Chinchuluun, A., Pardalos, P.M., Migdalas, A., Pitsoulis, L. (eds.) Pareto Optimality, Game Theory and Equilibria. SOIA, vol. 17, pp. 159–185. Springer, New York (2008). https://doi.org/10.1007/978-0-387-77247-9_7

10. Deng, X., Fang, Q., Sun, X.: Finding nucleolus of flow game. J. Comb. Optim. **18**(1), 64–86 (2009)
11. Deng, X., Ibaraki, T., Nagamochi, H.: Algorithmic aspects of the core of combinatorial optimization games. Math. Oper. Res. **24**(3), 751–766 (1999)
12. Edmonds, J.: Maximum matching and a polyhedron with 0, 1-vertices. J. Res. Natl. Bur. Stan. B **69**(125–130), 55–56 (1965)
13. Edmonds, J.: Paths, trees, and flowers. Can. J. Math. **17**(3), 449–467 (1965)
14. Elkind, E., Goldberg, L.A., Goldberg, P., Wooldridge, M.: Computational complexity of weighted threshold games. In: Proceedings of the National Conference on Artificial Intelligence, p. 718 (2007)
15. Faigle, U., Kern, W., Fekete, S.P., Hochstättler, W.: The nucleon of cooperative games and an algorithm for matching games. Math. Program. **83**(1–3), 195–211 (1998)
16. Faigle, U., Kern, W., Kuipers, J.: Note computing the nucleolus of min-cost spanning tree games is NP-hard. Int. J. Game Theory **27**(3), 443–450 (1998)
17. Faigle, U., Kern, W., Kuipers, J.: On the computation of the nucleolus of a cooperative game. Int. J. Game Theory **30**(1), 79–98 (2001)
18. Faigle, U., Kern, W., Kuipers, J.: Computing an element in the lexicographic kernel of a game. Math. Methods Oper. Res. **63**(3), 427–433 (2006)
19. Faigle, U., Kern, W., Paulusma, D.: Note on the computational complexity of least core concepts for min-cost spanning tree games. Math. Methods Oper. Res. **52**(1), 23–38 (2000)
20. Farczadi, L.: Matchings and games on networks. University of Waterloo (2015)
21. Gillies, D.B.: Solutions to general non-zero-sum games. Contrib. Theory Games **4**(40), 47–85 (1959)
22. Granot, D., Granot, F., Zhu, W.R.: Characterization sets for the nucleolus. Int. J. Game Theory **27**(3), 359–374 (1998)
23. Granot, D., Maschler, M., Owen, G., Zhu, W.R.: The kernel/nucleolus of a standard tree game. Int. J. Game Theory **25**(2), 219–244 (1996)
24. Grötschel, M., Lovász, L., Schrijver, A.: Geometric Algorithms and Combinatorial Optimization, Algorithms and Combinatorics: Study and Research Texts, vol. 2. Springer, Heidelberg (1988). https://doi.org/10.1007/978-3-642-97881-4
25. Grötschel, M., Lovász, L., Schrijver, A.: Geometric Algorithms and Combinatorial Optimization, vol. 2. Springer, New York (2012)
26. Kern, W., Paulusma, D.: Matching games: the least core and the nucleolus. Math. Oper. Res. **28**(2), 294–308 (2003)
27. Koenemann, J., Pashkovich, K., Toth, J.: Computing the nucleolus of weighted cooperative matching games in polynomial time. arXiv preprint arXiv:1803.03249 (2018)
28. Kopelowitz, A.: Computation of the kernels of simple games and the nucleolus of n-person games. Technical report, Hebrew Univ Jerusalem (Israel) Dept of Mathematics (1967)
29. Kuhn, H.W.: The Hungarian method for the assignment problem. Naval Res. Logist. Q. **2**(1–2), 83–97 (1955)
30. Kuipers, J., Solymosi, T., Aarts, H.: Computing the nucleolus of some combinatorially-structured games. Math. Program. **88**(3), 541–563 (2000)
31. Lau, L.C., Ravi, R., Singh, M.: Iterative Methods in Combinatorial Optimization, vol. 46. Cambridge University Press, Cambridge (2011)
32. Lemaire, J.: An application of game theory: cost allocation. ASTIN Bull. J. IAA **14**(1), 61–81 (1984)

33. Maschler, M., Peleg, B., Shapley, L.S.: Geometric properties of the kernel, nucleolus, and related solution concepts. Math. Oper. Res. **4**(4), 303–338 (1979)
34. Megiddo, N.: Computational complexity of the game theory approach to cost allocation for a tree. Math. Oper. Res. **3**(3), 189–196 (1978)
35. Paulusma, D.: Complexity Aspects of Cooperative Games. Twente University Press, Enschede (2001)
36. Potters, J., Reijnierse, H., Biswas, A.: The nucleolus of balanced simple flow networks. Games Econ. Behav. **54**(1), 205–225 (2006)
37. Rothvoß, T.: The matching polytope has exponential extension complexity. J. ACM (JACM) **64**(6), 41 (2017)
38. Schmeidler, D.: The nucleolus of a characteristic function game. SIAM J. Appl. Math. **17**(6), 1163–1170 (1969)
39. Shapley, L.S., Shubik, M.: The assignment game I: the core. Int. J. Game Theory **1**(1), 111–130 (1971)
40. Solymosi, T., Raghavan, T.E.: An algorithm for finding the nucleolus of assignment games. Int. J. Game Theory **23**(2), 119–143 (1994)
41. Stearns, R.E.: Convergent transfer schemes for n-person games. Trans. Am. Math. Soc. **134**(3), 449–459 (1968)
42. Ziegler, G.M.: Lectures on Polytopes, vol. 152. Springer, New York (2012)

Breaking Symmetries to Rescue Sum of Squares: The Case of Makespan Scheduling

Victor Verdugo[1,2] and José Verschae[1(✉)]

[1] Institute of Engineering Sciences, Universidad de O'Higgins, Rancagua, Chile
{victor.verdugo,jose.verschae}@uoh.cl
[2] Department of Mathematics, London School of Economics, London, UK

Abstract. The Sum of Squares (SoS) hierarchy gives an automatized technique to create a family of increasingly tighter convex relaxations for binary programs. There are several problems for which a constant number of rounds give integrality gaps matching the best known approximation algorithm. In many other, however, ad-hoc techniques give significantly better approximation factors. The lower bounds instances, in many cases, are invariant under the action of a large permutation group. The main purpose of this paper is to study how the presence of symmetries on a formulation degrades the performance of the relaxation obtained by the SoS hierarchy. We do so for the special case of the minimum makespan problem on identical machines. Our first result is to show that a linear number of rounds of SoS applied over the *configuration linear program* yields an integrality gap of at least 1.0009. This improves on the recent work by Kurpisz et al. [30] that shows an analogous result for the weaker LS_+ and SA hierarchies. Then, we consider the weaker *assignment linear program* and add a well chosen set of symmetry breaking inequalities that removes a subset of the machine permutation symmetries. We show that applying the SoS hierarchy for $O_\varepsilon(1)$ rounds to this linear program reduces the integrality gap to $(1 + \varepsilon)$. Our results suggest that the presence of symmetries were the main barrier preventing the SoS hierarchy to give tight relaxations.

1 Introduction

The lift-and-project *hierarchies*, as Sherali & Adams (SA), Lovász & Schrijver (LS), or Sum of Squares (SoS), are systematic methods for obtaining a family of increasingly tight relaxations, parameterized on the *round* of the hierarchy. Arguably, it is not well understood for which problems these hierarchies yield relaxations that match the best possible approximation algorithm. Indeed, there are some positive results, but many other strong negative results for algorithmically easy problems. This shows a natural limitation to the power of hierarchies as one-fit-all technique. Quite remarkably, the instances used for obtaining lower

This work has been partially funded by Project Fondecyt Nr. 1181527.

A. Lodi and V. Nagarajan (Eds.): IPCO 2019, LNCS 11480, pp. 427–441, 2019.
https://doi.org/10.1007/978-3-030-17953-3_32

bounds often have a very symmetric structure [13,30,31,44,46], which suggests a strong connection between the tightness of the relaxation given by these hierarchies and symmetries. The main purpose of this work is to study this connection for a specific relevant problem, namely, minimum *makespan scheduling on identical machines*. This problem is one of the first considered under the lens of approximation algorithms [12], and since then it has been studied extensively. The input of the problem consists of a set J of n jobs, each having an integral processing time $p_j > 0$, and a set $M = [m]$ of m identical machines. Given an assignment $\sigma : J \to M$, the load of a machine i is the total processing time of jobs assigned to i, that is, $\sum_{j \in \sigma^{-1}(i)} p_j$. The objective is to find an assignment of jobs to machines that minimizes the *makespan*, that is, the maximum load. The problem is strongly NP-hard and admits several *polynomial-time approximation schemes* based on different techniques, as dynamic programming and some tractable versions of integer programming [1,2,7,15,16,19,20].

Integrality Gaps. The minimum makespan problem has two natural linear relaxations, which have been widely studied in the literature. The *assignment linear program* uses binary variables x_{ij} which indicate if job j is assigned to machine i. It is easy to see that its integrality gap is 2. The stronger *configuration linear program*, uses an exponential number of variables y_{iC} which indicate whether the set of jobs assigned to i has C as a multiset of processing times. Kurpisz et al. [30] showed that the configuration linear program has an integrality gap of at least $1024/1023 \approx 1.0009$ even after a linear number of rounds of the LS_+ or SA hierarchies. Hence, the same lower bound holds when the ground formulation is the assignment linear program. On the other hand, Kurpisz et al. [30] leave open if the SoS hierarchy applied to the configuration linear program has a $(1 + \varepsilon)$ integrality gap after constantly many rounds for constant $\varepsilon > 0$, i.e., $O_\varepsilon(1)$ rounds. Our first main contribution is a negative answer to this question.

Theorem 1. *Consider the problem of scheduling identical machines to minimize the makespan. For each $n \in \mathbb{N}$ there exists an instance with n jobs such that, after $\Omega(n)$ rounds of the SoS hierarchy over the configuration linear program, the obtained semidefinite relaxation has an integrality gap of at least 1.0009.*

Since the configuration linear program is stronger than the assignment linear program, our result holds if we apply $\Omega(n)$ rounds of SoS over the assignment linear program. The proof of the lower bound relies on tools from representation theory of symmetric groups over polynomials rings and it is inspired on the recent work by Raymond et al. for invariant sums of squares over k-hypercubes [47]. On one hand, the lower bound is obtained by constructing high-degree *pseudo-expectations*, and by obtaining symmetry-reduced decompositions of the polynomial ideal defined by the configuration liner program, on the other hand. The machinery from representation theory allows to restrict attention to invariant polynomials, and we combine this with a strong *pseudoindependence* result for a well chosen polynomial spanning set. Our analysis is also connected to the work of Razborov on flag algebras and we believe it can be of independent interest for analyzing lower bounds in the context of SoS in presence of symmetries [48,49].

Symmetries and Hierarchies. It is natural to explore whether symmetry handling techniques might help overcoming the limitation given by Theorem 1. A natural source of symmetry for this problem comes from the fact that the machines are identical: Given a schedule, we obtain other with the same makespan by permuting the assignment over the machines. In other words, the assignment linear program is *invariant* under the action of the symmetric group on the set of machines. The question we study is the following: *Is it possible to obtain a polynomial size linear or semidefinite program with an integrality gap of at most $(1 + \varepsilon)$ that is not invariant under the machine symmetries?* That is, our goal is to understand if the group action is deteriorating the quality of the relaxations obtained from the SoS hierarchy. This time, we provide a positive answer.

Theorem 2. *Consider the problem of scheduling identical machines to minimize the makespan. After adding a set of linearly many symmetry breaking inequalities to the assignment linear program, $O_\varepsilon(1)$ rounds of the SoS hierarchy yields a convex relaxation with an integrality gap of at most $(1 + \varepsilon)$, for any $\varepsilon > 0$.*

The theorem is based on introducing a formulation that *breaks* the symmetries in the invariant assignment program by adding new constraints. This enforces that any integer feasible solution of the formulation should respect a *lexicographic* order over the machine configurations. On top of the linear program obtained from adding the aforementioned constraints, we apply the SoS hierarchy. Using the *decomposition theorem* [23], we can can construct a solution that is integral on a well chosen set of machines M' of size $O_\varepsilon(1)$. Our symmetry breaking inequalities imply that between two consecutive machines in M', our solution assigns approximately the same configurations, and thus we can construct an approximately optimal solution.[1]

Literature Review

Upper Bounds. The first application of semidefinite programming in the context of approximation algorithms was by the work of Goemans & Williamson for Max-Cut [11]. There are not many positive results in this line for other combinatorial optimization problems, but of particular interest to our work is the SoS based approximation scheme by Karlin et al. to the Max-Knapsack problem [23]. They use a structural *decomposition theorem* satisfied by the SoS hierarchy, which makes a difference with respect to other classic hierarchies. Recently, for a constant number of machines, Levey and Rothvoss design an approximation scheme with a sub-exponential number of rounds in the weaker SA hierarchy [34], which was recently improved to a QPTAS [8]. A lot of attention has received the SoS method in order to design algorithms for high-dimensional problems. Among them we find matrix and tensor completion [5,45], tensor decomposition [37] and clustering [26,46].

---

[1] The current version gives a high-level exposition of our work. The full version with examples, figures, explanatory comments and technical details can be found in [54].

Lower Bounds. The first was obtained in the context of *positivstellensatz certificates* was by Grigoriev [13], showing the necessity of a linear number of SoS rounds to refute an easy Knapsack instance. Similar results were obtained by Laurent [31] for Max-Cut and by Kurpisz et al. for unconstrained polynomial optimization [27]. The same authors show that for a certain polynomial-time single machine scheduling problem, the SoS hierarchy exhibits an unbounded integrality gap even in a high-degree regime [27,29]. Remarkable are the work of Grigoriev [14] and Schoenebeck [52] exhibiting the difficulty for SoS to certify the insatisfiability of random 3-SAT instances in subexponential time, and recently there have been efforts on unifying frameworks to show lower bounds on random CSP's [3,24,25]. For estimation and detection problems, lower bounds have been shown for the planted clique problem, k-densest subgraph and tensor PCA, among others [4,17].

Invariant Sum of Squares. Remarkable in this line is the work of Gatermann & Parrillo, that studied how to obtain reduced sums of squares certificates of non-negativity when the polynomial is invariant under the action of a group, using tools from representation theory [9]. Recently, Raymond et al. developed on the Gatermann & Parrillo method to construct symmetry-reduced sum of squares certificates for polynomials over k-subset hypercubes [47]. Furthermore, the authors make an interesting connection with the Razborov method and flag algebras [48,49]. Blekherman et al. provided degree bounds on rational representations for certificates over the hypercube, recovering as a corollary known lower bounds for combinatorial optimization problems like Max-Cut [6,32]. Kurpisz et al. provide a method for proving SoS lower bounds when the formulations exhibits a high degree of symmetry, even under the presence of symmetries [28].

Symmetry Handling in Integer Programming. The integer programming community have dealt with symmetries by either breaking them [18,22,35], or devising symmetry-aware exact algorithms as isomorphism pruning [38], orbital branching [42] and orbitopal fixing [21]. Ostrowski [41] uses the SA hierarchy and reduces the dimension of the lifted relaxation to obtain a faster algorithm. For an extensive treatment we refer to the surveys by Margot [39] and Liberti [36].

2   Preliminaries: Sum of Squares and Pseudoexpectations

In what follows we denote by $\mathbb{R}[x]$ the ring of polynomials with real coefficients. Binary integer programming belongs to a larger class of problems in *polynomial optimization*, where the constraints are defined by polynomials in the variables indeterminates. More specifically, for given index sets M, J and E, consider the set $\mathcal{K}$ defined as

$$\Big\{ x \in \mathbb{R}^E : g_i(x) \geq 0 \quad \forall i \in M, \ h_j(x) = 0 \quad \forall j \in J, x_e^2 - x_e = 0 \quad \forall e \in E \Big\}, \quad (1)$$

where $g_i, h_j \in \mathbb{R}[x]$ for all $i \in M$ and for all $j \in J$. In particular, for binary integer programming the equality and inequality constraints are affine functions. We denote by $\mathbf{I}_E$ the ideal of polynomials in $\mathbb{R}[x]$ generated by $\{x_e^2 - x_e : e \in E\}$, and let $\mathbb{R}[x]/\mathbf{I}_E$ be the quotient ring of polynomials that vanish in the ideal $\mathbf{I}_E$. That is, $f, g \in \mathbb{R}[x]$ are in the same equivalence class of the quotient ring if $f - g \in \mathbf{I}_E$, that we denote $f \equiv g \mod \mathbf{I}_E$. Alternatively, $f \equiv g \mod \mathbf{I}_E$ if and only if the polynomials evaluate to the same values on the vertices of the hypercube, that is, $f(x) = g(x)$ for all $x \in \{0,1\}^E$. Observe that the equivalence classes in the quotient ring are in bijection with the *square-free* polynomials in $\mathbb{R}[x]$, that is, polynomials where no variable appears squared. In what follows we identify elements of $\mathbb{R}[x]/\mathbf{I}_E$ in this way. Given $S \subseteq E$, we denote by x_S the square-free monomial that is obtained from the product of the variables indexed by the elements in S, that is, $x_S = \prod_{e \in S} x_e$. The degree of a polynomial $f \in \mathbb{R}[x]/\mathbf{I}_E$ is denoted by $\deg(f)$, and we say that f is a *sum of squares* polynomial, for short SoS, if there exist polynomials $\{s_\alpha\}_{\alpha \in \mathcal{A}}$ for a finite family $\mathcal{A}$ in the quotient ring such that $f \equiv \sum_{\alpha \in \mathcal{A}} s_\alpha^2 \mod \mathbf{I}_E$.

Certificates and SoS *Method.* The question of certifying the infeasibility of (1) is hard in general but sometimes it is possible to find *simple* certificates of infeasibility. We say that there exists a degree-ℓ SoS certificate of infeasibility for $\mathcal{K}$ if there exist SoS polynomials s_0 and $\{s_i\}_{i \in M}$, and polynomials $\{r_j\}_{j \in J}$ such that

$$-1 \equiv s_0 + \sum_{i \in M} s_i g_i + \sum_{j \in J} r_j h_j \mod \mathbf{I}_E, \qquad (2)$$

and the degree of every polynomial in the right hand side is at most ℓ. The SoS algorithm iteratively checks the existence of a SoS certificate, parameterized in the degree, and each step of the algorithm is called a *round*. Since $|E|$ is an upper bound on the certificate degree, the method is guaranteed to terminate [33,43]. Furthermore, the existence of a degree-ℓ SoS certificate can be decided by solving a semidefinite program, in time $|E|^{O(\ell)}$. In the case of binary integer programming, if $\mathcal{K}$ is infeasible there exists a degree-ℓ SoS certificate, with $\ell \leq |E|$ [33]. We say that a linear functional $\widetilde{\mathbb{E}} : \mathbb{R}[x]/\mathbf{I}_E \to \mathbb{R}$ is a degree-ℓ SoS *pseudoexpectation* for (1), if it satisfies the following properties:

1. $\widetilde{\mathbb{E}}(1) = 1$,
2. $\widetilde{\mathbb{E}}(f^2) \geq 0$ for all $f \in \mathbb{R}[x]/\mathbf{I}_E$ with $\deg(f) \leq \ell/2$,
3. $\widetilde{\mathbb{E}}(f^2 g_i) \geq 0$ for all $i \in M$, for all $f \in \mathbb{R}[x]/\mathbf{I}_E$ with $\deg(f^2 g_i) \leq \ell$,
4. $\widetilde{\mathbb{E}}(f h_j) = 0$ for all $j \in J$, for all $f \in \mathbb{R}[x]/\mathbf{I}_E$ with $\deg(f h_j) \leq \ell$,

where $\bar{p}$ is the square-free representation of p after polynomial division by the Gröbner basis $\{x_e^2 - x_e : e \in E\}$. That is, $p \equiv \bar{p} \mod \mathbf{I}_E$.[2]

Lemma 3. *Assume that $\mathcal{K}$ defined in (1) is empty. If there exists a degree-ℓ SoS pseudoexpectation for $\mathcal{K}$ then there is no degree-ℓ SoS certificate of infeasibility.*

This lemma provides a way of finding lower bounds on the minimum value of a certificate's degree, and it will used in Sect. 3. In the following we refer to *low-degree* when the degree of a certificate or the pseudoexpectation is $O(1)$.

---

[2] In what follows, every time we evaluate a polynomial in the pseudoexpectation we are doing it over the square-free representation. We omit the bar notation for simplicity.

3   Lower Bound: Symmetries Are Hard for SoS

In what follows we show that the SoS method fails to provide a low-degree certificate of infeasibility for a certain family of scheduling instances. The program
we analyze is known as the *configuration linear program*, that has proven to
be powerful for different scheduling and packing problems [10, 53]. Given a value
$T > 0$, a *configuration* corresponds to a multiset of processing times such that its
total sum does not exceed T. The multiplicity $m(p, C)$ indicates the number of
times that the processing time p appears in the multiset C. The *load* of a configuration C is just the total processing time, $\sum_{p \in \{p_j : j \in J\}} m(p, C) \cdot p$. Given T, let
$\mathcal{C}$ denote the set of all configurations with load at most T. For each combination
of $i \in M$ and a configuration $C \in \mathcal{C}$, the program has a variable y_{iC} that models
whether machine i is scheduled according to configuration C. Let n_p denote the
number of jobs in J with processing time p. Consider the formulation, clp(T),
given by (i) $\sum_{C \in \mathcal{C}} y_{iC} = 1$ for all $i \in M$, (ii) $\sum_{i \in M} \sum_{C \in \mathcal{C}} m(p, C) y_{iC} = n_p$ for
all $p \in \{p_j : j \in J\}$ and (iii) $y_{iC} \in \{0, 1\}$ for all $i \in M, C \in \mathcal{C}$. The configuration
linear program corresponds to the linear relaxation where the last constraint is
changed to $y_{iC} \geq 0$. We briefly describe the construction of a family of hard
instances $\{I_k\}_{k \in \mathbb{N}}$ for the configuration linear program in [30]. Let $T = 1023$,
and for each odd $k \in \mathbb{N}$ we have $n = 15k$ jobs and $3k$ machines. There are 15
different job-sizes with value $O(1)$, each one with multiplicity k. There exist a
set of special configurations $\{C_1, \ldots, C_6\}$, called *matching configurations*, such
that the program above is infeasible if and only if the program restricted to the
matching configurations is infeasible [30, Lemma 2].

Theorem 4 ([30]). *For each odd k, there exists a degree-$\lfloor k/2 \rfloor$ SA pseudoexpectation for the configuration linear program.*

3.1   A Symmetry-Reduced Decomposition of the Scheduling Ideal

Given $T > 0$, the variable set of the configuration linear program is $E = [m] \times
\mathcal{C}$, and the symmetric group S_m acts over the monomials in $\mathbb{R}[y]$ according to
$\sigma y_{iC} = y_{\sigma(i)C}$ for every $\sigma \in S_m$. The action extends linearly to $\mathbb{R}[y]/\mathbf{I}_E$, and
the configuration linear program is invariant under this action, that is, for every
$y \in$ clp(T) and every $\sigma \in S_m$ we have $\sigma y \in$ clp(T). We say that a polynomial
$f \in \mathbb{R}[y]/\mathbf{I}_E$ is S_m-invariant if $\sigma f = f$ for every $\sigma \in S_m$. In particular, if f is
invariant we have $f = 1/|S_m| \sum_{\sigma \in S_m} \sigma f := \text{sym}(f)$, which is the symmetrization
or Reynolds operator of the group action. We say that a linear function $\mathcal{L}$ over
the quotient ring is S_m-symmetric if for every polynomial $f \in \mathbb{R}[y]/\mathbf{I}_E$ we have
$\mathcal{L}(f) = \mathcal{L}(\text{sym}(f))$.

Lemma 5. *Let $\widetilde{\mathbb{E}}$ be a symmetric linear operator over $\mathbb{R}[y]/\mathbf{I}_E$ such that for
every invariant SoS polynomial g of degree at most ℓ we have $\widetilde{\mathbb{E}}(g) \geq 0$. Then,
$\widetilde{\mathbb{E}}(f^2) \geq 0$ for every $f \in \mathbb{R}[y]/\mathbf{I}_E$ with $\deg(f) \leq \ell/2$.*

Therefore, when $\widetilde{\mathbb{E}}$ is symmetric we restrict our attention to those polynomials that are invariant and SoS. To analyze the action of the symmetric

group over $\mathbb{R}[y]$ we introduce some tools from representation theory to characterize the invariant S_m-modules of the polynomial ring [51]. We say that a S_m-module V is *irreducible* if the only invariant subspaces are $\{0\}$ and V. Any S_m-module V can be decomposed into irreducible modules, and the decomposition is indexed by the *partitions* of m. A partition of m is a vector $(\lambda_1, \ldots, \lambda_t)$ such that $\lambda_1 \geq \lambda_2 \geq \cdots \lambda_t > 0$ and $\lambda_1 + \cdots + \lambda_t = m$. We denote by $\lambda \vdash m$ when λ is a partition of m. Then, V can be decomposed as $V = \bigoplus_{\lambda \vdash m} V_\lambda$, that is, a direct sum where each V_λ is an irreducible S_m-module of V [51]. Each of the subspaces in the direct sum is called an *isotypic component*. A *tableau* of shape λ is a bijective filling between $[m]$ and the cells of a grid with t rows, and every row $r \in [t]$ has length λ_r. In this case, the *shape* or *Young diagram* of the tableau is λ. For a tableau τ_λ of shape λ, we denote by $\mathrm{row}_r(\tau_\lambda)$ the subset of $[m]$ that fills row r in the tableau. The row group $\mathbf{R}_{\tau_\lambda}$ is the subgroup of S_m that stabilizes the rows of the tableau τ_λ, that is,

$$\mathbf{R}_{\tau_\lambda} = \Big\{ \sigma \in S_m : \sigma \cdot \mathrm{row}_r(\tau_\lambda) = \mathrm{row}_r(\tau_\lambda) \text{ for every } r \in [t] \Big\}.$$

Invariant SoS *Polynomials.* Let $\mathbf{Q}^\ell$ be the quotient ring $\mathbb{R}[y]/\mathbf{I}_E$ restricted to polynomials of degree at most ℓ and let $\mathbf{Q}^\ell = \bigoplus_{\lambda \vdash m} \mathbf{Q}^\ell_\lambda$ be its isotypic decomposition. Given a tableau τ_λ of shape λ, let $\mathbf{W}_{\tau_\lambda}$ be the *row subspace* of fixed points in $\mathbf{Q}^\ell$ for the row group $\mathbf{R}_{\tau_\lambda}$, that is, $\mathbf{W}_{\tau_\lambda} = \{q \in \mathbf{Q}^\ell_\lambda : \sigma q = q \text{ for all } \sigma \in \mathbf{R}_{\tau_\lambda}\}$. The following result follows from the work of Gaterman and Parrillo [9] and the recent work of Raymond et al. [47] in the context of symmetry reduction for invariant semidefinite programs. In what follows, $\langle A, B \rangle$ denotes the trace of AB. Given $\ell \in [m]$, let Λ_ℓ be the partitions of m that are lexicographically larger than $(m - \ell, 1, \ldots, 1)$.

Theorem 6. *Suppose that $g \in \mathbb{R}[y]/\mathbf{I}_E$ is a degree-ℓ SoS and S_m-invariant polynomial. For each partition $\lambda \in \Lambda_\ell$, let τ_λ be a tableau of shape λ and let $\mathcal{P}^\lambda = \{p_1^\lambda, \ldots, p_{\ell_\lambda}^\lambda\}$ be a set of polynomials such that $\mathrm{span}(\mathcal{P}^\lambda) \supseteq \mathbf{W}_{\tau_\lambda}$. Then, for each partition $\lambda \in \Lambda_\ell$ there exists a $\ell_\lambda \times \ell_\lambda$ positive semidefinite matrix M_λ such that $g = \sum_{\lambda \in \Lambda_\ell} \langle M_\lambda, Z^\lambda \rangle$, where $Z_{ij}^\lambda = \mathrm{sym}(p_i^\lambda p_j^\lambda)$.*

Together with Lemma 5, it is enough to study pseudoexpectations for each of the partitions in Λ_ℓ separately. Theorem 6 gives us flexibility in the spanning set that we use for describing the row subspaces. If A is a matrix with entries in $\mathbb{R}[y]/\mathbf{I}_E$, let $\widetilde{\mathbb{E}}(A)$ be the matrix obtained by applying $\widetilde{\mathbb{E}}$ to each entry of A.

Lemma 7. *Suppose that for each $\lambda \in \Lambda_\ell$, the spanning set $\mathcal{P}^\lambda$ of $\mathbf{W}_{\tau_\lambda}$ is such that $\widetilde{\mathbb{E}}(Z^\lambda)$ is positive semidefinite. Then, if $\deg(f) \leq \ell/2$ we have $\widetilde{\mathbb{E}}(f^2) \geq 0$.*

3.2 Spanning Sets of the Scheduling Ideal

In this section we show how to construct the spanning sets of the row subspaces in order to apply Lemma 7, which together with a particular linear operator provides the existence of a high-degree SoS pseudoexpectation. We say that $S \subseteq [m] \times \mathcal{C}$ is a *partial schedule* if for every $i \in [m]$ we have $\delta_S(i) \leq 1$, where δ_S

is the vertex degree in the (directed) bipartite graph G_S with vertex partition $[m]$ and $\mathcal{C}$, and edges S. We denote by $\mathcal{M}(S)$ the set of machines incident to a partial schedule S, that is, $\{i \in [m] : \delta_S(i) = 1\}$. Sometimes is convenient to see a partial schedule S as a function from $\mathcal{M}(S)$ to $\mathcal{C}$, so we say that S is partial schedule with domain $\mathcal{M}(S)$. Let **sched** be the ideal of polynomials in $\mathbb{R}[y]$ generated by $\left\{\sum_{C \in \mathcal{C}} y_{iC} - 1 : i \in [m]\right\} \cup \left\{y_{iC}^2 - y_{iC} : i \in [m], C \in \mathcal{C}\right\}$. Let $\mathbf{Q}_{\text{sched}}^{\ell}$ be the polynomials in **sched** with degree equal to ℓ that vanish in the ideal **sched**.

Theorem 8. $\mathbf{Q}_{\text{sched}}^{\ell}$ *is spanned by* $\{y_S : |S| = \ell \text{ and } S \text{ is a partial schedule}\}$.

3.3   Spanning Sets of the Invariant Row Subspace

In previous section we provided a reduced spanning set for the quotient ring vanishing in **sched**. In the following we construct spanning sets for the invariant row subspaces. Given a tableau τ_λ with shape λ, the hook(τ_λ) is the tableau with shape $(\lambda_1, 1, \ldots, 1) \in \mathbb{Z}^{m-\lambda_1+1}$, its first row it is equal to the first row of τ_λ and the remaining elements of τ_λ fill the rest of the cells in increasing order over the rows. That part is called the *tail* of the hook, and we denote by tail(τ_λ) the elements of $[m]$ in the tail of hook(τ_λ). We say that a partial schedule is in γ-*profile*, with $\gamma : \mathcal{C} \to \mathbb{Z}_+$, if for every $C \in \mathcal{C}$ we have $\delta_S(C) = \gamma(C)$. A partial schedule in γ-profile has size $\sum_{C \in \mathcal{C}} \gamma(C)$, a quantity that we denote by $\|\gamma\|$. We denote by supp(γ) the support of the vector γ, namely, $\{C \in \mathcal{C} : \gamma(C) > 0\}$.

Definition 9. *Given a partial schedule T, we say that a partial schedule A over $[m] \setminus \mathcal{M}(T)$ is a (T, γ)-extension if A is in γ-profile. We denote by $\mathcal{F}(T, \gamma)$ the set of (T, γ)-extensions. In particular, every (T, γ)-extension has size $\|\gamma\|$.*

Given a partial schedule T and a γ-profile, let $\mathcal{B}_{T,\gamma}$ be the polynomial defined by $\mathcal{B}_{T,\gamma} = \sum_{A \in \mathcal{F}(T,\gamma)} y_A$, if $\gamma \neq 0$, and 1 otherwise. In words, this polynomial corresponds to the sum over all those partial schedules in γ-profile that are not incident to $\mathcal{M}(T)$. The following theorem is the main result of this section.

Theorem 10. *Let $\lambda \in \Lambda_\ell$ and a tableau τ_λ of shape λ. Then, the row subspace W_{τ_λ} of $\mathbf{Q}_{\text{sched}}^{\ell}$ is spanned by $\mathcal{P}^\lambda$, defined as*

$$\bigcup_{\omega=0}^{\ell} \left\{y_T \mathcal{B}_{T,\gamma} : T \text{ is partial schedule with } \mathcal{M}(T) = \text{tail}(\tau_\lambda) \text{ and } \|\gamma\| = \omega\right\}. \quad (3)$$

3.4   High-Degree SoS Pseudoexpectation: Proof of Theorem 1

Recall that for every k odd, the hard instance I_k has $m = 3k$ machines. Also, the linear operators we consider are supported over partial schedules incident to a set of *matching configurations*, $\{C_1, \ldots, C_6\}$. Consider $\widetilde{\mathbb{E}} : \mathbb{R}[y]/\mathbf{I}_E \to \mathbb{R}$ such that for every partial schedule S of cardinality at most $k/2$ we have $\widetilde{\mathbb{E}}(y_S) = \frac{1}{(3k)_{|S|}} \prod_{j=1}^{6} (k/2)_{\delta_S(C_j)}$, where $(a)_b = a(a-1) \cdots (a-b+1)$, and $(a)_0 = 1$; $\widetilde{\mathbb{E}}$ is zero elsewhere. We state the main result that implies Theorem 1.

Theorem 11. *For every odd k, the operator $\widetilde{\mathbb{E}}$ is a degree-$\lfloor k/6 \rfloor$ SoS pseudoexpectation for the configuration linear program in instance I_k and $T = 1023$.*

Theorem 4 guarantees that for every k odd, $\widetilde{\mathbb{E}}$ is a degree-$\lfloor k/2 \rfloor$ SA pseudoexpectation, and therefore a degree-$\lfloor k/6 \rfloor$ SA pseudoexpectation as well. In particular, properties (1) and (4) are satisfied. Since the configuration linear program is constructed from equality constraints, it is enough to check property (2) for high enough degree, in this case $\ell = \lfloor k/6 \rfloor$. To check property (2) we require a notion of *conditional pseudoexpectations.* Given a partial schedule T, consider the operator $\widetilde{\mathbb{E}}_T : \mathbb{R}[y]/\mathbf{I}_E \to \mathbb{R}$ such that $\widetilde{\mathbb{E}}_T(y_S) = \frac{1}{(3k-|T|)!} \prod_{j=1}^{6}(k/2 - \delta_T(C_j))_{\delta_S(C_j)}$ for every partial schedule S over the machines $[m] \setminus \mathcal{M}(T)$ and zero otherwise.

Lemma 12. *If T, S are disjoint partial schedules, then $\widetilde{\mathbb{E}}(y_T y_S) = \widetilde{\mathbb{E}}_T(y_S) \widetilde{\mathbb{E}}(y_T)$.*

Lemma 13. *Let T be a partial schedule and γ, μ a pair of configuration profiles with $|T| + \|\gamma\| + \|\mu\| \le k/2$ and $\mathrm{supp}(\gamma)$, $\mathrm{supp}(\mu) \subseteq \{C_1, \ldots, C_6\}$. Then, $\widetilde{\mathbb{E}}_T(\mathcal{B}_{T,\gamma} \mathcal{B}_{T,\mu}) = \widetilde{\mathbb{E}}_T(\mathcal{B}_{T,\gamma}) \widetilde{\mathbb{E}}_T(\mathcal{B}_{T,\mu})$.*

Proof Idea of Theorem 11. By Lemma 7 it is enough to prove that $\widetilde{\mathbb{E}}(Z^\lambda)$ obtained from the spanning set in (3) is positive semidefinite for each partition $\lambda \in \lambda_\ell$. By Lemma 12, $\widetilde{\mathbb{E}}(Z^\lambda)$ is block diagonal, with one block for each partial schedule with domain tail(τ_λ). By Lemma 13 each block is a rank-1 matrix, and therefore positive semidefinite. The detailed proof can be found in the full version.

4   Breaking Symmetries to Approximate with SoS

In the previous section we showed that the action of the symmetric group is hard to tackle for the SoS method. In the following we show how to obtain almost optimal programs in terms of integrality gap if we apply the SoS hierarchy after breaking symmetries on the *assignment linear program.*[3] In this model, there are variables x_{ij} indicating whether job j is assigned to machine i. For a given guess on the optimal makespan T, consider the formulation assign(T) defined by: (i) $\sum_{i \in M} x_{ij} = 1$ for all $j \in J$, (ii) $\sum_{j \in J} x_{ij} p_j \le T$ for all $i \in M$, and (iii) $x_{ij} \in \{0, 1\}$ for all i, j. The assignment linear program corresponds to the linear relaxation where the last constraint is changed to $x_{ij} \ge 0$. The symmetric group S_m acts over the monomials in $\mathbb{R}[x]$ according to $\sigma x_{ij} = x_{\sigma(i)j}$, for every $\sigma \in S_m$. The action extends linearly to $\mathbb{R}[x]/\mathbf{I}_E$, and assign($T$) is invariant under this action, that is, for every $x \in$ assign(T) and $\sigma \in S_m$ we have $\sigma x \in$ assign(T).

---

[3] Recently, it was noted that the SoS hierarchy cannot necessarily be solved in polytime, even for a constant number of rounds [40]. As we will show in the journal version, this is not an issue as the Sherali-Adams hierarchy suffices for our purposes.

4.1   Symmetry Breaking Inequalities

In what follows we break symmetries by forcing a specific order on the configurations over the machines. Suppose we have a partitioning $\mathcal{J}$ of the jobs set J into s parts, $\mathcal{J} = \{J_1, \ldots, J_s\}$. For example, suppose the job sizes are ordered from largest to smallest, that is $p^1 > p^2 > \cdots > p^s$ where $s = |\{p_j : j \in J\}|$ is the number of different job sizes. For the case $J_q = \{j \in J : p_j = p^q\}$ we call $\mathcal{J}$ the *job-sizes partition*. Given a partitioning of the jobs, a *configuration C* is a multiset of elements in $\{1, \ldots, s\}$. Recall that for every $q \in \{1, \ldots, s\}$, the multiplicity of q in C, $m(q, C)$, is the number of times that q appears repeated in C. As we did before, we denote by $\mathcal{C}$ the set of all configurations. Observe that it coincides with the configuration notion introduced in the context of the configuration linear program if we consider the job-sizes partition. We say that a configuration C is *lexicographically* larger than S, and we denote $C >_{\text{lex}} S$, if there exists $q \in [s]$ such that $m(\ell, C) = m(\ell, S)$ for all $\ell < q$ and $m(q, C) > m(q, S)$. The relation $>_{\text{lex}}$ defines a total order over $\mathcal{C}$. Given a positive integer B and a partitioning $\mathcal{J}$ of the jobs, consider the program $\text{assign}(B, T)$ obtained by intersecting $\text{assign}(T)$ with a set of *symmetry breaking inequalities*,

$$\text{assign}(T) \cap \bigcap_{i=1}^{m-1} \left\{ x \in \mathbb{R}^{M \times J} : \sum_{q=1}^{s} B^{s-q} \sum_{j \in J_q} \left(x_{ij} - x_{(i+1)j} \right) \geq 0 \right\}.$$

We remark that the symmetry breaking constraints depend on the partitioning of J. In the following we show that for sufficiently large, but polynomially sized B, every solution in $\text{assign}(B, T)$ obeys the lexicographic order on configurations. Given a feasible integer solution $x \in \text{assign}(T)$ and a machine $i \in M$, let $\text{conf}_i(x) \in \mathcal{C}$ be the configuration defined by the number of jobs for each possible part that are scheduled in i according to x, that is, for every $q \in \{1, \ldots, s\}$, $m(q, \text{conf}_i(x)) = \sum_{j \in J_q} x_{ij}$.

Theorem 14. *There exists $B^* = O(|J|^2)$ such that for all (integral) solution $x \in \text{assign}(B, T)$ and for each $i \in M \setminus \{m\}$, we have $\text{conf}_i(x) \geq_{\text{lex}} \text{conf}_{i+1}(x)$.*

4.2   Balanced Partitionings

Now we will show that low-degree SoS pseudoexpectations of $\text{assign}(B, T)$ can be rounded to obtain integral solutions with almost optimal makespan, where the degree depends only on the number of configurations in $\mathcal{C}$ and the size of $\mathcal{J}$. A partitioning $\mathcal{J}$ is *α-balanced*, for $\alpha \geq 1$, if for every $K, H \subseteq J$ such that $\text{conf}(K) = \text{conf}(H)$, $\sum_{j \in K} p_j \leq \alpha \sum_{j \in H} p_j$. Observe that the job-sizes partitioning is 1-balanced. A key parameter is the maximum number of jobs that can be assigned to a machine with makespan T, that is, $\lambda = \max\{|K| : \sum_{j \in K} p_j \leq T\}$. Recall that $\mathcal{C}$ depends on the partitioning and let $\tau(\mathcal{C}) = 2\lambda|\mathcal{C}|$.

Theorem 15. *Consider a value $T > 0$ and an α-balanced partitioning of J. Suppose there exists a degree-$\tau(\mathcal{C})$ SoS pseudoexpectation for $\text{assign}(B^*, T)$. Then, we can find in polynomial time an integral solution $x^{\text{lex}} \in \text{assign}(B, \alpha T)$.*

For every $\varepsilon > 0$ we show how to construct a $(1 + \varepsilon)$-balanced partitioning, which combined with the previous theorem yields the approximation scheme. An overview can be found in the appendix and the full proof in the full version [54].

A   SDP Based Approximation Scheme: Proof Sketch of Theorem 2

The subset of long jobs is denoted by J_{long} and the short jobs are $J_{\text{short}} = J \setminus J_{\text{long}}$. We consider a partitioning obtained by grouping jobs with a *similar* processing time.

Lemma 16. *For every $\varepsilon > 0$, there exists a partitioning, called ε-partitioning, that is $(1 + \varepsilon)$-balanced.*

Consider $\text{assign}(B^*, T)$ obtained from the ε-partitionings and according to Theorem 14. Using binary search we look for the smallest T such that there exists a degree-$\tau(\mathcal{C})$ SoS pseudoexpectation for the long jobs. Theorem 15 constructs a schedule for the long jobs. The short jobs are scheduled greedily. Details can be found in the full version [54].

B   Pseudoexpectation Rounding: Proof Overview of Theorem 15

Observe that for the hard instances shown for the configuration linear program, for $T = 1023$ we have that $\tau(\mathcal{C}) = O(1)$. Recall that for $T = 1023$ there is no feasible schedule for the instance and the job-sizes partition is 1-balanced. Therefore, for every odd k, Theorem 15 guarantees that the degree of a SoS pseudoexpectation in $\text{assign}(B^*, T)$ is upper bounded by a constant, and therefore there is a low-degree SoS certificate of infeasiblity.

One of the key tools in our rounding algorithm is the notion of *pseudoexpectation conditioning*. Consider a degree-ℓ pseudoexpectation $\widetilde{\mathbb{E}}$, let $i \in M$ be a machine and $K \subseteq J$ such that $\widetilde{\mathbb{E}}\left(\prod_{j \in K} x_{ij} \prod_{j \in J \setminus K} (1 - x_{ij}) \right) > 0$. Observe that the polynomial in this expression is equal to 1 if and only if $x_{ij} = 1$ for every $j \in K$ and $x_{ij} = 0$ for every $j \in J \setminus K$. That is, machine i is scheduled integrally with the jobs in K. For simplicity, we call $\phi_{i,K} = \prod_{j \in K} x_{ij} \prod_{j \in J \setminus K} (1 - x_{ij})$. The (i, K)-*conditioning* of $\widetilde{\mathbb{E}}$ corresponds to the linear operator over $\mathbb{R}[x]/\mathbf{I}_E$ defined by $\widetilde{\mathbb{E}}_{i,K}(x_I) = \widetilde{\mathbb{E}}(x_I \phi_{i,K}) / \widetilde{\mathbb{E}}(\phi_{i,K})$, for every $I \subseteq M \times J$. Intuitively, the (i, K)-conditioning is the pseudoexpectation value obtained by conditioning on the event that machine i is scheduled integrally with the jobs in K.

Lemma 17. *Let $\widetilde{\mathbb{E}}$ be a degree-ℓ pseudoexpectation with $\ell \geq 2\lambda$ and consider a machine $i \in M$. Then, the following holds:*

(a) If $\widetilde{\mathbb{E}}(\phi_{i,K}) > 0$, then $|K| \leq \lambda$.
(b) $\widetilde{\mathbb{E}}_{i,K}(x_{ij}) = 1$ for every $j \in K$ and $\widetilde{\mathbb{E}}_{i,K}(x_{ij}) = 0$ for every $j \in J \setminus K$.

(c) If there exists $H \subseteq J$ such that $\widetilde{\mathbb{E}}(\phi_{i,H}) > 0$, then $\sum_{K \subseteq J} \widetilde{\mathbb{E}}(\phi_{i,K}) = 1$ and $\widetilde{\mathbb{E}} = \sum_{K \subseteq J : \widetilde{\mathbb{E}}(\phi_{i,K}) > 0} \widetilde{\mathbb{E}}(\phi_{i,K}) \cdot \widetilde{\mathbb{E}}_{i,K}$.

In particular, property (c) above justifies the intuition behind since it decomposes the pseudoexpectation as a convex combination of conditionings. We now state the Decomposition Theorem adapted to the assignment linear program in the language of pseudoexpectations. It was originally introduced using the moments approach, but they are equivalent and we refer to [50] for a proof and a detailed exposition of the SoS hierarchy.

Theorem 18 ([23]). *Let $\widetilde{\mathbb{E}}$ be a degree-ℓ SoS pseudoexpectation of* assign(B, T), *with $\ell \geq 2\lambda$. Then, for every machine $i \in M$ and a subset of jobs $K \subseteq J$ such that $\widetilde{\mathbb{E}}(\phi_{i,K}) > 0$, the operator $\widetilde{\mathbb{E}}_{i,K}$ is a degree-$(\ell - 2\lambda)$ SoS pseudoexpectation of* assign(B, T).

Lemma 17 (b) guarantees that a conditioning $\widetilde{\mathbb{E}}_{i,K}$ is integral for machine i and this machine is scheduled with exactly the jobs in K, when $\widetilde{\mathbb{E}}(\phi_{i,K}) > 0$. In our algorithm we iteratively decompose the current pseudoexpectation according to the above conditionings. Every time we perform this step we obtain an integral machine schedule, and therefore in order to progress we require that machine to remain integral along the execution.

Proposition 19. *Let $\widetilde{\mathbb{E}}$ be a degree-ℓ SoS pseudoexpectation of* assign(B^*, T), *with $\ell \geq 2\lambda$, and let $h \in M$ be such that $\widetilde{\mathbb{E}}(x_{hj}) \in \{0,1\}$ for each $j \in J$. Let $i \in M$ and $K \subseteq J$ be such that $\widetilde{\mathbb{E}}_{i,K}(\phi_{i,K}) > 0$. Then, $\widetilde{\mathbb{E}}_{i,K}(x_{hj}) = \widetilde{\mathbb{E}}(x_{hj}) \in \{0,1\}$ for each $j \in J$.*

Overview of the Rounding Algorithm. Consider a partitioning of the jobs that is α-balanced. If we start from a high enough level of the hierarchy, we get at the end of the procedure a solution that is feasible for assign(B^*, T), and therefore, the configurations of the integral machines have to obey the lexicographic order. The algorithm consist of two phases. In Phase 1, we use the solution obtained from high enough level of the hierarchy to find the last machine which is fractionally scheduled according to configuration C^1 using the Decomposition Theorem, and pick the corresponding conditioning pseudoexpectation. We then proceed by finding the last machine scheduled fractionally according to C^2 in the pseudoexpectation conditioning, and so on, for every configuration C^k. We end up with a pseudoexpectation that is integral for all these machines, and it respects the lexicographic order. The number of conditioning steps is upper bounded by the number of configurations. In Phase 2, we greedily construct the schedule for the rest of the machines and jobs that have not been assigned yet. The correctness of Phase 2 is guaranteed by certifying the feasibility of a particular transportation problem. We call the schedule obtained in this way the *lexicographic schedule*, x^{lex}. The detailed analysis of the rounding algorithm can be found in the full version.

References

1. Alon, N., Azar, Y., Woeginger, G.J., Yadid, T.: Approximation schemes for scheduling. In: Proceedings of the Eighth Annual ACM-SIAM Symposium on Discrete Algorithms, SODA 2018, pp. 493–500, ACM-SIAM (1997)
2. Alon, N., Azar, Y., Woeginger, G.J., Yadid, T.: Approximation schemes for scheduling on parallel machines. J. Sched. **1**(1), 55–66 (1998)
3. Barak, B., Chan, S.O., Kothari, P.K.: Sum of squares lower bounds from pair wise independence. In: Proceedings of the Forty-Seventh Annual ACM Symposium on Theory of Computing, STOC 2015, pp. 97–106. ACM (2015)
4. Barak, B., Hopkins, S.B., Kelner, J., Kothari, P., Moitra, A., Potechin, A.: A nearly tight sum-of-squares lower bound for the planted clique problem. In: Proceedings of Foundations of Computer Science, FOCS 2016, pp. 428–437. IEEE (2016)
5. Barak, B., Moitra, A.: Noisy tensor completion via the sum-of-squares hierarchy. In: Proceedings of the 29th Conference on Learning Theory, COLT 2016, pp. 417–445 (2016)
6. Blekherman, G., Gouveia, J., Pfeiffer, J.: Sums of squares on the hypercube. Mathematische Zeitschrift **284**(1–2), 41–54 (2016)
7. Eisenbrand, F., Weismantel, R.: Proximity results and faster algorithms for integer programming using the Steinitz lemma. In: Proceedings of the Twenty-Ninth Annual ACM-SIAM Symposium on Discrete Algorithms, SODA 2018, pp. 808–816 (2018)
8. Garg, S.: Quasi-PTAS for scheduling with precedences using LP hierarchies. In: Proceedings of the 45th International Colloquium on Automata, Languages, and Programming, ICALP 2018, pp. 59:1–59:13 (2018)
9. Gatermann, K., Parrilo, P.A.: Symmetry groups, semidefinite programs, and sums of squares. J. Pure Appl. Algebra **192**(1–3), 95–128 (2004)
10. Goemans, M., Rothvoß, T.: Polynomiality for bin packing with a constant number of item types. In: Proceedings of the Twenty-Fifth annual ACM-SIAM Symposium on Discrete Algorithms, pp. 830–839. SIAM (2014)
11. Goemans, M., Williamson, D.: Improved approximation algorithms for maximum cut and satisfiability problems using semidefinite programming. J. ACM **42**, 1115–1145 (1995)
12. Graham, R.L.: Bounds for certain multiprocessing anomalies. Bell Syst. Tech. J. **45**(9), 1563–1581 (1966)
13. Grigoriev, D.: Complexity of positivstellensatz proofs for the knapsack. Comput. Complex. **10**(2), 139–154 (2001)
14. Grigoriev, D.: Linear lower bound on degrees of positivstellensatz calculus proofs for the parity. Theoret. Comput. Sci. **259**(1–2), 613–622 (2001)
15. Hochbaum, D.: Approximation Algorithms for NP-Hard Problems. PWS Publishing Co., Boston (1996)
16. Hochbaum, D., Shmoys, D.: Using dual approximation algorithms for scheduling problems theoretical and practical results. J. ACM **34**, 144–162 (1987)
17. Hopkins, S.B., Kothari, P.K., Potechin, A., Raghavendra, P., Schramm, T., Steurer, D.: The power of sum-of-squares for detecting hidden structures. In: Proceedings of the 58th IEEE Annual Symposium on Foundations of Computer Science, FOCS 2017, pp. 720–731. IEEE (2017)
18. Jans, R.: Solving lot-sizing problems on parallel identical machines using symmetry-breaking constraints. INFORMS J. Comput. **21**(1), 123–136 (2009)

19. Jansen, K.: An EPTAS for scheduling jobs on uniform processors: using an MILP relaxation with a constant number of integral variables. SIAM J. Discrete Math. **24**(2), 457–485 (2010)
20. Jansen, K., Klein, K., Verschae, J.: Closing the gap for makespan scheduling via sparsification techniques. In: Proceedings of the 43rd International Colloquium on Automata, Languages, and Programming, ICALP 2016, pp. 72:1–72:13 (2016)
21. Kaibel, V., Peinhardt, M., Pfetsch, M.E.: Orbitopal fixing. Discrete Optim. **8**, 595–610 (2011)
22. Kaibel, V., Pfetsch, M.: Packing and partitioning orbitopes. Math. Program. **114**(1), 1–36 (2008)
23. Karlin, A.R., Mathieu, C., Nguyen, C.T.: Integrality gaps of linear and semi-definite programming relaxations for knapsack. In: Günlük, O., Woeginger, G.J. (eds.) IPCO 2011. LNCS, vol. 6655, pp. 301–314. Springer, Heidelberg (2011). https://doi.org/10.1007/978-3-642-20807-2_24
24. Kothari, P., O'Donnell, R., Schramm, T.: SOS lower bounds with hard constraints: think global, act local. arXiv preprint arXiv:1809.01207 (2018)
25. Kothari, P.K., Mori, R., O'Donnell, R., Witmer, D.: Sum of squares lower bounds for refuting any CSP. In: Proceedings of the 49th Annual ACM SIGACT Symposium on Theory of Computing, STOC 2017, pp. 132–145. ACM (2017)
26. Kothari, P.K., Steinhardt, J., Steurer, D.: Robust moment estimation and improved clustering via sum of squares. In: Proceedings of the 50th Annual ACM SIGACT Symposium on Theory of Computing, STOC 2018, pp. 1035–1046. ACM (2018)
27. Kurpisz, A., Leppänen, S., Mastrolilli, M.: On the hardest problem formulations for the 0/1 Lasserre hierarchy. Math. Oper. Res. **42**(1), 135–143 (2016)
28. Kurpisz, A., Leppänen, S., Mastrolilli, M.: Sum-of-squares hierarchy lower bounds for symmetric formulations. In: Louveaux, Q., Skutella, M. (eds.) IPCO 2016. LNCS, vol. 9682, pp. 362–374. Springer, Cham (2016). https://doi.org/10.1007/978-3-319-33461-5_30
29. Kurpisz, A., Leppänen, S., Mastrolilli, M.: An unbounded sum-of-squares hierarchy integrality gap for a polynomially solvable problem. Math. Program. **166**(1–2), 1–17 (2017)
30. Kurpisz, A., Mastrolilli, M., Mathieu, C., Mömke, T., Verdugo, V., Wiese, A.: Semidefinite and linear programming integrality gaps for scheduling identical machines. Math. Program. **172**(1–2), 231–248 (2018)
31. Laurent, M.: Lower bound for the number of iterations in semidefinite hierarchies for the cut polytope. Math. Oper. Res. **28**(4), 871–883 (2003)
32. Laurent, M.: Semidefinite representations for finite varieties. Math. Program. **109**(1), 1–26 (2007)
33. Laurent, M.: Sums of squares, moment matrices and optimization over polynomials. In: Putinar, M., Sullivant, S. (eds.) Emerging Applications of Algebraic Geometry, vol. 149, pp. 157–270. Springer, New York (2009). https://doi.org/10.1007/978-0-387-09686-5_7
34. Levey, E., Rothvoss, T.: A (1+epsilon)-approximation for makespan scheduling with precedence constraints using LP hierarchies. In:Proceedings of the 48th Annual ACM SIGACT Symposium on Theory of Computing, STOC 2016, pp. 168–177. ACM (2016)
35. Liberti, L.: Automatic generation of symmetry-breaking constraints. In: Yang, B., Du, D.-Z., Wang, C.A. (eds.) COCOA 2008. LNCS, vol. 5165, pp. 328–338. Springer, Heidelberg (2008). https://doi.org/10.1007/978-3-540-85097-7_31

36. Liberti, L.: Symmetry in mathematical programming. In: Lee, J., Leyffer, S. (eds.) Mixed Integer Nonlinear Programming. The IMA Volumes in Mathematics and its Applications, pp. 263–283. Springer, New York (2012). https://doi.org/10.1007/978-1-4614-1927-3_9

37. Ma, T., Shi, J., Steurer, D.: Polynomial-time tensor decompositions with sum-of-squares. In: Proceedings of the 48th Annual ACM SIGACT Symposium on Theory of Computing, STOC 2016, pp. 438–446. IEEE (2016)

38. Margot, F.: Pruning by isomorphism in branch-and-cut. Math. Program. **94**, 71–90 (2002)

39. Margot, F.: Symmetry in integer linear programming. In: Jünger, M., et al. (eds.) 50 Years of Integer Programming 1958–2008, pp. 647–686. Springer, Heidelberg (2010). https://doi.org/10.1007/978-3-540-68279-0_17

40. O'Donnell, R.: SOS is not obviously automatizable, even approximately. In: Proceedings of the 8th Innovations in Theoretical Computer Science Conference (ITCS 2017). Leibniz International Proceedings in Informatics (LIPIcs), vol. 67, pp. 59:1–59:10. Schloss Dagstuhl-Leibniz-Zentrum fuer Informatik (2017)

41. Ostrowski, J.: Using symmetry to optimize over the Sherali-Adams relaxation. Math. Program. Comput. **6**, 405–428 (2014)

42. Ostrowski, J., Linderoth, J., Rossi, F., Smriglio, S.: Orbital branching. Math. Program. **126**, 147–178 (2011)

43. Parrilo, P.: Semidefinite programming relaxations for semialgebraic problems. Math. program. **96**, 293–320 (2003)

44. Potechin, A.: Sum of squares lower bounds from symmetry and a good story. arXiv preprint arXiv:1711.11469 (2017)

45. Potechin, A., Steurer, D.: Exact tensor completion with sum-of-squares. arXiv preprint arXiv:1702.06237 (2017)

46. Raghavendra, P., Schramm, T., Steurer, D.: High-dimensional estimation via sum-of-squares proofs. arXiv preprint arXiv:1807.11419 (2018)

47. Raymond, A., Saunderson, J., Singh, M., Thomas, R.R.: Symmetric sums of squares over k-subset hypercubes. Math. Program. **167**(2), 315–354 (2018)

48. Razborov, A.A.: Flag algebras. J. Symbol. Logic **72**(4), 1239–1282 (2007)

49. Razborov, A.A.: On 3-hypergraphs with forbidden 4-vertex configurations. SIAM J. Discrete Math. **24**(3), 946–963 (2010)

50. Rothvoß, T.: The Lasserre hierarchy in approximation algorithms. Lecture notes for MAPSP (2013)

51. Sagan, B.E.: The Symmetric Group: Representations, Combinatorial Algorithms, and Symmetric Functions. GTM, vol. 203. Springer, New York (2001). https://doi.org/10.1007/978-1-4757-6804-6

52. Schoenebeck, G.: Linear level Lasserre lower bounds for certain k-CSPs. In: Proceedings of the 49th Annual IEEE Symposium on Foundations of Computer Science, FOCS 2008, pp. 593–602. IEEE (2008)

53. Svensson, O.: Santa claus schedules jobs on unrelated machines. SIAM J. Comput. **41**(5), 1318–1341 (2012)

54. Verdugo, V., Verschae, J.: Breaking symmetries to rescue sum of squares: the case of makespan scheduling. arXiv preprint arXiv:abs/1811.08539 (2018)

Random Projections for Quadratic Programs over a Euclidean Ball

Ky Vu[1], Pierre-Louis Poirion[2], Claudia D'Ambrosio[3], and Leo Liberti[3(✉)]

[1] Department of Mathematics, FPT University, Hanoi, Vietnam
vukhacky@gmail.com
[2] RIKEN Center for Advanced Intelligence Project, Tokyo, Japan
pierre-louis.poirion@riken.jp
[3] CNRS LIX Ecole Polytechnique, 91128 Palaiseau, France
{dambrosio,liberti}@lix.polytechnique.fr

Abstract. Random projections are used as dimensional reduction techniques in many situations. They project a set of points in a high dimensional space to a lower dimensional one while approximately preserving all pairwise Euclidean distances. Usually, random projections are applied to numerical data. In this paper, however, we present a successful application of random projections to quadratic programming problems subject to polyhedral and a Euclidean ball constraint. We derive approximate feasibility and optimality results for the lower dimensional problem. We then show the practical usefulness of this idea on many random instances, as well as on two portfolio optimization instances with over 25M nonzeros in the (quadratic) risk term.

1 Introduction

In this paper we show that Random Projections (RP) can be applied to Quadratic Programming (QP) problems subject to linear inequality constraints and a single Euclidean ball constraint. We consider the following pair of QP formulations:

$$\left.\begin{array}{r}\max_y y^\top \tilde{Q} y + \tilde{c}^\top y \\ \tilde{A} y \leq \tilde{b} \\ \|y\|_2 \leq R, \end{array}\right\} \quad (1) \qquad \left.\begin{array}{r}\max_x x^\top Q x + c^\top x \\ A x \leq b \\ \|x\|_2 \leq 1. \end{array}\right\} \quad (2)$$

In Eq. (1), y is a vector of n decision variables, $\tilde{Q}$ is a symmetric $n \times n$ matrix, $\tilde{c} \in \mathbb{R}^n$, $\tilde{A}$ is $m \times n$, $\tilde{b} \in \mathbb{R}^m$, and R is a positive scalar. We also assume that $\tilde{A} y \leq \tilde{b}$ defines a full dimensional polyhedron and that $\tilde{b} \geq 0$ (this can be relaxed by translation if a feasible point for $\tilde{A} y \leq \tilde{b}$ is known). No assumption is made on $\tilde{Q}$. Eq. (2) is a scaled version of Eq. (1), where $Q = R^2 \tilde{Q}$, $c = R\tilde{c}$, $A = \tilde{A}/\mu$

This paper has received funding from the European Union's Horizon 2020 research and innovation programme under the Marie Sklodowska-Curie grant agreement n. 764759 "MINOA".

© Springer Nature Switzerland AG 2019
A. Lodi and V. Nagarajan (Eds.): IPCO 2019, LNCS 11480, pp. 442–452, 2019.
https://doi.org/10.1007/978-3-030-17953-3_33

(where $\mu = \max_j \|\tilde{A}_j\|_2$ and $\tilde{A}_j$ is the j-th column of $\tilde{A}$), and $b = \tilde{b}/(R\mu)$. Given a solution x^* of Eq. (2), then $y^* = Rx^*$ is a solution of Eq. (1). Note that all the columns of A are vectors of norm ≤ 1.

QP is now a ripe field with many applications (e.g., portfolio optimization, constrained linear regression, stable set problem, maximum cut and many more). The significance of the ball constraint is technical, but it could simply be interpreted to mean "bounded", since for all bounded QPs we can find a large enough R (in Eq. (1)) so that all solutions fall within a ball of radius R. In practice, however, if R is too large it might lead to ill scaling of Eq. (2). Note, however, that Eq. (1) is interesting in its own right as it is the formulation of the well-known *trust region subproblem*.

If we assume that all the data are rational, then the decision version of Eq. (2) without the ball constraint is **NP**-complete [9]. Moreover, by [10,11], the decision version of Eq. (2) without the polyhedral constraints is in **P** (and hence also in **NP**). For Eq. (2), one of the following applies: (i) some of the linear inequalities are active at the optimum; (ii) the ball inequality is active at the optimum; (iii) a combination of (i) and (ii); (iv) the optimum is unconstrained. In the first two cases the results in [9–11] apply, and the problem is in **NP**. Case (iii) falls in both of the first two categories, and the problem is still in **NP**. For case (iv) we can tell apart optimality vs. unboundedness by testing whether Q has negative eigenvalues or not [10]. Hence, the decision versions of Eqs. (1) and (2) are in **NP**.

RPs are random matrices which are used to perform dimensionality reduction on a set of vectors while approximately preserving all pairwise Euclidean distances with high probability. The goal of this paper is the applicability of RPs to bounded QPs such as those of Eq. (2). Specifically, we will define a projected version of Eq. (2) and prove that it is likely to have approximately the same optima as the original QP. We also perform a computational verification of our claim and show that the theoretical results, which are asymptotic in nature, also apply in practice.

RPs are usually applied to numerical data in view of speeding up algorithms which are essentially based on Euclidean distances, such as k-means or k-nearest neighbours. Since RPs ensure approximations of Euclidean distances by definition, it is perhaps not so surprising that they should work well in those settings. The focus of the present work is the much more counter-intuitive statement that a Mathematical Programming formulation might be approximately invariant (as regards feasibility and optimality) w.r.t. randomly projecting the input parameters. Similarly in spirit to our previous work on Linear Programming [13], but using a different projection and proof techniques, the results of this paper are independent of any solution algorithm, and largely independent of Euclidean distances (barring the ℓ_2 ball bounding the feasible region, which is applied to decision variables rather than data). While RPs have already been applied to some optimization problems, these are usually unconstrained minimizations of ℓ_2 norms and/or assume small Gaussian or doubling dimension of the feasible set [8,15]: two assumptions we do not make.

The rest of this paper is organized as follows. In Sect. 2 we define RPs and the projected QP. In Sect. 3 we introduce some theoretical results about random projections. In Sect. 4 we prove the main theorems about RPs applied to QP. In Sect. 5 we discuss computational results.

2 Definitions

RPs are simple but powerful tools for dimension reduction [4,8,12,13,15]. They are often constructed as random matrices sampled from some given distribution classes. The simplest examples are suitably scaled matrices sampled componentwise from independently identically distributed (i.i.d.) random variables with Gaussian $N(0,1)$, uniform on $[-1,1]$, or Rademacher ± 1 distributions. One of the most important features of a RP is that it approximately preserves the norm of any given vector with high probability. In particular, let $P \in \mathbb{R}^{d \times n}$ be a RP, e.g. sample every component of P from $N(0,1/\sqrt{d})$. Then, for any $x \in \mathbb{R}^n$ and $\varepsilon \in (0,1)$, we have

$$\mathsf{Prob}\left[(1-\varepsilon)\|x\|_2^2 \le \|Px\|_2^2 \le (1+\varepsilon)\|x\|_2^2\right] \ge 1 - 2e^{-\mathcal{C}\varepsilon^2 d}, \tag{3}$$

where $\mathcal{C}$ is a *universal constant* (in fact a more precise statement should be existentially quantified by "there exists a constant $\mathcal{C}$ such that...").

Perhaps the most famous application of RPs is the *Johnson-Lindenstrauss lemma* [2]. It states that, for any $\varepsilon \in (0,1)$ and for any finite set $X \subseteq \mathbb{R}^n$, there is a mapping $F : \mathbb{R}^n \to \mathbb{R}^d$, in which $d = O(\frac{\ln|X|}{\varepsilon^2})$, such that

$$\forall x,y \in X \quad (1-\varepsilon)\|x-y\|_2^2 \le \|F(x)-F(y)\|_2^2 \le (1+\varepsilon)\|x-y\|_2^2.$$

Such a mapping F can be realized as the matrix P above. The existence of the correct mapping is shown (by the probabilistic method) using the union bound. Moreover, the probability of sampling a correct mapping can be made arbitrarily high. In practice, we found that there is often no need to re-sample P.

In the following, all norm symbols $\|\cdot\|$ will be assumed to refer to the ℓ_2 norm $\|\cdot\|_2$. We sample our RPs from Gaussian ensembles (in practice, we also specify their density, see Sect. 5).

2.1 The Randomly Projected QP

Let $P \in \mathbb{R}^{d \times n}$ be a RP. We want to "project" each vector $x \in \mathbb{R}^n$ to a lower dimensional vector $Px \in \mathbb{R}^d$. Consider the following *projected problem*:

$$\max\{x^\top(P^\top PQP^\top P)x + c^\top P^\top Px \mid AP^\top Px \le b, \|Px\| \le 1\}.$$

By setting $u = Px$, $\bar{c} = Pc$, $\bar{A} = AP^\top$, $\bar{Q} = PQP^\top$, we can rewrite it as

$$\max_{u \in \, \mathsf{Im}(P)} \{u^\top \bar{Q}u + \bar{c}^\top u \mid \bar{A}u \le b, \|u\| \le 1\}, \tag{4}$$

where $\mathsf{Im}(P)$ is the image space generated by P. Since P is (randomly) generated with full rank with probability 1, it is very likely to be a surjective mapping. Therefore, we assume it is safe to remove the constraint $u \in \mathsf{Im}(P)$ and study the smaller dimensional problem:

$$\max_{u \in \mathbb{R}^d} \{u^\top \bar{Q}u + \bar{c}^\top u \mid \bar{A}u \le b, \ \|u\| \le 1\}, \tag{5}$$

where u ranges in $\mathbb{R}^d$. As we will show later, Eq. (5) yields a good approximate solution of Eq. (2) with high probability.

3 Some Properties of Random Projections

It is known that singular values of random matrices often concentrate around their means. In the case when the RP is sampled from Gaussian ensembles, this phenomenon is well-understood due to many current research efforts. The following lemma, which is proved in [16], uses this phenomenon to show that, when $P \in \mathbb{R}^{d \times n}$ is a Gaussian random matrix (with the number of row significantly smaller than the number of columns), then $PP^\top$ is very close to an identity matrix. This gives an intuitive explanation as to why Eq. (5) has desirable approximate properties w.r.t. Eq. (2).

Lemma 3.1 ([16]). *Let $P \in \mathbb{R}^{d \times n}$ be a RP. Then for any $\delta > 0$ and $0 < \varepsilon < \frac{1}{2}$, with probability at least $1 - \delta$, we have $\|PP^\top - I\|_2 \le \varepsilon$ provided that*

$$n \ge (d+1)\ln(2d/\delta)/(\mathcal{C}_1 \varepsilon^2), \tag{6}$$

where $\| \cdot \|_2$ is the spectral norm of the matrix and $\mathcal{C}_1 > \frac{1}{4}$ is some universal constant.

This lemma also tells us that, when we go from low to high dimensions, with high probability we can ensure that the norms of all the points endure small distortions. Indeed, for any vector $u \in \mathbb{R}^d$, then

$$\|P^\top u\|^2 - \|u\|^2 = \langle P^\top u, P^\top u \rangle - \langle u, u \rangle = \langle (PP^\top - I)u, u \rangle \in [-\varepsilon\|u\|^2, \varepsilon\|u\|^2],$$

due to the Cauchy-Schwarz inequality. Moreover, it implies that $\|P^\top\|_2 \le (1+\varepsilon)$ with probability at least $1 - \delta$.

Condition (6) is not difficult to satisfy in practice, since d is often very small compared to n. On the other hand, n should be large enough to dominate the effect of $\frac{1}{\varepsilon^2}$.

Lemma 3.2. *Let $P \in \mathbb{R}^{d \times n}$ be a RP satisfying Eq. (3) and let $0 < \varepsilon < 1$. Then there is a universal constant $\mathcal{C}_0$ such that the following statements hold.*

(i) *For any $x, y \in \mathbb{R}^n$, $\langle x, y \rangle - \varepsilon\|x\| \, \|y\| \le \langle Px, Py \rangle \le \langle x, y \rangle + \varepsilon\|x\| \, \|y\|$ with probability at least $1 - 4e^{-\mathcal{C}_0 \varepsilon^2 d}$.*

(ii) Let $\mathbf{1}$ *be the all-one vector. For any* $x \in \mathbb{R}^n$ *and* $A \in \mathbb{R}^{m \times n}$ *having unit row vectors, we have* $Ax - \varepsilon\|x\|\mathbf{1} \le AP^{\top}Px \le Ax + \varepsilon\|x\|\mathbf{1}$ *with probability at least* $1 - 4\,m\,e^{-\mathcal{C}_0\varepsilon^2 d}$.

(iii) For any two vectors $x, y \in \mathbb{R}^n$ *and a square matrix* $Q \in \mathbb{R}^{n \times n}$, *then with probability at least* $1 - 8\,k\,e^{-\mathcal{C}_0\varepsilon^2 d}$, *we have:*

$$x^{\top}Qy - 3\varepsilon\|x\|\,\|y\|\,\|Q\|_* \le x^{\top}P^{\top}PQP^{\top}Py \le x^{\top}Qy + 3\varepsilon\|x\|\,\|y\|\,\|Q\|_*,$$

in which $\|Q\|_*$ *is the nuclear norm of* Q *and* k *is the rank of* Q.

4 Approximate Optimality

We now prove that the objective of the quadratic problem in Eq. (2) is approximately preserved under RPs. To do so, we study the relations between this and two other problems:

$(\mathrm{QP}_{\varepsilon}^{-}) \quad \max\{u^{\top}PQP^{\top}u + (Pc)^{\top}u \mid AP^{\top}u \le b, \quad \|u\| \le 1 - \varepsilon, u \in \mathbb{R}^d\}$

$(\mathrm{QP}_{\varepsilon}^{+}) \quad \max\{u^{\top}PQP^{\top}u + (Pc)^{\top}u \mid AP^{\top}u \le b + \varepsilon, \ \|u\| \le 1 + \varepsilon, u \in \mathbb{R}^d\}.$

We first state the following feasibility result.

Theorem 4.1. *Let* $P \in \mathbb{R}^{d \times n}$ *be a RP. Let* $\delta \in (0,1)$. *Assume further that Eq. (6) holds for some universal constant* $\mathcal{C}_1 > \frac{1}{4}$. *Then with probability at least* $1 - \delta$, *for any feasible solution* u *of the projected problem* $(\mathrm{QP}_{\varepsilon}^{-})$, $P^{\top}u$ *is also feasible for the original problem in Eq. (2).*

We remark the following universal property of Theorem 4.1: with a fixed probability, feasibility holds for all vectors u (instead of a given vector).

Proof. Let $\mathcal{C}_1$ be as in Lemma 3.1. Let u be any feasible solution for the projected problem $(\mathrm{QP}_{\varepsilon}^{-})$ and take $\hat{x} = P^{\top}u$. Then we have $A\hat{x} = AP^{\top}u \le b$ and

$$\|P^{\top}u\|^2 = \langle P^{\top}u, P^{\top}u \rangle = \langle u, u \rangle + \langle (PP^{\top} - I)u, u \rangle \le (1 + \varepsilon)\|u\|^2$$

with probability at least $1 - \delta$ (by Lemma 3.1). This implies that $\|\hat{x}\| \le (1 + \varepsilon/2)\|u\|$; and since $\|u\| \le 1 - \varepsilon$, we have $\|\hat{x}\| \le (1 + \varepsilon/2)(1 - \varepsilon) < 1$ with probability at least $1 - \delta$, which proves the theorem. $\qquad\square$

Let u_{ε}^{-} and u_{ε}^{+} be optimal solutions for these two problems, respectively. Denote by $x_{\varepsilon}^{-} = P^{\top}u_{\varepsilon}^{-}$ and $x_{\varepsilon}^{+} = P^{\top}u_{\varepsilon}^{+}$. Let x^* be an optimal solution for the original problem in Eq. (2). We will bound $x^{*\top}Qx^* + c^{\top}x^*$ between $x_{\varepsilon}^{-\top}Qx_{\varepsilon}^{-} + c^{\top}x_{\varepsilon}^{-}$ and $x_{\varepsilon}^{+\top}Qx_{\varepsilon}^{+} + c^{\top}x_{\varepsilon}^{+}$, the two values that are expected to be approximately close to each other.

Theorem 4.2. *Let* $P \in \mathbb{R}^{d \times n}$ *be a RP, and let* $\delta \in (0,1)$. *Let* x^* *be an optimal solution for the original problem Eq. (2). Then there are universal constants* $\mathcal{C}_0 > 1$ *and* $\mathcal{C}_1 > \frac{1}{4}$ *such that, if* $d \ge \ln(m/\delta)/(\mathcal{C}_0\varepsilon^2)$ *and Eq. (6) are satisfied, we will have the following two statements. (i) With probability at least* $1 - \delta$, *the solution* x_{ε}^{-} *is feasible for the original problem Eq. (2). (ii) With probability at least* $1 - \delta$,

$$x_{\varepsilon}^{-\top}Qx_{\varepsilon}^{-} + c^{\top}x_{\varepsilon}^{-} \le x^{*\top}Qx^* + c^{\top}x^* \le x_{\varepsilon}^{+\top}Qx_{\varepsilon}^{+} + c^{\top}x_{\varepsilon}^{+} + 3\varepsilon\|Q\|_* + \varepsilon\|c\|.$$

Proof. The constants $\mathcal{C}_0$ and $\mathcal{C}_1$ are chosen in the same way as before. (i) By Theorem 4.1, with probability at least $1 - \delta$, for any feasible point u of the projected problem $(\mathrm{QP}_\varepsilon^-)$, $P^\top u$ is also feasible for the original problem Eq. (2). Therefore, it must hold also for x_ε^-.

(ii) By Part (i) above, with probability at least $1 - \delta$, x_ε^- is feasible for the original problem Eq. (2). Therefore, we have $x_\varepsilon^{-\top} Q x_\varepsilon^- + c^\top x_\varepsilon^- \leq x^{*\top} Q x^* + c^\top x^*$ with probability at least $1 - \delta$. Moreover, due to Lemma 3.2, with probability at least $1 - 8(k+1)e^{-\mathcal{C}_0\varepsilon^2 d}$, where k is the rank of Q, we have

$$x^{*\top} Q x^* \leq x^{*\top} P^\top P Q P^\top P x^* + 3\varepsilon\|x^*\|^2\,\|Q\|_* \leq x^{*\top} P^\top P Q P^\top P x^* + 3\varepsilon\|Q\|_*$$
$$\text{and } c^\top x^* \leq \qquad c^\top P^\top P x^* + \varepsilon\|c\|\,\|x^*\| \qquad \leq c^\top P^\top P x^* + \varepsilon\|c\|,$$

since $\|x^*\| \leq 1$. Hence $x^{*\top} Q x^* + c^\top x^* \leq x^{*\top} P^\top P Q P^\top P x^* + c^\top P^\top P x^* + \varepsilon\|c\| + 3\varepsilon\|Q\|_*$. On the other hand, let $\hat{u} = P x^*$; due to Lemma 3.2, we have

$$A P^\top \hat{u} = A P^\top P x^* \leq A x^* + \varepsilon\|x^*\|\mathbf{1} \leq A x^* + \varepsilon\mathbf{1} \leq b + \varepsilon$$

with probability at least $1 - 4m\,e^{-\mathcal{C}_0\varepsilon^2 d}$ (the last inequality holds by the assumption $b \geq 0$), and $\|\hat{u}\| = \|P x^*\| \leq (1+\varepsilon)\|x^*\| \leq (1+\varepsilon)$ with probability at least $1 - 2e^{-\mathcal{C}_0\varepsilon^2 d}$ (by Lemma 3.2). Therefore, $\hat{u}$ is a feasible solution for the problem $(\mathrm{QP}_\varepsilon^+)$ with probability at least $1 - (4m+2)e^{-\mathcal{C}_0\varepsilon^2 d}$. Due to the optimality of u_ε^+ for the problem $(\mathrm{QP}_\varepsilon^+)$, it follows that

$$x^{*\top} Q x^* + c^\top x^* \leq x^{*\top} P^\top P Q P^\top P x^* + c^\top P^\top P x^* + \varepsilon\|c\| + 3\varepsilon\|Q\|_*$$
$$= \hat{u}^\top P Q P^\top \hat{u} + c^\top P^\top \hat{u} + \varepsilon\|c\| + 3\varepsilon\|Q\|_*$$
$$\leq u_\varepsilon^{+\top} P Q P^\top u_\varepsilon^+ + (Pc)^\top u_\varepsilon^+ + \varepsilon\|c\| + 3\varepsilon\|Q\|_*$$
$$= x_\varepsilon^{+\top} Q x_\varepsilon^+ + c^\top x_\varepsilon^+ + \varepsilon\|c\| + 3\varepsilon\|Q\|_*,$$

with probability at least $1 - (4m+6)e^{-\mathcal{C}_0\varepsilon^2 d}$, which is at least $1 - \delta$ for the chosen universal constant $\mathcal{C}_0$. Hence $x^{*\top} Q x^* + c^\top x^* \leq x_\varepsilon^{+\top} Q x_\varepsilon^+ + c^\top x_\varepsilon^+ + 3\varepsilon\|Q\|_* + \varepsilon\|c\|$, which concludes the proof. □

The above result implies that the value of $x^{*\top} Q x^* + c^\top x^*$ lies between $x_\varepsilon^{-\top} Q x_\varepsilon^- + c^\top x_\varepsilon^-$ and $x_\varepsilon^{+\top} Q x_\varepsilon^+ + c^\top x_\varepsilon^+$. It remains to prove that these two values are not so far from each other. Let $S^* = \{x \in \mathbb{R}^n \mid Ax \leq b, \|x\| \leq 1\}$. Since, by assumption, the feasible set of (2) is full dimensional, S^* is also full dimensional.

We associate with each set S a positive number $\mathrm{FULL}(S) > 0$, which is considered as a fullness measure of S and is defined as the maximum radius of any closed ball contained in S. Now, from our assumption, we have $\mathrm{FULL}(S^*) = r^* > 0$, where r^* is the radius of the greatest ball inscribed in S^* (see Fig. 1, left).

The following lemma characterizes the fullness of S_ε^+ with respect to r^*, where $S_\varepsilon^+ = \{u \in \mathbb{R}^d \mid A P^\top u \leq b + \varepsilon, \|u\| \leq 1 + \varepsilon\}$ is the feasible set of the problem $(\mathrm{QP}_\varepsilon^+)$.

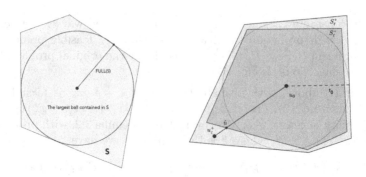

Fig. 1. Left: fullness of a set. Right: idea of the proof of Theorem 4.4.

Lemma 4.3. *Let S^* be full-dimensional with $\mathrm{FULL}(S^*) = r^*$. Then with probability at least $1 - 3\delta$, S_ε^+ is also full-dimensional with the fullness measure $\mathrm{FULL}(S_\varepsilon^+) \geq (1 - \varepsilon)r^*$.*

The proof of this lemma extensively uses the fact that, for any row vector $a \in \mathbb{R}^n$ $\sup_{\|u\| \leq r} a^\top u = r\|a\|$, which is actually the equality condition in the Cauchy-Schwartz inequality.

Now we will estimate the gap between the two objective functions of the problems $(\mathrm{QP}_\varepsilon^+)$ and $(\mathrm{QP}_\varepsilon^-)$ using the fullness measure. The theorem states that, as long as the fullness of the original polyhedron is large enough, the gap between them is small. Figure 1, right, gives the proof idea.

Theorem 4.4. *Let $0 < \varepsilon < 0.1$. Then with probability at least $1 - 4\delta$, we have*
$$x_\varepsilon^{-\top}Qx_\varepsilon^- + c^\top x_\varepsilon^- \leq x_\varepsilon^{+\top}Qx_\varepsilon^+ + c^\top x_\varepsilon^+ < \tfrac{(1+\varepsilon)^2}{(1-\varepsilon)^2}(x_\varepsilon^{-\top}Qx_\varepsilon^- + c^\top x_\varepsilon^-) + \tfrac{\varepsilon}{\mathrm{FULL}(S^*)}(36 + 18\|c\|).$$

Proof. Let $B(u_0, r_0)$ be a closed ball with maximum radius that is contained in S_ε^+. In order to establish the relation between u_ε^+ and u_ε^-, our idea is to move u_ε^+ closer to u_0, so that the new point is contained in S_ε^-. Therefore, its objective value will be less that the value of u_ε^-, but quite close to the objective value of u_ε^+. We define $\hat{u} = (1-\lambda)u_\varepsilon^+ + \lambda u_0$ for some $\lambda \in (0,1)$: we want to find λ such that $\hat{u}$ is feasible for $(\mathrm{QP}_\varepsilon^-)$ while its corresponding objective value is not so different from $x_\varepsilon^{+\top}Qx_\varepsilon^+ + c^\top x_\varepsilon^+$. Since for all $\|u\| \leq r_0$, $AP^\top(u_0 + u) = AP^\top u_0 + AP^\top u \leq b + \varepsilon$. Then $AP^\top u_0 \leq b + \varepsilon - r_0(\|A_1 P^\top\|, \ldots, \|A_m P^\top\|)^\top$. Therefore, we have, with probability at least $1 - \delta$,

$$AP^\top u_0 \leq b + \varepsilon - r_0(1-\varepsilon)(\|A_1\|, \ldots, \|A_m\|)^\top = b + \varepsilon - r_0(1-\varepsilon).$$

Hence $AP^\top \hat{u} = (1-\lambda)AP^\top u_\varepsilon^+ + \lambda AP^\top u_0 \leq b + \varepsilon - \lambda r_0(1-\varepsilon) \leq b + \varepsilon - \tfrac{1}{2}\lambda r_0$, since we can assume w.l.o.g. that $\varepsilon \leq \tfrac{1}{2}$. Hence, $AP^\top \hat{u} \leq b$ if we choose $\varepsilon \leq \lambda \tfrac{r_0}{2}$. Furthermore $\|\hat{u}\| \leq 1 + \varepsilon$ Hence, when we choose $\lambda = 2\tfrac{\varepsilon}{r_0}$, then $\tfrac{1-\varepsilon}{1+\varepsilon}\hat{u}$

is feasible for the problem $(\mathrm{QP}_\varepsilon^-)$ with probability at least $1 - \delta$. Therefore, $\frac{1+\varepsilon}{1-\varepsilon}u_\varepsilon^{-\top}PQP^\top u_\varepsilon^- + (Pc)^\top u_\varepsilon^-$ is greater than or equal to $\hat{u}^\top PQP^\top \hat{u} + (Pc)^\top \hat{u} =$

$$
\begin{aligned}
&= \left(u_\varepsilon^+ + \lambda(u_0 - u_\varepsilon^+)\right)^\top PQP^\top \left(u_\varepsilon^+ + \lambda(u_0 - u_\varepsilon^+)\right) + (Pc)^\top \hat{u} \\
&= u_\varepsilon^{+\top}PQP^\top u_\varepsilon^+ + \lambda u_\varepsilon^{+\top}PQP^\top \left(u_0 - u_\varepsilon^+\right) + \lambda(u_0 - u_\varepsilon^+)^\top PQP^\top u_\varepsilon^+ \\
&\quad + \lambda^2(u_0 - u_\varepsilon^+)^\top PQP^\top(u_0 - u_\varepsilon^+) + (Pc)^\top \hat{u}.
\end{aligned}
$$

However, from Lemma 3.2 and the Cauchy-Schwartz inequality, we have

$$
\begin{aligned}
|u_\varepsilon^{+\top}PQP^\top\left(u_0 - u_\varepsilon^+\right)| &\leq \|P^\top u_\varepsilon^+\|\,\|Q\|_2\,\|P^\top(u_0 - u_\varepsilon^+)\| \\
&\leq (1+\varepsilon)^2\|u_\varepsilon^+\|\,\|Q\|_2\,\|(u_0 - u_\varepsilon^+)\| \leq 2(1+\varepsilon)^4\,\|Q\|_2
\end{aligned}
$$

(since $\|u_\varepsilon^+\|$ and $\|u_\varepsilon^-\| \leq 1 + \varepsilon$), and similarly for other terms. We then have

$$
\hat{u}^\top PQP^\top \hat{u} \geq u_\varepsilon^{+\top}PQP^\top u_\varepsilon^+ - (4\lambda + 4\lambda^2)(1+\varepsilon)^4\,\|Q\|_2.
$$

Since $\varepsilon < 0.1$, we have $(1+\varepsilon)^4 < 2$ and we can assume that $\lambda < 1$. Then we have

$$
\begin{aligned}
\hat{u}^\top PQP^\top \hat{u} &> u_\varepsilon^{+\top}PQP^\top u_\varepsilon^+ - 16\lambda\|Q\|_2 \\
&= u_\varepsilon^{+\top}PQP^\top u_\varepsilon^+ - 32\varepsilon/r_0 \quad (\text{since } \|Q\|_2 = 1) \\
&\geq u_\varepsilon^{+\top}PQP^\top u_\varepsilon^+ - 32\varepsilon/((1-\varepsilon)\mathrm{FULL}(S^*)) \quad (\text{due to Lemma 4.3}) \\
&> u_\varepsilon^{+\top}PQP^\top u_\varepsilon^+ - 36\varepsilon/\mathrm{FULL}(S^*) \quad (\text{since } \varepsilon \leq 0.1),
\end{aligned}
$$

with probability at least $1 - 2\delta$. Furthermore, we have

$$
c^\top P^\top \hat{u} = c^\top P^\top u_\varepsilon^+ + \lambda c^\top P^\top(u_0 - u_\varepsilon^+) \geq c^\top P^\top u_\varepsilon^+ - \frac{4(1+\varepsilon)\varepsilon}{r_0}\|Pc\|.
$$

We know that $r_0 \geq (1 - \varepsilon)r^*$, hence

$$
c^\top P^\top \hat{u} \geq c^\top P^\top u_\varepsilon^+ - \frac{4(1+\varepsilon)\varepsilon}{r_0}\|Pc\| \geq c^\top P^\top u_\varepsilon^+ - \frac{4(1+\varepsilon)^2\varepsilon}{(1-\varepsilon)r^*}\|c\|,
$$

with probability at least $1 - \delta$. The results holds by $\varepsilon < 0.1$.   □

5   Computational Results

Although we developed our theory for dense Gaussian RPs, in practice one can decrease computational costs considerably by using sparsity [1,3]. All of the results of this paper, aside from Lemma 3.1, actually hold (unchanged) also for *sub-gaussian* RPs. Amongst sub-gaussian RPs, we elect to use $d \times n$ matrices where each component is sampled from $\mathrm{N}(0, \frac{1}{\sqrt{d}})$ with some given probability dens $\in (0,1)$. Although we are not going to include the generalization of Lemma 3.1 to sub-gaussian RPs here for lack of space, the proof exploits the fact that the largest and smallest singular values of sub-gaussian RPs are approximately the same.

All tests were carried out on a single core of a 4-CPU machine with 64 GB RAM, each CPU of which has 8 cores (Intel Xeon CPU E5-2620 v4@2.10 GHz).

5.1   Random Instances

Our first computational test is carried out of randomly generated feasible instances of Eq. (2) with Q negative semidefinite (we make this assumption in order to compute guaranteed global maxima in acceptable CPU times for comparison purposes: the projection technique is independent of the convexity of the objective function). We generate all instances varying the following parameters: number of constraints $m \in \{10, 100, 1000\}$, of variables $n \in \{2000, 3000\}$, random number generation distribution $\text{distr} \in \{0, 1\}$ (choosing between uniform distributions $\mathsf{U}(0, 1)$ and $\mathsf{U}(-1, 1)$) and density $\text{dens} \in \{0.1, 0.6\}$ for matrices A and Q. As mentioned above, our RPs are sparse random Gaussian matrices P with $\varepsilon \in \{0.10, 0.15, 0.20\}$ and density $\text{dens}_P \in \{0.2, 0.5, 1.0\}$. This yields a total of 216 solution logs (all instances with all RP generation methods) obtained using the IPOPT solver [14]. We benchmark means, standard deviations, maximum and minimum values for: (i) CPU time (solution of original vs. projected problem to optimality, where the projected CPU time also includes RP sampling, matrix multiplication and solution retrieval time); (ii) objective function ratio $\rho = \frac{|f^*_{\text{org}} - f^*_{\text{retr}}|}{\max(|f^*_{\text{org}}|, |f^*_{\text{retr}}|)}$, where f^*_{org} is the optimal objective function value of Eq. (2) and f^*_{retr} is the value of the objective function of Eq. (2) evaluated at the solution retrieved from the projected problem Eq. (5); (iii) average feasibility errors denoted ace, are, be for, respectively, $Ax \le b$, ranges $-1 \le x \le 1$ and $\|x\| \le 1$. All our coding was carried out in Julia+JuMP [5].

Table 1. Computational results for random instances.

	CPU_{org}	CPU_{proj}	ρ	ace	are	be
mean	37.691	**14.590**	0.103	0.0	0.0	2.237
stdev	49.984	**15.057**	0.070	0.0	0.0	0.916
min	8.750	**2.170**	0.000	0.0	0.0	0.921
max	198.350	**61.340**	0.485	0.0	0.0	3.886

Table 1 shows a consistent behaviour of our projection technique: projected formulations take considerably less time to solve (despite pre- and post-processing steps), and yield solutions having objective function values within around 10% of the optimum, with no feasibility error w.r.t. linear and range constraints. There is a large ball error, however, which we are unable to explain at this time—we are looking into it. Scaling the retrieved solution back to norm 1 yields a feasible point but increases ρ considerably (to around 0.4 on average).

Table 2 shows the trade-off between approximation quality and efficiency in function of the parameters ε and dens_P of the RP (blacker is better). The best compromise appears to be achieved for $\varepsilon = 0.15$ and $\text{dens}_P = 0.2$.

Table 2. Trade-off between approximation and efficiency for ε and dens$_P$.

dens$_P$	0.2			0.5			1.0		
ε	CPU$_{org}$	CPU$_{proj}$	ρ	CPU$_{org}$	CPU$_{proj}$	ρ	CPU$_{org}$	CPU$_{proj}$	ρ
0.10	37.57	19.91	**0.07**	37.90	21.02	**0.06**	37.76	24.36	**0.07**
0.15	37.56	11.26	0.10	37.81	11.95	0.09	37.78	12.16	0.11
0.20	37.50	**10.04**	0.14	37.71	**10.30**	0.15	37.62	**10.32**	0.14

5.2 Two Large Portfolio Instances

We consider two large-scale Markowitz portfolio [6,7] instances where the objective is a scalarized version of risk minimization (using correlation rather than covariance for better scaling) and return maximization. The system $Ax \leq b$ encodes the portfolio constraints $0 \leq x \leq 1$ (with $-x \leq 0$ being part of the inequality constraints) and $\sum_j x_j \leq 1$, which imply $\|x\|_2 \leq 1$. The stock price data were obtained from Kaggle (goo.gl/XHfhi2), and yielded fully dense Q matrices. We have used the $\varepsilon = 0.15$ and dens$_P = 0.2$ RP settings obtained from Table 2. Our results are presented in Table 3. The computational savings are remarkable, the optimal objective function values are within a reasonable approximation ratio, but the retrieved solutions are slightly infeasible w.r.t. linear and range constraints. Specifically, some of the components of the retrieved solutions are very slightly negative (0.001 for `etfs` and 0.0005 for `stocks` on average), which is an issue we had also observed in applying RPs to Linear Programs [13]. The ball errors are again high.

Table 3. Results for two large instances of Markowitz' portfolio problem.

Instance	n	nnz(Q)	CPU$_{org}$	CPU$_{proj}$	ρ	ace	are	be
etfs	1344	902,496	534.32	**11.38**	0.270	0.026	0.001	1.570
stocks	7163	25,650,703	266,713.40	**132.78**	0.007	0.023	0.001	3.927

6 Conclusion

We prove that random projections can be used to generate lower dimensional QPs, bounded by a Euclidean ball constraint, which have approximately the same global optimum with arbitrarily high probability as their original counterparts. Computational results are exhibited to substantiate our claim and show the applicability of our techniques.

As a corollary, we remark that our results are also applicable to reduce the number of variables of inequality constrained LPs (with a Euclidean ball constraint), since there is no assumption on Q (so $Q = 0$ is a possibility). We are in the process of deriving theorems which ensure better bounds given this specific structure.

References

1. Achlioptas, D.: Database-friendly random projections: Johnson-Lindenstrauss with binary coins. J. Comput. Syst. Sci. **66**, 671–687 (2003)
2. Johnson, W., Lindenstrauss, J.: Extensions of Lipschitz mappings into a Hilbert space. In: Hedlund, G. (ed.) Conference in Modern Analysis and Probability. Contemporary Mathematics, vol. 26, pp. 189–206. American Mathematical Society, Providence (1984)
3. Kane, D., Nelson, J.: Sparser Johnson-Lindenstrauss transforms. J. ACM **61**(1), 4 (2014)
4. Liberti, L., Vu, K.: Barvinok's Naive algorithm in distance geometry. Oper. Res. Lett. **46**, 476–481 (2018)
5. Lubin, M., Dunning, I.: Computing in operations research using Julia. INFORMS J. Comput. **27**(2), 238–248 (2015). https://doi.org/10.1287/ijoc.2014.0623
6. Markowitz, H.: Portfolio selection. J. Financ. **7**(1), 77–91 (1952)
7. Mencarelli, L., D'Ambrosio, C.: Complex portfolio selection via convex mixed-integer quadratic programming: a survey. Int. Trans. Oper. Res. **26**, 389–414 (2019)
8. Pilanci, M., Wainwright, M.: Randomized sketches of convex programs with sharp guarantees. In: International Symposium on Information Theory (ISIT), pp. 921–925. IEEE, Piscataway (2014)
9. Vavasis, S.: Quadratic programming is in NP. Inf. Process. Lett. **36**, 73–77 (1990)
10. Vavasis, S., Zippel, R.: Proving polynomial-time for sphere-constrained quadratic programming. Technical report 90–1182, Department of Computer Science, Cornell University (1990)
11. Vavasis, S.: Nonlinear Optimization: Complexity Issues. Oxford University Press, Oxford (1991)
12. Vu, K., Poirion, P.L., Liberti, L.: Gaussian random projections for Euclidean membership problems. Discrete Appl. Math. (2018). https://doi.org/10.1016/j.dam.2018.08.025
13. Vu, K., Poirion, P.L., Liberti, L.: Random projections for linear programming. Math. Oper. Res. **43**, 1051–1404 (2018). https://doi.org/10.1287/moor.2017.0894
14. Wächter, A., Biegler, L.: On the implementation of an interior-point filter line-search algorithm for large-scale nonlinear programming. Math. Program. **106**(1), 25–57 (2006)
15. Woodruff, D.: Sketching as a tool for linear algebra. Found. Trends Theoret. Comput. Sci. **10**(1–2), 1–157 (2014)
16. Zhang, L., Mahdavi, M., Jin, R., Yang, T., Zhu, S.: Recovering the optimal solution by dual random projection. In: Shalev-Shwartz, S., Steinwart, I. (eds.) Conference on Learning Theory (COLT), Proceedings of Machine Learning Research, vol. 30, pp. 135–157 (2013). http://www.jmlr.org

Author Index

Printed in the United States
By Bookmasters